Earth System Science

From Biogeochemical Cycles to Global Change

edited by

Michael C. Jacobson
Robert J. Charlson
Henning Rodhe
Gordon H. Orians

ACADEMIC PRESS

A Harcourt Science and Technology Company

SAN DIEGO   SAN FRANCISCO   NEW YORK   BOSTON
LONDON   SYDNEY   TOKYO

Academic Press
A Harcourt Science and Technology Company
32 Jamestown Road, London NW1 7BY
http://www.academicpress.com

Academic Press
525 B Street, Suite 1900, San Diego, California 92101-4495, USA
http://www.academicpress.com

ISBN 0-12-379370-X

A catalogue for this book is available from the British Library

Typeset by Paston PrePress Ltd, Beccles, Suffolk
Printed in Great Britain by The Bath Press, Bath

00 01 02 03 04 BP 9 8 7 6 5 4 3 2 1

Contents

Color Plates are located between pp. 194–195

Authors

Theodore L Anderson, Department of Atmospheric Sciences, University of Washington, Box 351640, Seattle, WA 98195–1640, USA

Sharon E. Anthony, Evergreen State College, 2700 Evergreen Parkway NW, Olympia, WA 98505, USA

Mark M. Benjamin, Department of Civil and Environmental Engineering, University of Washington, Box 352700, Seattle, WA 98195–2700, USA

Edward J. Brook, Department of Geology, Department of Environmental Science, Washington State University, Vancouver, WA 98686, USA

Donald E. Brownlee, Department of Astronomy, University of Washington, Box 351580, Seattle, WA 98195–1580, USA

Stephen J. Burges, Department of Civil and Environmental Engineering, University of Washington, Box 352700, Seattle, WA 98195–2700, USA

Samuel S. Butcher (retired), PO Box 54, Willow Creek, MT 59760, USA

Robert J. Charlson, Department of Atmospheric Sciences, University of Washington, Box 351640, Seattle, WA 98195–1640, USA

Kurt M. Cuffey, Department of Geography, 501 McCone Hall, University of California, Berkeley, CA 94720, USA

Steven Emerson, School of Oceanography, University of Washington, Box 357940, Seattle, WA 98195–7940, USA

Rolf O. Hallberg, Geologiska Institutionen, Stockholms Universitet, S-106 91 Stockholm, Sweden

Patricia C. Henshaw, Northwest Hydraulic Consultants, 16300 Christenson Road, Suite 350, Seattle, WA 98188, USA

Kim Holmén, Meteorologiska Institutionen, Stockholms Universitet, S-106 91 Stockholm, Sweden

Bruce D. Honeyman, Laboratory for Applied and Environmental Radiochemistry, Environmental Science and Engineering Division, Colorado School of Mines, Golden, CO 80401, USA

Michael C. Jacobson, Department of Laboratory Medicine, University of Washington, Box 357110, Seattle, WA 98195–7110, USA

Daniel A. Jaffe, University of Washington-Bothell, 22011 26th Avenue SE, Bothell, WA 98021, USA

Richard A. Jahnke, Skidaway Institute of Oceanography, 10 Ocean Science Circle, Savannah, GA 31411, USA

Russell E. McDuff, School of Oceanography, University of Washington, Box 357940, Seattle, WA 98195–7940, USA

David R. Montgomery, Department of Geological Sciences, University of Washington, Box 351310, Seattle, WA 98195–1310, USA

James W. Murray, School of Oceanography, University of Washington, Box 357940, Seattle, WA 98195–7940, USA

Gordon H. Orians, Department of Zoology, University of Washington, Box 351800, Seattle, WA 98195–1800, USA

Henning Rodhe, Meteorologiska Institutionen, Stockholms Universitet, S-106 91, Stockholm, Sweden

Henri Spaltenstein, University of Lausanne, Valentine 18, 1400 Yverdon, Switzerland

James T. Staley, Department of Microbiology, University of Washington, Box 357242, Seattle, WA 98195–7242, USA

Robert F. Stallard, US Geological Survey, 3215 Marine Street, Boulder, CO 80303, USA

Fiorenzo C. Ugolini, Dipartimento di Scienza

del Suolo e Nutrizione della Pianta, Università degli Studi, Piazzale delle Cascine 15, 50144 Firenze, Italy

Gordon V. Wolfe, College of Oceanic and Atmospheric Sciences, 104 Ocean Adminis-

tration Building, Corvallis, OR 97331–5503, USA

Darlene Zabowski, College of Forest Resources, University of Washington, Box 352100, Seattle, WA 98195–2100, USA

Preface to the Second Edition

Nearly 23 years have passed since Bert Bolin visited the University of Washington and lectured on the question, "Can mankind change the composition of the atmosphere?" and it has been seven years since the first edition of this book appeared. Between the mid 1970s and the early 1990s, the study of biogeochemical cycles emerged as a means to integrate large areas of environmental science. The first edition offered this approach in 1992, and has been used widely as both a text and reference book. Its success, measured in the distribution of over 5000 copies, indicates a widespread appreciation for this integrative approach.

The second edition contains virtually all of the material from the first edition in an updated and edited form. The second edition deliberately extends the integrative approach into three new chapters (16, 17, and 18) on the acid–base and redox balances of the Earth, the coupling of biogeochemical cycles and climate, and the paleorecords of environmental chemistry as deduced from studies of ice cores. This new, fourth section of the book thus gives examples of what we call the Earth system approach. Each of these chapters examines an integrative topic. A new chapter has also been added on water and the hydrologic cycle, which was never specifically treated in the first edition. Along with these new chapters, the original chapters covering the sediments and the pedosphere have been merged into a single chapter because of the strong connection between the two sub-jects. Besides the addition of several new co-authors, the list of editors has changed. All of the editors of the first edition – Samuel S. Butcher, Robert J. Charlson, Gordon Orians, and Gordon Wolfe – still appear as co-authors of their respective chapters.

As a consequence of the extension into integration, the title has been changed to *Earth System Science: From Biogeochemical Cycles to Global Change*. However, despite this new title and new chapters, this book is still about fundamental science; it is not issue oriented. This edition conveys the same philosophy as the earlier one, and the first edition preface (following) still conveys these basic principles around which the book is organized. This edition is more clearly divided into four sections, each with its own introductory summary. The reader is urged to read these summaries in order to gain the perspective that we have attempted to present.

Once again, this book would not have been possible without the contributions of the chapter authors and the very important work of those who prepared the manuscript and illustrations. Much of the typing was done again by Sheila Parker, and Kay Dewar prepared all of the new figures. Michele Kruegel and Monte Lapka provided additional graphics and computer support. A generous gift from the Ford Corporation allowed Michael Jacobson the needed time as a post-doctoral scientist to perform the tasks of chief editor.

Preface to the First Edition

Human activity is affecting the global environment in a profound way. Some of these changes are due to high rates of additions of materials to the environment. Other changes result from losses of habitat and the associated extinctions of species. Many of these human interventions now occur on a scale capable of changing the global biogeochemical cycles upon which life and the Earth's climate depend.

Biogeochemical cycles describe the transformation and movement of chemical substances in a global context. This text is designed for courses intended to present an integrated perspective on biogeochemical cycles. Courses focusing on this subject are offered at advanced undergraduate and graduate levels in many colleges and universities. These courses are usually presented by a person with a specialty in one of the conventional scientific disciplines, supplemented by guest lecturers and readings in other areas. Our goal has been to provide a comprehensive treatment under one cover so that the components are integrated and the need for additional reading is reduced.

The text has its roots in courses on biogeochemical cycles offered at the University of Washington and at the University of Stockholm. The course at the University of Washington was started by two of the authors (Charlson and Murray), and an essential part of the course has been visits by faculty from other disciplines. Many of the chapters in this text spring from materials prepared for those presentations. Some of the authors are former students in this course.

Much of the work important to the study of biogeochemical cycles is done in traditional disciplines – ranging from astronomy to zoology. Many disciplines that have developed fairly recently (such as chemical oceanography) also play very important roles. This is likely to continue to be the case. Nonetheless, given the nature of biogeochemistry, specialists need to understand what their disciplines can bring to the subject and what are the needs of the other disciplines. To fully comprehend these cycles, a person must also integrate material from several disciplines.

Although our goal has been to be comprehensive, adjustments have had to be made. In managing the compromise between depth of coverage and maintaining a reasonable size for a textbook, many topics are given only limited space. We hope that readers will nevertheless gain an appreciation of the scope of biogeochemical cycles and will be adequately prepared to understand the growing literature in the field.

This book is about fundamental aspects of the science of biogeochemistry. As such, and while it is relevant to the major issues of global change, it is not issue oriented. Not does this book attempt to review all of the research on these topics. It does, however, emphasize fundamental aspects of the physical, chemical, biological, and Earth sciences that are of lasting importance for integrative studies of the Earth.

We assume that our readers have a background in science attainable by completing a university level course in introductory chemistry. We also expect our readers to be involved in one of the disciplines integral to the study of biogeochemical cycles. This includes appropriate subdisciplines of chemistry, biology, and geology, and the sciences that deal with soils, atmospheres, and oceans.

Bert Bolin's visit to the University of Washington in 1976 provided a major stimulus for thinking about biogeochemical cycles at the university. Active work on this text began with a grant from the Rockefeller Foundation to the Institute for Environmental Studies at the University of Washington in 1978. Rockefeller assistance made it possible to bring several scientists to the University of Washington to discuss the role of their specialty in biogeo-chemical cycles. Visits from M. Alexander, P. L. Brezonik, P. J. Crutzen, R. A. Duce, R. O. Hall-berg, H. D. Holland, M. L. Jackson, G. E. Likens, F. T. Mackenzie, S. Odén, H. Rodhe, and H. J. Simpson played important roles in shaping our approach to the text.

Several individuals played important roles during the final preparation of the manuscript. Most of the typing was done by Sheila Parker. Drafting of figures was done by Kay Dewar and April Ryan. Last but not least, we owe thanks to the many students at the University of Washington and the University of Stockholm who explored this subject initially without a textbook and then with draft chapters. Their enthusiasm for the subject and their comments and criticism have helped maintain our interest in this manuscript over almost a decade.

Part One

Basic Concepts for Earth System Science

The first part of this book is designed to provide background information that is useful when studying the rest of the text.

Chapter 1 is an introduction to the emerging discipline of Earth system science, covering the history and philosophy of this field. The key to understanding this book and this field is to first realize that the Earth is materially a closed, dynamic system. Because the Earth is not a closed system with respect to energy (i.e., solar radiation), there is a constant cycling of the elements through the various parts of the planet. The movement and transformation of major elements (C, N, S, P, and trace metals) are described individually, using the concept of *biogeochemical cycles*, which describe the flux of material in and out of the various *geospheres* (atmosphere, hydrosphere, pedosphere, lithosphere) and the chemical and physical transformations that occur there. Each cycle is studied more or less individually, to allow a way of simplifying the Earth system into a series of smaller, more manageable subsystems. Humans have modified the natural cycling of these elements in a variety of ways, which has lead to chemical and physical changes on Earth on the planetary scale. Using biogeochemical cycles to study the Earth system is one of the major themes of this book.

In order to understand the Earth's character as a planet, it also is helpful to have an understanding of how the elements in our solar system were formed. Chapter 2 starts with the Big Bang theory and continues with how very small grains eventually came together and accreted to form the beginnings of what would eventually become the Earth and other planets, about 4.5×10^9 years ago (4.5 Gyr). The initial processes of the Earth's evolution involved heat generated from radioactive decay and kinetic energy of projectiles impacting the infant planet, causing chemical differentiation of the Earth's interior into a core, mantle, and crust. As we will see in Part Two of the book, the movement of tectonic plates that arose because of this differentiation is necessary for the continuation of life and the biogeochemical cycling of the elements.

Probably the most important characteristic of Earth from a human perspective is its abundant life. It is the only planet we know of that supports a *biosphere*. The distribution of the elements on the planet were initially controlled by the physical and chemical processes that are described in Chapter 2, but biological processes have been at work in affecting chemical dispersal ever since life first appeared about 3.5×10^9 years ago (3.5 Gyr). In Chapter 3 we investigate the beginnings and evolution of life on Earth, as well as the principal biochemical systems which affect distribution of the elements. The diversity of life on the planet is due to a large number of factors, especially evolutionary agents that can change a population of organisms over time. Evolution of species can alter the ways in which the organisms consume nutrients and energy, which eventually impacts the surrounding environment by altering the elemental cycles. These environmental changes can themselves act as evolutionary agents and affect the genetic code of future generations. It is important for us to have a basic understanding of these biological processes so we understand their effect on the rest of the system, and how the Earth came to be in the state it is in today. Since we, as humans, are also part of the biosphere, we also have a personal interest in how changes to the Earth system affect biological processes on the planet.

Earth System Science
ISBN 0-12-379370-X

The ultimate goal in studying Earth systems is to understand the system well enough that we can explain past changes (e.g., why did the ice ages start?) and predict the future of the system (e.g., what will the concentration of atmospheric CO_2 be in 100 years?). This is possible only if the system can be accurately modeled. Armed with the material in Chapter 4, the reader should be able to develop a feeling for how robust a model prediction presented in the popular press is likely to be. Modeling also provides a way of evaluating what remains poorly understood in a system. Since modeling is such an integral part of studying Earth systems, and biogeochemical cycles in particular, we have included Chapter 4 as a basic modeling primer. The simplest types of models are box models, which describe movement of material between compartments or reservoirs. These models are introduced in this chapter, and used again in Part Three, which covers the individual elemental cycles. This chapter also introduces basic concepts such as residence time, turnover time, response time, and steady-state conditions, which are fundamental in discussing a biogeochemical cycle. It also provides an inventory of the basic information needed about global systems in order to be useful. The recent development of fast computers has dramatically expanded our ability to use the modeling concepts presented in this chapter.

The last chapter in this introductory part covers the basic physical chemistry that is required for using the rest of the book. The main ideas of this chapter relate to basic thermodynamics and kinetics. The thermodynamic conditions determine whether a reaction will occur spontaneously, and if so whether the reaction releases energy and how much of the products are produced compared to the amount of reactants once the system reaches thermodynamic equilibrium. Kinetics, on the other hand, determine how fast a reaction occurs if it is thermodynamically favorable. In the natural environment, we have systems for which reactions would be thermodynamically favorable, but the kinetics are so slow that the system remains in a state of perpetual disequilibrium. A good example of one such system is our atmosphere, as is also covered later in Chapter 7. As part of the presentation of thermodynamics, a section on oxidation–reduction (redox) is included in this chapter. This is meant primarily as preparation for Chapter 16, but it is important to keep this material in mind for the rest of the book as well, since redox reactions are responsible for many of the elemental transitions in biogeochemical cycles.

1

Introduction: Biogeochemical Cycles as Fundamental Constructs for Studying Earth System Science and Global Change

Michael C. Jacobson, Robert J. Charlson, and Henning Rodhe

1.1 Introduction

The latter part of the 20th century has seen remarkable advances in science and technology. Accomplishments in biochemistry and medicine, computer technology, and telecommunications have benefited nearly everyone on Earth to one degree or another. Along with these advances that have improved our quality of life, scientific research into the study of the Earth has revealed a planetary system that is more complex and dynamic than anyone would have imagined even 50 years ago. The Earth and the environment have become one of society's greatest concerns, perhaps as the result of these discoveries combined with the quick dissemination of information that is now possible with modern telecommunications.

The basis of most environmental issues is pollution. But what is pollution? Keep in mind that with very minor exceptions, virtually *all* of the atoms in the solid, liquid, and gaseous parts of the Earth have been a part of the planet for *all* of its approximately 4.5 billion years of existence. Very few of these atoms have changed (i.e., by radioactive decay) or departed to space.

This includes all of the atoms in your own body and in all other living things, which have also been permanent residents of the Earth through the eons. This means that the Earth is an essentially closed system with respect to atomic matter, and is therefore governed by the law of conservation of mass. This law dictates that all of the Earth's molecules must be made of the same aggregation of atoms even though molecular forms may vary, evolve, and be transported within and around the planetary system. Pollution, therefore, is a human-induced change in the distribution of atoms from one place on Earth to another.

In order to understand the impact of pollution on Earth, we must realize that the planet itself is not stagnant, but continually moving material around the system naturally. Any human (*anthropogenic*) redistribution in the elements is superimposed on these continuous natural events. Energy from the sun and radioactive decay from the Earth's interior drive these processes, which are often *cyclic* in nature. As a result, almost all of the rocks composing the continents have been processed at least once through a chemical and physical cycle involving

Earth System Science
ISBN 0-12-379370-X

weathering, formation of sediments, and sub-duction, being subjected to great heat and pressure to produce new igneous rocks. The water in the oceans has been evaporated, rained out, and returned via rivers and groundwater flow many tens of thousands of times. The main gases in the atmosphere (nitrogen and oxygen) are cycled frequently through living organisms. The combined effect of these dynamic transports and transformations is a planet that is in a state of continual physical, chemical and biological evolution. A bird's eye, cartoon view of the dynamic Earth system is shown in Plate 1. This book is about putting together all of the different dynamic parts of this figure into an understandable, coordinated picture. In the last chapter of the book, we will revisit the topic of human modification of the system in detail.

1.1.1 Biogeochemical Cycles and Geospheres

Aside from the cyclic systems listed above, there is a complementary set of *chemical* cycles that we can describe for each of the most important biological elements (carbon, nitrogen, oxygen, sulfur, and the trace metals). These *biogeochemical cycles* are descriptions of the transport and transformation of the elements through various segments of the Earth system, called *geospheres*. We use these constructs to compartmentalize the larger Earth system into more manageable, chemically definable parts.

What are the geospheres? One of them is easily definable and requires no special introduction: the *atmosphere* is the gas-phase envelope surrounding the globe of the Earth. Another geosphere is the *hydrosphere*, which includes all of the oceans, and freshwater bodies of water on the planet. The *lithosphere* is the entirety of rocks on Earth, including rocks exposed to the atmosphere, under the waters of the hydrosphere, and the entire interior parts of the planet. The *pedosphere* (literally that upon which we walk) comprises the soils of the Earth. The geospheres listed thus far are more or less geographically definable, but there is a geosphere that can exist *within* all of the other geospheres: the *biosphere*, which is the collection

of the biota (all living things) on the planet. The interfaces between the geospheres are often fuzzy and difficult to define. For example, ocean sediments contain water as well as rock and organic material; it is difficult to say exactly where the hydrosphere ends and the lithosphere starts. Part of the hydrosphere exists in the atmosphere as rain and cloud droplets.

The constant transport of material within and through the geospheres is powered by the sun and by the heat of the Earth's interior. A simple diagram of these geospheric concepts and the energy that moves material within them is presented in Fig. 1-1. The result of the interactions shown in Plate 1 and Fig. 1-1 is an Earth system that is complex, coupled, and evolving.

In addition to the natural evolution of the interacting geospheres, human activities have brought about an entirely new set of perturbations to the system. Because many political and social issues surround the problem of human induced *global change*, there are both basic and applied scientific motivations to study biogeochemical cycles and their roles in the Earth system. The need for development and application of basic science to the broad policy issues of dealing with global change have inspired the formation of a new integrative scientific discipline, *Earth system science* (NASA, 1986).

The subject offers a number of challenges that are important for the scientific community to address. Probably the largest challenge is integrating knowledge and material from many disciplines. This is a major theme of this chapter and of this book, as will be seen in the sections that follow. If the scientific community is not able to integrate the science necessary to describe biogeochemical systems, it seems unlikely that it will be easy for society to derive solutions for the problems raised by global change.

The principal obstacles facing us as scientists studying Earth system science are the finite resources of most educational institutions. Development of this subject requires that we think of novel ways to do interdisciplinary work in a setting dominated by traditional disciplines. Although we can draw heavily on work being done in recently formed disciplines such as chemical oceanography, stable isotope geo-

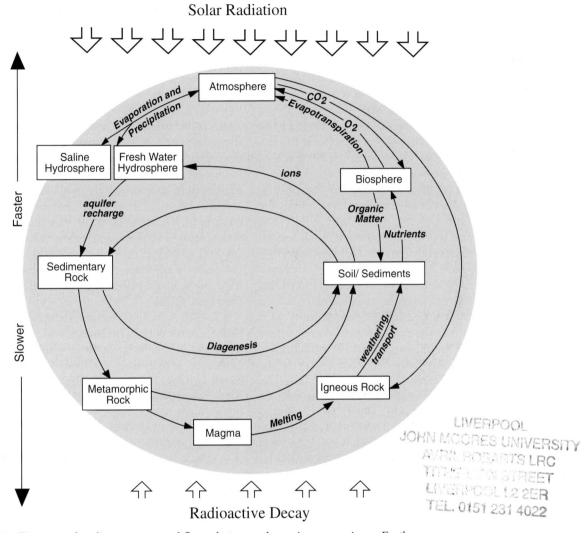

Fig. 1-1 Diagram of cyclic processes and fluxes between the major reservoirs on Earth.

chemistry, and atmospheric chemistry, we may be able to glean some clues in how to accomplish our goal of integrating the disciplines by examining the history of Earth system study. Indeed, the recognition that biogeochemical issues may be significant for mankind goes back several hundred years.

1.2 History

Some of the earliest work in the study of biogeochemical cycles and their role in the physical functioning of the planet was by James Hutton (1788), who viewed the Earth as a "superorganism, and that its proper study should be by physiology." More than 100 years later, a classic paper by Svante Arrhenius (1896) appeared, called "On the influence of carbonic acid in the air on the temperature of the ground." This work provided a paradigm for quantitatively connecting the greenhouse effect of carbon dioxide to climate, as well as to the global biogeochemical cycle of carbon. A truly meaningful study of these issues clearly requires input from nearly all of the natural sciences (chemistry, physics, biology, geology, meteorology, etc.). These scientific disciplines, which were still in their early formative stages when Arrhenius' pioneering work emerged, have since evolved over the past century into highly refined and useful, but largely separated entities. Accordingly, most scientists have adopted a *reductionist* approach to biogeochemistry (i.e., simplifying large scientific problems into smaller parts to be examined by an individual discipline). In contrast, as we stated above, one of the goals of Earth system science (referred to early on as *natural philosophy*) is to integrate the natural sciences to strive towards understanding the entire system.

In addition to the work of Arrhenius, several other scientists have made notable contributions that have helped to mold the approach and content of Earth system science (and of this book). The term *biosphere* was originally coined by the Austrian geologist Eduard Suess as a way of defining the parts of the Earth's surface that support life (Suess, 1875). The global view portrayed by Arrhenius' lengthy quote of the paper on the carbon cycle by Arvid Högbom set the tone for coupling the biosphere to geochemistry (Arrhenius, 1896). This coupling was strongly emphasized and promoted by the Russian mineralogist Vladimir Ivanovich Vernadsky, who published a series of lectures on the subject, first in Russian (1926), then in French (1929), and recently translated into English (1997). In these writings, Vernadsky developed an integrative definition of the *biosphere* as all living things and everything connected to them. He strongly believed that the Earth is a set of connected parts that can only be studied in an indivisible, holistic way. Vernadsky went so far as to suggest that geologic phenomena at the Earth's surface are inherently caused by life itself, so his definition of the biosphere includes all of the Earth's geologic features. Vernadsky's teachings remained relatively obscure until Hutchinson (1970) popularized them in a famous article in *Scientific American*. Vernadsky came to be called the father of modern biogeochemistry by James Lovelock (1972), who went on to suggest that feedbacks in the biosphere–climatic system lead to *homeostasis* of basic Earth processes. Lovelock called this new feed-

back-based integrative science *geophysiology* and the system itself *Gaia*, which is a Greek word meaning *Mother Earth*.

This view of the coupled nature of the Earth system has not dominated the historical development of the key disciplines of the natural sciences. Many evolutionary biologists have viewed the changing physical climate of the Earth as an externally imposed factor to which the biosphere must adapt. Likewise, many geologists and geophysicists have viewed the evolution of the planet as being governed primarily or exclusively by chemical and physical processes. Under this perspective, free oxygen in the atmosphere is viewed as a constant factor. The disparate views of the biological and geologic communities have coexisted for nearly a century, and (with few exceptions) their merging has been controversial (see, e.g., Dawkins, 1976).

The *Gaia hypothesis* was put forward by Lovelock together with Lynn Margulis (Lovelock and Margulis, 1974) to provide a single scientific basis for integrating all components of the Earth system. This theory suggests that Earth's biota as well as the planet itself are parts of a quasi-living entity that "has a capacity for homeostasis" (or self-regulation) (Lovelock, 1986). The earliest version of Lovelock's Gaia hypothesis contained phrases like "by and for the biosphere," which implied a sense of purposefulness on the part of the biota to evolve in ways that would suit its own continued existence. As the Gaia hypothesis has evolved, the interdependence of the evolution of biota and geophysical/geochemical systems is described in non-teleological terms. Lovelock (1991) himself recently stated what seems obvious: "In no way do organisms simply 'adapt' to a dead world determined by physics and chemistry alone. They live in a world that is the breath and bones of their ancestors and that they are now sustaining."

1.3 Evidence for the Coupled Nature of the Earth System

The biosphere is ultimately what ties the major systems of the Earth together. Studying the fundamental differences between a planet full of life (the Earth) versus one that lacks life (e.g., Mars) evidences this. Dead planets are in or near a state of perpetual thermodynamic equilibrium, but as pointed out by Lovelock (1972), the spheres of the Earth and the chemical constituents within them are far out of equilibrium in a thermodynamic sense, as evidenced by the following points:

- O_2, N_2, and H_2O are the main molecular forms coexisting in the atmosphere, but the condition of thermodynamic equilibrium would require that HNO_3 be formed from these gases and subsequently dissolve in the oceans.
- O_2 coexists with combustible biomass in plants.
- Acidic materials in the atmosphere (e.g. CO_2 and H_2CO_3, SO_2 and H_2SO_3, HNO_3 and so on) coexist with basic materials in both igneous and sedimentary rocks (e.g. FeS_2 and $CaCO_3$).

We cover each of these types of examples in separate chapters of this book, but there is a clear connection as well. In all of these examples, the main factor that maintains thermodynamic disequilibrium is the living biosphere. Without the biosphere, some abiotic photochemical reactions would proceed, as would reactions associated with volcanism. But without the continuous production of oxygen in photosynthesis, various oxidation processes (e.g., with reduced organic matter at the Earth's surface, reduced sulfur or iron compounds in rocks and sediments) would consume free O_2 and move the atmosphere towards thermodynamic equilibrium. The present-day chemical functioning of the planet is thus intimately tied to the biosphere.

All living organisms require at least one mobile phase (gas or liquid) in order to exist. Life on Earth as we know it would be impossible without the involvement of the liquid phase of water. The gas phase is necessary for life forms that consume gaseous substances or that produce gaseous waste products. Hence, the very functioning of the biosphere implicitly depends on the existence of the mobile atmosphere and hydrosphere, both of which are in intimate contact with the solid phases of the planet and exchange substances with them. Since the atmosphere and hydrosphere are distributed globally and because they each are mixed on large, often global spatial scales, the chemical influences of the biosphere are evident everywhere on Earth.

A rich base of empirical evidence convincingly illustrates the complexity of the couplings within the Earth system. One of the most important pieces of evidence of this interconnectedness is the chronology provided by chemical and isotopic analysis of ice cores from the deepest, oldest ice on the planet. The Vostok ice core (Lorius *et al.*, 1985; Saltzman, pers. comm., 1998) currently yields information covering more than 400 000 years, from the current interglacial time, through the most recent ice age, through the previous interglacial time and back to an earlier ice age. Figure 1-2 shows plots of several chemical variables against time. The data are derived from detailed analyses of the chemical composition of the ice. These data will be discussed in more detail in Chapter 17, but even a cursory examination reveals features that indicate regulation of the system by the biosphere:

- The isotopically inferred temperature shows two relatively stable climatic states: ice ages and interglacial periods. The temperature during the current interglacial is about the same as during the past one. Likewise, the temperature during the coldest part of the last ice age was about the same as the minimum temperature during the previous glacial period.
- CO_2 is higher in interglacial times and lower in ice ages, indicating significant differences in photosynthesis between these two climatic states.
- Sulfate and methane sulfonic acid (MSA) (both of which are produced primarily by photosynthetic algae in the ocean) are *low* during interglacials and *high* in ice ages, opposite to the CO_2 trend in Antarctica (although not in Greenland).

These chemical constituents all vary in synchrony and two climatic states as defined by temperature coincide with the "climatic states"

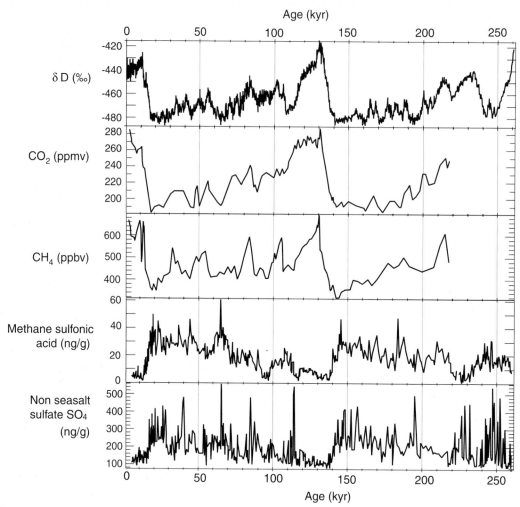

Fig. 1-2 Chemical data from the Vostok ice core. The graph of δD can be taken as a proxy for temperature changes, as described in Chapter 18. CO_2 and CH_4 are greenhouse gases and vary in the same direction as temperature. Non-seasalt sulfate and methane sulfonic acid are both sulfur species existing in the particle phase, and are positively correlated with each other, but negatively with T. Major variations for all of these variables seem to correlate either positively or negatively with each other, indicating a coupled system. δD, non-seasalt sulfate, and methane sulfonic acid data kindly provided by Dr Eric Saltzman. CO_2 data are from Barnola *et al.* (1987) and Jouzel *et al.* (1993). CH_4 data are from Chappellaz *et al.* (1990) and Jouzel *et al.* (1993). (ppmv = parts per million by volume; ppbv = parts per billion by volime)

as defined by variation in the chemical variables. Although these correlations and anti-correlations do not *prove* that the Earth functions as an integrated biogeochemical system, it does strongly suggest that the many subsystems involved are coupled. Furthermore, the presence of the quasi "set points" of the temperature and chemical variables suggests the possible existence of feedbacks that steer

the Earth system into one or another preferred climatic and biogeochemical state. Currently, no viable alternative hypothesis has been advanced to explain these correlations. To unravel the detailed functioning of the entire Earth system, we must first establish a basis for understanding the individual "spheres," and the individual biogeochemical cycles that are involved.

1.4 Philosophy of Using the Cycle Approach to Describe Natural Systems on Earth

Viewing the Earth system as a set of coupled biogeochemical cycles that both depend on and influence the climate allows us to conceptually simplify the movement of material on Earth and its couplings to climate. Much of this book is presented using this approach, giving *budgets* for the *flux* of material into and out of various *reservoirs*. Take for instance the circulation of water between the oceans, atmosphere, and continents. In this example, the reservoirs would be "the oceans," "the water in the atmosphere," "the ground water," etc. Using the most basic description of the cyclic processes that take place, we can mathematically and quantitatively model these cycles to describe and predict the distribution of the important chemical constituents of the planet. The fundamental concepts that govern modeling of biogeochemical cycles are covered in Chapter 4, but we introduce some of the main ideas here, since these concepts and definitions are useful for the entire text.

A basic goal of the cycle approach is to determine how the fluxes between the reservoirs depend on the content of the reservoirs and on other external factors. In many cases the details of the distribution of an element within each reservoir are disregarded, and for the most simplified calculations, the amounts of material in each reservoir are assumed to remain constant (i.e., there is a condition of *steady state*). This allows a chemical *budget* to be defined for the entire cycle.

A steady-state, flux-based approach to describing the physical–chemical environment on Earth has advantages as well as disadvantages. Some advantages are that:

- It provides an overview of fluxes and reservoir contents.
- It gives a basis for quantitative modeling.
- It helps to estimate the relative magnitudes of anthropogenic and natural fluxes.
- It stimulates questions such as, "Where is the material coming from? Where is it going next?"
- It helps to identify gaps in knowledge.

There are, however, some disadvantages:

- The analysis is by necessity superficial. It provides little or no insight into what goes on inside the reservoirs or into the nature of the fluxes between them.
- It gives a false impression of certainty. Typically at least one of the fluxes in a cycle is calculated by the imposed mathematical necessity of balancing a steady-state budget. Such estimates may erroneously be taken to represent solid knowledge.
- The analysis is based on averaged quantities that cannot always be easily measured because of spatial variation and other complicating factors.

Also, many important geophysical problems cannot be studied using a simplified cycle approach. Weather forecasting, for example, requires a detailed knowledge about the distribution of winds, temperature, etc. within the atmosphere. Weather cannot be forecast using a reservoir model with the atmosphere as one of the reservoirs. It would not be much better even if the atmosphere were divided into several reservoirs. A forecast model requires a resolution fine enough to resolve explicitly the structure of the most important weather phenomena such as cyclones, anticyclones, and wave patterns on spatial scales as small as a few hundred kilometers. Spatially explicit models are based either on a division of the physical space into a large but finite number of grid cubes (grid point models) or on a separation of the variables into different wave numbers (spectral models).

1.5 Reservoir Models and Cycles – Some Definitions

The models used to study biogeochemical cycles are described by a set of terms whose definitions must be clearly understood at the outset. We define them here as they are used throughout the book.

Reservoir (box, compartment). An amount of material defined by certain physical, chemical or biological characteristics that, for the purposes of analysis we consider to be reasonably homogeneous. Examples:

- Oxygen in the atmosphere.
- Carbon monoxide in the southern hemisphere.
- Carbon in living organic matter in the ocean surface layer.
- Amount of ocean water having a density between ρ_1 and ρ_2.
- Sulfur in sedimentary rocks.

If the reservoir is defined by its physical boundaries, the content of the specific element is called its *burden*. We will denote the content of a reservoir by M. The dimension of M would normally be mass, although it could also be, e.g., moles.

Flux. The amount of material transferred from one reservoir to another per unit time, in general denoted by F (mass per time). Examples:

- The rate of evaporation of water from the ocean surface to the atmosphere.
- The rate of oxidation of N_2O in the stratosphere (i.e., flux from the atmospheric N_2O–nitrogen reservoir to the stratospheric NO_x–nitrogen reservoir).
- The rate of deposition of phosphorus on marine sediments.

In more specific studies of transport processes, flux is normally defined as the amount of material transferred per unit area per unit time. To distinguish between these two conflicting definitions, we refer to the latter as "flux density."

Source. A flux (Q) of material *into* a reservoir.

Sink. A flux (S) of material *out of* a reservoir – very often this flux is assumed to be proportional to the content of the reservoir ($S = kM$). In such cases the sink flux is referred to as a first-order process. If the sink flux is constant, independent on the reservoir content, the process is of zero order. Higher-order fluxes, i.e. $S = kM^\alpha$ with $\alpha > 1$, also occur.

Budget. A balance sheet of all sources and sinks of a reservoir. If sources and sinks balance and do not change with time, the reservoir is in

steady state, i.e., M does not change with time. Usually some fluxes are better known than others. If steady state prevails, a flux that is unknown *a priori* can be estimated by difference from the other fluxes. If this is done, it should be made very clear in the presentation of the budget which of the fluxes is estimated as a difference.

Turnover time. The turnover time of a reservoir is the ratio of the content M of the reservoir to the sum of its sinks S or the ratio of M to the sources Q. The turnover time is the time it will take to empty the reservoir in the absence of sources if the sinks remain constant. It is also a measure of the average time spent by individual molecules or atoms in the reservoir (more about this will be presented in Chapter 4).

Cycle. A system consisting of two or more connected reservoirs, where a large part of the material is transferred through the system in a cyclic fashion. If all material cycles *within* the system, the system is *closed*. Many systems of connected reservoirs are not cyclic, but instead material flows unidirectionally. In such systems some reservoirs (at the end of the chain) may be *accumulative*, whereas others remain balanced (*nonaccumulative*); cf. Holland (1978).

Biogeochemical cycle. As discussed early in the chapter, this term describes the global or regional cycles of the "life elements" C, N, S, and P with reservoirs including the whole or part of the atmosphere, the ocean, the sediments, and the living organisms. The term can be applied to the corresponding cycles of other elements or compounds.

Budgets and cycles can be considered on very different *spatial scales*. In this book we concentrate on global, hemispheric and regional scales. The choice of a suitable scale (i.e. the size of the reservoirs), is determined by the goals of the analysis as well as by the homogeneity of the spatial distribution. For example, in carbon cycle models it is reasonable to consider the atmosphere as one reservoir (the concentration of CO_2 in the atmosphere is fairly uniform). On the other hand, oceanic carbon content and carbon exchange processes exhibit large spatial variations and it is reasonable to separate the surface layer from the deeper layers, the Atlantic from the Pacific, etc. Many sulfur and nitrogen compounds in the atmosphere occur in very different concentrations in different regions of the world. For these compounds regional budgets tell us more about local dynamics than do global budgets.

1.6 The Philosophy of Integration as a Basis for Understanding the Earth System

1.6.1 *Recognizing the Interconnected Nature of the Earth System*

Although it is clearly established that it is necessary to study all of the geospheres and all of the biogeochemical cycles in order to understand and predict the workings of the entire Earth system, it is not sufficient to stop there. The functioning of the biosphere and each of the individual physical spheres of the planet involves continuous and strong interactions, making all parts of the Earth system dependent to some degree on all the other parts. Earth system science attempts to understand current conditions by extending analyses backwards and forward in time – to include the earliest stages of the evolution of life on Earth, as well as projections of human-induced global change in the future.

The task of integrating the spheres and the biogeochemical cycles emerges as a necessary if daunting challenge. Disciplinary science has provided little in the way of precedent for us, although substantial guidance is provided by the pioneers like Arrhenius, Vernadsky, and Lovelock who presaged these global developments. Another major contributor to this kind of thinking is the field of *systems analysis* or *systems engineering*, and the field of cybernetics, though these lack any degree of focus on the Earth system.

1.6.2 *Examples of Integration of Global Systems*

In the chapters that follow, we cannot provide a complete and unified description for the integra-tion of the entire system; however, we can provide a set of examples of integrated subsystems that will illustrate the nature of the interactions that constitute major pieces of the whole. Key features of the process of integrating natural-science fundamentals are (1) some degree of globality or global applicability, (2) interaction of two or more biogeochemical cycles, and (3) coupling of the interacting geospheres. While there are many such subsystems, we will mention a few of them to which we will return in the closing chapters. These examples of integrated subsystems also introduce and maintain a three-way focus on one or more geosphere(s), the biogeochemical cycles, and on systems integration. While these topics might appear to be non-parallel, they share the key feature of an integrative global view.

1.6.2.1 *The hydrologic cycle*

Far from being just the processing of water on Earth, this cycle is the basis for a wide range of meteorologic, geochemical, and biological systems. Water is the transport medium for all nutrients in the biosphere. Water vapor condensed into clouds is the chief control on planetary albedo. The cycling of water is also one of the major mechanisms for the transportation of sensible heat (e.g. in oceanic circulation) and latent heat that is released when water falls from the air.

1.6.2.2 *Acid–base and oxidation–reduction systems*

It is often taken for granted that the oxygen content of the air is nearly constant at ca. 20% of the atmospheric volume, that most of the liquid water on the planet is aerobic (i.e. contains O_2), and that most water has pH values relatively close to "neutral" (close to 7). However, these circumstances are not mere coincidences but are in fact consequences of the interaction of key global biogeochemical cycles. For instance, the pH of rainwater is often determined by the relative amounts of ammonia and sulfuric acid cycled through the atmosphere, a clear example of interaction between the nitrogen and sulfur cycles.

1.6.2.3   Ice-age/interglacial climatic flip-flops

Although these are often viewed as being merely changes in physical climate as indexed by temperature, they really represent changes in a large range of biogeochemical phenomena. The rich and rapidly growing body of data from ice cores and sediments strongly supports the notion that the Earth functions as a coupled system.

1.6.2.4   The climate system with biogeochemical feedbacks

Climate is often viewed as the aggregate of all of the elements of weather, with quantitative definitions being purely physical. However, because of couplings of carbon dioxide and many other atmospheric species to both physical climate and to the biosphere, the stability of the climate system depends in principle on the nature of feedbacks involving the biosphere. For example, the notion that sulfate particles originating from the oxidation of dimethylsulfide emitted by marine phytoplankton can affect the *albedo* (reflectivity) of clouds (Charlson *et al.*, 1987). At this point these feedbacks are mostly unidentified, and poorly quantified.

1.6.2.5   Anthropogenic modification of the Earth system

Again, the myriad influences of human activity are usually viewed as separate effects (global warming, acid rain, ozone loss, urban pollution, etc.) However, these individual symptoms clearly have major interdependencies that must be understood if humans are to learn how to coexist with a stable Earth system.

1.7   The Limitations and Challenges of Understanding Earth Systems

Earth system science is a young science with great potential, but we must exercise caution in not overlooking important details of traditional, disciplinary science in our attempt to develop this new and integrative science. The foundation upon which we will proceed in this book is to provide the basic disciplinary components, starting with fundamental concepts of modeling, Earth science, biology, and chemistry. Having reviewed these basic scientific building blocks we move on to a survey the biogeochemical cycles of key elements. Following this, we present a set of integrative topics.

The three user groups of this book (students, teachers, and researchers) will all discover that the challenge of understanding Earth system science is at least as much of a problem of integration of well developed fundamental fields into a global context as it is to refine the disciplinary pieces themselves. Users who are practiced in a traditional field utilizing reductionism will need to expand their thinking into other disciplines as well as into the problem of how to combine their own field into the larger picture. In doing so, they will find an opportunity to extend the scope of their own discipline towards and into the global context.

References

Arrhenius, S. (1896). On the influence of carbonic acid in the air upon the temperature of the ground. *Philosophical Magazine and Journal of Science* S.5, Vol 4, No 251, p. 237 ff.

Barnola, J. M., Raynaud, D., Korotkevich, Y. S., and Lorius, C. (1987). Vostok ice core provides 160 000 year record of atmospheric CO_2. *Nature* **329**, 408–413.

Chappellaz, J., Barnola, J. M., Raynaud, D., Korotkevich, Y. S., and Lorius, C. (1990). Atmospheric methane record over the last climatic cycle revealed by the Vostok ice core. *Nature* **345**, 127–131.

Charlson, R. J., Lovelock, J. E., Andreae, M. O., and Warren, S. G. (1987). Oceanic phytoplankton, atmospheric sulphur, cloud albedo and climate. *Nature* **326**, 655–661.

Dawkins, R. (1976). "The Selfish Gene." Oxford University Press, New York.

Holland, H. D. (1978). "The Chemistry of the Atmosphere and Oceans." p. 351. Wiley-Interscience, New York.

Hutchinson, G. E. (1970). The Biosphere. *Scient. Am.* September, 45–53.

Hutton, J. (1788). "Theory of the Earth; or an investigation of the laws observable in the composition, dissolution and restoration of land upon the globe." *R. Soc. Edin. Trans.* **1**, 209–304.

Jouzel, J., Lorius, J. R., Petit, C. *et al.* (1993). Vostok ice-core – a continuous isotope temperature record over the last climatic cycle (160 000 years). *Nature* **329**, 403–408.

Lovelock, J. E. (1972). Gaia as seen through the atmosphere. *Atmos. Environ.* **6**, 452–453.

Lovelock, J. E. and Margulis, M. (1974). Atmospheric homeostasis by and for the biosphere. *Tellus* **26**, 1–10.

Lovelock, J. E. (1986). ''The Biosphere.'' *New Scient.* July 17, 51.

Lovelock, J. E. (1991). Geophysiology – the science of Gaia. *In* ''Scientists on Gaia'' (S. A. Schneider and P. J. Boston, eds), pp. 3–10. MIT Press, Cambridge, MA.

Lorius, C., Jouzel, J., Ritz, C., Merlivat, L., Barkov, N. I., Korotkevich, Y. S., and Kotlyakov, V. M. (1985). A 150 000-year climatic record from Antarctic ice. *Nature* **316**, 591–596.

NASA (1986). ''Earth System Science – Overview – A Program for Global Change.'' Earth System Sciences Committee, NASA Advisory Council, Washington, DC.

Suess, E. (1875). ''Die Enstehung der Alpen'' (The Origin of the Alps). W. Braunmüller, Vienna.

Vernadsky, V. I. (1997). ''The Biosphere.'' (D. B. Langmuir, transl.; revised and annotated by M. A. S. McMenamin), Copernicus Books, New York.

2

The Origin and Early Evolution of the Earth

D. E. Brownlee

2.1 Introduction

The subject of this book is the Earth, and its cycles and processes. The specific focus is the Earth in its present state, but to understand how it came to this point and predict how it may change in the future, it is important to examine the full history of the Earth. We need to examine its origin and evolution as a planetary body. Advances in planetary science, including models and abundant new data provide a valuable framework for understanding fundamental aspects of development of the Earth. The recent discovery of planets outside the solar system extends the scope of our understanding of planetary processes and properties. Many of the Earth's properties are taken for granted as simply being "natural," but comparison with other planets and models of planetary evolution indicates that the character of the Earth is actually quite odd. The unusual properties include life, an ocean, "moderate" surface temperature and the presence of plate tectonics. We will discuss the origin and evolution of the Earth to provide a planetary perspective and insight into some of the fundamental Earth processes. We will start with the origin of the elements that the Earth and solar system formed from.

2.2 Pre-Solar Evolution: The Origin of the Elements

The Earth is the end product of a series of evolutionary processes that began at the time the first elements were produced, 15 billion years ago. The Earth's composition, its evolution and many of its chemical and physical cycles described in this book are influenced to various degrees by processes that occurred during or before the formation of the solar system. These processes include nuclear reactions to produce the elements, gravitational collapse to produce stars and protoplanetary systems, condensation to produce solid grains, and accretion to accumulate grains into planets. The basic framework of this scheme is believed to be generally well understood, although many of the details of even the most fundamental processes such as condensation and accretion are highly uncertain.

The composition of the Earth was determined both by the chemical composition of the solar nebula, from which the sun and planets formed, and by the nature of the physical processes that concentrated materials to form planets. The bulk elemental and isotopic composition of the nebula is believed, or usually assumed to be identical to that of the sun. The few exceptions to this include elements and isotopes such as lithium and deuterium that are destroyed in the bulk of the sun's interior by nuclear reactions. The composition of the sun as determined by optical spectroscopy is similar to the majority of stars in our galaxy, and accordingly the relative abundances of the elements in the sun are referred to as "cosmic abundances." Although the cosmic abundance pattern is commonly seen in other stars there are dramatic exceptions, such as stars composed of iron or solid nuclear matter, as in the case with neutron stars. The

Earth System Science
ISBN 0-12-379370-X

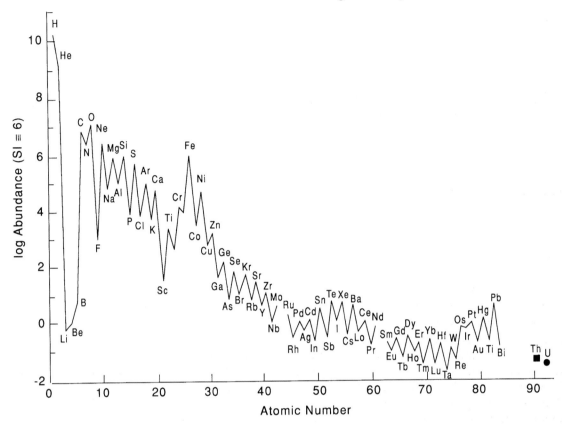

Fig. 2-1 Cosmic (solar) abundances of the elements, relative to Si, which is given the arbitrary value of 10^6.

best estimation of solar abundances is based on data from optical spectroscopy and meteorite studies, and in some cases extrapolation and nuclear theory. The measured solar abundances are listed in Fig. 2-1 and Table 2-1. It is believed to be accurate to about 10% for the majority of elements. The major features of the solar abundance distribution are a strong decrease in abundance for heavier elements, large deficiency of Li, Be, and B, and broad abundance peak centered near Fe. The factor of 10 higher abundance of even atomic number nuclei relative to their immediate odd atomic number neighbors is due to the higher binding energy of nuclei with even numbers of protons. The abundance curve of odd nuclei plotted against mass is a very smooth function. The cosmic abundance pattern is the net result of nuclear reactions that occurred during the origin of the universe and in the interiors of later generations of stars.

Over 99% of the atoms in the sun are H and He and are believed to have formed in the "Big Bang," the origin of the universe (Silk, 1989). Nucleosynthesis that occurred during the Big Bang produced basically the cosmic abundance of ^{1}H, and ^{4}He and small amounts of ^{2}H, ^{3}He, and ^{7}Li but essentially no heavier elements. The Li/H ratio produced by the Big Bang was about 10^{-9}. Element formation occurred over a quarter hour time period during which temperatures dropped from very high values to less than about 10^9 K. After this brief period of expansion and cooling the universe no longer contained matter that was hot and dense enough for nuclear reactions to occur. The fundamental nuclear reaction that occurred in the Big Bang was the fusion of hydrogen to form helium. Synthesis stopped at ^{7}Li because formation of the next abundant element, carbon, required higher densities than existed in the universe at the appropriate temperature range. Inside stars

Table 2-1 Solar abundances of the elements (atoms/10^6 atoms of Si)

Element	Abundance	Element	Abundance	Element	Abundance
1 H	2.72×10^{10}	29 Cu	514	58 Ce	1.16
2 He	2.18×10^9	30 Zn	1.26×10^3	59 Pr	0.174
3 Li	59.7	31 Ga	37.8	60 Nd	0.836
4 Be	0.78	32 Ge	118	62 Sm	0.261
5 B	24	33 As	6.79	63 Eu	0.0972
6 C	12.1×10^7	34 Se	62.1	64 Gd	0.331
7 N	2.48×10^6	35 Br	11.8	65 Tb	0.0589
8 O	2.01×10^7	36 Kr	45.3	66 Dy	0.398
9 F	843	37 Rb	7.09	67 Ho	0.0875
10 Ne	3.78×10^6	38 Sr	23.8	68 Er	0.253
11 Na	5.7×10^4	39 Y	4.64	69 Tm	0.0386
12 Mg	1.075×10^6	40 Zr	10.7	70 Yb	0.243
13 Al	8.49×10^4	41 Nb	0.71	71 Lu	0.0369
14 Si	1.00×10^6	42 Mo	2.52	72 Hf	0.176
15 P	1.04×10^4	44 Ru	1.86	73 Ta	0.0226
16 S	5.15×10^5	45 Rh	0.344	74 W	0.137
17 Cl	5.24×10^3	46 Pd	1.39	75 Re	0.0507
18 Ar	1.04×10^5	47 Ag	0.529	76 Os	0.717
19 K	3.77×10^3	48 Cd	1.69	77 Ir	0.660
20 Ca	6.11×10^4	49 In	0.184	78 Pt	1.37
21 Sc	33.8	50 Sn	3.82	79 Au	0.186
22 Ti	2.40×10^3	51 Sb	0.352	80 Hg	0.52
23 V	295	52 Te	4.91	81 Tl	0.184
24 Cr	1.34×10^4	53 I	0.90	82 Pb	3.15
25 Mn	9.51×10^3	54 Xe	4.35	83 Bi	0.144
26 Fe	9.0×10^5	55 Cs	0.372	90 Th	0.0335
27 Co	2.25×10^3	56 Ba	4.36	92 U	0.0090
28 Ni	4.93×10^4	57 La	0.448		

the temperatures and densities are sufficiently high for synthesis of heavy elements. Without future generations of hot dense matter, in the form of stars, there would have never have been any elements that were heavier or more chemically interesting than H, He, and Li (Penzias, 1979).

The H and He produced in the Big Bang served as "feed stock" from which all heavier elements were later created. Less than 1% of the H produced in the Big Bang has been consumed by subsequent element production and thus heavy elements are rare. Essentially all of the heavier elements now in the Earth were produced after the Big Bang inside stars. Following the Big Bang, the universe expanded to the point where instabilities formed galaxies, mass concentrations from which up to 10^{14} stars could develop.

The reaction that is the major energy source for stars similar to the sun is the fusion of H to He. The basic reactions (the proton–proton chain) in the sun are shown in Fig. 2-2. Although these reactions are the major source of energy in the solar system, they proceed at a remarkably slow and uniform rate. In the sun's core where the temperature is 14×10^6 K, the lifetime of a proton before it is fused to deuterium is 10^{10} years. The average energy generation rate for the entire sun is only 200 µW/kg. Hydrogen burning occurs in stars whose interior temperatures are in the 10^7 to 10^8 K range and while there are several reaction chains depending on the stellar mass, the general fusion reactions forming He are similar to those that occurred during the Big Bang.

For stars like the sun, the burning of H to He

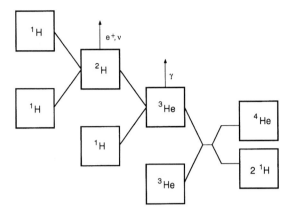

Fig. 2-2 Nuclear reactions of the H-to-He primary fusion sequence in the sun.

occurs deep in stellar interiors at a fairly stable rate for over 90% of the star's lifetime. For the sun this "main sequence" stage will last for about 10^{10} years before H depletion and rising core temperature initiate a set of more energetic nuclear reactions that occur in the final stages of its evolutionary lifetime. Throughout the geologic record of Earth the sun has been a main sequence star burning H to He. On the basis of theory and observations of similar stars it is expected that the total luminosity (total radiated power) emitted from the sun should not undergo large changes. Short- and long-term changes do occur, however, and they may have had significant effects on the Earth and its physical and chemical cycles. Short-term variability, at the level of 0.1% per year is observed by spacecraft. The most important change, and the one that is the most predictable, is the gradual increase in the solar luminosity over geologic time. As hydrogen is burned in the sun's core, the mean molecular weight of nuclei results in a slow but relentless increase in temperature of the core. To maintain the star in hydrostatic equilibrium the internal pressure must remain constant to support the weight of overlying matter. To maintain constant pressure as the mean molecular weight of the gas increases, the temperature must rise. The temperature rise results in increased energy generation and it has been estimated that over the past four billion years the "solar constant" (the intensity of sunlight at Earth) has increased by

about 30%. Over the next three billion years it will increase by more than an additional 30%. Astronomically such a change seems minor, but the stress that the increased brightness places on terrestrial processes is very large. In fact it is quite remarkable the surface temperature of the Earth has remained as constant as it has while the solar luminosity has increased. A 30% decrease in the sun's output should decrease the temperature of a 300 K airless body heated by sunlight by 20 K. In the case of the Earth, this lesser illumination would lower the mean surface temperature to the freezing point, if there were no other compensating processes occurring. There is no evidence that the Earth's oceans have ever frozen over. The remarkable stability of the Earth's temperature in light of solar changes and large changes in the composition of the atmosphere has led to the "Gaia" hypothesis (Lovelock, 1979). This theory suggests that biological organisms on the Earth collectively act to moderate the long-term atmospheric environment to the mutual benefit of terrestrial life. An alternative hypothesis, not involving organisms, is described by Kasting *et al.* (1988). The next chapter discusses the Gaia hypothesis in detail.

Following the long duration of hydrogen burning, stars enter the red giant phase where increasingly heavier elements are produced. Increasing temperatures in the stellar cores allow more massive, highly charged nuclei to collide with sufficient energy to penetrate coulomb barriers and initiate fusion reactions. The first major step is the fusion of He to form C, a reaction that takes place above 10^8 K. This occurs by the triple alpha process, an interaction that requires essentially a three-body collision between He nuclei. The nearly simultaneous collision of three particles requires high densities and is the reason why this reaction did not occur in the Big Bang. In the triple alpha sequence two He nuclei collide to form a highly unstable ^{8}Be nucleus. The ^{8}Be must then interact with a third He to form ^{12}C on a very short time scale because of the 10^{-16} s decay time of ^{8}Be. The reaction is very temperature dependent and the He burning phase is violent, unstable, and relatively short-lived. At temperatures above 10^8 K fusion reactions can produce elements up to Fe. From He to Fe the binding energy per nucleon increases with atomic number and fusion reactions are usually exothermic and provide an energy source. Beyond Fe the binding energy per nucleon decreases and exothermic reactions do not occur. Up to Fe many of the nuclei are products of alpha reactions, which involve fusion with a He nucleus. Because of this and the fact that there is high binding energy for nuclei that are multiples of ^{4}He all of the most abundant isotopes for elements up to Fe are multiples of ^{4}He (i.e. ^{12}C , ^{16}O, ^{32}S, ^{24}Mg, ^{28}Si, etc.); see Fig. 2-1.

During the red giant phase of stellar evolution, free neutrons are generated by reactions such as ^{13}C(α,n) ^{16}O and ^{22}Ne(α,n) ^{25}Mg. (The (α,n) notation signifies a nuclear reaction where an alpha particle combines with the first nucleus and a neutron is ejected to form the second nucleus.) The neutrons, having no charge, can interact with nuclei of any mass at the existing temperatures and can in principle build up the elements to Bi, the heaviest stable element. The steady source of neutrons in the interiors of stable, evolved stars produces what is known as the "s process," the buildup of heavy elements by the slow interaction with a low flux of neutrons. The more rapid "r process" occurs in

explosive environments where the neutron flux is high. The mechanism of the s process is illustrated in Fig. 2-3. Starting with a seed isotope, successive neutron captures build up increasingly neutron-rich isotopes of the same element until an unstable isotope is reached. The typical decay of this neutron-rich isotope is beta decay, which produces the next element in the periodic table. The new element will have one more proton and one less neutron than its radioactive parent. In beta decay a neutron disintegrates into a proton, a neutrino and an ejected electron (beta particle). The new element created in the s process will then add new neutrons until it reaches a neutron-rich isotope that undergoes beta decay to form yet the next element. The s process can produce the isotopes along the "valley of beta stability" in the chart of the nuclides, the chart of isotopes plotted on a graph of total neutrons in nuclei versus atomic number. The relative abundance of an isotope produced by the s process is proportional to its binding energy and inversely proportional to its neutron capture cross-section. Tightly bound nuclei have small cross-sections and are slower to absorb neutrons to form heavier isotopes. The s process cannot produce isotopes that are particularly neutron rich or neutron poor, but it can

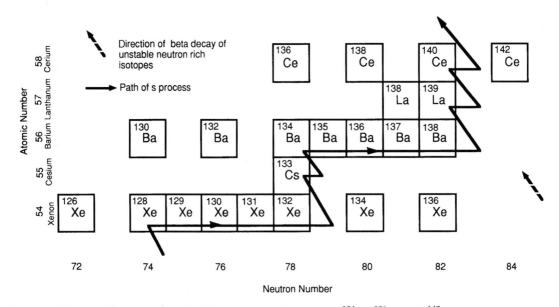

Fig. 2-3 Schematic showing the path of the s process. The isotopes ^{134}Xe, ^{136}Xe, and ^{142}Ce are beyond the reach of s process nucleosynthesis and are only produced by the r process.

produce most of the cosmically abundant elements between ^{56}Fe and ^{209}Bi.

Because the path of the s process is blocked by isotopes that undergo rapid beta decay, it cannot produce neutron-rich isotopes or elements beyond Bi, the heaviest stable element. These elements can be created by the r process, which is believed to occur in cataclysmic stellar explosions such as supernovae. In the r process the neutron flux is so high that the interaction time between nuclei and neutrons is shorter that the beta decay lifetime of the isotopes of interest. The s process chain stops at the first unstable isotope of an element because there is time for the isotope to decay, forming a new element. In the r process, the reaction rate with neutrons is shorter than beta decay times and very neutron-rich and highly unstable isotopes are created that ultimately beta decay to form stable elements. The paths of the r process are shown in Fig. 2-3. The r process can produce neutron-rich isotopes such as ^{134}Xe and ^{136}Xe that cannot be reached in the s process chain (Fig. 2-3).

Thus the origin of the elements began in the Big Bang, but the formation of most of the elements important for physical and chemical processes on the Earth occurred in stars. The element production process required cycles of star formation, element formation in stellar cores, and ejection of matter to produce a gas enriched in heavy elements from which new generations of stars could form. The atoms in the Earth are products of reactions in a large number of stars; typical atoms have been cycled through several generations of stars. The synthesis of material and subsequent mixing of dust and gas between stars produced the solar mix of elements in the proportions that are called *cosmic abundances*. Although some isotopes are exceedingly rare, nuclear processes in stars have produced every known stable isotope.

In addition to stable elements, radioactive elements are also produced in stars. The unstable but relatively long-lived isotopes ^{40}K, ^{232}Th, ^{235}U, and ^{238}U make up the internal heat source that drives volcanic activity and processes related to internal convection in the terrestrial planets. The short-lived transuranium elements such as Rn and Ra that are found on the Earth are all products of U and Th decay.

These isotopes are sometimes used as tracers of natural terrestrial processes and cycles. Long-lived isotopes, such as ^{87}Rb and ^{147}Sm are used for precise dating of geological samples. When the solar system formed it also contained several short-lived isotopes that have since decayed and are now extinct in natural systems. These include ^{26}Al, ^{60}Fe, ^{244}Pu, ^{107}Pd, and ^{129}I. ^{26}Al with a half-life of less than a million years is particularly important because it is a potentially powerful heat source for planetary bodies and because its existence in the early solar system places tight constraints on the early solar system chronology.

2.3   The Origin of the Solar System

The sun and planets formed 4.55×10^9 years ago from interstellar gas and dust. This interstellar material had a bulk elemental composition similar to that of the sun, but the elements in the initial material were highly fractionated between the solid and gaseous phases. Most of the condensable materials were in the form of submicrometer dust grains, while materials that do not condense under astrophysical conditions, such as H, He, and the noble gases, were gaseous. The grains are believed to be mixtures of silicates and carbonaceous matter. In popular models for grains they have silicate cores about 100 nm in diameter coated with 100 nm thick mantles of compounds composed primarily of H, C, N, and O. Irradiation by ultraviolet light and charged particles likely leads to the formation of complex cross-linked polymers from condensed low-atomic-weight compounds in space.

The formation of the solar system is believed to have begun when a cloud of gas and dust became unstable to gravitational collapse and started an essentially unconstrained freefall. During the freefall, the dimensions of the cloud decreased by a factor of nearly 1000 by the time a stable rotating lenticular nebula was established. The nebula was stable to further collapse because gravitational forces were countered by gas pressure and centrifugal forces. The condition for gravitational collapse of a cloud is that the internal gravitational potential energy must exceed twice the kinetic energy. This is expressed by

$$\frac{3GM^2}{5R} > \frac{3kTM}{\mu m_H}$$

where k is the Boltzmann constant, R the initial cloud radius, G the universal gravitational constant, M the cloud mass, T the temperature, μ the mean molecular mass, and m_H the mass of the proton. The collapse of interstellar material to form the solar nebula takes on the order of 10^5 to 10^6 years depending on the initial conditions. During most of the collapse the cloud is transparent to its own emitted infrared radiation and it cools to a temperature near 10 K until nebular densities are reached and the cloud becomes opaque. During this isothermal collapse phase, any condensable matter not originally on grains certainly condensed. The solar nebula that formed is believed to have been a stable rotating disk somewhat larger than the present planetary system. The sun formed in the center and the planets formed from materials that accumulated in the disk. The planets Jupiter and Saturn must have formed by some variant of gravitational collapse, because to a good approximation their elemental compositions match that of the bulk nebula. Most of their mass is H and He, elements that could only have been in gaseous form. The other planets apparently formed out of solids. The outer planets Uranus and Neptune formed from icy and rocky materials, while the terrestrial planets, Mars, Earth, Venus, and Mercury, formed exclusively from rocky and metallic particles.

The solar nebula was hot and dense near its center and became cooler and more diffuse with increasing radial distance. Modern theories of evolution in the nebula indicate that the major heat source was frictional viscosity within the rotating disk of gas. The viscosity and associated redistribution of energy and angular momentum was the result of convective movement of gas in a gas disk that had differential rotation with radial distance from its center. In the outer regions of the nebula, it is likely that heating was never sufficient to vaporize pre-existing interstellar dust grains. In the inner regions of the nebula the original grains were apparently vaporized or extensively altered and the inner (terrestrial) planets must have formed from second-generation solids that condensed from the nebular gas. This condensation process is temperature dependent and it influenced the composition of the planets that formed. The bulk of the mass of each planet appears to have formed largely from local material in a "feeding zone," an annular ring of nebular material. Grains condensed and then were eventually swept up to form planetary bodies. The composition of material in feeding zones depended on temperature and radial distance from the center of the nebula and this determined planetary compositions.

2.4   Condensation

The sequence of condensation of solids from a solar composition at a nebular pressure of 10 Pa (about 10^{-4} atm) is shown in Fig. 2-4. This sequence is calculated for what solids could exist in equilibrium with the solar nebula at various temperatures. Attainment of true equilibrium requires the lack of nucleation barriers, and efficient diffusion within solids so that grain interiors can maintain equilibrium with the gas phase. While strict equilibrium condensation may have occurred at higher temperatures, perfect equilibrium probably did not occur with grains larger than a few micrometers in size or at lower temperatures where diffusion is slow. At temperatures sufficiently above 1500 K all elements were in the gas phase. The first solid grains to condense would be the highly refractory but cosmically rare elements like Pt, Os, Ir, and Re. The first abundant solids to form are oxides and silicates of Ca, Al, and Ti. Inclusions in certain meteorites are rich in these elements, although it is still not clear whether they are actually preserved condensates or refractory residues of volatilized material. Around 1400 K compounds of the most abundant elements in the Earth, Mg, Si, and Fe, condense. At this high temperature Mg and Si condense as Fe-free silicates and Fe condenses as an FeNi metal alloy. At lower temperatures, silicates that maintain equilibrium with the gas can incorporate FeO. At 750 K, independent of pressure, Fe metal in contact with the nebula should react

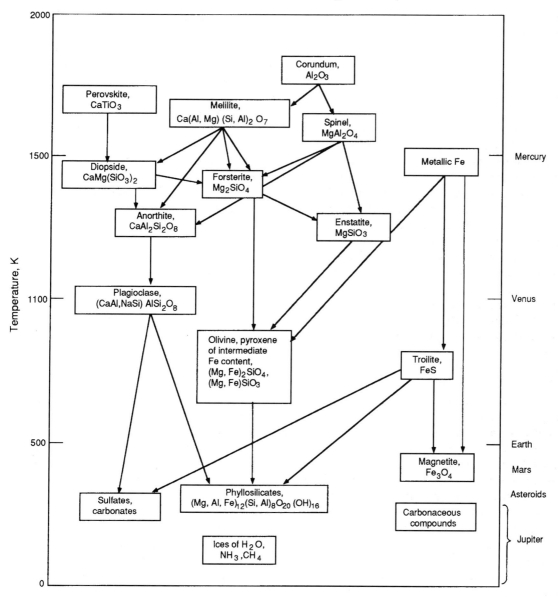

Fig. 2-4 The sequence of condensation of solids from a solar composition gas at a nebular pressure of 10 Pa (ca. 10^{-4} atm). (Modified with permission from J. A. Wood, "The Solar System," p. 162, Copyright © 1979, Prentice-Hall, Englewood Cliffs, NJ.)

with H_2S gas and form FeS, the first sulfur-bearing solid. Iron remaining in contact with gas below 450 K should react with H_2O forming magnetite, Fe_3O_4. Also at this temperature the first water can be incorporated into solids in the form of bound water in hydrated silicates. At temperatures below 250 K, water and clathrates of methane and ammonia can form. Ultimately,

if the temperature ever drops as low as 40 K pure methane ice can exist. The approximate temperatures at which materials condensed and accreted to form the planets are indicated on the right side of Fig. 2-4. While there certainly must have been complications, such as radial mixing of material and non-equilibrium condensation and grain destruction, the equilibrium condensation sequence is generally consistent with the observed properties of the planets and meteorites, at least for compounds that form at high and moderate temperatures. The reality of low-temperature gas–grain reactions such as hydration of silicates and formation of magnetite from metal is in question because these low-temperature equilibrium processes may not have occurred in the time scales available in the rather short-lived solar nebula.

Evidence for condensation is seen in the meteorites, fragments of the asteroids that formed in the region between Mars and Jupiter. The stony meteorites that have elemental compositions that closely match those of the sun (except for volatile elements such as H, He, N etc.) are called *chondrites* after the presence of small spherical particles called *chondrules*. Although the chondrites generally contain close to undifferentiated solar compositions, there are elemental fractions in these objects that are related to condensation processes. Different chondrite groups are distinguished by Fe/Si ratios that vary by 50%, ratios of Ca, Al, and Ti to Si that vary by about 40%, and abundances of volatile elements, such as Cd, Bi, In, and Pb, that vary by orders of magnitude. The depletion of volatile elements is believed to be due to incomplete condensation. The correlated depletion of Ca, Al, and Ti in all but the most primitive meteorites is most likely due to the loss of an early condensate that was composed of these elements. An early condensate of Ca, Al, Ti, and other highly refractory elements could have been separated by accretion from the region of the nebula where grains condensing later formed the Al-depleted chondrites. The Fe/Si fractionation may be the result of the different accretion efficiencies of metal and silicate grains. But even this process is affected by condensation temperature because it determines how Fe is divided between metal, sulfide, oxide, and silicate phases.

Effects of condensation are also seen in the bulk compositions of the planets and their satellites. The outer planets, Uranus and Neptune, have overall densities consistent with their formation from icy and stony solids. The satellites of Uranus have typical densities of 1.3 g/cm^3, which would tend to indicate a large ice com-

ponent. The inner planets, which are composed of silicates and metal have uncompressed densities ranging from 3.4 for Mars to 5.5 for Mercury. The range in densities among the terrestrial planets is largely due to differences in the oxidation state of Fe. Most of the mass of these planets is composed of Fe, Mg, and Si with sufficient oxygen to totally oxidize Si and Mg. The elements Fe, Mg, and Si appear to occur near solar atomic abundances of approximately 1:1:1. The oxidation state of iron ranges from almost completely oxidized in the case of Mars to completely reduced as in Mercury. This range in oxidation state is consistent with equilibration with nebular gas at low and high temperatures, respectively. The oxidation state of Fe in chondrites also ranges from completely reduced in the case of the enstatite chondrites to completely oxidized in the carbonaceous chondrites. Even though the meteorites probably formed beyond the region of the terrestrial planets, the range of oxidation state is undoubtedly the result of nebular pressures and temperatures.

Water and carbon play critical roles in many of the Earth's chemical and physical cycles and yet their origin on the Earth is somewhat mysterious. Carbon and water could easily form solid compounds in the outer regions of the solar nebula, and accordingly the outer planets and many of their satellites contain abundant water and carbon. The type I carbonaceous chondrites, meteorites that presumably formed in the asteroid belt between the terrestrial and outer planets, contain up to 5% (m/m) carbon and up to 20% (m/m) water of hydration. Comets may contain up to 50% water ice and 25% carbon. The terrestrial planets are comparatively depleted in carbon and water by orders of magnitude. The concentration of water for the whole Earth is less that 0.1 wt% and carbon is less than 500 ppm. Actually, it is remarkable that the Earth contains any of these compounds at all. As an example of how depleted in carbon and water the Earth could have been, consider the moon, where indigenous carbon and water are undetectable. Looking at Fig. 2-4 it can be seen that no water- or carbon-bearing solids should have condensed by equilibrium processes at the temperatures and pressures that probably were typical in the zone of the solar nebula that produced the Earth. Water of hydration does not occur in silicates until the temperature is below about 350 K, and ice could not exist in nebular conditions until the temperature was below 200 K. Temperatures low enough for even the formation of hydrated phases probably did not exist within the region of the terrestrial planets. The origin of carbonaceous materials is even more mysterious. In the nebula the distribution of carbon between CO and CH_4 should be controlled by the following equilibrium:

$$CO + 3H_2 \leftrightarrow CH_4 + H_2O$$

Above 650 K carbon in the nebula should be primarily in gaseous CO but below 650 K, if equilibrium persisted, nearly all carbon would be reduced to CH_4. The condensation temperature of methane is 50 K and if all carbon were in this form then it could not have been efficiently incorporated into solids except perhaps at the extreme outer edges of the solar nebula. It is likely, however, that equilibrium between CO and CH_4 did not occur on the available time scale and CO was probably an important reservoir of carbon throughout the nebula. With abundant CO, catalytic reactions on grain surfaces could form carbonaceous coatings. It has been suggested that Fischer–Tropsch reactions similar to the following, produced some of the solid carbonaceous matter in meteorites that have delivered organic solids to the Earth throughout its history:

$$10CO + 21H_2 \rightarrow C_{10}H_{22} + 10H_2O$$

Carbonaceous solids also reach Earth in the form of organic and icy materials that condensed in the cold outer regions of the solar nebula and also as organic materials preserved in interstellar grains.

As evidenced by their low abundances, carbon compounds, water, and other volatiles such as nitrogen compounds were probably not significantly abundant constituents of the bulk of the solids that formed near the Earth. Many of the carriers of these volatiles condensed in cooler, more distant regions and were then scattered into the region where the Earth was forming. Fragments of comets and asteroids formed in the outer solar system still fall to Earth at a rate of 1×10^7 kg/yr and early in the

Earth's history the rate must have been higher. Certainly some of Earth's volatile elements were accreted from comets and other bodies from the outer solar system but it is yet unknown if this was the major source of the more volatile elements that comprise the atmosphere and oceans. One possible tracer that could be used to determine the cometary input is the noble gas composition of comets and the atmosphere. Once outgassed, noble gases tend to remain in the atmosphere and are not influenced by subsequent planetary activity. Unfortunately, the noble gas composition of comets is not presently known. An argument against a common source such as comets for volatiles on the terrestrial planets is based on the differences in the ^{36}Ar contents of planetary atmospheres directly measured using spacecraft. The primordial ^{36}Ar content on Venus is nearly 100 times higher than that for Earth, which is in turn nearly 100 times the Mars value. If this volatile species were carried by the same material that brought water and carbon, then all three could not have been derived from a common comet source.

2.5   Accretion of the Planets

Condensed nebular solids, along with possible pre-solar solids, accumulated by the processes of accretion to form planetary bodies. The process began with low-velocity collisions of micrometer-sized dust and terminated with bodies as large as the Earth and Mars colliding with velocities of over 5 km/s. The first stage involved low-velocity collision of dust grains to form bodies of centimeter size. The movements of dust-sized objects were strongly coupled to gas motions but centimeter and larger particles could de-couple from local gas motions and settle to form a layer. The larger particles, with smaller surface area to mass ratios, are dominated by gravity, not nebular winds. Pulled by the vertical component of gravity, these rock-sized particles would attempt to settle to the central plane of the nebula, forming a thin disk with a relatively high concentration of dust, rocks and boulders. Energy dissipation in the disk by collisions and gas friction should produce an extremely thin disk system somewhat analogous to the rings of Saturn. The particles should initially have almost perfectly circular orbits about the sun at near zero inclinations so that the relative velocity between particles was quite low. It was from this "cold disk" of small bodies with similar orbits that the accretion of larger bodies, and ultimately planets, began.

As the bodies grew by collisions and accretion, gravitational perturbations from each other and from nearby forming planetesimals caused orbits to become less circular and more inclined to the plane of the solar system. As orbits become less similar to each other, their relative impact velocities rise, ultimately up to the km/s range. The increase in velocity dispersion is due to gravitational effects of the larger bodies stirring up the relative motions of all planetesimals and make their orbits less circular. Bodies that pass near another body are scattered, usually into a less circular orbit. In the higher-velocity regimes, collisions are often destructive. An impact at 4 km/s has the same energy per mass and release rate as the chemical energy in high explosives. The accretion process is complex and involves both destruction and net growth. The process is also highly competitive with the largest body in the growing swarm having a selective advantage because of its gravitational field. Gravity expands the capture cross-section to areas much larger than the geometrical cross-section and it helps in retaining rebounded material from high-velocity impacts. At one time, the material that formed the Earth was in the form of a million or so bodies a hundred kilometers in diameter. In the end all of these were incorporated into one object or were ejected out of the feeding zone. The final stages of accretion involved collision with very large bodies perhaps as large as Mars.

A collision with a Mars-sized object may have resulted in the formation of the Earth's moon. Our moon is by no means the largest satellite in the solar system, but it is unusual in that it and the moon of Pluto are the largest moons relative the mass of the planets they orbit. Geochemical studies of returned lunar samples have shown that close similarities exist between the bulk composition of the moon and the Earth's mantle. In particular, the abundances of siderophiles ("iron loving" elements such as Ir and Au that concentrate in planetary iron cores) in the moon are similar to those in the Earth's mantle. These observations have lead to the popular hypotheses that the moon formed from parts of a large impactor and Earth mantle material ejected into orbit by the impact. The collision must have occurred early in the Earth's history but after separation of its core.

The formation of the large moon has had a profound effect on Earth history. The massive moon has provided an important stabilizing effect on the Earth's obliquity, the tilt of its spin axis relative to the orbit plane. The obliquity has changed only a few degrees over geologic time but without the stabilizing effects of the moon, the obliquity could have varied as much as 90°. Such change would have caused substantial changes in the atmosphere, climate and ocean circulation. As an example, if the obliquity were over 50°, the equatorial regions would receive less solar energy over the year than the polar regions. It is believed that the obliquity of Mars has changed significantly over its history and this may have played a role in the apparent large degree of variability of its atmosphere. Mars cannot presently support liquid water at its surface but there is evidence that it could in the past. The moon also plays a large role in producing tides, an effect important to a variety of biological and chemical cycles. Early in the Earth's history the moon was closer and the tidal effects were much larger. The tidal interactions between the Earth and moon result in a slowing of the Earth's spin rate and an increase in the Earth–moon distance. Over the past 500 000 years the Earth's spin rate has decreased about 10% by this effect.

After planetary accretion was complete there remained two groups of surviving planetesimals, the comets and asteroids. These populations still exist and play an important role in the Earth's history. Asteroids from the belt between Mars and Jupiter and comets from reservoirs beyond the outer planets are stochastically perturbed into Earth-crossing orbits and they have collided with Earth throughout its entire history. The impact rate for 1 km diameter bodies is approximately three per million years and impacts of 10 km size bodies occur on a

100 Myr time scale. The collision of a 10 km asteroid or comet produces a crater nearly 200 km in diameter and the atmospheric and oceanic effects from shock processes and ejecta can produce severe stress on terrestrial organisms. Major global effects can include heavy particulate loading of the atmosphere, production of nitrogen oxides from shock effects and strong infrared radiation from ejected matter. The cretaceous–tertiary extinctions that occurred 65 Myr ago are marked with a global layer that contains shocked quartz grains and the trace element signature of meteoritic material. The total mass of the layer is consistent with ejecta from the hypervelocity impact of a 10 km comet or asteroid. The above impact rates have remained relatively constant over the past 3.9 Gyr, although they were much higher before this time. The surface of the moon provides an excellent record of this impact history. The smooth basalt-filled mare on the Earth's side of the moon provide a record of cratering over the past 3.5 Gyr while the bright highlands areas reveal the intense era of large impact events that occurred before this time.

2.6 Early Evolution of the Earth

When the Earth was in its early stages of accretion, it was presumably a cool object because the rate of accretional heat added to it was small. In the final stages, accretional energy was appreciable and must have heated the upper regions of the Earth to high temperatures. When the Earth approached its present mass, projectiles would impact at a minimum velocity of 11.2 km/s, the velocity of escape. The kinetic energy at this velocity is high and, at least for very large objects, a substantial fraction of the impact energy can be trapped within the Earth. The oldest rocks on Earth were formed 600 Myr after its formation and unfortunately there are no direct records of what the Earth was like at this stage. The moon, however, does preserve a record of its earliest history, and it provides a glimpse of what it might have been like. The moon actually melted down to a depth of about 400 km, forming a "magma ocean" from which its >60 km thick anorthositic crust crystallized.

It is likely that a major portion of the upper parts of the Earth also melted at some time during the final stages of accretion.

After the planet accreted, additional heat sources were radioactive decay and gravitational energy released by formation of its core. If the Earth contained amounts of ^{26}Al and ^{60}Fe, similar to those seen in primitive meteorites, the decay of these short-lived isotopes would provide an early and intense source of thermal energy within the first few million years of its history. The other radioactive heat sources (U, Th, and K) are longer lived and release heat on time scales of over 10^8 years. Partial melting of the upper layers leads to the formation of molten iron masses that sink towards the Earth's interior. The gravitational settling of dense metal releases heat that creates a single pulse of gravitational energy. All of these heat sources lead to the chemical differentiation of the Earth into a core, mantle and crust, and to outgassing of volatiles to form an atmosphere. Apparently all of the terrestrial planets underwent differentiation, although there may have been significant variations in the details of crustal development and subsequent evolution. Later evolution depends on the amount of internal heat available and the thickness of crust through which geologic activity must penetrate. The Earth is still a very active body with volcanic activity and widespread plate movement. Its closest neighbors, Mars and Venus, are less active and have had different evolutionary histories.

It now appears that the early evolution of the Earth may have been much more violent than commonly imagined in the past. The evidence and general agreement that the moon probably formed as the result of the impact of a Mars-sized object or larger, underscores calculations indicating that the Earth's accretion included impact with bodies of planetary size. These great impacts influenced volatile loss and core formation. It is likely that great disturbances and heat of large impacts caused core formation to occur simultaneously with accretion rather than after the planet was fully assembled, as is commonly assumed. The violence of these events may be responsible of the depletion of volatiles in the Earth. The volatile element

potassium, the most important source of radiogenic heat, is only one-fifth as abundant in the Earth as it is in primitive meteorites. This depletion may be related to the composition of the bodies that accreted to from Earth, but it might also be due to volatile loss during great impacts such as that the is believed to have formed the moon.

The evolution of the Earth's atmosphere and oceans are important for many of the cycles described in this book. Both of these evolved by outgassing but there is little information about the history of these processes. Again, comets could have been a main source of some of the volatiles (H_2O, CH_4, etc); however, proof of this source is still lacking (see Section 2.4 above). Unfortunately, bombardment of the inner planets by large projectiles pulverized planetary surfaces until about 3.9 billion years ago, erasing essentially all direct records of the Earth's early history. The Earth's ocean and its atmosphere are unique in comparison with Mars and Venus. Neither of these planets have oceans and their atmospheres are composed largely of CO_2 (Table 2-2). Mars has ice on and in its surface materials but Venus has essentially no water. With a surface temperature of 650 K, water near the surface of Venus could exist only as a gas. It has been suggested that Venus may have had as much water as the Earth but lost it in a catastrophic blow-off process driven by extreme greenhouse heating early in its history. (See Chapter 7 for a discussion of loss by escape from the upper atmosphere.) Another possibility is that Venus could have lost an early atmosphere and hydrosphere as a result of large impacts, a process that may also have "eroded" the atmospheres of Earth and Mars. The CO_2 inventory of Venus is similar to that on the Earth except that on Venus it is in the atmosphere; on the Earth, all but trace amounts are trapped in carbonate rocks. Eventually, with continually increasing solar output or changes in the atmosphere, the CO_2 content of the Earth's atmosphere may increase. Increased greenhouse heating and positive feedback would lead to higher surface temperatures. Mars, in contrast, is in a nearly permanent ice age with only transient periods when conditions are sufficient for the existence of liquid water. The most remarkable aspect of the Earth is its atmosphere. It is composed of oxygen and nitrogen in contact with liquid water in highly non-equilibrium proportions. Except for ^{40}Ar produced by decay of ^{40}K, the atmospheric composition is controlled by biological processes that act on time scales that are certainly faster than anywhere else in the solar system. Early in the Earth's history the atmosphere was probably dominated by CO or CO_2, like Mars and Venus. In time the CO_2 was incorporated into carbonate rocks and nitrogen and oxygen came under the control of biological processes. The rise of oxygen due to photosynthesis started early in the Earth's history and reached modern levels before the start of the Cambrian, 600 million years before the present.

Table 2-2 Atmospheric compositions of the terrestrial planets

Planet	Relative pressure	Principal gases, % v	Other gases, ppmv
Mercury	10^{-15}	He, ca. 98	
		H, ca. 2	
Venus	90	CO_2, 96	H_2O, ca. 100; SO_2, 150; Ar, 70
		N_2, 3.5	CO, 40; Ne, 5; HCl, 0.4; HF, 0.01
Earth	1	N_2, 78	CO_2, 365; Ne, 18; He, 5; Kr, 1.1
		O_2, 21	Xe, 0.087; CH_4, 1.7; H_2, 0.5; SO_2, 0.0001; O_3, 0.04;
			Ar, 0.92; NO_2, 0.00001; NH_3, 0.0001; N_2O, 0.3
		Ar, 0.93	CO, 0.12; NH_3, 0.01; NO_2, 0.001
Mars	0.007	CO_2, 95	O_2, 1300; CO, 700; H_2O, 300
		N_2, 2.7	Ne, 2.5; Kr, 0.3; Xe, 0.08; O_3, 0.1

2.7 Earth and the Development of Life

The Earth began with processes of great violence, impacts that may have physically torn the planet in two followed by gravitational reassembly. As previously discussed such impacts may have resulted in formation of the moon. The majority of accretion occurred rather quickly and within a 10^7 year time period, the era of planet distorting impacts was over. Following accretion, the planet entered an era where smaller bodies would occasionally impact, the largest of these being only a few hundred kilometers in diameter and they would produce impact features hundreds to thousands of kilometers in diameter. Typically they were smaller than the impacts that occurred earlier but they were much larger than those that would impact after this period, which is often called the "era of heavy bombardment." It was the time during which the large impact basins, such as Mare Imbrium, formed on the lunar surface. The era ended 3.9 Gyr ago when the supply of sufficiently large impactors in Earth-crossing orbits was effectively depleted. The largest impact basin known on the moon is South Aitken basin, which is over 2000 km in diameter. Although no record of this period survives for the Earth, extrapolation from the lunar record would suggest that the Earth would have experienced over a thousand basin-forming impacts. The impact rate during the "heavy bombardment era" would have actually been rather low, with tens of thousands to millions of years between major impacts.

It is unclear what the Earth was like during this heavy bombardment era but it is certain that it was punctuated with great impact events. The Earth's surface may have been warm due to a dense atmosphere and greenhouse heating or it may have been cool due the relative faintness of the early sun. It was largely a water- or ice-covered planet and it is possible there were no continents – only short-lived volcanic islands. The atmosphere would have been transparent to the ultraviolet and the Earth's surface would have been quite inhospitable for any early life. Life may have formed during this time but the impact of 100 km bodies provides enough global heating to essentially sterilize the planet down to depths, in some cases, of several kilometers. If life formed on the early Earth it may have been destroyed and reformed many times; hence this period has also been called the period of "impact frustration" (Maher and Stevenson, 1988). No long-term life was possible until after 3.9 Gyr ago when the great impacts had ceased.

It is significant that the earliest records of life on Earth start shortly after the period of impact frustration. Apparently life formed as soon as the conditions permitted it. Life originated from compounds produced by prebiotic organic chemistry. The source of the molecules included those produced on Earth by energetic processes such as impacts and electrical discharges as well as those that fell in from space. Whatever processes occurred, they would have had to happen either in the deep ocean or in what might have been rare regions of land and shallow water.

The Earth is a highly unusual planet because life did evolve on it and it thrived to the extent that the surface and atmosphere of the planet were greatly modified. The Earth is unique in this respect relative to all known astronomical bodies (Taylor, 1999). The Earth's location, composition, and evolutionary history are all significant factors in the planet's success in nurturing life. Critical factors include its temperature, its atmosphere, its oceans, its long-term stability and its "just right" abundance of water and other light element compounds.

One of the Earth's most important properties is its location; it lies within the habitable zone (HZ) of the sun. The HZ is the range of solar distances where an Earth-like planet can have liquid water on its surface. Besides its surface temperature, the Earth contains a mix of land and ocean that seem highly suitable for life. The presence of land is critical because of its role in removal of CO_2 from the atmosphere and also it ultimately serves as a habitat of the rich diversity of life that occurs on land and in the shallow water surrounding it. The silicate–CO_2-weathering cycle, which is discussed in Chapters 8 and 11, is probably the most biologically important chemical process on Earth because it provides a means of keeping the surface temperature in a region appropriate for life. This cycle requires land and shallow water. It is remarkable, though, that the planet contains land at all. If it were not for continents, the Earth would be essentially a water-covered planet with only small islands like Hawaii and Iceland providing contact with the atmosphere.

Overall, the Earth is unique compared to any other planet we know of in its remarkable perfection in supporting life. The integrated systems that allow this to happen started with events occurring billions of years ago. Keeping this historical perspective in mind while studying present-day Earth systems will help complete the picture of the planet.

References

Lovelock, J. E. (1979). "Gaia, a New Look at Life on Earth." Oxford University Press, Oxford.

Kasting, J. F., Toon, O. B., and Pollack, J. B. (1988). How climate evolved on the terrestrial planets. *Scient. Am.* **258**(2), 90–97.

Maher, K. A. and Stevenson, D. J. (1988). Impact frustration of the origin of life. *Nature* **331**, 612–614.

Penzias, A. A. (1979). The origin of the elements. *Science* **205**, 549–554.

Silk, J. (1989). "The Big Bang." W. H. Freeman, New York.

Taylor, S. R. (1999). On the difficulties of making Earth-like planets. *Meteoritics and Planet. Sci.* **34**, 317–329.

Zahnle, K. J. and Sleep, N. H. (1997). Impacts and the early evolution of life. *In* "Comets and the Origin and Evolution of Life" (P. J. Thomas, C. F. Chyba and C. P. McKay, eds), pp. 175–208. Springer, New York.

3

Evolution and the Biosphere

James T. Staley and Gordon H. Orians

Although physical and chemical factors were exclusively responsible for affecting the distributions of elements 4.5 billion years ago when Earth formed (see Chapter 2), this is no longer true. Ever since life originated on Earth more than 3.5 Gyr (Gyr = 10^9 years) ago, biological processes have become increasingly important in determining the distribution of elements and the compounds into which they are incorporated. To understand the basic features of the Earth system and the functioning of biogeochemical cycles, it is necessary to know something about the chemistry of living organisms, how organisms derive energy, and how they influence and are influenced by the physicochemical states of the environments in which they live. In this chapter we discuss how life originated in a non-living world, the biochemical machinery of living organisms, the mechanisms that have governed the evolution of life during the past 3.5 Gyr, the ecological organization of the living world, and some of the major impacts organisms have had on the development and continued functioning of Earth's processes. In addition, we present an overview of some of the important roles organisms play in the functioning of biogeochemical cycles.

3.1 The Origin of Life on Earth

People have sought to explain the origin of organisms since at least the beginning of historical times. At the time of the Renaissance most Europeans believed that living organisms developed from non-living materials, i.e. that maggots could be "generated" by allowing meat to decay. The careful experiments of the Italian naturalist Francesco Redi (1626–1698) on decaying meat temporarily set to rest this belief. But the theory of spontaneous generation of life gained many new adherents with the discovery of microbes by Anton van Leeuwenhoek (1632–1723), and by experiments showing that mixtures, such as boiled aqueous extracts of meat, developed microorganisms even after prolonged heating in sealed vessels. Scientists reasonably, but incorrectly, inferred that the microorganisms were developing spontaneously from these non-living materials rather than, as was actually the case, from heat-resistant bacterial spores that survived the boiling temperatures.

Scientists do not believe that life is arising from non-life on Earth today, but, if life originated on Earth, as it apparently did, it must have developed from non-living materials. Current scientific views of when and how life might have originated and evolved are based upon imaginative chemical experiments in the laboratory, combined with studies of the fossil record and ways of dating events in the remote past.

Early students of the origin of life were misled because they believed that Earth was very young, in part because no methods were available for dating ancient events. Today, suitable methods exist for determining the age of materials that are billions of years old, and the fossil record of ancient organisms has vastly improved. The evolution of living organisms

Earth System Science
ISBN 0-12-379370-X

was in part possible because of important changes in the physical environment, but living organisms in turn caused profound changes in the physical nature of the Earth and its atmosphere. Ever since the evolution of life, living organisms have been major participants in biogeochemical cycles. This biological activity, which continues today, is being increasingly influenced by human activities.

3.1.1   Synthesis of Organic Molecules on the Primeval Earth

We are uncertain just how life originated, but a reasonable first step in the development of organisms was the non-biological synthesis of compounds of carbon upon which all living systems are based. There is considerable evidence indicating that these compounds can be formed abiologically. For example, one group of meteorites, the carbonaceous chondrites, contains organic matter, which was presumably formed in intergalactic space. Some of these meteorites contain a variety of organic constituents, including sugars, organic acids, and amino acids. The fact that the amino acids in meteorites contain equal portions of the D and L stereoisomers (see Section 3.2) provides evidence that these organic materials were produced abiotically, or at least were not contaminants derived from Earth, where organisms produce primarily L-amino acids.

Stanley Miller (1953) conducted pioneering experiments that had a profound influence on subsequent thinking. He set out to determine if organic chemicals could be formed from water and the various gases that were presumed to have existed in the atmosphere 4 Gyr ago. He used an atmosphere containing methane, ammonia, and hydrogen as its principal constituents of biological interest. This atmosphere differed from the current one by lacking oxygen, carbon dioxide, and molecular nitrogen. The experiments were conducted in a glass-enclosed system (Fig. 3-1). Water, representing the sea, was contained in one flask which was connected to a condenser; the other part, representing the atmosphere, contained the gases. Electrodes inserted into the atmos-

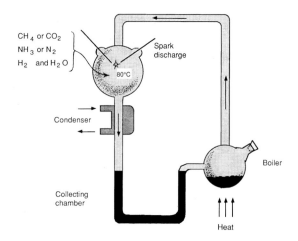

Fig. 3-1 Schematic of the apparatus Stanley Miller used to demonstrate formation of amino acids from simple inorganic compounds under conditions similar to those of the early Earth.

phere were used to produce sparks to simulate lightning. In a relatively short time of operation, amino acids and a variety of other organic substances were produced under these anaerobic conditions. The energy source for his experiments, electric discharges, is destructive of life as we know it, and is not used today by any biological systems. Other energy sources probably powered the evolution of life.

An alternative view of Earth's early atmosphere was advanced by Rubey (1951, 1955). He proposed that the early atmosphere was strongly affected by volcanic outgassing, which released carbon dioxide, nitrogen, sulfur compounds such as hydrogen sulfide, methane, hydrogen, and water. In his view, the early atmosphere was very similar to that of present-day Earth except that it was anaerobic. Rubey's view, which is supported by chemical analyses of the oceans and atmosphere, has gained wide acceptance among evolutionists and has stimulated research to determine whether organic compounds can be synthesized abiotically under such atmospheric conditions. The results from these experiments, which have been discussed by Chang *et al.* (1983), confirm that organic compounds can be synthesized by many combinations of these gases, again supporting the plausibility of prebiotic organic synthesis. Others are investigating the abiotic synthesis of organic compounds such as amino acids under high hydrostatic pressure and temperature (Amend and Shock, 1998). This approach is consistent with the view increasingly favored by some biologists, that life may have arisen in marine hydrothermal vents. Organic compounds can be produced under so many environmental conditions that the current question is – which of the many possible conditions were the ones actually found on the early Earth?

Although much is known about how the organic compounds might have been synthesized abiogenically, scientists still have a poor understanding of how those molecules were assembled to form the first living organisms. All organisms, even the simplest of them, are extremely complex structures that carry out a rich variety of chemical and physical processes. And they are able to precisely control exchanges of materials between themselves and the environment. Thus, it is not surprising that no one has produced living organisms or functioning prototypes of them in the laboratory.

3.2 The Machinery of Life

Even though we don't know how life arose, we can describe in great detail its physical and chemical structure. But simple descriptions of the structure of living organisms do not enable us to define exactly what life is. One way to define life is to state what appear to be the simplest requirements for life as we know it. They include: (a) the presence of a semipermeable membrane to control the passage of materials into and out of the compartments (cells) of which all living organisms are composed; (b) chemical machinery for the synthesis and degradation of essential molecules; (c) genetic material that encodes the synthesis of the molecules required to catalyze those syntheses and degradations; (d) sufficient structure to prevent unwanted reactions from occurring; and (e) machinery to duplicate all the above capabilities in the formation of a new organism (i.e. reproduction).

We do not know the minimum number of molecules required to perform these activities.

Viruses, on the border of life, are non-cellular, and are able to multiply only by utilizing the complex genetic and structural machinery of more complex cellular organisms, which they parasitize. The largest viruses contain enough genetic material to encode only about 100 different proteins. The smallest known organisms that are able to live independently are single-celled bacterial parasites called the Rickettsiae that contain enough genes to encode about 200–400 proteins. Whether smaller independently living organisms could have survived prior to the evolution of more complex forms of life is not known.

3.2.1 The Cellular Structure of Life

The unit of construction of all living organisms on Earth is the cell. Some organisms consist of a single cell; others contain many cells. Cells range in size from less than 1 μm (10^{-6} m) to more than 500 μm in diameter. All cells have the same basic structures: a bounding *cell membrane*, a *nucleus* or *nuclear material*, and *cytoplasm* in which most biochemical reactions take place.

3.2.2 The Chemical Basis of Life

To build their structures and to carry out the myriad biochemical reactions that take place within their cells, organisms need a source of energy. The needed energy is obtained via biochemical pathways driven either by sunlight or by energy contained in reduced chemical compounds.

Life is based on interactions among a set of large organic molecules, each of which is assembled from smaller molecules. In addition to carbon and hydrogen, most naturally occurring organic molecules contain one or more of four key elements, all of which are from the second and third period of the periodic table: N, O, P, and S. Carbon atoms form most of the skeleton of these molecules. Bonding among carbon atoms can lead to very large molecules with a variety of structures. Hydrogen may share one or more of the valence electrons of carbon and may also participate in *hydrogen bonds*. Hydrogen bonds also involve nitrogen and oxygen and are important in determining the structures of many molecules.

Most organophosphorus compounds are phosphates, $R-PO_4$. The bond energy of the phosphate to the rest of the molecule is central to the flow of energy in *all* metabolic processes. Sulfur, found largely as sulfide, is central to providing three-dimensional rigidity by uniting parts of a molecule by the disulfide ($-S-S-$) bridge.

Amino acids, the building blocks of giant protein molecules have a carboxyl group and an amino group attached to the same carbon atom. A *protein* is a linear polymer of amino acids combined by peptide linkages. Twenty different amino acids are common in proteins. Their side chains, which have a variety of chemical properties, control the shapes and functions of proteins. Some of these side chains are hydrophobic, others are hydrophilic, and still others occur either on the surface or the interiors of proteins.

Carbohydrates form a diverse group of compounds that share an approximate formula $(CH_2O)_n$. The major categories of carbohydrates are *monosaccharides* (simple sugars); *oligosaccharides* (small numbers of simple sugars linked together); and *polysaccharides* (very large molecules), among which are starches, glycogen, cellulose, and other important compounds. *Derivative carbohydrates*, such as sugar phosphates and amino sugars, contain additional elements. Important amino sugars are chitin, the principal structural carbohydrate in insect skeletons and cell walls of fungi, and cartilage.

Lipids are insoluble in water and release large amounts of energy when they are metabolized. The lipids of Eubacteria and Eucarya are composed of two principal building blocks, *fatty acids*, and *glycerol*. Three fatty acids (carboxylic acids with long hydrocarbon tails) combine with one molecule of glycerol by ester linkages to form a triglyceride. More complex lipids are formed by the addition of other groups, the most important of which contain phosphorus. The lipids of Archaea are quite different in that they do not contain fatty acids.

Deoxyribonucleic acid (DNA) is the genetic material of all organisms, including plants, animals, and microorganisms. (Some viruses lack DNA, but use RNA (ribonucleic acid) in its place.) DNA, which carries all the hereditary information of the organism, is replicated and passed from parent to offspring. RNA is formed on a DNA template in the nucleus of a cell. The RNA carries the genetic information to the cytoplasm where it is used to produce proteins on the ribosomes. The specific proteins formed include enzymes which carry out the characteristic activities of the organism. Both RNA and DNA are formed from monomers, called nucleotides, each of which consists of a simple sugar, a phosphate group, and a nitrogen-containing base. The complex structure of DNA is founded on only four bases. A tremendous wealth of information is contained in the precise ordering of these bases to form the genetic code.

The major classes of macromolecules and their subunits are described more fully in the appendix that follows this chapter.

3.2.3 The Energetics of Living Organisms

To maintain themselves, to grow, and to reproduce, all organisms must obtain raw materials and energy from the environment. The raw materials – chemicals – are digested, and the products are used to build large carbon-based molecules. The energy obtained from chemical digestion is used to power the synthetic reactions. These conversions of matter and energy are called *metabolism*. Organisms can be viewed as devices for capturing, processing, and converting matter and energy from one form to another.

The energy obtained from chemical digestion is stored in cells in the short term in the form of ATP (adenosine triphosphate) (Fig. 3-2). ATP contains high-energy bonds in its triphosphate group. When it is hydrolyzed to form ADP (adenosine diphosphate) and H_2O, 30 kJ/mol of energy is released. In cellular metabolism, this energy is used to carry out synthetic reactions for the production of DNA, proteins, carbohydrates, and all other cellular materials.

The energy stored in ATP is released during reactions catalyzed by proteins called enzymes. Enzymes typically have molecular weights of

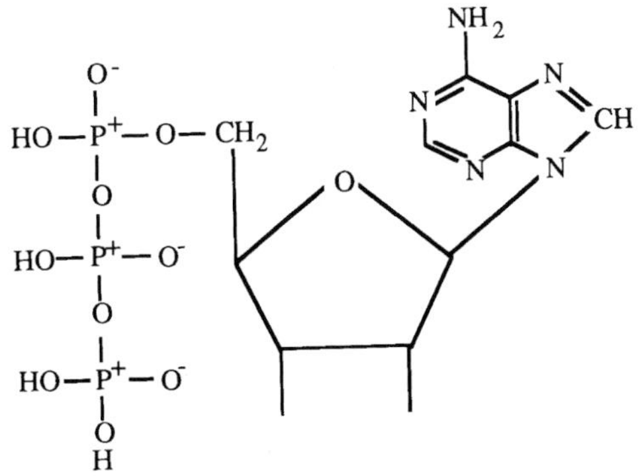

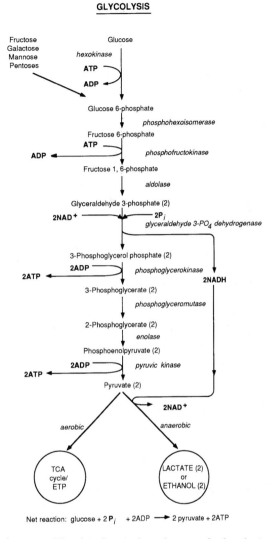

Fig. 3-2 Chemical diagram of ATP (adenosine triphosphate). The three functional groups are the base adenosine (upper right), a five-carbon ribose sugar (middle), and three molecules of phosphate (left). Lines at bottom of sugar ring indicate hydroxyl groups.

about 30 000, but they vary greatly in size. Enzymes function by lowering the activation energy for reactions by complexing with the substrate for the reactions. In the absence of enzymes, many vital cellular reactions would not take place at all or would proceed much more slowly than they do.

Organisms obtain their energy either by oxidizing preformed organic molecules (heterotrophs = "other feeders") or by using an external source of energy, such as sunlight or chemically reduced matter, to drive the synthesis of large, energy-rich molecules (autotrophs = "self-feeders").

3.2.3.1 Energy generation by heterotrophic organisms

Heterotrophic organisms obtain their energy by oxidizing reduced organic compounds. In the absence of oxygen, organisms may oxidize organic compounds either by fermentation or by anaerobic respiration. In contrast, aerobic organisms can function only in oxygenated environments. Anaerobic processes probably evolved early in Earth's history, before oxygen was readily available.

3.2.3.1.1 Fermentation. Many bacteria live in anaerobic environments where they ferment carbohydrates to produce substances such as

Fig. 3-3 The biochemical pathway of glycolysis, which obtains energy from the breakdown of 6-carbon sugars to a pair of 3-carbon pyruvate molecules. Enzymes at each reaction in the sequence are in italics. The energy generated is stored in the form of ATP (see Fig. 3-2).

ethanol, propionic acid, butyric acid, formate, hydrogen gas, and carbon dioxide, depending on the species and its enzymes, as well as the substrate (Perry and Staley, 1997). One example of this is the fermentation carried out by lactic acid bacteria. In this process sugars are first oxidized anaerobically to form pyruvic acid. In this process ATP is produced during one of the steps (Fig. 3-3).

Table 3-1 Electron acceptors that are used in the biodegradation of organic material in marine sediments. More on the chemistry of these processes is presented in Chapters 8 and 16

Electron acceptor	Process	Approximate depth in sediment (cm)
O_2	Aerobic respiration	0–0.5[a]
FeO(OH)	Iron respiration	0–0.5[a]
NO_3^-	Denitrification	0.5–5[a]
SO_4^{2-}	Sulfate reduction	5–100[b]
CO_2	Methanogenesis	>100[b]

[a] Shallow (possibly water column).
[b] Deeper (sediment).

3.2.3.1.2 Anaerobic respiration. Some microbes carry out other chemical transformations in anaerobic freshwater and marine sediments and in waterlogged soils. Anaerobic bacteria use substances such as nitrate, iron and manganese oxides, sulfate, and carbon dioxide as the ultimate electron acceptors in the absence of oxygen. In typical marine sediments (Table 3-1) a gradient exists in which aerobic respiration occurs in the surficial sediments. The oxygen is depleted in the underlying sediments so aerobic respiration is replaced by anaerobic respiration. These include nitrate reduction (denitrification), iron and manganese reduction, sulfate reduction, and carbon dioxide reduction (methanogenesis) in progressively deeper sediment layers. Some microorganisms are able to move vertically in the gradient to take advantage of local availability of electron acceptors.

3.2.3.1.3 Aerobic respiration. Many organisms carry out aerobic respiration in which enzymes remove electrons from organic compounds and pass them through a chain of carriers including flavoproteins and cytochromes located in intracellular membranes (Fig. 3-4) until finally they are used to reduce oxygen to produce water. ATP is produced by an enzyme called ATPase, that is located in the cell membrane, and the process is driven by a proton gradient across the membrane.

If aerobic respiration continues to completion, all of the organic material is oxidized to form carbon dioxide and water. Much more energy is

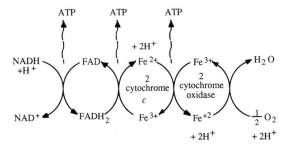

Fig. 3-4 Electron transport process schematic, showing coupled series of oxidation–reduction reactions that terminate with the reduction of molecular oxygen to water. The three molecules of ATP shown are generated by an enzyme called ATPase which is located in the cell membrane and forms ATP from a proton gradient created across the membrane.

potentially available in this process than in fermentations because the organic compound is completely oxidized to inorganic constituents. For example, in lactic acid fermentation, a net gain of only two ATP molecules results from each molecule of glucose degraded. In contrast, 36 ATP molecules can be generated per molecule of glucose during aerobic respiration.

Some bacteria generate energy by the oxidation of reduced inorganic compounds, using special enzymes that allow them to remove electrons from these materials. The electrons are passed through electron transport systems in which a proton gradient is formed and ATP is generated by membrane-bound ATPases. For example, nitrifying bacteria oxidize ammonia to nitrite or nitrite to nitrate; sulfur bacteria oxidize hydrogen sulfide or sulfur to sulfur or sulfate; iron bacteria oxidize reduced ferrous iron to form hematite; hydrogen bacteria use hydrogen gas as an energy source. Some of these bacteria are aerobic; others are anaerobic. For example, some hydrogen-oxidizing bacteria use sulfate as an electron acceptor.

3.2.3.2 *Energy generation by autotrophic organisms*

Autotrophic organisms derive their carbon from carbon dioxide. They use either light or an inorganic chemical as their energy source. The dominant form of autotrophy is photosynthesis, a process that uses light as an energy source. Photosynthetic organisms produce special pigments such as chlorophyll *a* that absorb light in internal cell membranes. When a phototroph is illuminated, photons cause electrons to be emitted from chlorophyll *a*. These electrons are then passed through an electron transport chain in a membrane. Protons are released across the membrane as in respiration, and they produce ATP via ATPases.

Photoautotrophic organisms, such as algae, cyanobacteria, and plants, all contain chlorophyll *a* and obtain energy by a process known as *oxygenic photosynthesis*. The overall chemical reaction of this process is:

$$CO_2 + H_2O \rightarrow (CH_2O)_n + O_2 \qquad (1)$$

where $(CH_2O)_n$ refers to organic material. In this reaction, the oxygen is derived from water. The sequence of reactions by which plants, algae, and cyanobacteria fix carbon dioxide is referred to as the Calvin–Benson cycle (Fig. 3-5).

However, many photosynthetic bacteria, such as purple sulfur and green sulfur bacteria contain special bacteriochlorophyll compounds (not chlorophyll *a*) and carry out *anoxygenic photosynthesis* without producing oxygen:

$$CO_2 + H_2S \rightarrow (CH_2O)_n + S^0 \qquad (2)$$

Purple sulfur bacteria fix carbon dioxide using the Calvin–Benson cycle, but green sulfur bacteria use a completely different pathway, the reverse tricarboxylic acid cycle. Other photosynthetic bacteria use still different pathways for CO_2 fixation (Perry and Staley, 1997).

Carbon dioxide can also be fixed by bacteria that use inorganic chemicals as an energy source, a process called *chemoautotrophy*. For example, sulfur-oxidizing bacteria obtain energy by oxidizing hydrogen sulfide or elemental sulfur as an energy source and produce sulfate as an end product. They fix carbon dioxide using the Calvin–Benson cycle. Methanogens fix CO_2 using an entirely different pathway. In addition, acetogenic bacteria fix carbon dioxide and produce acetic acid (as well as cell material). Both acetogens and chemoautotrophic methanogens use hydrogen gas as an energy source.

3.2.3.3 *Molecules synthesized by organisms*

Using the energy obtained from photosynthesis, chemoautotrophy, and heterotrophy, organisms synthesize an amazing variety of molecules. For the purpose of understanding the roles of biochemical syntheses carried out by organisms, we can group this enormous variety of molecules into a rather limited number of groups according to their functions. Basically molecules are used for (a) cell structure; (b) metabolism, energy storage, and energy transfer; (c) information storage and information transfer; (d) modifiers of other chemicals (enzymes); and (e) defense against predators, parasites, and competitors. A given molecule may, of course, serve more than one function. Because of the functional requirements of molecules for these different purposes, some types are important in biogeochemical cycles; others are not, as will be described later. The key factors influencing the biogeochemical significance of a molecule are its *per capita* rate of production, the abundance of its producers, its rate of chemical decomposition, the nature of its degradation products, and the mobility of the molecule in its original and transformed states.

3.3 Evolutionary Mechanisms

Biological evolution is a change over time in the genetic composition of members of a population of organisms that are mating with one another. Thus, evolution is a population process, not a

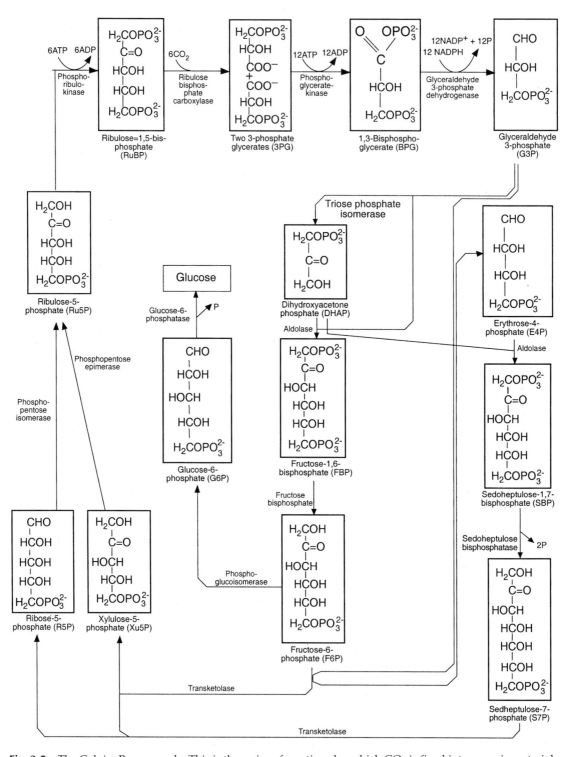

Fig. 3-5 The Calvin–Benson cycle. This is the series of reactions by which CO_2 is fixed into organic material.

change that happens to individuals; to understand evolutionary changes "population thinking" is required. Changes that happen over a small number of generations constitute *microevolution*. Changes that take centuries, millennia, or longer to be completed are called *macroevolution*.

Much of what we know about the history of life on Earth comes from *fossils*, the preserved remains of organisms or impressions of organisms in materials that eventually became rocks. Although fossils have been known since prehistoric times, their significance as the remains or traces of early life forms was not appreciated until the Renaissance. The careful studies of William Smith (1769–1839), an English geologist, led to the recognition that certain identifiable strata in sedimentary rocks always contained the same types of fossils. Paleontologists also noted that the lowest strata with fossils contained fewer types of organisms and ones that were simpler in structure than strata closer to the surface. In the most ancient strata there is no evidence at all for living organisms, but rocks dated at 3.5 Gyr of age contain spherical carbon-containing structures approximately the size of modern prokaryotic cells, fossils of filamentous bacteria, and finely stratified undulating sediments (Fig. 3-6), thought to be fossilized stromatolites (Hofmann and Schopf, 1983). Microbial fossils are now known from deposits older than 3.5 Gyr of age, indicating that the first microorganisms had evolved within 1 Gyr after the formation of Earth (Awramik *et al.*, 1983).

The major periods during Earth's evolution were originally demarcated by fossils in strata of different ages, even though the ages of the strata had not been determined (Table 3-2). The Hadean eon, which extended from the time of the origin of Earth to about 3.9 Gyr ago, was a period during which Earth's crust formed. So much debris hit the surface of Earth at that time, and so much crustal movement has occurred since then, that no rocks survive on the surface from this eon. During the Archaean eon, which extended from about 3.9 to 2.6 Gyr ago, more stable crustal features developed and life evolved. The major metabolic patterns of living organisms also evolved during this eon, including fermentation, photosynthesis, and the ability of cells to convert atmospheric nitrogen into a useful form (i.e., nitrogen fixation). The Proterozoic eon extended from 2.6 Gyr ago to about 0.6 Gyr ago. During this time new cell types and sexual reproduction evolved, leading to the evolution of plants and animals.

The fossil record illustrates that most changes in lineages of organisms were gradual. Even the most rapid ones required millions of years for their realization. Species persist, on average, for only a few million years, and the course of biological evolution has been interrupted many times by periods of mass extinction. After each of these episodes, the diversity of life has rebounded, but several million years or more were required before the original diversity of life was reestablished or exceeded. After periods of mass extinction, the groups of organisms that became dominant usually differed from those that dominated the biota prior to the extinction episode. Rates of evolutionary change have been very uneven. Many species experienced long periods of *stasis*, during which they changed very little. Periods of stasis were repeatedly interrupted by periods of rapid evolutionary change, but not all lineages underwent rapid evolution at the same time.

Because biological evolution is a change over time in the genetic composition of members of a population, evolutionary biologists attempt to measure of genetic variability and how it changes. The genetic constitution governing a heritable trait is called its *genotype*. A population evolves when individuals with different genotypes survive or reproduce at different rates, but agents of evolution do not act directly on a genotype. They act on the physical expression of an organism's genotype – its *phenotype*. Not all phenotypic variation is governed by genotypes. For example, the leaves on a tree or shrub are normally genetically identical, but they may differ dramatically in size and shape as a result of differences in the amount of wind and sunlight to which they are exposed when they expand. Such environmentally induced variation is ecologically important, but only traits that are, at least in part, heritable, determine the direction of evolution.

Individuals of most species have two sets of

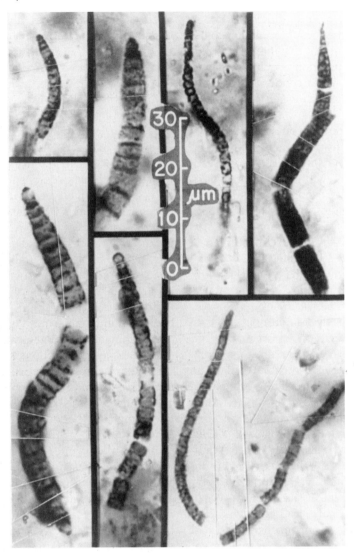

Fig. 3-6 Microfossils of several filamentous microorganisms. Bitter Springs Formation, Central Australia. Dated at 0.85 Gyr. (Courtesy William Schopf.)

chromosomes, the structures on which the genes are linearly arranged, in their cells. One chromosome is derived from their maternal parent and one from their paternal parent. A gene occupies a particular place (locus) on a chromosome, and different forms of a gene – called *alleles* – are present in most populations. An individual may have the same or different alleles for a particular gene on the chromosomes it inherited from its maternal and paternal parents. The alleles of a particular gene are designated by variations of a single label, i.e. A and a, or A^1, A^2, and A^3. The percentages of different alleles at each locus describe the genetic structure of a population.

An equilibrium population, one that is not changing genetically, persists under conditions that were discovered independently in 1908 by the British mathematician G. H. Hardy and the German physician W. Weinberg. The Hardy–Weinberg Rule, which is the foundation of population genetics, consists of three assumptions and two major mathematical results. The assumptions are that the population is very large, that mating is random, and that no

Table 3-2   Geological history of the Earth

Eon/Era	Period	Began[a]	Major physical and biological events
Hadean		4.5 Gyr	
Archean		3.8 Gyr	Origin of life; prokaryotes flourish. Photosynthetic cells liberate oxygen, O_2 first appears in the atmosphere
Proterozoic		2.5 Gyr	O_2 levels increase to >1% of current level. Eukaryotes evolve; several animal phyla appear
Phanerozoic		600 Myr	O_2 levels increase to >5% of current level
Paleozoic	Cambrian	600 Myr	Climate warms, O_2 levels approach current level; most animal phyla present, including some that failed to survive; algae and cyanobacteria diversify
	Ordovician	500 Myr	Diversification of echinoderms, other invertebrate phyla, jawless fishes. *Mass extinction* at end of period (ca. 85% of all species disappear)
	Silurian	440 Myr	Sea levels rise, two large continents form, hothouse climate. Diversification of jawless fishes, first bony fishes; invasion of land by vascular plants and arthropods
	Devonian	400 Myr	Continents collide at end of period. Asteroid probably collides with Earth. Diversification of bony and cartilaginous fishes; trilobites diversify; origin of ammonoids, amphibians, insects; first forests. *Mass extinction* at end of period (ca. 75% of species disappear)
	Carboniferous	345 Myr	Climate cools, marked latitudinal gradients. Extensive forests of early vascular plants, especially club mosses, horsetails, ferns. Coal beds form. Amphibians diversify; first reptiles appear. Radiation of early insect orders
	Permian	290 Myr	All land united in one large continent – Pangaea; large glaciers form. Reptiles, including mammal-like forms, radiate; amphibians decline; diverse orders of insects evolve. Conifers appear. *Mass extinction* at end of period (ca. 95% of all species disappear)
Mesozoic	Triassic	245 Myr	Continents begin to drift apart. Early dinosaurs; first mammals; gymnosperms become dominant; diversification of marine invertebrates. *Mass extinction* at end of period (ca. 75% of all species disappear)
	Jurassic	195 Myr	Two large continents form: Laurasia (north) and Gondwana (south). Dinosaurs diversify; first birds and mammals evolve; gymnosperms dominate terrestrial vegetation; ammonites radiate into diverse forms
	Cretaceous	138 Myr	Gondwana begins to break up. Continued radiation of flowering plants; mammals begin diversifying. Meteorite strikes Yucatan Peninsula at end of period causing *mass extinction* (ca. 75% of all species disappear)
Cenozoic	Tertiary	66 Myr	Climate Cools. Continents nearing modern positions. Drying trend in middle of period. Radiation of birds, mammals, flowering plants, pollinating insects
	Quaternary	3 Myr	Repeated glaciations, North and South America join; mass extinctions of large mammals, evolution of *Homo*; rise of civilizations, humans begin to modify biogeochemical cycles

[a] Time before present. 1 Gyr = 1 billion years.
[b] After Purves *et al.* (1995).

agents of evolution are acting on the population. If these conditions hold, the frequencies of alleles at a locus will remain constant from generation to generation, and, after one generation of random mating, the genotypic frequencies will remain in the proportions

$$p^{2(AA)} + 2pq^{(Aa)} + q^{2(aa)} = 1$$

where p is the frequency of allele A in the population and q is the frequency of allele a in the population, provided that the gene has only two allelic forms, A and a. The mathematical derivation of these results, which need not concern us here, is presented in all elementary biology texts.

It is clear that the conditions necessary for the Hardy–Weinberg Rule to apply are rarely met in nature. Populations are often small, mating is often nonrandom with respect to genotype (likes mate with likes), and evolutionary agents typically are acting on populations. The rule is nonetheless important because significant deviations of genotype frequencies in a population from Hardy–Weinberg expectations are evidence that an agent of evolution is in action.

Sexual reproduction, by bringing together the genetic material from two parents, greatly increases the amount of genetic variation present within a population. In a population that reproduces asexually, offspring are genetically identical to their parents unless genes in the parents have mutated. Sexual recombination, on the other hand, generates an enormous variety of genotypic combinations upon which the agents of evolution can act. Therefore, sexual recombination increases the evolutionary potential of a population, but sexually reproducing populations do not necessarily evolve rapidly.

3.3.1   Evolutionary Agents

Evolutionary agents are forces that change allele and genotype frequencies in populations, causing them to deviate from Hardy–Weinberg expectations. The known evolutionary agents are mutation, gene flow, genetic drift, nonrandom mating, and natural selection.

3.3.1.1   Mutation

Mutation is a stable, heritable change of a gene from one allele to another, which both creates and maintains genetic variability in populations. Most mutations adversely affect the survival and reproductive success of their bearers, but if the physical or biological environment changes, previously neutral or harmful alleles may become beneficial. Mutation rates typically are very low, but they are sufficient to create considerable genetic variation over many generations.

3.3.1.2   Gene flow

When individuals migrate to, and then breed in a new location, they may add new alleles to a population or may change the frequencies of alleles already present. Similarly, emigrants may remove alleles or may change the frequencies of alleles in the populations from which they departed.

3.3.1.3   Genetic drift

Genetic drift is alteration of allele frequencies in small populations by chance events. During times when large populations are reduced to small numbers of individuals, genetic variation is likely to be lost by chance. Genetic drift also occurs when a small number of individuals colonize a new region. The pioneers are unlikely to have all the alleles found in the source population and the frequencies of the ones they do have are likely to differ from those in the source population.

3.3.1.4   Nonrandom mating

This results when individuals with certain genotypes mate more often with individuals of either the same or different genotypes than would be expected on a random basis. Self-fertilization, which is common among many groups of organisms, particularly plants, is an extreme form of nonrandom mating. Another common type of nonrandom mating is preferential mating by females with certain males.

3.3.1.5   Natural selection

Because not all individuals in a population survive and reproduce equally well in a particular environment, some individuals contribute more offspring to subsequent generations than do other individuals. Such differential contribution of offspring resulting from variations in heritable traits was called natural selection by Charles Darwin. *Natural selection is* especially important because it is *the only evolutionary agent that adapts organisms to their environments.*

Natural selection can produce several different outcomes. It may (1) preserve the genetic characteristics of a population by favoring average individuals (*stabilizing selection*), (2) change the characteristics of a population by favoring individuals that vary in one direction from the population mean (*directional selection*), or (3) change the characteristics of a population by favoring individuals that vary in opposite directions from the population mean (*disruptive selection*). Stabilizing selection is the norm because most populations are not evolving rapidly most of the time, but evolutionary changes depend on directional or disruptive selection.

3.3.2   Catastrophic Events and the Course of Evolution

Although natural selection is the only evolutionary agent that adapts organisms to their environments, the course of evolution has been profoundly influenced by major environmental changes, some of which had catastrophic effects. Some of these events resulted from Earth's internal processes, such as the activity of volcanoes and the shifting and colliding of continents. Others were the result of external events, such as collision of meteorites with Earth.

The movement of Earth's crustal plates and the continents they contain – *continental drift* – has had enormous effects on climate, sea levels, and the distributions of organisms. Mass extinctions of organisms have usually accompanied major drops in sea levels. The collision of all the continents to form the gigantic landmass called Pangaea about 260 million years ago, triggered massive volcanic eruptions. The volcanoes ejected enough ash into Earth's atmosphere to significantly reduce sunlight penetration and trigger massive glaciation. The result was the extinction of about 90% of all species on Earth, both terrestrial and marine.

Through much of its history, Earth's climate was much warmer than it is today, and temperatures decreased more slowly toward the poles. At other times, Earth was colder than it is today, and massive glaciers formed at high latitudes. We live in one of the colder periods of Earth's history.

External events have also triggered important changes. At least 30 meteorites hit Earth each year, but collisions with very large meteorites are very rare. One, about 10 km in diameter that collided with Earth 65 million years ago, caused massive firestorms and tidal waves and triggered the extinction of many species of marine organisms and all terrestrial animals larger than about 25 kg in body weight.

3.3.3   The Biological Consequences of Evolution

All living organisms are descendants of a lineage of unicellular organisms that lived almost four billion years ago. Over the course of evolution, living organisms stored greater quantities of information and evolved increasingly complex mechanisms for using it. But if the evolution of complexity were the entire story, only one kind of organism might exist on Earth today. Instead, Earth is populated by many millions of genetically different kinds of organisms, called *species*, that rarely or never interbreed with one another.

For two species to form from a single one, the ancestral population must become divided into two or more separate populations among which gene exchange does not occur. Over time, genetic differences may accumulate as the separated groups adapt to the particular environments in which they live. If enough differences accumulate, individuals of one population may not be able to breed with individuals of the other if their ranges subsequently overlap. They will have become different species.

Gene flow among members of a population

may be interrupted in several ways. A common method is division of the population by a barrier, such as a water gap for terrestrial organisms, dry land for aquatic organisms, and mountains. Barriers can form when continents drift, sea levels rise and fall, or climates change. Populations separated in these ways are typically large initially. They evolve differences because the places in which they live are or become different. Alternatively, gene flow may be interrupted if some members of a population cross an existing barrier and form a new isolated population.

Gene flow among members of a population may also be interrupted in the absence of geographical separation. The most common means is *polyploidy*, a duplication of the number of chromosomes in the cells of its members. Polyploidy can arise in two ways. One way is the accidental production during cell division of cells having four (tetraploid) rather than the normal two (diploid) sets of chromosomes. Tetraploid individuals usually cannot produce fertile offspring by mating with diploids because their chromosomes do not pair properly during cell division, but they can mate with one another to form a new evolutionary lineage. A polyploid species can also be produced when individuals of two different species, whose chromosomes do not pair properly during cell division, interbreed. The resulting individuals are usually sterile, but they may be able to reproduce asexually. After many generations, their descendants may eventually become fertile as a result of further chromosome duplication. Speciation by means of polyploidy is very rare among animals but has been very important in the evolution of flowering plants. More than half of all species of flowering plants are polyploids.

Repeated splitting of lineages of organisms into separate species has resulted in the great richness and diversity of life that lived in the past and is found on Earth today. Earth would be very different if speciation had been a rare event during the life's history. Biogeochemical cycles are influenced by millions of species, each adapted to live under a particular range of conditions and to use environmental resources in a particular way.

3.4   The Diversity of Living Organisms

Based on the structure of their cells, organisms can be grouped into two broad categories. *Prokaryotic* (meaning "before-nucleus") organisms have structurally simple cells. Their DNA is not bound by a nuclear membrane (Fig. 3-7), and they typically have a single, circular chromosome. By contrast, the cells of *eukaryotic* organisms (meaning "true nucleus") have a nuclear membrane that surrounds their chromosomes (Fig. 3-8). Typically each eukaryotic cell has two sets of chromosomes, one set derived from each parent. Eukaryotic cells also contain additional membrane-bound structures outside the nucleus. The most important of these are mitochondria, the sites of aerobic respiration, and – among photosynthetic plants – chloroplasts, the centers of photosynthetic activity.

Although many prokaryotic organisms are single-celled (unicellular), some exist as multicellular filaments or collections of cells. Eukaryotic organisms may be unicellular or multicellular. Most eukaryotic cells are at least 5 μm in diameter, but many are much larger. The cells of most prokaryotes are small, ranging from 0.2 to 1 μm in diameter, but a few are much larger.

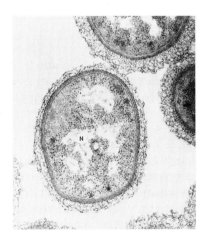

Fig. 3-7 A thin section through a prokaryotic cell. Note that the nuclear material (N) is not bound by a membrane, but is free in the cytoplasm. Mitochondria and other intracytoplasmic structures are absent. (Reprinted with permission from J. J. Cardamone, Jr., Univ. of Pittsburgh/Biological Photo Service.)

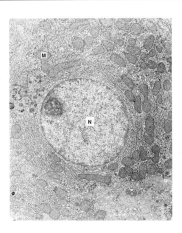

Fig. 3-8   A thin section showing a eukaryotic cell. Note the nucleus (N) is bound by a nuclear membrane. In the cytoplasm of the cell are many mitochondria (M) and intracytoplasmic membranes. (Reprinted with permission from Richard Rodewald, Univ. of Virginia/Biological Photo Service.)

Each cell of a prokaryotic organism typically divides by an asexual process, after duplication of its chromosome, to give rise to two essentially identical daughter cells. Gene exchange among prokaryotes occurs in some species when two cells of different mating types come into contact or by virus-mediated transfers. Typically, only part of the genetic material of a cell is transferred to a recipient cell. Most eukaryotic organisms undergo a sexual process during reproduction in which the genetic material is duplicated and a complete set is transferred to the recipient cell.

Because most complex organisms require oxygen for growth, multicellular organisms probably did not evolve until after the appearance of oxygen on Earth, i.e. not before some 2 Gyr ago. Indeed, even though oxygen might have been produced prior to 2 Gyr ago, considerable time would have been required before atmospheric concentrations of oxygen became substantial. Initially oxygen would have reacted with reduced compounds such as iron; only after these more reduced forms were fully oxidized would oxygen have accumulated to a significant level. Moreover, when oxygen first accumulated on Earth, it would have been toxic to nearly all organisms, just as O_2 kills most organisms that live in environments lacking oxygen today. This may have been the first major event of biologically caused pollution, but it provided conditions that favored both the ability to tolerate oxygen and the ability to use it in aerobic respiration. Prokaryotes predominated from about 3.5 Gyr ago to about 0.6 Gyr ago, or three-quarters of the time life has existed. The tremendous variety of eukaryotic organisms originated and evolved largely during the last 0.6 Gyr, after Earth's atmosphere became oxygenated.

3.4.1   The Classification of Organisms

To help us comprehend the incredible diversity of life, biologists have developed a system of classifying organisms. In that system, first proposed by the Swedish biologist Carolus Linnaeus in 1758, each species, for example, *Homo sapiens*, is given two names. The first name identifies the *genus* (plural, genera), a group of closely related species; the second designates the species. Because this system of binomial nomenclature has been universally adopted, scientists throughout the world are able to use a single name for a particular species of organism.

Species and genera are, in turn, grouped into larger units in a hierarchical system. The units in the system cluster organisms that are believed to share a common ancestor, that is, the classification system attempts to represent evolutionary relationships among organisms. Members of a genus share a relatively recent ancestor. The most recent common ancestor of members of larger taxonomic units lived in the more distant past.

For two reasons, determining evolutionary relationships among organisms is a difficult task. First, the fossil record is incomplete; only a small fraction of species that have ever lived on Earth have left a fossil record, and typically only hard body parts are preserved. Second, organisms are so variable in their morphological and physiological features that they share few traits with which they can be compared. However, an important evolutionary record is preserved in the sequences of certain macromolecules that are shared by all living organisms. The most important of these universally shared molecules is ribosomal RNA (rRNA). The RNA of ribosomes, which carry out protein synthesis, contains a small and a large subunit. A widely accepted phylogeny of all life on Earth has been erected by comparing the structure of RNA in the small subunit. In the classification system based on these data, organisms are grouped into three domains, the *Bacteria*, the *Archaea*, and the *Eucarya* (Table 3-3).

Members of the domains Bacteria and Archaea are prokaryotic. Their cells lack the complex, membrane-bound structures found in cells of the third domain. Approximately half the mass of living organisms on Earth consists of prokaryotic organisms (Whitman *et al.*, 1998). Bacteria are divided into more than a dozen major groups, such as the Gram positive bacteria, Spirochetes, Chlamydia, Proteobacteria, Green filamentous bacteria, and Planctomycetes, as well as others. These organisms, which are found in nearly all soils and aquatic environments, include the medically important pathogenic bacteria. Almost all contain a cell wall macromolecule – peptidoglycan – whose synthesis is inhibited by penicillin. Processes carried out by members of this group include nitrogen fixation, denitrification, nitrification, anoxygenic photosynthesis, sulfate reduction, sulfur and sulfide oxidation, acetogenesis, and methane oxidation. Members of the group also degrade toxic compounds such as aliphatic and polycyclic aromatic hydrocarbons and halogenated organic compounds such as chlorinated phenols.

The Archaea differ from the Bacteria in that they lack peptidoglycan and contain special membranes that lack the fatty acids found in Bacteria and Eucarya. The Archaea also are characterized by their unusual habitats and metabolic processes. Some of them, which can grow at temperatures in excess of 100°C, live in hot springs and marine hydrothermal vents. Others grow in saturated brines. Another group is noted for its unique metabolic capability to produce methane gas (methanogenic) from hydrogen and carbon dioxide as well as methanol or acetic acid. These methanogens are obligate anaerobes that live in reduced aquatic sediments, soils, and in the intestinal tracts of animals such as ruminants.

Members of the other domain – *Eucarya* –

have more complex cells with nuclei and elaborate cellular compartments. They are classified into four kingdoms – Protista, Plantae, Fungi, and Animalia.

Protists are a heterogeneous group of microorganisms that either have a single cell or are groups of similar cells joined together. Some are autotrophs, others are ingestive heterotrophs, and still others are heterotrophs. One phylum (Sporozoa) consists of nonmotile forms, but members of the others move by means of ameboid motion, ciliary action, or flagella. Many marine species (foraminifera) secrete skeletons of calcium carbonate and their remains are the major contributors to the formation of limestone. Others (radiolarians) secrete glassy siliceous skeletons that are the principal components of sediments under many tropical seas.

Fungi are mostly multicellular, heterotrophic organisms that absorb their food. Most species are saprophytes (living on dead matter) and they produce, at some time in their life cycles, characteristic and often complex reproductive structures that differentiate them from protists, plants, or animals. Some are parasites. Fungal cell walls contain a variety of polysaccharides, often including cellulose and, in some species, chitin. Lichens are composite organisms that consist of a meshwork of a fungus and some photosynthetic organism, either an alga or a cyanobacterium.

Plants range in size from single-celled forms to large trees and vines. The *Rhodophyta, Chlorophyta, Pyrrophyta, Chrysophyta*, and *Phaeophyta*, collectively known as algae, abound in fresh and marine waters and on moist terrestrial substrates. They account for more than one-fourth of the photosynthesis occurring on Earth. They use a variety of photosynthetic pigments, apparently adaptations to the very different light regimes found at different depths in aquatic environments. They synthesize a number of different molecules for the storage of their food reserves (starch, fats, and oils). The materials used in the construction of their cell walls are also highly varied (cellulose, pectin substances, silica, lignin, mucilage, and calcium carbonate). The largest species may attain lengths in excess of 35 m, but the cells of even

Table 3-3 Major phylogenetic groups of living organisms

Taxon	Representatives
Domain Bacteria	
phyla:	Proteobacteria, Cyanobacteria, Gram-positive Bacteria, Chlamydia, Spirochaetes, Planctomycetes, Green sulphur bacteria, Green filamentous bacteria
	Verrucomicrobia, Cytophaga-Flavobacterium, Deinococci
Domain Archaea	
phyla:	Methanogens, Extreme Halophiles, Thermoplasmas, Hyperthermophiles
Domain Eukarya	
Kingdom Protista	
Phylum Mastigophora	Flagellates
Phylum Sarcodina	Amebas and their relatives
Phylum Actinopodia	Actinopods
Phylum Foraminifera	Foraminiferans
Phylum Sporozoa	Ameboid parasites
Phylum Ciliophora	Ciliates
Phylum Apicomplexa	Apicomplexans
Phylum Pyrrophyta	Dinoglagellates
Phylum Chrysophyta	Diatoms
Phylum Phaeophyta	Brown algae
Phylum Rhodophyta	Red algae
Phylum Chlorophyta	Green algae
Kingdom Plantae	
Phylum Hepatophyta	Liverworts
Phylum Anthocerophyta	Hornworts
Phylum Broyphyta	Mosses
Phylum Lycophyta	Club mosses
Phylum Sphenophyta	Horsetails
Phylum Psilophyta	Whisk ferns
Phylum Pterophyta	Ferns
Phylum Cycadophyta	Cycads
Phylum Ginkgophyta	Ginkgos
Phylum Gnetophyta	Gnetum, Ephedras
Phylum Coniferophyta	Conifers
Phylum Angiospermae	Flowering plants
Kingdom Fungi	
Phylum Chytridiomycota	Water molds
Phylum Zygomycota	Bread molds
Phylum Basidiomycota	Mushrooms, rusts, smuts
Phylum Ascomycota	Yeasts, sac fungi
Kingdom Animalia (only the major phyla are listed)	
Phylum Porifera	Sponges
Phylum Cnidaria	Hydras, jellyfish, corals
Phylum Ctenophora	Comb jellies
Phylum Platyhelminthes	Flatworms
Phylum Rotifera	Rotifers
Phylum Nematoda	Round worms
Phylum Mollusca	Chitons, snails, clams, squids, octopi
Phylum Annelida	Segmented worms
Phylum Arthropoda	Scorpions, spiders, crabs, insects, millipedes, centipedes
Phylum Bryozoa	Moss animals
Phylum Brachiopoda	Brachiopods
Phylum Echinodermata	Sea lilies, seastars, sea urchins, sand dollars, sea cucumbers
Phylum Chordata	Tunicates, sharks, bony fishes, amphibians, reptiles, birds, mammals

the largest species are not differentiated into distinct types. The large blades of marine kelps, for example, consist primarily of thin sheets of identical cells.

Mosses and liverworts (*Bryophyta*) are more complex than algae. Some of the larger species have structures that superficially appear similar to roots, stems and leaves, but they lack the internal conducting systems present in the vascular plants (*Tracheophyta*). Internal transport systems (vascular systems) make possible the large sizes of terrestrial plants where the soil is the source of some requisites (water, mineral nutrients) and the air is the source of others (CO_2, sunlight). The different groups of vascular plants are characterized primarily by their methods of reproduction. Vascular plants are the source of all wood.

More than one million species of animals (kingdom *Animalia*) have been described by scientists. Estimates of the true number range to higher than 30 million because most species of insects and other arthropods are as yet not described. Because they are heterotrophs, animals represent less biomass than the autotrophs upon which they depend for their food, but because many of them construct sturdy skeletons that are durable and resistant to degradation, they are important contributors to biogeochemical cycles. And, of course, the species producing the largest contributions to and perturbations of biogeochemical cycles is an animal – *Homo sapiens*.

3.5 The Ecological Organization of the Living World

Ecology is the study of the distribution and abundance of organisms, their interrelationships, and the communities of which they are a part. Ecological investigations range from the study of the behavior of individuals in response to their environments, to the flow of energy and matter through organisms and the environment at regional to global scales. Here we focus on *ecosystems* – spatially explicit units that include all organisms that live there and relevant components of their abiotic environment – because the consequences of interactions of organisms

with the physical environment at large scales are the subject matter of this book.

The bodies of living organisms are excellent sources of energy-rich molecules that can be used to fuel the machinery of other organisms. Since early in the evolution of life, some organisms have obtained their energy and materials by consuming others. The major dynamics of the living world today are dominated by eating and being eaten. Even though most species interact strongly with only a small proportion of species living in the same area, a rich network of connections exists in all ecosystems. Ecologists devote much of their efforts to studying the webs of connections that form the bases of ecological communities.

3.5.1 Energy Flow in Ecosystems

The entry of materials and energy into the living world is largely via photosynthesis. Except for a few ecosystems (deep-sea thermal systems, deep-Earth microbial communities) in which solar energy is not the main energy source, almost all energy utilized by organisms comes from (or once came from) the sun. The fraction of solar energy falling on Earth's surface that is used in photosynthesis is small. Earth intercepts 5×10^{24} J of energy per year from the sun, only about 3×10^{20} J of which are captured by photosynthesis. Chlorophyll *a* and its associated pigments absorb only a small fraction of the total energy present in sunlight. Most incident solar radiation is reflected from Earth's surface. Much is converted to heat, some of which is used to evaporate water and, hence, to drive the global hydrologic cycle.

The organisms that obtain their energy from the same general source constitute a *trophic level*. The major trophic levels in ecological communities are photosynthesizers (primary producers), herbivores (eaters of plants or parts of them), primary carnivores (eaters of herbivores), secondary carnivores (eaters of primary carnivores), tertiary carnivores, and detritivores (eaters of the dead remains of once living organisms) (Fig. 3-9). A sequence of linkages in which a plant is eaten by an herbivore, which is in turn eaten by a primary carnivore, and so on, is

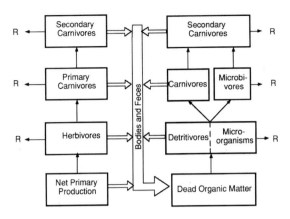

Fig. 3-9 Trophic levels in ecosystems. Thin arrows show flow of energy up the food chain (through living biomass) and the broad arrows show the complementary flow of dead organic matter (detritus) back down. R indicates respiration.

called a *food chain*. Because most species in a community eat and are eaten by more than one species, food chains are interconnected to form *food webs*.

Trophic levels are imprecise categories; many organisms, such as human beings, obtain their energy from two or more trophic levels.

The total amount of energy that plants assimilate by photosynthesis is called *gross primary production*. The production that remains after subtracting the energy that plants use for maintenance and building tissues is called *net primary production*. The global distribution of net primary production in Earth's major ecological zones is shown in Table 3-4. Oceans, despite their much larger surface area, contribute less than half of Earth's net primary production because surface waters, where photosynthesis can take place, are highly deficient in nutrients over most of the oceans. Oceanic production is concentrated in coastal areas, especially where upwelling of deep water brings nutrients to the surface. On land, photosynthesis is often limited by dryness, cold temperatures, and nutrient shortages, which is why subtropical and tropical areas contribute much more than their proportional share to global primary production. Productivity also changes in response to disturbances, such as fires, volcanic eruptions, and severe storms.

A shortage of any of some two dozen chemical elements that are essential for the growth of organisms can reduce ecosystem productivity, but phosphorus and nitrogen are often the most limiting nutrients, which is why these two elements are standard components of commercial fertilizers. Phosphorus is often limiting because it moves through soil pores and aquatic

Table 3-4 Net annual primary production of Earth's major ecological zones[a]

Zone	Area $(10^6\ km^2)$	Primary production of carbon $(Pg)^b$
Terrestrial	133.9	
Tropical evergreen forests		17.8
Broadleaf deciduous forests		1.5
Mixed broadleaf/needle forests		3.1
Coniferous forests		4.5
Savannahs		16.8
Grasslands		2.4
Deserts		1.5
Tundra		0.8
Cultivated areas		8.0
Total terrestrial		56.4
Marine	361.0	48.5
Global total	510.3	104.9

[a] After Field, Behrenfeld, Randerson and Falkowski, *Science* 281, 237–240, 1998.
[b] $1\ Pg = 10^{15}$ g.

ecosystems in the form of phosphate salts or organic phosphate compounds dissolved in water. If oxygen is present and pH is neutral or alkaline, phosphate complexes with calcium or iron and become immobile.

Much of the energy captured by an organism is used to support its basic metabolism and is dissipated as heat. Only the energy content of an organism's net production – its growth plus reproduction – is available to organisms at the next trophic level. Moreover, much of the energy potentially available to organisms in the next higher trophic level is not captured by them. Many organisms or parts of them die and are consumed by detritivores, which dominate the flow of energy from green plants and are the primary pathway for return of nutrients in a form useful for assimilation by green plants.

On average, only about 10% of the energy captured by one trophic level is taken in by the next level, but there are wide variations around this overall average. For example, the small, unicellular algae that dominate marine photosynthesis have high growth and cell division rates, possess easy-to-digest tissues, and can be eaten by very small animals. As a result, aquatic herbivores are able to consume about 51% of the primary production of oceans and lakes (Cyr and Pace, 1993). In many marine areas there may be, for parts of the year, a larger standing crop of herbivores than of the plants upon which they feed, due to the very high algal growth and reproduction rates. Similarly, in grasslands, where plants allocate most of their net production to easily digested tissues, mammals may consume 30 to 40% of the annual net aboveground net primary production; insects may consume an additional 5 to 15%. By contrast, in forested ecosystems, where plants allocate much of their energy to production of wood, which is difficult for most organisms to digest, less than 3% of net primary production may be consumed by herbivores.

The consequences of the massive "loss" of energy accompanying passage from one trophic level to another also include the fact that organisms low in the trophic ladder tend to dominate the cycling of elements through the biosphere. This is especially true on land where vascular plants dominate both the physical structure and the flow of energy in ecosystems. In the oceans, because the major photosynthesizers are small and structurally simple, the cycling of many elements is much more strongly influenced by herbivores and carnivores (Longhurst, 1998).

3.5.2 Elemental Cycles in Ecosystems

Energy is not recycled in ecosystems because at each transformation much of it is dissipated as heat, a form of energy that cannot be used by organisms to power their metabolism. Chemical elements, on the other hand, are not lost when they are transferred among organisms. Instead, they cycle through organisms and the environment. The quantities of carbon, nitrogen, phosphorus, calcium, sodium, sulfur, hydrogen, and oxygen, together with smaller amounts of other chemical elements, that are available to organisms are strongly influenced by how organisms obtain them, how long they hold on to them, and what they do with them while they have them.

The basic principles that govern the behavior of nutrients also govern the behavior of toxic materials in the environment, but some contaminants, such as metals and organic compounds, build up to very high concentrations in organisms via *biomagnification*. Any substance that is retained more efficiently than carbon or nutrients is likely to increase in concentration in an organism's tissues. At each step in a food chain, the concentration of such a contaminant increases in the consumers' tissues. The most serious contaminants whose concentrations have been greatly biomagnified are mercury and organochlorine compounds (pesticides such as DDT and chlordane, PCBs, chloromethanes). Microscopic photosynthetic aquatic organisms can concentrate PCBs up to 100 000 times more than in the surrounding water.

3.6 The Impact of Life on Biogeochemical Cycles

Organisms are involved in many transformations that have important environmental consequences. The activities of microbes are especially diverse and important. Pesticide degradation, cellulose decomposition, toxin production, deposition of limestone, rock and mineral weathering, formation of sulfur deposits, production of antibiotics, and methyl mercury formation are examples of the many processes of ecological and geochemical significance that are mediated, at least in part, by microorganisms (Table 3-5). Many further details of important

transformations are brought out in the chapters on the cycles of specific elements.

3.6.1 The Hydrologic Cycle

The major reservoirs of water on Earth are the oceans. The hydrologic cycle is driven primarily by evaporation of water from the oceans, lakes,

Table 3-5   Elemental cycles and selected important transformations in which organisms play roles

Cycle	Process (chemical transformation)	Biological group responsible
Carbon	Carbon dioxide fixation ($CO_2 \rightarrow$ organic material)	Photosynthetic organisms: plants, algae, bacteria
		Chemoautotrophic organisms: nitrifying bacteria, some sulfur oxidizers, iron oxidizers, hydrogen oxidizers
	Aerobic respiration (organic material $+ O_2 \rightarrow CO_2 + H_2O$)	All plants, animals, and strictly aerobic microbes
	Organic decomposition (or mineralization) (organic material $\pm O_2 \rightarrow \rightarrow$ inorganic material)	Microorganisms, especially fungi and bacteria
	Methane production [$CO_2 + H_2$ (or simple organic compound such as acetate) $\rightarrow CH_4 + H_2O$]	Methane-producing bacteria
Nitrogen	Nitrogen fixation [$N_2 \rightarrow \rightarrow RNH_2$ some amino group)]	Free living prokaryotes: *Azotobacter* (organic spp., some *Clostridium* spp., Cyanobacteria, photosynthetic bacteria
		Symbiotic prokaryotes: *Rhizobium* spp. and others
	Nitrification ($NH_3 \rightarrow NO_2^-$) ($NO_2^- \rightarrow NO_3^-$)	Chemoautotrophic nitrifying bacteria
	Dissimilatory denitrification[a] ($NO_3^- \rightarrow N_2, N_2O$)	Anaerobic respiring bacteria
	Ammonification (organic nitrogen $\rightarrow NH_3$)	Many microbes, especially bacteria
Sulfur	Sulfur oxidation ($H_2S \rightarrow \rightarrow S \rightarrow S_2O_3^{2-} \rightarrow \rightarrow SO_4^{2-}$)	Purple and green sulfur photosynthetic bacteria, some cyanobacteria
		Chemoautotrophic sulfur oxidizers
	Dissimilatory sulfate reduction[b] ($SO_4^{2-} \rightarrow \rightarrow H_2S$)	Sulfate-reducing bacteria
	Dimethyl sulfide production $SO_4^{2-} \rightarrow (CH_3)_2S$	Certain marine algae
Metal cycles	Iron and manganese oxidation and reduction	Iron bacteria and manganese bacteria

[a] Assimilatory denitrifiers reduce nitrate to the amino acid level where it is incorporated into protein. Many plants and bacteria can do this and, therefore, use nitrate as a nitrogen source.
[b] Assimilatory sulfate reducers reduce sulfate to the sulfhydryl level where it is incorporated into the sulfur amino acids of protein. Many plants and bacteria can do this.

and rivers. However, land plants also influence the hydrologic cycle by taking up water through their root systems. Impurities in the water provide nutrients for plant growth and water is a raw material for photosynthesis. Excess water escapes to the atmosphere through tiny openings in the leaves stomata, a process known as *evapotranspiration*.

The synthesis of 1 kg of dry plant biomass requires the evapotranspiration of about 300 L of water, although smaller amounts of water are needed by some plants such as desert cacti. Approximately one-third of the annual continental rainfall (100 cm/yr) is returned to the atmosphere by evapotranspiration. Although it accounts for only about 10–15% of global evaporation, plant evapotranspiration can play a major role in local climates. For example, a molecule of water falling on the upper Amazon Basin is recycled on average five times during its eventual return to the Atlantic Ocean.

3.6.2   The Carbon Cycle

As described in Chapter 11, fast fluxes in the carbon cycle are driven by removal of carbon dioxide from the environment during photosynthesis by plants, algae, and cyanobacteria, which are the major photosynthesizers in most environments. Chemoautotrophic bacteria, which are also important primary producers, play a major role in carbon fixation in environments lacking light. Carbon dioxide is returned to the environment as a byproduct of the respiration of all organisms, including photosynthesizers during night-time hours. Organic decomposition, which also produces carbon dioxide and water and consumes oxygen, is carried out by saprophytic fungi and bacteria that obtain the organic material from non-living sources, such as dead trees, and other non-living plant and animal material. Animals, too, are important decomposers of plant and animal organic material.

The methane-producing bacteria that derive their energy from the oxidation of simple organic compounds such as methanol and acetate, release large quantities of methane. They are responsible for swamp gas (methane) production in the sediments of aquatic habitats. Some reside as symbionts in the rumen of cattle and hindguts of termites, whose metabolism releases large quantities of methane to the atmosphere.

3.6.3   The Nitrogen Cycle

Nitrogen, despite its abundance on the atmosphere, is the element that most commonly limits ecosystem productivity (Vitousek and Howarth, 1991). Atmospheric nitrogen, N_2, cannot be used by most organisms. To enter an ecosystem, N_2 must be "fixed" as ammonia, nitrate, nitrite, or an organic compound of nitrogen. Small amounts of nitrogen are fixed by lightning and a small amount of ammonia is vented into the atmosphere by volcanoes, but most nitrogen in the biosphere is fixed by organisms – by cyanobacteria in aquatic systems and by symbiotic bacteria associated with the roots of certain plants in terrestrial ecosystems. Nitrogen is lost rapidly from ecosystems via groundwater, vaporization of ammonia, and denitrification, but it is released slowly from decaying organic matter. Thus, nitrogen is typically limiting because it is lost rapidly but is fixed and recycled slowly. Humans have roughly doubled the input of fixed nitrogen to the biosphere, primarily by combustion of fossil fuels and industrial fixation of nitrogen for use as fertilizer (Vitousek *et al.*, 1997). Managing the nitrogen cycle is certain to be a major problem for human society in the foreseeable future.

Dissimilatory denitrification occurs anaerobically and is mediated by bacteria that use nitrate in place of oxygen as an acceptor of electrons during respiration. The result is the formation of molecular nitrogen and nitrous oxide. The nitrous oxide plays a role in the chemistry of stratospheric ozone, and is, therefore, extremely important biogeochemically. These bacteria derive energy from the anaerobic oxidation of organic compounds. Many organisms, especially bacteria, decompose organic material with the release of ammonia, a process referred to as ammonification.

3.6.4 The Sulfur Cycle

Reduced sulfur compounds serve as hydrogen donors for anoxygenic photosynthetic bacteria such as the green and purple sulfur bacteria and some cyanobacteria. In contrast, chemoautotrophic sulfur bacteria obtain energy from the oxidation of reduced sulfur compounds including hydrogen sulfide, sulfur, and thiosulfate. As with the nitrifying bacteria, these bacteria are primarily aerobic and use carbon dioxide as their source of carbon. The ultimate product of their metabolism is sulfuric acid. These bacteria are responsible for the production of acid mine waters in areas where strip mining has exposed pyrite minerals to rainfall and oxygen. Some of these bacteria can grow at pH values as low as 1.0; pH values of 3.0 and 4.0 are common in runoff streams from mining areas. Fish cannot live in these waters and most plants cannot grow in such highly acidic soils.

Dissimilatory sulfate reducers such as *Desulfovibrio* derive their energy from the anaerobic oxidation of organic compounds such as lactic acid and acetic acid. Sulfate is reduced and large amounts of hydrogen sulfide are generated in this process. The black sediments of aquatic habitats that smell of sulfide are due to the activities of these bacteria. The black coloration is caused by the formation of metal sulfides, primarily iron sulfide. These bacteria are especially important in marine habitats because of the high concentrations of sulfate that exists there.

Dimethylsulfide (DMS) is the major volatile sulfur compound of biogenic origin emitted from the oceans into the atmosphere. It is estimated that the annual global sea-to-air flux is 15–40 million metric tons of sulfur per year. DMS is produced by the enzymatic cleavage of dimethylpropiothetin (DMPT). The function of DMPT in these algae is uncertain, but there is strong evidence that it may function as a very effective osmoregulator (Andreae and Bernard, 1984; Vairavamurthy *et al.*, 1985). The dipolar ionic nature of DMPT gives the molecule a very low membrane permeability. The osmotic role of DMPT is also suggested by the fact that most freshwater algae produce little or no DMS, although cyanobacteria do. The dimethylsulfide produced by marine algae reacts in the atmosphere to form sulfuric acid as well as ammonium sulfate, ammonium bisulfate, and methane sulfonic acid, all of which have low vapor pressures in the atmosphere and can condense to form aerosol particles. These particles can affect climate by changing the reflective properties of the marine atmosphere and by providing particles on which cloud droplets can nucleate. (See Chapters 7 and 17 for more details.)

3.6.5 The Phosphorus Cycle

Phosphorus is not an abundant constituent of the biosphere, but it is an essential component of living organisms as a component of nucleic acids and high-energy compounds such as ATP. Organisms have evolved mechanisms for concentrating phosphorus, which is often a limiting nutrient, from soil and water. In freshwaters, algal blooms are frequently controlled by the availability of phosphate. Microorganisms are able to store phosphate as a polymer inside their cells, which they use during periods of phosphorus limitation.

3.6.6 The Metal Cycles

Many bacteria can oxidize and reduce metals and metallic ions. Some can even derive energy from those oxidative processes. For example, iron oxidizers, such as species of *Thiobacillus*, can grow on reduced iron compounds and obtain energy from their oxidation if the pH is sufficiently low, as in pyrite oxidation in acid mine waters. No chemoautotrophs are known to be able to derive energy from the oxidation of manganese, but many heterotrophic bacteria can do so. Many bacteria are also involved in the deposition of oxidized iron and manganese compounds, thereby immobilizing these elements. Trace amounts of metals can be very important to organisms either through their toxic effects or the roles that metals play in enzymes and energy transfer compounds.

Table 3-6   Biological sources of Earth's major atmospheric gases[a]

Gas	Principal biological source	Residence time in the atmosphere
Nitrogen	Bacteria	10^7 to 10^9 years
Oxygen	Photosynthesis	Thousands of years
Carbon dioxide	Organism respiration, fuel combustion	About 100 years
Carbon monoxide	Bacterial processes, incomplete combustion	A few months
Methane	Bacteria in anaerobic environments	A few years
Nitrous oxide	Bacteria and fungi	About 100 years
Ammonia	Bacteria and fungi	A few days
NO_x	Reaction of pollutants in sunlight	A few days
Hydrogen sulfide	Anaerobic bacteria	A few days
Hydrogen	Photosynthetic bacteria, methane oxidation	A few years

[a]Modified from Lovelock and Margulis (1974).

3.6.7   Long-term Influences of Living Organisms on Earth's Atmosphere

The long-term influences of living organisms on Earth's atmosphere have been enormous. The biological inputs of major atmospheric gases is given in Table 3-6. Had Earth remained lifeless, concentrations of carbon dioxide in the atmosphere probably would have remained very high, and the temperature would be very different from that of today. Oxygen would have slowly increased due to the splitting of water by sunlight, but it would not have risen above 1% of its present concentration. The atmosphere of Earth also differs in the chemical interactions taking place in it. If life were to disappear today, nitrogen in the atmosphere would eventually be transformed into nitrate, which would be transferred to the oceans, lowering their pH considerably from present values (Hutchinson, 1944).

The mean temperature of Earth is a result of input of energy from the sun and loss of energy by emission of radiation. The input of energy is a function of the reflectivity (albedo) of Earth. Ice reflects 80–95% of incident light, dry grassland 30–40%, and a conifer forest 10–15%. Therefore seasonal changes in vegetation substantially alter the amount of radiation absorbed by the surface. Major changes in vegetation due to human activity have the same effect. The surface temperature of Earth also depends on atmospheric concentrations of carbon dioxide, nitrous oxide, and methane, greenhouse gases whose concentrations are influenced by the biosphere.

3.6.8   Evolutionary Perspectives on Gaia

It is now generally appreciated that the chemical disequilibrium of Earth's atmosphere is due to the activities of living organisms, but acceptance of this view is surprisingly recent. Indeed, until 1974, when James Lovelock and Lynn Margulis built on Vernadsky's theories and proposed the Gaia hypothesis, most scientists believed that living organisms were buffeted by powerful physical forces that they were unable to influence. The form in which the Gaia hypothesis was first advanced (Lovelock, 1979), proposed that living organisms evolved certain traits because those traits stabilized the composition of the atmosphere, thereby maintaining conditions favorable for life. In response to various criticisms, Lovelock subsequently modified his views. The modified form of Gaia postulates only that the activities of living organisms have had, and continue to have, major effects on the composition of the atmosphere. It does not specify why organisms evolved those features.

Debates over the validity of the Gaia hypothesis have focused on two very different issues. One concerns the degree to which the composition of the atmosphere and, hence, Earth's climate, have actually been precisely regulated. This is an empirical issue that can be and is being resolved by accumulating detailed data on temporal changes in Earth's atmosphere and climate. Fluctuations have occurred, but they have been less than would have been occurred without the involvement of life.

The second issue, which is more relevant to the focus of this chapter, concerns the evolutionary basis of the profound influences living organisms have on Earth's atmosphere. In part this debate continues because of a widespread failure to appreciate limitations on the mechanisms of evolutionary change. For adaptive evolutionary change to happen, an allele must confer benefits to its possessors when the allele is present in only a tiny fraction of members of a population. In other words, an allele must be able to increase in frequency when it is rare.

The severe limitation that this requirement imposes is made clear by considering one of Lovelock's most famous and provocative examples – Daisy World. Lovelock proposed that Earth's temperature could have been maintained at a relatively constant level despite a steady reduction in solar energy input over the eons by a change in Earth's albedo caused by a change from a landscape dominated by, say, white daisies to one dominated by dark daisies. Calculations of the potential consequences of changes in the reflective properties of dominant organisms on Earth's temperature are not controversial. The evolutionary problem is to identify what would favor dark daisies when they were rare. Clearly, when dark daisies are rare, they would have an utterly trivial influence of Earth's albedo.

An answer follows from noting that in a cooling world, a dark daisy might well benefit directly (increased photosynthetic rate) by being warmer. If so, dark daisies could outproduce their white associates and, over many generations, come to dominate the daisy population. The end result would be a significant change in Earth's albedo, but the change would be a byproduct of benefits accruing to individual daisies from their color. This example illustrates why most evolutionary biologists reject the view that organisms evolved traits because

of their effects on the atmosphere, while accepting that the activities of organisms do exert powerful influences on the maintenance of Earth's atmosphere.

3.7   How Biogeochemical Cycles Affect Life

Geochemical conditions on Earth have profound effects on living organisms. Our anthropocentric view knows that life flourishes at neutral pH, temperature ranges from freezing to 40°C, and pressures of about 1 atm. However, microbial life occurs at much more extreme conditions as well. In addition, organisms are influenced by changes in biogeochemical cycles.

3.7.1   *Extreme Environments*

Living organisms require water, an energy source, and the necessary nutrients to make cellular material for growth. Some places on Earth do not provide combinations of these resources necessary for organisms to survive. For example, some areas of the Atacama Desert in Peru have received no rain for more than 30 years (E. I. Friedmann, personal communication) and harbor no life. Likewise, living organisms are not known to grow at temperatures exceeding about 115°C and below about −10°C (Morita, 1975). Other physico-chemical constraints on life include high and low pH, high salinity, and high radiation. The pH range for microbial growth is remarkably broad, extending from about pH 1 to more than 12. Some Archaea are able to grow in saturated brines. Other microorganisms, such as the bacterial genus *Deinococcus,* can survive high doses of UV and radioisotopic radiation. Furthermore, some environments, such as acid hot springs, that have stressful combinations of these limiting factors, nonetheless support life. Thus, microorganisms have evolved to live in a wide range of environments on Earth, many of which are inhospitable to more complex organisms.

3.7.2 Response of Ecosystems to Changes in Biogeochemical Cycles and Climate

Perturbation of normal geochemical cycles is increasing as a result of human activities. One example is the ozone hole in the stratosphere centered over the North and South Poles. The ozone layer, which absorbs ultraviolet radiation in the stratosphere, is formed by the reaction of oxygen with sunlight. Stratospheric ozone concentrations are decreasing because chlorofluorocarbon (CFC) compounds used as refrigerants and air conditioners (such as Freon) are increasingly released to the atmosphere where they are converted into fluorine, chlorine, and carbon by ultraviolet light in the stratosphere. Chlorine interacts with ozone to convert it to oxygen. Chlorine is not degraded in this process, so it can cause the destruction of many additional ozone molecules.

The increase in fossil fuel burning since the industrial revolution, coupled with deforestation, has resulted in increased concentrations of carbon dioxide in the atmosphere. Increased concentrations of carbon dioxide affect plant and animal life both directly and indirectly. Directly, higher atmospheric carbon dioxide concentrations can, if other factors are not limiting, stimulate photosynthesis, and hence, plant growth rates (Oechel and Strain, 1985). This "fertilizing" effect is most pronounced in agricultural systems in which levels of water and mineral nutrients are highly favorable (Bazzaz, 1990; Mooney et al., 1991). Natural ecosystems in which rates of photosynthesis are limited by moisture, temperature, or low levels of soil nutrients, may experience only brief or minor increases in productivity (Field et al., 1992; Mellilo et al., 1993; Parton et al., 1995; Schimel, 1995). However, in combination with higher temperatures and longer growing seasons, carbon dioxide enrichment may result in increased rates of photosynthesis and increased rates of storage of carbon in forested ecosystems, as is currently happening in temperate and tropical forests in the New World (Fan et al., 1998; Phillips et al., 1998).

Increases in concentrations of carbon dioxide within projected ranges are not expected to have direct effects on animals. However, changes in plant productivity and chemical composition of plant tissues that are caused by carbon dioxide enrichment, alter the palatability of plant tissues, and, hence, population dynamics of herbivores and the carnivores that feed on them (Ayres, 1993; Fajer et al., 1992).

Increases in temperature can affect natural ecosystems by changing the spatial distribution of climates and, hence, ecosystems. For example, the species living in the major vegetation regions in eastern North America are predicted to be displaced northward by 100 to 1000 km (Overpeck et al., 1991). Thus, pines, oaks, birch, spruce, and non-grass prairie plants must either undergo major shifts in ranges or face extinction. Furthermore, the period of time over which climate warming is projected to occur is on the order of hundreds rather than thousands of years as occurred during the most recent postglacial period. Plant species differ in their abilities to adapt to and respond to rapid changes in climate and habitats. Species that have broad ranges, good dispersal mechanisms, and broad habitat tolerances are expected to handle the changes better than species with more restrictive requirements and poorer dispersal abilities. Similar arguments apply to animals. Animals that are judged to be most susceptible include local endemic species, rare species, top carnivores, migratory species, and species that depend on highly specific plant resources (Terborgh and Winter, 1980).

Species interactions and community structure will also be influenced by major climate changes, but details are very difficult to predict. The paleoecological record indicates that the species that composed the communities of organisms did not simply shift their ranges together during the cooling that accompanied glacial expansion or the warming that followed the retreat of the glaciers. Instead, species responded individualistically. As a result, past ecological communities were characterized by mixtures of species that differ from any found on Earth today. The composition of future ecological communities is certain to hold many surprises, not all of which are likely to be favorable from a human perspective.

References

Amend, J. P. and Shock, E. L. (1998). Energetics of amino acid synthesis in hydrothermal ecosystems. *Science* **281**, 1659–1662.

Andreae, M. O. and Bernard, W. R. (1984). The marine chemistry of dimethyl sulfide. *Marine Chem.* **14**, 267–269.

Awramik, S. M., Schopf, J. W., and Walter, M. R. (1983). Filamentous fossil bacteria 3.5 × 10^9 years old from the Archaen of Western Australia. *Precambrian Res.* **20**, 357–374.

Ayers, M. P. (1993). Plant defense, herbivory, and climate change. *In* "Biotic Interactions and Global Change" (P. M. Kareiva, J. G. Kingsolver, and R. B. Huey, eds), pp. 75–94. Sinauer Associates, Sunderland, MA.

Bazzaz, F. A. (1990). The response of natural ecosystems to the rising CO_2 levels. *Annu. Rev. Ecol. System.* **21**, 167–196.

Chang, S., Desmarias, D., Mack, R., Miller, S. L., and Strathern, G. E. (1983). Prebiotic organic synthesis and the origin of life. *In* "Earth's Earliest Biosphere, Its Origin and Evolution" (J. W. Schopf, ed.), pp. 53–88. Princeton University Press, Princeton, New Jersey.

Cyr, H. and Pace, M. L. (1993). Magnitude and patterns of herbivory in aquatic and terrestrial ecosystems. *Nature* **36**, 148–150.

Fajer, E. D., Bowers, M. D., and Bazzaz, F. A. (1992). The effects of nutrients and enriched CO_2 environments on production of carbon-based allelochemicals in *Plantago*: a test of the carbon/nutrient balance hypothesis. *Am. Natur.* **140**, 707–723.

Fan, S., Gloor, M., Mahlman, J., Pacala, S., Sarmiento, J., Takahashi, T., and Tans, P. (1998). A large terrestrial carbon sink in North America implied by atmospheric and oceanic carbon dioxide data and models. *Science* **282**, 442–446.

Field, C. B., Chapin III, F. S., Matson, P. A., and Mooney, H. A. (1992). Responses of terrestrial ecosystems to the changing atmosphere: a resource-based approach. *Annu. Rev. Ecol. System.* **23**, 201–235.

Hofmann, H. J. and Schopf, J. W. (1983). Early Proterozoic microfossils. *In* "Earth's Earliest Biosphere, Its Origin and Evolution" (J. W. Schopf, ed.), pp. 321–359. Princeton University Press, Princeton, New Jersey.

Hutchinson, G. E. (1944). Nitrogen in the biogeochemistry of the atmosphere. *Am. Scient.* **32**, 178–195.

Longhurst, A. (1998). "Ecological Geography of the Sea." Academic Press, London.

Lovelock, J. E. (1979). "Gaia. A New Look at Life on Earth." Oxford University Press, Oxford.

Lovelock, J. E. and Margulis, L. (1974). Atmospheric homoeostatis by and for the biosphere: the Gaia hypothesis. *Tellus* **26**, 1–10.

Melillo, J. M., McGuire, A. D., Kicklighter, D. W., Moore III, B., Vorosmarty, C. J., and Schloss, A. L. (1993). Global climate change and terrestrial net production. *Nature* **363**, 234–240.

Miller, S. L. (1953). A production of amino acids under possible primitive Earth conditions. *Science* **117**, 528–529.

Mooney, H. A., Drake, B. G., Luxmoore, R. J., Oechel, W. C., and Pitelka, L. F. (1991). Predicting ecosystem responses to elevated CO_2 concentrations. *Bioscience* **41**, 96–104.

Morita, R. Y. (1975). Psychrophilic bacteria. *Bacteriol. Rev.* **39**, 144–167.

Oechel, W. C. and Strain, B. R. (1985). Native species responses to increased carbon dioxide concentration. *In* "Direct Effects of Increasing Carbon Dioxide on Vegetation" (B. R. Strain and J. D. Cure, eds), pp. 117–154. Springfield, VA, U.S. Department of Energy.

Overpeck, J. T., Bartlein, P. J., and Webb III, T. (1991). Potential magnitude of future vegetation change in eastern North America: Comparisons with the past. *Science* **254**, 692–695.

Parton, W. J., Scurlock, J. M. O., Ojima, D. S., Schimel, D. S., Hall, D. O., and SCOPEGRAM group members (1995). Impact of climate change on grassland production and soil carbon worldwide. *Glob. Change Biol.* **1**, 13–22.

Perry, J. J. and Staley, J. T. (1997). "Microbiology: Dynamics and Diversity." Saunders College Publishing, Ft. Worth, TX.

Phillips, O. L., Malhi, Y., Higuchi, N., Laurance, W. F., Nuñez, P. V., Vasquez, R. M., Laurance, S. G., Ferreria, L. V., Stern, M., Brown, S., and Grace, J. (1998). Changes in the carbon balance of tropical forests: evidence from long-term plots. *Science* **282**, 4398–4442.

Rubey, W. W. (1951). Geological history of sea water. An attempt to state the problem. *Geol. Soc. Am. Bull.* **62**, 1111–1148.

Rubey, W. W. (1955). Development of the hydrosphere and atmosphere, with special reference to probable composition of the early atmosphere. *In* "Crust of the Earth" (A. Poldenvaart, ed.), pp. 631–650. Geological Society of America, New York.

Schimel, D. S. (1995). Terrestrial ecosystems and the carbon cycle. *Glob. Change Biol.* **1**, 77–91.

Terborgh, J. and Winter, B. (1980). Some causes of extinction. *In* "Conservation Biology: An Evolutionary-Ecological Perspective'' (M. E. Soulé and B. A. Wilcox, eds), pp. 119–133. Sinauer Associates, Sunderland, MA.

Vairavamurthy, A., Andreae, M. O., and Iverson, R. L. (1985). Biosynthesis of dimethylpropionthetin by *Hymenomnonas carterae* in relation to sulfur source and salinity variations. *Limnol. Oceanogr.* **30**, 59–70.

Vitousek, P. M. and Howarth, R. W. (1991). Nitrogen limitation on land and in the sea. How can it occur? *Biogeochemistry* **13**, 87–115.

Vitousek, P. M., Aber, J. D., Howarth, R. W., Likens, G. E., Matson, P. A., Schindler, D. W., Schlesinger, W. H., and Tilman, D. (1997). Human alteration of the global nitrogen cycle: Sources and consequences. *Ecol. Apps.* **7**, 737–750.

Whitman, W. B., Coleman, D. C., and Wiebe, W. J. (1998). Prokaryotes: the unseen majority. *Proc. Natl Acad. Sci. USA* **95**, 6578–6583.

Appendix

Many types of *organic molecules* are basic constituents of living organisms or are produced by them. Besides carbon and hydrogen, most naturally occurring organic molecules contain one or more of four key elements, all of which are from the second and third period of the periodic table: N, O, P, and S. Carbon atoms form most of the skeleton of these molecules. Bonding of carbon can lead to very large molecules with a variety of structures. Hydrogen may share one or more

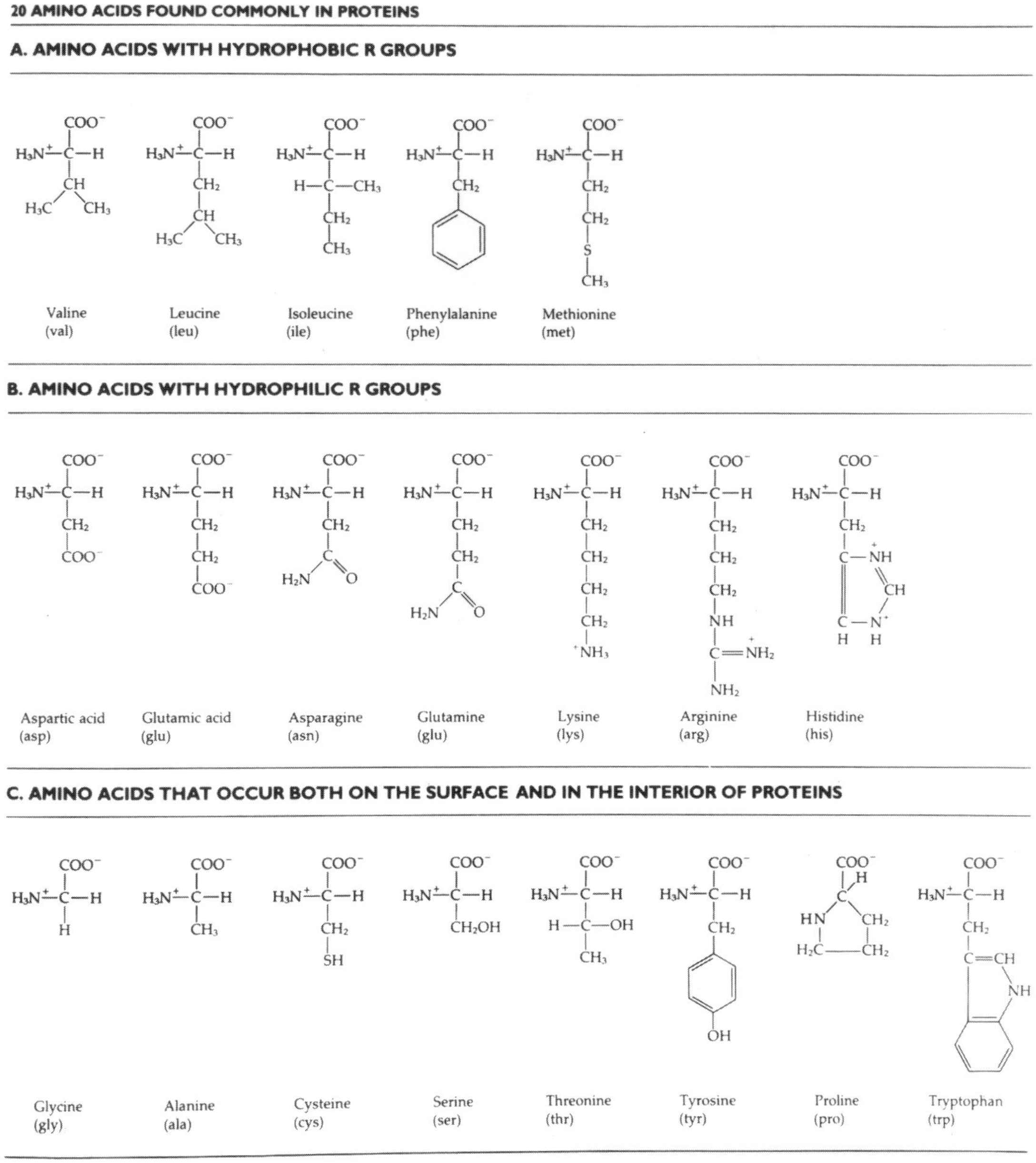

20 AMINO ACIDS FOUND COMMONLY IN PROTEINS

A. AMINO ACIDS WITH HYDROPHOBIC R GROUPS

Valine (val)   Leucine (leu)   Isoleucine (ile)   Phenylalanine (phe)   Methionine (met)

B. AMINO ACIDS WITH HYDROPHILIC R GROUPS

Aspartic acid (asp)   Glutamic acid (glu)   Asparagine (asn)   Glutamine (glu)   Lysine (lys)   Arginine (arg)   Histidine (his)

C. AMINO ACIDS THAT OCCUR BOTH ON THE SURFACE AND IN THE INTERIOR OF PROTEINS

Glycine (gly)   Alanine (ala)   Cysteine (cys)   Serine (ser)   Threonine (thr)   Tyrosine (tyr)   Proline (pro)   Tryptophan (trp)

Fig. 3-10   The 20 protein amino acids divided by R group character as (a) hydrophobic, (b) hydrophilic, and (c) mixed. (Reprinted with permission from W. K. Purves and G. H. Orians, "Life: The Science of Biology," pp. 63–81, Copyright © 1987 by Sinauer Associates, Inc., Sunderland, MA.)

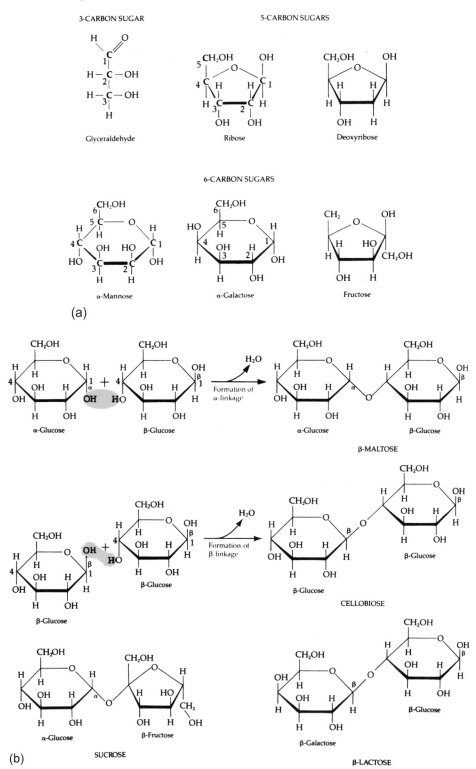

Fig. 3-11 Carbohydrates in (a) 3-, 5-, and 6-carbon sugars (monosaccharides), (b) oligosaccharides, and (c) polysaccharides. (Reprinted with permission from W. K. Purves and G. H. Orians, "Life: The Science of Biology," pp. 63–81, Copyright © 1987 by Sinauer Associates, Inc., Sunderland, MA.)

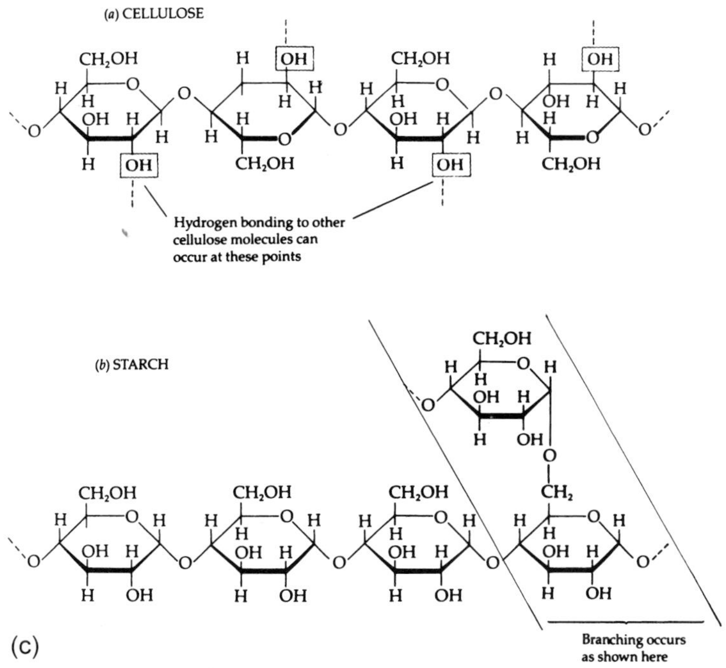

(*a*) CELLULOSE

Hydrogen bonding to other cellulose molecules can occur at these points

(*b*) STARCH

CH$_2$OH

Branching occurs as shown here

(c)

Fig. 3-11 (Continued)

of the valence electrons of carbon and may also participate in *hydrogen bonds*. Hydrogen bonds also involve nitrogen and oxygen and are important in determining the structures of DNA and many other molecules. Much of the nitrogen is found in amino (—NH$_2$) groups, which provide the possibility of basic behavior as well as polarity, hydrophilic behavior, and in some cases, water solubility. Oxygen, in carbonyl (C=O), carboxyl (—COOH), and hydroxyl groups (—OH), provides acidic and hydrophilic character.

Most organo-phosphorus compounds are phosphates, R–PO$_4$. The bond energy of the phosphate to the rest of the molecule is central to the flow of energy in *all* metabolic processes. Sulfur, found largely as sulfide, is central to providing three dimensional rigidity by uniting parts of a molecule by the disulfide (—S—S—) bridge.

Amino acids, the building blocks of giant protein molecules have a carboxyl group and an amino group attached to the same carbon atom (Fig. 3-10). A *protein* is a linear polymer of amino acids combined by peptide linkages. Twenty different amino acids are common in proteins.

Their side chains, which have a variety of chemical properties, control the shapes and functions of proteins. Some of these side chains are hydrophobic (Fig. 3-10a), others are hydrophilic (Fig. 3.10b), and still others (Figure 3.10c) occur either on the surface or the interiors of proteins.

Carbohydrates form a diverse group of compounds (Fig. 3-11) that share an approximate formula (CH$_2$O)$_n$. The major categories of carbohydrates are *monosaccharides* (simple sugars, Fig. 3-11a); *oligosaccharides* (small numbers of simple sugars linked together, Fig. 3-11b); and *polysaccharides* (very large molecules, Fig. 3-11c), among which are starches, glycogen, cellulose, and other important compounds. *Derivative carbohydrates*, such as sugar phosphates and amino sugars, contain additional elements. Important amino sugars are cartilage and chitin, the principal structural carbohydrate in insect skeletons and cell walls of fungi.

Lipids (Fig. 3-12) are insoluble in water and release large amounts of energy when they are metabolized. Lipids are composed of two principal building blocks, *fatty acids* (Fig. 3-12a), and *glycerol* (Fig. 3-12b). Three fatty acids (carboxylic

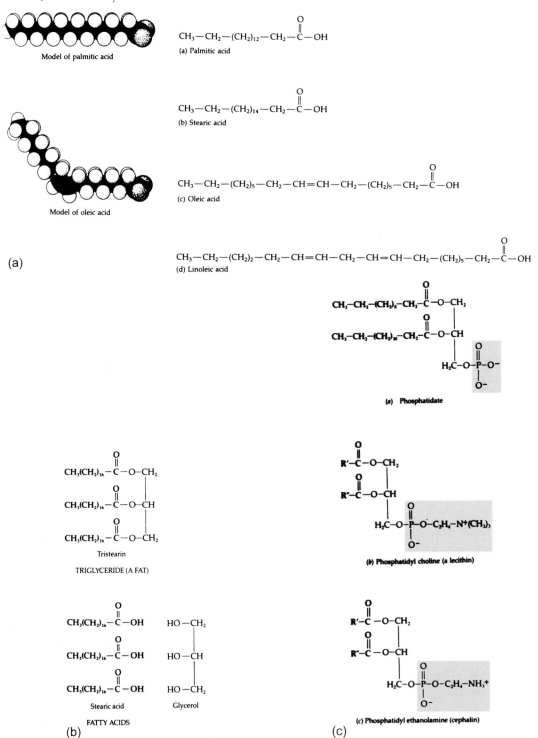

Fig. 3-12 Lipids consist of a triglyceride, three fatty acids such as those in (a) joined to glycerol (b). Other lipids include other functional groups such as phosphate derivatives (c). (Reprinted with permission from W. K. Purves and G. H. Orians, "Life: The Science of Biology," pp. 63–81, Copyright © 1987 by Sinauer Associates, Inc., Sunderland, MA.)

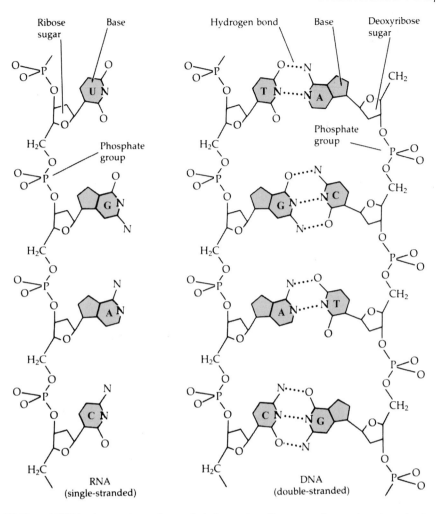

Fig. 3-13 RNA and DNA, the carriers of genetic information. Bases are denoted as A = adenine, T = thymine, G = guanine, C = cytosine, and U = uracil. Note that RNA contains U where DNA contains T. (Reprinted with permission from W. K. Purves and G. H. Orians, "Life: The Science of Biology," pp. 63–81, Copyright © 1987 by Sinauer Associates, Inc., Sunderland, MA.)

acids with long hydrocarbon tails) combine with one molecule of glycerol to form a triglyceride. More complex lipids are formed by the addition of other groups, the most important of which contain phosphorus (Fig. 3.12c).

Deoxyribonucleic acid (DNA, Fig. 3-13) is the genetic material of all organisms, including plants, animals, and microorganisms. (Some viruses lack DNA, but use RNA (ribonucleic acid) in its place.) DNA carries all the hereditary information of the organism and is therefore replicated and passed from parent to offspring. RNA is formed on DNA in the nucleus of the cell. The RNA carries the genetic information to the cytoplasm where it is used to produce proteins on the ribosomes. The specific proteins formed include enzymes which carry out the characteristic activities of the organism. Both RNA and DNA are formed from monomers, called nucleotides, each of which consists of a simple sugar, a phosphate group, and a nitrogen-containing base. The complex structure of DNA is founded on only four bases. A tremendous wealth of information is contained in the precise ordering of these bases to form the genetic code.

4

Modeling Biogeochemical Cycles

Henning Rodhe

4.1 Introductory Remarks

To formulate a model is to put together pieces of knowledge about a particular system into a consistent pattern that can form the basis for (1) interpretation of the past history of the system and (2) prediction of the future of the system. To be credible and useful, any model of a physical, chemical or biological system must rely on both scientific fundamentals and observations of the world around us. High-quality observational data are the basis upon which our understanding of the environment rests. However, observations themselves are not very useful unless the results can be interpreted in some kind of model. Thus observations and modeling go hand in hand.

This chapter focuses on types of models used to describe the functioning of biogeochemical cycles, i.e., reservoir or box models. Certain fundamental concepts are introduced and some examples are given of applications to biogeochemical cycles. Further examples can be found in the chapters devoted to the various cycles. The chapter also contains a brief discussion of the nature and mathematical description of exchange and transport processes that occur in the oceans and in the atmosphere. This chapter assumes familiarity with the definitions and basic concepts listed in Section 1.5 of the introduction such as reservoir, flux, cycle, etc.

Modeling biogeochemical cycles normally involves estimating the spatial and temporal averages for concentrations and fluxes in and out of reservoirs (i.e., reservoir modeling). The spatial average (i.e., the physical size of the reservoir itself) often has a horizontal size approaching that of a continent, or larger, i.e., >1000 km. The time scales corresponding to this spatial average are months, or longer. This means that day-to-day variations in weather and ocean currents are not generally considered explicitly when modeling biogeochemical cycles.

The advent of fast computers and the availability of detailed data on the occurrence of certain chemical species have made it possible to construct meaningful cycle models with a much smaller and faster spatial and temporal resolution. These spatial and time scales correspond to those in weather forecast models, i.e. down to 100 km and 1 h. Transport processes (e.g., for CO_2 and sulfur compounds) in the oceans and atmosphere can be explicitly described in such models. These are often referred to as "tracer transport models." This type of model will also be discussed briefly in this chapter.

4.2 Time Scales and Single Reservoir Systems

4.2.1 Turnover Time

Consider the reservoir shown in Fig. 4-1. The turnover time is the ratio between the content (M) of a substance (tracer) in the reservoir and the total flux out of it (S):

$$\tau_0 = \frac{M}{S} \tag{1}$$

Earth System Science
ISBN 0-12-379370-X

Fig. 4-1 Schematic illustration of a single reservoir with source flux Q, sink flux S, and content M.

The turnover time may be thought of as the time it would take to empty the reservoir if the sink (S) remained constant while the sources were zero ($\tau_0 S = M$). This time scale is also sometimes referred to as "renewal time" or "flushing time." In the common case when the sink is proportional to the reservoir content ($S = kM$), the turnover time is the inverse of the proportionality constant (k^{-1}), which is analogous to first-order chemical kinetics.

In fluid reservoirs like the atmosphere or the ocean, the turnover time of a tracer is also related to the spatial and temporal variability of its concentration within the reservoir; a long turnover time corresponds to a small variability and vice versa (Junge, 1974; Hamrud, 1983). Figure 4-2 shows a plot of measured trace gas variability in the atmosphere versus turnover time estimated by applying budget considerations as indicated by Equation (1). An inverse relation is obvious, but the scatter in the data

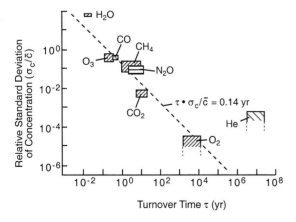

Fig. 4-2 Inverse relationship between relative standard deviation of atmospheric concentration and turnover time for important trace chemicals in the troposphere. (Modified from Junge (1974) with permission from the Swedish Geophysical Society.)

implies some departure from this simple relation.

If material is removed from the reservoir by two or more separate processes, each with a flux S_i, then turnover times with respect to each process can be defined as:

$$\tau_{0i} = \frac{M}{S_i} \qquad (2)$$

Since $\sum S_i = S$, these time scales are related to the turnover time of the reservoir, τ_0, by

$$\tau_0^{-1} = \sum \tau_{0i}^{-1} \qquad (3)$$

As an application of the turnover time concept, let us consider the model of the carbon cycle shown in Fig. 4-3. This diagram is different from the one used in the chapter on the carbon cycle (Chapter 11), because it serves our purposes better for this chapter. The values given for the various fluxes and burdens are very similar to the corresponding figure in Chapter 11 (Fig. 11-1).

The turnover time of carbon in biota in the ocean surface water is $3 \times 10^{15}/(4 + 36) \times 10^{15}$ yr ≈ 1 month. The turnover time with respect to settling of detritus to deeper layers is considerably longer: 9 months. Faster removal processes in this case must determine the turnover time: respiration and decomposition.

The equation describing the rate of change of the content of a reservoir can be written as

$$\frac{dM}{dt} = Q - S = Q - \frac{M}{\tau_0} \qquad (4)$$

If the reservoir is in a steady state ($dM/dt = 0$) then the sources (Q) and sinks (S) must balance. In this case Q can replace S in Equation (1).

4.2.2 Residence Time (Transit Time)

The *residence time* is the time spent in a reservoir by an *individual* atom or molecule. It is also the age of a molecule when it leaves the reservoir. If the pathway of a tracer from the source to the sink is characterized by a physical transport, the word *transit time* can also be used. Even for a single chemical substance, different atoms and molecules will have different residence times in a given reservoir. Let the probability density

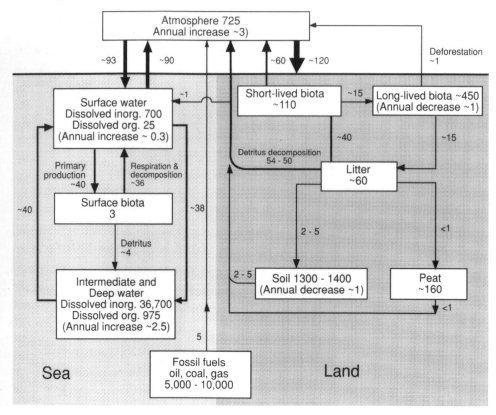

Fig. 4-3   Principal reservoirs and fluxes in the carbon cycle. Units are 10^{15} g (Pg) C (burdens) and Pg C/yr (fluxes). (From Bolin (1986) with permission from John Wiley and Sons.)

function of residence times be denoted by $\phi(\tau)$, where $\phi(\tau)d\tau$ describes the fraction of the tracer having a residence time in the interval to τ to $\tau + d\tau$. The *average residence time* (average transit time) τ_r is defined by:

$$\tau_r = \int_0^\infty \tau\phi(\tau)d\tau \qquad (5)$$

In many cases the word "average" is left out and this quantity is simply referred to as "residence time."

The shape of the probability density function, $\phi(\tau)$, depends on the system. Some examples are shown in Fig. 4-4. This figure also contains probability density of age (see Section 4.2.3). Figure 4-4a might correspond to a lake with inlet and outlet on opposite sides of the lake. Most water molecules will then have a residence time in the lake roughly equal to the time it takes for the mean current to carry the water from the

inlet to the outlet, i.e., τ_r. Very few molecules will have a residence time much greater than or much less than τ_r, as illustrated in the ϕ curve. Another example is a human population where most people live to attain mature age. The ϕ curve in this case can be interpreted as the frequency function for the age at which people die; few die very young and few survive the average age of death by more than 50%.

Figure 4-4b illustrates exponential decay. A simple example could be the reservoir of all ^{238}U on Earth. The half-life of this radionuclide is 4.5×10^9 years. Since the Earth is approximately 4.5×10^9 years old, this implies that the content of the ^{238}U reservoir today is about half of what it was when the Earth was formed. The probability density function of residence time of the uranium atoms originally present is an exponential decay function. The average residence time is 6.5×10^9 years. (The average value of

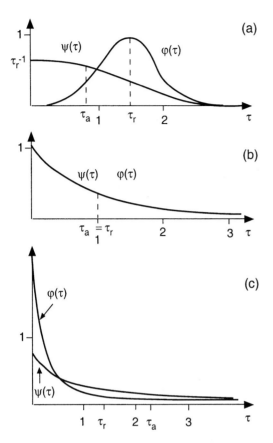

Fig. 4-4 The *age* frequency function $\psi(\tau)$ and the *residence time* frequency function $\phi(\tau)$ and the corresponding average values τ_a and τ_r for the three cases described in the text: (a) $\tau_a > \tau_r$; (b) $\tau_a = \tau_r$; (c) $\tau_a > \tau_r$. (Adapted from Bolin and Rodhe (1973) with permission from the Swedish Geophysical Society.)

time for an exponential decay function is the half-life divided by ln 2.) In a well-mixed reservoir all particles always have the same probability of being removed. In such situations the frequency function for the residence time is also exponential.

In the reservoir corresponding to Fig. 4-4c the removal is biased towards "young" particles. This might occur when the sink is located close to the source (the "short circuit" case).

4.2.3 Age

The *age* of an atom or molecule in a reservoir is the time since it entered the reservoir. As with residence times, the probability density function of ages, $\psi(\tau)$, has a shape that depends on the situation. In a steady-state reservoir, however, $\psi(\tau)$ is always a nonincreasing function. The shapes of $\psi(\tau)$ corresponding to the three residence time distributions discussed above are included in Fig. 4-4.

The *average age* of atoms in a reservoir is given by

$$\tau_a = \int_0^\infty \tau\psi(\tau)\mathrm{d}\tau \qquad (6)$$

4.2.4 Relations Between τ_0, τ_r and τ_a

It can be shown that for a reservoir in steady state, τ_0 is equal to τ_r, i.e. the turnover time is equal to the average residence time spent in the reservoir by individual particles (Eriksson, 1971; Bolin and Rodhe, 1973). This may seem to be a trivial result but it is actually of great significance. For example, if τ_0 can be estimated from budget considerations by comparing fluxes and burdens in Equation (1) and if the average transport velocity (V) within the reservoir is known, the average distance ($L = V\tau_r$) over which the transport takes place in the reservoir can be estimated.

The relation between τ_0 and τ_a is not as simple. τ_a may be larger or less than τ_0 depending on the shape of the age probability density function (as shown in Fig. 4-4). For a well-mixed reservoir, or one with a first-order removal process, $\tau_a = \tau_0$ (Fig. 4-4b).

In the case of a human population corresponding to Fig. 4-4a, τ_a is only about half of τ_0. This example applies to the average age of all Swedes, which is around 40 years, whereas the average residence time, i.e., the average length of life (average age at death) is almost 80 years.

In the situation where most atoms leave the reservoir soon and few of them remain very long (Fig. 4-4c), τ_a is larger than τ_0 (the "short circuit" case). Some further examples of age distributions and relations between τ_a and τ_0 are given in Lerman (1979).

When equating τ_0 and τ_r it must be made clear that the flux, S, which defines τ_0 (see Equation

(1)) is the *gross flux* and not a *net flux*. For example, removal of water from the atmosphere occurs both by precipitation and dry deposition (direct uptake by diffusion to the surface). Dry deposition is not normally explicitly evaluated but subtracted from the gross evaporation flux to yield the net evaporation from the surface. The turnover time of water in the atmosphere calculated as the ratio between the atmospheric content and the precipitation rate (about 10 days) is thus *not* equal to the average residence time of water molecules in the atmosphere. The actual value of the average residence time of individual water molecules is substantially shorter.

4.2.5   Response Time

The response time (relaxation time, adjustment time) of a reservoir is a time scale that characterizes the adjustment to equilibrium after a sudden change in the system. A precise definition is not easy to give except in special circumstances like in the following example.

Consider a single reservoir, like the one shown in Fig. 4-1, for which the sink is proportional to the content ($S = kM$) and which is initially in a steady state with fluxes $Q_0 = S_0$ and content M_0. The turnover time of this reservoir is

$$\tau_0 = \frac{M_0}{S_0} = \frac{1}{k} \tag{7}$$

Suppose now that the source strength is suddenly changed to a new value Q_1. How long would it take for the reservoir to reach a new steady state? The adjustment process is described by the differential equation

$$\frac{dM}{dt} = Q_1 - S = Q_1 - kM \tag{8}$$

with the initial condition $M(t = 0) = M_0$.

The solution

$$M(t) = M_1 - (M_1 - M_0)\exp(-kt) \tag{9}$$

approaches the new steady-state value ($M_1 = Q_1/k$) with a response time equal to k^{-1} or τ_0. The change of the reservoir mass from the initial value M_0 to the final value M_1 is illus-

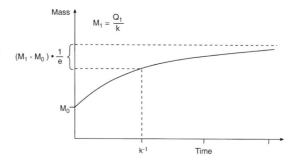

Fig. 4-5   Illustration of an exponential adjustment process. In this case, the response time is equal to k^{-1}.

trated in Fig. 4-5. In this case, with an exponential adjustment, the response time is defined as the time it takes to reduce the imbalance to $e^{-1} = 37\%$ of the initial imbalance. This time scale is sometimes referred to as "*e*-folding time." Thus, for a single reservoir with a sink proportional to its content, the response time equals the turnover time.

As a specific example, consider oceanic sulfate as the reservoir. Its main source is river runoff (pre-industrial value: 100 Tg S/yr) and the sink is probably incorporation into the lithosphere by hydrogeothermal circulation in mid-ocean ridges (100 Tg S/yr, McDuff and Morel, 1980). This is discussed more fully in Chapter 13. The content of sulfate in the oceans is about 1.3×10^9 Tg S. If we make the (unrealistic) assumption that the present runoff, which due to man-made activities has increased to 200 Tg S/yr, would continue indefinitely, how fast would the sulfate concentration in the ocean adjust to a new equilibrium value? The time scale characterizing the adjustment would be $\tau_0 \approx 1.3 \times 10^9$ Tg/$(10^2$ Tg/yr$) \approx 10^7$ years and the new equilibrium concentration eventually approached would be twice the original value. A more detailed treatment of a similar problem can be found in Southam and Hay (1976).

4.2.6   Reservoirs in Non-steady State

Let us analyze the situation when one observes a change in reservoir content and wants to draw conclusions regarding the sources and sinks. We rewrite Equation (8) as

$$\frac{1}{M}\frac{dM}{dt} = \frac{Q}{M} - \frac{1}{\tau_0}$$

where $\tau_0 = 1/k$ is the turnover time in the steady state situation. Let us denote the left side of the equation (the observed rate of change of the reservoir content) as τ_{obs}^{-1}. If the mass were observed to increase by, say, 1% per year, τ_{obs} would be 100 years. Two limiting cases can be singled out:

1. $\tau_{obs} \gg \tau_0$. In this case there has to be an approximate balance between the two terms on the right-hand side of the equation.

$$\frac{Q}{M} \approx \frac{1}{\tau_0} \quad (\text{or}) \quad Q \approx \frac{M}{\tau_0}$$

This means that the observed change in M mainly reflects a change in the source flux Q or the sink function. As an example we may take the methane concentration in the atmosphere, which in recent years has been increasing by about 0.5% per year. The turnover time is estimated to be about 10 years, i.e., much less than τ_{obs} (200 years). Consequently, the observed rate of increase in atmospheric methane is a direct consequence of a similar rate of increase of emissions into the atmosphere. (In fact, this is not quite true. A fraction of the observed increase is probably due to a decrease in sink strength caused by a decrease in the concentration of hydroxyl radicals responsible for the decomposition of methane in the atmosphere.)

2. $\tau_{obs} \ll \tau_0$. In this case $dM/dt \approx Q$ which means that there is an increase in reservoir content about equal to the source flux with little influence on the part of the sink. The reservoir is then in an accumulative stage and its mass is increasing with time largely as a function of Q, irrespective of whether Q itself is increasing, decreasing or constant. A good example of this situation is sulfur hexafluoride (SF_6) whose concentration in the atmosphere is currently increasing by about 0.5% per year ($\tau_{obs} = 200$ years) as a result of various industrial emissions (IPCC, 1996). Because of inefficient removal processes (SF_6

is a very stable molecule) the turnover time of SF_6 is as large as 3000 years. The rate of increase is thus a reflection of an imbalance between sources and sinks rather than an increase in the source flux Q.

In situations where τ_{obs} is comparable in magnitude to τ_0, a more complex relation prevails between Q, S, and M. Atmospheric CO_2 falls in this last category although its turnover time (3–4 years, cf. Fig. 4-3) is much shorter than τ_{obs} (about 300 years). This is because the atmospheric CO_2 reservoir is closely coupled to the carbon reservoir in the biota and in the surface layer of the oceans (Section 4.3). The effective turnover time of the combined system is actually several hundred years (Rodhe and Björkström, 1979).

4.3   Coupled Reservoirs

The treatment of time scales and dynamic behavior of single reservoirs given in the previous section can easily be generalized to systems of two or more reservoirs. While the simple system analyzed in the previous section illustrates many important characteristics of cycles, most natural cycles are more complex. The matrix method described in Section 4.3.1 provides an approach to systems with very large numbers of reservoirs that is at least simple in notation. The treatments in the preceding section and in Section 4.3.1 are still limited to linear systems. In many cases we assume linearity because our knowledge is not adequate to assume any other dependence and because the solution of linear systems is straightforward. There are, however, some important cases where non-linearities are reasonably well understood. In such cases analytical mathematical solutions, corresponding to those given in Section 4.3.1, normally do not exist and the mathematical expressions describing the dynamics of the cycle have to be replaced by finite difference expressions that may be solved by computer. A few of these cases are described in Section 4.3.2.

As important as coupled reservoirs and nonlinear systems are, the less mathematically inclined may want to read this section only for its qualitative material. The treatment described here is not essential for understanding the material later in the book. However, an excellent overview of solving differential equations by the eigenvalue–eigenvector method used in the next section appears in Section 3.6 of Braun (1983).

4.3.1   Linear Systems

A linear system of reservoirs is one where the fluxes between the reservoirs are linearly related to the reservoir contents. A special case, that is commonly assumed to apply, is one where the fluxes between reservoirs are proportional to the content of the reservoirs where they originate. Under this proportionality assumption the flux F_{ij} from reservoir i to reservoir j is given by

$$F_{ij} = k_{ij}M_i \qquad (10)$$

The rate of change of the amount M_i in reservoir i is thus

$$\frac{dM_i}{dt} = \sum_{j=1}^{n} k_{ji}M_j - M_i \sum_{j=1}^{n} k_{ij} \quad \text{for} \quad j \neq i \quad (11)$$

where n is the total number of reservoirs in the system.

This system of differential equations can be written in matrix form as

$$\frac{d\mathbf{M}}{dt} = \mathbf{kM} \qquad (12)$$

where the vector $\mathbf{M}$ is equal to $(M_1, M_2, \ldots, M_n)$ and the elements of matrix $\mathbf{k}$ are linear combinations of the coefficients k_{ij}. The solution to Equation (12) describes the adjustment of all reservoirs to a steady state by a finite sum of exponential decay functions (Lasaga, 1980; Chameides and Perdue, 1997). The time scales of the exponential decay factors correspond to the nonzero eigenvalues of the matrix $\mathbf{k}$. The response time of the system, τ_{cycle}, may be defined by

$$\tau_{\text{cycle}} = \frac{1}{|E_1|} \qquad (13)$$

where E_1 is the nonzero eigenvalue with smallest absolute value (Lasaga, 1980). The treatment

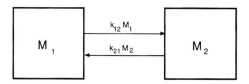

Fig. 4-6   A coupled two-reservoir system with fluxes proportional to the content of the emitting reservoirs.

can be generalized by adding an external forcing function on the right-hand side of Equations (11) and (12).

As an illustration of the concept introduced above, let us consider a coupled two-reservoir system with no external forcing (Fig. 4-6). The dynamic behavior of this system is governed by the two differential equations

$$\begin{aligned} \frac{dM_1}{dt} &= -k_{12}M_1 + k_{21}M_2 \\ \frac{dM_2}{dt} &= k_{12}M_1 - k_{21}M_2 \end{aligned} \qquad (14)$$

the expression of conservation of mass

$$M_1 + M_2 = M_T \qquad (15)$$

and the initial condition

$$\begin{aligned} M_1(t = 0) &= M_{10} \\ M_2(t = 0) &= M_T - M_{10} \end{aligned} \qquad (16)$$

Equations (14) can be written in matrix form as

$$\frac{d\mathbf{M}}{dt} = \mathbf{kM} \qquad (17)$$

where $\mathbf{M}$ is the vector (M_1, M_2) describing the contents of the two reservoirs and $\mathbf{k}$ the matrix:

$$\begin{pmatrix} -k_{12} & k_{21} \\ k_{12} & -k_{21} \end{pmatrix}$$

The eigenvalues of $\mathbf{k}$ are the solutions to the equation

$$\begin{vmatrix} -k_{12} - \lambda & k_{21} \\ k_{12} & -k_{21} - \lambda \end{vmatrix} =$$

$$(-k_{12} - \lambda)(-k_{21} - \lambda) - k_{12}k_{21} = 0 \qquad (18)$$

$\lambda_1 = 0$ and $\lambda_2 = -(k_{12} + k_{21})$. The general solution to Equation (17) can be written as

$$\mathbf{M}(t) = \psi_1 \exp(\lambda_1 t) + \psi_2 \exp(\lambda_2 t) \qquad (19)$$

where ψ_1 and ψ_2 are the eigenvectors of the matrix $\mathbf{k}$. In our case, we have

$$\mathbf{M}(t) = \psi_1 + \psi_2 \exp(-(k_{12} + k_{21})t) \qquad (20)$$

or, in component form and in terms of the initial conditions:

$$M_1(t) = \frac{k_{21}}{k_{12} + k_{21}} M_T$$
$$+ \left(M_{10} - \frac{k_{21}M_T}{k_{12} + k_{21}} \right) \exp[-(k_{12} + k_{21})t]$$
$$M_2(t) = \frac{k_{12}}{k_{12} + k_{21}} M_T \qquad (21)$$
$$+ \left(M_T - M_{10} - \frac{k_{12}M_T}{k_{12} + k_{21}} \right)$$
$$\times \exp[-k_{12} + k_{21})t]$$

It is seen that in the steady state the total mass is distributed between the two reservoirs in proportion to the sink coefficients (in reverse proportion to the turnover times), independent of the initial distribution.

In this simple case there is only one time scale characterizing the adjustment process, that is $(k_{12} + k_{21})^{-1}$. This is also the response time, τ_{cycle}, as defined by Equation (13).

$$\tau_{\text{cycle}} = \frac{1}{k_{12} + k_{21}} \qquad (22)$$

or, if expressed in terms of the turnover times of the two reservoirs:

$$\tau_{\text{cycle}}^{-1} = \tau_{01}^{-1} + \tau_{02}^{-1} \qquad (23)$$

The response time in this simple model will depend on the turnover times of both reservoirs and will always be shorter than the shortest of the two turnover times. If τ_{01} is equal to τ_{02}, then τ_{cycle} will be equal to half of this value.

An investigation of the dynamic behavior of a coupled three-reservoir system using the techniques described above is included in the problems listed at the end of the chapter.

It should be noted that the steady-state solution of Equation (12) is not necessarily unique. This can easily be seen in the case of the four-reservoir system shown in Fig. 4-7. In the steady state all material will end up in the two accumulating reservoirs at the bottom. However, the distribution between these two reservoirs will

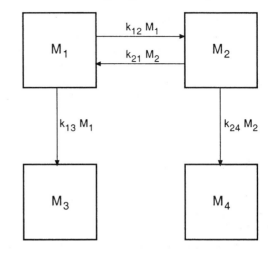

Fig. 4-7 Example of a coupled reservoir system where the steady-state distribution of mass is not uniquely determined by the parameters describing the fluxes within the system but also by the initial conditions (see text).

depend on the amount initially located in the two upper reservoirs.

Before turning to nonlinear situations, let us consider two specific examples of coupled linear systems. The first describes the dynamic behavior of a multireservoir system; the second represents a steady-state situation of an open two-reservoir system.

Example 1. As a specific example of a time-dependent linear system we may take the model of the phosphorus cycle shown in Fig. 4-8. This is a duplicate of the figure shown in the chapter on the phosphorous cycle (Fig. 14-7). The authors used a computer to solve the system of equations in Equation (11) with a time-dependent source term added to represent the transient situation with an exponentially increasing industrial mining input (7% increase per year). Lasaga (1980) studied the same situation in a more elegant way using matrix algebra. The evolution of the phosphorus content of the various reservoirs (except in sediments) during the first 70 years is shown in Table 4-1. Within this time frame, the only noticeable change is seen to occur in the land reservoir. Lasaga showed that the adjustment time scale of the system, τ_{cycle}, is 53 000 years. This is much

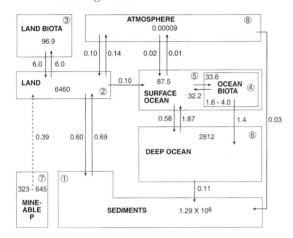

Fig. 4-8   The global phosphorus cycle. Values shown are in Tmol and Tmol/yr. (Adapted from Lerman *et al.* (1975) and modified to include atmospheric transfers. The mass of P in each reservoir and rates of exchange are taken from Jahnke (1992), MacKenzie *et al.* (1993) and Follmi (1996).)

shorter than the turnover time of the sediment reservoir (2×10^8 years) but much longer than the turnover times of all other reservoirs. This cycle is described in greater detail in Chapter 14.

Example 2. As a much simpler example let us consider a system consisting of two connected reservoirs as depicted in Fig. 4-9. Steady state is assumed to prevail. Material is introduced at a constant rate Q into reservoir 1. Some of this material is removed (S_1) and the rest (T) is transferred to reservoir 2, from which it is

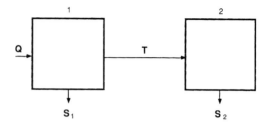

Figure 4-9   An open two-reservoir system.

removed at a rate S_2. The turnover times (average residence times) of the two reservoirs and of the combined reservoir (defined as the sum of the two reservoirs) are easily calculated to be

$$\tau_{01} = \frac{M_1}{T + S_1} = \frac{M_1}{Q} \tag{24}$$

$$\tau_{02} = \frac{M_2}{S_2} = \frac{M_2}{T} \tag{25}$$

$$\tau_0 = \frac{M_1 + M_2}{S_1 + S_2} = \frac{M_1 + M_2}{Q}$$
$$= \frac{M_1}{Q} + \frac{M_2}{S_2} \cdot \frac{S_2}{Q} = \frac{M_1}{Q} + \frac{M_2}{S_2} \cdot \frac{T}{Q} = \tau_{01} + \alpha\tau_{02} \tag{26}$$

where $\alpha = T/Q$ is the fraction of the material passing through reservoir 1 that is transferred to reservoir 2. An example application of this two-reservoir model is that it has been used to study the oxidation of sulfur in the atmosphere where SO_2–sulfur was treated as one reservoir and sulfate–sulfur as the other (Rodhe, 1978). In the

Table 4-1   Response of phosphorus cycle to mining output. Phosphorus amounts are given in Tg P ($1\,\text{Tg} = 10^{12}$ g). Initial contents and fluxes as in Fig. 4-7 (system at steady state). In addition, a perturbation is introduced by the flux from reservoir 7 (mineable phosphorus) to reservoir 2 (land phosphorus), which is given by $12\exp(0.07t)$ in units of Tg P/yr

Time (years)	Land	Land biota	Organic biota	Surface ocean	Deep ocean
0	200 000	3000	138	2710	87 100
10	200 173	3000	138	2710	87 100
20	200 522	3001	138	2710	87 100
30	201 224	3003	138	2710	87 100
40	202 636	3008	138	2710	87 100
50	205 481	3018	138	2710	87 100
60	211 208	3018	138	2711	87 100
70	222 741	3078	138	2712	87 101

special case where $S_1 = 0$, $T = Q$ and $\alpha = 1$ (all material introduced in reservoir 1 is transferred to reservoir 2) the turnover time of the combined reservoir equals the sum of the turnover times of the individual reservoirs.

4.3.2 Non-linear Systems

In many situations the assumption about linear relations between removal rates and reservoir contents is invalid and more complex relations must be assumed. No simple theory exists for treating the various non-linear situations that are possible. The following discussion will be limited to a few examples of non-linear reservoir/flux relations and cycles. For a more comprehensive discussion, see the review by Lasaga (1980).

Consider a single reservoir with a constant rate of supply and a removal rate proportional to the square of the reservoir content. The equation governing the rate of change of the reservoir content is

$$\frac{dM}{dt} = Q - BM^2 \qquad (27)$$

If $M(0) = 0$, the solution to this equation is

$$M = \sqrt{\frac{Q}{B} \cdot \frac{1 - \exp(-2\sqrt{QBt})}{1 + \exp(-2\sqrt{QBt})}} \qquad (28)$$

This is graphically illustrated in Fig. 4-10. Initially, the mass increases almost linearly with time. After the time $(2\sqrt{QB})^{-1}$ the removal term becomes effective and the mass begins to level

off. M eventually reaches a steady state equal to $\sqrt{Q/B}$ but the response time scale is not as easily defined as in the linear case. Relative to a simple exponential relaxation process the adjustment given by Equation (28) is more rapid initially, and slower as time progresses.

In general, if the removal flux is dependent upon the reservoir content raised to the power α ($\alpha \neq 1$), i.e., $S = BM^\alpha$, the adjustment process will be faster or slower than the steady-state turnover time depending on whether α is larger or smaller than unity (Rodhe and Björkström, 1979).

A similar simple non-linear adjustment process is described by the equation

$$\frac{dM}{dt} = AM - BM^2 \qquad (29)$$

which is a common model for the growth of biological systems (it is called logistical growth). The term AM represents exponential growth (unlimited supply of space and nutrients) and the term BM^2 is a removal term, a negative feedback effect of "crowdedness." Initially (where $AM_0 \gg BM_0^2$), the growth will be close to exponential and will then gradually level off to the equilibrium value A/B (Fig. 4-11).

A sink flux that has a weaker than proportional dependence on the content M of the emitting reservoir is often described by the Michaelis–Menten equation:

$$S = \frac{CM}{(M + D)} \qquad (30)$$

where C is a rate coefficient and D a term

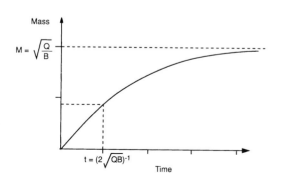

Fig. 4-10 The shape of the function given in Equation (28).

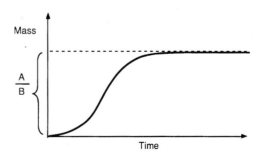

Fig. 4-11 Shape of "logistical growth." The rate of change increases slowly initially. The rate of growth reaches a maximum and eventually drops to zero as the mass levels off, approaching the value A/B.

(Michaelis constant) which determines the deviation from proportionality; a small D (compared to M) corresponds to a weak dependence of S on M and a large D corresponds to a near proportional dependence.

An important example of non-linearity in a biogeochemical cycle is the exchange of carbon dioxide between the ocean surface water and the atmosphere and between the atmosphere and the terrestrial system. To illustrate some effects of these non-linearities, let us consider the simplified model of the carbon cycle shown in Fig. 4-12. M_S represents the sum of all forms of dissolved carbon (CO_2, H_2CO_3, HCO_3^- and

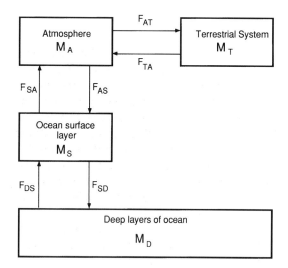

Fig. 4-12 Simplified model of the biogeochemical carbon cycle. (Adapted from Rodhe and Björkström (1979) with the permission of the Swedish Geophysical Society.)

CO_3^{2-}). The ocean to atmosphere flux, which is dependent on the concentration of the dissolved species $CO_2(aq)$, is related to the total carbon content in the surface layer (M_S) by

$$F_{SA} = k_{SA} M_S^{\alpha_{SA}} \qquad (31)$$

where the exponent α_{SA} (the buffer, or Revelle factor) is about 9. The buffer factor results from the equilibrium between $CO_2(g)$ and the more prevalent forms of dissolved carbon. This effect is discussed further in Chapter 11. As a consequence of this strong dependence of F_{SA} on M_S, a substantial increase in CO_2 in the atmosphere is balanced by a small increase of M_S.

Similarly, the flux from the atmosphere to the terrestrial system may be represented by the expression

$$F_{AT} = k_{AT} M_A^{\alpha_{AT}} \qquad (32)$$

The exponent α_{AT} is considerably less than unity owing to the fact that CO_2 generally is not the limiting factor for vegetation growth. This means that even a substantial increase in M_A does not produce a corresponding increase in F_{AT}.

Assuming that the carbon cycle of Fig. 4-12 will remain a closed system over several thousands of years, we can ask how the equilibrium distribution within the system would change after the introduction of a certain amount of fossil carbon. Table 4-2 contains the answer for two different assumptions about the total input. The first 1000 Pg corresponds to the total input from fossil fuel up to about the year 2000; the second (6000 Pg) is roughly equal to the now

Table 4-2 Steady-state carbon contents (unit: $Pg = 10^{15}$ g) for the four-reservoir model of Fig. 4-11: (a) during the unperturbed (pre-industrial) situation; (b) after the introduction of 1000 Pg carbon; and (c) after the introduction of 6000 Pg carbon

	Pre-industrial content (Pg)	After 1000 Pg		After 6000 Pg	
		Content (Pg)	% increase	Content (Pg)	% increase
Atmosphere	700	840	20	1880	170
Terrestrial system	3000	3110	4	3655	22
Ocean surface layer	1000	1020	2	1115	12
Deep ocean	35 000	35 730	2	39 050	12

known accessible reserves of fossil carbon (Keeling and Bacastow, 1977).

If all fluxes are proportional to the reservoir contents, the percentage change in reservoir content will be equal for all the reservoirs. The non-linear relations discussed above give rise to substantial variations between the reservoirs. Note that the atmospheric reservoir is much more significantly perturbed than any of the other three reservoirs. Even in the case with a 6000 Pg input, the carbon content of the oceans does not increase by more than 12% at steady state.

However, with "only" 1000 Pg emitted into the system, i.e. less than 3% of the total amount of carbon in the four reservoirs, the atmospheric reservoir would still remain significantly affected (20%) at steady state. In this case the change in oceanic carbon would be only 2% and hardly noticeable. The steady-state distributions are independent of where the addition occurs. If the CO_2 from fossil fuel combustion were collected and dumped into the ocean, the final distribution would still be the same.

If all fluxes were proportional to the reservoir content, i.e., if α_{SA} and α_{AT} were unity, all reservoirs would be equally affected; 15% in the 6000 Pg case and 2.5% in the 1000 Pg case.

4.4 Fluxes Influenced by the Receiving Reservoir

There are some important situations in which a flux between two reservoirs is determined not only by the mass of the emitting reservoir but also by the mass of the receptor. Uptake of CO_2, or indeed any other nutrient by a plant community depends also on the magnitude of its biomass because that determines the size of the surfaces where photosynthesis take place. Consider, for example, the uptake of atmospheric CO_2 by terrestrial biota. A reasonable parameterization of this flux would be

$$F_{AT} = \frac{k_{ATB} M_A M_{TB}}{M_A + D} \qquad (33)$$

where the notations follow those in Fig. 4-12 with M_{TB} being the content of carbon in terrestrial biota and D, a Michaelis constant. One

problem with an expression like this is that mathematically speaking, the flux F_{AT} and the mass M_{TB} may grow without bounds; the larger M_{TB}, the larger the flux to it, i.e., exponential growth. To avoid such a mathematical explosion, Williams (1987) suggested that the factor M_{TB} in Equation (33) be replaced by

$$\frac{M_{TB}(M_{TB_{max}} - M_{TB})}{M_{TB_{max}}}$$

where $M_{TB_{max}}$ is an upper limit to the size of M_{TB}. Once M_{TB} approaches the value $M_{TB_{max}}$, the flux diminishes to zero and M_{TB} is kept limited.

4.5 Coupled Cycles

An important class of cycles with non-linear behavior is represented by situations when coupling occurs between cycles of different elements. The behavior of coupled systems of this type has been studied in detail by Prigogine (1967) and others. In these systems, multiple equilibria are sometimes possible and oscillatory behavior can occur. There have been suggestions that atmospheric systems of chemical species, coupled by chemical reactions, could exhibit multiple equilibria under realistic ranges of concentration (Fox *et al.*, 1982; White, 1984). However, no such situations have been confirmed by measurements.

The cycles of carbon and the other main plant nutrients are coupled in a fundamental way by the involvement of these elements in photosynthetic assimilation and plant growth. Redfield (1934) and several others have shown that there are approximately constant proportions of C, N, S, and P in marine plankton and land plants ("Redfield ratios"); see Chapter 10. This implies that the exchange flux of one of these elements between the biota reservoir and the atmosphere – or ocean – must be strongly influenced by the flux of the others.

Williams (1987) has pointed out that there are two main approaches to the treatment of such couplings. The first is to apply flux expressions like the one described for CO_2 in Section 4.4, in Equation (33), and let both the rate coefficient (k) and the upper limit of the biota reservoir size

($M_{TB_{max}}$) be explicit functions of the available concentrations of the other nutrients. This approach allows for a pronounced interdependence between the fluxes of the different nutrients but it does not ensure that the Redfield ratios are maintained. In the second approach the contents of the nutrients in the biota reservoir are forced to remain close to the Redfield ratios. This method was used by Mackenzie *et al.* (1993) in their study of the global cycles of C, N, P, and S and their interactions. They were able to demonstrate how a human perturbation in one of these element cycles could influence the cycles of the other elements.

4.6   Forward and Inverse Modeling

In most cases models describing biogeochemical cycles are used to estimate the concentration (or total mass) in the various reservoirs based on information about source and sink processes, as in the examples given in Section 4.4. This is often called *forward modeling*. If direct measurements of the concentration are available, they can be compared to the model estimates. This process is referred to as *model testing*. If there are significant differences between observations and model simulations, improvements in the model are necessary. A natural step is then to reconsider the specification of the sources and/or the sinks and perform additional simulations.

Inverse modeling represents a situation when a model is used, in a systematic fashion, to estimate the magnitude of sources or sinks from observed concentrations (mass). In the simplest case with a single reservoir (Fig. 4-1) in steady state, the formal solution of the inverse problem follows directly from the relations $Q = kM$ or $k = Q/M$, depending on whether Q or k is the quantity being sought. In a situation when concentrations vary in space and time and the model consists of several reservoirs the inverse modeling becomes more complex and statistical methods, such as minimizing the squares of deviations between observations and simulations, have to be employed. For example, Prinn *et al.* (1992) used an eight-box model of the atmosphere (four latitude bands and two height layers) together with observations of

methyl chloroform (CH_3CCl_3) from a global network and information about its emission rate to estimate the removal rate of this gas in the atmosphere. Since the only removal process for methyl chloroform is the reaction with hydroxyl radical and the rate of this chemical reaction is well established, their result also provided an estimate of the concentration of hydroxyl radical in the atmosphere. This estimate of the atmospheric concentration of hydroxyl radical has been very useful in connection with later (forward) modeling of many other chemical compounds that are also removed by reaction with the hydroxyl radical, e.g., CO, CH_4, SO_2, etc.

4.7   High-Resolution Models

So far the focus of this chapter has been on relatively simple box models with each box representing a reservoir with well-defined boundaries in terms of its physical, chemical or biological characteristics, e.g., the ocean surface layer, the atmosphere, the terrestrial biota, etc. In many situations it is also important to model the distribution of a species within such a reservoir, especially the fluid reservoirs (atmosphere and water bodies) where transport processes are rapid. A common tool for modeling such processes is a gridpoint model in which the fluid space is divided into smaller boxes, each one of them represented mathematically in the model by a single gridpoint. The physical transport of the species in question, by mean motions as well as by turbulence, can then be described according to Equations (40) and (41) in Section 4.8.1 if the spatial derivatives are approximated by finite differences between adjacent gridpoints.

The simplest kind of gridpoint model is one where only one spatial dimension is considered, most often the vertical. Such one-dimensional models are particularly useful when the conditions are horizontally homogeneous and the main transport occurs in the vertical direction. Examples of such situations are the vertical distribution of CO_2 within the ocean (except for the downwelling regions in high latitudes, Siegenthaler, 1983) and the vertical distribution of ozone and other trace gases in the stratosphere (Ko *et al.*, 1989).

Two-dimensional gridpoint models enable more realistic descriptions of transport processes in that circulation cells like the Hadley cell in the tropical and subtropical troposphere (cf. Section 10.5.2) can be explicitly modeled. Models of this kind, with height vs. latitude dimensions, have been very useful for improving our understanding of the global-scale distributions of long-lived gaseous components in the atmosphere. For species with an atmospheric lifetime that is short compared to the characteristic time for mixing longitudinally around the globe, i.e., several weeks, two-dimensional (height/latitude) models are less useful. This is the case for reactive gases like SO_2 and NO_x as well as for species carried by aerosol particles, which have average atmospheric lifetimes of only a few days. For such species the concentration also exhibits large variations in the longitudinal direction, so three-dimensional models (height/latitude/longitude) are required.

In three-dimensional models many more oceanic and atmospheric motion systems can be explicitly described, including ocean currents and monsoon circulations and extratropical cyclones in the atmosphere. The drawback is that these models quickly become very complex and require large computer facilities. This limitation has recently become less of a problem because of the rapid development of computers and the increasing number of high-quality observations for model testing. Indeed, regional and global scale three-dimensional models have become a standard tool for studies of biogeochemical cycles, especially their atmospheric component.

Although many important features of oceanic and atmospheric circulation can be explicitly resolved in three-dimensional gridpoint models, there will always be many processes that occur on the sub-gridscale level that cannot. The effects of these sub-gridscale processes must be *parameterized*, i.e., summarized in a statistical fashion in a way related to the large-scale flow. The purpose of parameterization is to describe the combined effect of sub-gridscale processes on the larger-scale variables. For example, convective motion systems in the oceans and in the atmosphere occur on spatial scales of a few km or less and therefore cannot be resolved explicitly in large-scale models with a grid size of 100 km or more. However, the combined effect of such motions is of fundamental importance for vertical energy transport. Parameterization schemes therefore have to be used to describe how convection develops under certain conditions and how they influence the large-scale flow in the ocean and in the atmosphere.

Figure 4-13 shows an example from a three-dimensional model simulation of the global atmospheric sulfur balance (Feichter *et al.*, 1996). The model had a grid resolution of about 500 km in the horizontal and on average 1 km in the vertical. The chemical scheme of the model included emissions of dimethyl sulfide (DMS) from the oceans and SO_2 from industrial processes and volcanoes. Atmospheric DMS is oxidized by the hydroxyl radical to form SO_2, which, in turn, is further oxidized to sulfuric acid and sulfates by reaction with either hydroxyl radical in the gas phase or with hydrogen peroxide or ozone in cloud droplets. Both SO_2 and aerosol sulfate are removed from the atmosphere by dry and wet deposition processes. The reasonable agreement between the simulated and observed wet deposition of sulfate indicates that the most important processes affecting the atmospheric sulfur balance have been adequately treated in the model.

In gridpoint models, transport processes such as speed and direction of wind and ocean currents, and turbulent diffusivities (see Section 4.8.1) normally have to be prescribed. Information on these physical quantities may come from observations or from other (dynamic) models, which calculate the flow patterns from basic hydrodynamic equations. Tracer transport models, in which the transport processes are prescribed in this way, are often referred to as *off-line models*. An *on-line model*, on the other hand, is one where the tracers have been incorporated directly into a dynamic model such that the tracer concentrations and the motions are calculated simultaneously. A major advantage of an on-line model is that feedbacks of the tracer on the energy balance can be described

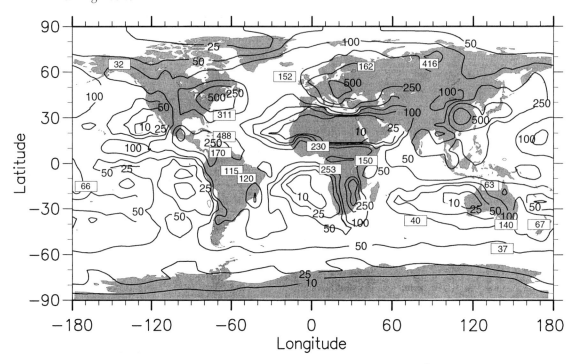

Fig. 4-13 Calculated and observed annual wet deposition of sulfur in mg S/m^2 per year. (Reprinted from "Atmospheric Environment," Volume 30, Feichter, J., Kjellström, E., Rodhe, H., Dentener, F., Lelieveld, J., and Roelofs, G.-J., Simulation of the tropospheric sulfur cycle in a global climate model, pp. 1693–1707, Copyright © 1996, with permission from Elsevier Science.)

realistically. Examples of such dynamically interactive tracers that are important for the climate system are CO_2 and the other greenhouse gases, as well as aerosols. Due to the great complexity of global models that combine in an interactive fashion tracer transport and explicit hydrodynamics, few such simulations have yet been carried out.

Dynamic models based on gridpoint or spectral discretization have been used by meteorologists to forecast weather for several decades. Today, the climate issue has stimulated rapid development of models designed for simulations of climate and its variations on time scales of decades and centuries. Unlike weather forecast models, global-scale climate models – general circulation models (GCMs) – must include descriptions of ocean circulation, the distribution of ice, snow and vegetation, and especially the exchange of heat, water and momentum (friction) between the atmosphere and the underlying surface. Figure 4-14 is a

schematic of the components considered in a GCM.

4.8 Transport Processes

So far we have not gone in-depth into the nature of the transport processes responsible for fluxes of material between and within reservoirs. This section includes a very brief discussion of some of the processes that are important in the context of global biogeochemical cycles. More comprehensive treatments can be found in textbooks on geology, oceanography and meteorology and in reviews such as Lerman (1979) and Liss and Slinn (1983).

4.8.1 Advection, Turbulent Flux and Molecular Diffusion

Let us consider a fluid in which a tracer *i* is

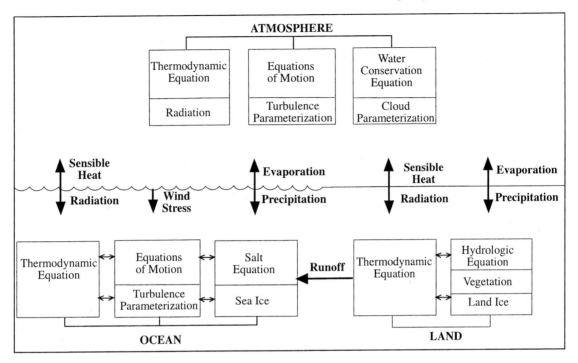

Fig. 4-14 Schematic diagram showing the components of a global climate model (GCM). (Reprinted from Hartmann (1994), with permission from Academic Press.)

mixed. A flux of the tracer within the fluid can be brought about either by organized fluid motion or by molecular diffusion. These two flux processes can be written as

$$\mathbf{F}_{i1} = \mathbf{V}q_i\rho = \mathbf{V}c_i \qquad (34)$$

and

$$\mathbf{F}_{i2} = -D_i\rho\nabla q_i \qquad (35)$$

where $\mathbf{F}_{i1}$ and $\mathbf{F}_{i2}$ denote the flux vectors of the tracer (dimension: $M/(L^2T)$), $\mathbf{V}$ the fluid velocity vector (L/T), ρ the density of the fluid (M/L^3), q_i the tracer mixing ratio (M/M), c_i the mass concentration of the tracer (M/L^3), D the molecular diffusivity (L^2/T) and ∇ the gradient operator (L^{-1}). The expression ∇q_i: denotes the vector ($\partial q_i/\partial x, \partial q_i/\partial y, \partial q_i/\partial z$).

The continuity of tracer mass is expressed by the equation

$$\frac{\partial c_i}{\partial t} = -\nabla \cdot \mathbf{F}_i + Q - S = -\nabla \cdot (\mathbf{F}_{i1} + \mathbf{F}_{i2}) + Q - S$$

$$= -\nabla \cdot (\mathbf{V}c_i) + \nabla \cdot (D_i\rho_p\nabla q_i) + Q - S \qquad (36)$$

where Q and S represent production and removal of the tracer ($M/(L^3T)$). Here $\nabla \cdot \mathbf{F}_i$ denotes the scalar quantity

$$\frac{\partial F_{ix}}{\partial x} + \frac{\partial F_{iy}}{\partial y} + \frac{\partial F_{iz}}{\partial z}$$

If variations in fluid density and diffusivity can be neglected we have

$$\frac{\partial c_i}{\partial t} = -\nabla \cdot (\mathbf{V}c_i) + D\nabla^2 c_i + Q - S \qquad (37)$$

In most situations a fluid would be turbulent implying that the velocity vector, as well as the concentration c_i, exhibits considerable variability on time scales smaller than those of prime interest. This situation can be described by writing these quantities as the sum of an average quantity (normally a time average) and a perturbation

$$\mathbf{V} = \bar{\mathbf{V}} + \mathbf{V}'$$
$$c_i = \bar{c}_i + c'_i$$

From Equation (34), the transport flux $\mathbf{F}_{i1}$, then becomes

$$\mathbf{F}_{i1} = (\bar{\mathbf{V}} + \mathbf{V}')(\bar{c} + c_i') = \bar{\mathbf{V}}\bar{c}_i + \bar{\mathbf{V}}c_i' + \mathbf{V}'\bar{c}_i + \mathbf{V}'c_i' \tag{38}$$

and its average value

$$\overline{\mathbf{F}_{i1}} = \bar{\mathbf{V}}\bar{c}_i + \overline{\mathbf{V}'c_i'} \tag{39}$$

Note that the averages of $\mathbf{V}'$ and c' are equal to zero. The continuity equation can now be written as

$$\frac{\partial \bar{c}_i}{\partial t} = -\nabla \cdot (\bar{\mathbf{V}}\bar{c}_i) - \nabla \cdot (\overline{\mathbf{V}'c_i'}) + D\nabla^2\bar{c}_i + Q - S \tag{40}$$

The first two terms on the right side of Equation (40) describe the contributions from transport by advection and by turbulent flux, respectively. The separation of the motion flux into advection and turbulent flux is somewhat arbitrary; depending upon the circumstances the averaging time can be anything from a few minutes to a year or even more.

Since in most situations the perturbation quantities ($\mathbf{V}'$ and c_i') are not explicitly resolved, it is not possible to evaluate the turbulent flux term directly. Instead, it must be related to the distribution of averaged quantities – a process referred to as parameterization. A common assumption is to relate the turbulent flux vector to the gradient of the averaged tracer distribution, which is analogous with the molecular diffusion expression, Equation (35).

$$(\mathbf{F}_{i2})_{\text{turb}} = \overline{\mathbf{V}'c'} = -k_{\text{turb}}\nabla\bar{c}_i \tag{41}$$

The coefficient k_{turb} introduced in Equation (41) (dimension: L^2/T) is called the turbulent, or eddy diffusivity. In the general case the eddy diffusivity is given separate values for the three spatial dimensions. It must be remembered that the eddy diffusivities are not constants in any real sense (like the molecular diffusivities) and that their numerical values are very uncertain. The assumption underlying Equation (41) is therefore open to question.

In most cases, the term expressing the divergence of the molecular flux in Equation (40) ($D\nabla^2\bar{c}_i$) can be neglected compared to the other two transport terms. Important excep-

tions occur, e.g. in a thin layer of the atmosphere close to the surface and in similar layers of the oceans close to the ocean floor and to the surface (viscous sublayers). Molecular diffusion is also an important transport process in the upper atmosphere, at heights above 100 km.

Order-of-magnitude values for the vertical eddy diffusivity in the atmosphere and the ocean are shown in Fig. 4-15. The values for the viscous layers represent molecular diffusivities of a typical air molecule like N_2.

Development in recent years of fast-response instruments able to measure rapid fluctuations of the wind velocity ($\mathbf{V}'$) and of the tracer concentration (c'), has made it possible to calculate the turbulent flux directly from the correlation expression in Equation (41), without having to resort to uncertain assumptions about eddy diffusivities. For example, Grelle and Lindroth (1996) used this *eddy-correlation technique* to calculate the vertical flux of CO_2 above a forest canopy in Sweden. Since the mean vertical velocity ($\bar{w}$) has to vanish above such a flat surface, the only contribution to the vertical flux of CO_2 comes from the eddy-correlation term ($\overline{c'w'}$). In order to capture the contributions from all important eddies, both the anemometer and the CO_2 instrument must be able to resolve fluctuations on time scales down to about 0.1 s.

A type of motion that is often very important in both the oceans and the atmosphere is *convection*. This is a vertical mixing process where parcels of water (or air) are rapidly transported in the vertical direction due to their buoyancy. In the oceans, this occurs when the surface water becomes denser than the underlying water – by cooling and/or increased salinity due to evaporation – and parcels of water sink down within days to depths of up to several km. In the atmosphere, convective motions occur when surface air is heated by conduction from the underlying surface. Air parcels having a horizontal dimension on the order of 1 km then rise and sometimes reach a height as high as 10–15 km within less than an hour, especially in tropical areas. Cumulus and cumulonimbus clouds are visible manifestations of convection in the atmosphere. In some circumstances, con-

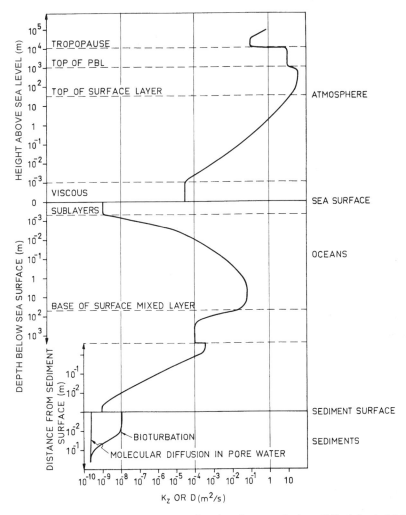

Fig. 4-15 Orders of magnitude of the average vertical molecular or turbulent diffusivity (whichever is largest) through the atmosphere, oceans, and uppermost layer of ocean sediments.

vection may contribute to a very rapid vertical mixing.

4.8.2 Other Transport Processes

Under some circumstances transport processes other than fluid motion and molecular diffusion are important. One important example is *sedimentation* due to gravity acting on particulate matter submerged in a fluid, e.g., removal of dissolved sulfur from the atmosphere by precipitation scavenging, or transport of organic carbon from the surface waters to the deep layers and to the sediment by settling detritus. The rate of transport by sedimentation is determined essentially by the size and density of the particles and by the counteracting drag exerted by the fluid.

Geochemically significant mixing and transport can sometimes be accomplished by biological processes. An interesting example is redistribution of sediment material caused by the movements of worms and other organisms (*bioturbation*).

Exchange processes between the atmosphere and oceans and between the oceans and the sediments are treated below in separate sections.

4.8.3   Air-Sea Exchange

4.8.3.1   Gas transfer

The magnitude and direction of the net flux density, F, of any gaseous species across an air–water interface is positive if the flux is directed from the atmosphere to the ocean. F is related to the difference in concentration (Δc), in the two phases by the relation

$$F = K\Delta c \qquad (42)$$

Here $\Delta c = c_a - K_H c_w$ with c_a and c_w representing the concentrations in the air and water respectively and K_H the Henry's law constant. The parameter K, linking the flux and the concentration difference, has the dimension of a velocity. It is often referred to as the transfer (or piston) velocity. The reciprocal of the transfer velocity corresponds to a resistance to transfer across the surface. The total resistance ($R = K^{-1}$) can be viewed as the sum of an air resistance (R_a) and a water resistance (R_w):

$$R = R_a + R_w = \frac{1}{k_a} + \frac{K_H}{\alpha k_l} \qquad (43)$$

The parameters k_a and k_l are the transfer velocities for chemically unreactive gases through the viscous sublayers in the air and water, respectively. They relate the flux density F to the concentration gradients across the viscous sublayers through expressions similar to Equation (42):

$$F = k_a(c_a - c_{a,i})$$
$$F = k_l(c_{w,i} - c_w) \qquad (44)$$

Here $c_{a,i}$ and $c_{w,i}$ are the concentrations right at the interface (cf. Fig. 4-16). They are related by $c_{a,i} = K_H c_{w,i}$.

The parameter α in Equation (43) quantifies any enhancement in the value of k_l due to chemical reactivity of the gas in the water. Its value is unity for an unreactive gas; for gases with rapid aqueous phase reactions (e.g., SO_2) much higher values can occur.

A comparison of the resistance in air and water for different gases shows that the resistance in the water dominates for gases with low solubility that are unreactive in the aqueous phase (e.g., O_2, N_2, CO_2, CH_4). For gases of

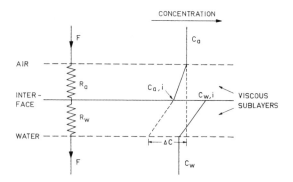

Fig. 4-16   A simplified model of flux resistances and concentration gradients in the viscous sublayers at the air–sea interface.

high solubility or rapid aqueous chemistry (e.g., H_2O, SO_2, NH_3) processes in the air control the interfacial transfer.

The numerical values of the transfer velocity K for the different gases are not well established. Its magnitude depends on such factors as wind speed, surface waves, bubbles and heat transfer. A globally averaged value of K often used for CO_2 is about 10 cm/h. Transport at the sea–air interface is also discussed in Chapter 10; for a review see Liss (1983).

4.8.3.2   Transfer of particles

Liquid water, including its soluble and insoluble constituents, is transferred from the oceans to the atmosphere when air bubbles in the water rise to the surface. These bubbles form from air trapped by breaking waves, "whitecaps." As the bubbles burst at the surface, water droplets are injected into the atmosphere. These water droplets are small enough to remain airborne for several hours. Whitecaps begin to form in winds common over the oceans, and a significant amount of seasalt made airborne in this way is transported to the continents and deposited in coastal areas.

The flux of particles in the other direction, deposition on the ocean surface, occurs intermittently in precipitation (wet deposition) and more continuously as a direct uptake by the surface (dry deposition). These flux densities may be represented by a product of the concentration of particulate matter in air close to the surface and parameters often referred to as deposition velocities:

$$F = F_w + F_d = (v_w + v_d)c_a \qquad (45)$$

The deposition velocities depend on the size distribution of the particulate matter, on the frequency of occurrence and intensity of precipitation, the chemical composition of the particles, the wind speed, nature of the surface, etc. Typical values of v_w and v_d for particles below about 1 μm in diameter are in the range 0.1 to 1 cm/s (Slinn, 1983). The average residence time in the atmosphere for such particles is a few days.

4.8.4   Sediment–Water Exchange

The sediment surface separates a mixture of solid sediment and interstitial water from the overlying water. Growth of the sediment results from accumulation of solid particles and inclusion of water in the pore space between the particles. The rates of sediment deposition vary from a few millimeters per 1000 years in the pelagic ocean up to centimeters per year in lakes and coastal areas. The resulting flux density of solid particles to the sediment surface is normally in the range 0.006 to 6 kg/m^2 per year (Lerman, 1979). The corresponding flux density of materials dissolved in the trapped water is 10^{-6} to 10^{-3} kg/m^2 per year. Chemical species may also be transported across the sediment surface by other transport processes. The main processes are (Lerman, 1979):

1. Sedimentation of solids (mineral, skeletal and organic materials).
2. Flux of dissolved material and water into sediment, contributing to the growth of the sediment column.
3. Upward flow of pore water and dissolved material caused by pressure gradients.
4. Molecular diffusional fluxes in pore water.
5. Mixing of sediment and water at the interface (bioturbation and water turbulence).

An estimate of the advective fluxes (processes 1, 2, and 3) requires knowledge of the concentration of the species in solutions and in the solid particles as well as of the rates of sedimentation and pore water flow. The diffusive type processes, 4 and 5, depend on vertical gradients of the concentrations of the species as well as on the diffusivities. In regions where bioturbation occurs, the effective diffusivity in the uppermost centimeters of the sediments can be more than that due to molecular diffusion in the pore water alone (cf. Fig. 4-15).

4.9   Time Scales of Mixing in the Atmosphere and Oceans

It is often important to know how long an element spends in one environment before it is transported somewhere else in the Earth system. For example, if a time scale characterizing a chemical or physical transformation process in a region has been estimated, a comparison with the time scale characterizing the transport away from the region will tell which process is likely to dominate.

The question of residence time and its definition in a steady-state reservoir was discussed earlier in this chapter. The average residence time in the reservoir was shown to be equal to the turnover time $\tau_0 = M/S$ where M is the mass of the reservoir and S the total flux out of it. It is important to note that if one considers the exchange between two reservoirs of different mass, the time scale of exchange will be different depending upon whether the perspective is from the small or the big reservoir. An interesting example is that of mixing between the troposphere and stratosphere in the atmosphere. Studies of radioactive nuclides injected into the lower stratosphere by bomb testing have shown that the time scale characterizing the exchange between the lower stratosphere and troposphere is one to a few years. This means that a "particle" injected in the lower stratosphere will stay for this time, on average, before entering the troposphere. On the other hand, a gas molecule like N$_2$O, which is chemically stable in the troposphere, will spend several decades in the troposphere before it is mixed up into the lower stratosphere, where it is decomposed by photochemical processes. So although the gross flux of air from the troposphere to the stratosphere, F, is equal to the gross flux of air from the stratosphere to the troposphere, the time scale of mixing between these two reservoirs is very different. From the tropospheric point of view, the time scale τ_T is several decades, but the time scale of mixing as seen from the stratosphere (τ_S) is only a few years. The reason for the difference is the small mass of the stratosphere, M_S as compared to the troposphere, M_T. Formally, we can write

$$\frac{\tau_S}{\tau_T} = \frac{M_S/F}{M_T/F} = \frac{M_S}{M_T} \approx 0.1$$

Time scales of transport can also be applied to situations when no well-defined reservoirs can be defined. If the dominant transport process is advection by mean flow or sedimentation by gravity, the time scale characterizing the transport between two places is simply $\tau_{adv} = L/V$ where L is the distance and V the transport velocity. Given a typical wind speed of 20 m/s in the mid-latitude tropospheric westerlies, the time of transport around the globe would be about 2 weeks.

In situations where the transport is governed by diffusive processes a time scale of transport can be defined as

$$\tau_{turb} = \frac{L^2}{D} \qquad (46)$$

where L is the distance and D is the diffusivity (molecular or turbulent). Applying this definition to the vertical mixing through the surface mixed layer of the ocean, assuming the depth of the layer to be 50 m and the turbulent diffusivity 0.1 m^2/s, we get

$$\tau_{turb} = \frac{(50)^2}{0.1} s \approx 7\,h$$

Some important time scales characterizing the transport within the oceanic and atmospheric environments are summarized in Fig. 4-17. In view of the somewhat ambiguous nature of the definitions of these time scales, the numbers should not be considered as more than indications of the magnitudes.

Acknowledgement. I wish to thank Anders Björkström and Michael Jacobson for valuable comments on the manuscript.

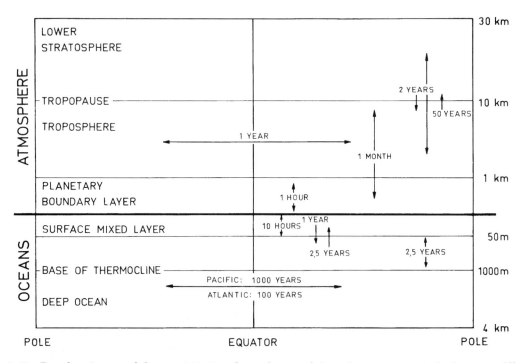

Fig. 4-17   Rough estimates of characteristic time for exchange of air and water respectively, between dfferent parts of the atmosphere and oceans.

Questions

4-1 Consider a reservoir with two separate sources Q_1 and Q_2 and a single sink S. The magnitudes of Q_1 and S and their uncertainties have been estimated to be 75 ± 20 and 100 ± 30 (arbitrary units). Assuming that there is no direct way of estimating Q_2, how would you derive its magnitude and uncertainty range from budget considerations? What assumption must be made regarding the reservoir?

4-2 Calculate the turnover time of carbon in the various reservoirs given in Fig. 4-3.

4-3 What is the relation between the turnover time τ_0 the average transit time τ_r, and the average age τ_a, in a reservoir where all "particles" spend an equal time in the reservoir?

4-4 Consider a reservoir with a source flux Q and two sink fluxes S_1 and S_2. S_1 and S_2 are proportional to the reservoir content M with proportionality constants k_1 and k_2. The values of k_1 and k_2 are (1 yr^{-1}) and (0.2 yr^{-1}), respectively. The system is initially in steady state with $M = M_0$ and $Q = S_{10} + S_{20}$. Describe the change in time of M if the source is suddenly reduced to half its initial value. What is the response time of the reservoir?

4-5 Consider the water balance of a lake with a constant source flux Q. The outlet is the "threshold" type where the sink is proportional to the mass of water above a threshold value M_1; $S = k(M - M_1)$. Calculate the turnover time of water at steady state and the response time relative to changes in Q.

4-6 For the more mathematically inclined: Investigate the dynamic behavior of a coupled linear three reservoir model using the technique outlined in Section 4.3.1.

Answers can be found on p. 509.

References

Bolin, B. (1986). How much CO_2 will remain in the atmosphere? *In* "The Greenhouse Effect, Climate Change, and Ecosystems," SCOPE Report 29 (B. Bolin, B. R. Döös, J. Jäger, and R. A. Warrick, eds). Wiley, Chichester.

Bolin, B. and Rodhe, H. (1973). A note on the concepts of age distribution and transit time in natural reservoirs. *Tellus* **25**, 58–62.

Braun, M. (1983). "Differential Equations and their Applications," 3rd edn (short version). Springer-Verlag, New York.

Chameides, W. L. and Perdue, E. M. (1997). "Biogeochemical Cycles." Oxford University Press, New York and Oxford.

Eriksson, E. (1971). Compartment models and reservoir theory. *Ann. Rev. Ecol. Syst.* **2**, 67–84.

Feichter, J., Kjellström, E., Rodhe, H., Dentener, F., Lelieveld, J., and Roelofs, G.-J. (1996). Simulation of the tropospheric sulfur cycle in a global climate model. *Atmos. Environ.* **30**, 1693–1707.

Follmi, K. B. (1996). The phosphorus cycle, phosphogenesis and marine phosphate-rich deposits. *Earth Sci. Rev.* **40**, 55–124.

Fox, J. L., Wofsy, S. C., McElroy, M. B., and Prather, M. J. (1982). A stratospheric chemical instability. *J. Geophys. Res.* **87**, 11 126–11 132.

Grelle, A. and Lindroth, A. (1996). Eddy-correlation system for long-term monitoring of fluxes of heat, water vapor and CO_2. *Global Change Biol.* **2**, 297–307.

Hamrud, M. (1983). Residence time and spatial variability for gases in the atmosphere. *Tellus* **35B**, 295–303.

Hartmann, D. J. (1994). "Global Physical Climatology." Academic Press, San Diego.

IPCC (1996). "Climate Change 1995, The Science of Climate Change." Intergovernmental Panel on Climate Change. Cambridge University Press.

Jahnke, R. A. (2000). Current edition.

Junge, C. E. (1974). Residence time and variability of tropospheric gases. *Tellus* **26**, 477–488.

Keeling, C. D. and Bacastow, R. B. (1977). Impact of individual gases on climate. *In* "Energy and Climate," pp. 72–95. US National Research Council, Washington, DC.

Ko, M. K. W., Sze, N. D., and Weisenstein, D.K. (1989). The roles of dynamical and chemical processes in determining the stratospheric concentration of ozone in one-dimensional and two-dimensional models. *J. Geophys. Res.* **94**, 9889–9896.

Lasaga, A. C. (1980). The kinetic treatment of geochemical cycles. *Geochim. Cosmochim. Acta* **44**, 815–828.

Lerman, A. (1979). "Geochemical Processes; Water and Sediment Environments." Wiley, New York.

Lerman, A., Mackenzie, F. T., and Garrels, R. M. (1975). Modeling of geochemical cycles: phosphorus as an example. *Geol. Soc. Am. Mem.* **142**, 205–218.

Liss, P. (1983). Gas transfer: experiments and geochemical implications. *In* "Air–Sea Exchange of Gases and Particles" (P. S. Liss and W. G. N. Slinn, eds), pp. 241–298. D. Reidel Publ. Co., Dordrecht.

Liss, P. S. and Slinn, W. G. N. (1983). "Air–Sea Exchange of Gases and Particles." D. Reidel, Dordrecht.

Mackenzie, F. T., Ver, L. M., Sabine, C., Lane, M., and Lerman, A. (1993). C, N, P, S global biogeochemical cycles and modeling of global change. *In* "Interactions of C, N, P and S Biogeochemical Cycles and Global Change" (Wollast *et al.*, eds). NATO ASI Series, Vol. 14. Springer-Verlag, Berlin.

McDuff, R. E. and Morel, F. M. M. (1980). The geochemical control of seawater (Sillén revisited). *Environ. Sci. Technol.* **14**, 1182–1186.

Prigogine, I. (1967). "Introduction to Thermodynamics of Irreversible Processes." Wiley-Interscience, New York.

Prinn, R., Cunnold, D., Simmonds, P., Alyea, F., Boldi, R., Crawford, A., Fraser, P., Gutzler, D., Hartley, D., Rosen, R., and Rasmussen, R. (1992). Global average concentration and trend for hydroxyl radicals deduced from ALE/GAGE trichloroethane (methyl chloroform) data for 1978–1990. *J. Geophys. Res.* **97**, 2445–2461.

Redfield, A. C. (1934). On the proportions of organic derivatives in sea water and their relation to the composition of plankton. *In* "James Johnstone Memorial Volume," pp. 176–192. Liverpool University Press, Liverpool, England.

Rodhe, H. (1978). Budgets and turnover times of atmospheric sulfur compounds. *Atmos. Environ.* **12**, 671–680.

Rodhe, H. and Björkström, A. (1979). Some consequences of non-proportionality between fluxes and reservoir contents in natural systems. *Tellus* **31**, 269–278.

Siegenthaler, U. (1983). Uptake of excess CO_2 by an outcrop-diffusion model of the ocean. *J. Geophys. Res.* **88**, 3599–3608.

Slinn, W. G. N. (1983). Air-sea transfer of particles. *In* "Air–Sea Exchange of Gases and Particles" (P. S. Liss and W. G. N. Slinn, eds), pp. 299–405. D. Reidel Publ. Co., Dordrecht.

Southam, J. R. and Hay, W. W. (1976). Dynamical formulations of Broecker's model for marine cycles of biologically incorporated elements. *Math. Geol.* **8**, 511–527.

White, W. H. (1984). Does the photochemistry of the troposphere admit more than one steady state? *Nature* **309**, 242–244.

Williams, G. R. (1987). The coupling of biogeochemical cycles of nutrients. *Biogeochemistry* **4**, 61–75.

5

Equilibrium, Rate, and Natural Systems

Samuel S. Butcher and Sharon E. Anthony

5.1 Introduction

This chapter applies the physical chemistry taught in the first year of undergraduate chemistry to chemical problems in the natural environment and introduces key chemical concepts to use and keep in mind for the rest of this book. The material in this chapter is especially important to consider when utilizing the modeling techniques presented in Chapter 4.

There are two principal chemical concepts we will cover that are important for studying the natural environment. The first is *thermodynamics*, which describes whether a system is at equilibrium or if it can spontaneously change by undergoing chemical reaction. We review the main first principles and extend the discussion to electrochemistry. The second main concept is how fast chemical reactions take place if they start. This study of the rate of chemical change is called *chemical kinetics*. We examine selected natural systems in which the rate of change helps determine the state of the system. Finally, we briefly go over some natural examples where both thermodynamic and kinetic factors are important. This brief chapter cannot provide the depth of treatment found in a textbook fully devoted to these physical chemical subjects. Those who wish a more detailed discussion of these concepts might turn to one of the following texts: Atkins (1994), Levine (1995), Alberty and Silbey (1997).

In many cases one can apply the first principles of thermodynamics and chemical kinetics to natural systems only with caution. The reason

has little to do with the shortcomings of the core chemical principles. Rather, the application of thermodynamics (including electrochemistry) can be restricted by the fact that some natural systems have reaction rates so slow that they exist for long periods under non-equilibrium conditions. Moreover, reaction rates can be difficult to characterize when they are sensitive to poorly understood catalytic (including enzymatic) effects or to surface effects. Applications of thermodynamics and kinetics, as they are presented in this chapter, require knowledge of many variables such as concentration, temperature, and pressure. However, in contrast to reactions in a stirred beaker, the natural world may have large gradients of these variables. The natural environment also differs because of the large number of chemical species that coexist and interact concurrently. Therefore, it is important that the limitations of chemical thermodynamics and kinetics be considered before they are applied to biogeochemical systems.

5.2 Thermodynamics

5.2.1 State Functions and Equilibrium Constraints

The laws of thermodynamics are the cornerstones of any description of a system at equilibrium. The First Law, also known as the Law of Conservation of Energy states that energy cannot be created or destroyed, i.e., the energy of the universe is constant. Thus if the internal

Earth System Science
ISBN 0-12-379370-X

energy of a system decreases in a reaction, the chemical change is accompanied by a release of energy, often in the form of heat. The *enthalpy*, *H*, is a property of the system that is equal to $E + PV$, where *E* is the internal energy of the system, *P* is pressure, and *V* is volume. At constant pressure, the change in *H* is the energy flow as heat. Enthalpy is a *state function* of the system, which means that it is a property that depends only on the state of the system and not how it managed to arrive at that state. Other state functions are temperature and pressure. Work and heat are examples of properties that are not state functions.

The Second Law of thermodynamics states that for a chemical process to be spontaneous, there must be an increase in entropy. Entropy (*S*) can be thought of as a measure of disorder.

Another property of the system, *G*, the free energy, or the Gibbs free energy, is related to enthalpy and entropy by:

$$G = H - TS \qquad (1)$$

When the temperature of reactants and products are equal, ΔG is given by

$$\Delta G = \Delta H - T\Delta S \qquad (2)$$

The second law also describes the equilibrium state of a system as one of *maximum entropy and minimum free energy*. For a system at constant temperature and pressure the *equilibrium condition* requires that the change in free energy is zero:

$$\Delta G = G_{\text{prod}} - G_{\text{react}} = 0 \qquad (3)$$

We represent a change in a quantity for any chemical reaction as the value of that quantity for the products of the reaction minus the value of that quantity for reactants. It is important to keep in mind that the terms reactant and product refer only to how the chemical equation is written and not to whether or not the substance is actually being formed or is disappearing.

Equation (3) defines the equilibrium condition under the constraint that temperature and pressure are constant. A related consequence of the Second Law is that if $\Delta G < 0$ the reaction of the reactant to product is *thermodynamically spontaneous*. Thermodynamic spontaneity means that

the system has the *potential* to react. The more negative ΔG, the more spontaneous the reaction. The actual time scale for the reaction to occur can vary greatly. A good example of a non-equilibrium system is the ambient pressure of nitrogen oxides in the troposphere, which can greatly exceed their equilibrium values for the equilibrium with N_2 and O_2. We discuss non-equilibrium systems later in the chapter.

Although ΔG is the overall determinant of spontaneity, it is convenient to examine the two thermodynamic components of ΔG. The ΔH term is largely a function of the strengths of the chemical bonds in reactant and product. To the extent that there are more strong bonds (and strong associations between molecules) in the product than in the reactant, ΔH will be negative. An examination of Equation (3) shows that a negative ΔH contributes to the overall spontaneity of the reaction.

The ΔS term is a measure of the relative degree of disorder in reactant and product. To the extent that the product has greater disorder than the reactant ΔS will be positive. A positive ΔS will contribute to the overall spontaneity of the reaction. ΔS for a reaction can be evaluated from tables of entropy data. Moreover, the sign of ΔS for gas phase reactions may often be determined without entropy tables. For gas phase systems in which the number of independent molecules and atoms is greater in the product, ΔS will be positive. If there is no change in the number of atoms and molecules for a gas phase reaction it is difficult to say (without evaluating ΔS from tables of data) whether ΔS is positive or negative. In the liquid phase these simple rules for ΔS are less easily applied because significant entropy effects occur in water and many other solvents. Ionic solutes can cause a relative lowering of entropy of the solution by forming highly ordered associations with water, whereas other solutes may increase the solution entropy by disrupting the structure of water and not replacing it with other low-entropy structures.

In practice, *G* and *H* for a substance are defined relative to the *G* and *H* for the constituent elements of that substance. These relative values are known as *free energy of formation* and *enthalpy of formation* for standard conditions and symbolized as ΔG^0 and ΔH^0. So where we indicated values of G and H in Equations (1) and (2), in practice we would use a free energy and enthalpy of formation, which are themselves a special kind of ΔG and ΔH. Values for these functions may be obtained from standard tables of thermodynamic data, usually for the reference temperature of 298.2 K and a pressure of 1.0 bar. The Chemical Rubber Company handbook (Lide, 1998) is one of the more commonly available sources.

More complete sources, including some with data for a range of temperatures, are listed in the references at the end of the chapter. Note that many tabulations (including many contemporary biological sources) still represent these energy functions in calories and that it may be necessary to make the conversion to joules (1 cal = 4.1840 J). Because of the definition of the energy of formation, elements in their standard state (carbon as graphite, chlorine as Cl_2 gas at one bar, bromine as Br_2 liquid, etc.) have free energies and enthalpies of formation equal to zero. If needed, the absolute entropies of substances (from which ΔS may be evaluated) are also available in standard sources.

For ions in the aqueous phase there is an additional complication in defining G and H. The change in G and H for the formation of an ion from its constituent element (e.g. $\frac{1}{2}Cl_2(g) + e^- \rightarrow Cl^-(aq)$) is defined *relative* to the change in G and H for the formation of H^+ from $H_2(g)$. This relative change in G and H is termed the *standard free energy or enthalpy of formation for ions*. As a result of this definition, ΔG^0 and ΔH^0 for the formation of H^+ are zero. In practice, the definitions given above lead to the following algorithm for determining ΔG^0 or ΔH^0 for a reaction. ΔG^0 for a reaction may be obtained by simply adding the ΔG^0 (formation) values for the product species and subtracting the sum of ΔG^0 (formation) for all reactant species. Where an element in its standard state or H^+ is involved, substitute zero.

5.2.2 Gas Phase Equilibria

For reactions in the gas phase, the free energy per mole of gas as a function of pressure is given by the following expression:

$$G = G^0 + RT \ln(P/P^0) \quad (4)$$

where P^0 is the standard reference pressure (commonly 1 bar or (1/1.01325) atmosphere), and P is the pressure of the gas, R is the *universal gas constant*, which has a value of 8.314 J/mol/K, and T is the temperature of the system in kelvins ($^\circ$C + 273). For the ideal systems considered here, we will treat the free energy per mole to be the same as the chemical potential. This treatment works best at very low pressures where gases approach ideal behavior.

For the gas phase equilibrium

$$N_2(g) + 2O_2(g) \leftrightarrow 2NO_2(g)$$

the change in free energy for this system at equilibrium is

$$\Delta G = 2G_{NO_2} - G_{N_2} - 2G_{O_2} = 0$$

substituting the expression for G^0 given in Equation (4), we derive the familiar equilibrium constant expression:

$$\Delta G^0 = -RT \ln \frac{\left(\frac{P_{NO_2}}{P^0}\right)^2}{\left(\frac{P_{N_2}}{P^0}\right)\left(\frac{P_{O_2}}{P^0}\right)^2} = -RT \ln K_P \quad (5)$$

We will see functions like the one occurring under the logarithm operator quite often. For efficiency, this is generally written as ln(Products)/(Reactants), where (Products) and (Reactants) denote the partial pressures of the species relative to the standard state pressure raised to a power that is equal to the stoichiometric coefficients. K_P is the equilibrium constant in terms of pressures. Since all pressures are in the same units, K_P is dimensionless. Note that in some literature there may be a combination of some power of P^0 with K_P to obtain an equilibrium constant with pressure units. In this case,

$$K_P' = \frac{K_P}{P^0} = \frac{P_{NO_2}^2}{P_{O_2}^2 P_{N_2}}$$

Although K_P for a gas phase reaction is defined in terms of the partial pressures of reactants and products, at times it is more convenient to express the equilibrium constant for

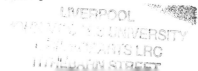

an equilibrium system in terms of moles or molecules per unit volume. The equilibrium constant in concentration terms is related to K_P by using the ideal gas equation by the following conversion:

$$K_C = K_P \left(\frac{P^0}{RT} \right)^{\Delta n} \qquad (6)$$

where K_C is the equilibrium constant in mole/volume units, and Δn is the change in the number of moles of gas for the reaction. In the case of the NO_2 equilibrium, $\Delta n = -1$.

5.2.3   Condensed Phase Equilibria

For reactions involving a nearly pure liquid or solid (often the case for water in aqueous solutions) G is given by the following expression, where X is the mole fraction of the liquid:

$$G = G^0 + RT \ln(X) \qquad (7)$$

However, for reactions occurring in dilute solutions, we use the following expression:

$$G = G^0 + RT \ln(c/c^0) \qquad (8)$$

which is parallel to the expression given for gases in Equation (4). Here c is the concentration of the solute (usually in units of moles per liter), and c^0 is the concentration in the standard state (defined as exactly 1 molar (M)). The ratio c/c^0 is related to the activity (for ideal solutions), which is defined by

$$G = G^0 + RT \ln(a) \qquad (9)$$

The activity is a measure of the tendency of a substance to react relative to its reacting tendency in the standard state. Here we relate activity to c/c^0 for ideal solutions. For ideal gases and ideal solvents, the activity approaches P/P^0 and X, respectively. Although c^0 is taken to be 1.0 M, Equation (8) works best when c is much less than 1.0 M.

One may combine the appropriate expressions for G for equilibria involving reactants in different phases to obtain a general expression, which relates the equilibrium constant to the change in free energy:

$$\Delta G^0 = -RT \ln(K_{eq}) \qquad (10)$$

Here K_{eq} may be a mixed expression involving pressures, mole fractions, and molar concentrations.

The van't Hoff equation describes the *temperature dependence* of the equilibrium constant.

$$\frac{d \ln K_{eq}}{dT} = \frac{\Delta H^0}{RT^2} \qquad (11)$$

Thus the degree of spontaneity at a given temperature depends on ΔG^0, but the *change* in spontaneity (as defined by the equilibrium constant) with temperature is given by Equation (11) and depends only on ΔH^0. If the change in temperature is small, one may assume that ΔH° is constant and integrate Equation (11) directly. For larger ranges in temperature one will need to know something about the temperature dependence of ΔH^0. This depends on the heat capacities of reactants and products.

5.2.4   Mixed Phases

As mentioned after Equation (10), the equilibrium constant may be expressed when the reactants are in several phases. As an example, the equilibrium between ammonia in a large cloud droplet and in the gas phase, $NH_3(aq)$ and $NH_3(g)$, is described by the equilibrium constant expression

$$K_{eq} = \frac{(P_{NH_3}/P^0)}{(c_{NH_3}/c^0)}$$

Here c_{NH_3} is the concentration of undissociated ammonia in water. The equilibrium constant for this class of equilibria is often defined in terms of a Henry's Law constant, K_H.

$$K_H = \frac{P_{N_3}}{c_{NH_3}}$$

The ΔG^0 value for $NH_3(aq)$ should be for dilute aqueous solutions. Note that K_H has units.

The solubility equilibrium for $CaCO_3$ (calcite) $\leftrightarrow Ca^{2+}(aq) + CO_3^{2-}(aq)$ is defined by

$$K_{eq} = \left(\frac{c_{Ca^{2+}}}{c^0} \right) \left(\frac{c_{CO_3^{2-}}}{c^0} \right) \quad \text{or} \quad K_{SP} = [Ca^{2+}][CO_3^{2-}]$$

where K_{SP} stands for *Solubility Product*. This equilibrium constant does not contain a term for $CaCO_3$ (calcite). To the extent that the calcite is a pure solid (does not contain dissolved impurities) and consists of particles large enough that surface effects are unimportant, this equilibrium does not depend on the "concentration" of calcite or particle size. Exceptions to this case are important in the formation of cloud droplets (where the particle size dependence is known as the Kelvin effect), which is discussed in Chapter 7, and in the solubility of finely divided solids.

5.2.5   Aggregate Variables

A knowledge of the concentrations of all reactants and products is necessary for a description of the equilibrium state. However, calculation of the concentrations can be a complex task because many compounds may be linked by chemical reactions. Changes in a variable such as pH or oxidation potential or light intensity can cause large shifts in the concentrations of these linked species. Aggregate variables may provide a means of simplifying the description of these complex systems. Here we look at two cases that involve acid–base reactions.

Al(III) is an example of an aquatic ion that forms a series of hydrated and protonated species. These include $AlOH^{2+}$, $Al(OH)_2^+$, $Al(OH)_3$, and other forms in addition to Al^{3+}. (For simplicity, we omit the H_2O molecules that complete the structures of these complexes.) Most of these species are amphoteric (able to act as an acid or a base). Thus the speciation of Al(III) and many other aquatic ions is sensitive to pH. In this case, an aggregate variable springs from the conservation of mass condition. In the case of dissolved aluminum, the total dissolved aluminum is given by

$$[Al_T] = [Al^{3+}] + [AlOH^{2+}] + [Al(OH)_2^+] + \ldots$$

For many problems Al_T is a constant or is known. The above condition thus constrains the concentrations of the individual species. A similar case occurs in carbonate equilibria, which leads to the formation of H_2CO_3,

HCO_3^-, and CO_3^{2-}. Total dissolved carbon is represented by

$$[C_T] = [CO_2(aq)] + [H_2CO_3] + [HCO_3^-] + [CO_3^{2-}]$$

In the case of CO_2 one must consider the hydrated form (H_2CO_3) and the non-hydrated form, $CO_2(aq)$. This latter form is in equilibrium with the atmosphere:

$$CO_2(g) \leftrightarrow CO_2(aq)$$

The $CO_2(aq)$ is in turn in equilibrium with $H_2CO_3(aq)$:

$$H_2CO_3(aq) \leftrightarrow CO_2(aq) + H_2O(l)$$

for which $K_{eq} \approx 650$. For many purposes, $CO_2(aq)$ and H_2CO_3 are lumped together as "carbonic acid," or $H_2CO_3^*$. The Henry's Law and acid–base equilibria are often written in terms of $H_2CO_3^*$ (Stumm and Morgan, 1981). Given the value of K_{eq}, most "carbonic acid" is in fact $CO_2(aq)$.

Alkalinity is helpful in describing the acid-neutralizing ability of an aqueous solution that may contain many ions. In any ionic equilibrium there is a conservation of charge condition.

$$\sum(+ \text{ charges}) = \sum(- \text{ charges})$$

We may further break this down to

$$\begin{aligned}
&\sum(+ \text{ charges for conservative ions}) \\
&+ \sum(+ \text{ charges for non-conservative ions}) \\
&= \sum(- \text{ charges for conservative ions}) \\
&+ \sum(- \text{ charges for non-conservative ions})
\end{aligned}$$
$$(12)$$

Conservative ions are ones that do *not* undergo acid–base reactions at the pH values of interest. These include Na^+, K^+, Ca^{2+}, SO_4^{2-}, Cl^-, etc. Non-conservative ions *do* undergo acid base reactions. These include H_3O^+, OH^-, HCO_3^-, and CO_3^{2-}. Alkalinity is defined as follows.

Alkalinity

$$
\begin{aligned}
&= \sum (+\text{charges for conservative ions}) \\
&- \sum (-\text{charges for conservative ions}) \\
&= \sum (-\text{charges for non-conservative ions}) \\
&- \sum (+\text{charges for non-conservative ions})
\end{aligned}
$$
(13)

If, for example, we make a 0.10 M solution of KOH, the only conservative ions would be K^+ at 0.10 M and the alkalinity would be 0.10 M. The alkalinity would be -0.10 M for a 0.10 M solution of HCl. Negative alkalinity is also known as acidity.

Solutions to complex ionic equilibrium problems may be obtained by a graphical log concentration method first used by Sillén (1959) and more recently described by Butler (1964) and Morel (1983). These types of problems are described further in Chapter 16 as they relate to natural systems. Computer-based numerical methods are also used to solve these problems (Morel, 1983).

5.2.6   Conditions Far-Removed from Standard Conditions

For many cases one needs to have values of thermodynamic variables for conditions very different from 298 K and 1.0 bar. These cases include reactions occurring above the tropopause, where pressures are several orders of magnitude less than 1.0 bar and temperatures are less than 200 K. The important reactions occurring in the high-temperature and high-pressure aqueous conditions of the mid-ocean rift zone, and the high-temperature and high-pressure conditions where important mineral transformations occur far below the Earth's surface are examples.

We offer two approaches for obtaining ΔH and ΔG values for these conditions. In the most straightforward approach, the functions $(G - H)/T$ and $(H - H_{298})/T$, are tabulated for extensive temperature ranges for many simple molecules. The advantage of these functions is

that they change in a very regular manner with respect to temperature and thus can easily be interpolated from tables. Once one has a value for these functions for a given T, then calculating ΔG_T or ΔH_T is straightforward.

The second approach to getting data relies on basic thermodynamic relationships between G and H, and T and P. For instance, the heat capacity at constant pressure (C_P) of a substance is defined by:

$$
C_P = \left(\frac{\partial H}{\partial T} \right)_P
$$
(14)

From this definition, we can obtain an expression for the temperature dependence of ΔH of a reaction, if the heat capacity at constant pressure is known. For the pressure dependence, the following fundamental relationship offers a good start:

$$
\left(\frac{\partial H}{\partial P} \right)_T = V(1 - \alpha T)
$$
(15)

where α is the coefficient of thermal expansion:

$$
\alpha = \left(\frac{1}{V} \right) \left(\frac{\partial V}{\partial T} \right)_P
$$
(16)

Thus, in this approach one will need tables (or functions) for C_P, α, and possibly the compressibility as functions of T and P for the substances in question.

5.2.7   Non-ideal Behavior

Departures from ideality have been studied extensively for gases and gas mixtures. For most conditions of interest in the Earth's atmosphere, the assumption of ideal behavior is a reasonable approximation. The two most prevalent gases (N_2 and O_2) are non-polar and have critical temperatures (126 K and 154 K) far below most temperatures of environmental interest. These gases behave fairly ideally even though their pressures are high. For other gases, the partial pressures common in the atmosphere are so low that ideal behavior is a good approximation.

Departures from ideality are much more common in the condensed phase. A significant departure from ideality results from the effect of the ionic strength of aqueous solutions on the energies of ions. This effect should be considered in any quantitative consideration of ionic equilibria in sea water (Stumm and Morgan, 1981).

5.2.8 Thermodynamic Description of Isotope Effects

Thermodynamic energy terms (and equilibrium constants) may differ for compounds containing different isotopic species of an element. This effect is described in theoretical detail by Urey (1947), and applications to geochemistry are discussed by Broecker and Oversby (1971) and Faure (1977). A good example is the case of the vapor/liquid equilibrium for water. The vapor pressure of a lighter isotopic species, $^1H_2^{16}O$, is higher relative to that of heavier species, $^1H^2H^{16}O$ (or $HD^{16}O$), $^1H_2^{18}O$, and others.

Small variations in isotopic composition are usually described by comparing the ratio of isotopes in the sample material to the ratio of isotopes in a reference material. The standard measure of isotopic composition is $\delta X'$ defined in parts per thousand (per mil or ‰) by

$$\delta X' = \left[\frac{\left(\dfrac{X'}{X}\right)_{sample}}{\left(\dfrac{X'}{X}\right)_{ref}} - 1 \right] \times 1000 \qquad (17)$$

The amounts of the standard isotopic species and the tracer isotopic species are represented by X and X' for the sample and the reference material. The reference substance is chosen arbitrarily, but is a substance that is homogeneous, available in reasonably large amounts, and measurable using standard analytical techniques for measuring isotopes (generally mass spectrometry). For instance, a sample of ocean water known as Standard Mean Ocean Water (SMOW) is used as a reference for ^2H and ^{18}O. Calcium carbonate from the Peedee sedimentary formation in North Carolina, USA (PDB) is used for ^{13}C. More information about using carbon isotopes is presented in Chapter 11.

If the sample has less of the tracer isotopic species than the reference material, δ is negative. Since many of the tracer species are heavier than the reference species (e.g., ^{14}C and ^{13}C vs. ^{12}C or ^{34}S vs. ^{32}S), substances having less of a tracer species than the standard species are said to be "light."

As an example, the ratio of the equilibrium vapor pressures for ^{16}O water, P_{16} and ^{18}O water, P_{18}, depends on temperature and is expressed by the following equation, derived from Faure (1977) (temperature is in kelvins):

$$\ln\left(\frac{P_{16}}{P_{18}}\right) = \frac{7.88}{T} - 0.0177 \qquad (18)$$

At 25°C the ratio of equilibrium pressures is 1.0088 (from the above equation). This means that pure $H_2^{16}O$ has a slightly greater vapor pressure than pure $H_2^{18}O$. The ratio of ^{18}O to ^{16}O in the atmosphere is equal to the ratio in the ocean times $1/1.0088$.

Water vapor formed at low latitudes by evaporation from the ocean is "lighter" than the ocean water from which it evaporated. As this water moves to higher latitudes and the first bit of vapor condenses and rains out, the rain is heavier than the vapor from which it forms, and the removal by rain makes the vapor remaining in the atmosphere even lighter. Further precipitation at still higher latitudes makes $\delta^{18}O$ for the remaining vapor even more negative. Precipitation at high latitudes is thus lighter than low-latitude precipitation.

As may be seen from the equation given above, the degree of fractionation of water increases as temperature decreases. $\delta^{18}O$ in ice cores from Greenland and Antarctica is more negative for those periods in which evaporation leading to precipitation occurred at a lower temperature. Measurements of $\delta^{18}O$ and δ^2H can thus be used as a means of tracking changes in mean temperatures (Faure, 1977; Saigne and Legrande, 1987). This and other information derived from ice cores is presented in Chapter 18. The kinetic basis for the isotope effect is discussed later in this chapter.

5.3 Oxidation and Reduction

Reactions in which electrons are transferred

from one reactant to another (oxidation–reduction or redox reactions) are a subset of systems described by thermodynamics. We focus on the transfer of electrons when examining conditions for spontaneity and equilibrium for oxidation reduction systems because the electrons are a common currency for comparing many different reactions. The potential for electrons to be transferred in one direction or another reflects the thermodynamic spontaneity of the reaction.

5.3.1   Half-Cell Conventions and the Nernst Equation

The elemental reaction used to describe a redox reaction is the half reaction, usually written as a reduction, as in the following case for the reduction of oxygen atoms in O_2 (oxidation state 0) to H_2O (oxidation state -2). The half-cell potential, E^0, is given in volts after the reaction:

$$O_2(g) + 4H^+ + 4e^- \rightarrow 2H_2O(liq) \quad E^0 = 1.2290 \text{ V}$$

E^0 represents the relative tendency for this reaction to occur. This can also be thought of as the driving force for the reaction. The half-cell free energy change, ΔG^0 is a related quantity that gives the free energy change for the given reaction at standard conditions relative to the reduction for the *standard hydrogen electrode*, for which the half reaction is

$$2H^+ + 2e^- \rightarrow H_2(gas) \quad \Delta G^0 \equiv 0$$

The half-cell potential and the half-cell free energy change are related by the following relationship for reversible conditions:

$$\Delta G^0 = -zFE^0 \tag{19}$$

where F is the Faraday constant ($= 94\,490$ C/mol or $94\,490$ J/V/mol) and z is the number of electrons appearing in the balanced reduction half cell. A negative value for ΔG^0 (or a positive value for E^0) means that the given reduction has a greater tendency to occur than does the reduction of H^+ to H_2. ΔG^0 values are easily obtained from tables of thermodynamic data. As mentioned above, ΔG^0 is zero by convention for elements in their standard states and for H^+.

The Nernst equation describes the dependence of the half-cell potential on concentration:

$$E = E^0 - \left(\frac{2.303RT}{zF}\right) \log \frac{[\text{Products}]}{[\text{Reactants}]} \tag{20}$$

The function [Products]/[Reactants] is the same as defined in Sections 5.2.2 and 5.2.3. The factor $2.303RT/F$ has the value 0.05916 at the common reference temperature of 298.2 K. The factor 2.303 results from the change from a natural logarithm (used to express the concentration dependence of thermodynamic functions) to the common logarithm (more often used in electrochemistry and in measurements of hydrogen ion activity). The symbolism E_h is often used to make clear that we are representing a half-cell potential relative to the standard hydrogen electrode. Since E^0 is listed above to be 1.2290 V for the reduction of oxygen, then assuming that the water is pure, the Nernst equation for reduction of oxygen comes out to be

$$E_{h_1} = 1.2290 - \frac{0.05916}{4} \log \left(\frac{1}{P_{O_2}[H^+]^4}\right) \tag{21}$$

5.3.2   Electron Activity and pε

As an alternative, the tendency for a reduction to occur may also be expressed in terms of a hypothetical electron activity based on the standard hydrogen electrode. Activity was functionally defined in Equation (9). The free energy of an electron is related to chemical activity of the electron by

$$G = G^0 + RT \ln(a_e) \tag{22}$$

where a_e is electron activity. The free energy change for a process in which z electrons move from a standard hydrogen electrode to some other electrode is therefore

$$\Delta G = [G^0 + RT \ln(a_e)] - [G^0 + RT \ln(a_{SHE})]$$
$$= zRT \ln(a_e/a_{SHE}) \tag{23}$$

where a_{SHE} is the standard hydrogen electrode electron activity.

We connected our earlier definition of activity to a standard state of 1.0 bar or 1.0 M or a mole fraction of unity. None of these make much sense for electrons, but we may define electron activity in terms of the standard hydrogen electrode. We define a_{SHE} to be unity, and we define a term related to electron activity, $p\varepsilon$:

$$p\varepsilon = -\log(a_e) = -\Delta G/(2.303zRT)$$
$$= E_h(F/2.303RT) \quad (24)$$

Hostettler (1984) discusses issues involved in associating $p\varepsilon$ with electron activity.

One of the conditions of spontaneity is that $\Delta G < 0$ at constant T and P. A new statement of this condition is that in a spontaneous process electrons (or any other substance) move from a state of higher activity to a state of lower activity. Because of the definition of $p\varepsilon$ as a negative logarithm of activity, the condition of spontaneity for $p\varepsilon$ is that electrons move from a more negative to a more positive $p\varepsilon$.

We can combine the definition of $p\varepsilon$ with the Nernst equation to obtain a relationship between $p\varepsilon$ and concentrations:

$$p\varepsilon = p\varepsilon^0 - \left(\frac{1}{z}\right)\log\left(\frac{[\text{Products}]}{[\text{Reactants}]}\right) \quad (25)$$

5.3.3 pε/pH Stability Diagrams

We can write $p\varepsilon$ expressions for the reduction of O_2 to H_2O and the reduction of H^+ to H_2. For the reaction

$$O_2 + 4H^+ + 4e^- \rightarrow 2H_2O$$

$$p\varepsilon = p\varepsilon^0 - \left(\frac{1}{4}\right)\log\left(\frac{1}{P_{O_2}[H^+]^4}\right)$$

If we use the relationship between $p\varepsilon^0$ and E^0 implied by Equation (22) and assume an oxygen pressure of 1.0 bar, then

$$p\varepsilon = +20.77 - pH$$

For the reduction of H^+ at a hydrogen pressure of 1.0 bar,

$$p\varepsilon = -pH.$$

We draw two lines representing these two expressions for $p\varepsilon$ as a function of pH in Fig. 5-1. We discuss five cases for this system:

Case 1. Along the lower line H_2 and H^+ are in equilibrium when $p\varepsilon = -pH$.

Case 2. If $p\varepsilon < -pH$ (perhaps as the result of

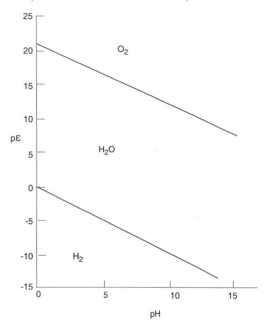

Fig. 5-1 Stability diagram for H_2O.

an applied voltage or the presence of another electrode), the electrons will have a greater activity than those in equilibrium with a standard hydrogen electrode, and H^+ will be reduced to H_2. This is the stability region for H_2.

Case 3. Along the upper line O_2 (at 1.0 bar) and H_2O are in equilibrium.

Case 4. Above the upper line, $p\varepsilon > +20.77 - pH$ and O_2 is stable.

Case 5. If $p\varepsilon$ lies between the two lines, H_2O is stable.

Consider for a moment what happens if we provide a path for electron flow between the O_2/H_2O electrode and the H^+/H_2O electrode. The activity of electrons in equilibrium with the O_2/H_2O electrode is $10^{-20.77}$ that of electrons associated with H^+/H_2O. Electrons will flow from the H^+/H_2O electrode to the O_2/H_2O electrode.

As a more complex example, we examine the stability of oxidation states of aqueous sulfur as a function of pH. This exercise will bring out the treatment of thermodynamically unstable species and the change of sulfur speciation with pH.

The sulfur species include those in the +6 oxidation state (or S(VI)) and also S(IV), S(0), and S(−II). The necessary reduction reactions are

1. $SO_4^{2-} + 2H^+ + 2e^- \rightarrow SO_3^{2-} + H_2O$,
   $\Delta G^0 = +20.82$ kJ
2. $SO_3^{2-} + 6H^+ + 4e^- \rightarrow S^0 + 3H_2O$,
   $\Delta G^0 = -224.94$ kJ
3. $S^0 + 2e^- \rightarrow S^{2-}$, $\Delta G^0 = +85.8$ kJ

Some of the thermodynamic data and equilibrium constants that are required follow.

Free energies of formation:

- H_2O(liq), -237.18 kJ/mol.
- SO_4^{2-}(aq), -744.6 kJ/mol.
- SO_3^{2-}(aq), -486.6 kJ/mol.
- S^{2-}(aq), $+85.8$ kJ/mol.

Acidity constants:

- H_2SO_4, $pK_{11} \sim -3$, $pK_{12} = 1.9$.
- H_2SO_3, $pK_{21} = 1.8$, $pK_{22} = 7.2$.
- H_2S, $pK_{31} = 7.1$, $pK_{32} = 14$.

We could write many more reactions involving these species, but, as will be seen, the ones above are sufficient. We will also assume a total sulfur concentration of 0.010 M, but we need to know how pH affects the distribution of sulfur-containing acids and bases. For any acid

$$HA \leftrightarrow H^+ + A^-$$

$$K_A = \frac{[H^+][A^-]}{[HA]}$$

which we rearrange to

$$\frac{K_A}{[H^+]} = \frac{[A^-]}{[HA]}$$

Looking at the second form of this equation, if $[H^+] \gg K_A$, then $[HA] \gg [A^-]$. Essentially all of the "A" species is present as HA under these conditions. When $[H^+] \ll K_A$, practically all of the "A" species are present as A^-.

We summarize below what this means for our sulfur species:

- S(VI) $-3 < pH < 1.9$, HSO_4^- is the dominant species;
  $pH > 1.9$, SO_4^{2-} is the dominant species.
- S(IV) $pH < 1.8$, H_2SO_3 is dominant;
  $1.8 < pH < 7.2$, HSO_3^- is dominant;
  $pH > 7.2$, SO_3^{2-} is dominant;

- S(−II) $pH < 7.1$, H_2S is dominant;
  $7.1 < pH < 14$, HS^- is dominant;
  $pH > 14$, S^{2-} is dominant.

We next write the expression for pε for each of our three reactions:

$$p\varepsilon_1 = -1.82 - \left(\frac{1}{2}\right) \log\left(\frac{[SO_3^{2-}]}{[SO_4^{2-}][H^+]^2}\right)$$

$$p\varepsilon_2 = +9.85 - \left(\frac{1}{4}\right) \log\left(\frac{1}{[SO_3^{2-}][H^+]^6}\right)$$

$$p\varepsilon_3 = -7.51 - \left(\frac{1}{2}\right) \log[S^{2-}]$$

In this section we will develop many different expressions for pε. At any given pH, only two or three of these equations will be relevant. We can begin by examining pε at any pH value, but let us begin at pH = 0. The dominant sulfur species at this pH are HSO_4^-, H_2SO_3, and H_2S. We express the sulfur concentrations in terms of these dominant ions by making the following substitutions:

$$[SO_4^{3-}] = K_{12}\frac{[HSO_4^-]}{[H^+]}$$

$$[SO_3^{2-}] = K_{21}K_{22}\frac{[H_2SO_3]}{[H^+]^2}$$

$$[S^{2-}] = K_{31}K_{32}\frac{[H_2S]}{[H^+]^2}$$

Using these substitutions with the numerical values for the constants, we obtain:

- for S(VI)/S(IV),

$$p\varepsilon_1 = +1.73 - \left(\frac{3}{2}\right)pH - \left(\frac{1}{2}\right)\log\left(\frac{[H_2SO_3]}{[HSO_4^-]}\right)$$

- for S(IV)/S(0),

$$p\varepsilon_2 = +7.60 - pH - \left(\frac{1}{4}\right)\log\left(\frac{1}{[H_2SO_3]}\right)$$

- for S(0)/S(−II),

$$p\varepsilon_3 = +3.04 - pH - \left(\frac{1}{2}\right)\log\left(\frac{1}{[H_2S]}\right)$$

These pε values actually represent electron activities for a new set of half-cell reactions derived from reactions 1, 2, and 3:

1. $HSO_4^- + 3H^+ + 2e^- \rightarrow H_2SO_3 + H_2O$

2. $H_2SO_3 + 4H^+ + 4e^- \rightarrow S^0 + 3H_2O$
3. $S^0 + 2H^+ + 2e^- \rightarrow H_2S$

The first thing to notice here is that $p\varepsilon_2 > p\varepsilon_1$ for all pH values and concentrations of interest to us. Given the role of S(IV) in these two half cells, this means that H_2SO_3 in reaction 1 can always give up electrons to H_2SO_3 in reaction 2. The overall reaction in this case is

$$3H_2SO_3 \rightarrow S^0 + 2HSO_4^- + H_2O + 2H^+$$

This is known as an *autoredox* reaction. S(IV) is thus not thermodynamically stable at all and need not be considered further in this problem. Instead of considering reactions 1 and 2, we consider the direction reduction of S(VI) to S(0),

4. $SO_4^{2-} + 8H^+ + 6e^- \rightarrow S^0 + 4H_2O$,
   $\Delta G^0 = -204.12$ kJ

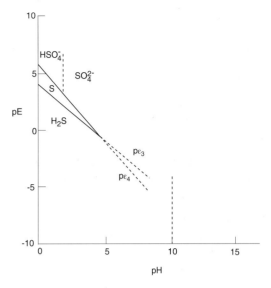

Fig. 5-2 Working diagram for sulfur.

The relevant equations are those for $p\varepsilon_3$ above and the following for $p\varepsilon_4$:

$$p\varepsilon_4 = +5.96 - \left(\frac{1}{6}\right)\log\left(\frac{1}{[SO_4^{2-}][H^+]^8}\right)$$

At pH near zero, this expression is

$$S(VI)/S(0) \quad p\varepsilon_4 = +5.64 - \left(\frac{7}{6}\right)pH$$
$$+ \left(\frac{1}{6}\right)\log[HSO_4^-]$$

We can now start to draw a figure representing $p\varepsilon$ as a function of pH, as shown in Fig. 5-2.

If $p\varepsilon$ lies below the $p\varepsilon_3$ line then only H_2S is stable; for $p\varepsilon$ values between the $p\varepsilon_3$ line and the $p\varepsilon_4$ line, S^0 is stable; for $p\varepsilon$ lying above the $p\varepsilon_4$ line, HSO_4^- is stable. At a pH of 1.9, the dominant S(VI) species changes to SO_4^{2-}. The slope of the $p\varepsilon_4$ curve changes slightly at $pH = 1.9$ reflecting a change in the number of protons in the balanced reaction:

$$S(VI)/S(0) \quad p\varepsilon_4 = +5.96 - \left(\frac{4}{3}\right)pH$$
$$+ \left(\frac{1}{6}\right)\log[SO_4^{2-}]$$

The lines for $p\varepsilon_3$ and $p\varepsilon_4$ cross at a pH of about 4.77. For pH > 4.77, $p\varepsilon_3 > p\varepsilon_4$ and S(0) is no longer thermodynamically stable. S(0) will undergo an autoredox reaction to form S(VI) and S(−II). The only sulfur half cell between

stable species is for the reduction of S(VI) to S(−II):

$$SO_4^{2-} + 8H^+ + 8e^- \rightarrow S^{2-} + 4H_2O$$
$$\Delta G^0 = -118.32 \text{ kJ}$$

$$S(VI)/S(-II) \quad p\varepsilon_4 = +5.64 - \left(\frac{7}{6}\right)pH$$
$$+ \left(\frac{1}{6}\right)\log[HSO_4^-]$$

For the pH region from 4.77 to 7.1, this becomes

$$p\varepsilon_5 = +5.23 - \left(\frac{5}{4}\right)pH - \left(\frac{1}{8}\right)\log\left(\frac{[H_2S]}{[SO_4^{2-}]}\right)$$

For 7.1 < pH < 14, this expression becomes

$$p\varepsilon_5 = +4.34 - \left(\frac{9}{8}\right)pH - \left(\frac{1}{8}\right)\log\left(\frac{[HS^-]}{[SO_4^{2-}]}\right)$$

For the case in which $[HS^-]$ and $[SO_4^{2-}]$ are equal, we have

$$p\varepsilon_5 = +4.34 - \left(\frac{9}{8}\right)pH$$

We present the $p\varepsilon$/pH diagram for the sulfur system in Fig. 5-3, which represents the sulfur system for a total sulfur concentration of 0.010 M. Note that as sulfur concentration decreases, the curve for $p\varepsilon_3$ will displace upward and the curve for $p\varepsilon_4$ will displace downward. The stability region for S(0) will be decreased and may even vanish.

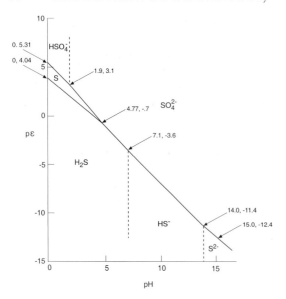

Fig. 5-3 Stability diagram for sulfur.

5.3.4   Natural Systems and the Nernst Equation

The extent to which natural systems are described by the Nernst equation depends on the relative rates at which electrons are transferred to and from various substances. These rates vary over several orders of magnitude. For example, the reduction of the hydrated ferric ion,

$$Fe^{3+} + e^- \rightarrow Fe^{2+}$$

may be a far simpler process then the reduction of nitrate,

$$2NO_3^- + 8H^+ + 8e^- \rightarrow N_2O + 5H_2O$$

The latter reaction must involve a large number of molecular steps and may be a much slower process. The mechanisms of a few inorganic electron transfer processes have been summarized by Taube (1968). The presence of very slow reactions when several redox couples are possible means that the E_h value measured with an instrument may not be related in a simple way to the concentrations of species present, and different redox couples may not be in equilibrium with one another. Lindberg and Runnells (1984) have presented data on the extent of disequilibrium

in ground waters. Bockris and Reddy (1970), Stumm and Morgan (1981), and Hostettler (1984) describe factors that determine the potential of non-equilibrium redox systems.

5.4   Chemical Kinetics

5.4.1   Reaction Rates

As described in the first part of this chapter, chemical thermodynamics can be used to predict whether a reaction will proceed spontaneously. However, thermodynamics does not provide any insight into how fast this reaction will proceed. This is an important consideration since time scales for spontaneous reactions can vary from nanoseconds to years. Chemical kinetics provides information on reaction rates that thermodynamics cannot. Used in concert, thermodynamics and kinetics can provide valuable insight into the chemical reactions involved in global biogeochemical cycles.

In chemical kinetics an empirical relationship is used to relate the overall rate of a process to the concentrations of various reactants. A common form for this expression is

$$\frac{dA}{dt} = -kA^m B^n C^p$$

This equation is known as the *rate law* for the reaction. The concentration of a reactant is described by A; dA/dt is the rate of change of A. The units of the *rate constant*, represented by k, depend on the units of the concentrations and on the values of m, n, and p. The parameters m, n, and p represent the order of the reaction with respect to A, B, and C, respectively. The exponents do not have to be integers in an empirical rate law. The order of the overall reaction is the sum of the exponents (m, n, and p) in the rate law. For non-reversible first-order reactions the scale time, tau, which was introduced in Chapter 4, is simply $1/k$. The scale time for second- and third-order reactions is a bit more difficult to assess in general terms because, among other reasons, it depends on what reactant is considered.

It is important to stress that the empirical rate law must be determined experimentally, with laboratory measurements of the rate constant and the coefficients m, n and p. A wide variety of laboratory experimental methods are used to determine these rate parameters. Further details on atmospheric and photochemical systems may be found in Calvert and Pitts (1966), Finlay-son-Pitts and Pitts (1986), Seinfeld (1986) and Wayne (1985).

The *temperature dependence* of a rate is often described by the temperature dependence of the rate constant, k. This dependence is often represented by the Arrhenius equation, $k = A \exp(-E_a/RT)$. For some reactions, the temperature relationship is instead written $k = AT^n \exp(-E_a/RT)$. The A term is the frequency factor for the reaction, which reflects the number of effective collisions producing a reaction. E_a is known as the *activation energy* for the reaction, and is a measure of the amount of energy input required to start a reaction (see also Benson, 1960; Moore and Pearson, 1981).

Rate constants for a large number of atmospheric reactions have been tabulated by Baulch *et al.* (1982, 1984) and Atkinson and Lloyd (1984). Reactions for the atmosphere as a whole and for cases involving aquatic systems, soils, and surface systems are often parameterized by the methods of Chapter 4. That is, the rate is taken to be a linear function or a power of some limiting reactant – often the compound of interest. As an example, the global uptake of CO_2 by photosynthesis is often represented in the empirical form $d[CO_2]/dt = -k[CO_2]^m$. Rates of reactions on solid surfaces tend to be much more complicated than gas phase reactions, but have been examined in selected cases for solids suspended in air, water, or in sediments.

5.4.2 Molecular Processes

Rather than always occurring in one step, reactions in the natural world often result from a series of simple processes between atoms and molecules resulting in a set of intermediate steps from reactants to products. The way multistep reactions occur can have a strong effect on the kinetics of the overall reaction. For instance, in the formation of NO from N_2 and O_2, the direct process in which a molecule of N_2 collides with a molecule of O_2 to produce two molecules of NO is extremely slow relative to other pathways for producing NO. Instead, the formation of NO at the high temperatures present during combustion and lightning, is thought to result from a multistep reaction known as the Zeldovich mechanism (Bagg, 1971), where M may be any of several molecules:

$$M + O_2 \rightarrow O + O + M$$
$$O + O + M \rightarrow O_2 + M$$
$$O + N_2 \rightarrow NO + N$$
$$N + O_2 \rightarrow NO + O$$

A *reaction mechanism* is a series of simple molecular processes, such as the Zeldovich mechanism, that lead to the formation of the product. As with the empirical rate law, the reaction mechanism must be determined experimentally. The process of assembling individual molecular steps to describe complex reactions has probably enjoyed its greatest success for gas phase reactions in the atmosphere. In the condensed phase, molecules spend a substantial fraction of the time in association with other molecules and it has proved difficult to characterize these associations. Once the mechanism is known, however, the rate law can be determined directly from the chemical equations for the individual molecular steps. Several examples are given below.

Three basic types of fundamental processes are recognized: unimolecular, bimolecular and termolecular. *Unimolecular* processes are reactions involving only one reactant molecule. Radioactive decay is an example of a unimolecular process:

$$^{14}C \rightarrow {}^{14}N + e^-$$

The rates of this process depend *only* on the concentration of reactant and the rate constant, which is generally temperature dependent.

$$\frac{d[^{14}C]}{dt} = \frac{d[^{14}N]}{dt} = -k[^{14}C]$$

The unit of the rate constant for a unimolecular process is $1/s$.

Photolytic reactions such as the decomposition of ozone by light are also unimolecular processes:

$$O_3 + hv \rightarrow O_2 + O$$

The rate of this photolytic reaction is given by

$$\frac{d[O_3]}{dt} = -\frac{d[O]}{dt} = -J[O_3]$$

Rate constants for photolytic reactions are commonly represented by the symbol J (unit 1/s). The first-order photolytic rate constant can be calculated using the formula:

$$J = \int \sigma(\lambda, T) I(\lambda) \phi(\lambda, T) d\lambda$$

The empirically determined absorption cross-section, $\sigma(\lambda, T)$ in units of cm^2/molecule, is a measure of the ability of a molecule to absorb light of a particular wavelength at a given temperature. The photon flux, $I(\lambda)$ in units of photons/cm s nm), represents the number of photons of a certain wavelength range arriving at a 1 cm^2 area per second. Therefore, $I(\lambda)$ and hence J increase with altitude, vary with time of day, and decrease to zero at night. The dimensionless quantum yield, $\phi(\lambda, T)$, describes the fraction of absorbed photons that results in the photolysis pathway of interest. For instance, O_3 can photolyze along several pathways including

$$O_3 + hv \rightarrow O_2 + O(^3P)$$
$$O_3 + hv \rightarrow O_2 + O(^1D)$$

$O(^3P)$ and $O(^1D)$ represent different electronic states of the oxygen atom. The quantum yield for $O(^1D)$ production, $\phi_{O(1D)}(\lambda, T)$, is the fraction of photons of wavelength λ absorbed by ozone at temperature T that result in the formation of the excited $O(^1D)$ atoms. Since photolysis can occur over a range of wavelengths, J is calculated over the integral from the shortest, λ_1, to the longest, λ_2, wavelength at which the photolytic reaction occurs.

Bimolecular processes are reactions in which two reactant molecules collide to form two or more product molecules. In most cases the reaction involves a rather simple rearrangement of bonds in the two molecules:

$$NO + O_3 \rightarrow NO_2 + O_2$$

Often, a single atom is transferred from one molecule to another and one bond is formed as another is broken. The rate of a bimolecular process depends on the product of concentrations of the two reactants. In this case

$$\frac{d[NO]}{dt} = -\frac{d[NO_2]}{dt} = -k[NO][O_3]$$

The units for the rate constant, k, for a bimolecular reaction are cm^3/molecule s.

Termolecular processes are common when two reactant molecules combine to form a single small molecule.

$$O_2 + O + M \rightarrow O_3 + M$$

Such reactions are often exothermic and the role of the third body is to carry away some of the energy released and thus stabilize the product molecule. In the absence of a collision with a third body, the highly vibrationally excited product molecule would usually decompose to its reactant molecules in the timescale of one vibrational period. Almost any molecule can act as a third body, although the rate constant may depend on the nature of the third body. In the Earth's atmosphere the most important third-body molecules are N_2 and O_2.

The rate of the reaction depends on the product of reactant concentrations, including the third body:

$$\frac{d[O_3]}{dt} = k[O_3][O][M]$$

In the atmosphere, $[M]$ is usually assumed to be atmospheric pressure at the altitude of interest. Therefore, unlike unimolecular and bimolecular processes, termolecular processes are pressure dependent. The units for the termolecular rate constant are cm^6/molecule s.

5.4.3   Catalytic Reactions

The chemistry of the stratospheric ozone will be sketched with a very broad brush in order to illustrate some of the characteristics of catalytic reactions. A model for the formation of ozone in the atmosphere was proposed by Chapman and may be represented by the following "oxygen only" mechanism (other aspects of stratospheric ozone are discussed in Chapters 7 and 12):

$$O_2 + hv \rightarrow O + O \qquad\qquad J_1$$
$$O + O_2 + M \rightarrow O_3 + M \qquad k_2$$
$$O_3 + hv \rightarrow O_2 + O \qquad\qquad J_3$$
$$O + O_3 \rightarrow 2O_2 \qquad\qquad k_4$$

Reactions 2 and 3 regulate the balance of O and O_3, but do not materially affect the O_3 concentration. Any ozone destroyed in the photolysis step (3) is quickly reformed in reaction 2. The amount of ozone present results from a balance between reaction 1, which generates the O atoms that rapidly form ozone, and reaction 4, which eliminates an oxygen atom and an ozone molecule. Under conditions of constant sunlight, which implies constant J_1 and J_3, the concentrations of O and O_3 remain constant with time and are said to correspond to the *steady state*. Under steady-state conditions the concentrations of O and O_3 are defined by the equations $d[O]/dt = 0$ and $d[O_3]/dt = 0$. Deriving the rate expressions for reactions 1–4 and applying the steady-state condition results in the equations given below that can be solved for $[O]$ and $[O_3]$.

$$0 = 2J_1[O_2] - k_2[O][O_2][M] + J_3[O_3] - k_4[O][O_3]$$
$$0 = k_2[O][O_2][M] - J_3[O_3] - k_4[O][O_3]$$

This simple oxygen-only mechanism consistently overestimates the O_3 concentration in the stratosphere as compared to measured values. This implies that there must be a mechanism for ozone destruction that the Chapman model does not account for. A series of *catalytic* ozone-destroying reactions causes the discrepancy. Shown below is an ozone-destroying mechanism with NO/NO_2 serving as a catalyst:

$$NO + O_3 \rightarrow NO_2 + O_2 \qquad k_5$$
$$NO_2 + O \rightarrow NO + O_2 \qquad k_6$$
$$O_3 + O \rightarrow 2O_2 \qquad\qquad \text{net reaction}$$

The net effect of reactions 5 and 6 produces the same end result as reaction 4 in the oxygen-only mechanism; O and O_3 are destroyed. The NO/NO_2 pair of compounds is referred to as a catalyst because it enhances the rate of the reaction ($O + O_3 \rightarrow 2O_2$) without being changed

in the process (Crutzen, 1971; Johnston, 1971). These and other catalysts have finite lifetimes. In this case it is thought that NO_2 is removed from the stratosphere as a nitric acid, $HONO_2$, a solute in precipitating polar stratospheric clouds. Catalysts are an "invisible hand" and can greatly increase the rate of a reaction. There are several ozone destroying catalysts including NO/NO_2, Cl/ClO, H/OH and OH/HO_2 which each undergo an analogous set of reactions. For instance, the catalytic cycle for the pair of chlorine radical species Cl and ClO is given below (Stolarski and Cicerone, 1974; Molina and Rowland, 1974):

$$Cl + O_3 \rightarrow ClO + O_2$$
$$ClO + O \rightarrow Cl + O_2$$

It is interesting to compare the rate constants of the oxygen-only ozone destruction reaction with those of the catalytic ozone destruction cycle. The rate constants for reactions 4–6 at 30 km are given below in units of cm^3 molecules^{-1} s^{-1}.

$$O_3 + O \rightarrow 2O_2 \qquad\qquad k_4 = 1.0 \times 10^{-15}$$
$$NO + O_3 \rightarrow NO_2 + O_2 \qquad k_5 = 4.7 \times 10^{-15}$$
$$NO_2 + O \rightarrow NO + O_2 \qquad k_6 = 9.3 \times 10^{-12}$$

The power of the catalyst may be seen in the relative rate constants. The rate constant for the loss of ozone due to reaction with NO is five times the rate constant for loss of ozone in the reaction with O atoms. The constant for loss of O atoms in the reaction with NO_2 is nearly four orders of magnitude faster than that due to reaction with ozone. These numbers are brought out to indicate the very large range in the values for rate constants. This range results mainly from differences in activation energies and the steric requirements for the reactions. (Steric requirements define the orientations of atoms necessary to form the new bonds.) Although the concentrations of the catalytic species in this example are lower than the ozone concentration by three or four orders of magnitude, their effect is magnified when each NO and NO_2 molecule goes through the catalytic cycle hundreds or thousands of times before being removed.

5.4.4 *Enzyme-Catalyzed Reactions*

Enzymes act as catalysts in biological systems to effectively regulate the rate of reactions and determine which products are formed. Enzyme-catalyzed reactions are often described by the Michaelis–Mentin mechanism that is represented by the following steps:

$$S + E \rightarrow ES^* \qquad k_1$$
$$ES^* \rightarrow E + S \qquad k_2$$
$$ES^* \rightarrow E + P \qquad k_3$$

In this case S represents the substrate, or the substance undergoing change and P represents the product molecule(s). E is the enzyme, which forms a complex, ES^*. This complex is capable of undergoing further reaction, with a lowered activation energy, to form a specific product molecule. The overall reaction is $S \rightarrow P$ with the enzyme being regenerated. The rate of product formation (or the negative of the rate of substrate loss) is given by

$$\frac{dP}{dt} = -\frac{dS}{dt} = \frac{k_3[E_0][S]}{\left(\dfrac{k_2 + k_3}{k_1} + [S]\right)}$$

where $[E_0]$ is the total enzyme concentration. This rather complex rate expression can be simplified in two limiting cases of substrate concentration:

$$[S] \ll \frac{k_2 + k_3}{k_1} \qquad \text{(Case I)}$$

$$[S] \gg \frac{k_2 + k_3}{k_1} \qquad \text{(Case II)}$$

In case I, $[S]$ can be neglected in the denominator of the rate expression, and the rate is now given by

$$\frac{dP}{dt} = \left(\frac{k_1 k_3}{k_2 + k_3}\right)[E]_0[S] = k'[E]_0[S]$$

Reaction 1 is the slowest step in this series of reactions leading to product formation. It is the *rate-limiting step*. Since this reaction involves bringing E and S together, it is a second-order reaction overall and first order with respect to the total enzyme concentration and the substrate concentration.

In case II, $(k_2 + k_3)/k_1$ can be disregarded in the denominator of the rate expression, and the rate law now becomes

$$\frac{dP}{dt} = k_3[E]_0$$

The rate is independent of the substrate concentration and first order with respect to enzyme concentration. In this case reaction (3), in which the complex decomposes to form the product, is the slowest step and is therefore rate limiting. Although this discussion has assumed that we have only an isolated enzyme reacting with a substrate, the same principles are applied to the more complex case when an entire organism, or a series of organisms consumes a substrate.

The case of bacterial reduction of sulfate to sulfide described by Berner (1984) provides a useful example. The dependence of sulfate reduction on sulfate concentration is shown in Fig. 5-4. Here we see that for $[SO_4^{2-}] < 5$ mM the rate is a linear function of sulfate concentration but for $[SO_4^{2-}] > 10$ mM the rate is reasonably independent of sulfate concentration. The sulfate concentration in the ocean is about 28 mM and thus in shallow marine sediments the reduction rate does not depend on sulfate concentration. (The rate *does* depend on the concentration of organisms and the concentration of other necessary reactants – organic carbon in this case.) In freshwaters the sulfate concentration is

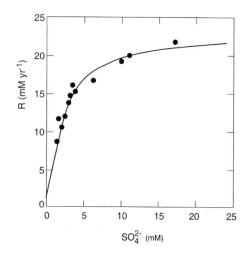

Fig. 5-4 The rate of bacterial sulfate reduction as a function of sulfate concentration. (Adapted from Berner (1984) with the permission of Pergamon Press.)

much less than 5 mM and the sulfate reduction rate does depend on the sulfate concentration (and may be independent of the concentration of organic carbon).

5.4.5   *Kinetic Isotope Effects*

Molecules containing different isotopic species of an element often react at different rates. The effect is noticeable if the rate-limiting step in a reaction is one in which a bond to the element in question is formed or broken. The primary kinetic isotope effect results from the lower vibrational frequencies of heavier isotopic species. The quantum mechanical zero point vibrational energies are also lower for heavier isotopes and thus slightly more energy is required to break a bond to ^{13}C, for instance, than to ^{12}C. Isotope ratios may be measured with high accuracy on small amounts of geochemical samples and thus one may infer something about the processes that form a compound from the observed isotope distribution.

For a kinetic isotope effect to exist, the reverse reaction must not occur to a significant extent (in which case we would have a thermodynamic isotope effect) and the molecules undergoing reaction must be drawn from a larger pool of molecules that do not react. If all the molecules are going to undergo reaction there will be no discrimination between isotopes and no kinetic isotopic effect. For these reasons, the kinetic isotope effect for ^{13}C in living matter occurs in the early stages of the photosynthetic process when a small fraction of the CO_2 (or HCO_3^-) available is added to organic substrates to form carboxylic acids.

Photosynthesis begins with the transfer of CO_2 into the cell from the atmosphere or water. Photosynthetic enzymes then transfer the inorganic carbon to a five-carbon organic compound to form two three-carbon carboxylic acid molecules. (This is the case for the C3 photosynthetic mechanism.) In these steps, reaction of $^{12}CO_2$ occurs slightly faster than reaction of $^{13}CO_2$ and the organism has a more negative $\delta^{13}C$ than the atmosphere or ocean from which it grows. Thus, while marine inorganic carbon has $\delta^{13}C$ of about 0‰ compared to the reference and atmospheric carbon has $\delta^{13}C$ of about -7‰, the $\delta^{13}C$ values for plants range from -10 to -30‰ (Schidlowski, 1988). The isotope distribution depends somewhat on the species of plant and it is a strong function of whether the plant fixes carbon by the C3 mechanism (most plants) or the C4 mechanism (a smaller group including corn, sugar cane, and some tropical plants).

Isotope effects also play an important role in the distribution of sulfur isotopes. The common state of sulfur in the oceans is sulfate and the most prevalent sulfur isotopes are ^{32}S (95.0%) and ^{34}S (4.2%). Sulfur is involved in a wide range of biologically driven and abiotic processes that include at least three oxidation states, S(VI), S(0), and S($-$II). Although sulfur isotope distributions are complex, it is possible to learn something of the processes that form sulfur compounds and the environment in which the compounds are formed by examining the isotopic ratios in sulfur compounds.

5.5   Non-Equilibrium Natural Systems

Thus far we have studied thermodynamics and kinetics under the assumption that the systems of interest are in equilibrium. However, some natural systems have reaction rates so slow that they exist for long periods under non-equilibrium conditions. The formation of nitric oxide serves as an interesting example.

The net reaction for NO formation is $N_2 + O_2 \rightarrow 2NO$, although the actual mechanism by which NO forms does not include the direct reaction of N_2 and O_2 to any significant extent. This direct reaction would involve the breaking of two strong bonds and the formation of two new bonds, an unlikely event. Rather, the oxidation of nitrogen begins with a simple reaction:

1:    $$M + O_2 \leftrightarrow O + O + M$$

This reaction occurs to only a small extent, but the oxygen atoms thus formed may form NO through the following catalytic cycle.

2:    $$O + N_2 \leftrightarrow NO + N$$

3:    $$N + O_2 \leftrightarrow NO + O$$

Net: $$N_2 + O_2 \leftrightarrow 2NO$$

The reverse reactions of 1, 2, and 3 are also important in establishing the equilibrium between N_2, O_2, and NO_2.

At low temperatures the rates of these reactions are very slow either because the rate constants are very small or because the concentrations of O and N are very small. For these reasons, equilibrium is not maintained at the low temperatures typical of the atmosphere. However, as the temperature rises, the rate constants for the critical steps increase rapidly because they each have large activation energies – E_a = 494 kJ/mol for reaction 1 and 316 kJ/mol for reaction 2. The larger rate constants contribute to a faster rate of NO production, and equilibrium is maintained at higher temperatures. The time scale for equilibrium for the overall reaction $N_2 + O_2 \rightarrow 2NO$ is less than a second for $T > 2000$ K.

It may seem unrealistic to consider such high temperatures, but in fact, many processes raise air to very high temperatures. Hydrocarbon or biomass combustion can produce temperatures of 1500–3000 K and lightning discharges can produce temperatures of the order of 30 000 K (Yung and McElroy, 1979). Other processes capable of producing high temperatures include shock waves from comet or meteorite impacts (Prinn and Fegley, 1987) and nuclear bomb explosions (Goldsmith *et al.*, 1973).

As the temperature of an N_2/O_2 mixture is increased above 2000 K the observed concentration of NO (as well as those for NO_2, N, O, and other species) will approach the equilibrium values appropriate for that temperature. As the temperature of the mixture of these gases decreases, the concentrations will follow the equilibrium values. Equilibrium will be maintained as long as the time scale for the chemical reaction is shorter than the time scale for the temperature change (that is, the chemical reaction is more rapid than the temperature change). The time scale for the chemical reaction increases rapidly as the temperature decreases because of the large activation energies. The concentrations of NO at ambient conditions reflect the lowest temperature at which the system was in equilibrium as it cooled.

The example described above for nitric oxide illustrates the interplay between thermodynamics and kinetics. At high temperatures, where the reaction rates are relatively high, the $N_2 + O_2 \rightarrow NO$ system is in equilibrium. At lower temperatures, however, the rate constants are so low that the system cannot achieve equilibrium and many of the thermodynamic principles described in this chapter would not apply.

The presence of a high concentration of oxygen in the contemporary atmosphere and the prevalence of substances that can react with oxygen in the atmosphere and on the surface of the Earth is another example of a non-equilibrium system.

Photosynthesis produces oxygen by the following redox reaction:

$$CO_2 + H_2O \rightarrow CH_2O(\text{fixed carbon}) + O_2$$

but most of the oxygen thus produced is removed by respiration and decomposition,

$$CH_2O + O_2 \rightarrow CO_2 + H_2O$$

A small fraction (less than 1%) of the fixed carbon produced by photosynthesis is buried and physically removed from any potential reaction with oxygen (until the buried material is brought to the surface – at a much later time). Thus, the oxygen in our contemporary atmosphere is the consequence of many millions of years of fixed carbon burial. More details on this topic can be found in Chapters 8 and 11.

The high concentration of oxygen in the atmosphere plays a central role in the photochemistry and chemical reactivity of the atmosphere. Atmospheric oxygen also defines the oxidation reduction potential of surface waters saturated with oxygen. The presence of oxygen defines the speciation of many other aquatic species in surface waters.

Still, a question arises as to why the high concentration of oxygen in the atmosphere does not react with the large amounts of reduced substances present. After all, the reaction between oxygen and fixed carbon is very exergonic; ΔG for the decomposition reaction above is about -480 kJ per mole of fixed carbon.

Oxygen concentrations can be high because the rates of many of the reactions of oxygen at ambient temperatures are very slow. The reaction of oxygen and fixed carbon by living systems involves control of the rate by a complex set of enzyme-mediated reactions. In fact, it's difficult to imagine an extensive range of life on a planet without the presence of reasonable concentrations of reactants that are thermodynamically unstable, but inhibited from undergoing abiotic reaction by high activation energies. Substances that undergo rapid reactions in the absence of enzymes could never reach high concentrations to provide energy for non-photosynthetic organisms. Lovelock (1979) described this connection between extensive life on a planet and the presence of something like oxygen while contemplating ways to recognize the presence of extraterrestrial life on Mars and other planets.

5.6 Summary

Thermodynamics establishes the boundaries of what is possible in the natural world. In many cases enough information is available to enable us to determine the thermodynamic spontaneity of processes. What reaction will occur if a reaction does take place? We often know much less about *how fast* the reaction is. Will the spontaneous reaction occur within our lifetime? Will the system reach equilibrium? The question of *rate* involves many issues that are much less well understood. We probably know more about gas phase rates than we do about rates in other media. Even so, it has taken many years to sort out the puzzle of stratospheric ozone depletion. Rates in the aqueous phase and other condensed media are much less well understood. Biological processes involving metabolic pathways in countless organisms, catalytic effects involving trace species, and processes occurring at phase boundaries are among the factors that complicate an understanding of rates in the natural world.

Although many natural systems are far from equilibrium, many localized regions of natural systems are well described in thermodynamic and equilibrium terms. As a general rule, if the reactions that redistribute compounds between the reactant and product states are fast, then equilibrium conditions may be applied. In some cases, part of a system can be described in equilibrium terms and part

cannot. As an example, ammonia is generally not in equilibrium with oxidized nitrogen in the atmosphere and surface waters and ammonia is not distributed between the oceans and atmosphere according to equilibrium expressions (Quinn *et al.*, 1988). Nonetheless, the distribution between NH_4^+ and $NH_3(aq)$ is described by the usual equilibrium expression because the proton exchange reaction is very fast. Generally speaking, one must be cautious in applying equilibrium relationships unless it is known that the underlying reactions are fast.

Questions

5-1 Consider the reaction $N_2(g) + 2O_2 \leftrightarrow 2NO_2(g)$, for which $\Delta G^0_{298} = 103.68 \text{ kJ/mol}$ and $\Delta H^0_{298} = 67.70 \text{ kJ/mol}$. (a) Evaluate K_P for this reaction at 298 K. (b) For the atmosphere at sea level, $P_{O_2} = 0.21$ bar and $P_{N_2} = 0.79$ bar. Evaluate the pressure of NO_2 in equilibrium with these pressures of O_2 and N_2 at 298 K. (c) Evaluate K_P at 500 K assuming that ΔH^0 is independent of temperature.

5-2 To illustrate an equilibrium involving different phases, consider the following important example. This could represent the equilibrium between a raindrop and atmospheric carbon dioxide:

$$H_2O(l) + CO_2(g) \rightarrow H_2CO_3(aq)$$

(a) Show that $K_{eq} = (c_{H_2CO_3}/c^0)/X_{H_2O}(P_{CO_2}/P^0)$. The equilibrium constant is related to the Henry's Law constant, $K_H = c_{H_2CO_3}/P_{CO_2} = 0.0334 \text{ M/bar}$ at 25°C. It is often a very good approximation to equate the mole fraction of water to one. (b) If ΔG^0 for $H_2O(l)$ is -237.18 kJ/mol, and ΔG^0 for $CO_2(g)$ is -394.37 kJ/mol, estimate ΔG^0 for $H_2CO_3(aq)$. (c) If ΔH^0 for the reaction is -20.4 kJ/mol, estimate the Henry's Law constant at 30°C. (The answer to (c) points to the connection between a negative enthalpy change and a decrease of the equilibrium constant with increasing temperature. See the discussion on the actual states of "CO_2" in water in Section 5.2.4.)

5-3 Evaluate the alkalinities of the following solutions: (a) 0.10 M $KHCO_3$. (b) 0.10 M K_2CO_3. (c) 0.10 M $Ca(OH)_2$. (d) 0.10 M NH_4Cl.

5-4 The ratio of ^{18}O to ^{16}O atoms is about 2045:1 000 000 in ocean water. (a) Suppose that

10^6 atoms of ^{16}O water evaporate in equilibrium with ocean water at 298 K. How many ^{18}O atoms will be associated with these ^{16}O atoms in the atmosphere? (b) Suppose now that the mixture of isotopes drifts to higher latitudes and is cooled to 293 K, whereupon half of the ^{16}O atoms condense to form raindrops. What number of ^{18}O atoms will condense and what number will remain in the atmosphere? (c) Evaluate $\delta^{18}O$ for the precipitation that forms in (b) and for the vapor that remains in the atmosphere. Describe how $\delta^{18}O$ will change as the moist air moves to yet higher latitudes and further condensation occurs.

5-5 Evaluate E^0 for the reduction of SO_4^{2-} to HSO_3^- $(3H^+ + SO_4^{2-} + 2e^- \rightarrow HSO_3^- + H_2O)$. ΔG^0 values (all in kJ/mol): SO_4^{2-} (-744.6), HSO_3^- (-527.8), H_2O (-237.18).

5-6 Write a balanced reaction for the spontaneous process when we allow a path for electron flow between the O_2/H_2O electrode and the H^+/H_2O electrode as described in Section 5.3.3.

5-7 In thinking about sulfur in the environment, we may need to include the pε lines for the $O_2/H_2O/H_2$ system. The line describing pε for O_2 at 1.0 bar in equilibrium with H_2O will lie above all of the sulfur diagram for the pH range 0 to 14. (a) If pε for electrons in equilibrium with O_2/H_2O is more positive than pε for electrons in equilibrium with SO_4^{2-}/S^0, which electrons will have greater activity? Will a spontaneous reaction occur involving O_2/H_2O and SO_4^{2-}/S^0? (b) Would you expect a sulfate solution to be stable in water saturated with O_2? (c) Would you expect a solution that is 0.10 M in HS^- and SO_4^{2-} in contact with 1.0 bar O_2 to be stable? What spontaneous reaction (if any) might occur?

5-8 Consider the NO/NO_2-catalyzed ozone destruction cycle, reactions 5 and 6 in Section 5.4.3. One could perform a calculation to determine which reaction is the *rate-limiting step* (i.e., the slowest step that determines the rate of the overall reaction) in this cycle. In this case, a theoretical doubling of k_5 reduces the ozone concentration by about 2%. On the other hand, doubling k_6 reduces the ozone concentration by nearly 50%. (a) Which reaction is the rate-limiting step in NO/NO_2-catalyzed ozone destruction? (b) The concentrations of NO and NO_2 are: [NO] = $2.9 \times 10^8/cm^3$ and $[NO_2] = 6.1 \times 10^9/cm^3$. How do these data support or refute your answer to (a)?

Answers can be found on p. 509.

References

Alberty, R. A. and Silbey, R. J. (1997). "Physical Chemistry." Wiley, New York.

Atkins, P. W. (1994). "Physical Chemistry," 5th edn. W. H. Freeman, New York.

Atkinson, R. and Lloyd, A. C. (1984). Evaluation of kinetic and mechanistic data for modeling of photochemical smog. *J. Phys. Chem. Ref. Data* **13**, 315–444.

Bagg, J. (1971). The Formation and Control of Oxides of Nitrogen in Air Pollution. *In* "Air Pollution Control," Part One (W. Strauss, ed.). Wiley, New York.

Baulch, D. L., Cox, R. A., Crutzen, P. J., Hampson, R. F., Kerr, J. A., Troe, J., and Watson, R. T. (1982). Evaluated kinetic and photochemical data for atmospheric chemistry: Supplement I. *J. Phys. Chem. Ref. Data 1982* **11**, 327–496.

Baulch, D. L., Cox, R. A., Hampson, R. F., Kerr, J. A., Troe, J., and Watson, R. T. (1984). Evaluated kinetic and photochemical data for atmospheric chemistry: Supplement II. *J. Phys. Chem. Ref. Data 1982* **13**, 1259–1380.

Benson, S. W. (1960). "The Foundations of Chemical Thermodynamics." McGraw-Hill, New York.

Berner, R. A. (1984). Sedimentary pyrite formation: An update. *Geochim. Cosmochim. Acta* **48**, 605–615.

Bockris, J. O. and Reddy, A. K. N. (1970). "Modern Electrochemistry." Plenum, New York.

Broecker, W. S. and Oversby, V. M. (1971). "Chemical Equilibrium in the Earth." McGraw-Hill, New York.

Butler, J. N. (1964). "Ionic Equilibrium." Addison-Wesley, Reading, MA.

Calvert, J. G. and Pitts, J. N. (1966). "Photochemistry." Wiley, New York.

Campbell, I. M. (1986). "Energy and the Atmosphere," 2nd edn. Wiley, New York.

Crutzen, P. J. (1971). Ozone production rates in an oxygen-hydrogen-nitrogen oxide atmosphere. *J. Geophys. Res.* **76**, 7311–7327.

Faure, G. (1977). "Principles of Isotope Geology." Wiley, New York.

Finlayson-Pitts, B. J. and Pitts, J. N. (1986). "Atmospheric Chemistry." Wiley, New York.

Goldsmith, P., Tuck, A. F., Foot, J. S., Simmons, E. L., and Newson, R. L. (1973). Nitrogen oxides, nuclear weapon testing, Concorde, and stratospheric ozone. *Nature* **244**, 545–551.

Hostettler, J. D. (1984). Electrode reaction, aqueous electrons, and redox potentials in natural waters. *Am. J. Sci.* **284**, 734–759.

Johnston, H. S. (1971). Reduction of stratospheric

ozone by nitrogen oxide catalysts from supersonic transports. *Science* **173**, 517–522.

Levine, I. N. (1995). "Physical Chemistry," 4th edn. McGraw-Hill, New York.

Lide, D. A., ed. (1998). "The Handbook of Chemistry and Physics." Chemical Rubber Co. Press, Cleveland, Ohio.

Lindberg, R. D. and Runnells, D. D. (1984). Groundwater redox reactions: An analysis of equilibrium state applied to Eh measurements and geochemical modeling. *Science* **225**, 925–927.

Lovelock, J. E. (1979) "Gaia: A New Look at Life on Earth." Oxford University Press, Oxford.

Molina, M. J. and Rowland, F. S. (1974). Stratospheric sink for chlorofluoromethanes: Chlorine atom-catalyzed destruction of ozone. *Nature* **249**, 810–812.

Moore, J. W. and Pearson, R. G. (1981). "Kinetics and Mechanism." Wiley, New York.

Morel, F. M. M. (1983). "Principles of Aquatic Chemistry." Wiley, New York.

Prinn, R. G. and Fegley, B. (1987). Bolide impacts, acid rain, and biospheric traumas at the Cretacious-Tertiary boundary. *Earth Planet. Sci. Lett.* **83**, 1–15.

Quinn, P. K., Charlson. R. J., and Bates, T. S. (1988). Simultaneous measurements of ammonia in the atmosphere and ocean. *Nature* **335**, 336–338.

Saigne, C. and Legrande, M. (1987). Measurements of methanesulfonic acid in Antarctic ice. *Nature* **330**, 240–242.

Schidlowski, M. (1988). A 3 800 million year isotopic record of life from carbon in sedimentary rocks. *Nature* **333**, 313–318.

Seinfeld, J. H. (1986). "Atmospheric Chemistry and Physics of Air Pollution." Wiley, New York.

Sillén, L. G. (1959). Graphical presentation of equilibrium data. *In* "Treatise in Analytical Chemistry," Part I, Vol. 1 (I. M. Kolthoff, P. J. Elving, and E. B. Sandell, eds), pp. 277–317. Interscience, New York.

Stolarski, R. S. and Cicerone, R. J. (1974). Stratospheric chlorine: A possible sink for ozone. *Can. J. Chem.* **52**, 1610–1615.

Stumm, W. and Morgan, J. J. (1981). "Aquatic Chemistry," 2nd edn. Wiley, New York.

Taube, H. (1968). Mechanisms of oxidation-reduction reactions. *J. Chem. Educ.* **45**, 453–461.

Urey, H. C. (1947). The thermodynamic properties of isotopic substances. *J. Chem. Soc.* **1947**, 562–581.

Wayne, R. P. (1985). "Chemistry of Atmospheres." Clarendon Press, Oxford.

Yung, Y. L. and McElroy, M. B. (1979). Fixation of nitrogen in the prebiotic atmosphere. *Science* **203**, 1002–1004.

Part Two

Properties of and Transfers between the Key Reservoirs

Part Two of this book focuses on the major "spheres" that deliver and receive the chemical constituents moving through the Earth system. This part is loosely organized in terms of the speed at which material is processed by a particular sphere. However, since the hydrologic cycle has traditionally served as a paradigm for considering mass balance in the elemental biogeochemical cycles, we begin by providing basic information on how the hydrosphere works in Chapter 6.

Earth is a unique planet in that it contains such a large amount of liquid water. The hydrosphere, which comprises this water, is both a reservoir (sphere) for material and a conduit of material between the various spheres (cycle). Its importance to the other cycles and spheres make it a logical starting point. Without this liquid phase and without its ability to evaporate into the gas phase, the Earth would be unlikely to have developed and maintained a biosphere. The polar nature of the water molecule, with its high boiling point determines the character of all aspects of the hydrosphere. We discuss its properties as a solvent, which allows it to participate in the fluxes of other material. Since water is necessary for life, we explore the variation of moisture on the planet, and the fluxes of water in and out of the reservoirs that contain it. The Earth's climate is also closely tied to the presence of water in different reservoirs and phases, especially in the control of the reflectivity of sunlight off of the planet. The hydrosphere and hydrologic cycle affect the cycling of all of the other biogeochemical cycles on Earth. Since the hydrologic cycle has itself undergone human

modification, it has indirectly been a vehicle for which the other cycles have been anthropogenically affected.

Closely connected to the hydrosphere and qualitatively similar to it (because it is highly mobile) is the atmosphere, covered in Chapter 7. The atmosphere is the least massive of the geospheres, the fastest moving and the one that is most sensitive to perturbations. Far from being a simple body of homogeneously mixed gases, the atmosphere contains a large amount of water in three different phases (vapor, liquid in the form of cloud droplets, and solid in the form of ice crystals in high clouds). In addition, the amounts of water in air are enormously variable in space and time. Other condensed-phase substances also exist (aerosol particles), ranging from supermicrometer dust particles down to molecular clusters of a few tens or hundreds of Ångström units in dimension. As might be expected from its small mass, the atmosphere presents rapid and extensive variability of composition.

Just as in the case for the hydrosphere, the atmosphere participates in all of the major biogeochemical cycles (except for phosphorus). In turn, the chemical composition of the atmosphere dictates its physical and optical properties, the latter being of great importance for the heat balance of Earth and its climate. Both major constituents (O_2, H_2O) and minor ones (CO_2, sulfur, nitrogen, and other carbon compounds) are involved in mediating the amounts and characteristics of both incoming solar and outgoing infrared radiation.

Chapter 8 considers the *pedosphere* (literally,

Earth System Science
ISBN 0-12-379370-X

that upon which we walk) which can be described as the interface where the *lithosphere* (rocks), the atmosphere, the hydrosphere, and the biosphere intersect. Like these three spheres, soils have a solid phase (made up of weathered rocks from the lithosphere and decayed matter from the biosphere), a liquid phase of pore water from the hydrosphere, and a gas phase of air trapped in spaces between the solid and liquid parts. The pedosphere deserves our attention because of its many roles in communicating with these other spheres, and the processes that provide living organisms soil nutrients for growing edible vegetation. Sometimes denoted as a separate reservoir, sediments formed from the chemical weathering and physical degradation of rocks exist both on land and under the hydrosphere. The solid phases existing in the pedosphere, soils and sediments have relatively low solubility and change much more slowly in composition and physical makeup than either the hydrosphere or atmosphere. However, the chemical reactions and transport of material that takes place in these parts of the Earth system assert long-term control on the chemical makeup of other systems.

The lithosphere – rocks – are the fundamental starting material for the biogeochemical development of the pedosphere. It is the slowest of the major spheres to change via tectonics and all forms of erosion. Chapter 9 surveys the factors – largely geological – that govern the biogeochemical functioning of this most massive of the geospheres. Although it is massive and slow moving, it plays extremely important roles in the Earth system. For example, in the bigger picture, the atmosphere and its condensed water yield acidic species (e.g., H_2CO_3, H_2SO_4, and HNO_3), the biosphere amplifies the acidity of carbon dioxide through its enhancement of dissolved CO_2 in groundwater, while the lithosphere provides the basic materials that react with the acidity via chemical weathering reactions. The upshot of this global acid–base interaction is the production and control of alkalinity in the global hydrosphere, mainly in the oceans.

To finish Part Two, Chapter 10 revisits the largest part of the hydrosphere – the oceans – again illustrating the central role of liquid water in the functioning of the Earth system. While water and the hydrosphere were the logical first topic of Part Two, they are also the logical last topic because the oceans are the sink for virtually all of the rapid portions of biogeochemical cycles. Some of the water evaporated from the oceans falls from the atmosphere as rain or snow, where it interacts with intervening and mediating entities – the lithosphere, biosphere, soils, and sediments. The consequence is a flow of dissolved and suspended matter to the oceans. These substances circulate within the marine biosphere, are carried around the globe in the great thermohaline circulation and in wind-driven currents, ultimately to be sedimented or rejoined into the tectonic cycle. The oceans therefore are both the beginning and the end of the hydrologic cycle, completing the loop for many of the elemental cycles as well.

6

Water and the Hydrosphere

Patricia C. Henshaw, Robert J. Charlson, and Stephen J. Burges

6.1 Introduction

It is hard to imagine any part of the Earth system that is more essential than or that has as many different functions as the water of the hydrosphere. In particular, the presence of a mobile liquid phase, with its long list of special chemical and physical properties, must be clearly identified as the main feature of Earth that separates it from the other terrestrial planets or from any known astronomical object. Close to home, the "terrestrial planets," Earth, Mars, and Venus are presumed to have accreted similar abundances of "excess volatiles" – H_2O, CO_2, etc. – but evolved very differently. Even in the earliest stages of planetary evolution, liquid water provided a medium in which chemical reactions occurred between atmospheric CO_2 and the minerals in primitive igneous rocks to allow the precipitation of carbonate minerals and to prevent a runaway greenhouse effect. While no exact chronology or quantification of this early chemical event can be given, it seems clear that some such process prevented the accumulation of all of the Earth's CO_2 and H_2O (as a vapor) in the atmosphere at the same time. This would have caused the Earth to be more or less like Venus – a condition from which there would appear to be no return to our present state. Before embarking on a description of this most important reservoir, it is useful – perhaps necessary – to reflect on the special properties of water itself. We can then proceed to a discussion of how the hydrosphere works.

6.1.1 Water as a Substance

The water molecule, H_2O, structurally

is a bent molecule with a very strong permanent dipole moment. This dipole is the result of the negatively charged O atom and the two positively charged H atoms (the whole molecule being neutral). The existence of this charge separation arises due to the near orthogonality of the orbitals of the bonding electrons of the central O atom, while its large magnitude arises comes from the lack of shielding of the bonding electron and the small size of the O atom.

The permanent dipole moment is so strong that it permits the function of what are called *hydrogen bonds* between the highly electronegative O atom of one molecule and a nearby hydrogen atom of another molecule (see Fig. 6-1). The hydrogen bond is not a chemical bond in the ordinary sense of the forces that hold molecules together, which can be deduced from its strength of ca. 20 kJ/mol. Ordinary molecular bonds have typical strengths (energy required to break them) of a few hundred kJ/mol.

It is the hydrogen bonds of water that give it unique physical and chemical properties, characteristics that set it apart from all of the other molecules formed from elements near the top of the periodic table. Table 6-1 compares several key properties of water to selected

Earth System Science
ISBN 0-12-379370-X

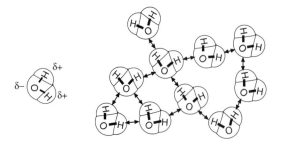

Fig. 6-1   Hydrogen bonds in liquid water.

simple compounds that might be expected to have similar properties to water but do not. In regard to melting and boiling point, water behaves like a much larger molecule, but it has low density like the low atomic number compound that it really is.

Likewise, liquid water has anomalously high molar *heat capacity* (75 J/mol K), meaning that liquid water can absorb relatively large amounts of heat from the sun by day and release it at night without much change of temperature. Owing to the large amount of liquid water at the surface of Earth, this large heat capacity is important in mediating temperatures and therefore climate.

Still further, water has large *latent heats of evaporation* and *freezing* (J/mol), all because of

the same hydrogen bonds. As one result, the solar heating of the planet (largely in the tropics and subtropics), which results primarily in evaporation, transfers latent heat to the atmosphere in the form of water vapor. Subsequent precipitation at colder temperatures (higher latitudes or altitudes) releases the latent heat, making water vapor an important heat-transport vehicle.

This latent heat of evaporation, L_e, also appears in the fundamental description of the dependence of the vapor pressure of water, p, on temperature, T – the Clausius–Clapeyron equation:

$$\frac{d(\ln p)}{dT} = \frac{L_e}{RT^2} \qquad (1)$$

or in integral form, between locations with p_1, T_1 and p_2, T_2:

$$\ln\frac{p_2}{p_1} = -\frac{L_e}{R}\left(\frac{1}{T_2} - \frac{1}{T_1}\right) \qquad (2)$$

where R is the universal gas constant in appropriate units.

The large value of L_e results in a very strong dependence of vapor pressure on temperature. As a result, the water vapor content of the air is extremely variable, from parts per million by volume in the coldest parts of the atmosphere to several percent in the warmest and wettest

Table 6-1   Anomalous properties of water

Property	Water	Comparison species
Boiling point	373 K	CH_4: 112 K (comparable size molecule) NH_3: 240 K H_2S: 211 K (dihydride) H_2Se: 231 K (dihydride) Ar: 87 K
Melting point	273 K	CH_4: 89 K NH_3: 195 K H_2S: 190 K H_2Se: 209 K Ar: 84 K
Heat capacity of liquid	4218 J/(K kg)	CH_4: 2170 J/K kg
Latent heat of vaporization (0°C)	2.5×10^6 J/kg	
Latent heat of freezing	3.3×10^5 J/kg	
Ratio of density frozen/density liquid (0°C)	0.92	
Surface tension	73 dyn/cm	CCl_4: 27 (nonpolar liquids) C_6H_6: 29

parts. The large latent heat of freezing of liquid water imposes a requirement for large transport of heat from bodies of water before the temperature can drop very much below 0°C, yet another type of thermostat.

A further unusual property of water is that it has a maximum density at around 4°C and expands upon freezing, again because of hydrogen bonds. There are more of these bonds in ice than in liquid water, creating a relatively open crystal structure in the solid phase. When ice melts (requiring the addition of a large amount of latent heat of freezing or fusion, L_f) some of the hydrogen bonds are broken, and a tighter packing of H_2O molecule results in the denser liquid.

The high latent heat of evaporation or vaporization – due to the hydrogen bonds causing attraction of water molecules to each other – also causes molecules at the surface of water to have cohesive forces. This results in water having anomalously high *surface tension*. In turn, this property plays a very strong role in the process of nucleation of cloud droplets, as one of the key factors involved in determining cloud droplet sizes and growth and coalescence rates. The latter is a significant factor in delivery of water to the continents by rain.

Very short wavelengths ($\lambda < 186$ nm) of UV radiation are required to dissociate the very strong O–H in water (bond strength = 456 kJ/mol). The large concentrations of O_2 and O_3 in air absorb incoming solar UV radiation high in the atmosphere, preventing much of this photodissociation. The strong bonds and the small size and mass of H_2O also give it a very complex infrared absorption spectrum that extends to shorter wavelengths (i.e. higher frequencies, ν, and higher energies, $h\nu$) than many other simple molecules (CO_2 for example). One absorption feature, the 6.3 μm band, is extremely strong, as can be seen in Fig. 6-2. The result of this strong IR absorption, the large amount of water vapor in the atmosphere, and the proximity of the 6.3 μm absorption band of water vapor to the peak of the Earth's black-body emission is that water vapor is by far the dominant greenhouse gas.

We can see the importance of water vapor as a greenhouse gas by comparing the greenhouse effect on Earth, a relatively humid planet, with

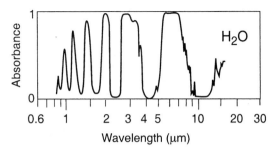

Fig. 6-2 Comparison of infrared absorbance of a vertical column of atmospheric CO_2 and H_2O vapor. The nearly total absorbance by H_2O between 5 and 7 μm, nearly coinciding with the peak of the wavelength-dependent emission of the surface, make H_2O a much more effective greenhouse gas. Liquid water (not shown) in clouds adds still more absorbance.

that on Mars, which is arid. Mars has an atmosphere that is ca. 95% CO_2 (by mass, about 50 times more than on Earth). The greenhouse effect of Martian CO_2 causes a temperature increase of only a few degrees. In contrast, the total natural greenhouse effect on Earth is ca. 33 K, the majority of which is due to water vapor and water clouds. Nonetheless, as will be seen in Chapter 17, the anthropogenic greenhouse effect due to enhanced CO_2, CH_4, N_2O etc. cannot be dismissed, as changes in the Earth's average temperature of even 1 K are significant.

Besides these special physical properties, hydrogen-bonded liquid water also has unique solvent and solution properties. One feature is high proton (H^+) mobility due to the ability of individual hydrogen nuclei to jump from one water molecule to the next. Recalling that at temperatures of about 300 K, the molar concentration in pure water of H_3O^+ ions is ca. 10^{-7} M, the "extra" proton can come from either of two water molecules. This freedom of H^+ to transfer from one to an adjacent "parent" molecule allows relatively high electrical conductivity. A proton added at one point in an aqueous solution causes a domino effect, because the initiating proton has only a short distance to travel to cause one to pop out somewhere else.

The existence of strongly polar water molecules and mobile protons also makes H_2O an excellent and almost universal solvent for ionic compounds and polar organic species. Compounds that will not significantly dissolve in water (i.e. saturated solutions with concentrations less than ca. 10^{-5} M) include aliphatic and aromatic hydrocarbons, as well as plastics and many other polymers.

6.1.2   The Right Abundance of Water to Support Life

As can be seen in Fig. 2–1 (abundance of elements), hydrogen and oxygen (along with carbon, magnesium, silicon, sulfur, and iron) are particularly abundant in the solar system, probably because the common isotopic forms of the latter six elements have nuclear masses that are multiples of the helium (He) nucleus. Oxygen is present in the Earth's crust in an abundance that exceeds the amount required to form oxides of silicon, sulfur, and iron in the crust; the excess oxygen occurs mostly as the volatiles CO_2 and H_2O. The CO_2 now resides primarily in carbonate rocks whereas the H_2O is almost all in the oceans.

While it is clear that the hydrosphere is a significant portion of the planet's mass, there is not, at least currently, so much water that the continents are submerged. Conversely, the oceans are large enough that their surface area would never become an important limiting factor in the hydrologic cycle. Although there have been many shifts in the balance of the hydrosphere, this condition has prevailed since the biosphere began to evolve. The presence of liquid water allowed the Earth to become and remain a living planet, and by the astronomical coincidence of the Earth's location, the planet received just the proper abundance to sustain and recycle this all-important resource, almost in perpetuity.

6.2   Global Water Balance

While the hydrosphere has long been appreciated as essential to life on Earth, only in the past couple of decades have scientists expanded their exploration of the global hydrologic cycle and its roles across the spectrum of Earth science disciplines. The Earth and its atmosphere, in the broadest view, are a complex, intimately coupled system of chemical, physical, and biological cycles, and water, with its myriad unique chemical and physical properties, plays a part in almost all of them.

To understand the role water plays in global cycles, it is necessary to first understand the mechanics of the water cycle. The hydrologic cycle is driven by solar radiation, which provides the energy necessary to overcome latent heat capacities involved in phase changes. Gravity plays a key role in returning condensed water to the surface as precipitation, and via runoff from the continents to the oceans. At the simplest level, the global water cycle results from imbalances between precipitation and evapotranspiration (ET) at the ocean and land surfaces. Globally, the oceans lose more water by evaporation than they gain by precipitation, whereas the land surface receives more precipitation than is lost through ET; runoff from the land surfaces then balances the ocean–atmosphere water deficit. The hydrologic cycle is significantly more complex than this simple description would suggest. In addition to the atmosphere, oceans, and rivers, significant amounts of water are stored in groundwater, glaciers and ice sheets, soil moisture, and, to a smaller extent, biomass. Figure 6-3 shows a schematic of the global hydrologic cycle, with storages in km^3 and fluxes in km^3/yr.

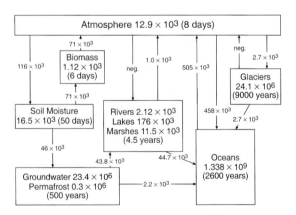

Fig. 6-3 Global water balance. Storages in km^3, fluxes in km^3/yr. Turnover times calculated as storage divided by total annual inflow. (Data from Shiklomanov and Sokolov, 1983.)

Table 6-2   Reservoir storage

Reservoir	UNESCO (1978) (millions of km^3)	Nace (1969) (millions of km^3)	Baumgartner and Reichel (1975) (millions of km^3)
Oceans	1338	1350	1348
Cryosphere	24.06	26.0	27.82
Groundwater	23.40	7.0	8.062
Fresh groundwater	10.53	—	7.956
Soil moisture	0.0165	0.150	0.0612
Permafrost	0.3000	—	—
Lakes	0.1764	0.230	0.224
Freshwater lakes	0.0910	0.125	0.126
Rivers	0.00212	0.0017	0.0011
Marshes	0.01147	—	—
Atmosphere	0.0129	0.013	0.013
Biosphere	0.00112	negligible	0.0011
Total	1386	1384	1384
Freshwater total	35.03	36.0	36.00

Calculation of the global water balance is a nontrivial problem. Gross storage volumes are calculated predominantly by multiplying surface areas by estimated average depths (UNESCO, 1978). While modern remote sensing technology has made it possible to determine areal extents of surface reservoirs, such as oceans, lakes, and ice sheets quite accurately, estimations of depths or thicknesses are still highly uncertain and often subjective. For example, the wide variations in groundwater estimates can often be attributed to differing interpretations of the extent of groundwater into the Earth's crust. Similarly, soil moisture depths vary from less than a meter to tens of meters or more, so global soil moisture averages are by necessity highly subjective. Even more well-defined reservoirs, such as oceans or lakes, cannot be accurately quantified without complete knowledge of their bathymetry and vertical temperature profiles, an impossible requirement on a global scale.

Although fluxes of precipitation and river discharge can be quite accurately determined on a local scale, large portions of the globe, especially the oceans and Antarctica, are essentially ungauged, requiring extensive extrapolation of existing data. Evaporation fluxes are even less well known, since calculation requires extensive knowledge of hydrologic and meteorologic parameters. Consequently, numerous estimates of storage and flux volumes exist for the hydrologic cycle; the representation in Fig. 6-3 was selected for its completeness. Some of the variability in estimates of the global water balance is reflected in the reservoir storage values shown in Table 6-2.

6.2.1   Reservoirs

Figure 6-3 shows the hydrologic cycle as seven primary reservoirs interconnected by a number of water fluxes. The role of each reservoir in the hydrologic cycle and its connections with other cycles is briefly summarized below, in order of storage volume.

6.2.1.1   Oceans

The oceans are by far the largest reservoir in the hydrologic cycle, containing more than 25 times as much water as the rest of the reservoirs combined. As another means of comparison, the volume of water in the oceans is four orders of magnitude larger than that in the next most visible reservoir, the world's lakes and rivers. The oceans are also one of the Earth's primary heat reservoirs and have absorbed approximately half of the CO_2 emitted to the atmosphere; consequently the oceans play an important role in climate. Coupled with atmospheric circulation, surface and thermohaline circulations in the oceans transfer heat from low to high latitudes and provide a modulating effect on global temperatures.

6.2.1.2   Glaciers and ice sheets

The cryosphere – the portion of the Earth's water frozen in ice caps, glaciers, and sea ice – contains the largest reserves of freshwater on Earth. Due to the remoteness of most of the planet's ice-covered areas, estimates of water stored in the cryosphere have a high degree of uncertainty. Current best estimates, based on improved remote sensing technology for determining ice extent, place ice storage at 3×10^7 km^3, of which the Antarctic ice sheet makes up about 90% (IPCC, 1996b). Since inflows and outflows are small, at least in the current climate mode, the cryosphere is a fairly static component of the hydrosphere. However, ice-covered portions of the planet do contribute significantly to albedo and can affect atmospheric circulation, making them important links in climatic feedbacks, as discussed in Chapter 17. The climatic record in ice sheets and glaciers is discussed in detail in Chapter 18.

6.2.1.3   Groundwater

The hydrologic cycle's lithospheric link, groundwater, consists of the water stored in underground aquifers, i.e. all subsurface storage below the subterranean water table. Groundwater reserves extend far down into the Earth's crust, although the active zone, which contains most of the fresh groundwater, is restricted to the upper reaches. This stratification results in a wide range of residence times for subsurface water, with some deep regions remaining essentially static for up to millions of years. For example, the "mining" of the Ogallala Aquifer underlying the Great Plains region of the United States is of particular concern because the aquifer contains substantial amounts of glacial melt-

water from the last ice age, which is essentially a nonrenewable supply.

Groundwater is fed through infiltration and percolation through the soil and is recycled via transpiration through plants, interflow into river networks, and some direct discharge to the ocean. This reservoir is extremely important in global water resources, though reserves in some areas are threatened by overdraft and pollution.

6.2.1.4   Lakes and rivers

Despite their small volume, lakes and rivers play a disproportionately large role in the cycling of water. River networks transport the majority of surface (rain and snowmelt) and subsurface runoff to the oceans to balance the hydrologic cycle. As such, rivers also provide a means of transport for eroded sediment and dissolved ions, nutrients, and organic matter, giving them a prominent role in tectonic and biogeochemical cycles. This is discussed more in Chapter 8. Nutrient cycling in aquatic systems, particularly lakes, is also an important link in particularly the phosphorus and nitrogen cycles. From a human standpoint, surface waters are the most critical reservoir of freshwater, as they provide water resources for most of the Earth's population (L'vovich, 1974).

6.2.1.5   Soil moisture and biomass

These two reservoirs, representing the land surface, link the atmosphere with other land-based hydrologic reservoirs and processes. Although storage in soil moisture is small, it plays a critical role in the cycling of water, acting as "a kind of intermediary between climate and meteorological factors on the one hand and the phenomena of the hydrologic regime (groundwater, rivers, and lakes) on the other" (L'vovich, 1974). Antecedent soil moisture conditions dictate how much precipitation can be infiltrated into the soil and how much is shed as surface runoff. The soil moisture reservoir feeds groundwater through percolation and plants through transpiration.

Biomass is not a particularly important storage reservoir, but it does play a large role in cycling, mainly via transpiration through land plants and, to a lesser extent, via photosynthesis. Soil–plant interactions are also the key determinants of land surface evaporation; in vegetated continental areas, most evaporation of soil moisture occurs by virtue of transpiration. In addition to evaporation, the land surface contributes to albedo, though surface effects have proven difficult to parameterize in global climate models. Albedo is a strong function of the availability of liquid water. Arid areas (deserts) have very high albedo compared to vegetated land. Consult Chapter 17 for more information on climate considerations.

6.2.1.6   Atmosphere

Although it is one of the smallest reservoirs in terms of water storage, the atmosphere is probably the second most important reservoir in the hydrosphere (after the oceans). The atmosphere has direct connections with all other reservoirs and the largest overall volume of fluxes. Water is present in the atmosphere in solid, liquid, and vapor forms, all of which are important components of the Earth's natural greenhouse effect. Cycling of water within the atmosphere, both physically (e.g. cloud formation) and chemically, is also integral to other biogeochemical cycles and climate. Consult Chapter 17 for more details.

6.2.2   Turnover Times

Average turnover time (defined as storage volume divided by annual inflow or outflow volume, assuming steady state) is a measure of how long it takes to replace the entire storage in a reservoir. In a steady-state system, which is a reasonable approximation for Earth cycles on geologic time scales, turnover times also provide a sense of how long it will take a reservoir to respond to perturbations in the cycle. Reservoirs with short turnover times are most sensitive, while those with longer turnover times respond more slowly and can often act as buffers on shorter time scales.

The enormous volume of the oceans results in an average turnover time of more than 2600 years, compared to less than 10 days for atmospheric water. Although the reservoir is much smaller than the oceans, the cryosphere has the longest turnover time due to the small input flux. Average turnover times for all seven reservoirs, calculated from the data in Fig. 6-3, are shown in Table 6-3.

Many hydrologic reservoirs can be further subdivided into smaller reservoirs, each with a characteristic turnover time. For example, water resides in the Pacific Ocean longer than in the Atlantic, and the oceans' surface waters cycle much more quickly than the deep ocean. Similarly, groundwater near the surface is much more active than deep reservoirs, which may cycle over thousands or millions of years, and water frozen in the soil as permafrost. Typical range in turnover times for hydrospheric reservoirs on a hillslope scale (10–10^3 m) are shown in Table 6-4 (estimates from Falkenmark and Chapman, 1989). Depths are estimated as typical volume averaged over the watershed area.

Global freshwater reserves (discounting pollution) are a small percentage of global water, accounting for only 35×10^6 km^3 of the total 1.386×10^9 km^3 global water supply (UNESCO,

Table 6-3   Reservoir turnover times

Reservoir	Volume (km^3)	Avg. turnover time
Oceans	1.338×10^9	2640 yrs
Cryosphere	24.1×10^6	8900 yrs
Groundwater/permafrost	23.7×10^6	515 yrs
Lakes/rivers	189 990	4.3 yrs
Soil moisture	16 500	52 days
Atmosphere	12 900	8.2 days
Biomass	1120	5.6 days

Table 6-4 Hillslope scale turnover times

Reservoir	Depth (mm)	Turnover times
Atmosphere	25[a]	8–10 days
Plants	5–50	Hours–days
Streams/rivers	3	Weeks
Lakes/reservoirs	—	Months–years
Soil moisture	$10–10^4$	Years
Groundwater	$10^4–10^5$	Days–10^6 years[b]

[a] Global average.
[b] Longer turnovers associated with large watershed areas.

1978). Assuming an input flux equal to oceanic evaporation, this would give a turnover time of about 750 years. The turnover time analysis is not strictly correct since freshwater resides in a number of interconnected reservoirs; however, since freshwater volume is essentially equal to non-marine storage and net evaporation is the only output flux from the oceans, this may be taken as a reasonable estimation. For practical purposes, freshwater resources available to humans cycle more rapidly, but since 97% of freshwater is stored in ice, the global average turnover is much longer.

6.2.3   Fluxes

Robert Horton, an influential pioneer in the field of hydrology, developed one of the first comprehensive representations of the hydrologic cycle in 1931. His original diagram, Fig. 6-4, illustrates the processes by which water moves between the Earth's hydrologic reservoirs. Hydrologic fluxes can be summed up in four

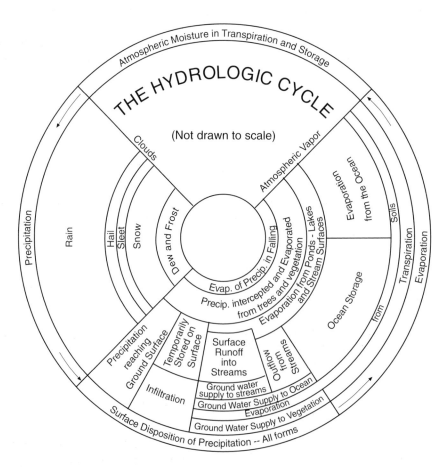

Fig. 6-4 The fluxes of the hydrologic cycle, developed by Robert Horton (1931).

processes, shown around the outside of Horton's wheel – precipitation, surface dissipation of precipitation, evaporation, and atmospheric moisture transport. This section will discuss precipitation, evaporation (and closely related transpiration), and runoff – the processes that link the oceans, atmosphere, and land surface. Atmospheric moisture is addressed in Chapter 7. These fluxes are highly variable over the Earth's surface in both space and time, which has extremely important implications for water resources; spatial and temporal variability is discussed in Section 6.3.

6.2.3.1 Precipitation

The flow of water from the atmosphere to the ocean and land surfaces as rain, snow, and ice constitutes the atmospheric efflux in the hydrologic cycle. Although most precipitation falls on the oceans (ca. 79% of the global total), precipitation onto land is much more hydrologically significant. On a global scale, nearly two-thirds of the land portion returns to the atmosphere via evapotranspiration (see below), while the remaining one-third contributes to groundwater and surface runoff. Precipitation is highly variable over the globe, with atmospheric circulation patterns concentrating it in the tropics and mid-latitudes.

An important component of precipitation on a regional scale comes from precipitation recycling; that is, a portion of the precipitation in a region comes from water vapor evaporated from within that region, with the remainder composed of atmospheric moisture advected into the region. The precipitation recycling ratio, the ratio of recycled precipitation to total precipitation, is then a function of evaporation and internal and external atmospheric moisture fluxes (Budyko, 1974; Eltahir and Bras, 1996). The settling of the Great Plains in the late 19th and early 20th centuries was in fact spurred by early (and largely unfounded) concepts of precipitation recycling. S. Aughey (cited in Holzman, 1937) wrote of Nebraska in 1880 that increased evaporation from cultivated land would increase moisture and rainfall – i.e., that "rain follows the plow."

6.2.3.2 Evapotranspiration

Water returns to the atmosphere via evaporation from the oceans and evapotranspiration from the land surface. Like precipitation, evaporation is largest over the oceans (88% of total) and is distributed non-uniformly around the globe. Evaporation requires a large input of energy to overcome the latent heat of vaporization, so global patterns are similar to radiation balance and temperature distributions, though anomalous local maxima and minima occur due to the effects of wind and water availability.

Evapotranspiration (ET) is the collective term for land surface evaporation and plant transpiration, which are difficult to isolate in practice. *Transpiration* refers to the process in which water is transported through plants and returned to the atmosphere through pores in the leaves called stomata, and is distinct from direct evaporation of intercepted precipitation from leaf surfaces. Some land surface processes and the roles of vegetation in the water and energy balances are illustrated in Fig. 6-5. Due to

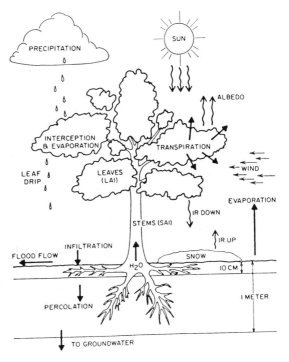

Fig. 6-5 Evaporation and transpiration from vegetation are among the complex land surface interactions in the hydrologic cycle. (From Dickinson, 1984.)

the number of variables involved, ET can be extremely difficult to measure and is often determined by closing the water or energy balance calculated from better-known components.

6.2.3.3   Runoff

The excess of evaporation from the oceans is made up for by runoff from the land. Although this flux is much smaller than precipitation and ET, it is a major link in many cycles and is of particular importance to humans in terms of water supply. Runoff can be broadly categorized into subsurface, or groundwater, flow and surface flow, consisting of overland runoff and river discharge.

6.2.3.3.1   Subsurface runoff. When precipitation hits the land surface, the vast majority does not go directly into the network of streams and rivers; in fact, it may be cycled several times before ever reaching a river and the ocean. Instead, most precipitation that is not intercepted by the vegetation canopy and re-evaporated infiltrates into the soil, where it may reside as soil moisture, percolate down to groundwater, or be transpired by plants.

Very little groundwater is discharged directly to the oceans, but groundwater does provide a significant contribution to stream discharge in most areas. Subsurface flow is generally much slower than surface runoff, allowing groundwater to provide perennial baseflow to streams far into a dry season, long after surface storm runoff has been discharged. Figure 6-6 shows a typical storm hydrograph, with baseflow and stormflow components indicated. Groundwater flow velocities have been found to follow Darcy's law:

$$v = \frac{K}{n} \cdot \frac{dh}{dL} \qquad (3)$$

where v is velocity, K is soil hydraulic conductivity with units of (length/time), n is the dynamic (or actively available) porosity, and dh/dL is the hydraulic gradient.

Hydraulic conductivities vary over a range of 10^{-12} cm/s for unfractured igneous and metamorphic rocks to 2 or 3 cm/s for porous (karst)

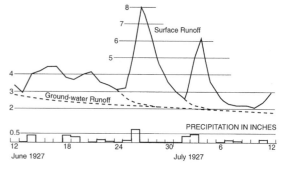

Fig. 6-6   Hydrograph showing the rapid contribution of surface runoff and more steady baseflow. Runoff in cubic feet per second, precipitaion in inches. (From Langbein and Wells, 1955.)

limestone and gravel; surface flow is typically on the order of a meter per second.

6.2.3.3.2   Surface runoff. Hydrologists have identified two processes for generating surface runoff over land. The first, *saturated overland flow* (SOF), is generated when precipitation (or snowmelt) occurs over a saturated soil; since water has nowhere to infiltrate, it then runs off over land. SOF typically occurs only in humid environments or where the water table rises to intersect with a stream. *Horton overland flow* (HOF or infiltration-limited overland flow) occurs when precipitation intensity exceeds the infiltration capacity of the soil in a non-saturated environment. In this case, only the excess precipitation (that exceeding the infiltration capacity) runs off over the surface. Both types of overland runoff generate relatively rapid flows that constitute the surface water contribution to the hydrograph (Fig. 6-6).

Figure 6-7 illustrates the runoff paths for HOF and SOF, as well as for subsurface stormflow and groundwater flow. Subsurface stormflow is a moderately rapid runoff process in which water flows to a stream through highly permeable surface soil layers (without reaching the water table). Note in Fig. 6-7 that while HOF and subsurface stormflow may occur over a large fraction of an infiltration-limited hillslope, SOF occurs over a smaller portion adjacent to the stream.

Hillslopes occupy about 99% of the landscape

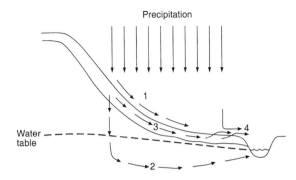

Precipitation

Water table

1

3

4

2

Fig. 6-7 Vertical cross-section showing pathways for surface and subsurface runoff. Path 1: HOF; path 2: groundwater flow; path 3: subsurface stormflow; path 4: SOF. (From Dunne and Leopold, 1978.)

and provide stream channels with water supply, making hillslope processes extremely important on a local scale. However, the much more visible component of surface runoff comes from river discharge. Globally, rivers discharge roughly 45 000 km^3 per year to the oceans (Shiklomanov and Sokolov, 1983). The 16 largest rivers account for more than one-third of total discharge, and over half of that contribution comes from the three largest. Table 6-5 lists the 10 largest rivers in the world in terms of average discharge rate (m^3/s) and annual discharge volume (km^3/yr) (Dingman, 1994).

While river discharge is the primary means of transferring water from the land to the oceans, its magnitude pales in comparison to circulation within the oceans themselves. The total average

Table 6-5 World's largest rivers

River	Discharge (m^3/s)	Discharge (km^3/yr)
Amazon	190 000	6000
Congo	42 000	1330
Yangtze	35 000	1100
Orinoco	29 000	915
Brahmaputra	20 000	630
La Plata	19 500	615
Yenesei	17 800	565
Mississippi	17 700	560
Lena	16 300	515
Mekong	15 900	500

discharge of the world's rivers is about 10^6 m^3/s or 1 Sverdrup (Sv); by comparison, the oceans' thermohaline circulation transports about 15 Sv (Broecker, 1997).

In addition to runoff, rivers transport products of upland weathering to the oceans, forming a key link in the tectonic cycle of uplift and erosion. This interaction will be explored further in Section 6.6.

6.3 Hydrologic Variability

Part of the difficulty in studying the hydrologic cycle arises from the huge variability in fluxes over time and space. Most of us have experienced the differences in rain and snowfall associated with changing seasons and observed the different precipitation patterns experienced by other areas of the country and the world. This hydrologic variability is present across virtually all spatial and temporal scales, from the smallest hillslope over a period of minutes to the entire globe over the geologic history of the Earth. Understanding hydrologic variability is particularly important for management of water resources. In that context, continental scale variations in precipitation and runoff within a year and over several years to decades are of the most interest.

Variability has traditionally been accounted for in hydrologic models in one of two ways. Stochastic models attempt to preserve statistical relationships between significant variables determined from historical records, while physically based models are designed to represent natural processes based on values of known variables and empirical parameters. While stochastic models tend to be simpler and less data intensive, they require long historical records, which are unavailable in many areas, and cannot represent conditions outside the range of historic values. Physically based models require extensive input data, but because they explicitly represent physical processes, they are more appropriate for studying effects of and responses to global change and can be used for limited extrapolation beyond the range of data used for their calibration and testing.

6.3.1 Precipitation

On a global scale, and as discussed in Chapter 7, precipitation patterns clearly reflect the convergence and divergence zones in the general atmospheric circulation. Rainfall peaks over the tropics, decreases in the subtropical latitudes, and exhibits more modest peaks at mid-latitudes before going to essentially zero at the poles. Figure 7-7 in the next chapter provides a more complete description of the overall latitude dependence. At the regional or continental scale, the precipitation patterns are complicated by many other factors, including mesoscale atmospheric circulations and orographic effects.

Figure 6-8 shows the average monthly precipitation for selected locations in the United States, illustrating both the spatial variability in total precipitation and the marked differences in distribution of precipitation through the year. The US exhibits four basic annual patterns for precipitation distribution, which can be used to classify climate (Thornthwaite, 1948). The West Coast experiences dry summers and winter precipitation maximums from storms coming in off the Pacific, while the interior of the continent tends to have late spring or early summer peaks associated with peaks in the soil moisture cycles. The northeast receives moderate precipitation throughout the year, while the southeast receives large amounts of rain from late summer thunderstorms fueled by the surrounding warm oceans. The rain shadow effects of the Sierra Nevada, Cascades, and Rocky Mountains, as well as the aridity of the desert southwest, can also be observed in Fig. 6-8.

6.3.2 Runoff

Because it depends on a number of conditions that are themselves inherently variable, runoff tends to vary even more than precipitation, particularly over time. Seasonal runoff patterns depend largely on latitude and altitude of the watershed, due to the importance of snowmelt in runoff peaks. In high-latitude basins or those with significant high-altitude contributing areas, peaks occur later than in low-latitude or low-altitude areas. Figure 6-9 shows the relative amount of runoff in each month for selected US rivers. While average flow in the southeastern rivers is nearly constant (approximately 8% per month), northern rivers and those fed by mountain snowpack show distinct peaks, with peak timing dependent on when snowmelt occurs.

Water resources decision making in many areas, particularly arid and semi-arid climates such as the American West, depends on interannual to decadal variations in surface water availability. In addition to more predictable seasonal differences, runoff tends to exhibit long-term trends alternating between flood and drought periods. Figure 6-10 shows historical wet and dry periods based on streamflow records for 50 world rivers. For the most part, these periods are consistent on a regional basis, though they appear to alternate on a hemispheric scale.

The importance of these long-term trends for water resources is illustrated by the case of the Colorado River Compact, which allocates the water of the Colorado River to the seven states in its drainage area. The agreement, signed in 1922, based allocations on annual discharge from the preceding decade, which remains the wettest period on record; subsequent drier years have given rise to numerous water rights disputes among the seven states and between the United States and Mexico.

6.3.3 Hydrologic Sensitivity

The ability to predict runoff and water availability is critical to water resources planners. However, the complex non-linearities of the hydrologic cycle make this an extremely difficult process. Even where precipitation is fairly well known, runoff prediction is a non-trivial problem, as land surface response depends as much (or more) on precipitation patterns and timing as on precipitation amount. The historical record of monthly rainfall and inflow at the Serpentine Dam, near Perth, Western Australia, provides an illustration of this sensitivity (Fig. 6-11a and b).

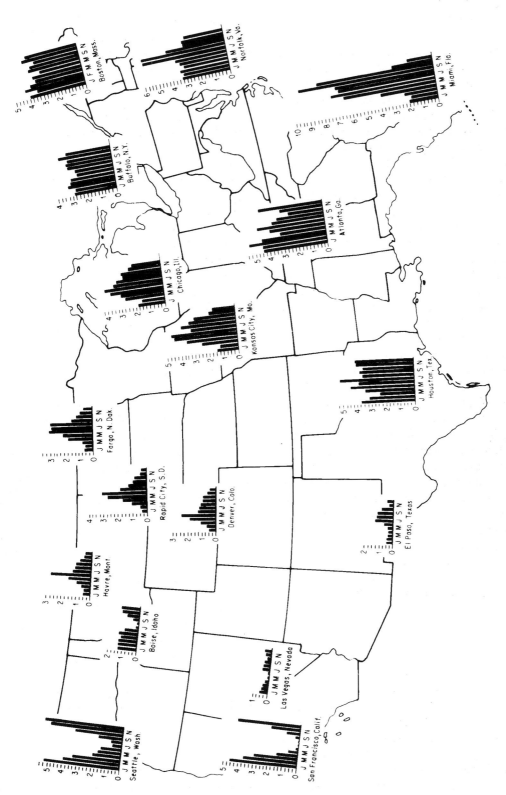

Fig. 6-8 Annual precipitation patterns for the United States. Precipitation in inches (1 inch = 25.4 mm). (From Linsley *et al.*, 1982.)

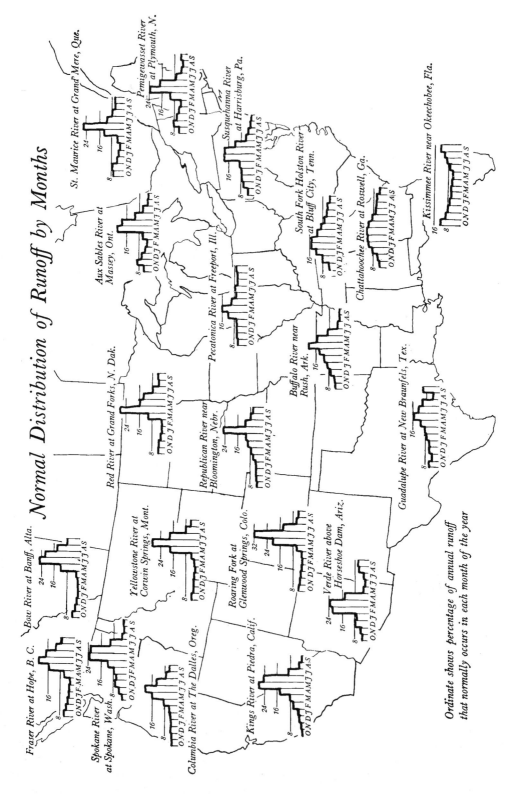

Fig. 6-9   Annual runoff patterns for the United States. Percent of normalized annual runoff in each month. (From Langbein and Wells, 1955.)

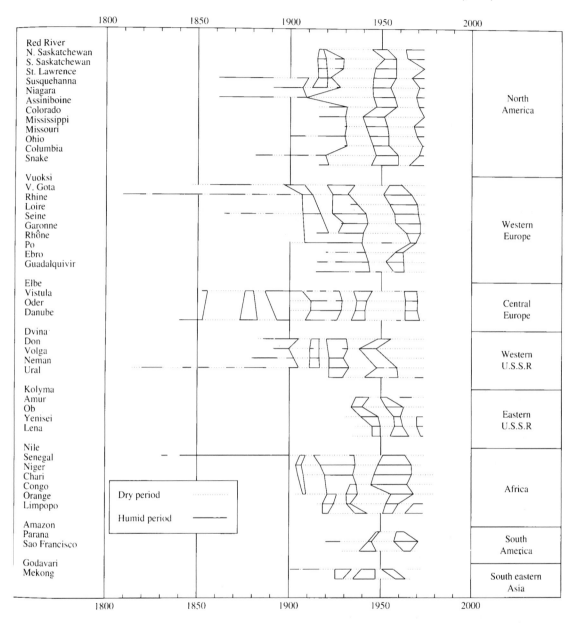

Fig. 6-10 Long-term global streamflow trends. Wet and dry periods from the historical record of 50 major rivers. (From Probst and Tardy, 1987.)

The precipitation record shows a mild decrease in rainfall for May, June, and July over the last 20 years of record. However, runoff decreased to less than half the historical average for May and June over the same period, and reduced runoff persisted into December, despite a return to normal or above normal precipitation levels over the latter half of the year.

Runoff sensitivity, particularly in arid and semi-arid climates, is largely a result of sensitivity in soil moisture response. If rainfall amount and frequency decrease, more soil moisture is lost to evapotranspiration, creating a soil moisture deficit that must be replaced before surface runoff or significant groundwater flow returns. The converse also tends to

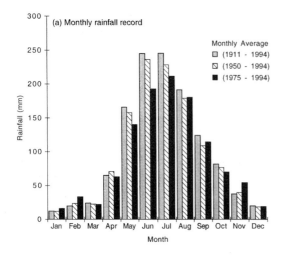

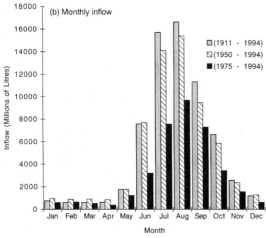

Fig. 6-11 (a) Monthly rainfall record. Serpentine dam, Perth, Western Australia. (From Burges, 1998.) (b) Monthly inflow. Serpentine Dam, Western Australia. (From Burges, 1998.)

be true: frequent storms can saturate the soil, generating more surface runoff and enhancing subsurface flow.

The complexities of land surface response and runoff generation have also presented a major obstacle to global climate modelers. Hydrologic response is linked to several important climate feedbacks (see Section 6.4.2), so until the hydrologic cycle, and in particular its land surface component, can be accurately represented, there is little hope for accurate assessments of global change.

6.4   Water and Climate

In addition to biogeochemical cycles (discussed in Section 6.5), the hydrosphere is a major component of many physical cycles, with climate among the most prominent. Water affects the solar radiation budget through albedo (primarily clouds and ice/snow), the terrestrial radiation budget as a strong absorber of terrestrial emissions, and global temperature distribution as the primary transporter of heat in the ocean and atmosphere.

6.4.1   Water and the Energy Balance

Water plays a crucial role in the redistribution of heat from the tropics to the high-latitude polar regions. Water transports heat in two forms. *Sensible heat* refers to the portion of the radiant energy budget that changes the temperature of the surface of the atmosphere. Sensible heat flux produces changes in temperature proportional to the product of density and specific heat (Shuttleworth, 1993). Since liquid water has one of the highest specific heats of any substance (4218 J/(kg K) at 0°C), the oceans can store and release vast amounts of sensible heat without large changes in their own temperatures. Surface ocean currents transport warm tropical water poleward and recirculate cooler surface waters toward the equator. These currents are complemented by the global thermohaline circulation, which circulates water through the depths of the oceans. Ice and water vapor can also store significant amounts of sensible heat, though their specific heat capacities are only one-half to one-third that of liquid water, respectively.

Latent heat is the energy associated with phase changes. Evaporation of water requires an energy input of 2.5×10^6 J per kilogram of water at 0°C, almost 600 times the specific heat. When water vapor is transported via atmospheric circulation and recondensed, latent heat energy is released at the new location. Atmospheric transport of water vapor thus transfers both latent and sensible heat from low to high latitudes.

Sensible and latent heat can be related

through the *Bowen ratio*, which is the ratio of sensible heat to latent heat flux at the surface. The Bowen ratio, R, can be estimated from atmospheric properties as follows:

$$R = \frac{SH}{LE} = 0.66 \frac{T_0 - T_a}{e_0 - e_a} \cdot \frac{p}{1000} \quad (4)$$

where SH = sensible heat (Energy/unit time and unit area); LE = latent energy flux (Energy/unit time and unit area); T_0 = surface temperature (°C); T_a = reference level air temperature (°C); e_0 = saturation vapor pressure (mbar); e_a = atmospheric vapor pressure (mbar); p = atmospheric pressure (mbar).

For saturated surfaces, the Bowen ratio can then be used to calculate evapotranspiration as a residual of the surface energy balance (Penman, 1948). Since direct measurement of ET is difficult and expensive, the energy balance method is fairly common.

6.4.2 Climate Feedbacks and Response to Global Warming

Five components of the hydrosphere play major roles in climate *feedbacks* – atmospheric moisture, clouds, snow and ice, land surface, and oceans. Changes to the hydrologic cycle, among other things, as a result of altered climate conditions are then referred to as *responses*. Interactions with climate can best be explored by examining potential response to a climate perturbation, in this case, predicted global warming.

Current debate regarding the role of the hydrologic cycle in climate focuses on potential responses to anthropogenically induced global warming through an enhanced greenhouse effect due to increased atmospheric CO_2 and other gases. Based on GCM simulation results, the Intergovernmental Panel on Climate Change (IPCC) has concluded that global mean surface atmospheric temperature will increase by 1.0 to 3.5°C in the next century, though this will be distributed unevenly over the Earth, with the most significant warming expected at high latitudes. Warmer temperatures are expected to accelerate the hydrologic cycle, though the magnitude and distribution of hydrologic changes,

particularly at the regional scale important for water resources, are much more speculative (IPCC, 1996a).

6.4.2.1 Climate feedbacks

In its assessment of climate change, the IPCC (1990) identified five hydrosphere-related feedback mechanisms in the climate system likely to be activated by increased greenhouse gas concentrations in the atmosphere. These feedbacks are briefly described below; for more detailed discussion of the climate system, refer to Chapter 17.

6.4.2.1.1 Atmospheric moisture. Short residence times and rapid phase changes for water in the atmosphere give it a disproportionately large influence on climate. Changes in atmospheric moisture affect cloud properties and are related to the cloud feedback mechanisms discussed below. In addition, water vapor is the most effective greenhouse gas due to its large heat capacity and absorption spectrum; thus expected increases in atmospheric vapor content (due to accelerated evaporation) would have a positive feedback effect. Increased evaporation and associated release of latent heat to the atmosphere (upon condensation) would warm the troposphere, increasing its moisture storage capacity. Additional water vapor then traps more terrestrial radiation, enhancing the greenhouse effect and further warming the troposphere. Although runaway warming would be prevented by changes in lapse rates to increase the flux of water vapor from the troposphere to higher altitudes, the predicted net effect of the atmospheric moisture feedback is surface and atmospheric warming (Ramanathan, 1988).

6.4.2.1.2 Clouds. Cloud feedback mechanisms are among the most complex in the climate system, due to the many disparate roles played by clouds, which control a large portion of the planetary albedo but also trap terrestrial radiation, reducing the energy escaping to space. To complicate matters further, different types of clouds behave differently in the same environment. In the present climate mode, clouds have a global mean cooling effect, but the sign of cloud feedback in climate models remains controversial.

Considerable uncertainty about the types of clouds that will increase due to greenhouse warming is the primary obstacle to predicting cloud feedback. Higher cloud top heights, assuming no change in cloud cover or water content, might result in surface warming due to greater capacity to absorb outgoing (terrestrial) radiation. Conversely, higher water content would increase cloud albedo and result in net surface cooling potentially capable of balancing additional greenhouse warming (Chahine, 1992). Ocean surface warming has also shown a moderating effect, increasing convective activity and the formation of high-albedo cirrus clouds (Chahine, 1992).

6.4.2.1.3 Snow–ice albedo.

Although cryospheric processes remain among the greatest sources of uncertainty in climate modeling, models have consistently shown enhanced warming at the poles, resulting in melting of sea ice and less snow and ice cover. Since snow and ice have higher albedos than the underlying land surfaces and oceans, this results in increased absorbed atmospheric radiation at the surface and further warming. Increased vegetative cover from the predicted poleward migration of currently more temperate biomes would also reduce surface albedo of the polar regions (IPCC, 1998). Significant melting of sea ice could also result in changes to ocean thermohaline circulation, as has been observed at several points in the paleoclimate record (see below).

6.4.2.1.4 Land surface/biosphere.

The complexity and high regional variability of land surface and biosphere effects make them extremely difficult to model, though some general feedbacks have been identified. Changes in precipitation (amount and temperature) and evaporation regimes will affect soil moisture storage and infiltration rates (which in turn influence runoff magnitude). Higher evaporation rates would be expected to reduce soil moisture and runoff, though this could be partially offset by reduced transpiration caused by elevated CO_2. Increased carbon dioxide increases stomatal

resistance in plants, with the potential to reduce transpiration loss per unit leaf area by up to 50%, but also increases plant growth. The net effect on transpiration is uncertain (Rosenberg *et al.*, 1990).

6.4.2.1.5 Ocean circulation.

Paleoclimate evidence has linked changes in thermohaline circulation and the formation of North Atlantic Deep Water (NADW) with several of the major climate shifts of the last glacial period. Therefore, ocean circulation feedback will probably play some role in determining the hydrologic and climatic response to greenhouse warming, though long ocean response times could make this a less important feedback in the short term. The mechanism for thermohaline response is the influx of low-density freshwater into the polar oceans due to precipitation and the melting of glaciers and icecaps. Since density is the dominant driving force for sinking, increased high-latitude precipitation and/or significant melting events could trigger a slowing or stoppage of deep-water formation. As a result, global circulation and poleward heat transfer would be greatly reduced, lowering surface temperatures and providing a negative feedback on global warming. Broecker (1997) hypothesized that a net increase in freshwater input to the North Atlantic of 50% – which is within predicted ranges – would disrupt the salt balance sufficiently to trigger an instability in and reorganization of the thermohaline circulation.

6.4.2.2 Hydrologic response

Accelerated hydrologic processes are predicted to result in an increase of global mean precipitation by 3 to 15%, though changes in regional precipitation would likely vary by $\pm 20\%$ (Schneider *et al.*, 1990). Like temperature, precipitation changes would be unevenly distributed, with high- and most mid-latitude areas receiving higher precipitation, while rainfall in the tropics may decrease. Changes in the amount and type of precipitation (rain versus snow) also have important implications for runoff magnitude and timing, with regional changes in runoff predicted to vary by as much as $\pm 50\%$ (Schneider *et al.*, 1990). Current studies of El Niño/southern oscillation and other circulation anomalies demonstrate the potential for significant effects of altered global circulation on precipitation patterns.

6.5   Water and Biogeochemical Cycles

Transport in water is an important mechanism for transfer of biogeochemical elements between the atmosphere, land, and oceans. In particular, rain is the primary means of removal from the atmosphere for many substances, and rivers (and to some extent groundwater) convey weathering products and runoff from the land surface to the oceans.

The following sections summarize only the most prominent interactions between the elemental cycles and the links in the hydrologic cycle. Water also plays a role in many chemical and biological reactions that are beyond the scope of this discussion. The carbon, nitrogen, sulfur, and phosphorus cycles are discussed in detail in Chapters 11, 12, 13, and 14, respectively.

6.5.1   Carbon

Rainwater and snowmelt water are primary factors determining the very nature of the terrestrial carbon cycle, with photosynthesis acting as the primary exchange mechanism from the atmosphere. Bicarbonate is the most prevalent ion in natural surface waters (rivers and lakes), which are extremely important in the carbon cycle, accounting for 90% of the carbon flux between the land surface and oceans (Holmén, Chapter 11). In addition, bicarbonate is a major component of soil water and a contributor to its natural acid–base balance. The carbonate equilibrium controls the pH of most natural waters, and high concentrations of bicarbonate provide a pH buffer in many systems. Other acid–base reactions (discussed in Chapter 16), particularly in the atmosphere, also influence pH (in both natural and polluted systems) but are generally less important than the carbonate system on a global basis.

6.5.2   Nitrogen

In the nitrogen cycle, precipitation is the primary mechanism for deposition from the atmosphere to the terrestrial–ocean system. Nitrate in rain is also a significant contributor to acid rain in the eastern US and Europe. River runoff is again the most significant flux between the terrestrial and ocean reservoirs in the nitrogen cycle. River loads are significantly increased by human activity, with fertilizers, agricultural and industrial runoffs, and acid rain contributing about one-sixth of river nitrogen in one study (Berner and Berner, 1987).

6.5.3   Sulfur

As with the nitrogen cycle, the sulfur cycle relies on rain and rivers for transport between the atmosphere, land surface, and oceans. The high solubility of H_2O_2 in water makes cloud droplets the locus of oxidation of SO_2. Rainout is the primary removal mechanism for sulfur (mainly as sulfate) from the atmosphere, although dry deposition can also be important, particularly for SO_2. Sulfur has an additional link with the atmospheric portion of the hydrologic cycle, as sulfate is the dominant component of cloud condensation nuclei in many environments (Bigg *et al.*, 1984). This is of particular concern for acid rain given the magnitude of anthropogenic sulfur emissions. Rivers are also an important transporter of sulfur, with sulfate representing the fourth most prevalent dissolved substance in global average river water (after bicarbonate, calcium ion, and silica).

6.5.4   Phosphorus

Unlike other biogeochemical elements, phosphorus does not have a significant atmospheric reservoir. Thus, while some amount of phosphorus is occasionally dissolved in rain, this does not represent an important link in the phosphorus cycle. River runoff is the primary means of transport between the land surface and oceans, and unlike the other elements discussed, phosphorus is transported in both dissolved and particulate form.

Phosphorus is extremely important in biological reactions and is thus cycled through biological systems many times before it ultimately reaches the ocean. Phosphorus cycling is particularly important in lake and wetland systems, where it is often temporarily stored in sediments. Consequently, the groundwater link is more important in the phosphorus cycle than in other elemental cycles. Concentrations of dissolved P in groundwater depend on biological and inorganic reactions.

6.6   Water and the Tectonic Cycles

In addition to transporting biogeochemical elements between the Earth's reservoirs, water serves as the primary change agent for the land surface itself, transporting the products of weathering and erosion through the river networks to the oceans. Weathering and erosion, primarily by water and glaciers, balance geologic uplift in mountain regions and redistribute sediments to the lowland floodplains and continental shelves to help maintain a constant recycling of the lithosphere.

The suspended sediment load carried by rivers, which discharge 13.5 Tg per year to the oceans, consists of particulate sediments and dissolved solids. Transport of larger bedload sediments is estimated at 1–2 Tg per year, although all of the bedload may not typically reach the ocean (Milliman and Meade, 1983). Since sediment load depends on erosion, yields tend to be highest for drainage basins with extensive geological activity. Consequently, glacial and southeast Asian rivers (due to high uplift rates in the Himalayas) have the highest sediment concentrations (sediment per unit runoff or drainage area), while desert rivers in Australia and Africa have the lowest (Milliman and Meade, 1983). The topic of sediment transport by rivers is a major focus of Chapter 8.

River water chemistry is determined by the relative concentrations of major dissolved components (bicarbonate, calcium ion, silica, and sulfate), which are in turn controlled by the environment. Rivers in precipitation-dominated environments (high precipitation and runoff) tend to have low concentrations of sodium-chloride-dominated salts. Evaporation-dominated rivers (low precipitation and runoff) are also NaCl dominated but with high total salt concentrations. Areas with moderate precipitation and runoff generate rock-dominated river chemistry, with moderate concentrations of calcium bicarbonate salts (Gibbs, 1970). Natural salt concentrations can be altered or overwhelmed by human activities, particularly by irrigated agriculture.

6.7   Anthropogenic Influences

Though the hydrosphere continues to operate in response to the same forces it always has, humans have had an unmistakable role in altering some of its balances. In general, these impacts have had relatively little effect on the overall global water balance, and there is little chance that direct manipulation of the hydrosphere will alter water storage and cycling on a global basis.

On regional scales, however, people have spent the last several thousand years trying to redistribute water resources temporally and spatially. Weirs, canals, and reservoirs have been built to control the timing of runoff and, more recently, to relocate surface water supplies, with the unintended result of greater evaporation losses from reservoir surfaces. Irrigated agriculture also diverts ocean-bound flow, much of which is then returned to the atmosphere through evapotranspiration. Thus, people pay for the privilege of redistributing water with greater losses to the atmosphere.

Dams and reservoirs represent some of the largest engineering projects of the 20th century, and they play a major role in the alteration of the hydrologic cycle on a regional scale. The Columbia River system in the northwestern United States and southwestern Canada is one of the most extensively dammed river systems in the world, with more than 50 dams providing irrigation, hydroelectric power, flood protection, and water supply for the Pacific Northwest. The dams have significantly altered the natural annual hydrograph, as shown in Fig. 6-12. The

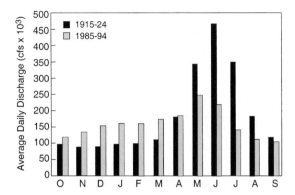

Fig. 6-12 Comparison of mean monthly averaged daily discharges for the Columbia River at The Dalles, Oregon for water years 1915–1924 (1 Oct. 1914–30 Sep. 1924) and 1985–1994. (Data from US Geological Survey, Station 14105700.)

May–June runoff peaks, 2 to 2.5 times the annual average runoff in the 1915–1924 period prior to dam construction, were barely 1.5 times the average flow between 1985 and 1994. Meanwhile, autumn low flows during the 1985–1994 period are close to the mean annual flow, compared to low flows at about half the mean prior to river regulation.

The consequences of this regulated system include significantly lower total discharge (beyond what would be expected from climate variability), ecological effects of altered freshwater inputs to the Pacific, and altered sediment budgets due to sediment trapping behind dams. One unintended result of the changed hydrograph has been reduced autumn and winter surface salinity from the mouth of the Columbia along the North American coast to the Aleutian Island chain, which has potentially negative ecological consequences for endangered salmon runs.

Regional water balances are also altered by agricultural and domestic water uses drawing on underground aquifers, increasingly at rates that exceed natural recharge capability and result in groundwater overdraft. Pollution of surface and groundwaters, though it has no physical effect on the water cycle itself, results in a loss of freshwater resources in addition to the effects on balances in other biogeochemical cycles.

One of the largest human influences on the hydrologic cycle results from changes in land use. Alteration of the land surface and natural vegetation disrupts the natural balance of precipitation, evapotranspiration, and runoff at a given location. This effect tends to be exaggerated by the fact that land use change (e.g. agriculture and urbanization) is often associated with the direct physical changes discussed above.

These and other direct human impacts on the hydrosphere are unlikely to affect the global hydrologic cycle, particularly since humans have not had a great deal of success in manipulating the water balances of the ocean and atmosphere, the largest and most sensitive reservoirs in the system, respectively. Significant anthropogenic effects on the hydrologic cycle are much more likely to arise from indirect changes, most notably human-induced climate change.

6.8 Conclusion

Cycling of water between the atmosphere, land surface, and oceans is important not only to humans and other organisms, which rely on water to live, but in maintaining balances in other cycles as well. Hydrologic fluxes, predominantly rain and rivers, transport significant amounts of carbon, nitrogen, sulfur, and phosphorus, among other elements, between reservoirs in their own biogeochemical cycles. Rivers are also a major link in the tectonic cycle, transporting sediment eroded from upland areas to inland basins and to the oceans. The heat capacities and physical properties of all phases of water also give the hydrosphere an important role in the global heat balance and climate.

To this point, direct human impacts on the hydrosphere have remained restricted to the regional scale. Although they can still be important, particularly in terms of water supply, these direct manipulations of the hydrologic cycle are unlikely to affect the global water balance significantly. However, this is not to suggest that the global water cycle is immune to human influence; its close ties to other physical and biogeochemical cycles subject the hydrologic cycle to indirect effects of human impacts on these cycles as well. The most immediate human threat to the existing global water balance, therefore, may not be the damming of rivers or mining of groundwater but more likely climate change induced by anthropogenic greenhouse gas emissions. Coupled with feedbacks linked to the sulfur cycle, carbon cycle, and biosphere, as well as internal feedbacks and responses, climate change has the potential to alter the hydrologic cycle more than the combined effects of thousands of years of hydraulic engineering.

The Earth's history, and its future, are shaped not by independent events but by an intricately linked series of feedbacks and responses spanning the spectrum of physical, chemical, and biological cycles, of which the hydrologic cycle is only a part, albeit a central one.

References

Baumgartner, A. and Reichel, E. (1975). "The World Water Balance: Mean Annual Global, Continental and Maritime Precipitation and Run-Off." Elsevier Scientific Publishers, Amsterdam.

Berner, E. K. and Berner, R. A. (1987). "The Global Water Cycle: Geochemistry and Environment." Prentice-Hall, Englewood Cliffs, NJ.

Bigg, E. K., Gras, J. L., and Evans, C. (1984). Origin of Aitken particles in remote regions of the Southern Hemisphere. *J. Atmos. Chem.* **1**, 203–214.

Broecker, W. S. (1997). Thermohaline circulation, the Achilles heel of our climate system: Will man-made CO_2 upset the current balance? *Science* **278**, 1582–1588.

Budyko, M. I. (1974). "Climate and Life." Academic Press, San Diego.

Burges, S. J. (1998). Streamflow prediction – capabilities, opportunities, and challenges. *In* "Hydrologic Science: Taking Stock and Looking Ahead" (National Research Council). National Academy Press, Washington, DC.

Chahine, M. T. (1992). The hydrologic cycle and its influence on climate. *Nature* **359**, 373–380.

Dickinson, R. E. (1984). Modeling evapotranspiration for three-dimensional global climate models. *Geophysical Monographs* (J. E. Hansen and T. Takahashi, eds.) **29**, 58–72. American Geophysical Union.

Dingman, S. L. (1994). "Physical Hydrology." Macmillan.

Dunne, T. and Leopold, L. B. (1978). "Water in Environmental Planning." W.H. Freeman, San Francisco.

Eltahir, E. A. B. and Bras, R. L. (1996). Precipitation recycling. *Rev. Geophys.* **34**, 367–378.

Falkenmark, M. and Chapman, T. (1989). "Comparative Hydrology: An Ecological Approach to Land and Water Resources." UNESCO, Paris.

Gibbs, R. J. (1970). Mechanisms controlling world water chemistry. *Science* **170**, 1088–1090.

Holzman, B. (1937). Sources of moisture for precipitation in the United States. *Technical Bulletin 589*, U.S. Department of Agriculture.

Horton, R. E. (1931). The field, scope, and status of the science of hydrology. *American Geophysical Union Transactions*, Reports and Papers, Hydrology, 189–202.

IPCC (1990). "Climate Change: The IPCC Scientific Assessment" (J. T. Houghton *et al.*, eds). Cambridge University Press, Cambridge, UK.

IPCC (1996a). "Climate Change 1995: The Science of Climate Change. Contribution of Working Group I to the Second Assessment Report of the Intergovernmental Panel on Climate Change" (J. T. Houghton *et al.*, eds.). Cambridge University Press, Cambridge, UK.

IPCC (1996b). "Climate Change 1995: The Science of Climate Change. Contribution of Working Group II to the Second Assessment Report of the Intergovernmental Panel on Climate Change" (R. T. Watson *et al.*, eds.). Cambridge University Press, Cambridge, UK.

IPCC (1998). "The Regional Impacts of Climate Change: An Assessment of Vulnerability" (R. T. Watson *et al.*, eds.). Cambridge University Press, Cambridge, UK.

Langbein, W. B. and Wells, J. V. B. (1955). The water in the rivers and creeks. *In* "Yearbook of Agriculture 1955," pp. 52–62. US Department of Agriculture.

Linsley, R. K., Kohler, M. A., and Paulhus, J. L. H. (1982). "Hydrology for Engineers," 3rd edn. McGraw-Hill, New York.

L'vovich, M. I. (1974). "World Water Resources and Their Future." Translated by R. L. Nace. American Geophysical Union, Washington, DC.

Milliman, J. D. and Meade, R. H. (1983). World-wide delivery of river sediment to the oceans. *J. Geol.* **91**, 1–21.

Nace, R. L. (1969). World water inventory and control. *In* "Water, Earth, and Man" (R. J. Chorley, ed.), pp. 31–42. Methuen and Co., Edinburgh, UK.

Penman, H. L. (1948). Natural evaporation from open water, bare soil and grass. *Proc. R. Soc. Lond.* **193**, 120–145.

Probst, J. L. and Tardy, Y. (1987). Long range stream-flow and world continental runoff fluctuations since the beginning of this century. *J. Hydrol.* **94**, 289–311.

Ramanathan, V. (1988). The greenhouse theory of climate change: A test by an inadvertent global experiment. *Science* **240**, 293–299.

Rosenberg, N. J., Kimball, B. A., Martin, P., and Cooper, C. F. (1990). From climate and CO_2 enrichment to evapotranspiration. *In* "Climate Change and US Water Resources" (P. E. Waggoner, ed.), pp. 151–175. Wiley, New York.

Schneider, S. H., Gleick, P. H., and Mearns, L. O. (1990). Prospects for climate change. *In* "Climate Change and US Water Resources" (P. E. Waggoner, ed.), pp. 41–73. Wiley, New York.

Shiklomanov, I. A. and Sokolov, A. A. (1983). Methodological basis of world water balance investigation and computation. *In* "New Approaches in Water Balance Computations." International Association for Hydrological Sciences Publication No. 148 (Proceedings of the Hamburg Symposium).

Shuttleworth, W. J. (1993). Evaporation. *In* "Handbook of Hydrology" (D. R. Maidment, ed.), Chapter 4. McGraw-Hill, New York.

Thornthwaite, C. W. (1948). An approach toward a rational classification of climate. *Am. Geogr. Rev.* **38**, 55–94.

UNESCO (1978). "World Water Balance and Water Resources of the Earth." UNESCO Press, Paris. (Translation of 1974 USSR publication.)

7

The Atmosphere

Robert J. Charlson

7.1 Definition

The atmosphere is a thin layer of gas uniformly covering the whole Earth. Its main constituents are nitrogen (N_2), oxygen (O_2), argon (Ar), water (H_2O gas, liquid and solid), and carbon dioxide (CO_2). The origins of these main constituents are discussed in Chapter 2. This chapter will first concentrate on the physical properties of the atmosphere and the ways in which these influence chemical composition – particularly through diffusion and transport. Then the chemical processes of the atmosphere are discussed with emphasis on minor constituents. As will become evident, most of the chemical functioning of the atmosphere involves substances other than N_2, O_2, Ar, H_2O, and CO_2. Even though some of these are often products of or participants in reactions, their abundance is so great that their concentrations are not perturbed. Because of the particular importance of CO_2, it is considered in detail in Chapter 11.

Five chemical features of the atmosphere are emphasized.

1. *Altitude dependence.* The composition varies with altitude. Part of that vertical structure is due to the physical behavior of the atmosphere while part is due to the influence of trace substances (notably ozone and condensed water) on thermal structure and mixing.

2. *Transport and diffusion.* With the exception of N_2, O_2, Ar, and numerous other long-lived species that are well-mixed in the bulk of the atmosphere, horizontal and vertical transport are closely coupled with chemical reactions in controlling atmospheric trace-substance concentrations.

3. *Composition.* Air is a mixture of a large number of species with concentrations varying in space and time. Of particular interest are ozone and compounds of sulfur, nitrogen, and carbon, and their chemical interactions.

4. *Role of composition in atmospheric physical process.* The composition of the atmosphere plays a distinct set of roles in controlling and affecting certain physical processes of the atmosphere, most notably the thermal structure.

5. *Processes that occur at the upper and lower boundaries of the atmosphere.* Many atmospheric constituents are formed, and many undergo a wide range of reactions at the lower boundary. At the upper boundary lighter elements are lost to space and some important substances are acquired.

Before setting out to discuss the vertical structure of the atmosphere, we note that it is useful to have access to conventional nomenclature. Figure 7-1, based on the thermal profile of the atmosphere, includes a number of commonly used definitions.

Earth System Science
ISBN 0-12-379370-X

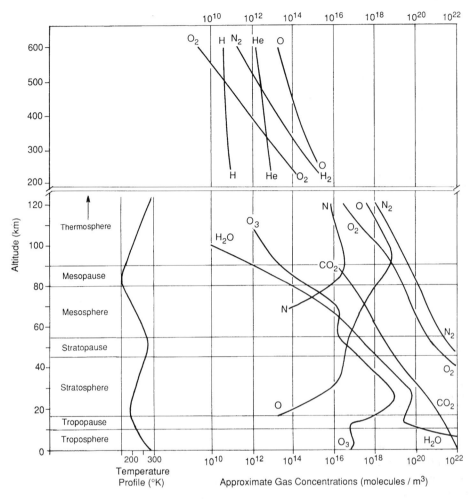

Fig. 7-1 Atmospheric vertical structure including temperature; composition and conventional names of atmospheric layers or altitude regions.

7.2 The Vertical Structure of the Atmosphere

7.2.1 Hydrostatic Equation

The atmosphere is very close to being in hydro-static equilibrium in the vertical dimension. This can be described by the hydrostatic equation:

$$\frac{dP}{dz} = -\rho g \qquad (1)$$

where P is pressure, ρ is density, g is the acceleration of gravity and z is the vertical coordinate. We have two choices for describing dP/dz. For an ideal gas, $PV = nRT$ where V is volume, n is number of moles, R is the gas constant and T is temperature. So,

$$P = \frac{n}{V}RT = \frac{mRT}{MV} \qquad (2)$$

where m is the mass and M is the molecular mass of the gas. Since $m/V = \rho$,

$$P = \frac{\rho RT}{M} \qquad (3)$$

The choices come in defining M for this mixture of gases. We might define M_i for each gas separately, or we might define a mean value $M = \sum_i X_i M_i$ where X_i is the mole fraction of component i. The use of M_i holds for P_i of any individual species in the absence of any physical mixing (e.g., by turbulence or Brownian motion), while M would be used in the case of perfect mixing.

7.2.2   Scale Height

If we define a molecular weight for each constituent, then we can rearrange Equation (1). Because $\rho = PM/RT$,

$$\frac{dP}{dz} = -\frac{PMg}{RT} \tag{4}$$

so

$$\frac{dP_i}{P_i} = -\left(\frac{M_i g}{RT}\right) dz = -\frac{dz}{H_i} \tag{5}$$

where $H_i \equiv RT/M_i g$ is called the scale height. In this situation, constituents with low M have large H so they tend to fall off in pressure (and concentration) slowly with altitude, while the opposite is true of constituents with high values of M. In such cases, each gas behaves as if no other substance were present. High-molecular-mass gases (e.g., Xe, Kr) would be concentrated in a layer at the bottom of the atmosphere and lighter gases (H_2, He) would extend to greater altitude. This diffusive separation is not generally significant at low altitudes but occurs increasingly at altitudes above 120 km. Turbulent mixing separates the atmosphere into two layers – the mixed layer at the bottom being called the homosphere and the upper layer, the heterosphere. The highest reaches of the atmosphere are thus dominated by H and He, and in the heterosphere heavy unreactive gases (40Ar, Xe, Kr, etc.) fall off rapidly with height. Figure 7-1 illustrates this compositional feature of the atmosphere at altitudes above ca. 120 km.

The abundance of light elements at high altitude leads to a finite flux of these substances escaping the Earth's gravitational field. This results from a combination of a very long mean free path and a few particles having the requisite escape velocity due to the high-velocity "tail" of the Boltzmann velocity distribution.

In terms of relevance to biogeochemical cycling, most of our emphasis is placed on the so-called homosphere (which really is homoge-neous *only* with respect to N_2, O_2, 40Ar, and other long-lived gases).

In the case of a mixed atmosphere, M cannot be defined precisely since the composition is variable (especially due to water vapor). If dry air is assumed (which is a good approximation most of the time at altitudes above about 5 km), then $M = 28.97$ g/mol. If the atmosphere is assumed to be roughly isothermal, then from Equation (5) pressure falls off with altitude as

$$P = P_0 \exp\left(-\frac{z}{H}\right) \tag{6}$$

Since H is constant if T and M are constant, $H \approx 8$ km if $T = 273$ K. H is the height the entire atmosphere would have if its density were constant at the sea level value throughout.

7.2.3   Lapse Rate

The atmosphere *is not* isothermal – largely due to the fact that it really is a compressible medium. In the simplest case of a dry atmosphere being mixed in the vertical direction with no addition or loss of energy, we might assume that an air parcel behaves adiabatically when it is not in contact with the ground. This implies that its enthalpy, H, is constant. If we define the geopotential at a given height, f, as the work needed to move a unit mass from sea level to that height, then

$$\phi = \int_0^z g\,dz \tag{7}$$

For one mole of an ideal gas, $dH = C_p\,dT + M d\phi$, where C_p is the constant pressure heat capacity per mole. Now the adiabatic condition implies that

$$d(C_p T + M\phi)$$

If we divide by dz and use Equation (1), we have

$$\frac{dT}{dz} = -\frac{Mg}{C_p} = -\Gamma_d$$

Γ_d is called the dry adiabatic lapse rate. For air, $C_p = 29.09$ J/(mol K), and on Earth $g = 9.81$ m/s, so $\Gamma_d = 9.8$ K/km.

Another way of expressing the way temperature varies in the vertical direction involves the concept of potential temperature, θ. Potential temperature is defined as the temperature a parcel of air would reach if brought adiabatically from its existing temperature and pressure to a standard pressure, P_0. Hence

$$\theta = T\left(\frac{P_0}{P}\right)^{R/C_P} \tag{8}$$

if T is the temperature at pressure P. This concept is widely used in meteorology, since in the absence of clouds and heat exchange it is convenient to assume that θ is constant for a parcel of air as it moves in the atmosphere.

7.2.4 Static Stability

Now, intuitively, we can consider a perfectly mixed cloudless atmosphere to have constant θ and thus it behaves in a sense like an isothermal body of water, i.e., no part of it is buoyant. If we allow for a layer of air of low θ to occur at the bottom of the atmosphere (like cold water at the bottom of a lake) it is stably stratified and an inversion is said to exist. This layer of relatively low θ acts as a barrier to vertical mixing and hence becomes a physical feature of the atmosphere that is dominant in controlling the dispersion of trace substances (see box).

Another widely used concept is that of a planetary boundary layer (PBL) in contact with the surface of the Earth above which lies the "free atmosphere." This PBL is to some degree a physically mixed layer due to the effects of shear-induced turbulence and convective overturning near the Earth's surface.

The PBL has different characteristics depending on wind speed and static stability; these can be roughly distributed between two extreme categories:

1. Cold air under warm air (inversion), such as warm air over snow or cold ocean water,

---

Static Stability

Fluids on the Earth's surface that are in hydrostatic equilibrium may be stable or unstable depending on their thermal structure. In the case of freshwater (an incompressible fluid), density decreases with temperature above ca. 4°C. Warm water lying over cold water is said to be stable. If warm water underlies cold, it is buoyant; it rises and is unstable. The buoyant force, F, on the parcel of fluid of unit volume and density ρ' is:

$$F = (\rho - \rho')g$$

where ρ is the density of the surrounding medium. Since the acceleration, a, is just F/ρ':

$$a = \left(\frac{\rho - \rho'}{\rho'}\right)g$$

by Archimedes' principle.

Now, in the case of an ideal gas, pressure, density, and temperature are related so that

$$a = g\left(\frac{(P/T)(P'/T')}{(P'/T')}\right)$$

where the prime again denotes the parcel of unit volume.

But in hydrostatic equilibrium and low acceleration, $P = P'$ so

$$a = g\left(\frac{T' - T}{T}\right)$$

If we let T' be represented by a simple Taylor series, i.e., $T' = T_0 + dT'/dz\Delta z$, and if the rest of the ideal atmosphere has a lapse rate dT/dz, so $T = T_0 + dT/dz\Delta z$, we have for small Δz

$$a = g\frac{\Delta z}{T_0}\left(\frac{dT'}{dz} - \frac{dT}{dz}\right)$$

If the acceleration is positive, our parcel is buoyant and spontaneous convection occurs. The atmospheric layer is said to be *unstable*. Negative acceleration implies that a small displacement, Δz, results in the parcel accelerating back toward its initial position and therefore indicates *stability*. If dT'/dz is that for an adiabatic test parcel $dT'/dz = -gM/C_P$ and dT/dz that of the existing layer, then for $dT/dz > -9.8$ K/km is stable and for $dT/dz < -9.8$ K/km is unstable. The 9.8 K/km figure then provides a simple benchmark for static stability of dry air.

coupled with low wind speed produces a thin PBL and thus a thin mixed layer. As an extreme example, in the Arctic winter the PBL may be only 100 m deep leading to the trapping of water and pollutants near the ground and the formation of ice fog. A less extreme but well-known example is the inversion in such cities as Los Angeles, London or Mexico City, again trapping trace substances and causing elevated concentration of pollutants.

2. The lapse rate in the PBL is unstable and vertical motion leads to the transport of significant amounts of energy upward, due to the buoyancy of air that has been in contact with the surface. A mixed layer forms up to a height where static stability of the air forms a barrier to thermally induced upward motion. This extreme occurs practically daily over the arid areas of the world and the barrier to upward mixing is often the tropopause itself. On the average in mid-latitudes, the unstable or mixed PBL is typically 1–2 km deep.

Figure 7-2 shows the vertical profiles of temperature, dew point, light scattering (a measure of aerosol concentration) and the concentrations of O_3 and SO_2. Here we see that up to about 1.5 km, the temperature, dew point, light scattering

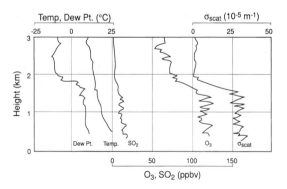

Fig. 7-2 Vertical profiles of physical (temperature, dew point, and backscatter coefficient) and chemical (ozone, sulfur dioxide) variables near Scranton, PA during the afternoon of 20 July 1978. (Modified with permission from P. K. Mueller and G. M. Hidy (1982). "The Sulfate Regional Documentation of SURE Sampling Sites", EPRI report EA-1901, v. 3, Electric Power Research Institute.)

(a measure of aerosol concentration) and the concentrations of O_3 and SO_2 are nearly constant. This indicates the presence of a mixed PBL. Above 1.5 km the profiles change dramatically.

In between the two extremes of stability and instability there are numerous near-neutral stability situations, resulting in varying degrees of vertical mixing. In this regime, the mixing depends on such factors as shear-induced turbulence and vertical mixing in and by clouds.

7.3 Vertical Motions, Relative Humidity, and Clouds

When air moves vertically, its temperature changes in response to the local pressure. Indeed, the amount of temperature change is quite large for small changes in height; one degree per 100 m, if the dry adiabatic lapse rate applies. Considering upward motion (and therefore cooling), an adiabatic decrease of pressure of only 10% due to an altitude increase from 100 to 1000 m results in a change in temperature of $-9°C$. This amount of cooling results in a major increase in the relative humidity (RH) due to the very strong dependence of the saturation vapor pressure of water on temperature. Details of this are discussed in Section 7.7 (below); however, the consequence of increasing RH due to upward vertical motion is that upward motions of more than a few tens to a few hundreds of meters often cause the air to reach RH = 100% and form clouds. It is important to realize that, even though the decrease in pressure causes a decrease in the amount of water in a fixed volume of air, the temperature decrease is more significant, causing an increase in RH.

Thus we see that vertical motions upward cause water clouds to develop; conversely air that descends becomes warm, causing the RH to decrease and clouds to disappear.

Vertical motions in the atmosphere are caused by a variety of factors:

1. Convection due to the solar heating of the Earth's surface. Upward velocities of 2–20 cm/s occur.

2. Upward motion associated with convergence

of horizontal motions (or vice versa, sinking due to divergence). This will be evident in the discussion of horizontal motions in Section 7.5. Again, vertical velocities of only cm/s usually are observed.

3. Horizontal motion over topographic features at the Earth's surface. A classic example of this is seen in the cap clouds associated with flow over mountains.

4. Buoyancy caused by the release of latent heat of condensation of water. As will be seen in Section 7.7, water releases a substantial amount of energy when it condenses.

Even though upward motion causes cooling of a parcel of air, the condensation of water vapor can maintain the temperature of a parcel of air above that of the surrounding air. When this happens, the parcel is buoyant and may accelerate further upwards. Indeed, this is an unstable situation which can result in violent updrafts at velocities of meters per second. Cumulus clouds are produced in this fashion, with other phenomena such as lightning, heavy precipitation and locally strong horizontal winds below the cloud (which provide the air needed to support the vertical motion).

On the average, the air over roughly half of the Earth's surface has an upward velocity and half has a downward velocity. This frontal activity (Section 7.5.3) and the interactions of marine air with the cold ocean surface result in about half of the Earth being covered by clouds and half being clear. As will be discussed in Chapter 17, this large fractional cloud cover is extremely important to the Earth's climate because it controls the planetary *albedo* (reflectivity).

7.4   The Ozone Layer and the Stratosphere

Another major feature of the vertical thermal structure of the atmosphere is due to the presence of ozone, O_3, in the stratosphere. This layer is caused by photochemical reactions involving oxygen. The absorption of solar UV radiation by O_3 causes the temperature in the stratosphere and mesosphere to be much higher than expected from an extension of the adiabatic temperature profile in the troposphere (see Fig. 7-1).

Briefly, oxygen can be photodissociated by solar UV of wavelength less than 242 nm:

$$O_2 + hv \rightarrow O + O \qquad (9)$$

Subsequently the following reactions occur:

$$O + O_2 + M \rightarrow O_3 + M \qquad (10)$$

$$O_3 + hv \rightarrow O + O_3 \qquad (11)$$

The dissociation of O_3 in Equation (11) occurs at longer wavelengths than for the dissociation of O_2. The progress of this reaction is halted by

$$O + O_3 \rightarrow 2O_2 \qquad (12)$$

In reality, many other chemical and photochemical processes take place leading to a sort of steady-state concentration of O_3 which is a sensitive function of height. To be accurate, it is necessary to include the reactions of nitrogen oxides, chlorine- and hydrogen-containing free radicals (molecules containing an unpaired electron). However, occurrence of a layer due to the altitude dependence of the photochemical processes is of fundamental geochemical importance and can be demonstrated simply by the approach of Chapman (1930).

The concentration of O_2 is approximately an exponential function of altitude:

$$\rho_{O_2} = \rho_{0,O_2} e^{-z/H} \qquad (13)$$

where ρ_{O_2} is the concentration of O_2, e.g. in molecules per unit volume, ρ_{0,O_2} is the concentration at $z = 0$, and H is the scale height. Now the intensity, I, of solar UV light falling on the atmosphere (in the direction of decreasing z) at an angle χ from the zenith will be attenuated as it penetrates into the atmosphere:

$$\frac{dI}{I} = A\rho_{O_2} \sec \chi dz \qquad (14)$$

or

$$\frac{dI}{I} = A\rho_{0,O_2} \sec \chi e^{-z/H} dz \qquad (14)$$

where A is the absorption cross-section for O_2. Integrating:

$$I = I_0 \exp(-A\rho_{0,O_2}H(\sec \chi)e^{-z/H}) \qquad (16)$$

Now, the rate of the production of O_3 via the

photolysis of O_2 is roughly given by the rate of photolysis of O_2 itself (the $O + O_2$ reaction is assumed to be fast). Thus, the rate of O_3 production as a function of altitude, $q(z)$, should be proportional to the rate of disappearance of photons as a function of altitude:

$$q(z) + \beta \left(\frac{dI}{dz} \right) \cos \chi \qquad (17)$$

where β denotes proportionality. Using Equation (16) we find that $q(z)$ has a maximum at a height

$$z_{max} = H \ln(A\rho_{0,O_2}H \sec \chi) \qquad (18)$$

The dependence of I on z results in a layer of O_3, the upper portion of the layer being controlled by the exponential decrease of ρ_{O_2} with altitude. The lower part of the layer is controlled by the fall-off of intensity of UV light as the solar beam penetrates into the increasingly dense atmosphere. More extensive treatments of this phenomenon can be found, e.g. in Wayne (1985, p. 117 ff.).

The resultant O_3 layer is critically important to life on Earth as a shield against UV radiation. It also is responsible for the thermal structure of the upper atmosphere and controls the lifetime of materials in the stratosphere. Many substances that are short-lived in the troposphere (e.g. aerosol particles) have lifetimes of a year or more in the stratosphere due to the near-zero removal by precipitation and the presence of the permanent thermal inversion and lack of vertical mixing that it causes.

Besides these features, the formation of a *layer* due to an interaction of a stratified fluid with light is itself noteworthy. Analogs to this phenomenon can be found in other media. Examples include photochemical reactions in the atmosphere near the Earth's surface, photochemical reactions in the surface water of the ocean and biological activity near the ocean surface.

7.5   Horizontal Motions, Atmospheric Transport, and Dispersion

The horizontal motion of the atmosphere (or wind) is characterized by four spatial scales. These, with their conventional names, are:

1. 0 to 10 km – the *micrometeorologic* scale, in which turbulent dispersion of materials is dominant.
2. 10 to hundreds of km – the *mesometeorologic* scale, in which both advection and turbulent dispersion are effective.
3. Hundreds to thousands of km – the *synoptic* scale, in which motions are those of whole weather systems. Advection is the dominant transport process.
4. $>5 \times 10^3$ km – the *global* scale.

Going along with these spatial scales, we can define temporal scales as well. Micrometeorologic processes tend to be important for times less than an hour, mesoscale processes, up to about a day, and synoptic scale, a few days or more.

7.5.1   Microscale Turbulent Diffusion

Accurate description of mixing processes on each of these scales is only possible in a few selected and idealized cases. One of the best understood cases is that of a turbulent PBL over flat terrain and a point source of a trace substance. In this case, the concentration downwind of the source is often described as a plume. Figure 7-3 shows such an idealized plume.

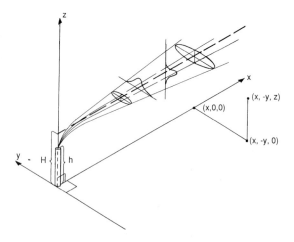

Fig. 7-3   Coordinate system showing the formation downwind from a source of Gaussian distributions of chemical concentrations in the horizontal and vertical. Ellipses denote the loci of two standard deviations.

Spreading in the downwind direction results from advection by the wind. Spreading at right angles to the wind results from turbulence. This description does not often hold for distances greater than a few tens of km. Mixing and transport over the mesoscale is extremely hard to describe and is often dominated by local topography, presence of organized vertical motions (e.g., into clouds or "thermals" due to convection) and stable layers that are embedded in the PBL.

7.5.2 Synoptic Scale Motion: The General Circulation

The motion of substances on the synoptic scale is often assumed to be pure advection. The flux through a unit area perpendicular to the wind is simply the product of wind velocity and concentration. If F is flux, V the velocity, and c concentration,

$$F = cV \qquad (19)$$

where such transport is called *advection*. (See also Chapter 4.)

The motions on the largest spatial scales amount to the aggregate of the world's synoptic weather systems, often called the general circulation. Both with respect to substances that have atmospheric lifetimes of a day or more and with regard to the advection of water, it is useful to depict the nature of this general circulation. The mean circulation is described to some extent in terms of the Hadley and Ferrell cells shown in Fig. 7-4. They describe a coupled circulation driven by the large input of solar radiation near the equator. While departures from the circulation in Fig. 7-4 are substantial, this average pattern does account for major aspects of the pattern of global precipitation. See also Fig. 17-1a and b, later in the book.

Three regions of the atmosphere are seen to have significant zonal components of flow and thus of advection. The *mid-latitude* troposphere at the surface tends to exhibit westerly flow (i.e., flow from west to east) on the average. This region contains the familiar high- and low-pressure systems that cause periodicity in mid-latitude weather. Depending on the lifetime of the substances of concern, the motion in these weather systems may be important.

The *tropical* regions of both of the hemispheres' troposphere exhibit easterly flow called the trade winds. Finally the *jet stream* – sometimes described as a river of air – flows at mid-latitude of both hemispheres with velocities of 25 to 50 m/sec from west to east, often carrying material completely around the Earth at its altitude close to the tropopause. It is in this flow that balloonists attempt to circle the globe.

7.5.3 Geostrophic Wind

Horizontal motion of the atmosphere, or wind, is a response of the air to the forces that are present. These include the force due to the pressure gradient, the Coriolis force associated with the rotation of the Earth, and frictional forces acting to retard any motion. If the acceleration of the air mass and frictional effects are small, the horizontal velocity is described by the following expression:

$$V_g \approx \frac{1}{\rho f} \cdot \frac{\partial P}{\partial x} \qquad (20)$$

This describes the *geostrophic wind* ($f = 2\omega \sin \phi$, where ω is the angular velocity due to the rotation of the Earth and ϕ is the latitude). The air moves parallel to the isobars (lines of constant pressure). The geostrophic wind blows counterclockwise around low-pressure systems in the northern hemisphere, clockwise in the southern.

At sea level, for $30°$ N and S, and a pressure gradient of 1 mbar per 100 km (or

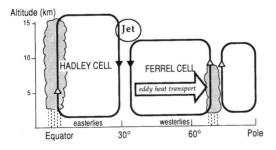

Fig. 7-4 Cross-section of the northern hemisphere atmosphere showing first-order circulation. The southern hemisphere is a mirror image.

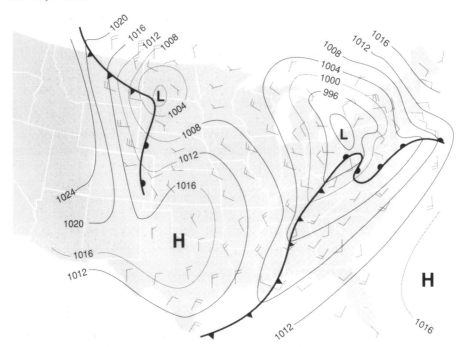

Fig. 7-5 Surface pressure map (millibars). Fronts are shown by heavy lines. H = high-pressure system; L = low-pressure system. Wind directions are shown by arrows; wind speeds correspond to the number of bars on the arrow tails.

1×10^{-3} Pa/m), $V_g \approx 15$ m/s. In many instances, the observed wind is indeed close to the geostrophic wind and it is often useful to have maps of isobars so that the transport trajectory can be approximated from V_g. For a complete derivation and explanation of the geostrophic wind, departures from it and related topics, the reader is referred to textbooks on meteorology (e.g., Wallace and Hobbs, 1977).

In the mid-latitude region depicted in Fig. 7-5, the motion is characterized by "large-scale eddy transport." Here the "eddies" are recognizable as ordinary high- and low-pressure weather systems, typically about 10^3 km in horizontal dimension. These eddies actually mix air from the polar regions with air from nearer the equator. At times, air parcels with different water content, different chemical composition and different thermodynamic characteristics are brought into contact. When cold dry air is mixed with warm moist air, clouds and precipitation occur. A *frontal system* is said to exist. Two such frontal systems are depicted in Fig. 7-5 (heavy lines in the midwest and southeast).

7.5.4　Meridional Transport of Water, the ITCZ

Among the consequences of this general circulation are convergent and divergent flows in the surface wind leading to systematic vertical motions, especially those of the Hadley cell in the tropics. Upward motion (such as near the equator in the Hadley cell) often results in the formation of clouds due to adiabatic cooling, while subsidence (downward motion) results in heating and the absence of clouds. Figure 7-6 shows composite satellite photographs depicting the mean brightness of the region from 40° N to 40° S. Clearly evident are the bands of clouds in the intertropical convergence zone (ITCZ) and the clear areas north and south of it. The influence of the land masses on this simplified picture is also apparent, clearly underscoring the difficulties in describing air motions in the vicinity of either topographic roughness or thermal discontinuity. Just at the top and bottom edges of the pictures at latitude 30–40° N or S, the cloudiness of the mid-latitude weather system is apparent. Plate 2 also shows the same features as detected by satellite-borne Lidar.

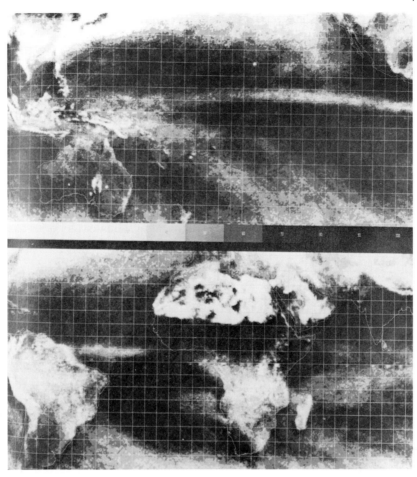

Fig. 7-6   Satellite observations of global reflectivity for January 1967–1970. White indicates areas of persistent cloudiness and relatively high precipitation, except for northern Africa where desert surface regions are highly reflective. (From US Air Force and US Department of Commerce (1971). "Global Atlas of Relative Cloud Cover," 1967–1970, Washington.)

Figure 7-7 depicts the transport of one substance – water – due to the general circulation. Here we see the overall consequence of the general circulation with its systematic pattern of vertical motions and weather systems. Water evaporates from the oceans and land surfaces at subtropical latitudes and is transported both toward the equator and the poles. Precipitation falls largely at the equator and in the mid-latitudes. Hence, the subtropics are arid, with evaporation exceeding precipitation. The polar regions likewise are arid due to water having been removed in mid-latitude weather systems prior to arrival in the Arctic and Antarctic. A more extensive discussion of factors influencing climate can be found in Chapter 17.

The turnover time of water vapor in the atmosphere obviously is a function of latitude and altitude. In the equatorial regions, its turnover time in the atmosphere is a few days, while water in the stratosphere has a turnover time of one year or more. Table 7-1 (Junge, 1963) provides an estimate of the average residence time for water vapor for various latitude ranges in the troposphere. Given this simple picture of vertical structure, motion, transport, and diffusion, we can proceed to examine the behavior of

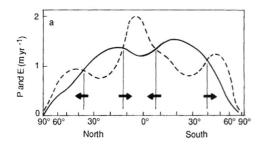

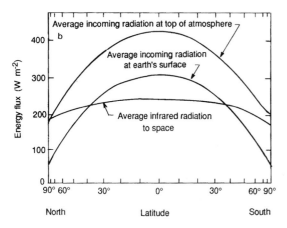

Fig. 7-7 (a) Average annual precipitation (P) and evaporation (E) per unit area versus latitude. Arrows represent the sense of the required water vapor flux in the atmosphere. (b) Incoming solar energy (top of atmosphere and surface) and outgoing terrestrial energy versus latitude.

reactive trace substances in this dynamic milieu. However, before we can do so, it is useful to briefly summarize the overall composition of the atmosphere.

7.6   Composition

Table 7-2 includes most of the main *gaseous* constituents of the troposphere with observed concentrations. In addition to gaseous species, the *condensed* phases of the atmosphere (i.e. aerosol particles and clouds) contain numerous other species. The physical characteristics and transformations of the aerosol state will be discussed later in Section 7.10. The list of major gaseous species can be organized in several different ways. In the table, it is in order of decreasing concentration. We can see that there are five approximate categories based simply on concentration:

1. The major gases – the concentration often given as a *percentage* (N_2, O_2, Ar, H_2O, CO_2).
2. Those gases having concentrations expressed in the *parts-per-million* range (Ne, He, CH_4, CO).
3. Gases expressed as *parts-per-billion* (O_3, NO, N_2O, SO_2).
4. Gases in the *parts-per-trillion* category (CCl_2F_2, CF_4, NH_3).
5. Gases expressed *in number of atoms or molecules per cubic centimeter* – notably radionuclides and free radicals like OH.

Alternatively, we could organize the list by variability in which we would see that N_2, O_2, and the noble gas concentrations are extremely stable, with increasing variability for substances of low concentration and for chemically reactive substances. Both the temporal and spatial variability are influenced by the same factors: source strength and its variability, sink mechanisms

Table 7-1   Average residence time of water vapor in the atmosphere as function of latitude

	Latitude range (degrees)								
	0–10	10–20	20–30	30–40	40–50	50–60	60–70	70–80	80–90
Average precipitable water (g/cm²)	4.1	3.5	2.7	2.1	1.6	1.3	1.0	0.7[a]	0.45[a]
Average precipitation (g/(cm² yr))	186	114	82	89	91	77	42	19	11
Residence time (days)	8.1	11.2	12.0	8.7	6.4	6.2	8.7	(13.4)	(15.0)

[a]Values extrapolated.

Table 7-2   Major gaseous constituents of dried air

		Average concentration (volume fraction)
N_2	Nitrogen	0.78084
O_2	Oxygen	0.20946
^{40}Ar	Argon	9.34×10^{-3}
CO_2	Carbon dioxide	3.7×10^{-4}
Ne	Neon	1.8×10^{-5}
He	Helium	5.24×10^{-6}
CH_4	Methane	1.7×10^{-6}
Kr	Krypton	1.13×10^{-6}
H_2	Hydrogen	5×10^{-7}
N_2O	Nitrous oxide	3×10^{-7}
Xe	Xenon	8.7×10^{-8}
CO	Carbon monoxide	$5–20 \times 10^{-8}$
OCS	Carbonyl sulfide	5×10^{-10}
O_3	Ozone	
	Troposphere (clean)	5×10^{-8}
	Troposphere (polluted)	4×10^{-7}
	Stratosphere	1×10^{-7} to 6×10^{-6}

and variability and atmospheric lifetime. Close to sources (such as in a polluted urban setting), variability is likely to be dominated by proximity to and variations of the source. Urban data, for example, often show clearly the influence of temporal features of human activities like automobile traffic. However, when observations are made in more remote settings, sink mechanisms or lifetimes tend to become more evident in determining variability. Junge (1974) posed a hypothesis relating variability to residence time, suggesting that there is a geometric and inverse relationship between the relative standard deviation of concentration and residence time as indicated in Fig. 7-8.

Focusing on the chemical reactivity, we could list the noble gases, N_2 and perhaps O_2 as the least reactive. Even though the reaction of N_2 and O_2 in the presence of H_2O is favored thermodynamically, the reaction rate is very slow so these two species do not end up as HNO_3 in the oceans. More will be said about this in Chapter 12, on the nitrogen cycle. Reactivity being a rather unspecific term, it seems logical to organize the composition on an element-by-element basis. However, before getting to the major elements (N, S, and C), it is useful to examine

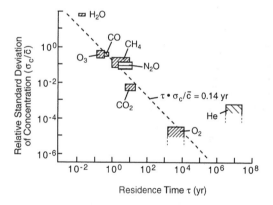

Fig. 7-8   Inverse relationship between relative standard deviation of concentration, σ_c/c, and residence time, τ, for important trace chemicals in the troposphere. (Modified with permission from C. E. Junge (1974). Residence variability of tropospheric trace gases, *Tellus* **26**, 477-488, Swedish Geophysical Society, Stockholm.)

H_2O as the most variable of the dominant species. In Table 7-2, we deliberately omitted water because of its variability. It can range from ppmv levels in the Antarctic and the stratosphere to several percent in moist tropical air. Thus, it is necessary to reference the concentrations in the table to dry air, or to devise another measure to get around the variability of water. Another scheme would be to present all the average concentrations relative to one of the more constant constituents, e.g., to nitrogen.

7.7 Atmospheric Water and Cloud Microphysics

As discussed in Chapter 6, water forms strong hydrogen bonds and these lead to a number of important features of its atmospheric behavior. All three phases of water exist in the atmosphere, and the condensed phases can exist in equilibrium with the gas phase. The equilibria between these phases is summarized by the phase diagram for water, Fig. 7-9.

We see from this diagram that partial pressures of H_2O at ordinary conditions range from very small values to perhaps 30 or 40 mbar. This corresponds to a mass concentration range up to about 25 g H_2O/m^3. In typical clouds, relatively little of this is in the condensed phase. Liquid water contents in the wettest of cumulus clouds are around a few grams per cubic meter; ordinary mid-latitude stratus clouds have 0.3–1 g/m³.

Water clouds play two key roles in biogeochemical cycles on the Earth:

1. They deliver water from the atmosphere to the Earth's surface as rain or snow, and are thus a key step in the hydrologic cycle.
2. They scavenge a variety of materials (e.g. nutrients and trace elements) from the air

and make them available for delivery in precipitation.

Thus we proceed to examine the physical–chemical nature of the cloud nucleation process.

Only two possibilities exist for explaining the existence of cloud formation in the atmosphere. If there were no particles to act as cloud condensation nuclei (CCN), water would condense into clouds at relative humidities (RH) of around 300%. That is, air can remain supersaturated below 300% with water vapor for long periods of time. If this were to occur, condensation would occur on surface objects and the hydrologic cycle would be very different from what is observed. Thus, a second possibility must be the case; particles are present in the air and act as CCN at much lower RH. These particles must be small enough to have small settling velocity, stay in the air for long periods of time and be lofted to the top of the troposphere by ordinary updrafts of cm/s velocity. Two further possibilities exist – the particles can either be water soluble or insoluble. In order to understand why it is likely that CCN are soluble, we examine the consequences of the effect of curvature on the saturation water pressure of water.

As a result of the high surface free energy of water, the vapor pressure of a water droplet increases with decreasing radius of curvature, r, as deduced by Kelvin:

$$RT \ln \frac{P}{P_\infty} = \frac{2M\sigma}{\rho r} \qquad (21)$$

where P_∞ is the water pressure over a flat surface and σ is the surface free energy (or surface tension). This, combined with vapor pressure depression due to dissolved substances (Raoult's law), results in a requirement for supersaturation (i.e., RH > 100%) as a condition for the formation of micrometer-sized droplets in clouds or fog. If the amount of soluble material is large, the supersaturation is small, while small soluble particles require higher supersaturation. Figure 7-10 shows the vapor pressure as a function of size for a variety of soluble particle masses.

These plots are called *Köhler curves* after their originator (Köhler, 1936). His assumptions that

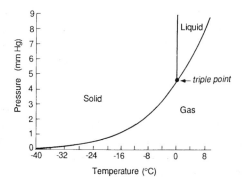

Fig. 7-9 *P–T* phase diagram for bulk water. (Based on data of the Smithsonian Meteorological Tables.)

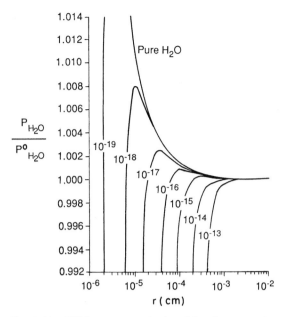

Fig. 7-10 Köhler curves calculated for the saturation ratio $P_{H_2O}/P^0_{H_2O}$ of a water droplet as a function of droplet radius r. The quantity im/M is given as a parameter for each line, where m = mass of dissolved salt, M = molecular mass of the salt, i = number of ions created by each salt molecule in the droplet.

cloud condensation nuclei (CCN) are water-soluble materials is now widely accepted. In the past, it was often thought that NaCl particles from the ocean were the main CCN; however, more recent studies have demonstrated the frequent dominance of sulfate particles with composition between H_2SO_4 and $(NH_4)_2SO_4$. Organic matter, for example the oxidation products of terpenes (from trees) also can act as CCN.

When a droplet reaches the peak of its appropriate curve, due to being in a region of RH greater than the RH for that critical size, it will continue to grow in an uncontrolled fashion. As it gets larger, the curvature effect decreases its vapor pressure and it enters a region of increased supersaturation relative to that at the peak of the Köhler curve. A particle that turns into a droplet and passes the critical size is said to be an *activated* CCN.

Following growth by condensation, droplets grow further by collision coalescence (colliding mainly due to different fall speeds). Some small amount of precipitation is produced in this fashion, recognizable as drizzle. Larger water particles and heavier precipitation occur when ice is present. Due to the ability of small droplets of liquid water to exist in a super-cooled state, most cloud water is liquid. Freezing is thought to occur due to the presence of an ice nucleating aerosol (IN), typically at temperatures of -5 to $-20°C$. Since ice has a lower vapor pressure than supercooled water at the same temperature, the ice particles grow at the expense of the droplets. Ice particles that are large enough to fall can subsequently collect larger amounts via collision with droplets with resulting graupel or hail (if the particle remains frozen) or rain (if the ice melts).

The overall rainfall rate and amount depend on these microphysical processes and even more greatly on the initial amount of water vapor present, and on the vertical motions that transport water upward, cool the air, and cause supersaturation to occur in the first place. Thus the delivery of water to the Earth's surface as one step in the hydrologic cycle is controlled by both microphysical and meteorologic processes. The global average precipitation amounts to about 75 cm/yr or 750 L/(m² yr).

Cloud nucleation also has *chemical* consequences. The soluble material of the CCN introduces solute into cloud droplets which, in many instances, is a major and even dominant ingredient of cloud and rainwater. A simple but useful expression for the amount of solute from CCN is

$$[X] = \frac{\varepsilon(X)_{air}}{M_X L} \qquad (22)$$

where $[X]$ = the average molarity of the solute X in cloud water; ε = fraction of aerosol particles of X that are activated CCN; M_X = molecular weight of X; $(X)_{air}$ = concentration of X in air entering the cloud (g/m³); L = liquid water content of the cloud (L/m²).

As an example, if 5 μg/m² of sulfate aerosol were present, and $\varepsilon \approx 1$, with $L = 1$ mL/m², then $[SO_4] = 5 \times 10^{-5}$ molar. This example is realistic for the industrialized areas such as eastern North America and Europe and for rain in clean marine areas.

In addition to solute from CCN, clouds

Table 7-3   Gaseous atmospheric sulfur compounds

Species	Concentration	Sources	Sinks
SO_2	0–0.5 ppmv (urban)	Oxidation of fossil fuel S	Direct reaction with earth surface, oxidation to sulfate
	20–200 pptv (remote)	Oxidation of S gases, volcanoes	
H_2S	0–40 pptv	Biological decay of protein in anaerobic water	Oxidation to SO_2
CH_3SH	Sub-ppbv	Plankton	Oxidation to SO_2
CH_3CH_2SH	Sub-ppbv		Oxidation to SO_2
OCS	500 pptv		Destruction in the stratosphere
CH_3SCH_3	20–200 pptv	Oceanic phytoplankton and algae	Oxidation to SO_2
CH_3SSCH_3	Small		Oxidation to SO_2
CS_2	10–20 pptv		Destruction in the stratosphere and tropospheric OH

contain dissolved gases (e.g., SO_2, NH_3, HCHO, H_2O_2, HNO_3 and many more). In turn some of these may react in the cloud droplets to form other substances which subsequently can appear in rainwater. Finally, falling raindrops can collect other materials (e.g., large dust particles) on their way to the Earth's surface. Thus, rainwater composition does not uniquely reflect the chemistry of the CCN.

7.8   Trace Atmospheric Constituents

7.8.1   *Sulfur Compounds*

There is a large variety of atmospheric sulfur compounds, in the gas, solid, and liquid phases. Table 7-3 lists a number of gaseous compounds, range of concentration, source, and sink (where known). As this list illustrates, a significant number of these gases contribute to the existence of oxidized sulfur in the forms of SO_2 and sulfate aerosol particles. Table 7-4 lists the oxyacids of sulfur and their ionized forms that could exist in the atmosphere. Of these the sulfates certainly are dominant, with H_2SO_4 and its products of neutralization with NH_3 as the most frequently reported forms.

As a result of the water solubility of both SO_2 and the sulfates, these compounds are frequently found in or associated with water in a condensed phase. The acidity of H_2SO_3 ($K_1 = 1.7 \times 10^{-2}$, $K_2 = 6.2 \times 10^{-8}$) can cause low pH in cloud and rainwater, although most measurements indicate that low rain pH is associated with SO_4^{2-} and NO^{3-}. In any case, SO_2 and SO_4^{2-} removal from air is dominated by precipitation. These points will be amplified and quantified later in Chapters 13 and 16.

There are two dominant stable isotopes of sulfur found in atmospheric sulfur compounds,

Table 7-4   Oxyacids, their salts, and ionized forms that could exist in atmospheric aerosol particles

Oxyacid	Formula	Salt/ionized form
Sulfuric	H_2SO_4	HSO_4^-, SO_4^{2-}
		NH_4HSO_4
		$(NH_4)_3H(SO_4)_2$
		$(NH_4)_2SO_4$
		$MgSO_4$
		$CaSO_4$
		Na_2SO_4
		$R-O-SO_3^-$
Sulfurous	$SO_2 \cdot H_2O$	HSO_3^-, SO_3^{2-}
Sulfonic	$R-SO_3-H$	$R-SO_3^-$
Hydroxymethane sulfonic	$CH_2(OH)SO_3H$	$CH_2(OH)SO_3^-$
Dithionic	$H_2S_2O_6$	$S_2O_6^{2-}$
Thiosulfuric	$H_2S_2O_3$	$S_2O_3^{2-}$

Table 7-5   Nitrogen atmospheric compounds

Species	Concentration	Source	Sink
NH_3	0–20 ppbv	Biological	Precipitation
$RNH_2 \ldots R_3N$	—[a]	Biological	Precipitation
N_2	78.084%	Primitive volatile, denitrification	Biological nitrification
N_2O	~ 0.3 ppmv	Biological	Photolysis in the stratosphere
(N_2O_3)	—[a]	Reaction intermediate	—[a]
NO	0–0.1 ppmv	Oxidation of N_2 in combustion	HNO_3
NO_2	0–0.1 ppmv	NO oxidation	HNO_3
HNO_2	—[a]	OH· + NO	Precipitation
HNO_3	—[a]	OH· + NO_2	Precipitation

[a] Global values not established.

^{32}S and ^{34}S. While it is attractive to utilize the ratio of these two for studies of atmospheric processes, source influences or sink mechanisms, no clearcut results have yet been demonstrated. The general features of the S isotope distributions will be summarized in Chapter 13.

7.8.2   Nitrogen Compounds

Like sulfur, nitrogen has stable compounds in a wide range of oxidation states and many of them are found in the atmosphere. Again, both gaseous and particulate forms exist as do a large number of water-soluble compounds. Table 7-5 lists the gaseous forms. The nitrogen cycle is discussed in Chapter 12.

Here we see a range of oxidation states from -3 to $+5$. The reduced forms are undoubtedly the most important gaseous bases in air, while the oxides tend to produce HNO_3 as one of the two dominant strong atmospheric acids (H_2SO_4 is the other one).

In particulate form, we find the condensed phase of some of these in the form of salts; as listed in Table 7-6.

7.8.3   Carbon Compounds

Unlike sulfur and nitrogen compounds in air where we found a wide range of oxidation states in relatively few compounds, carbon has a nearly unlimited number of compounds. These

Table 7-6   Nitrogenous aerosol constituents

Species	Concentration
$(NH_4)_2SO_4$	Up to 30 $\mu g/m^3$
NH_4HSO_4	
Amine sulfates	—
NH_4NO_3	Up to a few $\mu g/m^3$, but only in polluted air
Other nitrates	—
Organic N	—

compounds fall roughly into two categories: inorganic and organic compounds. The global carbon cycle is discussed in Chapter 11.

7.8.3.1   Elemental carbon

Common soot (also known as black carbon and refractory carbon) appears to contain significant amounts of elemental carbon in the molecular form of graphite, along with organic impurities. All forms of combustion of carbonaceous materials produce some soot, so its presence is ubiquitous in both pristine and polluted regions. The exact composition is variable and dependent on the nature of the source. All soots are, however, characterized by the presence of very small particles (submicrometer) and correspondingly a large ratio of surface area to mass. The graphitic component provides an exceedingly inert physical structure for such particles. Thus, soot particles can be transported large distances. They probably become coated with sulfates and other condensed material by Brownian coagulation, eventually come out of the atmosphere in rain and are eventually sequestered geologically in sediments.

The key features of soot are its chemical inertness, its physical and chemical adsorption properties, and its light absorption. The large surface area coupled with the presence of various organic functional groups allow the adsorption of many different materials onto the surfaces of the particles. This type of sorption occurs both in the aerosol phase and in the aqueous phase once particles are captured by cloud droplets. As a result, complex chemical processes occur on the surface of soot particles, and otherwise volatile species may be scavenged by the soot particles.

7.8.3.2   Carbon oxides

The oxides are gaseous and do not undergo reactions in the atmosphere that produce aerosol particles. Carbon monoxide is a relatively inert material with its main sinks in the atmosphere via reactions with free radicals, e.g.,

$$OH \cdot + CO \rightarrow CO_2 + H \cdot \qquad (23)$$

Other sinks – largely biological – probably exist at the Earth's surface. Seiler (1974) deduced a lifetime of ca. 0.5 year for CO, attesting to the lack of reactivity or water solubility in comparison to sulfur and nitrogen compounds.

Carbon dioxide is likewise an inert material. As a result, its only known sinks are photosynthesis and solubility in seawater. The cycle of carbon dioxide through the atmosphere will be a major focal point in Chapter 11.

Besides its inertness, CO_2 is modestly water soluble and in aqueous media forms carbonic acid which is a weak acid. Henry's Law for CO_2 states

$$P(CO_2) = K_H[H_2CO_3{}^*] \qquad (24)$$

where $K_H = 29$ atm/(mol L). ($H_2CO_3{}^*$ is the sum of H_2CO_3 and CO_2(aq). See Chapter 5.) Thus, 350 ppmv of CO_2 results in $[H_2CO_3{}^*] = 10^{-5}$ M in otherwise pure H_2O, as is often assumed for clouds and rain. Given that $K_1 = 4.5 \times 10^{-7}$ mol/L and $K_2 = 4.7 \times 10^{-11}$ mol/L for the first and second dissociations of H_2CO_3, cloud and rainwater affected only by H_2CO_3 would have a pH of about 5.6. However, other solutes (notably H_2SO_4) usually dominate the effect of H_2CO_3, even in pristine locations.

Most CO and CO_2 in the atmosphere contain the mass 12 isotope of carbon. However, due to the reaction of cosmic ray neutrons with nitrogen in the upper atmosphere, ^{14}C is produced. Nuclear bomb explosions also produce ^{14}C. The ^{14}C is oxidized, first to ^{14}CO and then to $^{14}CO_2$ by OH· radicals. As a result, all CO_2 in the atmosphere contains some ^{14}C, currently a fraction of ca. 10^{-12} of all CO_2. Since ^{14}C is radioactive (β-emitter, 0.156 MeV, half-life of 5770 years), all atmospheric CO_2 is slightly radioactive. Again, since atmospheric CO_2 is the carbon source for photosynthesis, all biomass contains ^{14}C and its level of radioactivity can be used to date the age of the biological material.

7.8.3.3   Organic carbon

The remaining carbon compounds fall into the category of organic molecules. The number of identified species is large – at least several hundred – so we cannot produce an exhaustive list here. Instead we will list molecular forms following conventional schemes for organic chemistry with a few selected samples.

Tables 7-7–7-11 give a sense of the range of organic molecules present in the atmosphere. Both natural sources and human activity contribute to the variety of organic molecules (Graedel, 1978). The sinks often involve *in-situ* reactions.

7.8.4   Other Trace Elements

The atmosphere may be an important transport medium for many other trace elements. Lead and other metals associated with industrial activity are found in remote ice caps and sediments. The transport of iron in wind-blown soil may provide this nutrient to remote marine areas. There may be phosphorus in the form of *phosphine*, PH_3, although the detection of volatile phosphorus has not been convincingly or extensively reported to date.

Table 7-7 Atmospheric hydrocarbons

Class	Compound	Typical source	Probable sink	Conc.
Alkanes	Methane	Microbes natural gas	OH	1.8 ppmv
	Ethane	Auto	OH	0–100 ppbv
	Hexane	Auto	OH	0–30 ppbv
	(Up to C_{37})	—	—	—
Alkenes	Ethene	Auto, microb., vegetation	OH, O_3	1–1000 ppbv
	Isoprene	Trees, auto	OH	0.2–30 ppbv
Alkynes	Acetylene	Auto, microbes	OH	0.2–200 ppbv
Terpenes	a-Pinene	Trees	O_3, OH	0–1 ppbv
	Limonene	Trees	O_3, OH	0–1 ppbv
Cyclic hydrocarbons	Cyclopentane	Auto	OH	0–10 ppbv
	Cyclohexane	Auto	OH	0–10 ppbv
	Cyclopentene	Auto	O_3, OH	0–10 ppbv
Aromatic hydrocarbons	Benzene	Auto	OH, O	—
	Toluene	Auto	OH	0–100 ppbv
Polynuclear aromatic hydrocarbons	Phenanthrene	Aluminum manufacture, combustion		0–300 ng/m^3
	Benzo(a)pyrene	Auto, wood combustion		0–100 ng/m^3

Table 7-8 Oxygenated organic compounds

Class	Example	Typical source	Probable sink	Conc.
Aliphatic aldehydes	Formaldehyde		hv, OH, rain	1–100 ppbv
	Acetaldehyde	Animal waste	hv, OH	1–10 ppbv
Olefinic aldehydes	Acrolein	Auto	O_3	0–1 ppbv
Aromatic aldehydes	4-methyl-benzaldehyde	Auto	—	0–300 pptv
Aliphatic ketones	Acetone	Animal waste	hv	0–10 ppbv
Cyclic ketones	1,2-cyclopentane	Known plant	—	No data
Aromatic ketones	Acetophone	Auto plant emissions	—	No data
Aliphatic acids	Formic acid	—	Rain	0–100 ppbv
	Pentanedioic acid	Cyclopentene	Rain	0–1 mg/m^3
Cyclic acids	Pinonic acid	a-Pinene + O_3	Rain	
Aromatic acids	Benzoic acid	Auto	Rain	0–400 ng/m^3
Aliphatic alcohols	Methanol CH_3OH	Animal waste	—	0–100 ppbv
Phenols	Phenol	Auto	—	0–3 ppbv

Table 7-9 Atmospheric nitrogen-containing organic compounds

Class	Example	Typical source	Probable sink	Conc.
Amines	Methylamine	Protein decay	H_2SO_4 aerosol	—
Nitrates	Peroxy acetyl nitrate	Photochemical	—	0–100 ppbv
Heterocyclic N compounds	Benzo[f]quinoline	Coal combustion	—	0–200 pg/m^3

Table 7-10 Atmospheric sulfur-containing organic compounds

Class	Example	Typical source	Probable sink	Conc.
Mercaptans	Methyl mercaptan	Microbiota	Oxidation to SO_2 by OH	0–4 ppbv
Sulfides	Dimethyl sulfide	Algae	Oxidation to SO_2 by OH	20–300 pptv

Table 7-11 Atmospheric halogen-containing organic molecules

Class	Example	Typical source	Probable sink	Conc.
Halogenated aliphatic compounds	CH3Cl	Biological	OH	620 pptv
	CH2Cl2	Industrial	OH	30 pptv
	CH_3CCl_3	Solvent	—	140 pptv
	CCl_4	Solvent	—	130 pptv
	$CFCl_3$ (CFC-11)	Propellant, refrigerant	Stratosphere	220 pptv
	CF_2Cl_2 (CFC-12)	Refrigerant	Stratosphere	375 pptv
Halogenated aromatic compounds	DDT	Pesticide	—	0.009–500 ng/m^3

7.9 Chemical Interactions of Trace Atmospheric Constituents

Unlike the chemistry of simple mixtures of small numbers of reactants as observed in the laboratory, the chemistry of the atmosphere involves complex interactions of large numbers of species. However, several key aspects of these interactions have been identified that account for major observable properties of the atmospheric chemical system. It is convenient to separate the description into gas phase and condensed phase interactions, not the least because different chemical and physical processes are involved in these two cases.

7.9.1 Gas Phase Interactions

Figure 7-11 and its caption (Crutzen, 1983) depict the most important of the gas phase and photochemical reactions in the atmosphere. Perhaps the single most important interaction involves the hydroxyl free radical, OH·. This extremely reactive radical is produced principally from the reactions of electronically excited atomic oxygen, $O(^1D)$, with water vapor. Photo-dissociation of ozone produces $O(^1D)$ and also the less reactive $O(^3P)$. In the troposphere, O_3 is produced largely by photochemical reactions involving still other free radicals, including the nitrogen oxides, NO and NO_2. OH· appears to be the dominant oxidant for CO, CH_4, SO_2, $(CH_3)_2S$ as well as the main source of HNO_3 and HNO_2.

7.9.2 Condensed Phase Interactions

Condensed phase interactions can be divided roughly into two further categories; chemical and physical. The latter involves all purely physical processes such as condensation of species of low volatility onto the surfaces of aerosol particles, adsorption, and absorption into liquid cloud and rainwater. Here, the interactions may be quite complex. For example, cloud droplets require a CCN, which in many instances is a particle of sulfate produced from SO_2 and gas–particle conversion. If this particle is strongly acidic (as is often the case) HNO_3 will not deposit on the aerosol particle; rather, it will be dissolved in liquid water in clouds and rain. Thus, even though HNO_3 is not very soluble in

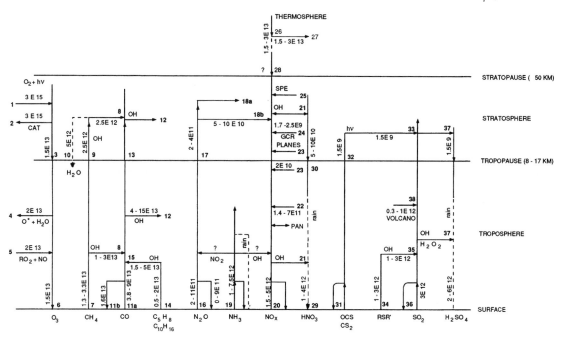

Fig. 7-11   Compilation of the most important photochemical processes in the atmosphere, including estimates of flux rates expressed in moles per year between the earth's surface and the atmosphere and within the atmosphere. (Modified with permission from P. J. Crutzen, Atmospheric interactions – homogeneous gas reactions of C, N, and S containing compounds. *In* B. Bolin and R. Cook (1983). "The Major Biogeochemical Cycles and Their Interactions," pp. 67–112, John Wiley, Chichester.)

1:
$$O_2 + h\nu \rightarrow 2O$$
$$O + O_2 + M \rightarrow O_3 + M \ (2x)$$

$$3O_2 \rightarrow 2O_3$$

2:
$$O_3 + h\nu \rightarrow O + O_2$$
$$XO + O \rightarrow X + O_2$$
$$X + O_3 \rightarrow XO + O_2$$

$$2O_3 \rightarrow 3O_2$$
$$X = NO, Cl, OH$$

3:   Downward flux to troposphere, small difference between 1 and 2

4:
$$O_3 + h\nu \rightarrow O(^1D) + O_2$$
$$O(^1D) + H_2O \rightarrow 2\,OH$$
$$HO_2 + O_3 \rightarrow OH + 2O_2$$

5:
$$RO_2 + NO \rightarrow RO\,NO_2$$
$$NO_2 + h\nu \rightarrow NO + O$$
$$O + O_2 + M \rightarrow O_3 + M$$

$$RO_2 + O_2 \rightarrow RO + O_3$$
$R = H$, CH_3, etc. From CO and hydrocarbon oxidation, e.g.
$$CO + OH \rightarrow H + CO_2$$
$$H + O_2 + M \rightarrow HO_2 + M$$

6:   Ozone destruction at ground; difference between 3, 4, and 5.

7:   Release of CH_4 at ground by variety of sources with range 1.3–3.3 $\times$ 10^{13} mol/yr. 2.5 $\times$ 10^{13} with average OH concentration of 6 $\times$ 10^5 molecules/cm^3.

8:
$$CH_4 + OH \rightarrow CH_3 + H_2O$$
$$CH_3 + O_2 + H \rightarrow CH_3O_2 + M$$
$$CH_3O_2 + NO \rightarrow CH_3O + NO_2$$
$$CH_3O + O_2 \rightarrow CH_2O + HO_2$$
$$CH_2O + h\nu \rightarrow CO + H_2$$
(and other oxidation routes)

9:   Flux of CH_4 to the stratosphere.

10:   Flux of H_2O to the troposphere from methane oxidation.

11a: Release of CO from a variety of sources, mostly man-made.

11b: Uptake of CO by microbiological processes in soils.

12:
$$CO + OH \rightarrow H + CO_2$$
Global loss of CO of 4–15 $\times$ 10^{13} mol/yr; 7 $\times$ 10^{13} calculated with [OH] of 6 $\times$ 10^5 molecules cm^{-3}.

Continued on p. 152

15: Isoprene and terpene oxidation to CO following reaction with OH.

16, 17: Release of N_2O to atmosphere by variety of sources; no significant sinks of N_2O in the troposphere have been discovered; stratospheric loss estimated by model calculations.

18a:
$$N_2O + hv \rightarrow N_2 + O$$
$$N_2O + O(^1D) \rightarrow N_2 + O_2$$

18b:
$$N_2O + O(^1D) \rightarrow 2NO$$

19: Release of NH_3 by variety of sources to the atmosphere; redeposition at the ground; most ammonia is removed by rain, but some NO_x loss and N_2O formation may be possible by gas phase reactions.

20: Release of NO_x at the ground by a variety of sources – redeposition at the ground.

21:
$$NO_2 + OH \rightarrow HNO_3$$
$$HNO_3 + hv \rightarrow OH\ NO_2$$
$$HNO_3 + OH \rightarrow H_2O + NO_3$$

22: NO_x produced from lightning.

23: NO_x produced by subsonic aircraft.

24: NO_x produced from galactic cosmic rays.

25: NO_x from sporadic solar proton events; maximum production recorded in August, 1972: 1×10^{10} moles produced.

26: NO production by fast photoelectrons in the thermosphere and by auroral activity.

27:
$$NO + N \rightarrow N_2 + O$$

28: Downward flux of NO to the stratosphere; a small difference between 26 and 27 may be important for this.

31, 32: COS destruction in the stratosphere calculated with a model; uptake of COS in the oceans and hydrolysis may imply an atmospheric lifetime of only a few years and a source of a few tens of billions (10^{10}) of moles per year.

33:
$$COS + hv \rightarrow S + CO$$
$$S + O_2 \rightarrow SO + O$$
$$SO + O_2 \rightarrow SO_2 + O$$

34: Release of H_2S, CH_3SCH_3 (DMS), and CH_3SH by biological processes in soils and waters.

35: Oxidation of H_2S, DMS, and CH_3SH to SO_2 after initial attack by OH.

36: Industrial release of SO_2.

37: SO_2 oxidation to H_2SO_4 on aerosols, in cloud droplets, and by gas phase reactions following attack by OH.

38. Volcanic injections of SO_2, averaged over past centuries.

---

the concentrated H_2SO_4 of the aerosol particle, the atmospheric residence time of HNO_3 is in part determined by the physical role of H_2SO_4 particles as CCN.

Chemical interactions also occur in the condensed phases. Some of these are expected to be quite complex, e.g., the reactions of free radicals on the surfaces of or within aerosol particles. Simpler sorts of interactions also exist. Perhaps the best understood is the acid–base relationship of NH_3 with strong acids in aerosol particles and in liquid water (see Chapter 16). Often, the main strong acid in the atmosphere is H_2SO_4, and one may consider the nature of the system consisting of H_2O (liquid), NH_3, H_2SO_4, and CO_2 under realistic atmospheric conditions. Carbon dioxide is not usually important to the acidity of atmospheric liquid water (Charlson and Rodhe, 1982); the dominant effects are due to NH_3 and H_2SO_4. The sensitivity the pH of cloud (or rainwater produced from it) to NH_3 and SO_4^{2-}

aerosol acting as CCN in a cloud are discussed further in Chapter 16.

7.10   Physical Transformations of Trace Substances in the Atmosphere

Perhaps because the unpolluted atmosphere can appear to be perfectly free of turbidity, it is not immediately obvious that it is a mixture of solid, gaseous, and liquid phases – even in the absence of clouds. Particles in the *aerosol** state constitute only a miniscule portion of the mass of the atmosphere – perhaps 10^{-9} or 10^{-10} in unpolluted cases. However, the condensed phases are important intermediates in the cycles of numerous elements, notably ammonia-N, sul-

* *Aerosols* are solid or liquid particles, suspended in the liquid state, that have stability to gravitational separation over a period of observation. Slow coagulation by Brownian motion is implied.

fate-S, and organic C. They are also absolutely necessary participants in the hydrologic cycle (see Sections 7.7 and 7.11).

Figure 7-12 depicts the main physical pathways by which aerosol particles are introduced into and removed from the air. Processes that occur within the atmosphere also transform particles as they age and are transported. This form of distribution of mass with size was originally discovered in polluted air in Los Angeles, but it is now known to hold for remote unpolluted locations as well (Whitby and Sverdrup, 1980). In the latter case, the

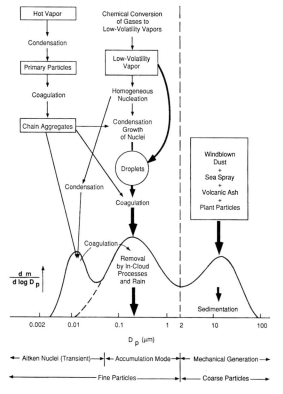

Fig. 7-12 Schematic of an atmospheric aerosol size distribution. This shows the three mass modes, the main sources of mass for each mode, and the principal processes involved in inserting mass into and removing mass from each mode (m = mass concentration, D_p = particle diameter). (Reproduced with permission from K. T. Whitby and G. M. Sverdrup (1983). California aerosols: their physical and chemical characteristics. *In* "The Character and Origin of Smog Aerosols" (G. M. Hidy, P. K. Mueller, D. Grosjean, B. R. Appel, and J. J. Wesolowski, eds), p. 483, John Wiley, New York.)

particle sizes of the accumulation mode (see Fig. 7-13) are probably somewhat smaller, perhaps by a factor of two.

Much of the fine particle aerosol is produced in the atmosphere by chemical reactions of gaseous precursors. Following the formation of very small *nuclei* (diameter less than about 0.1 μm) by chemical processes (e.g. the oxidation of SO_2 to H_2SO_4), the physical process of Brownian coagulation and deposition of additional reaction condensates cause the mass of nuclei to move up to dry diameters of 0.1–1.0 μm. The fine particle aerosol is thus composed of nuclei and larger conglomerates of material that has been accumulated. This is commonly called the *accumulation mode*. In addition, a *coarse* particle aerosol also exists, largely comprising seasalt and soil dust of surface derivation, with diameters larger than ca. 1 μm. Except in cloud droplets, there is limited chemical contact between the coarse mode and the accumulation mode.

Because the particles in the accumulation mode are very small (most of them have diameters less than 1 μm when dry), they have very small fall speeds (a 1 μm sphere of unit density has a fall speed of about 10^{-2} cm/s). Thus, they are only removed in any quantity by the formation of clouds with subsequent precipitation.

This brief description leads to Fig. 7-13 which depicts the physical transformations of trace substances that occur in the atmosphere. These physical transformations can be compared to the respective chemical transformations within the context of the individual elemental cycles (e.g., sulfur). This comparison suggests that the overall lifetime of some species in the atmosphere can be governed by the chemical reaction rates, while others are governed by these physical processes.

7.11 Influence of Atmospheric Composition on Climate

7.11.1 CO₂

Climate may be defined as the aggregate of all physical atmospheric properties and conditions. As such, it is absolutely clear that the chemical

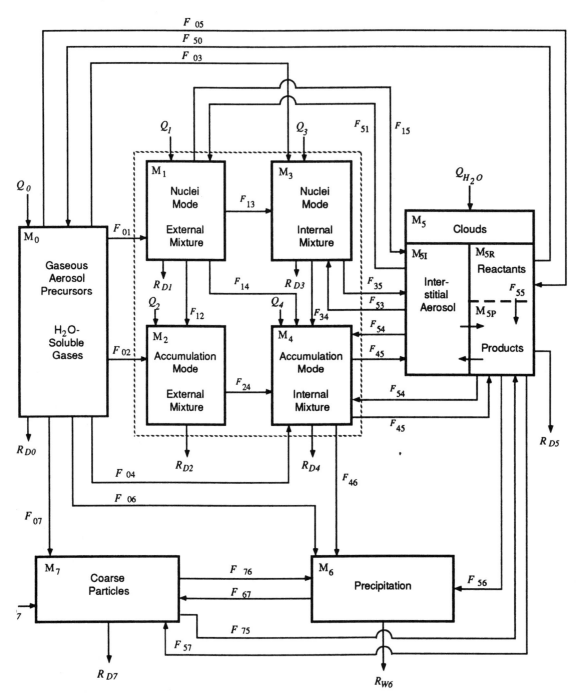

Fig. 7-13 Physical transformations of trace substances in the atmosphere. Each box represents a physically and chemically definable entity. The transformations are given in F_{ij} (from the ith to the jth box). Q_i represents sources contributing to the mass or burden, M_i, in the ith box. R_{Di} and R_{Wi} are dry and wet removals from M_i. The dashed box represents what may be called the fine-particle aerosol and could be a single box instead of the set of four sub-boxes ($i = 1, 2, 3, 4$). The physical transformations are as follows:

Continued on p. 155

F_{01}: Production of new nuclei-mode particles; F_{02}: Growth of existing accumulation-mode particles by the deposition of products of chemical reactions; F_{03}: Growth of pre-existing nuclei-mode particles, as in F_{02}; F_{04}: Growth of internally mixed accumulation mode, as in F_{02}; F_{05}: Dissolution of gaseous reactants in cloud drops; F_{50}: Reverse of F_{05} evaporation or gaseous exchange or both; F_{06}: Below-cloud scavenging of gaseous reactants or reactant products; F_{07}: Interaction of gases with coarse particles, e.g., $HNO(g) + seasalt \rightarrow coarse\ mode\ NO_3^-$; F_{12}: Brownian coagulation of nuclei-mode particles with themselves to produce accumulation-size (chemically) externally mixed particles; F_{13}: Adsorption, condensation; F_{14}: Coagulation; F_{15}: Cloud formation yielding nuclei-mode interstitial aerosol, coagulation with cloud droplets, or (unlikely) activation as a cloud condensation nuclei (CCN); F_{51}: Cloud evaporation releasing interstitial aerosol; F_{57}: Cloud evaporation releasing coarse particles; F_{24}: Adsorption, coagulation, condensation; F_{34}: Adsorption, coagulation, condensation; F_{35}: Cloud formation, as in F_{15}; F_{53}: Cloud evaporation, as in F_{51}; F_{45}: Cloud formation, likely to be CCN; F_{54}: Cloud evaporation, releasing CCN; F_{55}: Reactants in cloud water producing changes in solute mass; F_{46}: Below-cloud scavenging; F_{56}: Formation of precipitation; F_{67}: Evaporation of precipitation particles (raindrops) before reaching ground; F_{75}: Coarse particles acting as CCN; F_{76}: Below-cloud scavenging of coarse-mode particles; R_{D5}: Occult precipitation (deposition of cloud droplets directly to the Earth's surface, trees, etc.)

(Reproduced with permission from R. J. Charlson, W. L. Chameides, and D. Kley (1985). The transformations of sulfur and nitrogen in the remote atmosphere. *In* "The Biogeochemical Cycling of Sulfur and Nitrogen in the Remote Atmosphere" (J. N. Galloway, R. J. Charlson, M. O. Andreae and H. Rodhe, eds), pp. 67–80, D. Reidel Publishing Company, Dordrecht.)

composition of the atmosphere as well as the physical characteristics of condensed phase trace species are of leading importance as determinants of climate. A well-known example is the increase in the temperature of the Earth's surface due to the absorption of infrared radiation from the Earth's surface by CO_2 in the air. (See the box describing the greenhouse effect.) Without CO_2 the Earth's surface would be several degrees cooler than at present, depending on cloud cover, water vapor, and other controlling factors. Of course, there is substantial concern over the secular increase of CO_2, which will double from its pre-industrial level by the early to mid-21st century.

7.11.2 Other "Greenhouse Gases"

Carbon dioxide is not the only gas that can influence terrestrial infrared radiation, and infrared absorption is not the only way that composition influences climate. Other gases that are important for their infrared absorption, sometimes known as "greenhouse gases," include CH_4, CCl_2F_2 (CFC-12), $CFCl_3$ (CFC-11), N_2O, and O_3. Taken together these other species are about of equal importance to CO_2. That

some of these have increased dramatically due to human activities (the chlorofluorocarbons are only synthesized by man) or are increasing due to unknown causes (e.g., CH_4) suggests that the overall problem involves much more than just understanding or predicting CO_2.

7.11.3 Particles and Clouds

The condensed phases also are important to the physical processes of the atmosphere; however, their role in climate poses an almost entirely open set of scientific questions. The highest sensitivity of physical processes to atmospheric composition lies within the process of cloud nucleation. In turn, the albedo (or reflectivity for solar light) of clouds is sensitive to the number population and properties of CCN (Twomey, 1977). At this time, it appears impossible to predict how much the temperature of the Earth might be expected to increase (or decrease in some places) due to known changes in the concentrations of gases because aerosol and cloud effects cannot yet be predicted. In addition, since secular trends in the appropriate aerosol properties are not monitored very extensively there is no way to know

The Greenhouse Effect

Although greenhouses actually don't work the same way, this effect is so named because the result is the same; that is, the surface temperature of the Earth is increased because of the presence in air of gases that absorb infrared radiation. The primary "greenhouse gas" is water vapor, which accounts for much of the observed increase of the average surface temperature above that expected for this planet without an atmosphere. If the albedo (reflectivity) of Earth were 0.17 (the same as Mars), and without any infrared absorption in the atmosphere, the temperature would be ca. 260 K. Instead, it averages about 283 K. Increasing the CO_2 level from a 1980 level of 339 ppmv to an expected level in 2030 of 450 ppmv is calculated to result in an increase of only ca. $+0.7$ K. However, climate as we know it is a very sensitive function of temperature such that even a 1 K temperature increase would cause large changes if it persisted for a few decades. Predicted effects include melting of substantial amounts of continental ice, thermal expansion of the surface seawater and increases in sea level. Thus, it is important to understand the influence of all contributors to the "greenhouse effect."

Figure 7-14 shows one calculation of the expected temperature change profiles for 2030 as a function of altitude. Here, another key feature of the infrared interactions is evident. In the troposphere, the temperature increase is expected to be relatively uniform with height, due largely to the vertical mixing of the troposphere by convective motions. In the stratosphere the effect is just the opposite. The gases that absorb IR also emit energy to space, such that the stratosphere becomes cooler as the concentration of greenhouse gases increases.

Finally, we might ask why it is possible for a gas like $CFCl_3$ (CFC-11) to have any effect at all when its concentration is only ca. 0.2 ppbv? CO_2 is much more abundant (365 ppmv in 1996) and water vapor is dominant at the percent level. The reason for the sensitivity of the "greenhouse effect" to such gases lies in the details of their infrared absorption spectrum. Specifically, gases that absorb strongly within that part of the infrared region where water vapor and CO_2 do not absorb strongly are the ones that can have the biggest effect. The so-called "window" region between about 7 and 12 μm wavelength is of particular importance because (a) water vapor and CO_2 do not absorb there and (b) the Earth's surface emission is a maximum at around 10 μm. Further details on the greenhouse effect may be found in Goody and Yung (1989), Goody and Walker (1972) and Ramanathan *et al.* (1985). An excellent set of reviews can be found in the assessments of the Intergovernmental Panel on Climate Change, IPCC (1995). More details on the role of the greenhouse effect on climate can be found in Chapter 17.

---

the degree to which changes have occurred, e.g., due to human activity.

7.12 Chemical Processes and Exchanges at the Lower and Upper Boundaries of the Atmosphere

7.12.1 *CO₂, Photosynthesis, and Nutrient Exchange*

The atmosphere, as a single body of gas, is in physical contact with the entire surface of the Earth. It extends upward toward space, with density decreasing by roughly a factor of 10 every 16 km. Some processes of importance to geochemistry occur at the lower and upper boundaries. Evaporation of water from the Earth's surface is the first big step in the hydrologic cycle of the atmosphere. Indeed, this flux of water into and out of the atmosphere represents its largest flux and most massive cycle, as discussed in Chapter 6. Next is the exchange of CO_2 from the atmosphere to the biosphere via photosynthesis, with the return flow of CO_2 to the atmosphere via respiration, decay, and combustion. Many other trace substances also exchange through the biosphere in natural pro-

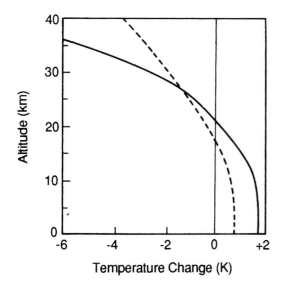

Fig. 7-14   Change in the vertical distribution of temperature due to an increase in CO_2 alone, and CO_2 along with other radiatively important trace gases. (Reproduced with permission from Ramanathan *et al.* (1985), with the permission of the American Geophysical Union.)

cesses that occur at the Earth's surface, notably nitrogen and sulfur species.

7.12.2   Reactions at the Surface

Another major process at the Earth's surface not involving rapid exchange is the chemical weathering of rocks and dissolution of exposed minerals. In some instances the key weathering reactant is H_3O^+ in rainwater (often associated with the atmospheric sulfur cycle), while in other cases H_3O^+ comes from high concentrations of CO_2, e.g., in vegetated soils.

Numerous atmospheric species react with the Earth's surface, mostly in ways that are not yet chemically described. The dissolution and reaction of SO_2 with the sea surface, with the aqueous phase inside of living organisms or with basic soils is one example. Removal of this sort from the atmosphere usually is called *dry removal* to distinguish it from removal by rain or snow. In this case, the removal flux is often empirically described by a *deposition velocity*, V_{dep}:

$$V_{dep} = \frac{F}{C} \qquad (25)$$

where F = the flux per unit area and time in appropriate units; C = concentration in the atmosphere, in units to match that of the flux.

At the Earth's surface, V_{dep} for many reactive species (e.g. SO_2, NO_2, O_3, HNO_3, etc.) is of the order of 1 cm/s.

7.12.3   Cosmic-Ray-Induced Nuclear Reactions

At the top of the atmosphere, more properly at altitudes where the density is sufficiently low, high-energy cosmic ray particles cause nuclear chemical reactions with important products. The production of radioactive ^{14}C (or radiocarbon) already has been mentioned.

Other radioisotopes known to be produced by cosmic rays include ^{10}Be, ^{3}H, ^{22}Na, ^{35}S, ^{7}Be, ^{33}P, and ^{32}P. Of these ^{35}S, ^{7}Be, ^{32}P, and ^{33}P have activities that are high enough to be measured in rainwater. In several instances, notably ^{14}C and ^{7}Be, these radioactive elements are useful as tracers.

7.12.4   Escape of H and He

As mentioned at the beginning of this chapter, diffusive separation of low atomic or molecular weight species into space causes them to be permanently lost from the Earth. Thus, the Earth is deficient in He and H_2 relative to the best estimates of initial terrestrial composition. Some species might be accreted from space; certainly, micrometeorites represent a small but identifiable flux. Published speculations exist regarding other substances, notably water. However, these would appear to be relatively unimportant at present.

References

Chapman, S. (1930). On ozone and atomic oxygen in the upper atmosphere. *Phil. Mag. S.7.* **10** (64), 369–383.

Charlson, R. J. and Rodhe, H. (1982). Factors influencing the natural acidity of rainwater. *Nature* **295**, 683.

Crutzen, P. J. (1983). Atmospheric interactions – homogeneous gas reactions of C, N and S containing compounds. *In* "The Major Biogeochemical Cycles and Their Interactions" (B. Bolin and R. Cook, eds). Scope 21, Wiley, Chichester.

Goody, R. M. and Walker, J. C. G. (1972). "Atmospheres." Prentice-Hall, Englewood Cliffs, NJ.

Goody, R. M. and Yung, Y. L. (1989). "Atmospheric Radiation." Oxford University Press, New York.

Graedel, T. E. (1978). "Chemical Compounds in the Atmosphere." Academic Press, New York.

Intergovernmental Panel on Climate Change (IPCC) (1995). "Climate Change 1995. The Science of Climate Change" (J. T. Houghton, L. G. M. Filho, B. A. Calander, N. Harris, A. Kattenberg, and K. Maskell, eds). Cambridge University Press, Cambridge.

Junge, C. E. (1963). "Air Chemistry and Radioactivity." Academic Press, New York.

Junge, C. E. (1974). Residence variability of tropospheric trace gases. *Tellus* **26**, 477–488.

Kent, G. S., Osborn, M. T., Trepte, C. R., and Skeens, K. M. (1997). LITE measurements of aerosols in the stratosphere and upper troposphere. *In* "Advances in Atmosphereic Remote Sensing with Lidar, Selected Papers of the 18th International Laser Radar Conference (ILRC), Berlin, 22–26 July, 1996" (A. Ansmann, R. Neuber, P. Rairoux, and U. Wandinger, eds), pp. 157–160.

Köhler, H. (1936). The nucleus in and the growth of hygroscopic droplets. *Trans. Faraday Soc.* **32**, 1152.

Ramanathan, V., Cicerone, R. J., Singh, H. B., and Kiehl, J. T. (1985). Trace gas trends and their potential role in climate change. *J. Geophys. Res.* **90**, 5547–5566.

Seiler, W. (1974). The cycle of atmospheric CO. *Tellus* **26**, 116–135.

Twomey, S. (1977). "Atmospheric Aerosols." Elsevier, Amsterdam.

Wallace, J. M. and Hobbs, P. V. (1977). "Atmospheric Sciences: An Introductory Survey." Academic Press, New York.

Wayne, R. P. (1985). "Chemistry of Atmospheres." Clarendon Press, Oxford.

Whitby, K. T. and Sverdrup, G. M. (1980). California aerosols: their physical and chemical characteristics. In "The Character and Origins of Smog Aerosols" (G. M. Hidy *et al.*, eds). Wiley, New York.

8

Soils, Watershed Processes, and Marine Sediments

David R. Montgomery, Darlene Zabowski, Fiorenzo C. Ugolini, Rolf O. Hallberg, and Henri Spaltenstein

8.1 Introduction

The surface of the Earth is a dynamic place. Over geologic time, rocks uplifted above sea level break down and are converted into soils by weathering processes. Soils release soluble components into rivers and can be eroded and transported across landmasses until both soluble and particulate components are eventually deposited in marine sedimentary basins. Once buried, high pressures and temperatures gradually convert the sediment back to rock. Tectonic processes can then uplift the new rock and expose it again at the Earth's surface, resulting in a cycle of uplift, erosion, deposition, burial, and renewed uplift called the *rock cycle*. The rock cycle continually modifies the Earth's surface (Fig. 8-1). Soils are an especially reactive component of the Earth's surface. They not only provide nutrients and water for terrestrial ecosystems, but they also store and exchange gases with the atmosphere, and affect the movement of surface and groundwater. Thus, soils affect and are affected by the biosphere, atmosphere, and hydrosphere.

Soil is a key component of the rock cycle because weathering and soil formation processes transform rock into more readily erodible material. Rates of soil formation may even limit the overall erosion rate of a landscape. Erosion processes are also a key linkage in the rock cycle

between soil production and the filling of sedimentary basins. When water falls onto the Earth's surface it can seep into the ground and percolate down to the water table or it can run off downslope to collect into streams that ultimately combine to form rivers. Flowing water transports eroded material until it is either deposited in local depositional environments or delivered to the oceans. If there was no sink for these sediments, the oceans would fill up in less than 100 Myr, and if there was no source for uplift of rocks on the continents they would be

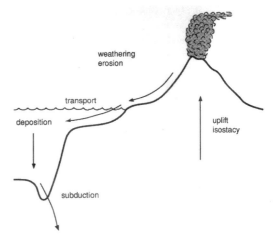

Fig. 8-1 Generalized cycle of sediments and sedimentary rocks.

Earth System Science
ISBN 0-12-379370-X

degraded to ocean level in less than 50 Myr (Holland, 1978, p. 146). Global tectonic activity prevents these scenarios through rock uplift and the return of sediments accumulated on the ocean floor back to the continents either by accretion or subduction at plate margins.

The material transformations and interactions that occur between soil, sediment, rock, water and the atmosphere during geological cycles of uplift and erosion are important in global biogeochemical cycles. Over geologic time, for example, material incorporated into sedimentary rocks can be sequestered in long-term storage in marine basins. In contrast, weathering products are vented to the atmosphere over much shorter time scales. The development of soils, erosional processes, and deposition in marine basins all play key roles in global biogeochemical cycles.

8.2   Weathering

Weathering occurs because rocks and minerals become exposed to physical and chemical conditions that differ from conditions under which they formed. Rocks form at higher temperatures and pressures than that of the surficial environment so they are unstable at the temperatures and pressure of the Earth's surface and are thus susceptible to weathering. The inorganic solid phase of any soil consists of a number of minerals displaying different degrees of weathering susceptibility. The extent of weathering of these minerals depends on the stabilities of the minerals and the physical and chemical environment to which the minerals are exposed in the soil or at surface conditions, including the supply of water and the removal or transport of weathering products (Garrels and Christ, 1965; Rai and Kittrick, 1989; Colman and Dethier, 1986).

Weathering can be separated into two types: physical and chemical. Physical weathering involves changes in the degree of consolidation with little or no chemical and mineralogical changes of rocks and minerals. Chemical weathering involves changes in chemical and mineralogical composition that generally act on the surfaces of rocks or minerals. Physical weathering increases the surface area of rocks and

minerals such that chemical weathering can proceed at a faster rate. In nature these two processes occur concurrently and are difficult to separate (Jackson and Sherman, 1953; Birkeland, 1999).

8.2.1   Physical Weathering

Rocks and minerals break when stressed above their tensile strength. Commonly, rocks fracture along joints, fissures, or planes that have developed during cooling, tectonism, and sedimentary processes or along lines of weakness at the boundaries between mineral grains. When previously buried rock masses are exposed at the Earth's surface, the lowering of the overburden pressure, or unloading, allows the rocks to expand. This expansion induces fracturing that aids in the conversion of rock to soil. Physical weathering processes expand these fractures or cause the development of new ones.

Frost wedging is the prying apart of materials by expansion of water when it freezes. The pressure produced by freezing water is well above the tensile strength of many rocks; however, this pressure may not be commonly attained in nature because rocks are not completely saturated but contain air gaps. Hydration shattering, the ordering and disordering of water molecules adsorbed at the surface of rocks, may be responsible for processes ascribed to frost wedging (Dunn and Hudec, 1972; Hudec, 1974). Nonetheless, the presence of shattered bedrock, and generally angular rock debris in cold environments provides sufficient evidence that frost wedging is at work. Laboratory experiments suggest that repeated freeze–thaw cycles can even produce clay-sized particles (Lautridou and Ozouf, 1982).

In arid environments, where the soluble products of weathering are not completely removed from the soil, saline solutions may circulate in the soil as well as in rock fractures. If upon evaporation the salt concentration increases above its saturation point, salt crystals form and grow (Goudie *et al.*, 1970). The growth of salt crystals in crevices can force open fractures. Salt weathering occurs in cold or hot deserts or areas where salts accumulate. Boulders, blocks, and cliffs affected by salt weathering display cavities and holes and sometimes acquire grotesque forms, as observed in the cold desert of Antarctica (Ugolini, 1986). Frost and salt weathering combined have a synergistic effect that could be more effective at breaking down rocks than salt or frost alone (Williams and Robinson, 1981).

Thermal expansion induced by insolation may be important in desert areas where rocky outcrops and soil surfaces are barren. In a desert, daily temperature excursions are wide and rocks are heated and cooled rapidly. Each type of mineral in a rock has a different coefficient of thermal expansion. Consequently, when a rock is heated or cooled, its minerals differentially expand and contract, thereby inducing stresses and strains in the rock and causing fractures. Ollier (1969) discussed examples of rock weathering due to insolation. Fire can develop temperatures far in excess of insolation and be quite effective in fracturing rocks (Blackwelder, 1927).

Plants and animals disrupt and disaggregate rocks and fracture or abrade individual grains or minerals. Endolithic algae growing in deserts may be capable of disintegrating rocks through shrinking and swelling (Friedman, 1971). Lichens are effective agents in physical weathering by extending fungal hyphae into rocks and by expansion and contraction of the thalli (Syers and Iskandar, 1973). Higher plants grow roots in rock crevices and eventually the increased pressure breaks and disrupts the substratum. In addition, the physical mixing of rock and soil that occurs from tree throw is a primary process in the conversion of bedrock into soil in forested regions. Earthworms, as discussed by Darwin (1896), digest and abrade a considerable amount of soil. Mammals, such as moles, gophers, and ground squirrels tunnel and excavate a substantial amount of soil when they build dens (Black and Montgomery, 1991; Butler, 1995). Similarly, rodents break down rocks and create fine particles (Ugolini and Edmonds, 1983).

8.2.2   Chemical Weathering

Chemical weathering involves chemical changes of rocks and minerals under near-surface conditions. Mineral grains in soils (see Table 8-1) are bathed in a film of water and the dissolution of these minerals depends on a number of factors. First, the solubility of the mineral affects the potential of a mineral to be weathered; this is determined largely by the number and strength of chemical bonds within the crystal lattice. Second, temperature affects the rate of weathering reactions. Third, the composition of the soil solution surrounding the mineral grains will determine weathering rates; solution pH, organic acids, carbonic acid, concentration of other ions already in solution, redox, and complexing ligands can all affect how readily the ions released by weathering can go into solution. And last, water; water is not only the universal solvent in the weathering environment, but it is also the vehicle for the redistribution of products of weathering. The amount of contact between the soil solution and the mineral surface in conjunction with the frequency of removal of soil solution containing ions released by weathering (and its replacement with new soil solution) will all determine how readily a mineral weathers. Taking these factors into consideration it is possible to determine the thermodynamic stability of minerals, and predict the weathering sequence of minerals in an environment (Garrels and Christ, 1965). There are six fundamental processes that chemically weather minerals. These are *dissolution, hydration, hydrolysis, acidolysis, chelation,* and *oxidation/reduction*.

8.2.2.1   Dissolution

Dissolution of a mineral occurs when the crystal lattice breaks down and it separates into its component ions in water. Minerals most affected are salts, sulfates, and carbonates. For example, calcite dissolution is described by

$$CaCO_3 \rightarrow Ca^{2+} + CO_3^{2-} \quad pK = 8.4 \qquad (1)$$

In this case the two ions, Ca^{2+} and CO_3^{2-}, are released into the soil solution and are able to react with water (to form bicarbonate or carbonic acid) or other solution components, or be removed from the soil by leaching. The

Table 8-1 Primary and secondary minerals commonly found in soils

Primary minerals	Approximate composition	Weatherability
Quartz	SiO_2	—
K-Feldspar	$KAlSi_3O_8$	+
Ca, Na-plagioclase	$CaAl_2Si_2O_8$ to $NaAlSi_3O_8$	+ to (+)
Muscovite	$KAl_3Si_3O_{10}(OH)_2$	+(+)
Amphibole	$Ca_2Al_2Mg_2Fe_3Si_6O_{22}(OH)_2$	+(+)
Biotite	$KAl(Mg,Fe)_3Si_3O_{10}(OH)_2$	++
Pyroxene	$Ca_2(Al,Fe)_4(Mg,Fe)_4Si_6O_{24}$	++
Apatite	$[3Ca_3(PO_4)_2]\cdot CaO$	++
Volcanic glass	Variable	++
Calcite	$CaCO_3$	+++
Dolomite	$(Ca,Mg)CO_3$	+++
Gypsum	$CaSO_4\cdot 2H_2O$	+++

Secondary minerals	Approximate composition	Type
Kaolinite	$Al_2Si_2O_5(OH)_4$	1:1 layer-silicate
Vermiculite	$(Al_{1.7}Mg_{0.3})Si_{3.6}Al_{0.4}O_{10}(OH)_2$	2:1 layer-silicate
Montmorillonite	$(Al_{1.7}Mg_{0.3})Si_{3.9}Al_{0.1}O_{10}(OH)_2$	2:1 layer-silicate
Chlorite	$(Mg_{2.6}Fe_{0.4})Si_{2.5}(Al,Fe)_{1.5}O_{10}(OH)_2$	2:1:1 layer-silicate
Allophane	$(SiO_2)_{1-2}Al_2O_5\cdot 2.5-3(H_2O)$	Pseudocrystalline, spherical
Imogolite	$SiO_2Al_2O_3\cdot 2.5H_2O$	Pseudocrystalline, strands
Hallyosite	$Al_2Si_2O_5(OH)_4\cdot 2H_2O$	Pseudocrystalline, tubular
Gibbsite	$Al(OH)_3$	Hydroxide
Goethite	$FeOOH$	Oxyhydroxide
Hematite	Fe_2O_3	Oxide
Ferrihydrite	$5Fe_2O_5\cdot 9H_2O$	Oxide

dissolution of $CaCO_3$ is regulated by the following reactions:

$$H_2O + CO_2(g) \Leftrightarrow H_2CO_3 \quad pK = 1.46 \quad (2)$$

$$H_2CO_3 \Leftrightarrow HCO_3^- + H^+ \quad pK = 6.35 \quad (3)$$

$$H^+ + CO_3^{2-} \Leftrightarrow HCO_3^- \quad pK = -10.33 \quad (4)$$

Overall:

$$CaCO_3(s) + H_2O + CO_2 \Leftrightarrow Ca^{2+} + 2HCO_3^-$$
$$pK = 5.8 \quad (5)$$

Dissolution of $CaCO_3$ is a congruent reaction; the entire mineral is weathered and results completely in soluble products. The above reaction is driven to the right by an increase of CO_2 partial pressure and by the removal of the Ca and/or bicarbonate. Any impurities present in the calcareous rock, such as silicates, oxides, organic compounds, and others, are left as residue. As the calcium and bicarbonate leach out over time, this residue becomes the substratum upon which soils develop in karst terrain found in areas of readily dissolved limestone. This terrain is characterized predominantly by underground drainage and marked by numerous abrupt ridges, fissures, sinkholes, and caverns.

8.2.2.2 Hydration and hydrolysis

Hydration is the incorporation of water molecule(s) into a mineral, which results in a structural as well as chemical change. This can drastically weaken the stability of a mineral, and make it very susceptible to other forms of chemical weathering. For example, hydration of anhydrite results in the formation of gypsum:

$$CaSO_4 + 2H_2O \Leftrightarrow CaSO_4\cdot 2H_2O \quad (6)$$
(anhydrite) $\qquad$ (gypsum)

Gypsum is a relatively soluble mineral and can undergo dissolution whereas anhydrite is less soluble.

Hydrolysis is the incorporation of either H^+ or OH^-, the components of water, into a mineral. Although water has a low dissociation constant, it is abundant in most environments. Even though little H^+ or OH^- may be provided by dissociation of water, the sheer volume of water moving through a soil over time makes hydrolysis an extremely important reaction.

As in dissolution, a chemical and structural change can occur from hydrolysis as the ions replaced by H^+ or OH^- may be of a different size so that the crystal structure is stressed and weakened. An example of this is the weathering of feldspar or goethite by H^+:

$$KAlSi_3O_8 + H^+ \rightarrow HAlSi_3O_8 + K^+ \qquad (7)$$
$$\text{(feldspar)}$$

$$FeOOH + 3H^+ \rightarrow Fe^{3+} + 2H_2O \qquad (8)$$
$$\text{(goethite)}$$

In Equation (7), an altered solid phase is produced by the weathering of feldspar with a K^+ ion released – an example of incongruent weathering (not everything is weathered into solution). In Equation (8), the goethite goes completely into solution—another example of congruent weathering. Both of these examples demonstrate a critical property of weathering, namely that almost all weathering reactions consume H^+. Thus, as long as weatherable minerals are present, weathering reactions can help counteract the natural tendency of soils to become acidic or neutralize the effects of acid rain.

8.2.2.3 Acidolysis

Acidolysis is a similar weathering reaction to hydrolysis in that H^+ is used to weather minerals, but in this case the source of H^+ is not water but organic or inorganic acids. Humic and fulvic acids (discussed in Section 8.3.2), carbonic acid, nitric or sulfuric acid, and low-molecular-weight organic acids such as oxalic acid can all provide H^+ to weather minerals. All of these acids occur naturally in soils; in addition nitric and sulfuric acid can be added to soil by acid pollution. The organic acids are prevalent in the upper soil where they cause intense weathering. Carbonic acid and bicarbonate are more important to weathering in young soils, or deep in the soil profile where organic acids are not prevalent.

8.2.2.4 Chelation

Besides attacking minerals by providing H^+, organic acids can also cause weathering by chelation. A chelator is a ligand capable of forming multiple bonds with a metal ion such as Fe, Al, or Ca, resulting in a ring-type structure with the metal incorporated into the complex. The large, complex organic acids formed in soils can act as chelators, and are capable of stripping metal ions from some primary minerals (Huang, 1989). Artificial chelators such as EDTA (ethylene diamine tetraacetic acid) are often used to test soils for the availability of micronutrients. Some low-molecular-weight organic acids are also capable of chelating metals.

8.2.2.5 Oxidation and reduction

Oxidation and reduction reactions weather minerals by the transfer of electrons. Minerals containing elements that can have multiple valence states such as Fe, Mn, S, or even N, are susceptible to redox reactions. A common reaction that occurs in soil involves both the oxidation and reduction of iron, which when present in a mineral is usually in the Fe(II) form. Fe(II) in the parent material oxidizes slowly in well-drained and aerated soils (Bohn *et al.*, 1985; Birkeland, 1999). In this oxidizing environment an electron may be removed from Fe^{2+} at mineral edges causing disruption in the crystal due to charge imbalance, triggering disintegration of mineral edges or making the mineral more susceptible to other forms of weathering. The Fe^{3+} released by weathering is very insoluble and readily combines with oxygen and water to form goethite

$$Fe^{3+} + 2H_2O \rightarrow FeOOH + 3H^+ \qquad (9)$$
$$\text{(goethite)}$$

This occurs in well-drained temperate soils, and is the reverse of Equation (8). In *very* well-drained warm soils, hematite can form. For example, a primary Fe-bearing mineral such as a pyroxene or amphibole weathers through oxidation to release Fe^{3+} into solution, which then precipitates out as goethite or hematite, depending on the environment. An important aspect of this process is that ultimately H^+ is released and able to weather other minerals. Minerals containing oxidized elements can undergo reduction reactions in anaerobic soils causing them to weather. Weathering of goethite in an anaerobic soil will release Fe^{2+} into solution.

Overall, weathering controls the chemistry of material that is transported into the sediment and that which stays behind in the soil. As an example, consider a general weathering reaction for an aluminosilicate (Stumm and Morgan, 1995):

$$\text{cation-Al-silicate} + H_2CO_3 + H_2O$$
$$\rightarrow HCO_3^- + \text{cation} + H_2SiO_4 + \text{layer-silicate clay} \tag{10}$$

A specific example of this would be the weathering of K-feldspar and the formation of kaolinite (see Table 8-1 for mineral definitions), a layer-silicate clay:

$$\underset{\text{(K-feldspar)}}{KAlSi_3O_8} + H_2CO_3 + H_2O$$
$$\rightarrow \underset{\text{(kaolinite)}}{Al_2Si_2O_5(OH)_4} + 2K^+ + 4H_4SiO_4 + HCO_3^- \tag{11}$$

Minerals formed in the soil can weather further with increasing quantities of H_2CO_3 and H_2O, or other sources of H^+ (Pedro, 1982). The cations released into solution (in this case K^+) can be moved out of the soil into groundwater, rivers,

and ultimately to the oceans and marine sediments. Al and Fe, however, tend to persist in the soil. Since they are not readily lost from soil and other cations are, soils ultimately become richer in Fe and Al oxides such as goethite, hematite, or gibbsite. See Section 8.3.2 for more information about clay formation in soil.

The weathering reactions given above show the key effects of weathering: the breakdown of the original rock minerals, the consumption of H^+, and the release of cations and silica into solution which can then be used to make new minerals or be lost from the soil into the groundwater and rivers.

8.3   Soils

Soil is a multi-phase system consisting of solids, liquids, and gases. In a typical soil, solids, liquids and gases compose about 50%, 20–30% and 20–30% respectively of the total soil volume (Brady and Weil, 1999). The solid phase can be broken down into two components: inorganic and organic matter, with organic matter ranging from 1 to 5% of the soil.

The inorganic component of soil is dominated by four elements: O, Si, Al, and Fe (Jackson, 1964). Together with Mg, Ca, Na, and K they constitute 99% of the soil mineral matter (see Table 8-2). Minerals in soil are divided into primary and secondary minerals. Primary minerals, which occur in igneous, metamorphic, and sedimentary rocks, are inherited by soil

Table 8-2   Elemental composition by atoms of the Earth's crust[a]

Element	%	Cumulative %	Element	%	Cumulative %
O	60.4%		Ti	0.19%	
Si	20.5	80.9%	P	0.07	
Al	6.3		F	0.07	
H	2.9	90.0	Mn	0.04	
Na	2.6		C	0.04	
Ca	1.9		S	0.02	
Fe	1.9		Sr	0.009	
Mg	1.8		Ba	0.006	99.90%
K	1.4	99.5			

[a] From Mason and Moore (1982).

from the parent material. Secondary minerals form in soils and include layer-silicate clays, amorphous (or non-crystalline) minerals, carbonates, phosphates, sulfides and sulfates, oxides, hydroxides, and oxyhydroxides. The most common primary and secondary minerals in soils are given in Table 8-1. Primary minerals are typically larger than secondary minerals. Primary minerals are seldom clay-sized ($\leqslant 2\ \mu m$), whereas secondary minerals are rarely larger than clay-sized. The small size of secondary minerals gives them a very high specific surface area, making them extremely reactive in soils.

Organic matter is incorporated into the soil from the detritus of organisms living on and within the soil. Plant litter is composed largely of C (44%), H (8%), and O (40%), with N, S, P, and other nutrients making up the remaining 8%. Soil organic matter consists of three fractions: recognizable litter, humus and colloidal organics. Humus forms from litter that has undergone decomposition and synthesis into a new amorphous organic compound with a brown to black color. Colloidal organics are soluble organic acids and other organic compounds that can stay in suspension or coat particle surfaces. Soils store substantial amounts of carbon – approximately 1500 Pg C is stored in the upper 1 m of soil in the world (Schlesinger, 1997). Living organisms are also a vital part of the soil as they facilitate many soil processes.

Soil air differs in composition from the atmosphere due to the activities of organisms and their interactions with the soil. Soil air is generally higher in CO_2 (frequently about 1%, but may be as high as 10%) and lower in oxygen (5 to 20%). The higher CO_2 and lower O_2 of soil air results from decomposition and root and microbial respiration that releases CO_2 and consumes oxygen. The exact composition of the soil air depends on the porosity of the soil. A soil with a high total porosity and large pores will allow faster diffusion of soil air to the atmosphere and have lower concentrations of CO_2 and higher concentrations of O_2. A soil with a low porosity or a soil that has its pores largely filled with water may have very high CO_2 and little O_2.

The liquid phase of the soil system is the soil water, or the *soil solution* as it is more appropriately called. Water can enter the soil from at the surface by rainfall or snowmelt, by upward movement of groundwater, or by lateral flow through soil. When water is in intimate contact with soil minerals, organics, organisms, and air, it acts as an interface between these different soil phases, providing a transport mechanism for elements from one location in the soil to another. Water is an excellent solvent; it dissolves many ions and contains organic and inorganic colloids in suspension. It also serves to remove elements from the soil by leaching, and can physically remove soil by erosion.

8.3.1   Soil Formation

Soil genesis is the result of four fundamental types of processes simultaneously operating at any part of the Earth's surface. As a soil develops, matter and energy enter the soil, can be transformed or translocated, and can leave the soil. The nature and magnitude of inputs, outputs, transformations, and translocations can vary widely from one site to another and result in numerous different types of soils.

8.3.1.1   Inputs and outputs

One reason soils form is because of the endless migration of ions, molecules, and particles into the soil from *meteoric inputs*. Examples of meteoric inputs include H_2O, CO_2, O_2, nitrogenous compounds, pollutants, salts, and dust. These molecules and compounds come from space, from the atmosphere and the oceans, and from other terrestrial systems.

Litter inputs come from vegetation and animals. These may be above ground litter as in tree leaves falling in autumn, or below ground as a plant dies and its roots become part of the soil organic matter. The composition of the meteoric inputs can be dramatically changed as inputs pass through the vegetative canopy (see Cronan (1984) and Johnson and Lindberg (1992) for a discussion of canopy processes and development of throughfall). Solar energy is another critical input to soils. Most soil reactions are driven by the energy released during decomposition of organic matter and by solar energy.

The release of ions through weathering is also considered an input to soils. Elements that were bound in mineral crystals are released into the soil solution. These ions can be involved in soil processes and the formation of new organic or inorganic materials, or leached from the soil into the groundwater.

Leaching is an important output from soils through which dissolved elements or suspended materials are carried downward by the soil solution and enter the groundwater. Most soils act as very good filters, retaining nutrients and organic matter, and even most metals and pesticides added as pollutants. Thus leaching losses to groundwater are normally minimal. However, the capacity of the soil to sorb ions or filter particulates can be exceeded in highly polluted, overfertilized, or eroded soils allowing passage of materials which then pollute the groundwater.

Erosion and gases released from the soil can also remove material from soils. Erosive outputs are usually from the upper portion of the soil, and may adversely impact soil processes and soil fertility by removing large quantities of the soil organic matter. Such losses of organic matter may remove substantial amounts of nutrients needed for plant growth. Gaseous losses are a normal efflux from soils to the atmosphere. Carbon is usually in the form of CO_2 when it effluxes from the soil to the atmosphere, but can be released as methane (CH_4) in wetlands where anaerobic processes are dominant. Nitrogen can evolve from the soil to the atmosphere as N_2, N_2O, or NH_3 depending on whether the soil is aerobic or anaerobic. Under anaerobic conditions, NO_3^- can be converted to N_2 or N_2O. Gaseous losses of S are typically as H_2S from wetlands. Volatile losses of C, N, and S through wildfire can be important in some environments.

8.3.1.2   *Transformations*

Organic matter and rocks are the building materials of soils, which both undergo extensive transformations within soil. These transformations include changes in physical as well as chemical properties and result in unique new soil characteristics. Weathering is one type of transformation of inorganic matter that occurs in soil (see discussion above). The formation of secondary minerals and the development of cation exchange capacity are others. The development of humus from fresh litter is another transformation that affects soil organic matter.

8.3.1.2.1   Secondary minerals. As weathering of primary minerals proceeds, ions are released into solution, and new minerals are formed. These new minerals, called secondary minerals, include layer silicate clay minerals, carbonates, phosphates, sulfates and sulfides, different hydroxides and oxyhydroxides of Al, Fe, Mn, Ti, and Si, and non-crystalline minerals such as allophane and imogolite. Secondary minerals, such as the clay minerals, may have a specific surface area in the range of 20–800 m^2/g and up to 1000 m^2/g in the case of imogolite (Wada, 1985). Surface area is very important because most chemical reactions in soil are surface reactions occurring at the interface of solids and the soil solution. Layer-silicate clays, oxides, and carbonates are the most widespread secondary minerals.

Layer-silicate structure, as in other silicate minerals, is dominated by the strong Si–O bond, which accounts for the relative insolubility of these minerals. Other elements involved in the building of layer silicates are Al, Mg, or Fe coordinated with O and OH. The spatial arrangement of Si and these metals with O and OH results in the formation of tetrahedral and octahedral sheets (see Fig. 8-2). The combination of the tetrahedral and octahedral sheets in different groupings, and in conjunction with different metal oxide sheets, generates a number of different layer silicate clays (see Table 8-1).

Once a layer-silicate clay forms, it does not necessarily remain in the soil forever. As conditions change it too may weather and a new mineral may form that is more in equilibrium with the new conditions. For example, it is common in young soils for the concentrations of cations such as K, Ca, or Mg in the soil solution to be high, but as primary minerals are weathered and disappear, cation concentrations will decrease. With a decrease in solution cations, a layer-silicate such as vermiculite will no longer be stable and can weather. In its place,

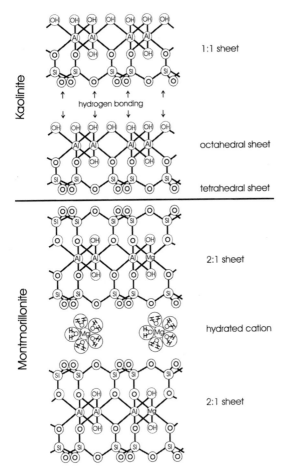

Fig. 8-2 Structure of a 1:1 (kaolinite) and a 2:1 (montmorillonite) layer-silicate clay mineral.

montmorillonite or another clay can form. Ultimately, kaolinite or Fe and Al oxides are most stable and, therefore most common in the oldest, most highly developed and leached soils.

8.3.1.2.2 Ion exchange capacity. A characteristic common to all the layer silicates is the electric charge present at their surfaces. This charge can develop when a clay is forming. With an abundance of Al^{3+} in solution when the clay is forming, some of the sites where Si^{4+} normally occurs in the crystal can be filled by Al^{3+} in place of Si^{4+} due to its similar ionic radius. This process is called isomorphic substitution. While the crystalline structure can persist with this substitution, the leftover negative charge from the O^{2-} and OH^- making up the rest of the clay

molecule must be satisfied. This is achieved by attracting cations to the surface of the clay or in between the layers of the clay mineral. Clays such as the smectites, which have little substitution and low charge (~ 0.6 to 0.25 charge per unit cell) can expand and allow water and cations to move in between the clay plates using the cations to balance the negative charge. The cations are electrostatically held; this allows one cation to easily take the place of another cation so that the cations are exchangeable. Thus, if a plant root releases H^+ into the soil solution, the H^+ can replace an NH_4^+ ion exchangeably held by a clay so that it goes into solution and can be taken up by the root. The quantity of cations that soils can exchangeably store is called its *cation exchange capacity* (CEC). It is fortunate for humans that the phenomenon of cation exchange capacity exists. Without this property, the nutrients released by weathering and decomposition would be easily lost to rivers and the ocean, leaving an infertile land. According to Jackson (1969), life that had originated in the seas was able to move onto the land because the clay produced by weathering had the capacity to hold cations and make them available to plants.

In a fertile agricultural soil, the exchangeable cations include most of the macronutrients needed by plants (NH_4^+, K^+, Ca^{2+}, and Mg^{2+}) along with Na^+. In an acid soil, many of the cation exchange sites are filled by Al^{3+} or H^+ ions, neither of which is essential for plants. When a soil has many of its exchange sites ($\geqslant 50\%$) filled by nutrient cations or Na^+, it has a high base saturation. Conversely, when a soil has mostly Al^{3+} and H^+ on its exchange sites, it has a low base saturation. Other layer silicates such as mica or chlorite (~ 0.9 to 1.0 charge per unit cell) have high negative charges and tend to fix cations (K and Mg, in particular) between their plates so that they do not expand and have little CEC. On the other hand, kaolinite has a charge of zero and thus has a CEC of almost zero, but kaolinite does not expand because hydrogen bonds hold the sheets together.

Oxides, non-crystalline minerals, and humified organic matter can also develop charges at their surfaces by reactions with the soil solution. In this case, the surface can have positive (CEC)

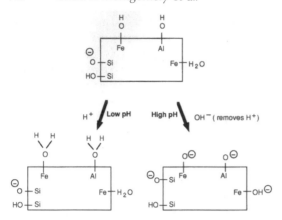

Fig. 8-3 Diagram illustrating the development of positively and negatively charged sites on surfaces of soil constituents, at low and high pH. (Reproduced with permission from R. L. Parfitt (1980). Chemical properties of variable charge soils. *In* "Soils with Variable Charge" (B. K. G. Theng, ed.), p. 168. New Zealand Society of Soil Science Offset Publications.)

or negative charges (*anion exchange capacity*, AEC) depending on the pH of the solution surrounding the particles. Figure 8-3 shows how this occurs. A pH-dependent charge can also occur at the incomplete bonds at the ends of layer silicate clay minerals. Ions held by pH-dependent charges are also completely exchangeable. The pH-dependent charges can aid in keeping either cations or anions (such as NO_3^-) in the soil and prevent them from leaching. An acidic soil has a much lower CEC than a basic soil, but may have AEC.

8.3.1.2.3 *Humus fractionation.* Another extremely important transformation that takes place in soils is the conversion of fresh organic litter into humus. Organic matter in soils is made up of partially decomposed plant and animal remains, substances synthesized during decomposition, and microbial bodies. Organic matter is derived directly or indirectly from photosynthesis. Decomposition processes incompletely break down the complex organic molecules, so that organic matter tends to accumulate in soils. Decomposition is an extremely important process in terms of releasing energy and nutrients to the soil system.

Plant litter consists mainly of sugars, cellulose, hemicellulose, lignin, waxes, and polyphe-

nols, and to a lesser extent proteins and cations (Paul and Clark, 1996). The rate of litter decomposition is a function of climate and the composition of the litter. The nitrogen content is often a limiting factor in decomposition, but lignin content may also control decomposition rates (Edmonds, 1979). In general, decomposition is most rapid in well-aerated, moist, mesic, and nearly-neutral pH soils. Cold, humid environments with high water tables and acidic conditions favor the accumulation of undecomposed organic matter and carbon storage in soils. When C is converted from CO_2 to an organic molecule during photosynthesis, it is reduced (gains electrons). Under aerobic conditions, decomposition oxidizes the C back to the 4+ valence state in CO_2 by using O as an electron acceptor. Anaerobic decomposition requires that a different electron acceptor be used, as little or no O is available. Nitrogen, Mn, Fe, S, and C can act as electron acceptors when O is not present. If C is used as an electron acceptor, methane is produced. Using C as an electron acceptor gives decomposers less energy than using any other element, thus decomposition will proceed extremely slowly under such a strongly reducing anaerobic environment.

Complete decomposition of organic matter releases CO_2, water, nutrients, and energy. Partial decomposition resynthesizes fresh organic matter into new organic compounds that are collectively referred to as humus or humic substances. Humic substances are most usefully described in terms of their solubility. Humic substances are "amorphous, dark-colored, hydrophilic, acidic, partly aromatic, chemically complex organic substances that range in molecular weight from a few hundred to several thousand" (Oades, 1989). Humic substances are fractionated into three main groups based on their solubility in acidic and basic extracts. Humic acids are soluble in base, but precipitate with acidification of the extract to pH 2. Fulvic acids are soluble in both alkaline and acidic water. Humin, the last humic group, is insoluble in either acid or base. Whereas these three humic fractions are structurally similar, their molecular weight and functional groups differ. Humic acid and humin contain more H, N, and S, but less O than fulvic acid. The total acidity and the number of carboxylic functional groups of fulvic acid are greater than those of humic acid and humin (Oades, 1989).

Humic substances are involved in many soil reactions, largely due to their high surface area (800–900 m^2/g) which can give humic material a high exchange capacity and a high absorptive capacity. This makes humic substances important for water retention, aggregation of soil particles, nutrient supply, and buffering of soil pH (Bohn *et al.*, 1985). Specific humic groups are also essential to certain soil processes. For example, fulvic acids with their high acidity and solubility, and high number of complexing functional groups, can weather Fe^{3+} and Al^{3+} from minerals, chelate them, and keep them in solution, thereby allowing their movement down the soil profile where they may deposit Fe and Al in lower horizons. On the other hand, humic acids, with their high molecular weight and lower solubility, help to create organic-rich surface soils; humic acids can complex with Ca, become very insoluble and create a deep, highly fertile, surface soil.

8.3.1.3   Translocations

Movement of raw and transformed materials can take place within the soil and results in zones of accumulation, depletion, or mixing. Formation, migration, and accumulation of different elements, clays, oxides, and organic matter can occur in different parts of the soil. These different zones or layers in soil that are approximately parallel to the surface are called *soil horizons*. Depleted or enriched soil horizons result in different depths in the soil having different chemical and physical properties. Translocations are caused by a combination of physical, chemical, and biological processes.

As an example, the migration of clay from the surface of a soil to a lower horizon results from several processes occurring when certain soil and environmental properties exist. First, clay-sized minerals must form, usually requiring weathering to have occurred. Clay minerals formed in the surface soil can then go into suspension when salt concentrations in solution are low. Seasonal rains can move the clay down

as water percolates through the soil and deposit the clay when the wetting front stops or soil pores become plugged. This process is called *lessivage* and results in a clay-enriched horizon (illuviated) beneath a coarser textured horizon (Duchaufour, 1982).

Movement of carbonates and salts can also occur in a similar fashion. As these minerals are weathered in the upper soil profile, their component ions go into solution and are moved down through the soil by rainfall entering the soil. As the water moves down the soil there may not be enough water to move the ions out of the soil, so they precipitate in a lower horizon where they accumulate. Such accumulations are common in arid environments with limited rainfall. In high rainfall areas, carbonates and salts are usually completely removed from the soil through leaching.

The kind of vegetation and environment in which a soil develops also can influence the mobility of organic and inorganic substances. One particular example occurs commonly in areas that have *Ericaceous* plants or coniferous forests, and cold, wet environments in which Fe, Al, and humic substances are translocated. This process, called *podzolization*, is particularly favored by a coarse soil texture so that water can move easily through the upper soil. In podzolization, the vegetation drops litter that is low in nutrients and favors the production of organic acids by incomplete decomposition. The soluble organic acids (fulvic acids in particular) migrate into the mineral soil where they can weather minerals by release of H^+ or chelation. The Fe and Al can then migrate down the soil profile along with the organic acids creating a zone of soil depleted in sesquioxides and organics (eluviated). Subsequently, the Fe, Al, and organic matter can precipitate out at a lower depth, creating a zone enriched in these materials. Further decomposition of this enriched horizon (illuviated) may release Al and Fe which may move further down the soil to create another zone enriched only in Fe and Al and where amorphous minerals can form (Lundstrom, 1994).

Some translocations can disrupt the concentration of materials and thereby provide mixing of different horizons rather than segregating them. Worms and other soil organisms do this by carrying down litter and organic rich soil into lower horizons and then tunneling up to the surface again carrying soil with them – thereby mixing horizon material in the process. This process is called bioturbation. Worms also contribute to the formation of humic substances with the passage of fresh litter through their gut and the release of partially decomposed humus. Some soil organisms such as worms are important for the creation of a deep, highly fertile, organic-rich surface soil because they aid in moving organic matter deeper into the soil. Mixing that can disrupt different accumulation zones in soils can also occur due to shrinking and swelling of clays in a soil with highly seasonal rainfall. Shrinking and cracking of the soil can allow surface soil to slough into the cracks and be moved lower in the soil, while lower soil swells upward in the wet season. In this case, organisms are not responsible for this type of soil mixing, called *pedoturbation*. Frost action and windthrow are other examples of pedoturbation.

8.3.1.4   *Soil horizons and profile development*

Soil horizons differ from one another in composition (e.g. clay or organic matter content), physical properties (e.g. color or particle size), or chemical properties such as pH or CEC. Five different soil horizons can form. An all-organic horizon (*O horizon*), typically occurs in wetlands or at the surface of forest soils. A mineral horizon (*A horizon*) that is rich in humified organic matter, may be found in prairies. A loss of Fe, Al, organic matter, or clay will create an *E horizon* (eluviated horizon). *B horizons* are zones of soil that have accumulated material from above or well-weathered soil that shows evidence of pedogenic processes through changes in color or development of soil aggregates. The lowest soil horizon is typically a *C horizon*, which is the least weathered zone of soil and is most similar to the original material the soil is forming from.

To provide even more information about a soil, subordinate horizon designations are used. These are modifiers of the master horizons that indicate specific properties of a horizon. Table 8-3 gives a summary of master horizon characteristics and some subordinate horizon designations used to modify them.

A *soil profile* is a vertical cross-section of the soil showing all of its constituent horizons. Different soils will have different soil profiles. Soil profiles may differ in the types of horizons they contain, the location of the horizons, or the depth of the horizons. Figure 8-4 gives examples of two different types of soil profiles. The dominant soil-forming processes at a site will determine what type of profile forms. In Fig. 8-4, the presence of a coniferous forest with a cold climate and high rainfall results in O, E, Bhs, Bs, and C horizons (podzolization). Fulvic acids are important in the creation of the E, Bhs, and Bs horizons as described above. Within a prairie, grasses input large quantities of organic matter directly into the soil from root turnover. In conjunction with deep mixing by organisms such as worms and prairie dogs, and with migration of carbonates to the lower profile due to seasonal rains, a deep A horizon develops which is rich in humic acids and humin, often with a Bk horizon below. Prairie soils may also have a Bw or Bt horizon below the A horizon. Other environments and organisms result in many other types of soil profiles that combine to create the pedosphere.

8.3.2   *The Pedosphere*

The pedosphere is the envelope of the Earth where soils occur and where soil-forming processes are active. Although the biosphere overlaps the pedosphere, the two are not coincident. Soil can develop in areas that are virtually abiotic such as in the high Arctic or in the ice-free areas of Antarctica (Ugolini and Jackson, 1982; Claridge and Campbell, 1984; Campbell and Claridge, 1987). Nevertheless, most soil formation does occur in the presence of life, and soil formation proceeds at a faster rate when biota are abundant (Ugolini and Edmonds, 1983). The rate of development of soils and the type of soil that forms are directly related to the soil-forming factors.

Table 8-3 Master soil horizons and some common subordinate horizon designations[a]

Master horizons

O   Organic horizon

A   Surface mineral horizon mixed with humus, typically darker-colored

E   Horizon showing a loss of Fe, Al, organic matter, or clay and showing a concentration of minerals resistant to weathering such as quartz; usually a pale gray or whitish color

B   Zone of accumulation of Fe, Al, organic matter, clay, salts, or carbonates; may also just show a pedogenic development by changes in color or structure from the parent material or C horizon

C   Unconsolidated material that shows some pedogenic effects on the parent material, but overall has little change; horizon least affected by soil-forming processes

Subordinate horizon designations

Used with O horizon

i   Slightly decomposed organic matter; most is recognizable litter

e   Intermediately decomposed organic matter; some is recognizable

a   Highly decomposed organic matter; almost none is recognizable

Used with A horizon

p   Plowed soil

Used with B horizon

h   Illuvial accumulation of organic matter

o   Residual accumulation of sesquioxides

s   Illuvial accumulation of sesquioxides

t   Accumulation of clay

w   Development of color or structure

Used with B or C horizon

g   Gleyed or waterlogged soil

f   Frozen

k   Accumulation of carbonates

[a] See "Keys to Soil Taxonomy" (Soil Survey Staff, 1998) for a complete listing.

8.3.2.1 Soil-forming factors

In the late 19th century, five factors were found to determine the rate and dominant processes that lead to the development of an individual soil. These are: climate, organisms, topography, parent materials, and time. The effects of climate on soils are numerous. For example, climate determines leaching rates by the quantity of water moving through the soil and determines reaction rates by temperature. It can also affect translocation in a soil by rainfall occurring constantly throughout the year (resulting in increased leaching) or rains occurring seasonally, which may be insufficient to leach ions, and only move material from one horizon to another. Organisms can influence soil by altering the quality and quantity of organic matter inputs and how rapidly they are decomposed or converted to humic substances. Topographic position can alter soil by changing whether a soil is well-aerated or waterlogged. Parent materials, the geologic materials or organic materials that the soil develops from, influence soil texture, acidity, weathering rates, and clay formation. Time determines how well developed a soil is relative to its environment. By continuing soil development for thousands to hundreds of thousands of years, soils become deeper, more highly weathered, contain more clay and, ultimately, under a strong leaching regime become rich in oxides.

8.3.2.2 Soil orders

Soils that have developed in different environments, or with different parent materials or ages

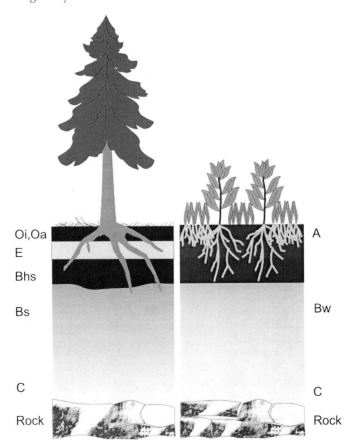

Oi,Oa
E
Bhs
Bs
C
Rock

A
Bw
C
Rock

Fig. 8-4 Typical soil horizon sequence for a Spodosol developed under a coniferous forest (left) and a Mollisol developed under grasses and herbaceous plants.

result in different combinations of soil horizons and different soil properties. Twelve major types of soil are recognized by the USDA Soil Taxonomy. These 12 soil types are called soil orders, the highest level of soil classification used in the US. Other classification systems use different but similar designations for their highest level of classification (e.g., FAO, 1971–81, Soil Survey Staff, 1998). Soil orders are distinguished by unique diagnostic horizons, unique climatic factors that control soil formation, or unique processes that create specific horizons within a soil. Table 8-4 shows the different soil orders and some unique characteristics and environmental factors that distinguish one from another. The table is arranged so that the soils with the least pedogenic development are at the top and those that are most highly developed are at the bottom. For example, Entisols are soils that have little horizon development, and usually do not even have a B horizon; they may have only a C horizon, or an A–C profile. In contrast, Oxisols are extremely old soils that have lost all their weatherable minerals, most cations, and residual Fe and Al oxides persist in the soil; normally these soils only develop in tropical areas where both temperatures and rainfall are high. Many other soils develop in other environments throughout the world and will be the "climax" soil for that environment.

8.3.2.3 Global soil patterns

Jenny (1941) attempted to quantitatively relate the factors of soil formation to soil properties such as N, C, or clay content, depth of leaching

Table 8-4   USDA soil orders with common ecosystems or environments in which they occur and shown in approximate order from low to high pedogenic development[a]

Soil order	Unique features	Ecosystem/environment
Entisols	Young, normally no B horizon	Variable
Gelisols	Cryoturbation; permafrost	Cold soils; polar areas
Inceptisols	Young, but have a Bw horizon	Variable
Andisols	Andic soil properties (Al humates, low pH, darkly colored A)	Volcanic parent materials
Aridisols	Variable	Dry soils; deserts
Histosols	All organic soil	Wetlands, riparian areas
Mollisols	Thick, dark A horizon with high base saturation	Temperate grasslands
Vertisols	>30% shrink–swell clays, cracking in dry season, deep A	Temperate to subtropical grasslands
Alfisols	Bt horizon with moderate base saturation	Temperate deciduous forests
Spodosols	Bhs and/or Bs horizon	Cool, wet, coniferous forests
Ultisols	Low base saturation and Bt horizon	Tropical to temperate, often forested
Oxisols	Highly weathered and leached, residual Fe and Al oxides	Tropical areas with a stable landscape

[a] See "Keys to Soil Taxonomy" (Soil Survey Staff, 1998) for complete criteria for each soil order.

of carbonates, and others. Determining the role of each environmental factor in influencing the development of any particular soil is difficult. Many soils have developed on the landscape over a long time, such soils are therefore polygenetic as they have acquired some of their properties under a constellation of soil-forming factors different from those currently in operation. The soil-forming factors are seldom independent. This is particularly true for organisms, especially vegetation, which is both influenced by climate and can alter microclimate. Many other interactions between soil-forming factors occur, such as between relief or topography and parent materials or time. Nevertheless, knowledge of the soil-forming factors can allow prediction of a general global soil pattern.

Figure 8-5 shows soil orders that form in relation to climate and vegetation along a transect from the poles to the equator along with processes occurring in the soil. In this conceptual diagram, it is assumed that there is a uniform parent material, similar topography and

equivalent time for soil formation. The effects of a cold climate and little vegetation result in slow, limited weathering and little organic inputs to soil. In slightly warmer climates, there is more vegetation and more organic inputs to soils. However, due to the cold climate, partial decomposition of litter results in formation of organic acids and intense leaching in the upper part of the soil, with accumulation of Fe and Al in the lower part of the soil – podzolization is dominant here, producing Spodosols. Under deciduous temperate forests, soils can undergo lessivage, resulting in clay accumulations in the lower profile. In such environments deeper soil profiles are common. These soils belong to the Alfisols and Ulfisols. In grasslands and deserts low rainfall may prevent carbonates from leaching out of the soil and keep cations in the soil so that base saturation is high. Grasses can help develop deep A horizons and humic acids can aid in weathering and clay formation (Mollisols). Under a tropical rainforest, old soils can be extremely weathered, cations will have been

Fig. 8-5   Soil and soil forming processes – a global view. The moisture and temperature regimes are generalized and intended only to show major pedoclimatic environments. Spodosols, for example, can also occur in a cryic regime and even in equatorial regions. Other orders could also occur in more than one moisture and temperature environment. _(overleaf)_

ZONE		COLD DESERT	POLAR DESERT	TUNDRA	BOREAL FOREST	TEMPERATE FOREST CONIFEROUS	TEMPERATE FOREST DECIDUOUS
Soil Order		Entisols	Gelisols	Inceptisols	Spodosols	Inceptisols	Alfisols
Climate	Moisture	ultra xeric	ultra xeric	aridic	udic	udic	
	Temperature	ultra pergelic	pergelic	cryic	frigid	mesic	
Examples of climate and vegetation effects on horizons	O, A,E, Bh			under-saturated non-mobile organic acids	very under-saturated mobile organic acids	under-saturated non-mobile organic acids	
	B		Fe-hydroxide neoformation	smectite, vermiculite formation	non-crystalline aluminosilicates and Fe-hydroxide neoformation	smectite,vermiculite formation by transformation of preexisting phyllosilicates	
	C	$CaCO_3$ precipitation	$CaCO_3$ precipitation	Fe-hydroxide neoformation	Fe-hydroxide neoformation	Fe-hydroxide neoformation	
	R	nil	nil	nil	nil	nil	
Speed of pedogenesis		extremely slow	very slow	slow	moderate	moderate	

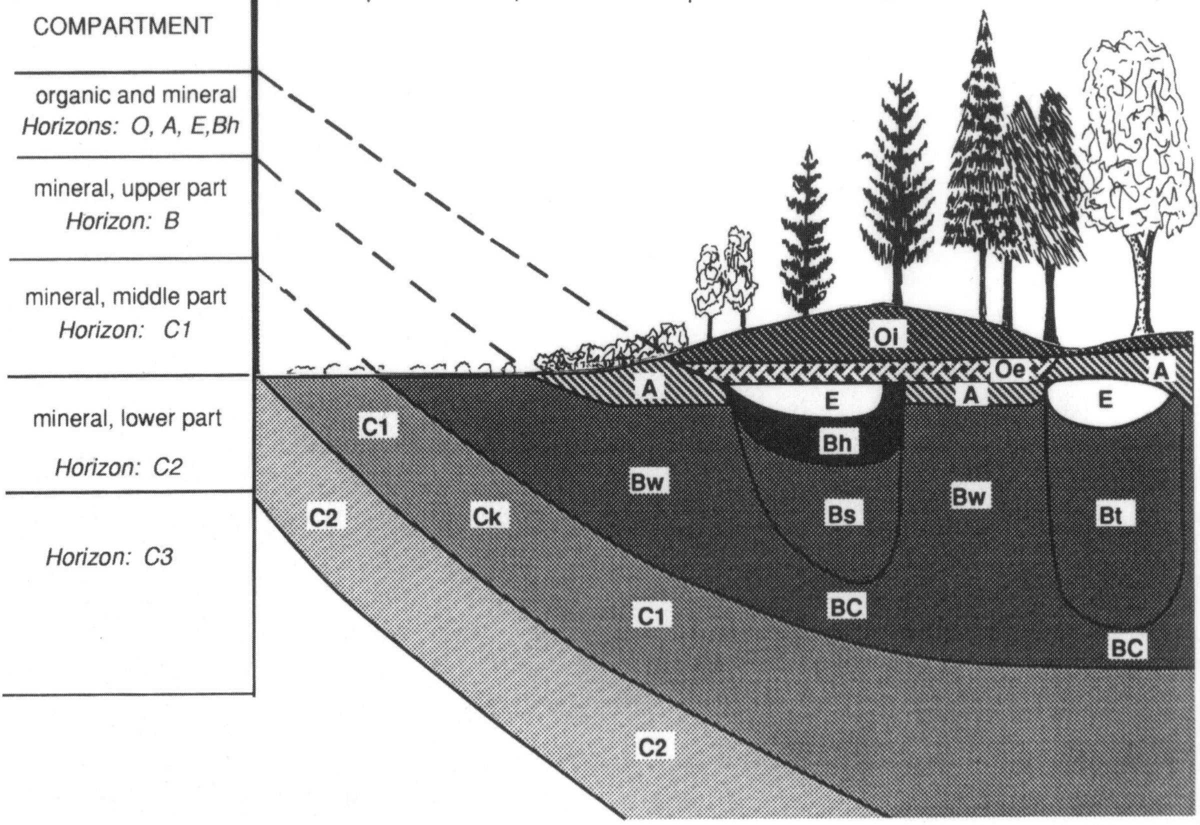

COMPARTMENT
organic and mineral
Horizons: O, A, E,Bh
mineral, upper part
Horizon: B
mineral, middle part
Horizon: C1
mineral, lower part
Horizon: C2
Horizon: C3
Oi
Oe
A
E
Bh
Bs
Bw
Bt
BC
C1
C2
Ck

Fig. 8.5

| GRASSLAND | DESERT | SAVANNA | | TROPICAL RAINFOREST |
		TREELESS	ARBOREAL	
Mollisols	Aridisols	Vertisols	Ultisols	Oxisols
ustic	aridic	ustic	udic	perudic
mesic	thermic	isothermic		isohyperthermic
saturated non-mobile organic acids		saturated non-mobile organic acids	under-saturated organic acids	very under-saturated mobile organic acids
Fe-hydroxide neoformation $CaCO_3$ precipitation	Fe-hydroxide neoformation $CaCO_3$ precipitation	smectite neoformation	kaolinite neoformation	Fe- and Al-hydroxide neoformation
$CaCO_3$ precipitation	$CaCO_3$ precipitation	$CaCO_3$ precipitation	Fe-hydroxide neoformation	kaolinite neoformation
nil	nil	nil		Fe-hydroxide neoformation
moderate	slow	fast		very fast

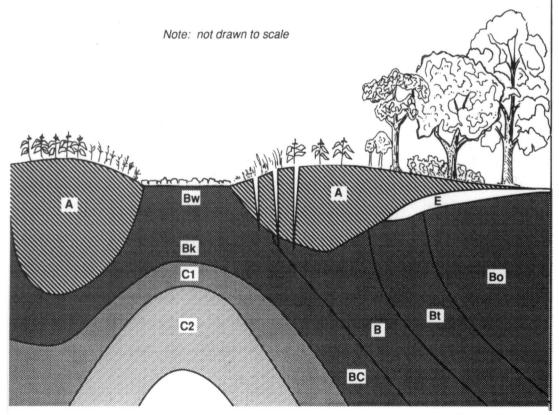

Note: not drawn to scale

leached, primary minerals are mostly gone, and only Fe and Al oxides remain in the soil along with kaolinite (Oxisols).

Plate 3 shows a map of dominant soil orders for the entire world. Although this map necessarily lacks detail due to its scale, the relationship between soils and the biosphere is evident. Different terrestrial ecosystems are correlated with climatic conditions and different soils are correlated with both. For example, Mollisols are common in areas where there are prairies or steppes; a result of grasses as the dominant vegetation and low, seasonal rainfall. Spodosols occur where coniferous forests dominate and the climate is cold and wet. Comparing Fig. 8-5 and Plate 3 carefully will show how strong this correlation is for the entire Earth.

8.3.2.4   Biogeochemical cycling in soils

One of the most important functions of the pedosphere is the cycling of elements that occurs within soils and the transfers that occur between the atmosphere, lithosphere, biosphere, and hydrosphere through soils. Soil is an interface between the atmosphere and lithosphere, between the biosphere and lithosphere, and between roots and soil organisms and the atmosphere. In many ways, soil acts as a "membrane" covering the continents and regulating the flow of elements between these other systems of the Earth.

This soil "membrane" has inputs and outputs, and can transform the elements entering it before these elements leave. Consider the simple cycle of potassium shown in Fig. 8-6. Inputs to the surface of the soil come from atmospheric deposition of particulates, fertilizer applications, and litterfall. Potassium that is released from a crystalline matrix by weathering is also considered an input to soils from rocks even though the rocks are contained within the pedosphere (Zabowski, 1990). Likewise, roots that die and begin to decompose provide inputs of K to the soil – the conversion of organically bound elements to an inorganic form is called mineralization. Within the soil, K can be used to form new clays, attached or released by cation exchange sites, released by dissolution of clays, or taken up by soil organisms (immobilized)

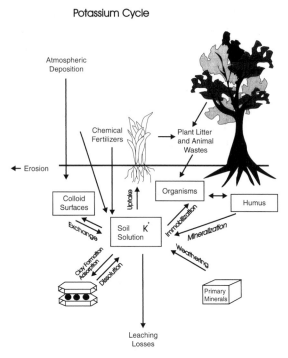

Potassium Cycle

Fig. 8-6   A cycle of potassium in soil.

where eventually they may be mineralized again when these organisms die. Note how the soil solution functions as the transfer mechanism. Potassium may also be removed from the soil by uptake into the biosphere (in which case it may eventually return to the soil through litterfall), erosion of soil, or leaching to groundwater which flows away from the soil to rivers. By knowing inputs and the quantity of an element in the soil, mean residence times (τ_r) and turnover times (τ_0) can be calculated. The basis for these calculations is presented in Chapter 4. Biogeochemical cycling in soils is further complicated by the different soil processes occurring in the different soil orders (see Fig. 8-6 and Table 8-4) which all cycle K and other elements at different rates.

The cycle of potassium is quite simple, as K does not change valence states or have a gaseous phase. In contrast, elements such as carbon and nitrogen both change phase and undergo redox reactions and undergo much more complicated cycling. For example, carbon captured by plants from the atmosphere is reduced to form organic matter, which is then oxidized by either organisms in the soil, or in roots to provide energy with a release of CO_2 if the soil is aerobic. Methane may be released if the soil is anaerobic. Other conversions of carbon to humic substances can make it very resistant to further decomposition. The τ_r of C in various soil fractions can range from a few months to thousands of years (Schlesinger, 1997). Nitrogen cycling is also very intricate, because N can exist as N_2, N_2O, NO_3^-, NO_2^-, NH_4^+, and NH_3 in soils. Although nitrogen is rarely input to soils by weathering, nitrogen-fixing organisms capture atmospheric N_2 and thereby act as a source of the element to soils. The rates at which nitrogen converts from one form to another in soil affect the rate of transfer of N from one "sphere" to another, owing to the fact that soils are at the interface of the atmosphere, biosphere, and hydrosphere, where nitrogen compounds reside. A complicating factor is that nitrogen is the nutrient most commonly limiting to plant growth. Carbon and nitrogen cycling are discussed in more detail in Chapters 11 and 12, respectively.

8.4 Watershed Processes

Watersheds, also known as drainage basins, define a natural context for the study of relationships among soils, geology, terrestrial ecosystems, and the hydrologic system because water and sediment travel downslope under the influence of gravity. This material is a continuation of some of what was presented in Chapter 6.

The concept of a drainage basin has no inherent scale, since watersheds range from small headwater valleys to the catchments of huge rivers that drain continents. The physical and chemical load carried in a river is produced by watershed processes, including weathering and exchange processes in the soil, how the water that runs off of a landscape as streamflow (often simply called runoff) is generated, and the relative importance of different geomorphological processes, climate, and the lithology of the bedrock. Climate and topography are two of the strongest influences on the sediment load of rivers, although sediment routing processes

and long-term storage of sediment in floodplains and structural valleys can create substantial time lags and differences in net sediment delivery to the oceans.

8.4.1 Runoff Processes

The processes through which rainfall is turned into runoff, together with the nature of the material through which water moves, control the chemical characteristics of streamflow. Specific runoff mechanisms operating in a landscape control the flowpaths by which water moves through the landscape. Flowpath-dependent differences, such as the total time that water spends in contact with different soil horizons or bedrock (residence time), can strongly influence runoff amounts and timing, the relative contribution of event (new) versus stored (old) water, and runoff chemistry.

8.4.1.1 Runoff mechanisms

The translation of rainfall into runoff occurs by a variety of mechanisms associated with different environments (refer back to Fig. 6-7). *Horton overland flow* (HOF) occurs primarily in arid or disturbed landscapes where rainfall intensity exceeds the infiltration rate of the ground surface long enough for ponding to occur (Horton, 1933). Observations that most rainfall infiltrates into the soil in humid, soil-mantled landscapes, and therefore that HOF is rare in such environments, led to the recognition of subsurface flow as a major mechanism of storm runoff (Loudermilk, 1934; Hursh, 1936). Subsurface stormflow (SSSF) dominates runoff generation in steep soil-mantled terrain where precipitation infiltrates and flows laterally either through macropores, or over a lower conductivity zone, such as at the base of a root mat or at the soil–bedrock boundary. Saturation overland flow (SOF) occurs in soil-mantled landscapes when an initially shallow water table rises enough to intersect the ground surface over a portion of the catchment (Hewlett and Hibbert, 1967), which then causes runoff by either return flow or direct precipitation onto saturated areas (Dunne and Black, 1970). Topographically driven patterns of soil

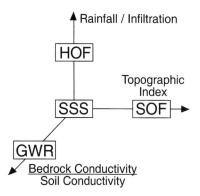

Fig. 8-7 Three principal ratios control the style of runoff generation prevalent in a landscape: (1) ratio of rainfall intensity to the infiltration capacity of the soil; (2) ratio of bedrock conductivity to soil conductivity; and (3) the topographic index defined by the ratio of the upslope drainage area to the ground slope. HOF = Horton overland flow; SOF = saturation overland flow; SSS = subsurface stormflow; GWR = groundwater flow.

moisture favor development of SOF in low-gradient, convergent topography where flow is concentrated. Water that infiltrates through the soil to the regional groundwater table contributes to groundwater flow that maintains low flows between storm events. The relative importance of these runoff generation mechanisms depends on several key ratios: rainfall intensity to infiltration rate; drainage area to local slope; and bedrock to soil conductivity (Fig. 8-7).

8.4.1.2   Runoff chemistry

The soil is the connecting link between rainfall and river flows, and the interaction of runoff processes and soils regulates the dissolved load in the hydrologic system. The soil plays a complex role in biogeochemical cycles, and the elements found in rivers as soluble salts or as particles are derived from weathering of rocks (Si, Al, Fe, Mg, Ca, K, P) or atmospheric fallout (S, N). One could predict the quality and quantity of this load to depend strongly upon the soils surrounding the river, but this is not always the case. In fact, the chemistry of rivers flowing on granitic terrains in western Europe and tropical Africa appear rather similar, although the soils in these two regions are dramatically different (Tardy, 1969). Such counter-intuitive results indicate that the dissolved load of rivers reflect not only soil properties, but also runoff pathways, lithological influences, and other environmental factors.

Most soil orders can be divided into two major compartments: an organo-mineral compartment on top, referred to as the biogeochemical compartment, and an underlying mineral compartment, referred to as the geochemical compartment. The character of runoff from the biogeochemical compartment depends strongly on the complex interactions among the climate, biota, parent material, topography and time, whereas in the geochemical compartment these interactions are less important. The chemistry of a river depends mostly on whether runoff flows overland, and thereby has little interaction with the soil, or on the level at which the water leaves the soil, as the soil solution can change dramatically as water moves through the different soil horizons.

8.4.1.2.1   Overland flow. Runoff by overland flow carries material to the river as particles of sand, silt, clay, and organic matter. This material consists of plant debris, more or less decomposed, and aggregates of humified organic material and mineral particles. Runoff from different soil orders developed on similar bedrock will also reflect soil properties; Spodosols developed on granite, for example, produce overland flow runoff that consists of very acidic organic matter, whereas overland flow from a Mollisol developed on a similar lithology yields aggregates of neutral humic acids and clay. In most environments, however, the load carried to the rivers by overland flow generally has a high nutrient content.

8.4.1.2.2   Biogeochemical compartment. The biogeochemical compartment consists of the O, A, E, and B horizons of soils, in which substantial organic matter can be an integral component of the soil. The load of the water leaving the soil from the biogeochemical compartment consists mainly of soluble compounds because the matrix of the soil acts as a filter that retains particulate matter. In some cases, however, clay and particulate humic substances are also carried away. The total dissolved load strongly reflects whether the soil experiences intense or mild weathering. Intense weathering is due to very acidic conditions or a very weatherable parent material which results in a soil solution that contains relatively high amounts of silica as H_4SiO_4. This silica cannot react with Al or Fe, because the latter elements are effectively complexed by humic substances. In addition, different soil orders yield different types of solution from the biogeochemical compartment. For example, Spodosols yield a very acidic and yellow solution containing fulvic acids partially saturated with Al and Fe, and containing some N and P, whereas Mollisols yield a neutral solution that contains relatively small amounts of organic matter. In general, the water exiting the biogeochemical compartment is likely to contain low concentrations of nutrients such as N and P, and higher concentrations of Ca, Mg, K, and Na in solution, accompanied by HCO_3^- and NO_3^-, small amounts of phosphate, and very little Fe and Al.

8.4.1.2.3 Geochemical compartment. The geochemical compartment consists of some lower B horizons, C horizons, and bedrock flowpaths. The solution entering streamflow from the geochemical compartment generally contains very little organic matter. For this reason, organic C and N, as well as Al and Fe, are present only in very small amounts, if at all. In this mineral environment, pH is a critical factor affecting the mobility of P because it is often too alkaline or too acid for P to exist as a soluble anion. The load carried to rivers from the geochemical compartment consists mostly of alkaline cations (Mg^{2+}, Ca^{2+}, Na^+, K^+), and the anion HCO_3^-. Silicic acid (H_4SiO_4) losses are strongly attenuated under mild weathering conditions, and therefore the Si concentration of the water reaching a river from the geochemical compartment is relatively low.

The combined influences of runoff generation mechanisms, runoff flowpaths, and soil properties together control runoff chemistry. In spite of the wide range of interactions that characterize terrestrial environments, a few broad generalities can be offered, as the chemical composition of streamflow typically contains little H_4SiO_4, variable amounts of alkaline cations (Ca, Mg, Na, K), and the HCO_3^- anion. There is little organic C or N, and a virtual absence of Al, Fe, and P. Prediction of runoff chemistry is complicated because the specific runoff pathways operating in a particular environment can have a greater impact on the composition of the river water than the composition of the soil itself.

8.4.2 Sediment Load of Rivers

The material transported by rivers consists of dissolved ions (dissolved load), sediment suspended in the flow (suspended load), and sediment transported along the bed of the river (bedload). The total load and the proportion of the load represented by these phases varies widely among rivers in different environments. In particular, climate, topography, and erosion influence the amount and composition of riverine sediment loads.

A typical watershed can be considered to be composed of a source area in its uplands, where sediment is produced from rock and introduced into the fluvial system, transport zones in which there is little storage of material, and downstream depositional areas where substantial sediment storage may occur (Schumm, 1977). The specific characteristics of a watershed control the connections and time lags between initial erosion of a sediment and its ultimate delivery to the marine environment. Sediment storage in floodplains can substantially delay sediment delivery as material may be stored there as floodplain deposits for decades to millenia before it is reintroduced to the river. The time required to route sediment out of a drainage basin is important for interpreting the cause and timing of erosional events that are recorded in marine sediments. Church and Slaymaker (1989), for example, showed that reworking of late Quaternary deposits dominates the contemporary sediment yield of the Fraser River in British Columbia, Canada.

The sediment load of a channel and the sediment yield of its drainage basin are expressed in different ways. The sediment load is the total mass of material moved by the river in a specified period of time, whereas the sediment yield of a drainage basin is the sediment delivered from the basin divided by the basin area and typically expressed in tonnes/km^2 per year. Hence, the sediment yield for a basin may decrease downstream, due to sediment storage, even though the total sediment load of the river draining it increases downstream. Comparison of sediment yields allows assessment of relative erosion rates among drainage basins on an equal area basis, and thereby normalizes for the effect of basin size on sediment load.

8.4.3   Dissolved Load

The dissolved load of rivers consists of material leached from the soil. In most natural ecosystems, the water leaving the soil and entering rivers contains mainly the bicarbonate anion HCO_3^-, a primary product of mineral weathering, and alkaline cations (mostly Ca^{2+} and Na^+). Nutrients such as N and P are added to the rivers in small amounts by other processes – soil erosion at the edge of the river, runoff, and from direct inputs such as leaves and other organic debris. Therefore, the nutrient content of rivers and lakes is generally low. In a heavily forested and still pristine watershed of southeast Alaska, Stednick (1981) measured the output of N and P to be about 4.5 and 0.8 kg/ha per year, (1 ha = 10^3 m^3) while it reaches 185 kg/ha per year for HCO_3^-, 275 kg/ha per year for Ca, 40 kg/ha per year for Na and 80 kg/ha per year for Si (measured as H_4SiO_4). One of the most important anthropogenic impacts for temperate ecosystems is the huge input of nutrients such as N and P to rivers that can occur from clear cutting of forests on steep slopes, drainage of wetlands, excessive fertilizer application, and agricultural practices that induce topsoil erosion.

In arid areas, runoff is often the main source of water reaching the valley bottom. The rivers carry a high nutrient load consisting mainly of N and P. Throughout the world, estuaries of rivers draining arid lands, or the lakes they empty into, are incredibly rich in aquatic life. Flood plains located downstream of arid areas are also known for their rich soils. The Yellow River in

China, the Colorado in the United States, and the Nile in Africa, are only some of the most famous examples. The construction of dams along these rivers allows flood control, and water for irrigation and power, but the retention of sediment behind the dams severely impacts some of the richest and most productive ecosystems in the world.

In humid areas, the concentration of soil-derived elements in river water is generally low. With humid and tropical climates, the concentration of H_4SiO_4 in soil solution is dictated by the equilibrium with kaolinite and tends to be low (Tardy, 1969; Kittrick, 1977; Velbel, 1985; Clayton, 1986; Pavich, 1986). In parent materials that are well drained and coarse textured, the H_4SiO_4 concentration in the soil solution going through the lower mineral compartment can be low enough to lead to gibbsite formation, even in temperate climates (Green and Eden, 1971; Dejou et al., 1972; Macias-Vaquez et al., 1987). In wetlands, river loads originate within the organo-mineral compartment and the rivers are rich in nutrients; however, the acidity of the water inhibits development of aquatic biomass and wildlife, especially in cold climates.

8.4.4   Bedload and Suspended Load

The physical transport of particles in a river occurs by two primary modes: bedload and suspended load. Bedload consists of material moved along the bed of the river by the tractive force exerted by flowing water. Bedload may roll or hop along the bottom, and individual particles may remain stationary for long periods of time between episodes of movement. Suspended load consists of material suspended within the flow and that is consequently advected by flowing water. Rivers and streams are naturally turbulent, and if the upward component of turbulence is sufficient to overcome the settling velocity of a particle, then it will tend to remain in suspension because the particles become resuspended before they can settle to the bottom of the flow. Suspended load consists of the finest particles transported by a river, and in general is composed of clay- and silt-sized

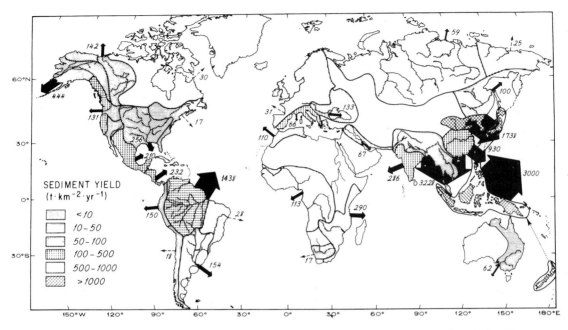

Figure 8-8   Annual suspended sediment discharge. Relative discharge is indicated by the width of the arrows. (From Milliman and Meade, 1983, reproduced with permission from University of Chicago Press.)

particles, but larger material can be suspended in especially fast water. The finest component of the suspended load is transported as washload that never settles and is therefore rapidly transported from a drainage basin. Suspended load constitutes the majority of sediment transported by most rivers, usually more than 80%, but there is substantial variability in the ratio of these two types of transport.

The travel time for suspended load is controlled by the flow velocity and the distance to the basin outlet. Flow velocities do not change much downstream in a typical river system (Leopold, 1953) and typically range from 0.1 to several m/s. Hence, suspended load should be able to travel at least 10 to 100 km per day and the travel time for suspended sediment to traverse even the longest rivers in the world should be less than a season. Although some of the suspended load will be deposited in floodplains, the component of the suspended load that does not get sequestered in terrestrial depositional environments is delivered almost as fast as the water that it flows in. Bedload travels much more slowly. In mountain drainage basins, the velocity of individual bedload clasts is on the

order of kilometers per year in steep mountain channels and may be on the order of only from one gravel bar to the next (tens to hundreds of meters) per year in lower-gradient pool-riffle type channels. Hence, we can think of a decoupling of transport rates, with transport of fine sediment involving relatively rapid response and coarser sediment potentially having much greater time lags between erosion and ultimate deposition in the ocean. At present, the combined global total for suspended load discharge into the oceans is ca. 20 Pg/yr (Milliman and Syvitski, 1992). A map showing the distribution of this discharge for rivers around the world is given in Fig. 8-8.

8.4.5   Climate, Topography, and Erosion

A fundamental distinction among landscapes is whether the net sediment flux (the total load carried by the river) is limited by the ability of erosional processes to carry sediment (transport-limited environments) or the availability of erodible material (weathering-limited environments). In general, soil-mantled landscapes can be considered transport limited, whereas landscapes with bedrock hillslopes are weathering limited. While the thickness of the soil mantle on hillslopes is dictated by the ratio of soil production to soil erosion rates (bedrock hillslopes result where erosion rates chronically exceed soil production rates), climate and vegetation are primary controls on both soil production and erosion. Wet climates with abundant vegetation tend to have soil-mantled landscapes, whereas dry climates with little vegetation tend to have relatively greater exposure of bedrock. Soil production increases with both greater precipitation and more extensive vegetation, whereas soil erosion rates also increase with greater precipitation but are inversely related to the extent of vegetation cover. The combined influence of vegetation acting to protect the ground surface and the tendency for higher rainfall to cause greater erosion leads us to expect a relation between mean annual precipitation and erosion rates in which erosion rates increase with increasing precipitation in arid areas to a maximum in semi-arid lands and then decrease in temperate and tropical areas due to the binding effect of vegetation counteracting the erosive potential of rainfall in areas with a complete ground cover of vegetation (Fig. 8–9).

The proportion of the total load of a river that is composed of dissolved load varies globally from close to 90% in the St. Lawrence river that drains low-gradient portions of Canada where numerous lakes act as effective traps for suspended and bedload material, to less than 10% for the Ganges river which receives a huge load of mechanical debris from the very high erosion rates in the Himalaya (Table 8-5). There are also general climatic controls on the relative importance of chemical and mechanical erosion on the composition of the load of rivers. The weathering history of a landscape is an important control on sediment yield (McLennan, 1993). Tropical areas with high rainfall and high temperatures tend to have high chemical loads, whereas cold environments tend to have relatively high mechanical loads due to relatively slow chemical weathering in such environments. However, watershed-specific geologic factors superim-

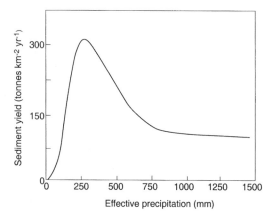

Fig. 8-9 Generalized variation of sediment yield with precipitation. (Modified from Langbein and Schumm, 1958.)

Table 8-5 Proportion of total load represented by dissolved load for large rivers. (Data from Summerfield, 1991)

River	Load (%)
Ganges	8
Amazon	18
Mississippi	20
Yukon	28
Zaire	42
Volga	64
St. Lawrence	89

pose substantial local variability onto such general global patterns.

Climate also exerts a primary control on the rates of mechanical erosion processes. While there has long been controversy over whether rivers or glaciers produce higher erosion rates, the consistently highest contemporary sediment yields come from glaciated catchments (Hallet *et al.*, 1996). Whether a glacier is frozen to its bed is also a very important control on the erosion rates, since such "cold bed" glaciers cause far less abrasion of their beds than do "warm bed" sliding glaciers. Hence, mountain drainage basins with valley glaciers in temperate environments typically have higher sediment yields than ice caps in polar landscapes.

Erosion rates by fluvial, hillslope, or glacial processes generally increase with increasing slope, and bedrock outcrops are found on the steepest slopes in most landscapes. Local relief (the difference between the highest and lowest points in an area of interest) is also correlated with erosion rates (Fig. 8-10) (Ruxton and McDougall, 1967; Ahnert, 1970), as areas with steep slopes tend to characterize areas of high relief and result in higher erosion. Deeply incised topography with steep slopes tends to produce the highest erosion rates.

Human or natural actions that significantly alter the erosion resistance of the ground surface can lead to dramatic increases in erosion rates. Channel networks dissect natural landscapes down to a fine-scale limit controlled by a threshold of channel initiation (Montgomery and Dietrich, 1992). In effect, channels begin where sufficient discharge collects to overcome the erosion resistance of the ground surface (Horton, 1945; Montgomery and Dietrich, 1988). Consequently, human or natural actions

that alter the critical shear stress of the ground surface can trigger expansion of the channel network to generate a network of rills, small channels and gullies that can rapidly dissect formerly undissected hillslopes, leading to very high sediment yields. Different styles of land use can result in very different erosion rates. For example, Dunne (1979) studied erosion rates from undisturbed forest, cut forest, and agricultural lands in Kenya and showed that sediment yields were substantially higher in managed landscapes where ground distur-bance was common (as shown in Fig. 8-11). Human-induced acceleration of erosion can result in sediment yields that greatly exceed the range of natural variability in all but the most catastrophically disturbed environments. Accelerated erosion from human activity is thought to have doubled the sediment yield to oceans over the past several thousand years, although recent dam construction has reduced this effect (Hay, 1998).

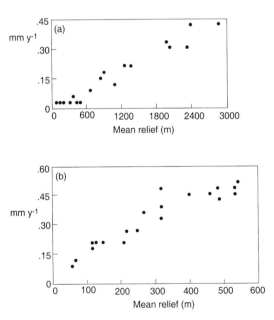

Fig. 8-10 Erosion rate as a function of relief for (a) mid-latitude medium-sized drainage basins (modi-fied from Ahnert, 1970), and (b) for Hydrographer's Volcano, Papua New Guinea (data source: Ruxton and McDougall, 1967).

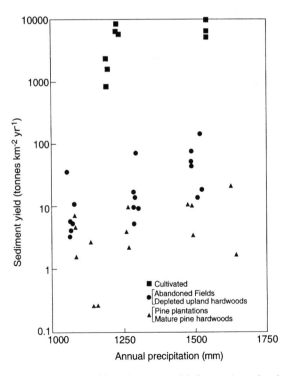

Fig. 8-11 Annual sediment yield for various land uses in northern Mississippi vs. annual precipitation. (After Ursic and Dendy, 1965.)

8.4.5.1   Sediment delivery

Sediment transport rates by fluvial processes are proportional to the product of flow depth and the energy gradient, or water surface slope (Richards, 1982). Hence, sediment transport increases with either deeper flow or steeper slopes, and conversely sediment transport rates decrease in shallower flow or on gentler slopes. Consequently, the transport capacity of a river decreases where flow spreads out across wide valley bottoms (such as in floodplains), or where channel gradients decrease abruptly (such as in alluvial fans at the foot of mountain ranges). The downstream decrease in transport capacity in turn leads to deposition of sediment in these locations. The proportion of the sediment eroded from a drainage basin that actually makes it to the mouth of the basin defines the sediment delivery ratio for the basin. This ratio tends to decrease with increasing basin size due to storage effects along floodplains or in structural valleys, and ranges from virtually complete delivery for short, steep catchments to less than half of the sediment load for large, continent-spanning rivers. In large floodplain river systems, such as the Amazon and the Ganges-Brahmaputra, roughly a third of the sediment load can be sequestered into long-term storage in floodplain sediments (Dunne *et al.*, 1998; Goodbred and Kuehl, 1998).

8.5   Marine Sediments

Sediments record aspects of the environments from which they were eroded and into which they were deposited. Consequently, a sequence of sedimentary layers contains information about environmental changes over time. Sediments are therefore a library of Earth history that can be read through an understanding of the processes behind their formation and post-depositional evolution. The recent sedimentary record reveals cultural impacts on the environment during the industrial era, while ancient sedimentary rocks provide insight into the evolution of free oxygen in the atmosphere during the Precambrian Era. But sediments are not simply inert records of past conditions; during formation and diagenesis they take an active part in biogeochemical cycles.

8.5.1   Formation of Sediments

Disintegration of rocks by weathering processes produces three types of weathering products: detrital material from the rock and vegetation; solutes from the dissolution of minerals and organic matter; and new minerals from chemical reactions between solutes and minerals. Marine sediments are formed by the deposition of these particles once they are transported to the ocean by rivers, glaciers and wind. Episodic delivery processes result in marine sediments that consist of depositional *layers* that record discrete depositional events, whereas a soil exhibits weathering *horizons* that record developmental processes, as discussed in Section 8.3.2. The mineral composition of detrital material depends on the type of source rock, and also on the duration and intensity of weathering and transport.

8.5.2   Composition of Sediments

Like soil, sediment contains three phases, and interactions among these phases can alter the nature of sediments through time.

8.5.2.1   The solid phase

The solid components of sediments are classified in a number of ways. Grain size conveys some basic information for exchange processes and correlates to a certain extent with the minerogenic distribution. Clay minerals, for example, occur in the $<2\,\mu m$ size fraction. For biogeochemical purposes, however, a granulometric description is far from sufficient. A more satisfactory classification is based on the material and genetic differences of marine sediments (Table 8-6). The arrangement of the particles is of even greater importance as it governs the permeability and porosity of the sediments. Marine sediments (with few exceptions) are composed of relatively porous clay aggregates and ran-

Table 8-6   Material-genetic classification of marine sediments[a]

Sediment type	% CaCO	% clastic and clayey material	% amorphous silica	% pelagic sed.	Composition
Detrital or epiclastic	<30	>50			Denudation products of continental rocks
Biogenic					
A. Calcareous	>30			~48	Foraminifera, coccoliths, calcareous algae, molluscs, bryozoa, and corals
B. Siliceous			>30	~14	Diatoms and radiolaria
Chemogenic					Iron-manganese nodules, glauconite, phosphorite, nodules, phillipsite, palagonite, celestobarite, and evaporites
Volcanogenic					Pyroclastic material
Polygenic	<10	>50	<10	~38	Red clay

[a] Modified from Lisitzin (1972) and Sverdrup *et al.* (1942).

domly distributed coarser grains of detrital material (Fig. 8-12).

The aggregates of sedimentary particles are usually arranged before deposition. Within these aggregates, forces are set up between atoms, molecules, and ions that depend on the ionic strength of the electrolyte. As a result, the arrangements of the particles are different in brackish and marine environments. Clays deposited in fresh water form relatively porous aggregates with small voids, while marine clays form large dense aggregates separated by large voids. Salinity variations in interstitial water are most significant in coastal areas and different degrees of permeability and porosity therefore may be expected in sediments of the continental margins as compared to deep sea sediments of the same grain size.

8.5.2.2   The liquid phase

The liquid phase of sediment consists of three different types of aqueous electrolytic solutions: (1) "normal" water of random ionic ordering at some distance from a solid surface; (2) adsorbed

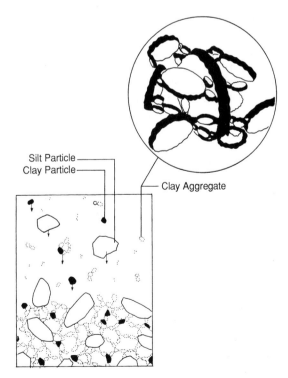

Fig. 8-12   Hypothetical particle arrangement in a sediment and a clay aggregate. (Adapted from Casagrande, 1940.)

water, essentially free from ions, on a solid surface; and (3) "structured" water with a layering of ionic ordering of types (1) and (2) near a solid surface. Because of the close relationship to the mineral particles in the sediment, interlamellar water is usually of types (2) and (3). The various types of water in sediments strongly influence other processes taking place in the aqueous phase.

8.5.2.3 The gaseous phase

The gaseous phase of the sediment–water interface includes oxygen, nitrogen, hydrogen sulfide, methane, carbon dioxide, and ammonia. The last-named occurs principally as the ammonium ion at pH values in the range 5 to 8. Oxygen and hydrogen sulfide react with each other spontaneously, and consequently these two gases are usually regarded as non-coexistent. In a sediment, however, heterogeneity of organic matter and permeability give rise to microenvironments of different chemical composition that can exist close to one another as more or less isolated chemical and biological systems. The presence of hydrogen sulfide is not therefore a reliable indicator of entirely anoxic conditions within the sediment. Fecal pellets may serve as such microniches at the uppermost part of the sediment, while burrowing organisms may distribute oxygen to the reduced part of the sediment.

Methane is produced by bacteria in the strongly reduced part of the sediment. Methane-producing bacteria are obligatory anaerobes and are most active below the zone of hydrogen sulfide production. Methane is used as a carbon source by heterotrophic microorganisms in the overlaying sediments. However, methane concentrations decrease rapidly below the sediment–water interface, and its importance for the biogeochemical cycles of carbon in the marine system is negligible.

Carbon dioxide is produced as a result of metabolism of all heterotrophic organisms. The concentrations of CO_2 in pore water of reduced sediments are therefore high. Autotrophic microorganisms consume CO_2 in the oxidized part of the sediment, which can vary in depth from a meter in deep sea sediments to a few mm

in organic-rich sediments of the continental margins. The buffer capacity of ocean water is very low and an addition of 0.5 mmol/kg of CO_2 will reduce the pH of seawater by more than one pH unit. It is therefore surprising that the concentration of CO_2 in pore water of marine sediments may reach values of 60 mmol/kg (Presley, 1969), and yet show pH shifts less than one unit. The increase in CO_2 therefore must be counterbalanced by other processes that tend to increase the pH (such as production of ammonia and increase in total alkalinity) so that the result is a fairly constant pH.

8.5.3   Diagenesis

Physico-chemical reactions within the sea–sediment interface tend to reach equilibrium. Those reactions that are so rapid that they occur prior to burial are referred to as halmyrolysis (e.g. formation of clay aggregates) while those that take place in the sediment are termed diagenesis. The diagenetic processes include cementation, compaction, diffusion, redox reactions, transformation of organic and inorganic material, and ion exchange phenomena. Only a short survey of some diagenetic processes is presented here; for a more detailed presentation of diagenetic processes the reader is referred to Berner (1980).

8.5.3.1   Cementation

Cementation is the precipitation of a binding material around grains, thereby filling the pores of a sediment. Berner (1971, p. 97) states that "cementation by silica must be predominantly a phenomenon of later diagenesis because almost no examples are found in recent marine sediments." In contrast, cementation by calcium carbonate may occur rapidly after deposition. A good example is beachrock, a mix of beach and intertidal sand (usually carbonate and skeletal fragments) cemented by $CaCO_3$ in subtropical to tropical climates. The cement forms so rapidly that human artifacts only a few decades old are commonly found cemented into the beachrock. Other examples of early diagenetic $CaCO_3$ cementation are provided by subtidal lithified calcarenites, reefs, and pelagic oozes (Mackenzie *et al.*, 1969). In all cases the cements are aragonite or high magnesium calcite indicating formation from seawater or other magnesium-rich solutions.

8.5.3.2 Compaction

Compaction is the decrease in volume of a sediment resulting primarily from an expulsion of water due to compression from deposition of overlying sediment. Rosenqvist (1962) stated on the basis of a study of approximately 100 stereoscopic micrographs of undisturbed marine clays that the particle arrangements within the clay aggregates were of the "corner/plane cardhouse" type (Fig. 8-12) suggested by Tan (1957) and Lambe (1958). From microstructural investigations of these "cardhouse" aggregates (Pusch, 1970) it can be concluded that during the compaction process, the aggregates approach one another and assume new, more stable orientations.

Thus the initial state of compaction results in a collapse of the more unstable original structures and a rearrangement of the particles into a tighter packing. More efficient packing will expel interstitial water from the sediment and decrease the porosity. As different clay minerals exhibit different packing characteristics, which in turn depend upon the chemical composition of the interstitial water, relationships become too complex for a simple generalization (Meade, 1966). The total compaction that occurs during accumulation of a sediment layer can be calculated from the following formula with the assumption that the rate of burial of sediment and the diagenetic processes, including bioturbation, are constant:

$$\text{Compaction} = \frac{h_0 - h}{h_0} = \frac{n_0 - n}{1 - n} \quad (12)$$

where h_0 = thickness of the layer at deposition; h = observed thickness of the layer; n_0 = porosity of the top-most sedimentary layer; n = porosity of the observed layer.

For the upper part of the sediment, where the sea–sediment interaction is most predominant, steady-state compaction is a reasonable assumption, as porosity and compaction undergo a linear change during burial.

8.5.4 Diffusion

Diffusion as referred to here is molecular diffusion in interstitial water. During early diagenesis the chemical transformation in a sediment depends on the reactivity and concentration of the components taking part in the reaction. Chemical transformations deplete the original concentration of these compounds, thereby setting up a gradient in the interstitial water. This gradient drives molecular diffusion. Diffusional transport and the kinetics of the transformation reactions determine the net effectiveness of the chemical reaction.

The presence of particles in the fluid medium complicates diffusion in a sediment due to the effects of porosity, represented by n, and tortuosity. Since tortuosity of natural sediments is seldom known it is more convenient to use the term "formation factor" or "lithological factor," denoted L, which takes into account everything but porosity. Fick's diffusion constant D is replaced by the whole sediment diffusion constant D_s, where $D_s < D$.

$$D_s = \frac{Dn}{L} \quad (13)$$

Manheim (1970) presents a critical review of the confusing variety of D_s data found in the literature, but according to Horne *et al.* (1969), pressures up to several thousand atmospheres seem to have no significant effect on D_s. Values for D_s are sometimes not available or are difficult to estimate but can be obtained indirectly by means of electrical conductivity measurements (Klinkenberg, 1951). Table 8-7 provides a list of D_s values found in the literature for the most

Table 8-7 Diffusion constants of sediments (D_s) for some common chemical species[a]

Na^+, K^+, NH_4^+, Cl^-, NO_3^-	5–7×10^{-6} cm^2/s
Ca^{2+}, Mg^{2+}, Mn^{2+}, Fe^{2+}, SO_4^{2-}	3–5×10^{-6} cm^2/s
Fe^{3+}, PO_4^{3-}, amorphous silica	1–3×10^{-6} cm^2/s

[a] The data given should serve only as reference values following the rule, the higher the ionic potential, the thicker the hydration layer of the water molecules around the ion, and the slower the ionic diffusion. Cations generally diffuse more rapidly than anions.

common dissolved chemical species in sediment.

Interactions between diffusion and chemical transformation determine the performance of a transformation process. Weisz (1973) described an approach to the mathematical description of the diffusion–transformation interaction for catalytic reactions, and a similar approach can be applied to sediments. The Weisz dimensionless factor compares the time scales of diffusion and chemical reaction:

$$\phi = \frac{\tau_{\text{diffusion}}}{\tau_{\text{reaction}}} = \frac{\left(\dfrac{R^2}{D}\right)}{\left(\dfrac{c}{dc/dt)}\right)} \qquad (14)$$

where R is a characteristic size of sediment particles. Using this approach, we can identify the following cases: (1) when $\phi < 1$ the time scale for diffusion is short compared with that for chemical reaction and reactivity will determine the rate of change of concentration; and (2) for $\phi > 1$ the time scale for diffusion is long and diffusion will play an important role in determining the reaction rate.

Finally, it must be stressed that diffusion of dissolved species in solutions is a key physicochemical process for the sea/sediment interaction and energy exchange at the sediment–water interface. The reader is referred to Cussler (1984) for a more comprehensive presentation of diffusion in fluid systems.

8.5.5   Redox Conditions

Reducing and oxidizing conditions in a sediment determine the chemical stability of the solid compounds and the direction of spontaneous reactions. The redox state can be recognized as a voltage potential measured with a platinum electrode. This voltage potential is usually referred to as E or E_h defined by the Nernst equation, which was introduced in Chapter 5, Section 5.3.1:

$$E_h = E_h^0 - \left(\frac{RT}{nF}\right) \ln \frac{[\text{Products}]}{[\text{Reactants}]} \qquad (15)$$

The Nernst equation is applicable only if the redox reaction is reversible. Not all reactions are completely reversible in natural systems; activities of reacting components may be too low or equilibrium may be reached very slowly. In a sediment, the biotic microenvironment may create a redox potential that is different from the surrounding macroenvironment. For this reason, measurements of E_h in natural systems must be cautiously evaluated and not used strictly for calculations of chemical equilibria. Calculations of redox equilibria are in some cases valuable, in the sense that they will give information about the direction of chemical reactions.

If a system is not at equilibrium, which is common for natural systems, each reaction has its own E_h value and the observed electrode potential is a mixed potential depending on the kinetics of several reactions. A redox pair with relatively high ion activity and whose electron exchange process is fast tends to dominate the registered E_h. Thus, measurements in a natural environment may not reveal information about all redox reactions but only from those reactions that are active enough to create a measurable potential difference on the electrode surface.

In marine sediments, usually only the uppermost layer of the sediment exhibits oxidizing conditions while the rest is reduced. The thickness of the oxidized layer and the reducing capacity of the sediment below depend on:

1. concentration of oxygen in the ambient bottom water;
2. the rate of oxygen penetration into the sediment;
3. the accessibility of utilizable organic matter for the bacterial activity.

The depositional rate of organic matter is higher in the coastal areas than in the open sea. Even though the exchange of water is higher in shallow areas of the ocean the deep water is not deficient in oxygen. We should thus expect to find a general trend of thicker oxidized sediments with increasing distance from the shoreline. Lynn and Bonatti (1965), for example, found that the thickness of the oxidized upper sediment in the Pacific Ocean between $15° \text{S}$ and $20° \text{N}$ increased with distance from the continent. Thicknesses were <1 cm near the shore and 8–15 cm at distances of 800 km offshore.

The inflection point of the redox gradient, constituting the boundary between oxidizing and reducing environments in sediments lies at around +250 mV. This boundary, the redox-cline (Hallberg, 1972), is recognized as comparable to other natural boundaries, such as the halocline and thermocline. The redoxcline is usually situated close to the sediment–water interface resulting in a redox turnover from oxidizing to reducing conditions during early diagenesis. The redox turnover will, in turn, produce disequilibrium within the sediment, causing dissolution of certain minerals and compounds and increased ion exchange across the redoxcline. The change of redox potential is a useful tool for describing the sedimentary environment.

8.5.6 Bioturbation

In oceanic sediments macro- (>1 mm) and microorganisms play an important role in the mixing of surface sediment layers. Burrowing by organisms in marine sediments is so common that it is the preservation of depositional structures that requires explanation, not their destruction (Arrhenius, 1952, p. 86). Bioturbation is the mixing of sediments by *in-situ* fauna, and is usually attributed to macro- and meso-fauna (0.1–1 mm of sieved sediments). Very little is known about the role of microfauna. The mixing of sediments by the former can easily be observed in coastal areas where their abundance is high and the resuspension and cycling of the annual sediment influx can be as high as 99% (Young, 1971). In anaerobic sediments, however, where macro- and mesofauna are absent due to lack of oxygen, there is still a large community of microorganisms; 10^7 individuals/cm^3 of sediments is not an uncommon value. Several types are motile and have been observed to swim 3 mm/day (Oppenheimer, 1960). If all bacteria in a cubic centimeter of sediment moved 1 mm/day they would cover a total distance of 10 km/day. These organisms cannot move particles but may have a significant effect on the mixing of the interstitial water, and thus also on the exchange between water and sediments.

8.6 Soils, Weathering, and Global Biogeochemical Cycles

The soil may represent a thin film on the surface of the Earth, but the importance of soils in global biogeochemical cycles arises from their role as the interface between the Earth, its atmosphere, and the biosphere. All terrestrial biological activity is founded upon soil productivity, and the weathering of rocks that helps to maintain atmospheric equilibrium occurs within soils. Soils provide the foundation for key aspects of global biogeochemical cycles.

The soil, an open system in the context of biogeochemical cycles, receives inputs and outputs of C and N, and mineral elements and is the foundation of the primary production of terrestrial ecosystems. The amount of carbon present in the soil is closely connected to the CO_2 concentration of the atmosphere, but atmospheric CO_2 is regulated mainly by the ocean rather than by the soil (see Chapters 10 and 11). The amount of N in the soil also does not influence the N in the atmosphere because the atmosphere is a huge reservoir regulated mainly by the ocean (see Chapter 12). Nevertheless, the soil has a tremendous influence on the nitrate load of rivers. Biogeochemical processes that occur in soils and the processes controlling the delivery of the products of these reactions to the oceans exert profound influences on global biogeochemical cycles.

Weathering processes take an active part in the cycling of oxygen and carbon, but does chemical weathering affect these cycles to a significant extent? Consider the following examples.

Oxygen is formed by photosynthesis according to the reaction

$$CO_2 + H_2O \rightarrow O_2 + CH_2O \tag{16}$$

The annual primary production of organic carbon through photosynthesis is on the order of 70 Pg/yr. The major part of this carbon is decomposed or respired in a process that also involves the biogeochemical transformation of nitrogen, sulfur, and many other elements. Only a small part of the annual primary production of organic carbon escapes decomposition and is buried in marine sediments. On average, sediments and sedimentary rocks contain 3% carbon, which corresponds to a net production of oxygen of about 300 Tg/yr. This production has the potential to change atmospheric oxygen over a time scale of 4 Myr. In contrast, the fossil record indicates that the partial pressure of oxygen in the atmosphere has fluctuated very little, at least during the last 600 Myr (Conway, 1943). Obviously there must be a sink that can accommodate the net annual production of oxygen. This sink is the annual weathering of rocks, which can be estimated from the approximately 18×10^{23} g of sediments formed during the last 600 Myr (Gregor, 1968). During this period, soil formation and sediment formation have fluctuated greatly, but the average weathering rate is 30×10^{14} g/yr. The average continental rock contains ferrous iron and sulfide sulfur (mainly as pyrite) and organic carbon. During the weathering processes these constituents are oxidized and consume oxygen. The amounts of oxygen necessary for each of these reactions are given in Table 8-8.

The total annual consumption of oxygen by weathering can be estimated (Holland, 1978) as follows:

$$(20 \pm 6 \text{ g O}_2/\text{kg rock}) \times (2 \times 10^{13} \text{ kg rock})$$
$$\approx 4 \times 10^{14} \text{ g O}_2 \qquad (17)$$

Note that this estimate of the annual O_2 loss to weathering processes is approximately equal to the estimated annual production of oxygen estimated above. Hence, the weathering of rocks and burial of organic carbon in sediments during their formation are important processes for the oxygen content of the atmosphere.

Weathering of rocks is also a sink for CO_2. Garrels and Mackenzie (1971) have estimated that the formation of 1500 g of sedimentary rocks requires 100 g of CO_2. If we use the same

number as above for the annual formation of sedimentary rocks, we have a CO_2 sink of 200 Tg/yr. About half of that comes from the oxidation of organic carbon in the weathered rock. The rest is from the atmosphere. The burning of fossil fuel increases the partial pressure of CO_2 in the atmosphere, but increased CO_2 levels increase the weathering rate. Thus, over geologic time, the weathering rate and sediment formation ultimately control the carbon dioxide content of the atmosphere.

Questions

8-1 Discuss how physical weathering operates in each of the following environments: (1) sea shore, (2) hot desert, (3) temperate forest.

8-2 Rewrite the weathering reactions shown in Section 8.3.2.2 using HNO_3 in place of H_2CO_3.

8-3 Why do we speak in terms of soil horizons and sediment layers?

8-4 Discuss the significance of clay minerals in a description of the solid phase of a sediment.

8-5 Give examples of early diagenetic processes.

8-6 Why do continental margins play a dominant role on the biogeochemical cycling of elements?

8-7 Give examples of bacterial transformations in a sediment that are of special importance for biogeochemical cycles.

8-8 Write a balanced reaction for formation of hematite from Fe^{2+}.

8-9 Explain the difference between a Mollisol and a Spodosol. How would cycling of Ca differ in each? N?

8-10 Track the possible fate of a K^+ ion from the moment it is released by weathering in a soil to its burial at sea.

References

Ahnert, F. (1970). Functional relationships between denudation, relief, and uplift in large mid-latitude drainage basins. *Am. J. Sci.* **268**, 243–263.

Arrhenius, G. (1952). *Rep. Swedish Deep-Sea Expedition (1947–1948)* **5**, 227.

Berner, R. A. (1971). "Principles of Chemical Sedimentology." McGraw-Hill, New York.

Berner, R. A. (1980). "Early diagenesis – a Theoretical Approach." Princeton University Press, Princeton, NJ.

Table 8-8 Average oxygen consumption during weathering of rocks[a]

$C + O_2 \rightarrow CO_2$	12 ± 3 g O_2/kg
$S^{2-} + 2O_2 \rightarrow SO_4^{2-}$	6 ± 2 g O_2/kg
$4 \text{ ''FeO''} + O_2 \rightarrow 2Fe_2O_3$	2 ± 1 g O_2/kg

[a] After Holland (1978, p. 285).

Birkeland, P. W. (1999). "Soils and Geomorphology." 2nd edn. Oxford University Press, New York.

Black, T. A. and Montgomery, D. R. (1991) Sediment transport by burrowing mammals, Marin County, California. *Earth Surf. Process. Landforms* **16**, 163–172.

Blackwelder, E. B. (1927). Fire as an agent in rock weathering. *J. Geol.* **35**, 134–140.

Bohn, H. L., McNeal, B. L., and O'Conner, G. A. (1985). "Soil Chemistry," 2nd Edn. John Wiley, New York.

Brady, N. C. and Weil, R. (1999). "The Nature and Properties of Soils," 12th edn. Prentice Hall.

Butler, D. R. (1995) "Zoogeomorphology: Animals as Geomorphic Agents", Cambridge University Press, Cambridge.

Campbell, I. B. and Claridge, G. G. C. (1987). "Antarctica: Soils, Weathering Processes and Environment." Developments in Soil Science, Vol 16. Elsevier, Amsterdam, The Netherlands.

Casagrande, A. (1940). The structure of clay and its importance in foundation engineering. *J. Boston Soc. Civ. Engrs* **19**(4), 168–209.

Church, M. and Slaymaker, O. (1989). Disequilibrium of Holocene sediment yield in glaciated British Columbia. *Nature* **337**, 452–454.

Claridge, G. G. C. and Campbell, I. B. (1984). Mineral transformations during weathering of dolerite under cold arid conditions. *N.Z. J Geol. Geophys.* **25**, 537–545.

Clayton, J. L. (1986). An estimate of plagioclase weathering rate in the Idaho batholith based upon geochemical transport rates. *In* "Rates of Chemical Weathering of Rocks and Minerals" (S. M. Coleman and D. P. Dethier, eds), Chap. 19, pp. 453–466. Academic Press, New York.

Colman, S. M. and Dethier, D. P. (eds) (1986). "Rates of Chemical Weathering of Rocks and Minerals." Academic Press, New York.

Conway, E. J. (1943). The chemical evolution of the ocean. *Proc. R.I.A.* **48B**, 161–212.

Cronan, C. S. (1984). Biogeochemical responses of forest canopies to acid precipitation. *In* "Direct and Indirect Effects of Acidic Deposition on Vegetation" (R. A. Linthurst, ed.), pp. 65–79. Butterworth, Boston, MA.

Cussler, E. L. (1984). "Diffusion-Mass Transfer in Fluid Systems." Cambridge University Press, Cambridge.

Darwin, C. (1896). "The Formation of Vegetable Mould Through the Action of Worms with Observations of their Habitats," pp. 305–313. D. Appleton, New York.

Dejou, J., Guyot, J., and Chaumont, C. (1972). La gibbsite, minéral banal d'altération des formations superficielles et des sols dévelopés sur roches cristallines dans les zones tempérées humides. *Int. Geol. Cong., 24th*, Section 10, pp. 417–425.

Dochaufour, P. (1982). "Pedology" (translated by T. R. Patan). George Allen and Unwin, London.

Dunn, J. R. and Hudec, P. P. (1972). Frost and sorbtion effects in argillaceous rocks. *In* "Frost action in Soils." Highway Research Board, *Natl. Acad. Sci.– Natl. Acad. Eng.* **393**, 65–78.

Dunne, T. (1979). Sediment yield and land use in tropical catchments. *J. Hydrol.* **42**, 281–300.

Dunne, T. and Black, R. D. (1970) Partial area contributions to storm runoff in a small New England watershed, *Water Resour. Res.* **6**, 1296–1311.

Dunne, T. and Leopold, L. B. (1978) "Water in Environmental Planning." W. H. Freeman, San Francisco.

Dunne, T., Mertes, L. A. K., Meade, R. H., Richey, J. E., and Forsberg, B. R. (1998). Exchanges of sediment between the flood plain and channel of the Amazon River in Brazil. *Geol. Soc. Am. Bull.* **110**, 450–467.

Edmonds, R. L. (1979). Decomposition and nutrient release in Douglas-fir needle litter in relation to stand development. *Can. J. For. Res.* **9**, 132–140.

FAO (1971–1981). "The FAO–UNESCO Soil Map of the World." Legend and 9 volumes. UNESCO, Paris.

Friedman, E. I. (1971). Light and scanning electron microscopy of the endolithic desert algal habitat. *Phycologia* **10**, 411–428.

Garrels, R. M. and Christ, C. L. (1965). "Solutions, Minerals, and Equilibria." Harper and Row, New York.

Garrels, R. M. and Mackenzie, F. T. (1971). "Evolution of Sedimentary Rocks." W. W. Norton, New York.

Goodbred, S. L., Jr and Kuehl, S. A. (1998). Floodplain process in the Bengal Basin and the storage of Ganges-Brahmaputra River sediment: an accretion study using ^{137}Cs and ^{210}Pb geochronology. *Sediment. Geol.* **121**, 239–258.

Goudie, A., Cooke, R., and Evans, I. (1970). Experimental investigation of rock weathering by salts. *Area* **4**, 42–48.

Green, C. P. and Eden, M. J. (1971). Gibbsite in weathered Dartmoor granite. *Geoderma* **6**, 315–317.

Gregor, C. B. (1968). The rate of denudation in Post-Algonkian time. *Koninkl. Ned. Akad. Wetenschap. Proc.* **71**, 22.

Hallberg, R. O. (1972). Sedimentary sulfide mineral formation – an energy circuit system approach. *Mineral. Deposit.* **7**, 189–201.

Hallet, B., Hunter, L., and Bogen, J. (1996). Rates of erosion and sediment yield by glaciers: A review of field data and their implications. *Glob. Planet. Change* **12**, 213–235.

Hay, W. W. (1998). Detrital sediment fluxes from continents to oceans. *Chem. Geol.* **145**, 287–323.

Hewlett, J. D. and Hibbert, A. R. (1967) Factors affecting the response of small watersheds to precipitation in humid areas. *In* "International Symposium on Forest Hydrology" (W. E. Sopper and W. H. Lull, eds), pp. 275–290. Pergamon, New York.

Holland, H. D. (1978). "The Chemistry of the Atmosphere and Oceans." Wiley-Interscience, New York.

Horne, R. A., Day, A. F., and Young, R. P. (1969). Ionic diffusion under high pressure in porous solid materials permeated with aqueous, electrolytic solution. *J. Phys. Chem.* **73**, 2782–2783.

Horton, R. H. (1933) The role of infiltration in the hydrologic cycle, EOS. *Trans. Am. Geoph. U.* **14**, 446–460.

Horton, R. E. (1945) Erosional development of streams and their drainage basins: Hydrophysical approach to quantitative morphology. *Bull. Geol. Soc. Am.* **56**, 275–370.

Hudec, P. P. (1974). Weathering of rocks in arctic and subarctic environments. *In* "Canadian Arctic Geology: Symposium on the Geology of the Canadian Arctic, Saskatoon 1973" (J. D. Aitken and D. J. Glass, eds), pp. 313–335. Geological Association of Canada/Canadian Society for Petrology and Geology, Calgary, Alberta.

Huang, P. M. (1989). Feldspars, olivines, pyroxenes, and amphiboles. *In* "Minerals in Soil Environments," 2nd edn (J. B. Dixon and S. B. Weed, eds), pp. 975–1050. Soil Sci. Soc. Am., Madison, WI.

Hursh, C. R. (1936) Storm-water and adsorption, EOS. *Trans. Am. Geoph. U.* **17**, 301–302.

Jackson, M. L. (1964). Chemical composition of soils. *In* "Chemistry of Soil" (F. E. Bear, ed.), pp. 71–151. Reinholt, New York.

Jackson, M. L. (1969). "Soil Chemical Analysis." Advanced Course. 2nd edn, 1973. Dep. of Soil Sci., Univ. of Wisconsin, Madison, WI.

Jackson, M. L. and Sherman, G. D. (1953). Chemical weathering of minerals in soils. *Adv. Agron.* **5**, 219–318.

Jenny, H. (1941). "Factors of Soil Formation: A System of Quantitative Pedology." McGraw-Hill, New York.

Johnson, D. W. and Lindberg, S. E. (1992). "Atmospheric deposition and forest nutrient cycling: a synthesis of the integrated forest study." Springer-Verlag, New York.

Kittrick, J. A. (1977). Mineral equilibria and the soil system. *In* "Minerals in Soil Environments" (J. B.

Dixon and S. B. Weed, eds), pp. 1–25. Soil Sci. Soc. Am., Madison, WI.

Klinkenberg, L. J. (1951). Analogy between diffusion and electrical conductivity in porous rocks. *Bull. Geol. Soc. Am.* **62**, 559–563.

Lambe, T. W. (1958). The structure of compacted clay. J. Soil Mech. Found. Div., Proc. ASCE, SM 2; 84, paper 1654.

Langbein, W. B. and Schumm, S. A. (1958) Yield of sediment in relation to mean annual precipitation, *Trans. Am. Geoph. U.* **39**, 1076–1084.

Lautridou, J. P. and Ozouf, J. C. (1982). Experimental frost shattering: 15 years of research at the Center de Geomorphologie du CNRS." *Prog. Phys. Georg.* **6**, 215–232.

Leopold, L. B. (1953) Downstream change of velocity in rivers. *Am. J. Sci.* **251**, 606–624.

Lisitzin, A. P. (1972). Sedimentation in the world ocean. *Soc. Econ. Paleontol. Mineral. Spec. Publ.* **17**, 1–218.

Loudermilk, W. C. (1934) The role of vegetation in erosion control and water conservation. *J. Forest.* **32**, 529–536.

Lundstrom, U. S. (1994) Significance of organic acids for weathering and the podzolization process. *Environ. Int.* **20**, 21–30.

Lynn, D. C. and Bonatti, E. (1965). Mobility of manganese in diagenesis of deep-sea sediments. *Marine Geol.* **3**, 457–474.

Macias-Vasquez, F. M., Fernandez-Marcos, L., and Chesworth, W. (1987). Transformations mineralogique dans les podzols et les sols podzolique de Galice (NW Espagna). *In* "Podzols et Podzolisation" (D. Righi and A. Chauvel, eds), pp. 163–177. Inst. Nat. de la Recher. Agronomique, Plaisir et Paris.

Mackenzie, F. T., Ginsburg, R. N., Land, L. S. and Bricker, O. P. (1969). Carbonate cements. *Bermuda Biological Station for Research Special Publication No. 3*, 325.

Manheim, F. T. (1970). The diffusion of ions in unconsolidated sediments. *Earth Planet. Sci. Lett.* **9**, 307–309.

Mason, B. and Moore, C. B. (1982). "Principles of Geochemistry," 4th edn. John Wiley, New York.

McLennan, S. M. (1993). Weathering and global denudation. *J. Geol.* **101**, 295–303.

Meade, R. H. (1966). Factors influencing the early stages of the compaction of clays and sands – review. *J. Sediment. Petrol.* **36**, 1085–1101.

Milliman, J. D. and Meade, R. H. (1983). World-wide delivery of river sediment to the oceans. *J. Geol.* **91**, 1–21.

Milliman, J. D. and Syvitski, J. P. M. (1992). Geomorphic/tectonic control of sediment to the ocean: The importance of small mountainous rivers. *J. Geol.* **100**, 525–544.

Montgomery, D. R. and Dietrich, W. E. (1988). Where do channels begin? *Nature* **336**, 232–234.

Montgomery, D. R. and Dietrich, W. E. (1992). Channel initiation and the problem of landscape scale. *Science* **255**, 826–830.

Oades, J. M. (1989). An introduction to organic matter in mineral soils. *In* "Minerals in Soil Environments," 2nd edn (J. B. Dixon and S. B. Weed, eds), pp. 89–159. Soil Sci. Soc. Am. Madison, WI.

Ollier, C. D. (1969). "Weathering." Oliver and Boyd, Edinburgh.

Oppenheimer, C. H. (1960). Bacterial activity in sediments of shallow marine bays. *Geochim. Cosmochim. Acta* **19**, 244–260.

Parfitt, R. L. (1980). Chemical properties of variable charge soils. *In* "Soils with Variable Charge" (B. K. G. Theng, ed.), pp. 167–194. New Zealand Society of Soil Science Offset Publications, Palmerston North, New Zealand.

Paul, E. A. and Clark, F. E. 1996. "Soil Microbiology and Biochemistry". Academic Press, San Diego.

Pavich, M. J. (1986). Processes and rates of saprolite production and erosion on a foliated granitic rock of the Virginia Piedmont. *In* "Rates of Chemical Weathering of Rocks and Minerals" (S. M. Coleman and D. P. Dethier, eds), pp. 551–590. Academic Press, New York.

Pedro, G. (1982). The conditions of formation of secondary constituents. *In* "Constituents and Properties of Soils" (M. Bonneau and B. Souchier, eds), pp. 63–81. Academic Press, New York.

Presley, B. J. (1969). Chemistry of interstitial water from marine sediments. Ph.D. thesis, Univ. Calif., Los Angeles.

Pusch, R. (1970). Clay microstructure. National Swedish Building Research. Document D8.

Rai, D. and Kittrick, J. A. (1989). Mineral equilibria and the soil system. *In* "Minerals in Soil Environments," 2nd edn (J. B. Dixon and S. B. Weed, eds), pp. 161–198. Soil Sci. Soc. Am., Madison, WI.

Richards, K. (1982) "Rivers: Form and Process in Alluvial Channels," Methuen & Co., London.

Rosenqvist, I. T. (1962). The influence of physico-chemical factors upon the mechanical properties of clays. *Proc. 9th Nat. Conf. on Clays and Clay Minerals*, pp. 12–27.

Ruxton, B. P. and McDougall, I. (1967) Denudation rates in northeast Papua from potassium-argon dating of lavas. *Am. J. Sci.* **265**, 545–561.

Schlesinger, W. H. (1997). "Biogeochemistry: an Analysis of Global Change." Academic Press, San Diego.

Schumm, S. A. (1977). "The Fluvial System." Wiley & Sons, New York.

Soil Survey Staff (1998). "Keys to Soil Taxonomy." 8th edn. USDA Natural Resource Conservation Service.

Stednick, J. D. (1981). Precipitation and streamwater chemistry in an undisturbed watershed in southeast Alaska. *Res. Pap. PNW-291*, US Department of Agriculture, Forest Service, Pacific Northwest and Range Experiment Station, Portland, Oregon.

Stumm, W. and Morgan, J. J. (1995). "Aquatic Chemistry," 3rd edn. John Wiley and Sons, New York.

Summerfield, M. A. (1991). "Global Geomorphology." Longman Scientific & Technical, Essex.

Sverdrup, H. U., Johnson, M. W., and Fleming, R. H. (1942). "The Oceans." Prentice-Hall, Englewood Cliffs, New Jersey.

Syers, J. K. and Iskandar, I. K. (1973). Pedogenic significance of lichens. *In* "The Lichens" (V. Ahmadjian and M. Hare, eds), pp. 225–248. Academic Press, New York.

Tan, T. K. (1957). Discussion on: Soil properties and their measurement. *Proc. 4th Int. Soil Mech. and Found. Engng.* **3**, 87–89.

Tardy, Y. (1969). Geochimie des alterations. Etude des arenes et des eaux de quelques massifs cristallins d'Europe et d'Afrique. These Doc. Etat, Univ. Strasbourg.

Ugolini, F.C. (1986). Processes and rates of weathering in cold and polar desert environments. In "Rates of Chemical Weathering of Rocks and Minerals" (S. M. Colman and D. P. Dethier, eds), pp. 193–235. Academic Press, New York.

Ugolini, F. C. and Edmonds, R. L. (1983). Soil biology. *In* "Pedogenesis and Soil Taxonomy 1. Concepts and Interactions" (L. P. Walding, N. E. Smeck and G. F. Hall, eds), pp. 193–231. Elsevier, Amsterdam, The Netherlands.

Ugolini, F. C. and Jackson, M. L. (1982). Weathering and mineral synthesis in Antarctic soils. In "Antarctic Geoscience" (C. Craddock, ed.), pp. 1101–1108. University of Wisconsin Press, Madison, WI.

Ursic, S. J. and Dendy, F. E. (1965). Sediment yields from small watersheds under various land uses and forest covers. Proceedings, Federal Inter-Agency Sedimentation Conference 1963. *US Dept. of Agriculture Misc. Publ.* **970**, 47–52.

Velbel, M. A. (1985). Hydrogeochemical constraints on mass balances in forested watersheds of the southern Appalachians. *In* "The Chemistry of Weathering" (J. I. Drever, ed.), pp. 231–247. G. Reidel , Dordrecht, Holland.

Wada, K. (1985). The distinctive properties of Ando-sols. *Adv. Soil Sci.* **2**, 173–229.

Weisz, P. B. (1973). Diffusion and chemical transformation. *Science* **179**, 433–440.

Williams, R. B. G. and Robinson, D. A. (1981). Weathering of sandstone by the combined action of frost and salt. *Earth Surf. Process. Landforms* **6**, 1–9.

Young, D. K. (1971). *Vie et Milieu*, Suppl. **22**, 557–571.

Zabowski, D. (1990). The role of mineral weathering in long-term site productivity. *In* "Impact of Intensive Harvesting on Forest Site Productivity" (W. J. Dyck and C. A. Mees, eds). *Proceedings, IEA/BE A3 Workshop*, South Island, NZ, March 1989. *IEA/BE T6/A6 Report No. 2. Forest Research Inst. Bulletin No. 159*, pp. 55–71.

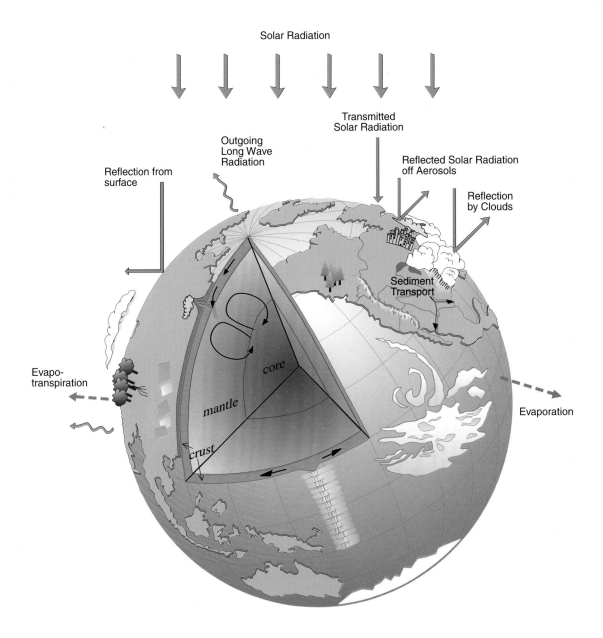

Plate 1. Overview of the cyclic processes on Earth. The placements of the geological and geographic features are not meant to represent their true placement on Earth, but rather are included to provide a view of many of the processes discussed in this book in one pictorial figure.

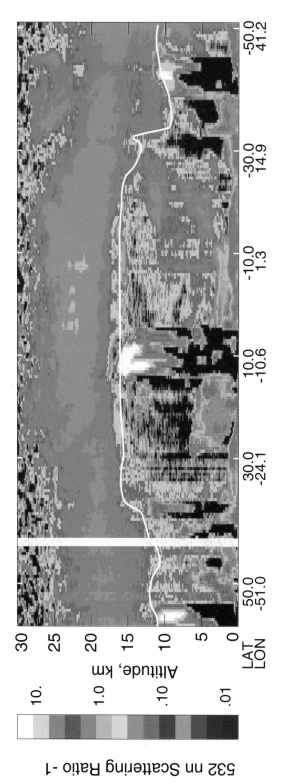

Plate 2. Altitude–distance image of the atmosphere obtained by LIDAR aboard the NASA Space Shuttle on 17th September 1994, across the Atlantic Ocean from approximately Newfoundland to South Africa. (Lidar stands for Light Detection And Ranging, the instrumental analog of RADAR with 532 nm radiation.) Approximately 30 minutes elapsed in going from the northern to southern hemisphere. The white and red tones are bright objects–clouds and thick dust, while the green and yellow tones are thin aerosols. Blue is clear air and black is no return. The horizonal white line is the tropopause obtained from metrological data. The vertical white stripe is due to momentary data loss. Visible are: the stratospheric aerosol left over from the emissions of Mt. Pinatubo (June 1991); the clouds of the ITCZ; Sahara dust over the Atlantic being carried by the trade winds; and the mid-latitude clouds associated with ordinary weather systems. Compare to Fig. 7-4 to see the consequences of the general circulation. (Figure courtesy of NASA, and is explained in detail Kent *et al.*, 1997.)

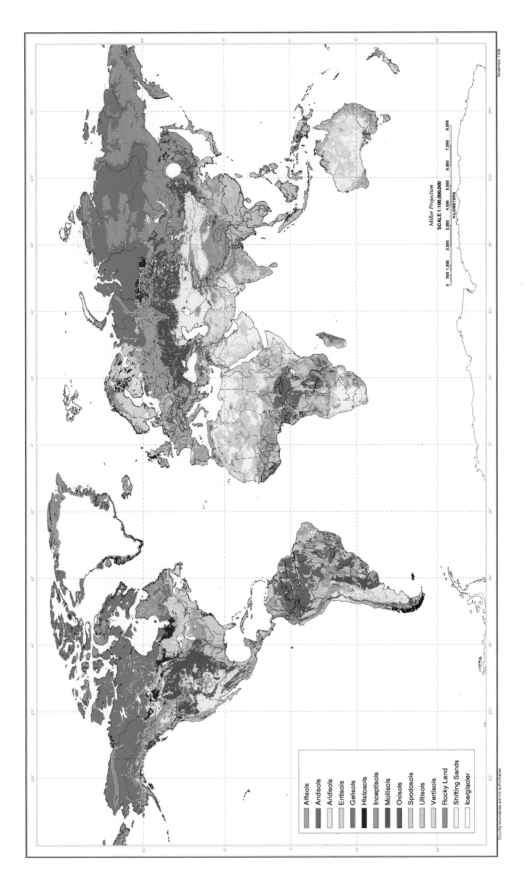

Legend:
Alfisols
Andisols
Aridisols
Entisols
Gelisols
Histosols
Inceptisols
Mollisols
Oxisols
Spodosols
Ultisols
Vertisols
Rocky Land
Shifting Sands
Ice/glacier

Miller Projection
SCALE 1:100,000,000
KILOMETERS
0 500 1,000 2,000 3,000 4,000 5,000 6,000 7,000 8,000

Country boundaries are not authoritative

November 1998

Plate 3. Global soil regions. (Courtesy of US Department of Agriculture, Natural Resources Conservation Service, Soil Survey Division, World Soil Resources.)

Plate 4. The global sea floor topography from satellite altimetry and ship depth soundings, as described in Smith and Sandwell, 1997. (Kindly provided by Dr Walter Smith of NOAA.)

9

Tectonic Processes and Erosion

Robert F. Stallard

9.1 Introduction

It is now widely accepted that the compositions of the atmosphere and world ocean are dynamically controlled. The atmosphere and the ocean are nearly homogeneous with respect to most major chemical constituents. Each can be viewed as a reservoir for which processes add material, remove material, and alter the compositions of substances internally. The history of the relative rates of these processes determines the concentrations of substances within a reservoir and the rate at which concentrations change. Commonly, only a few processes predominate in determining the flux of a substance between reservoirs. In turn, particular features of a predominant process are often critical in controlling the flux of a phase through that process. These are rate-controlling steps.

Weathering and erosion of bedrock are fundamental to the geochemical cycles that control the composition of the atmosphere, the oceans, and sedimentary rock. Consequently, identification of rate-controlling aspects of the erosion process is crucial to the analysis of global biogeochemical cycles. This chapter sketches how rates of erosion are controlled at different spatial and temporal scales, starting with the surfaces of mineral grains and expanding to whole continents. Most of the discussion focuses on how tectonic processes, continental freeboard, climate and the susceptibility of various lithologies to erosion influence the rate of continental denudation. Furthermore, to understand the development of the Earth through time, atmospheric properties that might affect erosion rates, such as temperature, moisture availability, oxygen and carbon dioxide partial pressure, etc., are examined.

The world today is in a state of exceptionally rapid transition. Two phenomena are involved. One is the rapid oscillations between a glacial and an interglacial climate mode over roughly 100 000-year intervals with transitions perhaps as short as decades. The other is the arrival of humans and the development of a technology fueled by geologic and biosphere energy sources. Both phenomena are responsible for tremendous shifts in geochemical cycles away from a dynamic equilibrium. In understanding rate-controlling steps we can better understand how rapid transitions affect the compositions of the ocean and atmospheric reservoirs.

Most of the examples in the following discussion are from the humid tropics. There are several fundamental reasons for focusing on the humid tropics as a study environment. Weathering and erosion are important to global geochemical cycles because of the chemical reactions that occur during weathering. Most of these reactions involve water, and although some are simple inorganic reactions, many are biologically mediated. The high temperatures, moist conditions, and luxuriant vegetation of this zone are ideal for rapid chemical weathering. Moreover, erosional processes in the humid tropics are particularly important on a global scale. According to Meybeck (1979), the humid tropics presently occupy about 25% of the Earth's land surface and supply about 65%

Earth System Science
ISBN 0-12-379370-X

of the dissolved silica and 38% of the ionic load delivered by rivers to the ocean. Data compiled by Milliman and Meade (1983) and Milliman and Syvitski (1992) indicate that the same region contributes about 50% of the total river solid load to the ocean. These studies demonstrate that the bulk of this material is derived from active orogenic belts and island arcs. Finally, the equatorial region was least affected by the climatic fluctuations of the ice ages.

The study of chemical weathering in drier and cooler regions of the Earth is beset by numerous complications and ambiguities. In dry regions, chemical weathering is very slow, and commonly many of the characteristics of landforms and soils seem to have been inherited from earlier, moister times. Moreover, materials derived from chemical weathering and atmospheric deposition often precipitate out as mineral deposits in soils. Cooler climates have been strongly affected by the repeated glaciations of the Pleistocene. In glaciated areas, soils are exceedingly young, and soils of adjacent nonglaciated regions were affected by a variety of periglacial processes. Still wider regions near ice sheets were veneered with loess.

This chapter examines climatic and tectonic controls on erosion in the tropics and the implications of these observations regarding the composition of erosion products in general. The role of glaciations in continental denudation will then be examined and contrasted with tropical conditions. Finally, we will briefly examine human effects.

9.2   Erosion, a Capsule Summary

The energy that powers terrestrial processes is derived primarily from the sun and from the Earth's internal heat production (mostly radioactive decay). Solar energy drives atmospheric motions, ocean circulation (tidal energy is minor), the hydrologic cycle, and photosynthesis. The Earth's internal heat drives convection that is largely manifested at the Earth's surface by the characteristic deformation and volcanism associated with plate tectonics, and by the hotspot volcanism associated with rising plumes of especially hot mantle material.

Erosion is the process that tears down the subaerial landforms constructed by crustal deformation and volcanism. Matter derived from the earth's crust generally moves from high elevation on land to low elevations in the ocean along pathways (rivers, glaciers, wind etc.) that are often long and complex. There are many pauses on the way (e.g., colluvial, alluvial, aeolian, lacustrine, and glacial deposits and lake and groundwaters), during which compositions can change. Erosion is typically most rapid in mountainous regions and coastal areas, and is slowest in flatlands. Matter is transported largely by rivers and to a lesser extent by winds and glaciers. During erosion, crustal material is initially mobilized by weathering.

9.2.1   Weathering and Crustal Breakdown

Weathering involves the chemical and physical breakdown of bedrock through its interaction with the hydrologic and atmospheric cycles. Chemical weathering is the breakdown of bedrock by chemical reactions. The products may include dissolved solids and new minerals (usually clays). Physical weathering is the physical breakdown of unweathered or partially chemically weathered bedrock; some old (primary) minerals remain intact. During erosion, these two types of weathering frequently operate in tandem. Chemical weathering weakens the rock; physical weathering finishes it off, making it available for transport processes. There are a few "rules of thumb":

1. Chemical weathering is more important in warm moist regions, whereas physical weathering is more important in cold dry areas.
2. Contributions made by chemical weathering are greater in regions where there is much vegetation.
3. Contributions made by physical weathering are much greater in steep terrains (i.e., more primary minerals remain), and overall weathering rates are higher.
4. Equivalent igneous and metamorphic lithologies appear to weather about twice as rapidly in island-arc and younger volcanic

terrains as compared to old cratonic settings (Stallard, 1995b).

The compositions of dissolved and solid erosion products are initially determined by the stability of the bedrock minerals at the site of weathering. The composition will change during transport to the ocean as the result of further weathering during storage in sedimentary deposits or during authigenic mineral formation in lakes, alluvial soils, and ground waters.

All minerals can be weathered chemically. The susceptibility of minerals varies considerably, however. This is illustrated nicely by the relative mineral stability for weathering under tropical conditions (Table 9-1). Note how stability appears to be closely related to composition. Similar mineral groups cluster together. For igneous and metamorphic minerals, stability is almost the reverse of the Bowen's reaction series (Goldich, 1938); the first minerals to crystallize out of a magma are the most susceptible to weathering. Note also that the vulnerability to chemical weathering of ionically bonded chemical sediments is in reverse order to the typical marine evaporite sequence (see Holland, 1974).

Minerals can weather congruently to produce only dissolved weathering products or incongruently to produce both dissolved cations (Na^+, K^+, Mg^{2+}, Ca^{2+}), silica ($Si(OH)_4$), and sometimes other constituents along with solid products that are cation-depleted and usually silica-depleted. Common minerals that always weather congruently are halite, anhydrite, gypsum, aragonite, calcite, dolomite, and quartz. Halite, anhydrite, and gypsum are so unstable that they almost never occur in the solid load of rivers. Calcite and dolomite sometimes occur in alkaline rivers that are supersaturated with respect to these minerals, for example, the Yellow River of China. Quartz is the most persistent of all common primary minerals and occurs in most river sediments. Minerals that contain iron and aluminum usually weather incongruently to produce clays and iron/aluminum sesquioxides (oxides and hydroxides of Fe^{3+} and Al^{3+}). Magnesium, potassium, and to a lesser extent calcium and sodium are retained in cation-rich clays such as the smectites, vermiculites, illites, and chlorites.

The most stable minerals are often physically eroded before they have a chance to chemically decompose. Minerals that decompose contribute to the dissolved load in rivers, and their solid chemical-weathering products contribute to the secondary minerals in the solid load. The secondary minerals and the more stable primary minerals are the most important constituents of clastic sedimentary rocks. Consequently, the secondary minerals of one cycle of erosion are

Table 9-1 Mineral stability in tropical soils[a]

MOST STABLE	
Quartz >	Pure silica
K-Feldspar, micas > Na-Feldspar > Ca-Feldspar, amphiboles > Pyroxenes, chlorite >	Igneous and metamorphic alumino-silicates
Dolomite > Calcite >	Carbonate minerals
Gypsum, anhydrite $\gg$ Halite	Evaporite minerals
LEAST STABLE	

[a] Adapted from Stallard (1985).
Na-feldspar + Ca-feldspar = plagioclase.

often the primary minerals of a subsequent cycle. When a weakly cemented sedimentary rock consisting of chemically stable minerals is exposed to weathering, it often breaks down physically, and although erosion might be very rapid, the solid products undergo little chemical alteration. If physical erosion processes are not very intense, most of the erosion products are dissolved, and stable primary minerals and secondary minerals will accumulate over the bedrock to form soil. Additional details on mineral transformations by weathering are provided in Chapter 7.

9.2.2   Weathering and the Atmospheric Gases

It is chemical weathering that makes continental denudation so important to geochemical cycles. Solution transport by rivers into the ocean is the largest single flux of many elements into the seawater reservoir. Carbon dioxide and molecular oxygen, compounds of fundamental biogeochemical importance, are consumed by weathering reactions involving primary minerals. The rate of consumption by weathering reactions, although much slower than the rate of the cycling through organisms, may be important in controlling the long-term concentrations of CO_2 and O_2 in the atmosphere and ocean (Garrels and Lerman, 1981; Berner *et al.*, 1983).

The consumption of CO_2 during weathering is indirect. Most members of two important classes of minerals, the silicates, and the carbonates, consume hydrogen ions and release alkali (Na^+, K^+) and alkaline-earth (Ca^{2+}, Mg^{2+}) cations during weathering. The primary proton donor is carbonic acid (H_2CO_3), formed by the hydrolysis of carbon dioxide. An accumulation of carbonate and bicarbonate ions in solution results. Organic acids, which are photosynthetic derivatives of CO_2 are often important in shallow soil horizons, and mineral acids (sulfuric, nitric, hydrochloric) from atmospheric, biological, or rock-weathering sources also can be important. The net effect of the reactions is to consume protons and, when weak acids are the proton donors, to produce alkalinity. Carbon dioxide is returned to the atmosphere primarily by carbonate deposition, metamorphic reactions, volcanism, and mid-ocean ridge hydrothermal circulation.

Oxygen is consumed largely by the oxidation of reduced iron in silicates and sulfides, of sulfur in sulfides, and of organic carbon in sedimentary rocks. Oxygen is returned to the atmosphere as a result of biological fixation of reduced carbon, sulfur, and iron and the subsequent burial of these compounds. Not all chemical weathering reactions involve CO_2 or O_2. Certain minerals, such as quartz (SiO_2) and the major evaporite minerals ($NaCl$, $CaSO_4$, and $CaSO_4 \cdot XH_2O$) dissolve without reacting with anything except water.

Atmospheric CO_2 and O_2 are also coupled through the formation and burial of and exhumation and weathering of organic carbon. When carbon dioxide is converted by photosynthesis to organic matter oxygen is released. If the organic matter is buried, then O_2 cannot recombine with the carbon to make CO_2, and the process is a net source of O_2 to the atmosphere. When buried carbon is reoxidized through weathering or burning, then O_2 is consumed and CO_2 is released. The burning of fossil fuels represents a tremendous artificial acceleration of this release.

9.3   Soils and the Local Weathering Environment

The initial partitioning of elements into dissolved and solid phases during continental denudation is controlled by the chemical and physical processes associated with soil genesis. The description of weathering processes as they occur in soil profiles is discussed in Chapter 7. A vast body of observational data, not to be reviewed here, has shown that soils are remarkably complex on the local scale. Soil systems are neither physically nor chemically homogeneous. Both solid and dissolved matter is washed down through soil profiles; capillary action operating in tandem with evaporation can lead to upward transport of dissolved material; frost action, root growth, tree falls, and burrowing can mix material within a profile; transport of fluids in soils on non-porous substrates is largely lateral; soil porosity is controlled by both chemical and physical processes; freezing can have profound effects on the structure of soil and the composition of residual fluids; biochemical processes generate many thermodynamically unstable reactants; the chemical activities of many chemical species (e.g., H_2O, CO_2, O_2) change tremendously on time scales that are infinitesimally short geologically (minutes to days); and reactions frequently occur in thin films and on grain surfaces.

9.3.1   *Weathering Reaction Kinetics*

It was once said that one of the reasons why weathering reaction kinetics has not been studied in more detail is that the reactions are too slow to be studied by graduate students. In recent years, great advances have been made in modeling, experimental design, and in examining mineral surfaces (White and Brantley, 1995). Quantum tunneling microscopy has permitted the examination of the placement of atoms on reactive surfaces permitting the testing of molecular-scale models of reaction chemistry. The resolution of electron microscopy has also improved as have the type and quality of tandem chemical measurements. Physically and chemically based models can now represent aspects of hydrology and chemical weathering, but as yet, integrated models that capture weathering processes in a small watershed do not exist. One factor is the incredible complexity of soil, in part imposed upon it by biological processes and season changes in temperature and hydrology.

Several simple experimental systems that simulate some aspect of the groundwater environment have been used to study the breakdown of individual minerals. These kinetics studies have encompassed quartz (Brantley *et al.*, 1986; Dove, 1995), feldspars (Blum and Stillings, 1995), pyroxenes and amphiboles (Brantley and Chen, 1995), carbonates (Berner, 1978), and glasses (White, 1983). Simultaneously, theoretical studies of weathering processes have advanced considerably (Casey and Ludwig, 1995; Lasaga, 1995; Nagy, 1995). Relative stability observed in laboratory weathering is consistent with field-based observations; however, experimental rates appear to be faster than those in natural systems. *In situ* studies (White, 1995) and complex models (Sverdrup and Warfvinge, 1995) have tried to bridge the differences between field and laboratory.

Studies of mineral grains from laboratory experiments and from natural soils, using electron microscopy and the recently developed atomic force microscopy, provide some interesting observations about reaction mechanisms and rate controls during weathering. A continuum of rate-limiting mechanisms can be defined between reactions that are transport (diffusion) controlled and those that are surface reaction controlled (Berner, 1978). In the former case, reactions are limited by diffusion of solution products away from the crystal surfaces. This situation occurs where reactions are sufficiently rapid that fluid concentrations near the primary mineral are close to equilibrium values for the driving reaction, which need not be the final equilibrium for the system. Under a microscope, grain surfaces appear smooth with rounded edges (Fig. 9-1). In the latter case, reaction rates are controlled by local surface energy on the crystals. Surface reaction control is important under circumstances where surrounding fluids are decidedly undersaturated relative to the equilibria that are driving the weathering reaction. Surfaces appear rough and pitted, commonly on zones of obvious crystallographic defects (Fig. 9-1, see also Gilkes *et al.*, 1980; Brantley and Chen, 1995; Hochella and Banfield, 1995). Most studies of primary minerals in natural soils show extensive pitting which is suggestive of surface reaction control.

There are several important caveats to the use of this observational model. There is, for example, a considerable range of undersaturation, down to 50% in the case of quartz, where surface defects are not especially reactive, and crystals appear smooth (Brantley *et al.*, 1986). The formation of weathering rinds and the build-up of clays and other secondary weathering products in the immediate vicinity of the weathering grains can strongly affect the transport processes. Finally, in soils, fluid compositions change rapidly and a wide range of reactions

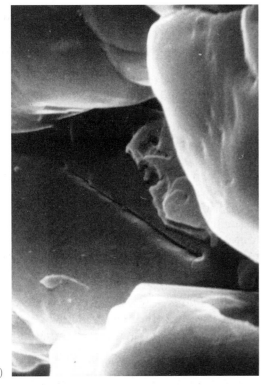

(a)   (b)

Fig. 9-1 Electron micrographs of quartz grains from a soil profile described in (a) Stallard (1985) and (b) Brantley *et al.* (1986). The quartz grains in (a) are from a sample close to fresh bedrock near the bottom of the profile. The sample in (b) is from near the top of the profile. Note that the surfaces of the grains in the left-hand photograph are smooth, whereas the right-hand grains are deeply etched. Presumably the etched grains are dissolving in soil waters that are strongly undersaturated with respect to quartz. Such etching indicates that reaction rates are affected by variations in surface energy. The smooth grains are probably dissolving at or close to equilibrium conditions.

may be important during weathering (Drever and Smith, 1978; Eberl *et al.*, 1986; Herbillon and Nahon, 1988; Trolard and Tardy, 1989; van Breeman, 1988a,b).

9.3.2 *Variable Chemistry of Soil Fluids, a Complicating Factor*

Weathering rates are most sensitive to the throughput of water. In soils, this is a decidedly discontinuous process. Typically, water flows through soil following rainfall or snowmelt. Once saturated, the flux of water is largely dependent on the physical properties of the soil and not on the rate of supply. Water that cannot be accommodated by flow through the soil,

because of soil saturation or high rate of supply, is rerouted into overland runoff and interacts minimally with primary minerals. Following wetting, typical soils drain rapidly. At most times, soils are drying. This scenario is evident in discharge curves for small river catchments, where precipitation and snowmelt events show up as spikes over a very long background discharge which reflects slow groundwater inputs.

Groundwater environments can be represented as a simple flow-through system. For the situation where chemical weathering of mineral grains is transport controlled, the weathering rate of a mineral should be directly dependent on the rate of throughput of water. For the situation where rates are controlled by surface reactions, weathering rate should be independent of the rate of throughput of water.

Soil water flow is decidedly episodic. During dry times the water solutions in the soil are probably fairly concentrated and not very reactive. Time-averaged reaction rates should be roughly proportional to the fraction of time reacting minerals are in contact with thermodynamically undersaturated (and reactive) water. In a study of the relationship between denudation rate and runoff for rivers draining igneous and metamorphic rock in Kenya, Dunne (1978) obtains the relationship of (denudation rate in tons/km^2 per year) = 0.28 (runoff in mm/year)$^{0.66}$.

The fact that soils dry between episodes of water flow complicates weathering reaction scenarios. During drying, solute concentrations in water films increase, and the areal extent of the films decrease. The chemical activity of water drops. As a result, chemical equilibria that might be important for controlling weathering reactions within wet soils are replaced by new equilibria reflecting the elevated concentration of solutes. A different suite of clays and sesquioxides might become stable and silica (opal), calcite, and various evaporite minerals can precipitate. During subsequent wetting, both primary minerals and secondary minerals formed under drying conditions may react. Features, such as etch pits may form during episodes of wetting and thermodynamic instability, even if they would not normally form under average or typical conditions. Moreover, secondary minerals that formed under dry conditions may persist through wetting cycles. The formation of calcium carbonate nodules in soils, caliche, is an example of this. The mineral constituents of such soils may not at equilibrium. The overall rate of chemical and physical weathering and the composition of weathering products changes with increasing runoff (see models in Stallard, 1995a,b). At low runoff, eroded solids are cation-rich and the ratio of silicon to cations in solutes is low. At high runoff, eroded solids are cation-depleted and silicon forms a large fraction of the solute load.

Freezing, which also produces residual fluids with elevated concentrations of dissolved solutes, presumably does not have as significant an effect as drying because lower temperatures are involved. Freeze–thaw cycles, however, can break apart rocks and expose fresh mineral surfaces.

9.3.3 Local Responses to Atmospheric Variables

The influence that variations of temperature and levels of atmospheric CO_2 and O_2 have on chemical weathering are more subtle. Temperature appears to have a direct effect on weathering rate (White and Blum, 1995). The silica concentration of rivers (Meybeck, 1979, 1987) and the alkalinity of ground waters in carbonate terrains (Harmon *et al.*, 1975) are both positively correlated with temperature variations. It is not clear, however, whether temperature-related variations in weathering rates are largely due to variations in vegetational activity that parallel temperature variations.

Partial pressures of O_2 and CO_2 in soils are controlled largely by soil biology. Oxygen is consumed and CO_2 is produced in soils by decay and root respiration. Plausible variations of the atmospheric concentration of these gases have little effect on their partial pressures in soils. The rate of the hydrologic cycle, however, is thought to respond directly to global mean temperature and is therefore indirectly sensitive to the partial pressure of atmospheric CO_2, which as a "greenhouse" gas, can affect global mean temperatures (see Berner *et al.*, 1983).

In summary, of all the local variables that can affect weathering rate, the supply of water is clearly the most important. Biology is very important as the supplier of proton donors and complexing agents for weathering reactions, as a mediator in the moisture budget for the soil, and as a controller of soil structure. Beyond suggesting that surface reactions are important in controlling the weathering rate for most minerals and confirming mineral stability sequences, laboratory models of weathering chemistry are just beginning to recognize the importance of soil structure, coatings, and biological processes in controlling weathering rates.

9.4 Slope Processes and the Susceptibility of Lithologies to Erosion

Weathering, atmospheric deposition, and the fixation of atmospheric gases are the ultimate sources of the material transported by rivers. These processes operate over the surface of the river catchment, and the resultant water and weathering products must be transported downslope before arriving in a channel. Examination of erosion processes on hillslopes provides insight as to how chemical weathering rates are controlled at an intermediate spatial scale and how chemical elements are partitioned between the dissolved and solid loads of rivers. Excellent models of watershed hydrology, such as TOPMODEL (Bevin and Kirkby, 1979; Bevin et al., 1995), have developed, and interfacing these with studies of physical, geochemical, and biological processes in soils and on hillslopes should be a major research direction in the future.

The erosion process on slopes can be envisioned as a continuum between the weathering-limited and transport-limited extremes (Carson and Kirkby, 1972; Stallard, 1985, 1995a). Erosion is classified as transport limited when the rate of supply of material by weathering exceeds the capacity of transport processes to remove the material. Erosion is weathering limited when the capacity of the transport process exceeds the rate at which material is generated by weathering. These two styles of erosion represent an interesting parallel to controls of weathering reaction rates on mineral surfaces, discussed earlier, wherein a similar continuum was defined between surface-reaction control and transport (diffusion) control (Stallard, 1988).

The weathering and transport processes associated with either end of the continuum are quite different. Where erosion is weathering limited, erosion rate is controlled by the rate at which chemical and physical weathering can supply dissolved or loose particulate material. In essence, erosion rates are controlled by susceptibility to weathering. Soils are thin, because loose material moves rapidly downslope. Much of this material is only partially weathered, because most rocks lose their structural integrity before they are completely chemically decomposed. Processes characteristic of weathering-limited regions include rockfalls, landslides, or anything that tends to maintain a fresh or slightly weathered rock surface (Table 9-2). These processes often operate at a threshold slope angle. In humid climates, weathering-limited conditions are associated with thin soils and steep straight slopes which often undergo parallel retreat at a threshold angle (Fig. 9-2).

In contrast, under transport-limited conditions, weathering rates are ultimately limited by the formation of soils that are sufficiently thick or impermeable to restrict free access by water to unweathered material. Erosion rates

Table 9-2 Erosion regimes, features, and processes associated with transport-limited and weathering-limited erosion

Transport limited	Weathering limited
Thick soils	Thin or no soils
Slight slopes that are convexo-concave	Steep slopes that are straight and at a threshold angle
Erosion rates	Erosion rates
independent of lithology	depend on bedrock susceptability
Processes:	Processes:
soil creep	rock slides
removal of dissolved phases	strong sheet wash
	soil avalanches
	removal of dissolved phases

Weathering limited: potential transport processes greater than weathering supply. Transport limited: supply by weathering greater than the capacity of transport processes to remove material (until weathering is slowed by feedback).

(a)   (b)

Fig. 9-2 Photographs illustrating (a) long straight slopes in the Andes and (b) convexo-concave slopes transitional into very flat terrain. The view of the Andes is taken from Huayna Picchu, the small peak next to Machu Picchu. The view is up the valley of a small tributary to the Urubamba River. Note that the V-shaped valley formed by fluvial erosion becomes a U-shaped glacial valley at its highest end. The glacier was active during the last ice age. (b) The Guayana Shield in southern Venezuela, taken from an airplane. In the foreground is the Orinoco River downriver from the Casiquiare Canal, a natural channel that connects the Orinoco and the Amazon River systems. The hills in the foreground are granite and about 200 m high.

are low, and soils and solid weathering products are cation deficient. In regions where transport-limited erosion predominates, soils are thick and slopes are slight and convexo-concave (Fig. 9-2). With time, these slopes tend towards increasing flatness. Soil creep is a process typical of transport-limited situations. Most soil mass movement and wash processes, however, are intermediate between weathering limited and transport limited in character.

Erosion associated with chemical weathering caused by circulating soil/ground water is intermediate between being transport limited and weathering limited in character. Chemical erosion would be transport limited whenever reactions at the mineral-grain level are also transport limited. As discussed earlier, this occurs when the flushing rate for water is sufficiently slow that equilibrium is reached with respect to some controlling reaction. Chemical weathering of carbonate rocks is probably transport limited as soil and ground waters are nearly saturated with respect to carbonate minerals (Holland *et al.*, 1964; Langmuir, 1971). In the case of silicate weathering, transport-limited erosion would occur where the silicates are particularly unstable or where water movement is restricted by low porosities or lack of hydraulic head in soils or bedrock.

9.4.1 Soils, Slopes, Vegetation, and Weathering Rate

For a given set of conditions (lithology, climate, slope, etc.), there is presumably an optimum soil thickness that maximizes the rate of bedrock weathering (Fig. 9-3) (Carson and Kirkby, 1972; Stallard, 1985). For less than optimum soil thicknesses, there is insufficient pore volume in the soil to accept all the water supplied by precipitation and downhill flow. Excess water runs off and does not interact with the subsurface soil and bedrock. In contrast, water infiltrates and circulates slowly through thick soils (especially where forested). If profile thicknesses greatly

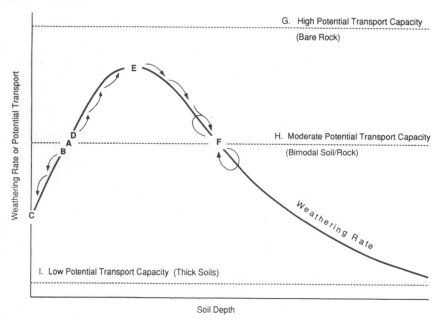

Fig. 9-3 Conceptual model to describe the interaction between chemical weathering of bedrock and down-slope transport of solid erosion products. It is assumed that chemical weathering is required to generate loose solid erosion products of the bedrock. Solid curve portrays a hypothetical relationship between soil thickness and rate of chemical weathering of bedrock. Dotted lines correspond to different potential transport capacities. Low potential transport capacity is expected on a flat terrain, whereas high transport is expected on steep terrain. For moderate capacity, C and F are equilibrium points. (Modified with permission from R. F. Stallard, River chemistry, geology, geomorphology, and soils in the Amazon and Orinoco basins. *In* J. I. Drever, ed. (1985), "The Chemistry of Weathering," D. Reidel Publishing Co., Dordrecht, The Netherlands.)

exceed the optimum, long residence times for water at the base of the profile lead to a reduction of weathering rate as a result of equilibration with respect to secondary phases. With increasing thickness, soil profiles also become more structured with definite horizons. Some of these can be rather impermeable. Water is routed through weathered material, and weathering rates are thereby reduced.

Figure 9-3 portrays a hypothetical model of how chemical weathering and transport processes interact to control soil thicknesses. The relationship between soil thickness and rate at which chemical weathering can generate loose solid material is indicated by the solid curve. The rate at which transport processes can potentially remove loose solid weathering products is indicated by horizontal dotted lines. The rate of generation by chemical weathering initially increases as more water has the opportunity to interact with bedrock in the soil. As soil thick-

ens, the optimum thickness for maximum rates of chemical weathering is exceeded, and weathering rates decrease with increasing soil thickness (E–F). The decrease presumably occurs because the rate of chemical weathering is limited by diminished access of reactive waters to unweathered materials. Recent work by Heimsath *et al.* (1997) indicates that an exponential decrease in weathering rate with increasing soil thickness is reasonable. If the potential transport rate exceeds weathering rate, loose soil material should not remain on the slope. If weathering rate exceeds potential transport rate, soils develop. As soils develop, weathering rates vary, leading to an evolution of the soil profile.

For soil profiles that are less than the optimum thickness, there is a destabilizing feedback between soil thickness and weathering rate. Assume that a thin soil is in a dynamic equilibrium such that weathering inputs balance transport losses (A in Fig. 9-3). Weathering rate can be reduced relative to transport rate by either increasing the strength of transport processes or by thinning the soil (B). Either way, transport removal would exceed weathering supply, and the soil would continue to thin. Eventually only a hard cohesive saprolite or bedrock would remain (C). If the soil is thickened, or if the capacity of transport processes is reduced, the soil would tend to accumulate (D–E). Finally, a situation involving stabilizing feedback occurs (F), and a thick soil forms such that transport of loose material balances weathering inputs. If soil thickens beyond point F, weathering rates decrease, and transport processes restore the soil thickness back to the stable value. This model suggests that soil distributions should be distinctly bimodal: either thin, weathering limited (G), or thick, transport limited (I). For some intermediate values of potential transport rate (H), either hard rock or a moderately thick soil could exist.

When soil thickness is at the stable value (F), erosion is transport limited. Chemical weathering is also transport limited. This is, however, not because of reaction kinetics; instead this limitation is primarily controlled by physical factors, most probably, restricted access of water to the primary minerals.

The effects of vegetation are complex. Vegetation reduces short-term physical erosion by sheltering and anchoring soils. This effect is equivalent to reducing potential transport in Fig. 9-3. A cover of vegetation does not necessarily reduce denudation rates. Vegetation can maintain a layer of soil on steep slopes, particularly under wet conditions. As the soil thickens, it often becomes unstable, detaches, and slides down the slope (Garwood *et al.*, 1979; Pain, 1972; Scott, 1975a; Scott and Street, 1976; Stallard, 1985; Wentworth, 1943; Larsen and Simon, 1993; Larsen and Torres Sánchez, 1998) (Fig. 9-4). Under such circumstances, weathering rates can be exceptionally high because of the extra moisture and bioacids; likewise, denudation rates are very high because of the continuous resupply of fresh rock. The effect of erosion following fires, tree falls, and land clearing on slopes is similar to that of slides (McNabb and Swanson, 1990; Meyer *et al.*, 1995; Scott, 1975b; Stallard, 1985,

Fig. 9-4 Photograph of landslides (soil avalanches) that occurred following earthquakes in Panama on July 17, 1976, near Jaque. In the background is a bay of the Pacific Ocean. The effects of this earthquake are described by Garwood *et al.* (1979), who estimated that about 42 km^2 (about 10%) of the region near the epicenter of the earthquake was devegetated. The bedrock is mostly island-arc basalts and andesites. (Photography by N. C. Garwood.)

1995a, 1998). On slight slopes, over extended periods of time (up to millions of years), vegetation may reduce weathering rates by allowing very thick soils to accumulate. For a given soil thickness, however, weathering should be faster because of the supply of bioacids.

9.4.2 Elemental Partitioning: The Role of Slope Processes

There are two principal ways to selectively partition different elements between the dissolved and solid loads in rivers: by selective chemical weathering of particular primary minerals, and by the formation of secondary phases that are enriched or depleted in certain elements, relative to bedrock (Stallard, 1985, 1995a).

Different styles of erosion are associated with different degrees of partitioning of elements between dissolved and solid load. As rocks weather chemically, they lose their structural integrity. When only kaolinite, gibbsite, or other cation-depleted phases form, as commonly happens with transport-limited erosion, cation ratios in solution should match those in bedrock. During weathering-limited erosion, unstable primary minerals are selectively removed, causing elemental partitioning. For example, in moist vegetated areas on crystalline rocks or indurated sediments, solifluction, soil avalanching, and sheet runoff remove weakly cohesive material (solum and soft saprolite), leaving a cohesive hard saprolite behind (Simon et al., 1990; Stallard, 1985; Stallard and Edmond, 1983). Where transport processes are sufficiently intense, more stable minerals such as zircon, quartz, potassium feldspars, and micas survive chemical weathering and are eroded. Some of these resistant minerals contain substantial K and Mg, but none contain much Na or Ca. Thus, K and Mg are enriched relative to Na and Ca in solid erosion products; Na and Ca are enriched in solution relative to K and Mg (Fig. 9-5). If Mg is incorporated into the lattice of many secondary clays, as often happens, this further accentuates its retention in bulk solids (Stallard et al., 1991).

Dissolved phases are assumed to best reflect the weathering processes occurring at the erosion site, because water is not usually stored for long periods (many years) in soils or during fluvial transport. Sediment moves slowly through river systems; the larger the river system the longer sediment takes to move through (Church and Slaymaker, 1989; Meade et al., 1990). Weathering products, however, can weather chemically during transport downslope and through fluvial systems. Solids degrade when they accumulate at the base of slopes (colluvium) or during storage on floodplains (alluvium); this obviously affects dissolved components (Johnsson et al., 1991; Stallard et al., 1991). Soil solutions also evolve as they flow

downslope. Contact with fresher materials is prolonged, and evapotranspiration can concentrate solutions (Carson and Kirkby, 1972; Stallard, 1988; Tardy et al., 1973). Both of these phenomena can cause the formation of new suites of clays and sesquioxides, and the precipitation of carbonates.

Aggradation or degradation of biomass or soil reservoirs may also produce effects that appear to be fractionation. This is because the elemental ratios in vegetation or soil reservoirs can be very different from those of bedrock. Sufficiently large and rapid changes in these reservoirs are sometimes evident in river chemistry. For example, the uptake and release of potassium in association with the seasonal growth and loss of leaves can affect the composition of streams that drain temperate deciduous forests (Likens et al., 1977; Vitousek, 1977).

9.5 Landforms, Tectonism, Sea Level, and Erosion

On a larger scale, landscape development reflects those mechanisms that expose bedrock, weather it, and transport the weathering products away. Present and past tectonism, geology, climate, soils, and vegetation are all important to landscape evolution. These factors often operate in tandem to produce characteristic landforms that presumably integrate the effects of both episodic and continuous processes over considerable periods of time.

Many important erosion-related phenomena are episodic and infrequent, such as flash floods, landslides, and glaciations, while others such as orogenesis and soil formation involve time scales that exceed those of major climate fluctuations. In either case, the time scale of human existence is too short to make adequate observations. Consequently, it is difficult to directly estimate the rates or characterize the effects of such phenomena on erosion products. The key to understanding weathering and erosion, on a continental scale, is to decipher the relationship between landforms, the processes that produce them, and the chemistry and discharge of riverborne materials.

To a first approximation, landscape formation

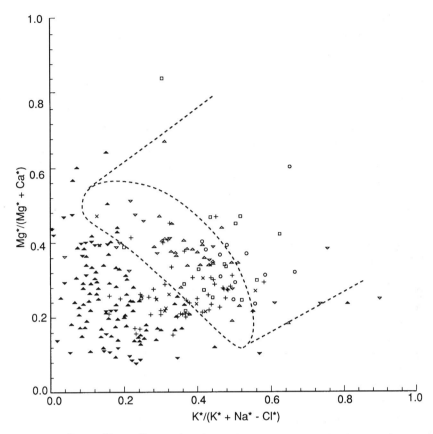

Fig. 9-5 Relation of $Mg^{2+}/(Mg^{2+}+Ca^{2+})$ to $K^+/(K^++Na^+-Cl^-)$ for dissolved material in surface waters of the Orinoco River and Amazon River basins, the Isthmus of Panama, and the Island of Taiwan and for rock types representative of bedrock in these basins. Symbols: ▲, samples from rivers that drain mostly mountain belts developed on felsic continental crust; ▼, samples from rivers that drain mostly island-arc mountain belts developed on mafic oceanic crust; +, samples from rivers that originate in continental mountain belts, but that drain large areas of craton; ×, rivers that drain alluvial and nonalluvial sedimentary rocks in foreland and intracratonic basins; □, rivers that drain only alluvial sediments in foreland and intracratonic basins; △, rivers that drain hilly to mountainous shield on felsic crust; ▽, rivers that drain hilly to mountainous shield on mafic crust or deeply eroded island arc; ○, rivers that drain peneplaned shield. The dashed oval represents the range of analyses for common igneous rocks; the dashed lines that extend away from the igneous-rock field represent the composition field for common sedimentary rocks. (Reproduced with permission from R. F. Stallard, Weathering and erosion in the humid tropics. *In* A. Lerman and M. Meybeck (1988). "Physical and Chemical Weathering in Geochemical Cycles", pp. 225–246, Kluwer Academic Publishers, Dordrecht, The Netherlands.)

processes can be viewed as being controlled by climatic and tectonic factors. Climate delimits where different types of weathering can occur, while tectonics controls how the effects of weathering are expressed. The role of tectonic setting is emphasized in this discussion because tectonic history is often critical to controlling denudation rates and the composition of erosion products. The object of the discussion is not to focus on the particulars of landscape development. Instead, landforms are primarily used to distinguish between different erosional regimes and to identify features that are useful for characterizing denudation and uplift rates, especially indicators of past sea level. The linkage between tectonic processes and landforms is embodied in the concept of a morphotectonic region.

9.5.1   The Effects of Erosional Regime

Many landforms have a convex upper slope, a straight main slope, and a concave lower slope. Carson and Kirkby (1972) argue that the main slope is dominated by weathering-limited processes; whereas, the upper and lower slopes are primarily areas of transport-limited erosion or even deposition in the case of the lower slope. If the overall slope is largely transport limited, it will undergo parallel retreat at the threshold angle. With parallel retreat, topographic form is maintained and characteristic landscapes are thereby generated.

The weathering regime exerts a major control on the production rate and the composition of erosion products from different lithologies (Stallard, 1985, 1988). For weathering-limited conditions, erosion rate is controlled by the susceptibility of the bedrock to weathering. The tectonic history, physical properties (porosity, shear strength, jointing, etc.), and chemical properties of the bedrock exert a major influence. For a given climate, within a specific catchment, each rock type should contribute an amount of material to river transport that is proportional to both its extent of exposure and its susceptibility to weathering. (Deep groundwater transport important in karst terrains would contribute additionally.) Solid erosion products should include abundant partially weathered (cation-rich) rock and mineral fragments. If the bedrock is especially physically unstable, it will contribute strongly to the solid load of rivers; likewise, chemically unstable bedrock will contribute strongly to the dissolved load.

In transport-limited conditions, however, susceptibility is not so important due to the isolating effect of thick soils. In the extreme situation of a flat landscape lowering at a uniform rate, the erosional contribution by a particular rock type should be related only to the area exposed. Solids would be cation-deficient.

9.5.2   The Effects of Deposition, Storage, and Burial

Sediment storage and remobilization involves huge mass fluxes and can respond rapidly to changes in climate and human activities. Accordingly, sedimentation on land may play a major role in the carbon cycle (Stallard, 1998). Sediment-laden rivers flowing over flat terrain commonly develop extensive floodplains. At times, floodplains coalesce into broad depositional alluvial plains such as the Llanos of South America. The sediments in those deposits weather chemically. Less stable minerals in the sediment are broken down and alluvial soils develop. Eventually only the most stable minerals, such as quartz, remain, and the clays are transformed into cation-deficient varieties. Sediment in such rivers, especially the sand, may go through many cycles of deposition, weathering, and erosion before it is transported out of the system. Compositionally, this sediment resembles that derived from transport-limited erosion. Elemental fractionation between the original bedrock and erosion products still occurs because of the permanent burial of some cation-rich material and the uninterrupted transport of much of the fine-grained suspended sediment out of the system (Johnsson *et al.*, 1988, 1991; Stallard, 1985, 1988).

9.5.3   The Effects of Tectonic Processes

Tectonic processes provide the material of which landscapes are made. Tectonism produces uplift followed by erosion, or subsidence followed by deposition. Sediment storage makes the calculation of denudation rates rather time-scale dependent. Sediment deposited at one time can be re-eroded through increases in river discharge caused by wetter climates or lowering of base level, perhaps brought about by uplift. What might be considered eroding bedrock on one scale can be just stored sediment in transit to the ocean on another, longer, scale. In the following sections, tectonically active and tectonically quiet areas are discussed separately to highlight the differences in erosional style between these two types of settings. An important aspect of this difference is the degree to which erosion is influenced by eustatic sea level fluctuations.

To a first approximation, fast tectonic processes occur at plate boundaries (Fig. 9-6). The

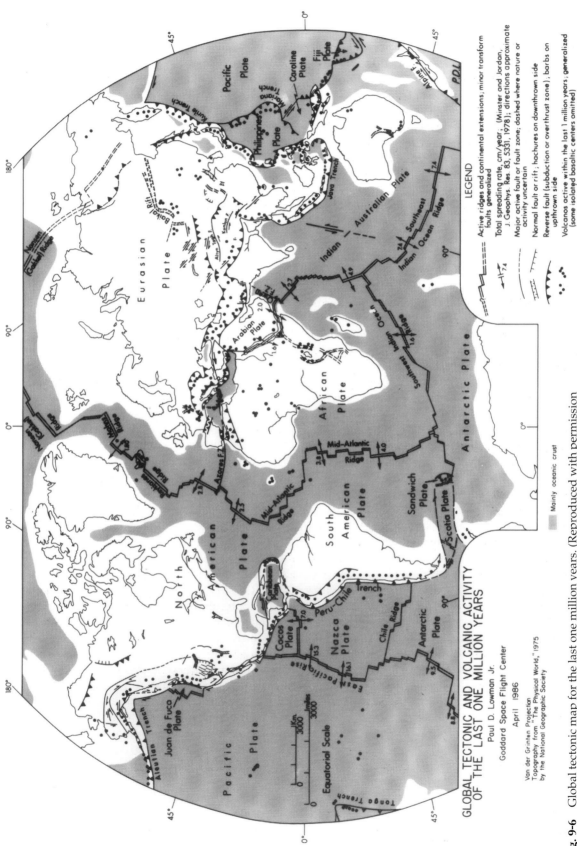

Fig. 9-6 Global tectonic map for the last one million years. (Reproduced with permission from P. D. Lowman, Jr. (1981). *Bull. Internat. Assoc. Engineer. Geol.* **23**, 37–49.)

primary exceptions to this are incipient rifts and the so-called "hot spots" or "mantle plumes," both of which are associated with uplift and volcanic activity in plate interiors. Collisional plate boundaries are of greatest interest, because these are the sites of the intense faulting, folding, and volcanism associated with the world's great mountain belts. Most divergent plate boundaries are underwater mid-ocean ridges, and are of no direct concern. Incipient divergent boundaries on land are associated with rift zones and their characteristic volcanism and block-faulted mountains. Transcurrent shearing (neither convergent nor divergent) plate boundaries are characterized by intense seismic activity, but less impressive mountain building. Hot spots are evidenced by local, sometimes intense volcanism, apparently fixed with respect to the mantle and not with the plate upon which it is occurring.

The term "craton" is generally used to refer to tectonically quiet or stable continental areas. Major subdivisions include "shields" where long-term erosion has exposed extensive areas of old crystalline basement, "platforms" where shields have a flat-lying sedimentary veneer, intracratonic basins where slow long-term subsidence has led to thicker sedimentary deposits on the craton, and "passive continental margins" where continental crust has rifted, separated, cooled, and subsided (Pitman, 1978; Sloss and Speed, 1974). Between cratons and mountain belts there is often a "foreland basin" where basement has slowly subsided as a result of sedimentary loading and tectonic downwarping. Intracratonic basins, passive margins, and foreland basins, represent loci of long-term sediment accumulation and storage (Ronov *et al.*, 1969; Sloss, 1963, 1979; Soares *et al.*, 1978; Vail *et al.*, 1977; Vail and Herdenbol, 1979). Intracratonic basins subside through especially long stretches of geologic time. The Amazon Trough, for example, has been active for all of the Phanerozoic (Soares *et al.*, 1978).

9.5.4   *Effects of Sea-Level Change on Erosion*

The ocean surface represents the master base level for continental erosion and sedimentation.

Given a sufficient period of time, in the absence of tectonic processes, continents would presumably be eroded flat to about sea level. It is not surprising, therefore, that most tectonically quiet areas on continents tend to have low elevations and are often flat, whereas tectonically active areas, mostly mountain belts, have high elevations and steep slopes (Figs 9-2 and 9-4).

High and low stands of sea level are directly recorded as sedimentary coastal onlap sequences and as erosional terraces. These records are complicated in regions of crustal instability, and the rate and nature of crustal deformation determines whether evidence of short-term or long-term sea-level fluctuations are preserved and how easily this evidence is interpreted. Because continental basement warps and fractures through time, and because evidence of sea level is erased by erosion, the interpretation of this evidence to produce sea-level curves for the Phanerozoic has been a subject of considerable debate.

Eustatic sea-level fluctuations occur on a wide range of time scales (Vail *et al.*, 1977; Vail and Herdenbol, 1979). The shortest duration changes appear to be associated with cyclic glacio-eustatic sea-level changes. Amplitudes are on the order of 100 m and appear to involve the superposition of several cycles ranging in period from 20 to 400 kyr. These cycles are related to changes of glacial ice and seawater volume apparently driven by climatic processes sensitive to Earth's orbital and rotation motions. On a thousand-fold longer time scale are sea-level fluctuations associated with the development of coastal onlap sequences and depositional sequences in intracratonic basins of North America, the Russian Platform, and Brazil. These fluctuations have amplitudes of 100 to 200 m and periods ranging from 10 to 80 Myr. Within regions of sluggish tectonics, the effects of sea-level change on these time scales seem to be especially important in affecting both landscape morphology and the nature of sedimentary deposits. All of the above variations are superimposed on two very long-term sea-level fluctuations that occurred over the last 700 Myr (Fischer, 1983). These have amplitudes of 300–400 m with low stands in Eocambrian and Triassic–Jurassic times. Lows apparently coincide with major episodes of continental break up. Sea level again seems to be approaching such a minimum.

The clearest sedimentary records of sea level change occur where there is a good and steady supply of sediment and where the land is slowly and steadily sinking relative to mean sea level (Pitman, 1978). Areas that commonly satisfy this requirement are passive margins, intracratonic basins, and the foreland basins of major orogenic belts. When sea level rises, thicker sedimentary units are deposited more inland, generating a coastal onlap. With sufficiently high sea level, epeiric seas begin to flood the cratons, often filling foreland and intracratonic basins before spilling out over normally high ground; this is a major transgression. At low sea level, very little deposition occurs on the cratons. Instead, most of the previously deposited sediments are removed by erosion and redeposited along the continental margins. Passive margins seem to have the most easily interpreted subsidence history, but the longest records are found in intracratonic basins. In the tectonically active areas, deformation and subsidence are typically too rapid to preserve information about eustatic sea level.

Several investigators have argued that there is a major tectonic component to the subsidence histories of passive margins and intracratonic basins. Sloss and Speed (1974) and Sloss (1979) note that subsidence episodes appear to be globally synchronous and coincident with high sea levels. They argue that simple sedimentary loading caused by deposition at times of high sea levels is not an adequate explanation and suggests that some common deep driving mechanism controls both sea level and subsidence.

In regions where land is steadily rising relative to mean sea level, the effects of sea-level fluctuations are sometimes recorded as erosional features on land. Whenever the rate of sea-level rise matches the rate of uplift, there is an apparent sea level still stand. Both deposition and erosion are controlled by this almost fixed base level, and a terrace may form. If sea level falls and again rises, the terrace will have risen sufficiently so that it is preserved upslope. Episodic uplift can produce terraces, but these should not be synchronous with similar terraces developed in other regions, unless episodes of uplift were being controlled by some deep-seated process. The best-formed terraces from the late Pleistocene–Holocene sea-level fluctuations occur on coasts where uplift has been rapid (Fig. 9-7). These take the form of combined wave-cut benches, beach ridges, and carbonate banks. Deformation and erosion in these areas is often so rapid, however, that no evidence of longer-term (even early Pleistocene) base-level changes is preserved. The erosional effects of long-term sea-level fluctuations, however, are spectacularly recorded on some cratons.

9.5.5 A General Concordance Between Erosion Rates and Uplift Rates

Figure 9-7 displays uplift curves, representing the elevation of various sea-level indicators versus time, for tectonically active (A, B, C) and quiet (D, E, F) regions. Note that uplift rates for Taiwan, a fold and thrust belt, exceed estimates for shield regions in South America by almost three orders of magnitude. If we assume that the hypsographic curves for any of these regions have remained fixed through time, which implies that the region has had a steady-state appearance, then uplift rates equate with denudation rates. However, because the climatic and tectonic processes that govern landform development do not operate constantly and continuously, current denudation rates may differ significantly from these values.

Dissolved solids concentrations (dissolved phases derived from the weathering of bedrock) can be used to estimate ranges of denudation rates for particular regions. Histograms of dissolved solids concentrations for different morphotectonic regions of the Amazon and Orinoco basins are presented in Fig. 9-8. Solution chemistry of rivers provides the best gage of current weathering for a particular terrain because sediment compositions are more difficult to interpret because of sediment storage as alluvium or colluvium. These denudation rates are in general agreement with more rigorous calculations (Edmond *et al.*, 1995; Gibbs, 1967; Lewis *et al.*,

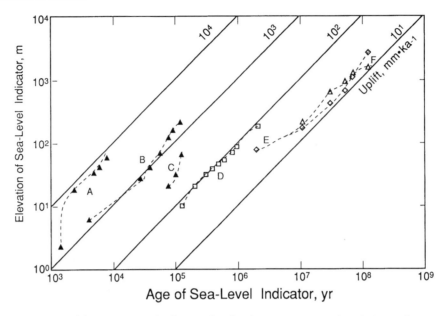

Fig. 9-7 The elevation of the terraces and other sea level indicators compared to their age for several locations (Stallard, 1988). To derive actual uplift rates, one needs to know the age of the terrace and the eustatic sea level at that time. (A) Elevation data from the Hengchun Peninsula, Tainan, and Eastern Coast Range of Taiwan, corrected for glacial-eustatic effects and grouped by intervals of about 1500 yr. Uplift seems to have been discontinuous, but has averaged about 5000 m/Myr (Peng et al., 1977). (B) Calculated composite terrace elevations for the Huon Peninsula, New Guinea, using data from Bloom et al. (1974). Uplift has been continuous and has ranged from 940 to 2560 m/Myr; the calculation used 1620 m/Myr. (C) terrace elevations for the Island of Barbados (Clermont traverse) from Matthews (1973); uplift rate has been about 400 m/Myr. (D) Depositional terraces in the Amazon Trough from Klammer (1984). Ages are estimated from those of interglacials on the standard oxygen isotope curve. Uplift is calculated to have been about 80 m/Myr. Klammer argues that the apparent change in elevation has been caused by drop in sea level since the Pliocene rather than by uplift. (E, F) Erosion surface from the Guayana and Brazilian Shields, respectively (King, 1957; McConnel, 1968; Aleva, 1984). (Reproduced with permission from R. F. Stallard (1988). Weathering and erosion in the humid tropics. *In* A. Lerman and M. Meybeck, "Physical and Chemical Weathering in Geochemical Cycles," pp. 225–246, Kluwer Academic Publishers, Dordrecht, The Netherlands.)

1987; Paolini, 1986; Stallard, 1980). Comparison with Fig. 9-7 shows a reasonable match between denudation rates and uplift rates for a particular type of terrain. The most concentrated water samples and highest denudation rates are observed in river basins in tectonically active areas.

9.6 Erosion in Tectonically Active Areas

In many active mountain belts, both rapid long-term uplift and erosion are sustained by tectonic recycling of sediments. For example, the island

of Taiwan (Fig. 9-7A), which is a fold and thrust belt, has been formed over the last several million years during an ongoing collision between the Luzon island arc and the Eurasian continental margin (Chi et al., 1981; Suppe, 1981). During this orogeny, metamorphosed and diagenetically altered sediments have been uplifted, eroded, and deposited, only to be reincorporated into the cycle once more (Manias et al., 1985). The central range of Taiwan has some of the highest denudation rates (13 000 tonne/km^2 per year solid + 650 tonne/km^2 per year dissolved = 5150 mm/kyr) measured for river catchments anywhere on

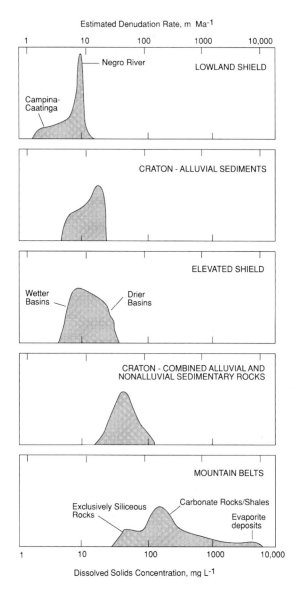

Fig. 9-8 Histogram of dissolved solids of samples from the Orinoco and Amazon River basins and corresponding denudation rates for morpho-tectonic regions in the humid tropics of South America (Stallard, 1985). The approximate denudation scale is calculated as the product of dissolved solids concentrations, mean annual runoff (1 m/yr), and a correction factor to account for large ratios of suspended load in rivers that drain mountain belts and for the greater than average annual precipitation in the lowlands close to the equator. The correction factor was treated as a linear function of dissolved solids and ranged from 2 for the most dilute rivers (dissolved solids less than 10 mg/L) to 4 for the most concentrated rivers (dissolved solids more than 1000 mg/L). Bedrock density is assumed to be 2.65 g/cm^3. (Reproduced with permission from R. F. Stallard (1988). Weathering and erosion in the humid tropics. *In* A. Lerman and M. Meybeck, "Physical and Chemical Weathering in Geochemical Cycles," pp. 225–246, Kluwer Academic Publishers, Dordrecht, The Netherlands.)

Earth (Li, 1976). The high denudation rate is a reflection of the poorly lithified, highly tectonized nature of the sedimentary rocks that compose the island. Sediment-yield data compiled by Milliman and Meade (1983) and Milliman and Syvitski (1992) indicate that island arcs and mountain belts in the tropical and subtropical west Pacific may contribute as more than 22% of all solid material discharged by rivers into the ocean. Furthermore, the tropical mountainous areas in southeast Asia and India may contribute another 33%.

9.6.1 Erosion and Orogeny

Mountain building usually involves compressional deformation of the crust. Studies of the physics of orogeny suggest that there is a feedback between the nature of the building process and denudation rates. Suppe (1981), Davis *et al.* (1983), and Dahlen *et al.* (1984) have modeled the effects of brittle deformation in accretionary fold-thrust mountain belts such as Taiwan and the Andes. The basis of their model is the hypothesis that rock deformation is governed by pressure-dependent, time-dependent brittle fracture or frictional sliding – "Coulomb behavior." In such belts, sediments are scraped off subducting lithosphere, and a large wedge of deformed sediment, separated from underriding crust by a basal decollement, is built. The process resembles the behavior of snow being pushed in front of a bulldozer blade. The mountain belt develops a regional profile that tapers towards the subducting plate. If the angle of the taper (0–6°) is less than a critical value, the wedge sticks to the basal decollement. Compression, due to the addition of material at the toe of the wedge, is taken up by the internal deformation of the wedge such that the wedge steepens. Once the angle of regional taper reaches the critical value, the entire wedge can slide over the basal decollement, deforming as material is added so as to maintain the critical taper. If too much material ($\sim 15\,km$) piles up, the deepest material starts to deform, greatly reducing basal resistance to sliding. Flat high plateaus, such as the Andean Altiplano and the Tibetan Plateau, may develop over the areas that are deforming in this manner.

Models indicate that the overall topographic profile of such belts evolves into a stable form such that erosional outputs balance accretional inputs (Suppe, 1981; Davis *et al.*, 1983). A mountain belt cannot reach unlimited height as more material is added. Two factors mitigate against this: (1) erosion and (2) changes in rock deformation from brittle to ductile with increasing depth. There should be a continuous supply of easily eroded material so long as accretion continues. Since erosion thins the wedge and reduces the taper, the intensity of deformation should increase with increasing erosion rates. It seems reasonable, therefore, that the lithological susceptibility to erosion and climate-related erosional intensity may in turn influence the form of the entire mountain range, not just the form of the slopes. Suppe (1981) argues that the width and regional slope of mountain belts is susceptible to the climate regime. For example, mountain belts might tend to be wider where erosion rates are reduced, such as in a region of dry climates. There must be some minimum regional denudation rate for a steady-state profile to form. Under the limiting case of no erosion, an accretionary mountain belt should continue to widen so long as accretion continues.

9.6.2 River Chemistry and Bedrock Susceptibility in Mountainous Regions

The composition of dissolved and solid material transported by rivers in mountain belts of the humid tropics is consistent with weathering-limited erosion. In the Andes, sediments constitute the principal basement lithology, and the river chemistry correlates with catchment geology (Stallard, 1980, 1985, 1988, 1995,a,b; Stallard and Edmond, 1983). For example, black shales have particularly high Mg:Ca ratios, and Bolivian rivers that drain black shales are exceptionally magnesium-rich. Rivers that drain evaporites have the highest total cation (TZ+) concentration, followed by those that drain carbonates, and finally by those that drain only siliceous rocks. This is illustrated by the trimodal nature of the histogram of dissolved solids from Andean rivers in Fig. 9-8 and by the ternary diagram in Fig. 9-9. Here, $Si(OH)_4$, alkalinity, and $(Cl^- + SO_4^{2-})$ are used as input markers for the respective lithologies, bearing in mind that alkalinity is also produced by weathering of silicate minerals, and SO_4^{2-} by weathering of pyrite and other sulfides. Total cations increase systematically from silica to alkalinity to (chloride + sulfate). Many Andean rivers are slightly supersaturated with respect to calcite (Stallard, 1988, 1995b; Stallard and Edmond, 1987); this saturation ultimately limits the supply of alkalinity to the rivers. Similar limits do not apply to Na, K, Mg, and Ca because of additional inputs from silicates and evaporites. Erosional contributions by the weathering of carbonates and especially by evaporite minerals greatly exceed those expected if inputs are assumed to be proportional to the fraction of the catchment area in which these lithologies are exposed. Much of the evaporite input is sustained by actively extruding salt diapirs (Benavides, 1968; Stallard, 1980; Stallard and Edmond, 1983).

The presence of unstable and cation-rich minerals in the suspended load and bed material of rivers that drain the Andes indicates that

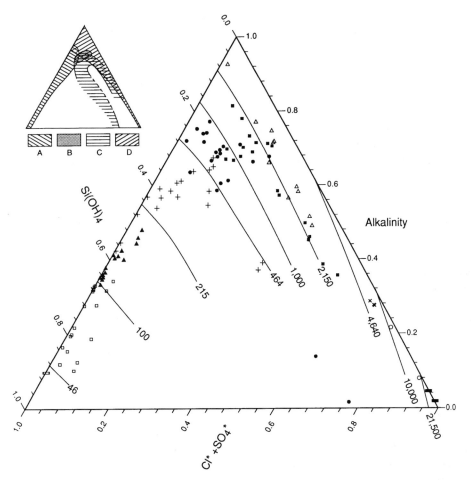

Fig. 9-9 Ternary diagram showing proportions of dissolved $Si(OH)_4$, carbonate alkalinity ($HCO_3^- + CO_3^{2-}$), and ($Cl^- + SO_4^{2-}$) in the Orinoco River and Amazon River basins. Charged species are in equivalents; $Si(OH)_4$ is in mole units. Curves, in the large figure, are numbered in total cation concentration (mequiv/L). Unlike previous figures, symbols represent the total cation concentration interval that includes the sample's concentration. The predominant symbol within each interval corresponds to samples whose concentrations plot within that interval. In the small figure, the patterned areas correspond to the predominant source of samples whose concentrations plot within the areas: A, streams that drain cratonic areas; B, streams that originate in mountain belts, but that drain large areas of cratons; C, streams that drain mountain belts with extensive black shales; D, streams that drain mountain belts with extensive carbonate rocks and evaporite deposits. (Reproduced with permission from R. F. Stallard (1988). Weathering and erosion in the humid tropics. *In* A. Lerman and M. Meybeck, "Physical and Chemical Weathering in Geochemical Cycles," pp. 225–246. Kluwer Academic Publishers, Dordrecht, The Netherlands.)

erosion is extraordinarily rapid. In these rivers, the solution enrichment of Na^+ relative to K^+ and Ca^{2+} relative to Mg^{2+}, when compared to bedrock ratios (Fig. 9-5) indicates that K- and Mg-rich solid phases are being eroded. Moreover, the chemical and mineralogical compositions of the sediment correlate with geology. The sands in tributaries that have their head-

waters in the mountain belts are commonly litharenites (DeCelles and Hertel, 1989; Franzinelli and Potter, 1983; Johnsson *et al.*, 1988, 1991; Potter, 1978, 1994; Stallard *et al.*, 1991). Especially unstable minerals such as calcite, amphiboles, and pyroxenes are present in samples from some of these rivers. Micas and 2:1 clays, including illites, vermiculites, and smectites are abundant in the fine-grained fraction (Gibbs, 1967; Irion, 1976, 1984; Koehnken, 1990; Stallard *et al.*, 1991). Chlorites (2:1:1 clays), quartz, feldspars, and amphiboles are also common in the silt-size fraction. Pyrophyllite occurs in some rivers. When compared to average igneous rocks, fine-grained sediments typically are enriched in Al relative to soluble cations, in Mg relative to Ca, and in K relative to Na (Stallard, 1985; Stallard *et al.*, 1991). Andean rivers often acquire an intense red color only after crossing regions with red beds. Furthermore, the suspended load of rivers in the Peruvian Andes that have not yet crossed the red beds is enriched in vermiculite and mica, whereas in the lower courses of the same rivers, it is enriched in smectite and kaolinite. Finally, the marine shales, which are abundant in the Bolivian, Colombian, and Venezuelan Andes, are very micaceous, and illite is particularly abundant in rivers that drain these regions.

Weathering products in island-arc and young volcanic terrains present an interesting contrast to those from tectonically assembled, continental mountain belts. When solutes from rivers draining siliceous rocks are compared, concentrations in the island-arc and younger volcanic settings are about twice those in continental mountain belts (Stallard, 1995b). The rocks in island-arc and volcanic settings are much younger than most continental rocks. One possible cause of higher concentrations is that rocks in the former settings have more volatiles in the mineral lattices and in fluid inclusions. These volatiles, in turn, make the rocks more susceptible to chemical weathering. Many volcanic rocks are undersaturated; they do not contain quartz. When these rocks weather, coarse saprolite pellets form. These are transported like sand and gravel. Within the pellets are grains of unweathered minerals (Johnsson and Stallard, 1989). When buried and compressed these pellets should transform into a sedimentary rock consisting of sand-size mineral grains suspended in a fine matrix – a greywacke.

9.7   Erosion of the Cratons

Erosion on cratons has been difficult to describe.

This is in part because denudation rates are very low and because sea-level fluctuation may be important to the erosion process. Cratons seem to undergo major episodes of erosion following drops of sea level. When the level drops to a stable stand of several million years, much of the landscape is eroded down to the new level and an erosion surface or a planation surface forms.

Two principal models have been put forward to describe the development of extremely eroded topography following a drop in base level in a region that had once been planed flat (James, 1959). One is the classic Davis cycle (Davis, 1932), the other is a model developed by King after Penck (King, 1953, 1967). These differ in their prediction of how the raised topography is dissected to form hills and valleys. According to Davis, rivers first cut steep valleys into the landscape, then their valleys broaden, and regional slopes are flattened until the last remnants of the original surface are worn away at interfluves. At this point the landscape is said to be mature. Slopes continue to flatten until formerly elevated interfluves become low swells. This is a peneplain. In King's model, initial slopes are formed along the edges of the uplift and the sides of large penetrating river valleys. Slopes then evolve into steepened equilibrium forms that separate the old erosion surface from an incipient new surface. As time proceeds, these slopes undergo parallel retreat into the older surface that is consumed while the younger surface is extended. The end product is a pediplain. The steep slopes are the locus of most of the erosion and the remaining terrain is almost flat, being drained by rivers having little erosive capacity. In the Davis model, no remnants of previous surfaces can exist. In King's model, however, successive erosional levels can remain stacked and can even actively expand through parallel retreat into the older surfaces long after the initiating change in base level has been superseded.

A search for stacked erosion surfaces became the centerpiece of work by King (1967), who compiled spectacular continental-scale examples from South America, Africa, India, and Australia. In a sense, these surfaces are rather like super-terraces, occurring within a narrow range of altitudes for a particular region. Surfaces nearest sea level are well defined and undissected, whereas the most elevated surfaces are usually remnants of small extent or are simply delineated by a large number of hills having peaks of similar height topped by deeply weathered soils (Fig. 9-10). In South America (Fig. 9-7E and F), Africa, and other tropical cratons at least five levels seem to be identifiable. King argued that to a first approximation the surfaces are globally synchronous (within about 10 Myr) and that the oldest ones are of great age, perhaps predating the rifting of the South Atlantic (late Jurassic to early Cretaceous).

Subsequent geomorphic studies and work related to bauxite exploration have produced better dating and descriptions of relationships among erosion surfaces in South America (Aleva, 1979, 1984; Krook, 1979; McConnell, 1968; Menendez and Sarmentearo, 1984; Zonneveld, 1969). The most economic bauxite deposits occur on the Neogene (55 Myr) surfaces.

Although bauxite from older surfaces is of high grade, they are of small extent and very dissected. The bauxite becomes progressively more ferruginous on younger surfaces. Where the substrate is quartzose, quartz persists through the entire weathering profile, which may be as much as 50 m thick on the 55 Myr surface (Menendez and Sarmentearo, 1984).

These surfaces are dated by classical stratigraphic techniques. Deposits of fluvial, lacustrine, or eolian sediments that contain dateable pollen or vertebrates are sometimes preserved on the surface. This is not common and dating usually is accomplished by tracing a surface into an area of subsidence such as a passive margin or an intracratonic basin. Occasionally, the surfaces have been sufficiently downwarped to be buried under marine sediments. More often, such as off the northeastern South American coast, erosion surfaces coincide with hiatuses in the sedimentary section; these are frequently overlain by deposits of quartz and bauxite gravels and sub-arkosic sands. This is thought to be

Fig. 9-10 Radar image of erosion surface in Venezuela (Petroleos de Venezuela S. A., 1977). The region bounded by this photograph is approximately 50 km by 100 km. The Orinoco crosses the left side of the image. The confluence of the Meta is just off the lower left corner of the image. To the west of the Orinoco are the Llanos of the Andean Foreland Basin; to the east of the Orinoco is the Guayana Shield. The erosion surfaces appear as areas of slightly dissected, raised topography in the Guayana Shield. An irregular NNW–SSE trending escarpment that starts to the left of center and runs to the lower edge separates the two surfaces.

indicative of the early dissection of thick and highly weathered soils following a drop in base level. Such hiatuses are not unlike those associated with sea level high stands seen in coastal onlap sequences. Accordingly, Aleva (1984) notes that the major surfaces appear to coincide with major high stands on the late Mesozoic–Cenozoic sea level curve. The most recent and youngest surfaces are ten or so glacio-eustatic terraces from the Amazon valley described by Klammer (1984) (Fig. 9-7D).

The Brazilian and Guayana shields are among the best exposures of elevated crystalline basement in the humid tropics. The topography can be spectacular with steep slopes, either of bare rock or thinly vegetated, often topped by high plateaus, some of which are quite expansive. The highest landforms are 2000–3000 m table-like mountains called "Tepuis" (Fig. 9-11). These are topped by thick orthoquartzites. Inselbergs are locally common. Great talus piles do not accumulate below cliffs and at the base of slopes. Much of this material probably just dissolves. A karst-like topography developed on the orthoquartzite, gneisses, and granites is found on some of the higher and presumably most ancient areas (Blancaneaux and Pouyllau, 1977; Szczerban, 1976; Urbani, 1986). The karst on quartzite is established to such a degree that stalactites consisting of opal, cristobalite, tridymite, and quartz have formed locally (Chalcraft and Pye, 1984). Aerial and radar photographs show that all but the youngest surfaces are dissected to various degrees, and that this dissection is geologically controlled. Many rivers appear to follow fault systems and dike swarms (Fig. 9-10, Schubert *et al.*, 1986; Stallard *et al.*, 1991). This would indicate that the older surfaces are no longer expanding in the manner described by King, since the dissected surfaces can no longer serve as fixed base levels for overlying scarps. Especially resistant lithologies such as quartzites and some intrusives persist as high elevations. Thus, even though weathering rates are very slow (Fig. 9-8C), the observation that lithology plays an important role in landscape morphology indicates that weathering-limited erosion is occurring.

Fig. 9-11 Photograph of Ayun Tepui, taken from Canaima, Venezuela. The cliffs, which are about 2000 m high, consist of orthoquartzite. The irregular top is a karst topography formed by the slow dissolution of the quartz. It is estimated that the top of this tepui, and similar ones elsewhere on the Guayana Shield, have been exposed since the Jurassic.

9.7.1   Erosion and the Slow Uplift of the Shields

The development of topography consisting of stacked erosion surfaces on many shields indicates that vast areas of cratons are undergoing a slow but sustained rise relative to sea level. The remnants of erosion surfaces are evidence of the former thickness of the cratons. It is interesting to speculate that this tendency to rise has held true for many shield areas throughout much of geologic time, thereby explaining the high metamorphic grade of rocks exposed on the surface of most shield areas. In this sense, shields would be dynamical opposites to intracratonic basins. In this context, continental platforms are areas that neither rise nor sink dramatically. Tectonic modes change with time. For example, the thick quartzites that top many of the highest peaks of the Guayana shield are evidence that this area once behaved as either a platform or a basin.

Part of the rising and sinking of cratons may be due to isostatic adjustments in response to erosion and to sediment accumulation. Part, however, may be due to deep-seated processes. Burke and Wilson (1972), Crough (1979), and Morgan (1983) have argued that uplift and erosion of shields is caused by random repeated passages of continental crust over mantle plumes. As a segment of crust passes over a mantle plume, it is heated and becomes more buoyant. This event produces a fairly rapid uplift followed by a slow reduction of elevation as a result of combined erosion and cooling. If erosion is sufficiently intense, the formerly elevated segment of crust becomes a depositional basin once it has completely cooled. Morgan (1983) used this idea to examine the relationship between hot spots and regional geology in the continents bracketing the Atlantic. He hypothesized the recent passage of a hot plume under the Guayana shield to explain its presently elevated nature.

The globally synchronous development of erosion surfaces and slow uplift rates of shields indicates that models based on localized heating of the crust and mantle are not adequate to explain the uplift process. To produce near simultaneous surfaces on separate continents, King (1967) invoked global epeirogeny – uplift driven by some sort of deep-seated process. The change in base level does not have to act continuously once scarps are initiated. They may become self sustaining to some degree, for as they retreat, isostatic adjustment would raise the topography (King, 1956), thereby stabilizing the effects of the initial base level change. The isostatic uplift would be cumulative, thus explaining the stacked topography. Global sea-level drop seems to be a more reasonable initiating event, but it is still a matter of speculation as to whether the uplift is continuous or episodic and isostatically self sustaining or driven by deep-seated, global, tectonic processes. There is, however, an interesting parallel between King's view of erosion surfaces and the idea of Sloss and Speed (1974) and Sloss (1979) that there is a tectonic component to the subsidence of sedimentary basins at times of high sea level.

9.7.2   Weathering-Limited Erosion on the Elevated Shield

Rivers from the shields transport very little dissolved or solid load, and the solids tend to be more resistant mineral phases (Franzinelli and Potter, 1983; Gibbs, 1967; Irion, 1976; Johnsson *et al.*, 1988, 1991; Koehnken, 1990; Potter, 1978; Stallard, 1995a,b; Stallard *et al.*, 1991). The major cations of the dissolved load are in bedrock proportions (Fig. 9-5), indicating that bedrock is strongly leached during weathering and most of the cation load is in solution. River sands include more potassic feldspars, as might be expected if weathering-limited erosion is occurring on the deep slopes. Erosion rates in these terrains are much lower than in active mountain belts like the Andes, even though elevations are as much as 3000 m. Both lithology and basement structure contribute to the low erosion rates in elevated crystalline-basement terrains (Stallard, 1988; Stallard *et al.*, 1991). Rocks are frequently massive and composed of minerals like quartz, potassium and sodium feldspars, and micas, which are somewhat resistant to weathering. Uplift apparently involved ductile (plastic) deformation and was followed by a long history of weathering that may have removed all unstable lithologies. The basement of active orogenic belts is not so massive owing to faulting, brittle deformation, and volcanism. Furthermore, active uplift provides a continuous supply of fresh unstable lithologies.

9.7.3 Transport-Limited Erosion on the Lowlands

Vast tracts of the South American and west African lowlands have predominantly transport-limited denudation regimes (Stallard, 1988; Stallard *et al.*, 1991). These regions represent the flattest and youngest erosional/depositional surfaces of the late Neogene. On the shield, the substrate is crystalline basement, while in the intracratonic basins, the Andean foreland basin, and the coastal plains, much of the sedimentary substrate consists of strongly "preweathered" fluvio-lacustrine sediment (Stallard, 1985, 1988; Stallard and Edmond, 1983; Stallard *et al.*, 1991). Soils and solid loads of rivers are rich in quartz, kaolinite, and iron sesquioxides, all of which are cation-depleted phases. River sands are quartzose (Franzinelli and Potter, 1983; Johnsson *et al.*, 1988, 1991) and the suspended load is largely kaolinite plus quartz. Aluminum sesquioxides (gibbsite) are much less common. In rivers with pH > 5 particles have iron sesquioxide coatings; at lower pH values coatings are absent (Stallard *et al.*, 1991). Dissolved major cations are in bedrock proportions. Dissolved solid concentrations, and by inference denudation rates, are exceedingly low (Fig. 9-8). This agrees with the diminution of weathering rates in conjunction with the development of thick soils. Dissolved loads tend to be lower on the shields than on the sediments, even though rocks of the shield are composed of less stable minerals; however, the low permeabilities of the rocks and many of the soils of the shields compared to those of the sediments apparently counterbalances this.

Where easily weathered lithologies such as carbonates and evaporites are near the surface, such as in the lower Amazon valley, their contribution to the rivers appears minor, probably because thick residual soil covers have devel-oped. This indicates that susceptibility to weathering is indeed less important in controlling lowland erosion rates. Carbonate platforms, such as southern Florida and the Yucatan in North America, are exceptions because of deep water circulation throughout the carbonate karst.

9.8   The Effects of Transients: Continental Ice Sheets and Human Technology

As seen in the previous discussion, susceptibility of bedrock to erosion is the primary factor controlling erosion rate in steep terrains, whereas, supply of reactants to reactive minerals limits erosion in flat terrains. Geochemists are frequently called upon to assess the impact of some human-induced environmental perturbation. Would pulverized granite make a good fertilizer in areas of thick, cation-depleted, tropical soils? What affect does permanent deforestation of a mountain slope have on erosion on a short time scale? On a time scale that is 10 times longer? What about acid rain? What about long-term climatic warming or increased atmospheric carbon dioxide? What happens when a formerly moist region dries out? Humans move as much solid matter about as do all other geomorphic processes (Hooke, 1994). Human modifications of soil genesis, erosion, plant productivity, and terrestrial sedimentation may be enough to account for the burial of a billion tonnes of carbon annually (Stallard, 1998). A billion tonnes of carbon is a significant fraction of the carbon released from the burning of fossil fuels. The Earth has recently undergone a remarkable perturbation, continental glaciation. Glaciations are an especially interesting context within which to evaluate the concepts of erosion regime.

9.8.1   Glacial Weathering and Erosion

During the past 2.4 million years the Earth has gone through several episodes of continental glaciation (see Hughes, 1985). Each of the last seven of these episodes (600 000 years) shows a characteristic pattern of waxing and waning

(Broecker and Denton, 1989). Each episode started slowly. Extensive ice sheets gradually build up in Canada (the Laurentide ice sheet), northern Europe (the European or Fenno-Scandian ice sheet), and across northern Asia. Smaller ice sheets form in Iceland and southern South America. Temperatures cool globally, and the altitude of mountain glacial formation descends about 1000 m. The ice caps of Greenland and Antarctica became more active. The glacial buildup lasts about 100 000 years and ends abruptly, over a few hundred years. Within the buildups and terminations there are short episodes of glacial advance and recession. The last ice age ended about 13 500 years ago.

Glaciers are powerful agents of physical erosion. In a detailed geomorphological study of the glaciated Canadian Shield, Sugden (1978) concluded that erosional style was related primarily to the basal thermal regime of the ice. From the center of divergence on an ice cap,

Sugden identified five idealized zones. Erosion is minimal under the cold-based ice at the center of the cap – shear from ice flow is minimal and the ice is anchored to the bedrock. Surrounding this was a zone of basal melting then a zone of basal freezing, and finally a zone of melting. Various processes drive erosion in these zones (Fig. 9-12). Sugden concludes that the excavation of debris is an important factor influencing the amount of erosion accomplished by an ice sheet. The excavation process contrasts strongly with fluvial systems or alpine glaciers, for which downslope transport is important. Sugden argues that landforms on the Canadian Shield are equilibrium forms with respect to the thermal regime of the ice sheet at maximum glaciation.

The estimation of depth of erosion by continental ice sheets has been a controversial endeavor. In mountainous regions such as the North American Rockies, alpine glaciers have

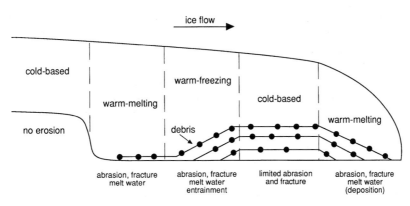

Fig. 9-12 Idealized model of the relationship between styles of glacial erosion and the basal thermal regime of ice, adapted from Sugden (1978). Many factors affect the basal thermal regime. The most important are surface temperature, ice thickness (pressure melting), and the rate of ice deformation (frictional heating). Cold-based ice is generally protective and has little erosive power unless it contains debris inherited from an up-glacial basal freezing zone. In warm-melted zones, basal slip between ice and rock promotes fracture, plucking, and abrasion. If ice subsequently passes into a freezing zone, bedrock particles can be entrained during the freezing process. This further enhances excavation of bedrock. During a glaciation, patterns of erosion may shift as ice sheets develop and retreat, but peak erosion seems to occur near glacial maxima. The top surface of an ice sheet is relatively flat. Erosion is more pronounced at basement topographic lows (either valleys or lowlands), relative to adjacent basement highs, because the thicker ice over the lows promotes pressure melting, and because low areas tend to be zones of ice convergence. Less porous rocks, such as crystalline shield may be more susceptible to glacial erosion than porous rocks such as sediments, because diversion of melt water throught the substrate may promote binding of the ice to the bedrock. Such ice would act more like cold-based ice. (Adapted with permission from D. E. Sugden (1978). Glacial erosion by the Laurentide ice sheet. *J. Glaciol.* **83**, 367–391, International Glaciological Society.)

scoured deep valleys into a topography that was, prior to glaciation, much smoother and more rolling. For continental ice sheets, it is much harder to find a convenient frame of reference, such as pre-existing topography, from which depth of erosion can be judged. White (1972) argued that the ice sheets had removed much of the sedimentary cover of the Canadian Shield and drew a cross-section indicating that there may have been as much as 1000 m of erosion. This conflicted with geologic, pedological, and geomorphological evidence that some areas of the Canadian Shield had been minimally eroded during the ice ages and that the oldest glacial deposits are rich in material from crystalline basement (Gravenor, 1975; Sugden, 1976). More recently, careful compilations of the quantities of glacial sediment deposited in the oceans around North America and in depositional basins on North America by Laine (1980) and Bell and Laine (1985) indicate that about 120 m or more of physical erosion were caused by continental ice sheets. If present river yields are used, about 20 m of chemical and 20 m of physical erosion would have occurred over the same time period (3 Myr assumed). Clearly, continental ice sheets are potent agents of erosion – on shields, ice evidently is more powerful than the processes associated with the humid tropical weathering.

The above-mentioned river yields are undoubtedly affected by the recent glacial scouring transient. Throughout the glaciated region there is still an abundance of loose glacial debris, much of which is fresh bedrock and mineral grains. This great abundance of fresh bedrock minerals at the ground surface contrasts markedly with the situation in flat regions undergoing transport-limited erosion where soil minerals are cation-deficient. Gravenor (1975) presents data that indicate that prior to glaciation the shield may once have had a soil mantle that even included bauxites. Presumably, both chemical and physical erosion must be proceeding much faster now than under steady-state conditions involving a thick soil. Thus, Bell and Laine (1985) used worst-case values – generally a good practice. What might be a more realistic estimate of weathering rates without glaciations?

9.8.2 Late-Glacial/Deglaciation Geochemical Anomalies

Ice ages end much more quickly than they develop, and the marine chemical record indicates that this was an unusual time. Late in the last glaciation and during glacial retreat, the oceans had $^{87}Sr/^{86}Sr$ ratios (Dia et al., 1992) and Ge/Si ratios greater than those of today (Froelich et al., 1992). The source of excess ^{87}Sr is radioactive decay of ^{87}Rb in potassium-bearing minerals; thus, an especially strong source of ^{87}Sr is the weathering of potassium-bearing minerals in old continental crust such as the Canadian Shield. In today's world, water that has high Ge/Si ratios are either draining terrains undergoing transport-limited erosion (Murnane and Stallard, 1990) or rivers contaminated by ash from coal burning, where the Ge is associated with organic matter and sulfides (Froelich et al., 1985). If we use today's world as a reference frame, the late-glacial data indicate a strong contribution of ^{87}Sr and Ge from a shield terrain undergoing intense tropical-style weathering (Froelich et al., 1992; Gibbs and Kump, 1994). This contradicts all that we know about global climate of the time (see Broecker and Denton, 1989).

Investigators studying warm-based mountain glaciers and glaciated areas have noted that even in granitic areas, waters are particularly high in Ca^{2+}, K^+, and SO_4^{2-} ions relative to other major soluble cations (Axtmann and Stallard, 1995; Drever and Hurcomb, 1986; Mast et al., 1990; Stallard, 1995b). One source of Ca^{2+} appears to be intergranular calcite, one of the last minerals to crystallize out of the cooling granite melt. Volumetrically, this calcite is a minor constituent. Calcite is also far more susceptible than other minerals in granite (Table 9-1). A primary reason for the Ca^{2+} abundance may be that the calcite is making a disproportionate contribution to the steam waters because of the recent exposure of fresh granite by glacial activity. Similarly, the excess SO_4^{2-} may be from the weathering of disseminated trace sulfides. The K^+ appears to come from the physical breakdown of micas in general and the oxidative weathering of vermiculite. Calcium-bearing minerals release strontium with a low $^{87}Sr/^{86}Sr$

ratio, while the ratio derived from the micas is exceedingly high, especially in old rocks. The weathering sources of Ca^{2+} and K^+ during soil development persist well after glaciers have departed (Blum *et al.*, 1994; Mast *et al.*, 1990; Stallard, 1995b), but are relatively weak compared to sub-glacial sources (Axtmann and Stallard, 1995). In the case of the Laurentide region, soil development took several thousand years and still continues (Harden *et al.*, 1992). Micas and sulfides are potential sources of excess Ge/Si. Thus, the peculiarities of late Ice Age seawater may be cause by the different styles of chemical weathering associated with glaciers (Stallard, 1995b). This points to the need to examine biogeochemical processes in detail before drawing analogies between different processes in various terrestrial environments (erosion regime) and the data that need to be explained (marine chemistry).

9.9   Conclusion

Weathering processes exert a major control on many aspects of the chemistry of the Earth's surface environment. Considerations of temporal and spatial scale are very important in evaluating how the weathering processes affect the chemical response of the atmosphere and ocean reservoirs. Continental denudation, the transfer of material from the land surface to the ocean, is a particularly important aspect of this response. In small stream basins, solution chemistry is controlled by processes that act on a hillslope scale. There is commonly little storage of water in most large river systems; thus for large river systems, hillslope processes remain important. Storage on a time scale of months to millions of years must be considered for solid material. Long-term burial results in a net loss to the river, and chemical weathering during storage can be important for large river systems.

In regions where the erosion regime is weathering limited, susceptibility of the bedrock to chemical and physical weathering controls erosion rates. This susceptibility relates directly to the chemical and physical properties of the rock. Susceptibility also depends on local climate. Moreover, weathering rates are affected by the

soils that form on the rock, and the nature of the vegetation that grows on the soil. Vegetation, by supplying bioacids that chemically degrade the rock, can increase the rate of erosion of rocks that are resistant to physical erosion.

Susceptibility of bedrock to erosion is not necessarily controlled by the reactivity of the mineral grains that make up the bedrock. In the case of limestones, the weathering reactions proceed until equilibrium is reached. The rate is controlled by the supply of reactants (protons). Rocks composed of silicate minerals are different. For example, the massive, poorly jointed granites of the Guayana Shield erode more slowly than the tectonized, highly cracked and sheared granites of the Andes. The difference is not caused by contrasting mineralogies, but by differences in the permeability of water and stability of steep slopes. Again, the supply of reactants is very important, but because the weathering does not proceed to equilibrium or to completion, the reactivity of individual mineral grains is also important. Thus, when compared to the bedrock proportions, K is enriched relative to Na and Mg relative to Ca in solid erosion products. Likewise, Na^+ and Ca^{2+} are enriched in solution relative to K^+ and Mg^{2+}. In flat terrains, these enrichments do not occur because, the mobilization of major soluble cations from bedrock is complete. The presences of volatiles and glasses may cause rocks from island arc terrains and younger volcanics to be more susceptible than equivalent older rocks from cratons.

In regions where erosion is transport limited, weathering rates are controlled by the supply of reactive fluids to unstable minerals. This is controlled by soil properties, regional base level, and ultimately, sea level.

The concept of weathering regime may be useful for interpreting the effects of important phenomena in Earth history. These include the evolution of life, changes in patterns of plate tectonic interactions, major meteorite impacts, glaciations, and so on. Continental chemical denudation depends on the proportions of the Earth's surface that are eroding under the two types of regimes: weathering-limitation associated with steeper slopes; transport-limitation associated with flat areas. Susceptibility has to be evaluated for each climate regime. In the case of glacial and periglacial conditions, chemical weathering processes are sufficiently different that they may be misinterpreted if one uses models developed for warmer parts of the world.

On the time scale of the entire history of the Earth, denudation rate is controlled by the tectonic processes that supply fresh bedrock to the subaerial Earth-surface environment. There are two important aspects to this. One is the uplift of mountains and volcanism to produce steep terrains that undergo weathering-limited erosion. The other is the rise and fall of global sea level that affects erosion and sedimentation on the cratons. Low sea level equates with erosion of the cratons. High sea level equates with cessation of erosion and even sedimentation on the cratons. Glaciers are the most potent agents of physical erosion on cratons and perhaps in mountain terrains were the bedrock is resistant to chemical weathering. The role of continental ice sheets in excavating shield over the history of the Earth deserves further consideration.

Questions

1. Provide some simple reasons for the rules of thumb for physical and chemical weathering given in Section 9.2.1.
2. Describe how vegetation can increase and decrease the weathering rate.
3. What is the distinction between physical and chemical weathering?
4. Describe the factors that limit mountain height. How does this relate to the areas of the Earth that deliver the largest dissolved and suspended loads to the oceans?
5. How might chemical and physical weathering have differed from today's world before the advent of rooted land plants? What about before any land plants, assuming that there were once green films?
6. Ice ages ended quickly. Processes that promote the increase of carbon dioxide in the atmosphere are positive feedbacks in ending an ice age. Is subglacial chemical weathering a positive or a negative feedback? What circumstances would allow one to give the opposite answer? (Hint:

look at the weathering reactions in Appendices 1 and 2 in Drever and Hurcomb (1986).)

References

Aleva, G. J. J. (1979). Bauxite and other duricrusts in Surinam: A review. *Geol. Mijnbouw* **58**, 321–336.

Aleva, G. J. J. (1984). Laterization, bauxitization and cyclic landscape development in the Guiana Shield. *In* "Bauxite: Proceedings of the 1984 Bauxite Symposium," Los Angeles, California, February 27, 1984–March 1, 1984 (L. Jacob, Jr., ed.), pp. 297–318. American Institute of Mining, Metallurgical and Petroleum Engineers, New York.

Axtmann, E. V. and Stallard, R. F. (1995). Chemical weathering in the South Cascade Glacier basin, comparison of subglacial and extra-glacial weathering. *In* "Biogeochemistry of Seasonally Snow-Covered Catchments," Boulder, Colorado, USA, July 3, 1995–July 14, 1995 (K. A. Tonnessen, M. W. Williams, and M. Tranter, eds), pp. 431–439. International Association of Hydrological Sciences: Washington, DC, International Association of Hydrological Sciences Publication **228**.

Bell, M. and Laine, E. P. (1985). Erosion of the Laurentide region of North America by glacial and glaciofluvial processes. *Quatern. Res.* **23**, 154–174.

Benavides, V. (1968). Saline deposits of South America. *Geological Society of America Special Paper* **88**, pp. 249–290.

Berner, R. A. (1978). Rate control of mineral dissolution under earth surface conditions. *Am. J. Sci.* **278**, 1235–1252.

Berner, R. A., Lasaga, A. C., and Garrels, R. M. (1983). The carbonate-silicate geochemical cycle and its effect on atmospheric carbon dioxide over the past 100 million years. *Am. J. Sci.* **283**, 641–683.

Bevin, K. J. and Kirkby, M. J. (1979). A physically based variable contributing area model of basin hydrology. *Hydrolog. Sci. Bull.* **23**, 419–437.

Bevin, K. J., Lamb, R., Quinn, P. F., Romanowicz, R., and Freer, J. (1995). TOPMODEL. *In* "Computer Models of Watershed Hydrology," Highlands Ranch, Colorado, 1994 (V. P. Singh, ed.), pp. 627–668. Water Resource Publications.

Blancaneaux, P. and Pouyllau, M. (1977). Formes d'altération pseudokarstiques en relation avec la géomorphologie des granites précambriens du type Rapakivi dans le territoire Fédéral de l'Amazone, Vénézuéla. *Cah. ORSTOM Sér. Pédol.* **15**, 131–142.

Blum, A. E. and Stillings, L. L. (1995). Feldspar dissolution kinetics. *In* "Chemical Weathering Rates of Silicate Minerals" (A. F. White and S. L.

Brantley, eds), Reviews in Mineralogy **31**, pp. 291–351. Mineralogical Society of America, Washington, DC.

Blum, J. D., Erel, Y., and Brown, K. (1994). $^{87}Sr/^{86}Sr$ ratios of Sierra Nevada stream waters: Implications for relative mineral weathering rates. *Geochim. Cosmochim. Acta* **58**, 5019–5025.

Brantley, S. L. and Chen, Y. (1995). Chemical weathering rates of pyroxenes and amphiboles. *In* "Chemical Weathering Rates of Silicate Minerals" (A. F. White and S. L. Brantley, eds), Mineralogical Society of America: Washington, DC, Reviews in Mineralogy **31**, 119–172.

Brantley, S. L., Crane, S. R., Crerar, D. A., Hellmann, R., and Stallard, R. F. (1986). Dissolution at dislocation etch pits in quartz. *Geochim. Cosmochim. Acta* **50**, 2349–2361.

Broecker, W. S. and Denton, G. H. (1989). The role of ocean-atmosphere reorganization in glacial cycles. *Geochim. Cosmochim. Acta* **53**, 2465–2510.

Burke, K. and Wilson, J. T. (1972). Is the African Plate stationary? *Nature* **239**, 387–390.

Carson, M. A. and Kirkby, M. J. (1972). "Hillslope, Form and Process." Cambridge University Press, Cambridge, England, p. 475.

Casey, W. H. and Ludwig, C. (1995). Silicate mineral dissolution as a ligand-exchange reaction. *In* "Chemical Weathering Rates of Silicate Minerals" (A. F. White and S. L. Brantley, eds), Mineralogical Society of America: Washington, DC, Reviews in Mineralogy **31**, 87–117.

Chalcraft, D. and Pye, K. (1984). Humid tropical weathering of quartzite in southeastern Venezuela. *Z. Geomorphol., N. F.* **28**, 321–332.

Chi, W. R., Namson, J., and Suppe, J. (1981). Stratigraphic record of plate interactions in the coastal range of eastern Taiwan. *Geol. Soc. China Mem.*, 491–530.

Church, M. and Slaymaker, O. (1989). Disequilibrium of Holocene sediment yield in glaciated British Columbia. *Nature* **337**, 452–454.

Crough, S. T. (1979). Hotspot epeirogeny. *Tectonophysics* **61**, 321–333.

Dahlen, F. A., Suppe, J., and Davis, D. (1984). Mechanics of fold-and-thrust belts and accretionary wedges: Cohesive Coulomb theory. *J. Geophys. Res.* **89**, 10087–10101.

Davis, D., Suppe, J., and Dahlen, F. A. (1983). Mechanics of fold-and-thrust belts and accretionary wedges. *J. Geophys. Res.* **88**, 1153–1172.

Davis, W. M. (1932). Piedmont Benchlands and Primärrümpfe. *Geol. Soc. Am. Bull.* **43**, 399–440.

DeCelles, P. G. and Hertel, F. (1989). Petrology of fluvial sands from the Amazonian foreland basin, Peru and Bolivia. *Geol. Soc. Am. Bull.* **101**, 1552–1562.

Dia, A. N., Cohen, A. S., O'Nions, R. K., and Shackleton, N. J. (1992). Seawater Sr-isotope variations over the past 300 ka and global climate cycles. *Nature* **356**, 786–788.

Dove, P. (1995). Kinetic and thermodynamic controls on silica reactivity in weathering environments. *In* "Chemical Weathering Rates of Silicate Minerals" (A. F. White and S. L. Brantley, eds), Mineralogical Society of America: Washington, DC, Reviews in Mineralogy **31**, 235–290.

Drever, J. I. and Hurcomb, D. R. (1986). Neutralization of atmospheric acidity by chemical weathering in an alpine drainage basin in the North Cascade Mountains. *Geology* **14**, 221–224.

Drever, J. I. and Smith, C. L. (1978). Cyclic wetting and drying of the soil zone as an influence on the chemistry of ground water in arid terrains. *Am. J. Sci.* **278**, 1448–1454.

Dunne, T. (1978). Rates of chemical denudation of silicate rocks in tropical catchments. *Nature* **274**, 244–246.

Eberl, D. D., Srodon, J., and Northrop, H. R. (1986). Potassium fixation in smectite by wetting and drying. *In* "Geochemical Processes at Mineral Surfaces," pp. 296–325. American Chemical Society, Washington, DC, ACS Symposium Series **323**.

Edmond, J. M., Palmer, M. R., Measures, C. I. Grant, B., and Stallard, R. F. (1995). The fluvial geochemistry and denudation rate of the Guayana Shield in Venezuela, Colombia, and Brazil. *Geochim. Cosmochim. Acta* **59**, 3301–3325.

Fischer, A. G. (1983). The two Phanerozoic supercycles. *In* "Catastrophies in Earth History: The New Uniformitarianism" (W. Berggren and J. Van Couvering, eds), pp. 129–150. Princeton University Press, Princeton, New Jersey.

Franzinelli, E. and Potter, P. E. (1983). Petrology, chemistry, and texture of modern river sands, Amazon River system. *J. Geol.* **91**, 23–39.

Froelich, P. N., Hambrick, G. A., Andreae, M. O., Mortlock, R. A., and Edmond, J. M. (1985). The geochemistry of inorganic germanium in natural waters. *J. Geophys. Res.* **90**, 1133–1141.

Froelich, P. N., Blanc, V., Mortlock, R. A., Chillrud, S. N., Dunstan, W., Udomkit, A., and Peng, T. H. (1992). River fluxes of dissolved silica to the ocean were higher during glacials: Ge/Si in diatoms, rivers, and oceans. *Paleoceanography* **7**, 739–767.

Garrels, R. M. and Lerman, A. (1981). Phanerozoic cycles of sedimentary carbon and sulfur. *Proc. Nat. Acad. Sci. USA* **78**, 4652–4656.

Garwood, N. C., Janos, D. P., and Brokaw, N. V. L.

(1979). Earthquake-caused landslides: A major disturbance to tropical forests. *Science* 205, 997–999.

Gibbs, M. T. and Kump, L. R. (1994). Global chemical erosion during the last glacial maximum and the present: Sensitivity to changes in lithology and hydrology. *Paleoceanography* 9, 529–543.

Gibbs, R. J. (1967). The geochemistry of the Amazon River system: Part I, The factors that control the salinity and composition and concentration of suspended solids. *Geol. Soc. Am. Bull.* 78, 1203–1232.

Gilkes, R. J., Suddhiprakarn, A. and Armitage, T. M. (1980). Scanning electron microscope morphology of deeply weathered granite. *Clays Clay Miner.* 28, 29–34.

Goldich, S. S. (1938). A study in rock weathering. *J. Geol.* 46, 17–58.

Gravenor, C. P. (1975). Erosion by continental ice sheets. *Am. J. Sci.* 275, 594–604.

Harden, J. W., Sundquist, E. T., Stallard, R. F., and Mark, R. K. (1992). Dynamics of soil carbon during deglaciation of the Laurentide Ice Sheet. *Science* 258, 1921–1924.

Harmon, R. S., White, W. B., Drake, J. J. and Hess, J. W. (1975). Regional hydrochemistry of North American carbonate terrains. *Water Resour. Res.* 11, 963–967.

Heimsath, A. M., Dietrich, W. E., Nishiizumi, K., and Finkel, R. C. (1997). The soil production function and landscape equilibrium. *Nature* 388, 358–361.

Herbillon, A. J. and Nahon, D. (1988). Laterites and laterization processes. *In* "Iron in Soils and Clay Minerals" (J. W. Stucki, B. A. Goodman, and U. Schertmann, eds), pp. 267–308. Kluwer Academic Publishers: Dordrecht, The Netherlands, NATO ASI Series C: Mathematical and Physical Sciences 217.

Hochella, M. F. Jr. and Banfield, J. F. (1995). Chemical weathering of silicates in nature: A microscopic perspective with theoretical considerations. *In* "Chemical Weathering Rates of Silicate Minerals" (A. F. White and S. L. Brantley, eds), pp. 353–406. Mineralogical Society of America: Washington, DC, Reviews in Mineralogy 31.

Holland, H. D. (1974). Marine evaporites and the composition of sea water during the Phanerozoic. *In* "Studies in Paleo-oceanography" (W. W. Hay, ed.), pp. 187–192. Society of Economic Paleontologists and Mineralogists, Tulsa, Oklahoma, Special Publication 20.

Holland, H. D., Kirsipu, T. V., Huebner, J. S., and Oxburgh, U. N. (1964). On some aspects of the chemical evolution of cave waters. *J. Geol.* 72, 36–67.

Hooke, R. L. (1994). On the efficacy of humans as geomorphic agents. *GSA Today* 4, 218, 224–225.

Hughes, T. J. (1985). The great Cenozoic ice sheet. *Palaeogeogr. Palaeoclim. Palaeoecol.* 50, 9–43.

Irion, G. (1976). Mineralogisch-geochemische Unterschungen an der pelitischen Fraktion amazonischer Oberboden und Sedimente. *Biogeographica* 7, 7–25.

Irion, G. (1984). Sedimentation and sediments of Amazonian rivers and evolution of the Amazonian landscape since Pliocene times. *In* "The Amazon, Limnology and Landscape Ecology of a Mighty Tropical River and its Basin" (H. Sioli, ed.), pp. 201–214. Dr. W. Junk Publishers, The Hague, The Netherlands.

James, P. E. (1959). The geomorphology of eastern Brazil as interpreted by Lester C. King. *Geogr. Rev.* 49, 240–246.

Johnsson, M. J. and Stallard, R. F. (1989). Physiographic controls on the composition of sediments derived from volcanic and sedimentary terrains on Barro Colorado Island, Panama. *J. Sediment. Petrol.* 59, 768–781.

Johnsson, M. J., Stallard, R. F., and Lundberg, N. (1991). Controls on the composition of fluvial sands from a tropical weathering environment: Sands of the Orinoco River drainage basin, Venezuela and Colombia. *Geol. Soc. Am. Bull.* 103, 1622–1647.

Johnsson, M. J., Stallard, R. F., and Meade, R. H. (1988). First-cycle quartz arenites in the Orinoco River basin, Venezuela and Colombia. *J. Geol.* 96, 263–277.

King, L. C. (1953). Canons of landscape evolution. *Geol. Soc. Am. Bull.* 64, 721–751.

King, L. C. (1956). A geomorfologia do Brasil oriental. *Rev. Brasil. Geograf.* 18, 147–265.

King, L. C. (1967). "The Morphology of the Earth," 762. Oliver and Boyd, Edinburgh.

Klammer, G. (1984). The relief of the extra-Andean Amazon basin. *In* "The Amazon, Limnology and Landscape Ecology of a Mighty Tropical River and its Basin" (H. Sioli, ed.), pp. 47–83. Dr. W. Junk Publishers, Dordrecht, The Netherlands.

Koehnken, L. (1990). The composition of fine-grained weathering products in a large tropical river system, and the transport of metals in fine-grained sediments in a temperate estuary, Ph.D. thesis, Princeton University, Department of Geological and Geophysical Sciences.

Krook, L. (1979). Sediment petrographical studies in northern Surinam, Ph.D. thesis, Free University.

Laine, E. P. (1980). New evidence from beneath the western North Atlantic for the depth of glacial erosion in Greenland and North America. *Quatern. Res.* 14, 188–198.

Langmuir, D. (1971). The geochemistry of some carbonate ground waters in central Pennsylvania. *Geochim. Cosmochim. Acta* **35**, 1023–1045.

Larsen, M. C. and Simon, A (1993). A rainfall intensity-duration threshold for landslides in a humid-tropical environment, Puerto Rico. *Geograf. Annal.* **75A**, 13–23.

Larsen, M. C. and Torres Sánchez, A. J. (1998). The frequency and distribution of recent landslides in three montane tropical regions of Puerto Rico. *Geomorphology* **24**, 309–331.

Lasaga, A. C. (1995). Fundamental approaches in describing mineral dissolution and precipitation rates. *In* "Chemical Weathering Rates of Silicate Minerals" (A. F. White and S. L. Brantley, eds), Mineralogical Society of America, Washington, DC, Reviews in Mineralogy **31**, 23–86.

Lewis, W. M. Jr., Hamilton, S. K., Jones, S. L. and Runnels, D. D. (1987). Major element chemistry, weathering, and element yields for the Caura River drainage, Venezuela. *Biogeochemistry* **4**, 159–181.

Li, Y. H. (1976). Denudation of Taiwan Island since the Pliocene Epoch. *Geology* **4**, 105–107.

Likens, G. E., Bormann, F. H., Pierce, R. S., Eaton, S., and Johnson, N. M. (1977). "Biogeochemistry of a Forested Ecosystem." Springer-Verlag, New York.

Manias, W. G., Covey, M., and Stallard, R. F. (1985). The effects of provenance and diagenesis on clay content and crystallinity in Miocene through Pleistocene deposits, southwestern Taiwan. *Petrol. Geol. Taiwan* 173–185.

Mast, M. A., Drever, J. I., and Baron, J. (1990). Chemical weathering in the Loch Vale watershed, Rocky Mountain National Park, Colorado. *Water Resour. Res.* **26**, 2971–2978.

McConnell, R. B. (1968). Planation surfaces in Guyana. *Geogr. J.* **134**, 506–520.

McNabb, D. H. and Swanson, F. J. (1990). Effects of fire on soil erosion. *In* "Natural and Prescribed Fire in the Pacific Northwest" (J. D. Walstad, S. L. Radosevich and D. V. Sandberg, eds), pp. 159–176. Oregon State University Press, Corvallis, Oregon.

Meade, R. H., Yuzyk, T. R., and Day, T. J. (1990). Movement and storage of sediment in rivers of the United States and Canada. *In* "Surface Water Hydrology" (M. G. Wolman and H. C. Riggs, eds), pp. 255–280. Geological Society of America, Boulder, Colorado, The Geology of North America **O-1**.

Menendez, A. and Sarmentearo, A. (1984). Geology of the Los Pijiguaos bauxite deposits, Venezuela. *In* "Bauxite, Proceedings of the 1984 Bauxite Symposium," Feb. 27–March 1, 1984 (L. Jacob, Jr., ed.),

pp. 387–406. American Institute of Mining, Metallurgical, and Petroleum Engineers, New York.

Meybeck, M. (1979). Concentrations des eaux fluviales en éléments majeurs et apports en solution aux océans. *Rev. Géol. Dynam. Géogr. Phys.* **21**, 215–246.

Meybeck, M. (1987). Global chemical weathering of superficial rocks estimated from river dissolved loads. *Am. J. Sci.* **287**, 401–428.

Meyer, G. A., Wells, S. G., and Jull, A. J. T. (1995). Fire and alluvial chronology in Yellowstone National Park: Climatic and intrinsic controls on Holocene geomorphic processes. *Geol. Soc. Am. Bull.* **107**, 1211–1230.

Milliman, J. D. and Meade, R. H. (1983). World-wide delivery of river sediment to the oceans. *J. Geol.* **91**, 1–21.

Milliman, J. D. and Syvitski, J. P. M. (1992). Geomorphic/tectonic control of sediment discharge to the ocean: The importance of small mountainous rivers. *J. Geol.* **100**, 525–544.

Morgan, W. J. (1983). Hotspot tracks and the early rifting of the Atlantic. *Tectonophysics* **94**, 123–139.

Murnane, R. J. and Stallard, R. F. (1990). Germanium and silicon in rivers of the Orinoco drainage basin, Venezuela and Colombia. *Nature* **344**, 749–752.

Nagy, K. L. (1995). Dissolution and precipitation kinetics of sheet silicates. *In* "Chemical Weathering Rates of Silicate Minerals" (A. F. White and S. L. Brantley, eds), Mineralogical Society of America, Washington, DC, Reviews in Mineralogy **31**, 173–233.

Pain, C. F. (1972). Characteristics and geomorphic effects of earthquake initiated landslides in the Albert Range of Papua New Guinea. *Eng. Geol.* **6**, 261–274.

Paolini, J. (1986). Transporte de carbono y minerales en el río Caroní. *Interciencia* **11**, 295–297.

Petroleos de Venezuela, S. A. (MARAVEN) (1977). Mosaico de Imágenes de Radar de Visión Lateral. Ministereo de Energía y Minas, Dirección General Sectoral de Geología y Minas, Caracas, Venezuela, Scale 1:2,500,000, Sheet No. NB19–8.

Pitman, W. C. III (1978). Relationship between eustacy and stratigraphic sequences of passive margins. *Geol. Soc. Am. Bull.* **89**, 1389–1403.

Potter, P. E. (1978). Petrology and chemistry of modern big river sands. *J. Geol.* **86**, 423–449.

Potter, P. E. (1994). Modern sands of South America: composition, provenance and global significance. *Geol. Rundschau* **83**, 212–232.

Ronov, A. B., Migdisov, A. A., and Barskaya, N. V. (1969). Tectonic cycles and regularities in the development of sedimentary rocks and paleogeographic environments of sedimentation of the Russian Platform (an approach to a quantitative study). *Sedimentology* **13**, 179–212.

Schubert, C., Briceño, H. O., and Fritz, P. (1986). Paleoenvironmental aspects of the Caroní-Paragua river basin (southeastern Venezuela). *Interciencia* **11**, 278–289.

Scott, G. A. J. (1975a). Relationships between vegetation cover and soil avalanching in Hawaii. *Proc. Assoc. Am. Geogr.* **7**, 208–212.

Scott, G. A. J. (1975b). Soil profile changes resulting from the conversion of forest to grassland in the montaña of Peru. *Great Plains-Rocky Mount. Geogr. J.* **4**, 124–130.

Scott, G. A. J. and Street, J. M. (1976). The role of chemical weathering in the formation of Hawaiian Amphitheatre-headed Valleys. *Z. Geomorph. N. F.* **20**, 171–189.

Simon, A., Larsen, M. C., and Hupp, C. R. (1990). The role of soil processes in determining mechanisms of slope failure and hillslope development in a humid-tropical forest, eastern Puerto Rico. *Geomorphology* **3**, 263–286.

Sloss, L. L. (1963). Sequences on the cratonic interior of North America. *Geol. Soc. Am. Bull.* **74**, 93–114.

Sloss, L. L. (1979). Global sea level changes: A view from the craton. *In* "Geological and Geophysical Investigations of Continental Margins" (J. S. Watkins, L. Montardert, and P. W. Dickerson, eds), American Association of Petroleum Geologists, Tulsa, Oklahoma, American Association of Petroleum Geologists Memoir **29**.

Sloss, L. L. and Speed, R. C. (1974). Relationships of cratonic and continental-margin tectonic episodes. *In* "Tectonics and Sedimentation" (W. R. Dickenson, ed.), pp. 98–119. Society of Economic Paleontologists and Mineralogists, Tulsa, Oklahoma, SEPM Special Publication **22**.

Soares, P. C., Landim, P. M. B., and Fulfaro, V. J. (1978). Tectonic cycles and sedimentary sequences in the Brazilian intracratonic basins. *Geol. Soc. Am. Bull.* **89**, 181–191.

Stallard, R. F. (1980). Major element geochemistry of the Amazon River system. Ph.D. Dissertation, Massachusetts Institute of Technology/Woods Hole Oceanographic Institution, Joint Program in Oceanography, Woods Hole Oceanographic Institution: Woods Hole, MA **WHOI-80-29**.

Stallard, R. F. (1985). River chemistry, geology, geomorphology, and soils in the Amazon and Orinoco basins. *In* "The Chemistry of Weathering" (J. I. Drever, ed.), pp. 293–316. D. Reidel Publishing Co., Dordrecht, Holland, NATO ASI Series C: Mathematical and Physical Sciences **149**.

Stallard, R. F. (1988). Weathering and erosion in the humid tropics. *In* "Physical and Chemical Weathering in Geochemical Cycles" (A. Lerman and M. Meybeck, eds), pp. 225–246. Kluwer Academic Publishers, Dordrecht, Holland, NATO ASI Series C: Mathematical and Physical Sciences **251**.

Stallard, R. F. (1995a). Relating chemical and physical erosion. *In* "Chemical Weathering Rates of Silicate Minerals" (A. F. White and S. L. Brantley, eds), Mineralogical Society of America, Washington, DC, Reviews in Mineralogy **31**, 543–564.

Stallard, R. F. (1995b). Tectonic, environmental, and human aspects of weathering and erosion: A global review using a steady-state perspective. *Ann. Rev. Earth Plan. Sci.* **12**, 11–39.

Stallard, R. F. (1998). Terrestrial sedimentation and the carbon cycle: Coupling weathering and erosion to carbon burial. *Glob. Biogeochem. Cycles* **12**, 231–252.

Stallard, R. F. and Edmond, J. M. (1983). Geochemistry of the Amazon 2: The influence of the geology and weathering environment on the dissolved load. *J. Geophys. Res.* **88**, 9671–9688, microfiche supplement.

Stallard, R. F. and Edmond, J. M. (1987). Geochemistry of the Amazon 3. Weathering chemistry and limits to dissolved inputs. *J. Geophys. Res.* **92**, 8293–8302.

Stallard, R. F., Koehnken, L., and Johnsson, M. J. (1991). Weathering processes and the composition of inorganic material transported through the Orinoco River system, Venezuela and Colombia. *Geoderma* **51**, 133–165.

Sugden, D. E. (1976). A case against deep erosion of shields by ice sheets. *Geology* **4**, 580–582.

Sugden, D. E. (1978). Glacial erosion by the Laurentide Ice Sheet. *J. Glaciol.* **20**, 367–391.

Suppe, J. (1981). Mechanics of mountain building in Taiwan. *Geol. Soc. China Mem.* 67–89.

Sverdrup, H. and Warfvinge, P. (1995). Estimating field weathering rates using laboratory kinetics. *In* "Chemical Weathering Rates of Silicate Minerals" (A. F. White and S. L. Brantley, eds), Mineralogical Society of America, Washington, DC, Reviews in Mineralogy **31**, 485–541.

Szczerban, E. (1976). Cavernas y simas en areniscas precámbricas del Territorio Federal Amazonas y Estado Bolívar. *Venezolana Direc. Geol. Bol. Geol. Public. Esp.* **7**, 1055–1072.

Tardy, Y., Bocquier, G., Paquet, H., and Millot, G. (1973). Formation of clay from granite and its distribution in relation to climate and topography. *Geoderma* **10**, 271–284.

Trolard, F. and Tardy, Y. (1989). An ideal solid

solution model for calculating solubility of clay minerals. *Clay Miner.* **24**, 1–21.

Urbani P. F. (1986). Notas sobre el origen de las cavidades en rocas cuarcíferas precámbricas del Grupo Roraima, Venezuela. *Interciencia* **11**, 298–300.

Vail, P. R. and Herdenbol, J. (1979). Sea-level changes during the Tertiary. *Oceanus* **22**, 71–79.

Vail, P. R., Mitchum, R. M., Todd, R. G., Widmier, J. M., Thompson III, S., Scngree, J. B., Bubb, J. N., and Hatlelid, W. G. (1977). Seismic stratigraphy and global changes of sea-level. *In* "Seismic-stratigraphy Applications to Hydrocarbon Exploration," American Association of Petroleum Geologists, Tulsa, Oklahoma, American Association of Petroleum Geologists Memoir **26**.

van Breeman, N. (1988a). Effects of seasonal redox processes involving iron on the chemistry of periodically reduced soils. *In* "Iron in Soils and Clay Minerals" (J. W. Stucki, B. A. Goodman, and U. Schwertmann, eds). Kluwer Academic Publishers, Dordrecht, The Netherlands, NATO ASI Series C: Mathematical and Physical Sciences **217**.

van Breeman, N. (1988b). Long-term chemical, mineralogical, and morphological effects of iron-redox processes in periodically flooded soils. *In* "Iron in Soils and Clay Minerals" (J. W. Stucki, B. A. Goodman, and U. Schwertmann, eds). Kluwer Academic Publishers, Dordrecht, The Netherlands, NATO ASI Series C: Mathematical and Physical Sciences **217**.

Vitousek, P. M. (1977). The regulation of element concentrations in mountain streams in the northeastern United States. *Ecol. Monogr.* **47**, 65–87.

Wentworth, C. K. (1943). Soil avalanches on Oahu, Hawaii. *Geol. Soc. Am. Bull.* **54**, 53–64.

White, A. F. (1983). Surface chemistry and dissolution kinetics of glassy rocks at 25°. *Geochim. Cosmochim. Acta* **47**, 805–815.

White, A. F. (1995). Chemical weathering rates of silicate minerals in soils. *In* "Chemical Weathering Rates of Silicate Minerals" (A. F. White and S. L. Brantley, eds), Mineralogical Society of America, Washington, DC, Reviews in Mineralogy **31**, 407–461.

White, A. F. and Blum, A. E. (1995). Effects of climate on chemical weathering in watersheds. *Geochim. Cosmochim. Acta* **59**, 1729–1747.

White, A. F. and Brantley, S. L. (1995). Chemical weathering rates of silicate minerals: An overview. *In* "Chemical Weathering Rates of Silicate Minerals" (A. F. White and S. L. Brantley, eds), Mineralogical Society of America, Washington, DC, Reviews in Mineralogy **31**, 1–22.

White, W. A. (1972). Deep erosion by continental ice sheets. *Geol. Soc. Am. Bull.* **83**, 1037–1096.

Zonneveld, J. I. S. (1969). Preliminary remarks on summit levels and the evolution of the relief in Surinam (S. America). *Verhand. Kon. Ned. Geol. Mijnbouwk. Gen.* **27**, 53–60.

10

The Oceans

James W. Murray

The oceans play a major role in the global cycles of most elements. As is evident in images from space, most of the Earth's surface is ocean. When viewed from space we see the oceans cover 71% of the Earth's surface. The oceans are in interactive contact with the lithosphere, atmosphere, and biosphere, and virtually all elements pass through the ocean at some point in their cycles. Given sufficient time, the water and sediments of the ocean are the receptacle of most natural and anthropogenic elements and compounds. Transport processes across the ocean boundaries and within the ocean are central to studies of the global cycles. Such processes as air/sea exchange of gases and aerosols, biological production of particles within the sea, and sedimentation need to be considered. The productivity of the ocean and climate are influenced by wind-generated surface currents and thermohaline circulation in the deep ocean. The complicated and diverse processes in estuaries influence how much material of riverine origin reaches the sea. There is currently a great deal of concern about how people are affecting the ocean and climate processes. In order to make sound predictions for the future we need a solid understanding of present and past conditions and especially how the ocean–climate system responds to natural and anthropogenic changes in forcing.

The ocean is also by far the largest reservoir for most of the elements in the atmosphere–biosphere–ocean system. Perturbations caused by our increased population and industrialization are passing through the ocean, and because the time scale for ocean circulation is long (about 2000 years) relative to the time scale of modern society, a new steady state or quasi-equilibrium will be established slowly. Until that time, local concentrations of toxic chemicals, especially in estuaries and bays with restricted circulation, will be the major concern for mankind.

In this chapter we first review some of the basic descriptive aspects of the ocean and its physiographic domains and show briefly how the ocean fits into the global water balance. We then present a brief review of surface and abyssal ocean circulation. The superposition of the biological cycle on ocean circulation is what controls the distribution of a large number of elements within the ocean, so spatial variations and the stoichiometry of biological productivity and the transport of biologically produced particles are reviewed. Ocean sediments are the main site of deposition for most elements and thus they record the course of events over geological time. Sediments are considered in Chapter 8 and are not considered in detail here. Finally, we review the basic properties of ocean chemistry and attempt to classify the elements into groups according to the mechanisms that control their distribution.

10.1 What is the Ocean?

The topography and structure of the ocean floor are highly variable from place to place and reflect tectonic processes within the Earth's

Earth System Science
ISBN 0-12-379370-X

interior. These features have varied in the past so that the ocean bottom of today is undoubtedly not like the ocean bottom of 50 million years ago. The major topographic systems, common to all oceans, are continental margins, ocean-basin floors, and oceanic ridge systems. Tectonic features such as fracture zones, plateaus, trenches, and mid-ocean ridges act to subdivide the main oceans into a larger number of smaller basins.

Mapping the sea floor using ships is a tedious process. The newest bathometric maps of the global oceans with horizontal resolution of 1 to 12 km have been derived by combining available depth soundings with high-resolution marine gravity data from the Geosat and ERS-1 spacecraft (Smith and Sandwell, 1997). Marine gravity anomalies are caused primarily by topographic variations on the ocean floor. This remote-sensing approach reveals all the intermediate and large-scale structures of the ocean basins (Plate 4) including incised canyons on continental margins, spreading ridges, fracture

zones, and seamounts. This approach has led to discovery of previously unknown features in remote locations.

The continental margin regions are the transition zones between continents and ocean basins. The major features at ocean margins are shown schematically in Fig. 10-1. Though the features may vary, the general features shown occur in all ocean basins in the form of either two sequences: shelf–slope–rise–basin or shelf–slope–trench–basin. The continental shelf is the submerged continuation of the adjacent land, modified in part by marine erosion or sediment deposition. The seaward edge of the continental shelf can frequently be clearly seen and it is called the shelf break. The shelf break tends to occur at a depth of about 200 m over most of the ocean. Sea level was 121 ± 5 m lower during Pleistocene glacial maxima (Fairbanks, 1989). At those times the shoreline was at the edge of present continental shelf, which was then a coastal plain. On average, the continental shelf is about 70 km wide, although it can vary

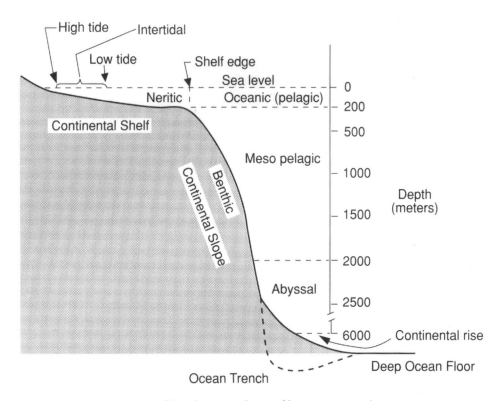

Fig. 10-1 Schematic representation of the physiographic profile at ocean margins.

widely (compare the east coast of China with the west coast of Peru). The Arctic Ocean has the largest proportion of shelf to total area of all the world's oceans. The continental slope is characterized as the region where the gradient of the topography changes from 1:1000 on the shelf to greater than 1:40. Thus continental slopes are the relatively narrow, steeply inclined submerged edges of the continents. The continental slope may form one side of an ocean trench as it does off the west coast of Mexico or Peru or it may grade into the continental rise as it does off the east coast of the US. The ocean trenches are the topographic reflection of subduction of oceanic plates beneath the continents. The greatest ocean depths occur in such trenches. The deepest is the Challenger Deep which descends to 11 035 m in the Marianas Trench. The continental rises are mainly depositional features that are the result of coalescing of thick wedges of sedimentary deposits carried by turbidity currents down the slope and along the margin by boundary currents. Deposition is caused by reduction in current speed when it flows out onto the gently sloping rise. Gradually the continental rise grades into ocean basins and abyssal plains.

The relationships between ocean depths and land elevations are shown in Fig. 10-2. On the average continents are 840 m above sea level, while the average depth of the oceans is 3730 m.

If the Earth were a smooth sphere with the land planed off to fill the ocean basins the earth would be uniformly covered by water to a depth of 2430 m.

The area, volume and average depth of the ocean basins and some marginal seas are given in Table 10-1. The Pacific Ocean is the largest and contains more than one-half of the Earth's water. It also receives the least river water per area of the major oceans (Table 10-2). Paradoxically it is also the least salty (Table 10-3). The land area of the entire Earth is strongly skewed toward the northern hemisphere.

The ocean contains the bulk of the Earth's water (1.37×10^{24} g) and moderates the global water cycle. The distribution of the mass of water is about 80% in the ocean and about 20% as pore water in sediments and sedimentary rocks. The reservoir of water in rivers, lakes, and the atmosphere is trivial (0.003%). Disregarding the pore water because it is not in free circulation, we find that 97% of the world's cycling water is in the ocean (Table 10-4). The average residence time of water in the atmosphere with respect to net transfer (evaporation minus precipitation over oceans) from the oceans to the continents is about one third of a year (0.13×10^{20} g/$(3.83 - 3.47 \times 10^{20}$ g/yr) = 0.33 yr). The ocean's role in controlling the water content of the atmosphere has important implications for past and present climates of the Earth. One of the possible positive feedbacks of global warming will be increased atmospheric water content resulting from warming of the sea surface.

10.2   Ocean Circulation

The chemistry and biology of the ocean are superimposed on the ocean's circulation, thus it is important to review briefly the forces driving this circulation and give some estimates of the transport rates. There are many reasons why it is important to understand the basics of the circulation. Four examples are given as an illustration.

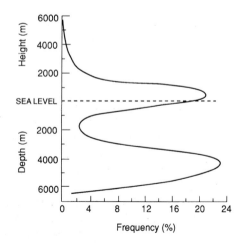

Fig. 10-2 A frequency distribution of elevation intervals of the Earth's surface.

1. Poleward-flowing, warm, surface, western boundary currents such as the Gulf Stream

Table 10-1   Area, mean depth, and volume of oceans and seas

Region	Area $(10^6 \, km^2)$	Mean depth (m)	Volume $(10^6 \, km^3)$
Pacific Ocean	165.25	4282	707.56
Atlantic Ocean	82.44	3926	323.61
Indian Ocean	73.44	3963	291.03
Three oceans only	321.13	4117	1322.20
Arctic Mediterranean	14.09	1205	16.98
American Mediterranean	4.32	2216	9.57
Mediterranean Sea and Black Sea	2.97	1429	4.24
Asiatic Mediterranean	8.14	1212	9.87
Baltic Sea	0.42	55	0.02
Hudson Bay	1.23	128	0.16
Red Sea	0.44	491	0.21
Persian Gulf	0.24	25	0.01
Marginal seas	8.08	874	7.06
Three Oceans plus adjacent seas	361.06	3795	1370.32
Pacific Ocean ⎫ including	179.68	4028	723.70
Atlantic Ocean ⎬ adjacent	106.46	3332	354.68
Indian Ocean ⎭ seas	74.92	3897	291.94

Table 10-2   A breakdown of the water balance for the four main ocean basins (cm/yr for the area of the respective basins)[a]

Ocean	Precipitation	Runoff from adjoining land areas	Evaporation	Water exchange with other oceans
Atlantic	78	20	104	6
Arctic	24	23	12	35
Indian	101	7	138	30
Pacific	121	6	114	13

[a] From M. I. Budyko (1958). "The Heat Balance of the Earth's Surface" (Trans. N. A. Stepanova). Office of Technical Services, Department of Commerce, Washington.

and the Kuroshiro have a profound effect on the sea-surface temperature (SST) and the climate of land areas bordering the oceans. For example, the Gulf Stream transports approximately 3.2 PW of heat to the North Atlantic (Hartmann, 1994), moderating the climate of northern Europe.

2. The El-Niño southern oscillation (ENSO) phenomenon is an interannual perturbation of the climate system characterized by a periodic weakening of the trade winds and warming of the surface water in the central and eastern equatorial Pacific Ocean. The impacts of ENSO are felt worldwide through disruption of atmospheric circulation and weather patterns (McPhaden, 1993; Wallace *et al.*, 1998).

3. Atmospheric testing of nuclear bombs resulted in contamination of the surface of the ocean with various isotopes including ^{14}C, ^{3}H, ^{90}Sr, ^{239}Pu, and ^{240}Pu. These isotopes are slowly being mixed through the ocean

Table 10-3 Average temperatures and salinity of the oceans, excluding adjacent seas[a]

	Temperature (°C)	Salinity (parts per thousand)
Pacific (total)	3.14	34.60
North Pacific	3.13	34.57
South Pacific	3.50	34.63
Indian (total)	3.88	34.78
Atlantic (total)	3.99	34.92
North Atlantic	5.08	35.09
South Atlantic	3.81	34.84
Southern Ocean[b]	0.71	34.65
World ocean (total)	3.51	34.72

[a] After L. V. Worthington (1981). The water masses of the world ocean: some results of a fine-scale census. *In* "Evolution of Physical Oceanography" (B. A. Warren and C. Wunsch, eds), pp. 42–69. MIT Press, Cambridge, MA.
[b] Ocean area surrounding Antarctica, south of 55°S.

Table 10-4 A detailed breakdown of the water volume in various reservoirs[a]

Environment	Water volume (10^3 km^3)	Percentage of total
Surface water		
Freshwater lakes	125	0.009
Saline lakes and inland seas	104	0.008
Rivers and streams	1.3	0.0001
Total	230	0.017
Subsurface water		
Soil moisture	67	0.005
Ground water	8000	0.62
Total	8067	0.625
Ice caps and glaciers	29,000	2.15
Atmosphere	13	0.001
Oceans	1,330,000	97.2
Totals (approx.)	1,364,000	100

[a] Data from Berner and Berner (1987).

and can be used as radioactive dyes and clocks (e.g. Broecker and Peng, 1982).

4. The atmospheric CO_2 concentration has been increasing since the beginning of the industrial age, but the increase (~ 3.2 Gt C/yr) is less than anthropogenic emissions and deforestation (~ 7.0 Gt C/yr) (Siegenthaler and Sarmiento, 1993). Some of the CO_2 has gone into the ocean (~ 2 Gt C/yr). All CO_2 taken up by the ocean is by the process of gas exchange. Some of the excess CO_2 has been transported into the intermediate and deep water by the subduction of water masses. Circulation replenishes the surface with water undersaturated with respect to the anthropogenically perturbed CO_2 levels.

In this section we briefly review what controls the density of seawater and the vertical density stratification of the ocean. Surface currents, abyssal circulation, and thermocline circulation are considered individually.

10.2.1   Density Stratification in the Ocean

The density of seawater is controlled by its salt content or salinity and its temperature. Salinity is historically defined as the total salt content of seawater and the units were given as grams of salt per kilogram of seawater or parts per thousand (‰). Salinity was expressed on a mass of seawater basis because mass, rather than volume, is conserved as temperature and pressure change. In modern oceanography salinity is determined as a conductivity ratio on the practical salinity scale and has no units (Millero, 1993). For more details on the formal definition of salinity and on the preparation of very accurate standards see UNESCO (1981). The salinity of surface seawater is controlled primarily by the balance between evaporation and precipitation. As a result the highest salinities are found in the subtropical central gyre regions centered at about 20° North and 20° South, where evaporation is extensive but rainfall is minimal. Surprisingly, they are not found at the equator where evaporation is large, but so is rainfall.

The temperature of seawater is fixed at the sea surface by heat exchange with the atmosphere. The average incoming energy from the sun at the Earth's surface is about four times higher at the equator than at the poles. The average infrared radiation heat loss to space is more constant with latitude. As a result there is a net input of heat into the tropical regions and this is where we find the warmest surface seawater. Heat is transferred from low to high latitudes by winds in the atmosphere and by currents in the ocean. The geothermal heat flux from the interior of the Earth is generally insignificant except in the vicinity of hydrothermal vents at spreading ridges and in relatively stagnant locations like the abyssal northern North Pacific (Joyce *et al.*, 1986) and the Black Sea (Murray *et al.*, 1991).

Because seawater signatures of temperature and salinity are acquired by processes occurring at the air–sea interface we can also state that the density characteristics of a parcel of seawater are determined when it is at the sea surface. This density signature is locked into the water when it sinks. The density will be modified by mixing with other parcels of water but if the density signatures of all the end member water masses are known, this mixing can be unraveled and the proportions of the different source waters to a given parcel can be determined.

To a first approximation the vertical density distribution of the ocean can be described as a three-layered structure. The surface layer is the region from the sea surface to the depth having a temperature of about 10°C. The transition region where the temperature decreases from 10°C to 4°C is called the thermocline. The deep sea is the region below the thermocline.

Because temperature (T) and salinity (S) are the main factors controlling density, oceanographers use T–S diagrams to describe the features of the different water masses. The average temperature and salinity of the world ocean and various parts of the ocean are given in Fig. 10-3 and Table 10-3. The North Atlantic contains the warmest and saltiest water of the major oceans. The Southern Ocean (the region around Antarctica) is the coldest and the North Pacific has the lowest average salinity.

Conventional T–S diagrams for specific locations in the individual oceans are shown in Fig. 10-4. The inflections in the curves reflect the inputs of water from different sources. The linear regions represent mixing intervals between these core sources. For example, in the Atlantic Ocean the curves reflect input from Antarctic Bottom Water (AABW), North Atlantic Deep Water (NADW), Antarctic Intermediate Water (AIW), Mediterranean Water (MW), and Warm Surface Water (WSW).

10.2.2 Surface Currents

Surface ocean currents respond primarily to the climatic wind field. The prevailing winds supply much of the energy that drives surface water movements. This becomes clear when charts of the surface winds and ocean surface currents are superimposed. The *wind-driven circulation* occurs principally in the upper few hundred meters and is therefore primarily a horizontal circulation although vertical motions can be induced when the geometry of surface circulation results in convergences (downwelling) or divergences (upwelling). The depth to which the surface circulation penetrates is dependent on the water column stratification. In the equatorial region the currents extend to 300–500 m while in the circumpolar region where stratification is weak the surface circulation can extend to the sea floor.

The net direction of motion of water is not always the same as the wind, because other

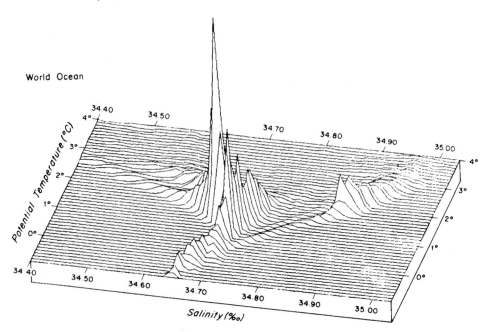

Fig. 10-3 Simulated three-dimensional *T–S* diagram of the water masses of the world ocean. Apparent elevation is proportional to volume. Elevation of highest peak corresponds to $26.0 \times 10^6 \, km^3$ per bivariate class $0.1°C \times 0.01‰$. (Reproduced with permission from L. V. Worthington, The water masses of the world ocean: some results of a fine-scale census. *In* B. A. Warren and C. Wunsch (1981). "Evaluation of Physical Oceanography," MIT Press, Cambridge, MA.)

factors come into play. These are shown schematically in Fig. 10-5. The wind blowing across the sea surface drags the surface along and sets this thin layer in motion. The surface drags the next layer and the process continues downward, involving successively deeper layers. As a result of friction between the layers each deeper layer moves more slowly than the one above and its motion is deflected to the right (clockwise) in the northern hemisphere by the Coriolis force (see Chapter 7). If this effect is represented by arrows (vectors) whose direction indicates current direction and length indicates speed, the change in current direction and speed with depth forms a spiral. This feature is called the Ekman spiral. If the wind blew continuously in one direction for a few days a well-developed Ekman spiral would develop. Under these conditions the integrated net transport over the entire depth of the Ekman spiral would be at 90° to the right of the wind direction (right in the northern hemisphere and left in the southern hemisphere). Normally the wind direction is

variable so that the actual net transport is some angle less than 90°.

As a result of Ekman transport, changes in sea-surface topography and the Coriolis force combine to form geostrophic currents. Take the North Pacific for example. The westerlies at ~40°N and the northeast trades (~10°N) set the North Pacific Current and North Equatorial Current in motion as a circular gyre. Because of the Ekman drift, surface water is pushed toward the center of the gyre (~25°N) and piles up to form a sea-surface "topographic high." As a result of the elevated sea surface, water tends to flow "downhill" in response to gravity. As it flows, however, the Coriolis force deflects the water to the right (in the northern hemisphere). When the current is constant and results from balance between the pressure gradient force due to the elevated seasurface and the Coriolis force, the flow is said to be in geostrophic balance. The actual flow is then nearly parallel to the contours of the elevated seasurface and clockwise. The seasurface topography of the Pacific Ocean was

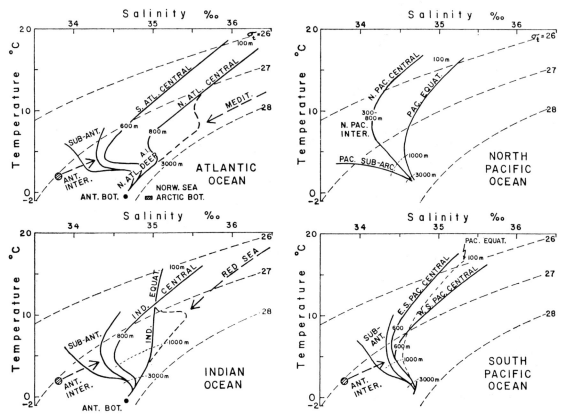

Fig. 10-4   Average temperature/salinity diagrams for the main water masses of the Atlantic, Indian, and Pacific Oceans. (Reproduced with permission from G. L. Pickard and W. J. Emery (1982). "Descriptive Physical Oceanography," pp. 138–139, Pergamon Press.)

determined by Tai and Wunsch (1983) from satellite altimetry. The absolute elevation of the subtropical gyre can be clearly seen and fits the schematic description given above.

As a result of these factors (wind, Ekman transport, Coriolis force) the surface ocean circulation in the mid-latitudes is characterized by clockwise gyres in the northern hemisphere and the counterclockwise gyres in the southern hemisphere. The main surface currents around these gyres for the world's oceans are shown in Fig. 10-6. The regions where Ekman transport tends to push water together are called convergences. Divergences result when surface waters are pushed apart.

Total transport by the surface currents varies greatly and reflects the mean currents and cross-sectional area. Some representative examples will illustrate the scale. The transport around

the subtropical gyre in the North Pacific is about 70 Sv (1 Sverdrup or Sv = 1 × 10⁶ m³/s). The Gulf Stream, which is a major northward flow off the east coast of North America, increases from 30 Sv in the Florida Straits to 150 Sv at 64°30'W, or 2000 km downstream.

10.2.3   El Niño Southern Oscillation (ENSO)

The equatorial Pacific is one of the best-studied regions of ocean divergence (Philander, 1990; McPhaden *et al.*, 1998). This is because of the ENSO phenomenon. The region around the equator normally experiences strong easterly trade winds that result in divergence from the equator and upwelling of colder, nutrient-rich water from below (Fig. 10-7). This "cold-tongue" typically extends from the coast of

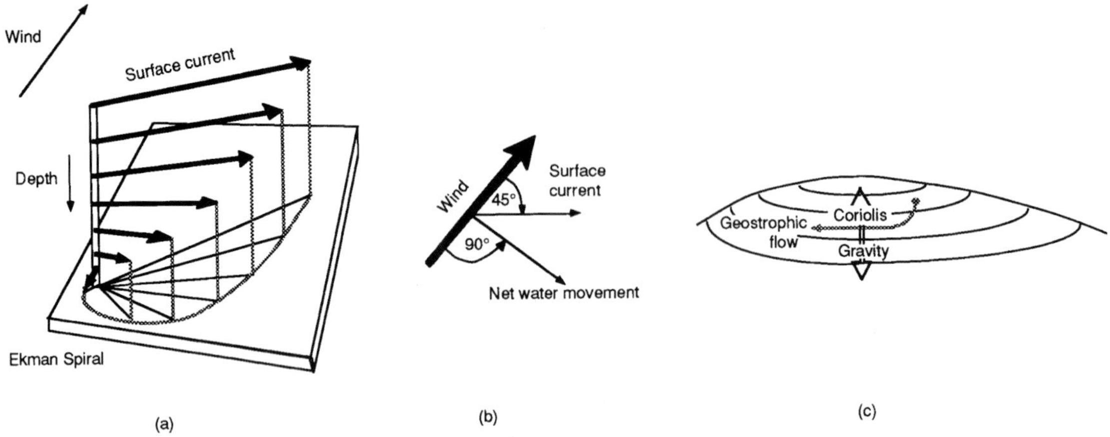

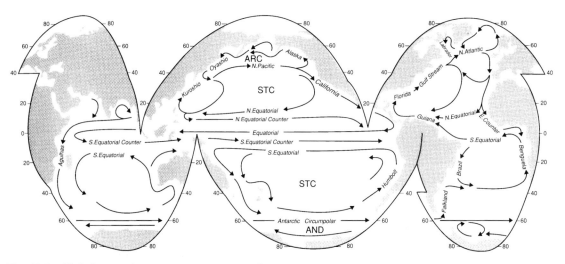

Fig. 10-5 Sketch of (a) current vectors with depth characteristic of an Eckman spiral; (b) relationship between wind, surface current, and net water movement vectors; and (c) production of circular gyres from the net interaction of the Coriolis force and Eckman transport.

Fig. 10-6 Global map of major ocean currents. AND = Antarctic Divergence; STC = subtropical convergence; ARC = Arctic convergence.

South America to about the date line (180°). These trade winds also drive near-equatorial surface flow westward as the South Equatorial Current (SEC). This piles up warm surface water in the western Pacific to create a deep warm pool and results in depression of the depth of the thermocline from east to west. The westward flow in the surface SEC is partly compensated by a return flow to the east in the thermocline (~150 m) called the Equatorial Undercurrent (EUC).

There is a zonal atmospheric circulation system associated with this normal ocean condition called the Walker Cell (Fig. 10-7). Evaporation rates are high over the warm pool and warm moist air ascends to great heights (deep convection) producing extensive cloud systems and rain. The Walker Cell is closed by westerly winds aloft and subsidence in the high-pressure zone of the eastern Pacific.

During El Niño (Fig. 10-7) the trade winds weaken, and even reverse, in the central and

Normal Conditions

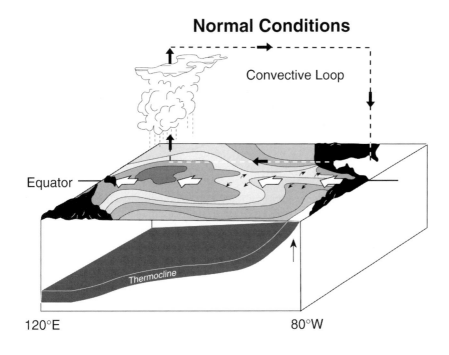

El Niño Conditions

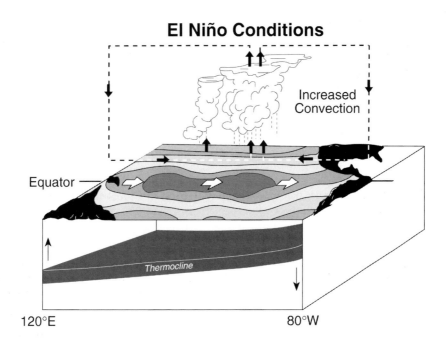

Fig. 10-7   Schematic of normal and El Niño conditions in the equatorial Pacific. (Figure kindly provided by Dr Michael McPhaden of NOAA.)

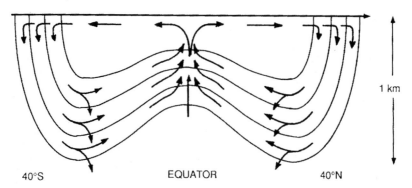

Fig. 10-8   The pathways followed by the water ventilating the main oceanic thermocline. (Reproduced with permission from W. S. Broecker and T.-H. Peng (1982). "Tracers in the Sea," p. 440, Eldigio Press, Palisades, NY.)

western Pacific resulting in a local eastward acceleration of the surface currents. Westerly wind events in the western Pacific excite downwelling equatorial Kelvin waves which propagate into the eastern equatorial Pacific where they depress the thermocline (Kessler and McPhaden, 1995) (Fig. 10-7). The winds in the eastern Pacific are usually still easterly favoring upwelling, but because the thermocline is depressed, warmer water is upwelled. The net result is migration of the "warm pool" and its associated atmospheric deep convection from the western Pacific to east of the date line (Fig. 10-7). Anomalously warm sea-surface temperatures occur from the coast of South America to the date line.

Deep convergence in the atmosphere is the main driving force for atmospheric circulation through the release of latent heat at mid-tropospheric levels. The zonal shift in the site of deep convection during El Niño affects atmospheric circulation and climate on a global basis (Wallace *et al.*, 1998). The variations in the upwelling also influence the flux of CO_2 from the ocean to the atmosphere (Feely *et al.*, 1997) and the biological characteristics of the region (Murray *et al.*, 1994).

10.2.4   Thermocline Circulation

The transition region between the surface and deep ocean is referred to as the thermocline. This is also a pycnocline zone where the density increases appreciably with increasing depth. Most of the density change results from the decrease in temperature (hence thermocline).

A simple but physically realistic model based on lateral transport has evolved to explain the origin of the thermocline. According to this view, the interior of the ocean is ventilated by rapid mixing and advection *along* isopycnal surfaces (Jenkins, 1980). The density surfaces that lie in the thermocline at 200 to 1000 m in the equatorial region shoal and outcrop at high latitudes. The argument is that water acquires its *T* and *S* (and chemical tracer) signature while at the sea surface and then sinks and is transported horizontally as shown in Fig. 10-8. A map showing the winter outcrops of isopycnal surfaces in the North Atlantic Ocean is shown in Fig. 10-9. Characteristic values of the horizontal eddy diffusion coefficient (*K*) are of the order of $10^7 \, \text{cm}^2/\text{s}$. Assuming a distance (*L*) of the order of 2000 km (30°) and assuming the characteristic time is $\tau = L^2/K$, we obtain a characteristic ventilation time for the main thermocline of about 130 years.

The horizontal isopycnal thermocline model is important for the problem of determining the fate of the excess atmospheric CO_2. The increase of CO_2 in the atmosphere is modulated by transport of excess CO_2 from the atmosphere into the interior of the ocean. The direct ventilation of the thermocline in its outcropping regions at high latitudes plays an important

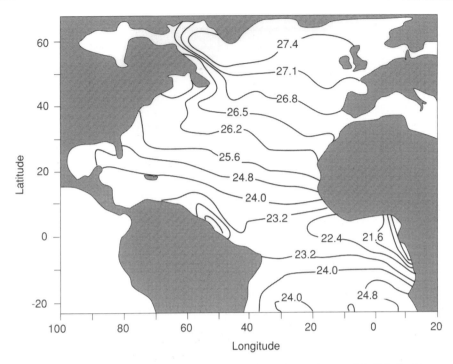

Fig. 10-9 Map of winter outcrops of isopycnal surfaces in the Atlantic Ocean. (Modified with permission from W. S. Broecker and T.-H. Peng (1982). "Tracers in the Sea," p. 394, Eldigio Press, Palisades, NY.)

role in removing CO_2 from the atmosphere (Brewer, 1978; Siegenthaler and Sarmiento, 1993).

Nuclear bomb produced $^{14}CO_2$ and ^3H (as HTO) have been used to describe and model this rapid thermocline ventilation (Ostlund *et al.*, 1974; Sarmiento *et al.*, 1982; Fine *et al.*, 1983). For example, changes in the distributions of tritium (Rooth and Ostlund, 1972) in the western Atlantic between 1972 (GEOSECS) and 1981 (TTO) are shown in Fig. 10-10 (Ostlund and Fine, 1979; Baes and Mulholland, 1985). In the 10 years following the atmospheric bomb tests of the early 1960s, a massive penetration of ^3H (tritium) into the thermocline has occurred at all depths. Comparison of the GEOSECS and TTO data, which have a 9 year time difference, clearly shows the rapid ventilation of the North Atlantic and the value of such "transient" tracers. A similar transient effect can be seen in the penetrative distribution of manmade chlorofluorocarbons, which have been released over a longer period (40 years) (Gammon *et al.*, 1982).

10.2.5 Abyssal Circulation

The circulation of the deep ocean below the thermocline is referred to as abyssal circulation. The currents are slow and difficult to measure but the pattern of circulation can be clearly seen in the properties of the abyssal water. For example, the water of lowest temperature in the water column is usually the densest and lies deepest. As a result, charts of the bottom water temperature have been useful in describing the pattern of the abyssal circulation (e.g., Mantyla, 1975; Mantyla and Reid, 1983). The topography of the sea floor plays an important role in constraining the circulation and much of the abyssal flow is funneled through passages such as the Denmark Straight, Gibbs Fracture Zone, Vema Channel, Samoan Passage, and Drake Passage.

A requirement of the *heat balance* for a steady-state ocean is that the input of new cold abyssal water (Antarctic Bottom Water and North Atlantic Deep Water) sinking in the high-latitude regions must be balanced by input of

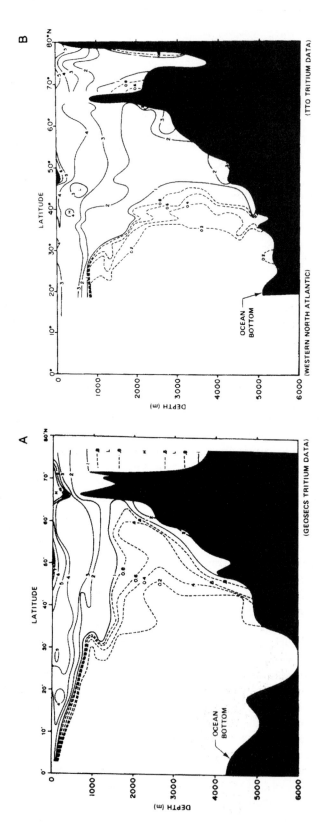

Fig. 10-10 Tritium section of the western Atlantic from 80°N to the equator versus depth (m). Vertical exaggeration is 2000:1. Horizontal scale is proportional to cruise track. (Reproduced with permission from H. G. Ostland and R. A. Fine (1979). Oceanic distribution and transport of tritium. *In* "Behaviour of Tritium in the Environment" (Proceedings of a Symposium, San Francisco, 16–20 October 1978, IAEA-SM-232/67, pp. 303–314. International Atomic Energy Agency, Vienna.)

heat by geothermal heating (heat flow from the Earth), downward convection of relatively warm water (e.g., from the Mediterranean) and downward diffusion of heat across the thermocline. A general mass balance of the world's oceans requires that the water sinking in the polar regions must be exactly balanced by upwelling of water from the abyssal ocean to the surface water. A combination of the mass and heat balances together with the forcing of the wind and the effect of a rotating Earth determine the nature of the abyssal circulation.

10.2.6   The Ocean Conveyor Belt

The ocean conveyor belt is one of the major elements of today's ocean circulation system (Broecker, 1997). A key feature is that it delivers an enormous amount of heat to the North Atlantic and this has profound implications for past, present, and probably future climates.

The conveyor belt is shown schematically in Fig. 10-11. Warm and salty surface currents in the western North Atlantic (e.g., the Gulf Stream) transport heat to the Norwegian–Greenland Seas where it is transferred to the atmosphere. This heat helps moderate the climate of northern Europe. The cooling increases the density resulting in formation of the now cold and salty North Atlantic Deep Water (NADW) (Worthington, 1970). The NADW travels south through the North and South Atlantic and then joins the Circumpolar Current that travels virtually unimpeded in a clockwise direction around the Antarctic Continent.

Deep water also forms along the margins of Antarctica and feeds the Circumpolar Current. The Weddell Sea, because of its very low temperature, is the main source of Antarctic Bottom Water (AABW) which flows northward at the very bottom into the South Atlantic, and then through the Vema Channel in the Rio Grande Rise into the North Atlantic. It ultimately returns southward as part of the NADW.

The Circumpolar Current is a blend of waters of North Atlantic ($\sim 47\%$) and Antarctic margin ($\sim 53\%$) origin (Broecker, 1997). This current is referred to as the Pacific Common Water and is the source of deep water to the Indian and Pacific Oceans. Deep water does not form in a similar way in the North Pacific because the salinity is too low (Warren, 1983). Pacific Common Water enters the Pacific in the southwest corner and flows north along the western boundary of the Tonga Trench. The abyssal circulation model of Stommel (1958) and Stommel and Arons (1960) predicted that deep waters flow most intensely along the western boundaries in all oceans and gradually circulate into the interior as allowed by topography. Most of the northward abyssal flow passes from the southwest Pacific to the north central Pacific through the Samoan Passage, located west of Samoa. In the North Pacific, the abyssal flow splits and goes west and east of the Hawaiian Islands. These flows meet again north of Hawaii where they mix, upwell and flow back to the South Pacific at mid-depths.

The conveyor belt is completed by return flow of surface water from the Pacific to the Atlantic. There are two main paths of this return flow, which amounts to about 19 Sv. Some passes through the Indonesian Archipelago, the Indian Ocean and around the tip of South Africa via the Agulhas Current (Gordon, 1985). Some enters the South Atlantic via the Drake Passage. Based on the salt budget, Broecker (1991) argued that the Drake Passage route transports about 50% more than the Agulhas Current. Finally there is a small transport (about 1 Sv) from the Pacific to the Atlantic through the Bering Strait.

The salt budget for the Atlantic, which is determined in part by the flux of freshwater through the atmosphere, drives the conveyor belt and can explain how it has varied in the past. At present there appears to be a net water vapor loss of about 0.32 Sv (greater than the flow of the Amazon) from the Atlantic to the Pacific. The NADW transports about 16.3 Sv of water with a salinity of 34.91. This is produced from 15 Sv of Gulf Stream water with a salinity of 35.8, 1 Sv of transport from the Bering Straits with $S = 32$ and a net excess of river inflow and rainfall over evaporation of about 0.3 Sv (Zauker and Broecker, 1992). It is easy to show that small changes in the freshwater budget can have a significant impact. For example, if the excess of precipitation plus runoff over evaporation

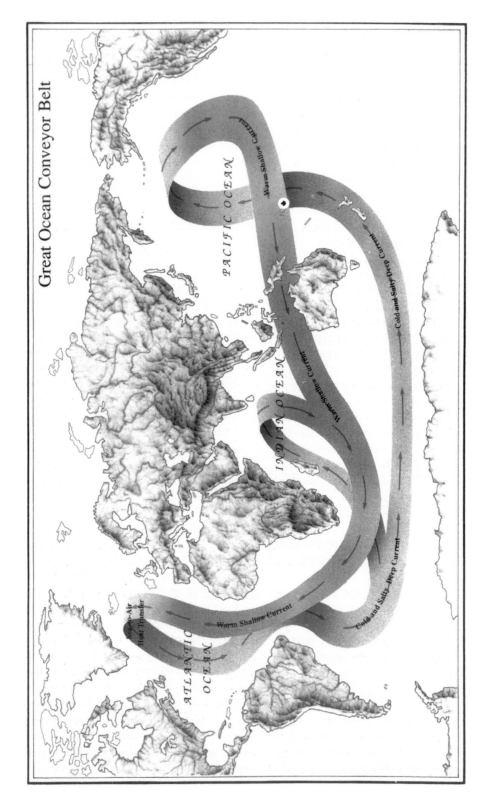

Fig. 10-11 Schematic of the ocean "conveyor belt" from Broecker (1991). (Reproduced with permission of the illustrator, Joe Le Monnier.)

increased by 50% to 0.45 Sv, the salt content of the NADW would decrease to 34.59. In order to compensate for the resulting reduction in density the water would have to be cooled by an additional 1.4°C and the conveyor would have to more than double its flow to restore the salt balance (Broecker, 1997). Model simulations have also shown that the oceans' thermohaline circulation is extremely sensitive to freshwater input (Manabe and Stouffer, 1995).

Although the general circulation patterns are fairly well known, it is difficult to quantify the rates of the various flows. Abyssal circulation is generally quite slow and variable on short time scales. The calculation of the rate of formation of abyssal water is also fraught with uncertainty. Probably the most promising means of assigning the time dimension to oceanic processes is through the study of the distribution of radioactive chemical tracers. Difficulties associated with the interpretation of radioactive tracer distributions lie both in the models used, non-conservative interactions, and the difference between the time scale of the physical transport phenomenon and the mean life of the tracer.

An example of the power of such tracers is in the "dating" of abyssal water using ^{14}C. ^{14}C has an atmospheric source and a half-life of 5720 years. Stuiver *et al.* (1983) measured the ^{14}C distribution in dissolved inorganic carbon in deep samples from major ocean basins (Fig. 10-12). These data were used to calibrate a box model which indicated that the replacement times for Atlantic, Indian, and Pacific Ocean deep waters (depths > 1500 m) are 275, 250, and 510 years respectively.

The present form of the conveyor belt appears to have been initiated by closure of the Panamanian seaway between the North and South American continents (Keigwin, 1982; Maier-Reimer *et al.*, 1990). Geologic evidence indicates that gradual closing of the Isthmus of Panama lasted from 13 to 1.9 Myr ago. Paleoceanographic records indicate that closure was sufficient by 4.6 Myr ago to cause a marked reorganization of ocean circulation (Burton *et al.*, 1997; Haug and Tiedemann, 1998). At this time the Gulf Stream intensified resulting in the transport of warm water to high latitudes. As a result NADW formation intensified and

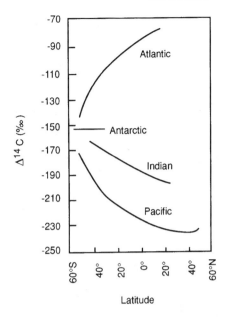

Fig. 10-12 The Δ^{14}C values of the cores of the North Atlantic, Pacific, and Indian Ocean deep waters. The oldest waters are encountered near 40°N in the Pacific Ocean. (Modified with permission from M. Stuiver *et al.* (1983). Abyssal water carbon-14 distribution and the age of the world oceans, *Science* **219**: 849–851, the AAAS.)

increased atmospheric moisture input to high latitudes helped trigger the growth of northern hemispheric ice sheets.

There is strong evidence that the conveyor belt has switched regularly from one mode of operation to another in the past. The associated changes in climate have been large, abrupt and global (Denton and Hendy, 1994). The changes appear to be driven by factors controlling the density of high-latitude North Atlantic surface water. These events appear to have been triggered by an increase in iceberg input, mainly from Canada (Bond *et al.*, 1992). These icebergs transport terrigenous debris across the North Atlantic. When they melt they deposit a layer of ice rafted material on the sea floor. These periodic events in the geologic record are called Heinrich events (Broecker, 1994; Bond *et al.*, 1997). The input of freshwater reduces the density of surface seawater and reduces production of NADW thus shutting down the present mode of the conveyor belt. The timing of these events has been perfectly preserved in the sediments of the North Atlantic from as far away as the Santa Barbara Basin (Behl and Kennett, 1996). At the time of these events the climate cools both at high latitudes and globally.

The climate records in Greenland ice reveal that over the past 60 000 years conditions switched back and forth between intense cold and moderate cold on a time scale of a few thousand years. These so-called Dansgaard–Oeschger cycles are characterized by abrupt changes in temperature, dust content, ice accumulation rate, methane concentration, and CO_2 content. The onset of these cold events occurred on time scales as short as a few decades to a few years (Alley et al., 1993). Each period of intense cold has been matched by an ice-rafting event in North Atlantic sediments. As a result of the switch to a colder climate, iceberg production slows and the salinity of the north Atlantic surface water slowly increases enabling NADW formation to occur again. The return to the warm phase occurs much more slowly, over a 1000 year time frame. These cyclic events appear to have continued in the Holocene, although with much muted amplitude (Alley et al., 1997).

There is great concern that one of the effects of global warming could be reduction in formation of NADW and associated reorganization of the conveyor-belt circulation (Manabe and Stouffer, 1995). The consequences of global warming will be to warm surface seawater and to intensify the hydrologic cycle. Both factors will make it more difficult to form deep water and could lead to an "anthropogenically" induced climate shift. Paradoxically, global warming could result in climate cooling for northern Europe.

10.3   Biological Processes

Almost all elements in the periodic table are involved in at least one way or another in the biological cycle of the ocean. Many elements are essential or required nutrients. Others are carried along as passive participants. In either case the rates of biological processes need to be known.

10.3.1   The RKR Model

Essentially all organic matter in the ocean is ultimately derived from inorganic starting materials (nutrients) converted by photosynthetic algae into biomass. A generalized model for the production of plankton biomass from nutrients in seawater was presented by Redfield, Ketchum and Richards (1963). The schematic "RKR" equation is given below:

$$106\ CO_2 + 16\ HNO_3 + H_3PO_4 + 122\ H_2O\ (Light) \rightarrow$$
$$(CH_2O)_{106}(NH_3)_{16}(H_3PO_4)\ (Plankton\ protoplasm)$$
$$+ 138\ O_2 \qquad (1)$$

This equation was originally proposed for "average" plankton, a category that included both zooplankton and phytoplankton. This mean elemental ratio of $C/N/P = 106/16/1$ by atoms is highly conserved (Falkowski et al., 1998) and reflects the average biochemical composition of marine phytoplankton and their early degradation products.

The following characteristics of the RKR reaction should be noted:

1. This is an organic redox reaction. Carbon in CO_2 and nitrogen in HNO_3 are reduced by oxygen from water as the oxygen in these compounds is oxidized to O_2. Only phosphorus does not undergo any change in oxidation state.

2. The reaction is endothermic. Energy from sunlight is stored in the form of high-energy C–C bonds (e.g., organic biomass) and O_2, the raw materials for the support of heterotrophic organisms dependent upon the food source.

3. This is not a reversible reaction in the strict sense and does not spontaneously seek equilibrium between products and reactants. The exothermic reverse reaction, respiration, occurs in a different part of phytoplankton cells or is mediated by heterotrophic organisms.

4. Inasmuch as the RKR model is a generalization, specific exceptions should be expected. The most important exceptions relate to growth conditions that can affect the stoichiometry of nutrient incorporation into plankton biomass. During respiration, the

reverse reaction occurs and nutrients are regenerated. Phosphorus tends to be regenerated preferentially relative to nitrogen which is preferentially regenerated relative to carbon. Recent interpretation of data along constant density surfaces in the Atlantic suggests that the regeneration ratios of $P:N:C:O_2$ are about 1:16:117:170, slightly different than the RKR ratios (Takahashi *et al.*, 1985; Anderson and Sarmiento, 1994).

10.3.2 Food Web Concepts

A schematic representation of biological processes in the marine euphotic zone is shown in Fig. 10-13. The links between the cycling of C, N, and O_2 are indicated. Total primary production is composed of two parts. The production driven by new nutrient input to the euphotic zone is called *new production* (Dugdale and Goering, 1967). New production is mainly in the form of the upward flux of nitrate from below but river and atmospheric input and nitrogen fixation (Karl *et al.*, 1997) are other possible sources. Other forms of nitrogen such as nitrite, ammonia, and urea may also be important under certain situations. The "new" nitrate is used to produce plankton protoplasm and oxygen according to the RKR equation. Some of the plant material produced is respired in the euphotic zone due to the combined efforts

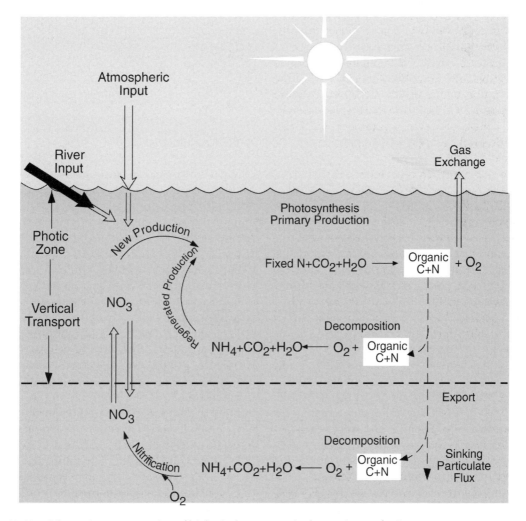

Fig. 10-13 Schematic representation of biological processes in the marine euphotic zone.

of zooplankton grazing and bacterial respiration. Oxygen is consumed and organic-N is released as ammonia according to

$$(CH_2O)_{106}(NH_3)_{16}(H_3PO_4) + 106O_2 \rightarrow$$

$$106CO_2 + 16NH_3 + H_3PO_4 + 106H_2O \quad (2)$$

The released ammonia is preferentially taken up by phytoplankton relative to nitrate (Dortch, 1990) to drive *regenerated production*. The *f*-ratio is used to describe the relative amounts of new and regenerated production (Dugdale and Goering, 1967) where

$$f = \frac{NO_3 \text{ uptake}}{NO_3 + NH_4 \text{ uptake}} \quad (3)$$

The *f*-ratio typically varies from values as low as 0.03 to 0.30 in the open ocean (e.g., McCarthy *et al.*, 1996) to values greater than 0.80 in the coastal ocean (Platt and Harrison, 1985).

If we define the sum of new plus regenerated production as *gross production* (*P*) and the difference of gross minus regenerated production as *net production* (*P* − *R*) then we can also express

$$f = (P - R)/P \quad (4)$$

Specific examples of marine ecosystem models can be seen in Frost (1987), Fasham *et al.* (1990), Frost and Franzen (1992), and Loukos *et al.* (1997).

As shown in Fig. 10-13, there is also a flux of O_2 produced during net photosynthesis from the ocean to the atmosphere and an export flux of particulate and dissolved organic matter out of the euphotic zone. For a steady-state system, new production should equal the flux of O_2 to the atmosphere and the export of organic carbon (Eppley and Peterson, 1979) (when all are expressed in the same units, e.g., moles of carbon). Such an ideal state probably rarely exists because the euphotic zone is a dynamic place. Unfortunately, there have been no studies where all three fluxes were measured at the same time. Part of the difficulty is that each flux needs to be integrated over different time scales. The oxygen flux approach has been applied in the subarctic north Pacific (Emerson *et al.*, 1991) and subtropical Pacific (Emerson *et al.*, 1995, 1997) and Atlantic (Jenkins and Goldman, 1985). The organic carbon export approach has

been evaluated in the equatorial Pacific (Murray *et al.*, 1996), subtropical Atlantic (Michaels *et al.*, 1994) and several locations by Buesseler (1998) and Hansell and Carlson (1998).

Integrated, interdisciplinary studies of elemental cycling in the euphotic zone have been one main focus of the Joint Global Ocean Flux Study (JGOFS) and these have contributed greatly to our understanding of carbon cycling in specific ocean regions. Multi-investigator JGOFS process studies have been conducted in the North Atlantic (North Atlantic Bloom Experiment; NABE) (Ducklow and Harris, 1993), Equatorial Pacific (EqPac) (Murray, 1995, 1996; Murray *et al.*, 1994, 1997), Subtropical Atlantic and Pacific (HOT and BATS) (Karl and Michaels, 1996), Arabian Sea (van Weering *et al.*, 1997; Smith, 1998), and Southern Ocean (Turner *et al.*, 1995; Gaillard and Treguer, 1997; Smetacek *et al.*, 1997).

10.3.3   Factors Affecting the Rate of Plankton Productivity

10.3.3.1   Nutrients

Liebig's Law of the Minimum states that under equal conditions of temperature and light, the nutrient available in the smallest quantity relative to the requirement of a plant will limit productivity. The "classic" approach for evaluating nutrient limitation is to compare the requirements of "average" marine plankton with "average" seawater. For example, plots of PO_4 versus NO_3 in deep ocean seawater show a very tight correlation with a slope of slightly less than 6.0 and a small but significant intercept on the PO_4 axis (e.g., Fanning, 1992; Gruber and Sarmiento, 1997). When these waters are upwelled nutrient uptake takes place with RKR proportions and NO_3 will run out first and become the limiting nutrient. Biological oceanographers have repeatedly demonstrated through enrichment experiments and observations of nutrient distribution that throughout the most of the coastal and open oceans phytoplankton productivity is most often limited by the availability of fixed inorganic N (Falkowski *et al.*, 1998). There is approximately a ten-fold excess of inorganic carbon (largely as HCO_3^-) in deep seawater relative to the availability of phosphorus and nitrogen availability which implies that carbon is never limiting in the ocean. In addition, except under very intense bloom conditions, the carbon fixed by plankton is provided by upwelling and not from the atmosphere. The reason deep ocean sea water is slightly depleted in N relative to RKR probably reflects nitrogen loss due to denitrification, which occurs primarily in the intense oxygen minimum zones of the eastern tropical north and south Pacific, the Arabian Sea and continental margin sediments throughout the world's oceans (Christensen *et al.*, 1987).

In the subtropical ocean gyres the situation is more complicated (Perry, 1976). These regions are considered to be the marine analogs of terrestrial deserts because all nutrients are greatly depleted and biological biomass is small. A recent time-series study (The Hawaiian Ocean Time-series or HOT) has revealed that the ecosystem of the north Pacific subtropical gyre is temporally and spatially variable (Karl *et al.*, 1995). This variability appears tied to the El Niño southern oscillation (ENSO) cycle. Increased stratification and decreased upper-ocean mixing during the 1991–92 El Niño event resulted in increased abundance and growth of nitrogen-fixing blue-green microorganisms called *Trichodesmium*. This resulted in a shift from the primarily nitrogen-limited regime that existed in 1981–90 to a phosphorus-limited condition in 1991–92. Growth of *Trichodesmium* spp. in subtropical habitats is favored under calm ocean conditions. Their ability to reduce N_2 can remove the fixed-nitrogen limitation. N_2 fixation may contribute up to half of the N required to sustain total annual organic matter export in this region (Karl *et al.*, 1997). N_2 fixation appears to be an important source of "new" nitrogen, especially under El Niño conditions. Thus the ecosystem in the subtropical gyres may switch periodically between N-limitation and P-limitation.

It has been argued that phosphorus limits oceanic productivity on the million year time scale (Broecker, 1971). The reason is that essentially all phosphorus in the ocean is introduced by rivers and thus ultimately from the weathering of continental rocks. This flux is, in effect, fixed by the rate of chemical weathering of the continents. By comparison fixed nitrogen can be derived from atmospheric N_2 (via nitrogen fixation by *Trichodesmium*) as well as by weathering of rocks. The reservoir of atmospheric N_2 is so large that nitrogen fixation can, over long time periods, adjust the overall supply of fixed nitrogen in seawater to the ratio needed by "average" plankton without significantly depleting the N_2 source.

Silicic acid (H_4SiO_4) is a necessary nutrient for diatoms, who build their shells from opal ($SiO_2 \cdot nH_2O$). Whether silicic acid becomes limiting for diatoms in seawater depends on the availability of Si relative to N and P. Estimates of diatom uptake of Si relative to P range from 16:1 to 23:1. Dugdale and Wilkerson (1998) and Dunne *et al.* (1999) have shown that much of the variability in new production in the equatorial Pacific may be tied to variability in diatom production. Diatom control is most important at times of very high nutrient concentrations and during non-steady-state times, perhaps because more iron is available at those times.

Over 20% of the world's open ocean surface waters are replete in light and major nutrients (nitrate, phosphate, and silicate), yet chlorophyll and productivity values remain low. These so-called "high-nitrate low-chlorophyll" or HNLC regimes (Chisholm and Morel, 1991) include the sub-arctic North Pacific (Martin and Fitzwater, 1988; Martin *et al.*, 1989; Miller *et al.*, 1991), the equatorial Pacific (Murray *et al.*, 1994; Fitzwater *et al.*, 1996) and the southern Ocean (Martin *et al.*, 1990). Iron concentrations are extremely low in these regions (Johnson *et al.*, 1997). The main source is probably particulate iron associated with atmospheric dust (Duce and Tindale, 1991). The equatorial undercurrent appears to be an additional source of the equatorial Pacific.

The results of two successful iron-fertilization experiments in the eastern equatorial Pacific have clearly shown that phytoplankton growth rate is limited by iron at that location (Martin *et al.*, 1994; Coale *et al.*, 1996). The species composition and size distributions of the ecosystem are influenced by iron availability (Landry *et al.*, 1997). In particular, large diatoms do not grow at optimum rates in the absence of sufficient iron. Loukos *et al.* (1997) used a simple ecosystem model with iron limitation to show that the main process causing the persistence of high surface nutrients was not the low specific growth rate of the phytoplankton assemblage but the efficient recycling of nitrogen as a consequence of the food web structure imposed by iron limitation. The first-order process responsible for low phytoplankton biomass is efficient grazing of the small cells by micrograzers, which is also an indirect consequence of iron limitation (Landry et al., 1997). Grazing balances primary production and controls phytoplankton biomass. Nevertheless, because of its impact on the food web, iron deficiency is the ultimate control of the HNLC condition.

Fluxes of continental dust preserved in ice cores of Greenland and Antarctica suggest a 30-fold increase in dust flux during the last Glacial Maximum. Dramatic increases in new biological production in the HNLC regions may have resulted, resulting in the draw-down of atmospheric CO_2 (Martin, 1990).

10.3.3.2 Light

Light is always necessary for photosynthesis (Raven and Johnston, 1991; Falkowski et al., 1992) and becomes limiting in the winter at high latitudes. In addition, the depth profiles of productivity and light energy correlate well at locations undergoing bloom conditions (Cullen et al., 1992). This suggests that the decline in productivity with depth reflects light penetration. Chemical constituents of seawater that can inhibit light penetration include dissolved humic substances (Gelbstuff) and suspended particulate matter. Both factors can become important factors in estuaries and other nearshore environments.

10.3.3.3 Availability of trace metals

Trace metals can serve as essential nutrients and as toxic substances (Sunda et al., 1991; Frausto da Silva and Williams, 1991). For example, cobalt is a component of vitamin B-12. This vitamin is essential for nitrogen fixing algae. In contrast, copper is toxic to marine phytoplankton at free ion concentrations similar to those found in seawater (Sunda and

Guillard, 1976). The possibility that iron availability may limit primary productivity was discussed earlier.

Nickel is required by plants when urea is the source of nitrogen (Price and Morel, 1991). Bicarbonate uptake by cells may be limited by Zn as HCO_3^- transport involves the zinc metalloenzyme carbonic anhydrase (Morel et al., 1994). Cadmium is not known to be required by organisms but because it can substitute for Zn in some metalloenzymes it can promote the growth of Zn-limited phytoplankton (Price and Morel, 1990). Cobalt can also substitute for Zn but less efficiently than Cd.

10.3.4   The Geographic Distribution of Primary Productivity

The geographic distribution of primary productivity in the ocean is shown in Fig. 10-14. High productivity is characteristic of marine zones where surface water is replenished with deeper water either by upwelling (as on western continental margins) or by deep mixing (as at high latitudes where stratification is less pronounced). Although upwelling regions are characterized by very high productivities ($\sim 300 \, g \, C/m^2$ per year) they together contribute less than 1% of the total ocean production (Table 10-5). Coastal regions have mean productivities of about $100 \, g \, C/m^2$ per year, but account for approximately 100 times the surface area of upwelling zones. These coastal regions contribute about 25% of the total primary production with the remaining 75% coming from the wide expanses (90% of total area) of low production ($50 \, g \, C/m^2$ per year) open ocean.

Recently, the ocean-basin distribution of marine biomass and productivity has been estimated by satellite remote sensing. Ocean color at different wavelengths is determined and used to estimate near-surface phytoplankton chlorophyll concentration. Production is then estimated from chlorophyll using either in situ calibration relationships or from empirical functional algorithms (e.g., Platt and Sathyendranth, 1988; Field et al., 1998). Such studies reveal a tremendous amount of temporal and spatial variability in ocean biological production.

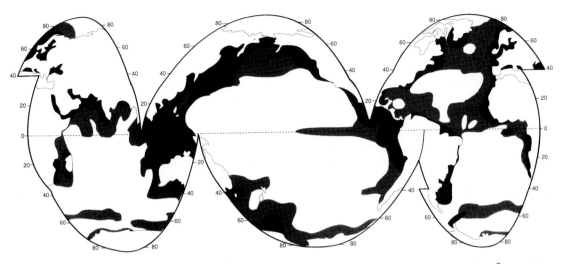

Fig. 10-14   Approximate geographical distibution of primary productivity in the oceans (g C/m² per year).

Table 10-5   Distribution of ocean productivity

Province	Percentage of ocean	Area $(10^6 \, \text{km}^2)$	Mean productivity[a] $(\text{g C/m}^2/\text{yr})$	Total productivity[a] $(10^{15} \, \text{g C/yr})$
Open ocean	90	326	50	16.3
Coastal zone[b]	9.9	36	100	3.6
Upwelling areas	0.1	0.36	300	0.1
Total				20

[a] From Ryther (1969).
[b] Includes offshore areas of high productivity.

10.3.5 Forms of Organic Matter in Seawater

To this point, organic matter in the ocean has been treated primarily as RKR average plankton. We now need to focus on the fate of this biologically produced organic carbon.

There is an exceedingly broad range of size of organic material in seawater, ranging from simple organic molecules, such as dissolved glucose (scale $\sim 10^{-9}$ m) to the blue whale ($\sim 10^2$ m). Although the distribution curve of organic particles is smooth over the 10^{-3} to 10^{-9} m size interval, it has become customary to divide these particles into "dissolved" and "particulate" categories on the basis of filtration through a 0.45 μm pore size filter. By operational definition, "dissolved" particles pass the filter whereas "particulate" materials are retained.

Although somewhat arbitrary, the 0.45 μm "cutoff" between dissolved and particulate organic is for the most part convenient. For example, particles above about 1.0 μm are observable with a microscope and tend to settle in seawater. Particles less than 1.0 μm are submicroscopic and generally sink very slowly and disperse as a result of Brownian motion. In addition, particles less than 0.45 μm fall below the range of most living organisms (except for some viruses and small bacteria).

Essentially all the dissolved organic matter in seawater can be assumed to be non-living. However, particulate organic matter can be either living or dead, with the latter often referred to as "detritus."

10.3.6 Organic Carbon Pathways in the Ocean

To understand the distribution and pathways of organic material in the ocean the key question is: "What happens to that 99% of the phytoplankton biomass that is remineralized between photosynthesis and burial?"

A major advance in our understanding of the processes controlling the vertical transport of organic matter in the ocean was the realization that most of the vertical flux of particulate material in the ocean water column is provided by large particles (> 128 μm) which account for less than 5% of the total mass concentration. These larger particles are predominantly zooplankton fecal pellets or marine snow particles. About 90% of all phytoplankton are eaten by zooplankton and encapsulated into large (100–300 μm), fast-sinking (50–500 m/day) fecal pellets. This result is particularly significant because the large, fast-sinking particles are less likely to be collected using conventional water samplers. As a result different types of "sediment traps" have been developed in order to collect a representative sample of the vertical particle flux.

Sediment trap studies in the open ocean show that the flux of organic carbon at any depth is directly proportional to the rate of primary productivity in the surface water and inversely proportional to the depth of the water column (Suess, 1980):

$$C_{flux} = \frac{C_{prod}}{0.0238Z + 0.212} \quad (5)$$

where C_{prod} is the primary production (g/m² per year), and C_{flux} is the flux at depth Z (m). The original data used to calibrate this equation are shown in Fig. 10-15 as a plot of the ratio of carbon flux/primary production versus water depth. As can be seen, about 10% of the primary production falls to a depth of 400 m, whereas only about 1% reaches 5000 m.

This general relationship has other implications and applications. If depth in Fig. 10-15 is transformed into time, then the slope of the plot represents a rate constant for in situ organic carbon loss from the sinking particles. Assum-

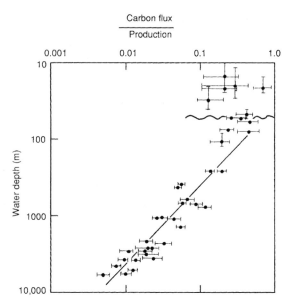

Fig. 10-15 Organic carbon fluxes with depth in the water column normalized to mean annual primary production rates at the sites of sediment trap deployment. The undulating line indicates the base of the euphotic zone; the horizontal error bars reflect variations in mean annual productivity as well as replicate flux measurements during the same season or over several seasons; vertical error bars are depth ranges of several sediment trap deployments and uncertainities in the exact depth location. (Reproduced with permission from E. Suess (1980). Particulate organic carbon flux in the oceans – surface productivity and oxygen utilization, *Nature* **288**: 260–263, Macmillan Magazines.)

ing an average settling rate of 100 m/day the previous equation becomes

$$2.38t = \frac{C_{prod}}{C_{flux} - 0.212} \quad (6)$$

where $t = Z/(dZ/dt)$. Thus, the estimated time for degradation of 90% of the primary production from falling particles is 4.1 days and after this time the particles would be at about 400 m.

This discussion suggests a rapid and relatively direct transport of organic material vertically through the ocean water column. However, this transport is not efficient and under "average" ocean conditions (primary productivity = 100 g C/m² per year and water

Table 10-6   Organic carbon reservoirs in the ocean

Depth (m)	Dissolved (10^6 tonnes C)	Approximate concentration (μg C/L)	Particulate (10^6 tonnes C)	Approximate concentration (μg C/L)	Living fraction (10^6 tonnes C)	Approximate % of particulate
$0–300^a$	110 000	1000–1500	11 000	100	550^f	5^f
$300–3800^b$	630 000	500–800	13 000	3–10	$<400^g$	$<3^g$
Totals	$740\,000^c$ $670\,000^d$ $1\,000\,000^e$		$24\,000^c$ $14\,000^d$ $30\,000^e$			

^{a}Concentrations vary widely with geographical area and with season. The depth at which the concentration tends to approach a more or less constant level varies widely from 100 to 500 m.
^{b}Concentrations more constant.
^{c}Williams (1971).
^{d}Menzel (1974).
^{e}Williams (1975).
^{f}Phytoplankton component only, see Table 10-4 and Williams (1975).
^{g}Total living matter, Parsons *et al.* (1977).

depth of 4000 m) only 1% of the production can be expected to reach the ocean floor.

10.3.7   Oceanic Reservoirs of Organic Carbon

The distribution of dissolved, total particulate, and living particulate organic carbon in the surface (0–300 m) and deep ocean (>300 m) are summarized in Table 10-6. Recent analytical advances have greatly improved our understanding of the distributions of DOC in the ocean (Hedges and Lee, 1993). The important aspects of this compilation are:

1. The organic carbon (and thus the organic matter) in seawater is predominantly in dissolved form (DOC) with an average for the whole ocean being 97%. In general, surface ocean concentrations are about 80 μM and decrease to about 40 μM below 500 m (Peltzer and Hayward, 1996; Hansell and Peltzer, 1998).
2. DOC in the deep ocean gradually decreases from 48 μM in the North Atlantic to 34 μM in the North Pacific (Hansell and Carlson, 1998).
3. The C/N ratio of dissolved organic matter (~ 17) is greater than the RKR ratio of 6.6. DON has a similar distribution as DOC and

appears to be derived from degradation-resistant biomolecules (McCarthy *et al.*, 1997).
4. Of the remaining particulate organic matter, very little is living (> 95% detritus).

10.3.8   An Oceanic Budget for Organic Carbon

A simple budget for the global carbon cycle is given in Fig. 10-16 (Siegenthaler and Sarmiento, 1993). In this version net primary production is 10 Gt C/yr (1 Gt = 10^{15} gC). This is in close agreement with the estimate of 7.2 Gt C/yr by Chavez and Toggweiler (1995). Falkowski *et al.* (1998) concluded that ocean primary production is about 45 Gt C/yr, which is about 43% of the global total (Field *et al.*, 1998). Thus the global average ocean *f*-ratio is about 20%. On average ocean biota (3 Gt C) turns over about every 1/3 year. According to this version 60% of the exported C is as DOC. More recent estimates suggest that the fraction as DOC is more like 20% (Hansell *et al.*, 1997; Murray *et al.*, 1996; Archer *et al.*, 1998).

The main mechanism for removal of organic carbon from the ocean is burial in sediments. This flux is equal to the average global sedimentation rate for marine sediments times their weight percent organic carbon. The total sink

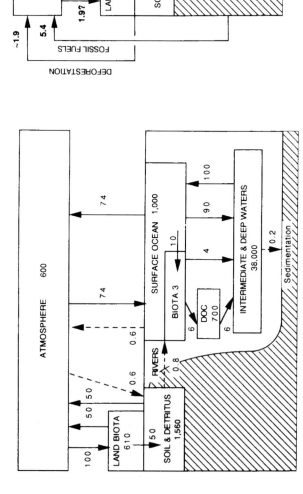

Fig. 10-16 A simple schematic of the carbon cycle. Part (a) is the pre-industrial case, and part (b) shows the contemporary reservoirs and fluxes, in Pg C and Pg C/yr, respectively (Pg C = 10^{15} g C). This diagram of the carbon cycle is similar to those presented in Chapters 4 and 11. (Reprinted by permission from *Nature* (1993). **365**: 119–125, Macmillan.)

by burial $(2 \times 10^{14} \text{g C/yr})$ does not nearly approach the production terms. If the reservoir of total organic matter in seawater is to remain constant (i.e., steady state) then over 98% of the input must be remineralized by respiration and decomposition either in the water column or the surface sediments. The closest estimate that can be made for the mean residence time of organic matter in the ocean is 10 000 years. This is obtained by dividing the reservoir mass by the sedimentary removal rate. This crude estimate is, however, slanted toward the refractory organic components and says little about organic carbon cycling rates in the ocean prior to sedimentation.

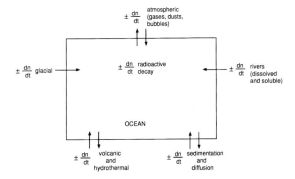

Fig. 10-17   A simplified box model of the ocean.

10.4   Chemistry of the Oceans

Chemical oceanography came to be identified as a discipline in its own right during the 1960s. The significance of chemical oceanography grew when it was realized that many of the stable and radioactive components of the ocean vary significantly in concentration and that knowledge of these variations could provide important information about natural processes. During the 1970s the distributions of most elements and isotopes became fairly well understood. In the process, the subdiscipline of marine chemistry emerged. This field focuses on the chemical reactions and mechanisms in the ocean and at its boundaries. A summary of the observed concentrations and are given in Table 10-7 (Quinby-Hunt and Turekian, 1983).

10.4.1   Residence Time

As a starting point we can view the ocean as one large reservoir to which materials are continuously added and removed (Fig. 10-17). The major sources of material include rivers and winds, which carry dissolved and particulate materials from the continents to the sea. The major removal process is the formation of marine sediments both by settling of particles through the water column as well as by precipitation of insoluble solid phases. For many ele-

ments hydrothermal circulation through the ocean crust may be important.

The concept of average residence time, or turnover time, provides a simple macroscopic approach for relating the concentrations in ocean reservoirs and the fluxes between them. For the single box ocean in Fig. 10-17 the rate of change of the concentration of component n can be expressed as

$$\left(\frac{\mathrm{d}n}{\mathrm{d}t}\right)_{ocean} = \sum_i \left(\frac{\mathrm{d}n_i}{\mathrm{d}t}\right) \tag{7}$$

If $(\mathrm{d}n/\mathrm{d}t) = 0$ we have a steady state. This is also referred to as a state of dynamic equilibrium where the rate of input equals the rate of removal.

The concept of residence time was first introduced by Barth (1952) and given by the following expression (see also Chapter 4 and Li, 1977):

$$\tau_0 = \frac{\text{Mass of element in the sea}}{\text{Mass supplied (or removed) per year}}$$

$$= \frac{M}{S} = \frac{M}{Q} \tag{8}$$

where Q and S represent the mean total input and removal rates, respectively, and M represents the total mass of an element dissolved in the sea. For most elements it appears that the removal rate is proportional to the total amount present or $S = kM$, where k is a first-order rate constant. At steady state, $\tau_0 = 1/k$. This relation predicts an inverse correlation between residence time and the removal rate constant, which must be a measure of the chemical

Table 10-7 Mean oceanic concentrations

Atomic number	Element	Species	Mean water concentration
1	Hydrogen	H_2O	
2	Helium	He (gas)	1.9 nmol/kg
3	Lithium	Li^+	178 µg/kg
4	Beryllium	$BeOH^+$	0.2 ng/kg
5	Boron	$B(OH)_3$	4.4 mg/kg
		$B(OH)_4^-$	
6	Carbon	ΣCO_2	2200 µmol/kg
7	Nitrogen	N_2, NO_2^-	590 µmol/kg
		NO_3, NH_4^+	30 µmol/kg
8	Oxygen	Dissolved O_2	150 µmol/kg
9	Fluorine	F^-, MgF^+	1.3 mg/kg
10	Neon	Ne (gas)	8 nmol/kg
11	Sodium	Na^+	10.781 g/kg
12	Magnesium	Mg^{2+}	1.28 g/kg
13	Aluminum	$Al(OH)_4^-$	1 µg/kg
14	Silicon	Silicate $Si(OH)_4$	110 µmol/kg
15	Phosphorus	Reactive phosphate	2 µmol/kg
16	Sulfur	Sulfate SO_4^{2-}	2.712 g/kg
17	Chlorine	Chloride Cl^-	19.353 g/kg
18	Argon	Ar (gas)	15.6 µmol/kg
19	Potassium	K^+	399 mg/kg
20	Calcium	Ca^{2+}	415 mg/kg (C.A.)
21	Scandium	$Sc(OH)_3^0$	<1 ng/kg
22	Titanium	$Ti(OH)_4^0$	<1 ng/kg
28	Vanadium	$H_2VO_4^-$, HVO_4^{2-}	<1 µg/kg
24	Chromium	Cr(tot)	330 ng/kg (Si)
		$Cr(OH)_3(s)$	330 ng/kg (Si, PO)
		CrO_4^{2-}	350 mg/kg (Si, NO)
25	Manganese	Mn^{3+}, $MnCl^+$	10 ng/kg
26	Iron	$Fe(OH)_2^+$, $Fe(OH)_4^-$	40 ng/kg
27	Cobalt	Co^{2+}	2 ng/kg
28	Nickel	Ni^{2+}	480 ng/kg
29	Copper	$CuCO_3^0$, $CuOH^+$	120 ng/kg
30	Zinc	$ZnOH^+$, Zn^{2+}, $ZnCO$	390 ng/kg
31	Gallium	$Ga(OH)_4^-$	10–20 ng/kg
32	Germanium	$Ge(OH)_4^0$	5 ng/kg
33	Arsenic	$HAsO_4^{2-}$, $H_2AsO_4^-$	2 µg/kg
		Dimethylarsenate	
34	Selenium	[Se(tot)]	170 ng/kg
		SeO_3^{2-}	
		[SeIV]	—
		[SeVI]	—
35	Bromine	Br^-	67 mg/kg
37	Rubidium	Rb^+	124 µg/kg
38	Strontium	Sr^{2+}	7.8 mg/kg (PO_4)
			7.7 mg/kg (Sr/Cl)
39	Yttrium	$Y(OH)_3^0$	13 ng/kg
40	Zirconium	$Zr(OH)_4^0$	<1 µg/kg
41	Niobium	—	1 ng/kg

continued opposite

Table 10-7   (*continued*)

Atomic number	Element	Species	Mean water concentration
42	Molybdenum	MoO_4^{2-}	$11\,\mu g/kg$
44	Ruthenium	—	$0.5\,ng/kg$
45	Rhodium	—	—
46	Palladium	—	—
47	Silver	$AgCl_2^-$	$3\,ng/kg$
48	Cadmium	$CdCl_2^0$	$70\,ng/kg$
49	Indium	$In(OH)_2^+$	$0.2\,ng/kg$
50	Tin	$SnO(OH)_3^-$	$0.5\,ng/kg$
51	Antimony	$Sb(OH)_6^-$	$0.2\,\mu g/kg$
52	Tellurium	$HTeO_3^-$	—
53	Iodine	IO_3^-, I^-	$59\,\mu g/kg$ (PO_4 corr.)
			$60\,\mu g/kg$ (NO_3 corr.)
54	Xenon	Xe(gas)	$0.5\,nmol/kg$
55	Cesium	Cs^+	$0.3\,ng/kg$
56	Barium	Ba^{2+}	$11.7\,\mu g/kg$
57	Lanthanum	$La(OH)_3^0$	$4\,ng/kg$
58	Cerium	$Ce(OH)_3$	$4\,ng/kg$
59	Praseodymium	$Pr(OH)_3$	$0.6\,ng/kg$
60	Neodymium	$Nd(OH)_3$	$4\,ng/kg$
61	Promethium	$Pm(OH)_3$	—
62	Samarium	$Sm(OH)_3$	$0.6\,ng/kg$
63	Europeum	$Eu(OH)_3$	$0.1\,ng/kg$
64	Gadolinium	$Gd(OH)_3$	$0.8\,ng/kg$
65	Terbium	$Tb(OH)_3$	$0.1\,ng/kg$
66	Dysprosium	$Dy(OH)_3$	$1\,ng/kg$
67	Holmium	$Ho(OH)_3$	$0.2\,ng/kg$
68	Erbium	$Er(OH)_3$	$0.9\,ng/kg$
69	Thulium	$Tm(OH)_3$	$0.2\,ng/kg$
70	Ytterbium	$Yb(OH)_3$	$0.9\,ng/kg$
71	Lutetium	$Lu(OH)_3^0$	$0.2\,ng/kg$
72	Hafnium	—	$<8\,ng/kg$
73	Tantalum	—	$<2.5\,ng/kg$
74	Tungsten	WO_4^{2-}	$<1\,ng/kg$
75	Rhenium	ReO_4^-	$4\,ng/kg$
76	Osmium	—	—
77	Iridium	—	—
78	Platinum	—	—
79	Gold	$AuCl_2^-$	$11\,ng/kg$
80	Mercury	$HgCl_4^{2-}$, $HgCl_3^0$	$6\,ng/kg$
81	Thallium	Tl^+	$12\,ng/kg$
82	Lead	$PbCO_3^0$	$1\,ng/kg$
83	Bismuth	BiO^+, $Bi(OH)_2^+$	$10\,ng/kg$
84	Polonium	PoO_3^3, $PoO(OH)_2^0$	—
86	Radon	Rn (gas)	—
88	Radium	Ra^{2+}	—
89	Actinium	—	—
90	Thorium	$Th(OH)_4^0$	$<0.7\,ng/kg$
91	Protactinium	—	—
92	Uranium	$UO_2(CO_3)_2^{4-}$	$3.2\,\mu g/kg$

reactivity. The approach to equilibrium for this problem is discussed in Chapter 4 and by Lasaga (1980, 1981).

Chemically reactive elements should have a short residence time in seawater and a low concentration. A positive correlation exists between the mean ocean residence time and the mean oceanic concentration; however, the scatter is too great for the plot to be used for predictive purposes. Whitfield and Turner (1979) and Whitfield (1979) have shown that a more important correlation exists between residence time and a measure of the partitioning of the elements between the ocean and crustal rocks. The rationale behind this approach is that the oceanic concentrations have been roughly constant, while the elements in crustal rocks have cycled through the oceans. This partitioning of the elements may reflect the long-term chemical controls. The relationship can be summarized by an equation of the form

$$\log \tau_0 = a \log K_D + b \qquad (9)$$

where K_D is the ratio of the mean concentration in seawater/mean concentration in the crust. Appropriate values for 40 elements have been tabulated by Whitfield (1979).

Li (1981) proposed that the distribution co-efficients reflect adsorption–desorption reactions at the surface of mineral grains. To emphasize this point, Li plotted a slightly different distribution coefficient, $\log C_{op}/C_{sw}$, where C_{op} and C_{sw} are the concentrations in oceanic pelagic clay sediments and seawater respectively) versus the first hydrolysis constants of the metals or the dissociation constant of the oxyanion acids. The argument is that those elements that hydrolyze the strongest will adsorb the strongest and thus have a larger preference for the solid phase as represented by pelagic clay. For oxyanions the larger the acidity constant the weaker the adsorption on solid phases.

Thus, the chemical reactivity of the elements in seawater is reflected by the residence time. It is important to note, however, that while residence times tell us something about the relative reactivities, they also tell us nothing about the nature of the reactions. The best source of clues for understanding these reactions is to study the shape of dissolved profiles of the different elements. When we do this we find that there are six main characteristic types of profiles as described in Table 10-8. Notice that most of these reactions occur at the phase discontinuities between the atmosphere, biosphere, hydrosphere, and lithosphere.

Table 10-8 Characteristic types of profiles of elements in the ocean with example elements and probable controlling mechanisms

Type of profile	Example elements	Mechanisms
1. Similar to salinity	Na, K, Mg, SO_4, F, Br	Conservative elements of very low reactivity
2. Sea-surface enrichment		Atmospheric input
	^{210}Pb, Mn	– natural
	^{90}Sr, Pb	– pollution, bomb tests
	NO	– photochemistry
	$As(CH_3)_2$, H_2, NO_2	Biological production
3. Photic-zone depletion with deep-ocean enrichment	Ca, Si, ΣCO_2, NO_3, PO_4, Cu, Ni	Biological uptake and regeneration
4. Mid-water maxima		
3000 m	^{3}He, Mn, CH_4, ^{222}Rn	Hydrothermal input
200 to 1000 m	^{3}H	Isopycnal transport
	Mn, NO_2	Redox chemistry in oxygen minimum
5. Bottom-water enrichment	^{222}Rn, ^{228}Ra, Mn, Si	Flux out of the sediments
6. Deep-ocean depletion	^{210}Pb, ^{230}Th, Cu	Scavenging by settling particles

10.4.2   Composition of seawater

10.4.2.1   Major ions

The salinity of seawater is defined as the grams of dissolved salt per kg of seawater. Using good technique, salinity can be reported to 0.001‰ or 1 ppm(m). By tradition the major ions have been defined as those that make a significant contribution to the salinity. Thus, major ions are those with concentrations greater than 1 mg/kg or 1 ppm(m). By this definition there are 11 major ions in seawater (Table 10-9).

The elements Na, K, Cl, SO, Br, B, and F are the most conservative major elements. No significant variations in the ratios of these elements to chlorine have been demonstrated. Strontium has a small (< 0.5%) depletion in the euphotic zone (Brass and Turekian, 1974) possibly due to the plankton *Acantharia*, which makes its shell from $SrSO_4$ (celestite). Calcium has been known since the 19th century to be about 0.5% enriched in the deep sea relative to surface waters. Alkalinity (HCO_3^-) also shows a deep enrichment. These elements are controlled by the formation and dissolution of $CaCO_3$ and are linked by the following reaction:

$$CaCO_3 + CO_2 + H_2O \rightarrow Ca^{2+} + 2\,HCO_3^- \quad (10)$$

Low-temperature circulation of seawater through mid-ocean ridge systems creates a deficiency in Mg and an excess in Ca at mid-depths (de Villiers and Nelson, 1999).

10.4.2.2   Minor elements

By definition a minor element in seawater is one has a concentration less than 1 ppm(m). It is experimentally challenging to determine the total concentrations, much less their major chemical forms. Development of new analytical techniques has greatly extended our knowledge (Johnson *et al.*, 1992). Because early data (prior to about 1975) was so erratic, the principle of oceanographic consistency was proposed as a test for the data (Boyle and Edmond, 1975). According to this principle the analyses of minor elements should:

1. Form smooth vertical profiles.
2. Have correlations with other elements that share the same controlling mechanism.

The concentrations of the minor elements are given in Table 10-7. There is at least one oceanographically consistent profile for most of the elements. Refer to Quinby-Hunt and Turekian (1983) for references and representative profiles for individual elements. The main mechanisms that control the distribution of minor elements are given in Table 10-8.

The chemical reactivity of minor elements in seawater is strongly influenced by their speciation (see Stumm and Brauner, 1975). For example, the Cu^{2+} ion is toxic to phytoplankton (Sunda and Guillard, 1976). Uranium (VI) forms the soluble carbonate complex, $UO_2(CO_3)_3^{4-}$, and as a result uranium behaves like an unreactive conservative element in seawater (Ku *et al.*, 1977).

Although the speciation of some minor elements has been determined directly by experimental means (e.g., ion selective electrodes, polarography, electron spin resonance) most of our thinking about speciation is based on equilibrium calculations. Garrels and Thompson

Table 10-9 The major ions of seawater: concentration at 35‰, ratio to chlorinity, and molar concentration[a]

	The major ions of seawater		
Ion	g/kg at $S = 35$‰	g/kg (chlorinity ‰)	mol/L
Cl^-	19.354	0.9989	5.46×10^{-1}
SO_4^{2-}	2.712	0.1400	2.82×10^{-2}
Br^-	0.0673	0.00347	8.4×10^{-4}
F^-	0.0013	0.000067	6.8×10^{-5}
B	0.0045	0.000232	4.1×10^{-4}
Na^+	10.77	0.5560	4.68×10^{-1}
Mg^{2+}	1.290	0.0665[b]	5.32×10^{-3}
Ca^{2+}	0.4121	0.02127	1.02×10^{-2}
K^+	0.399	0.0206	1.02×10^{-2}
Sr^{2+}	0.0079	0.00041	9.1×10^{-5}
HCO_3^-	0.142	—	2.387×10^{-3}

[a] From Wilson (1975).
[b] Recent reported values lie between 0.06612 and 0.06692.

(1962) conducted speciation calculations for the major elements of seawater. They showed that the major cations (Na, K, Ca, Mg) and Cl are mostly (> 90%) uncomplexed in seawater. The anions SO_4^{2-}, CO_3^{2-} and HCO_3^- are significantly complexed. When similar calculations are done for the minor elements in seawater we find a different story. Most of the minor elements exist as complex ions or ion pairs. In particular, the metals form complexes with anions (ligands) such as CO_3^{2-}, Cl^-, and especially OH^-. The best estimates of the speciation of the elements in seawater are given in Table 10-7.

Stumm and Brauner (1975) conducted a Garrels and Thompson type calculation for some major and minor elements in seawater. These results are shown in Table 10-10. There are actually two calculations shown. Table 10a is for an inorganic seawater model. In Table 10b the calculations are repeated using organic compounds with functional groups similar to those found in nature. The metals tested are listed in the left column followed by a column of their concentration in seawater (as best known at that time). The inorganic and organic ligands tested are listed across the top row. Of the metals studied only the major ions (Ca, Mg, Na, K) and Ni and Co occur mostly as the free metal ion. Complex species predominate for all other metals. When organic ligands are added to the model, only the speciation of Cu is seriously affected. Direct measurements by differential pulse anodic stripping voltammetry have shown that more than 99.7% of the total dissolved copper in surface seawater is associated with organic complexes (Coale and Bruland, 1988). More than 99.97% of dissolved Fe(III) is chelated by organic ligands (Rue and Bruland, 1995).

10.4.2.3   Dissolved gases

All deep waters of the ocean were once in contact with the atmosphere. Since over 95% of the total of all gases (except radon) reside in the atmosphere, the atmosphere dictates the ocean's gas contents. CO_2 is also a special case because the ocean has high total CO_2. As discussed in Chapter 7, the composition of the atmosphere is nearly constant horizontally.

The solubility of many gases depends mainly on molecular weight. The heavier the molecule the greater the solubility. Thus He is less soluble than Xe. The gases CO_2 and N_2O are exceptions to this general trend because they interact more strongly with water. Solubility also increases with decreasing temperature. Thus, high-latitude surface seawater has higher gas concentrations than seawater at mid- or low latitudes.

The equilibrium concentration in seawater is described by Henry's Law, which relates the partial pressure of the gas to its concentration (see Chapter 5 and Waser, 1966). Using the appropriate values of Henry's Law constant, K_H, and the partial pressures of gases in the atmosphere, the equilibrium concentrations of several gases are given in Table 10-11 for 0°C and 24°C.

In the ocean, inert gas concentrations tend to follow the temperature solubility dependence closely. This suggests that water parcels obtain their gas signatures when they are at the seasurface close to equilibrium with the atmosphere at ambient temperature.

The process of equilibration of the atmosphere with the ocean is called gas exchange. Several models are available, however, the simplest model for most practical problems is the one-layer stagnant boundary-layer model (Fig. 10-18). This model assumes that a well-mixed atmosphere and a well-mixed surface ocean are

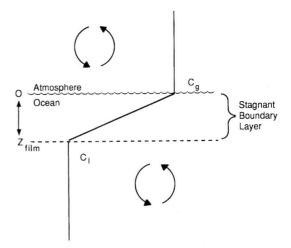

Fig. 10-18   A schematic of a stagnant boundary-layer gas exchange model. C_g = gas concentration at the liquid side of the interface; C_l = gas concentration at the base of the stagnant boundary layer; Z_{film} = stagnant boundary layer thickness.

Table 10-10 Equilibrium model: effect of complex formation on distribution of metals (all concentrations are given as $-\log(M)$). pH = 8.0, T = 25°C. Ligands: pSO$_4$ 1.95; pHCO$_3$ 2.76; pCO$_3$ 4.86; pCl 0.25.

A. Inorganic seawater

M	M$_T^a$	Free M	Major species			
Ca	1.97	2.03	CaSO$_4$	2.94	CaCO$_3$	3.50
Mg	1.26	1.31	MgSO$_4$	2.25	MgCO$_3$	3.3
Na	0.32	0.33	NaSO$_4$	1.97	NaHCO$_3$	3.3
K	1.97	1.98	KSO$_4$	3.93	—	—
Fe(III)	8.0	18.9	Fe(OH)$_2$	8.3	FeSO$_4$	18.5
Mn(II)	7.5	8.1	MnCl	7.8^b	MnCl	8.3^b
Cu(II)	7.7	9.2	CuCO$_3$	7.7	Cu(CO$_3$)$_2$	9.1
Cd	8.5	10.9	CdCl$_2$	8.7	CdCl	9.2
Ni	7.7	7.9	NiCl	8.3	NiSO$_4$	8.7
Pb	8.2	9.9	PbCO$_3$	8.6	PbOH	8.7
Co(II)	8.3	8.5	CoCl	9.0	CoSO$_4$	9.1
Ag	8.7	13.1	AgCl$_2$	8.7	AgCl	10.0
Zn	7.2	7.8	ZnOH	7.4	ZnCl	8.0

B. Inorganic seawater plus soluble organic matter (2.3 mg C/L)^c

Total concentration same as above						Organic complexes with ligands^d and free ligand concentration					
M	Free M	Major inorganic species				Acet. 5.21	Citr. 4.7	Tartr. 5.41	Glyc. 696	Glut. 6.89	Phthal. 5.2
Ca	2.03	CaSO$_4$	2.94	CaCO$_3$	3.50	7.41	5.90	6.41	9.06	8.19	6.28
Mg	1.31	MgSO$_4$	2.25	MgCO$_3$	3.3	6.06	5.25	5.56	7.31	6.34	—^{e}
Na	0.33	NaSO$_4$	1.97	NaHCO$_3$	3.3	—	—	—	—	—	—
K	1.98	KSO$_4$	3.93	—	—	—	—	—	—	—	—
Fe(III)	18.9	Fe(OH)$_2$	8.3	FeSO$_4$	18.5	20.7	8.6	—	15.9	13.7	—
Mn(II)	8.1	MnCl	7.8	MnCl$_2$	8.3	12.8	11.4	—	13.1	12.2	—
Cu(II)	10.8	CuCO$_3$	9.4	Cu(CO$_3$)$_2$	10.5	14.3	7.7	16.7	9.6	10.6	13.0
Cd	10.9	CdCl$_2$	8.7	CdCl	8.7	15.1	13.1	13.5	13.5	13.4	13.6
Ni	8.0	NiCl	8.5	NiSO$_4$	8.8	12.5	8.4	—	9.2	9.4	11.1
Pb	9.9	PbCO$_3$	8.6	PbOH	8.7	13.2	11.34	11.5	11.8	—	11.7
Co(II)	8.5	CoCl	9.0	CoSO$_4$	9.1	12.7	26.5	11.9	10.8	10.8	14.9
Ag	13.1	AgCl$_2$	8.7	AgCl	10.0	17.9	26.5	—	16.7	—	—
Zn	7.8	ZnOH	7.4	ZnCl	8.0	11.7	11.3	10.9	8.8	9.7	10.9
					%^{f}	13.0	98.6	44.9	0.7	6.6	7.5

^{a} Total concentration of metal species: note that Fe(III) is slightly oversaturated with respect to Fe(OH)$_3$(s); Cu(II) is oversaturated with respect to malachite but because formation is slow, precipitation of the solid has not been allowed. All other metals are thermodynamically soluble at the concentrations specified.

^{b} There is some uncertainty regarding the validity of the stability constants of chloro complexes of Mn^{2+}; according to other computations Mn^{2+} is a major inorganic species.

^{c} Organic matter of approximate composition C$_{13}$H$_{17}$O$_{12}$N consists of a mixture of acetate, citrate, tartrate, glycine, glutamic acid, and phthalate, each present at 7×10^{-6} M (11 mmol donor groups per liter).

^{d} The concentrations given refer to the sum of all complexes, e.g., CuCit, CuHCit, CuCit$_2$.

^{e} No stability constants for such complexes are available.

^{f} Percentage of total ligand bound to metal ions.

Table 10-11 Solubilities of various gases in surface ocean water

Gas	Partial pressure in dry air (atm)	Equilibrium concentration in surface seawater (cm^3/L)	
		0°C	24°C
H_2	5×10^{-7}	—	—
He	5.2×10^{-6}	4.1×10^{-5}	3.4×10^{-5}
Ne	1.8×10^{-5}	1.7×10^{-4}	1.5×10^{-4}
N_2	0.781	14	9
O	0.209	8.8	5.5
Ar	9.3×10^{-3}	0.36	0.22
CO_2	3.2×10^{-4}	0.47	0.23
Kr	1.1×10^{-6}	8.1×10^{-5}	4.9×10^{-5}
Xe	8.6×10^{-8}	1.2×10^{-5}	0.6×10^{-5}

separated by a film on the liquid side of the air–water interface through which gas transport is controlled by molecular diffusion. (A similar layer exists on the air side of the interface that can be neglected for most gases. SO_2 is a notable exception (Liss and Slater, 1974)).

If transport across this film is controlled by diffusion then from Fick's Second Law of Diffusion,

$$\frac{\partial C}{\partial t} = D\left(\frac{\partial^2 C}{\partial x^2}\right) \tag{11}$$

The boundary conditions are $C = C_g$ at $x = 0$ and $C = C_1$ at $x = z$. The steady-state concentration profile across the boundary layer is given by

$$C = \frac{C_1 - C_g}{z} x + C_g \tag{12}$$

The steady-state flux from the atmosphere to the ocean across the layer is given by Fick's First Law:

$$F = -D\left(\frac{\partial C}{\partial x}\right) = D\frac{C_g - C_1}{z} \tag{13}$$

This treatment may be compared with that given in Chapter 4. The top of the stagnant film is assumed to have a gas concentration in equilibrium with the overlying air (i.e., $C_g = K_H P_g$). The unknown values are the flux and the thickness of the diffusive layer z. The thickness z has been determined by analyses of isotopes (^{14}C and ^{222}Rn) that can be used to obtain the flux (Broecker and Peng, 1974; Peng *et al.*, 1979). The

thickness averaged over the entire ocean has been estimated from a ^{14}C balance to be $17\,\mu m$.

Since the units of D/z are the same as velocity we can think of this ratio as the velocity of two imaginary pistons: one moving up through the water pushing ahead of it a column of gas with the concentration of the gas in surface water (C_1) and one moving down into the sea carrying a column of gas with the concentration of the gas in the upper few molecular layers (C_g). For a hypothetical example with a film thickness of $17\,\mu m$ and a diffusion coefficient of $1 \times 10^{-5}\,cm^2/s$ the piston velocity is $5\,m/day$. Thus in each day a column of seawater $5\,m$ thick will exchange its gas with the atmosphere.

The piston velocity, or gas transfer velocity, is a function of wind speed. There are large differences in the relationships between piston velocity and wind speed, especially at higher wind speeds (e.g., Liss and Merlivat, 1986; Wanninkhof, 1992). This is the limiting factor for these calculations.

Example: Obtain a relationship for the residence time of gases in the atmosphere with respect to gas exchange:

$$\tau_{atm} = \frac{\text{Mass in atmosphere}}{\text{Flux into the ocean}} = \frac{M_{atm}}{F}$$

$$= \frac{P \cdot 3 \times 10^5 (atm)(mol/(m^2\ atm))}{P K_H (D/z)(atm)(mol/(m^3\ atm))(m/yr)}$$

$$= \frac{3 \times 10^5}{K_H (D/z)}$$

If we assume that $Z_{film} = 17\,\mu m$ and $D = 1 \times$

$10^{-5} \, cm^2/s$, then the piston velocity is about 1800 m/yr. The solubility of N_2, O_2, CO, H_2, NO, Ar, CH_4 etc. is about 1 mol/(m^3 atm), while it is about 30 mol/m^3 atm) for CO_2. Thus the residence time for most inert gases in the atmosphere is about 160 years. For CO_2 it is about 5.4 years or a factor of 30 faster than for the other gases.

10.4.2.4 Nutrients

Oceanic surface waters are efficiently stripped of nutrients by phytoplankton. If phytoplankton biomass was not reconverted into simple dissolved nutrients, the entire marine water column would be depleted in nutrients and growth would stop. But as we saw from the carbon balance presented earlier, more than 90% of the primary productivity is released back to the water column as a reverse RKR equation. This reverse reaction is called remineralization and is due to respiration. An important point is that while production via photosynthesis can only occur in surface waters, the remineralization by heterotrophic organisms can occur over the entire water column and in the underlying sediments.

It follows that deep seawater contains nutrients from two sources. First, it may contain nutrients that were present with the water when it sank from the surface. These are called "preformed nutrients." Second, it may contain nutrients derived by the *in situ* remineralization of organic particles. These are called oxidative nutrients.

The oxidative nutrients can be estimated from the RKR equation. From this model we might expect the four dissolved chemical species (O_2, CO_2, NO_3, PO_4) to vary in seawater according to the proportions predicted.

The key to understanding these remineralization reactions is the parameter Apparent Oxygen Utilization (AOU), defined as

$$AOU = O_2' - O_2 \qquad (14)$$

Where O_2' is the saturation value at the salinity and potential temperature of the sample. O_2 is the measured O_2 at the time of sampling. From the respiration form of the RKR equation it follows that for every 138 mol O_2 consumed *in situ* one gets 106 mol CO_2, 16 mol NH_3 and 1 mol H_3PO_4 (or using 170:117:16:1 using the more recent analysis by Anderson and Sarmiento, 1994). As can be seen, the slopes of the regression lines of AOU versus phosphate and nitrate closely correspond to the complete remineralization of average plankton.

The preformed nutrients are obtained by subtracting the oxidized nutrient from total nutrient (for N and P).

$$P_{total} = P_{preformed} + P_{oxidized} \quad P_{oxidized} = \tfrac{1}{138} AOU$$

$$N_{total} = N_{preformed} + N_{oxidized} \quad N_{oxidized} = \tfrac{16}{138} AOU$$

For silica an additional component of silicic acid generated by inorganic dissolution also occurs. One can estimate "oxidized" Si as being 23/138 AOU, but discrimination between preformed Si and inorganic Si is not possible. Representative profiles of PO_4, NO_3, and Si are given in Fig. 10-19. Note that after subtracting the oxidative nutrients the preformed values are relatively constant.

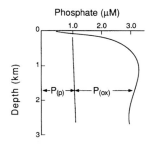
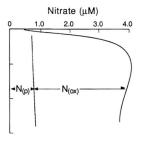
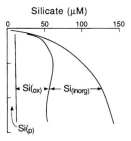

Fig. 10-19 Vertical distributions of various nutrients and their components.

10.4.2.5   Carbonate system

The carbonate system plays a pivotal role in most global cycles. For example, gas exchange of CO_2 is the exchange mechanism between the ocean and atmosphere. In the deep sea, the concentration of CO_3^{2-} ion determines the depth at which $CaCO_3$ is preserved in marine sediments.

Dissolved inorganic carbon is present as three main species which are H_2CO_3, HCO_3^- and CO_3^{2-}. Analytically we have to approach the carbonate system through measurements of pH, total CO_2 or DIC, alkalinity (Alk), and P_{CO_2}. In an open carbonate system there are six unknown species: H^+, OH^-, P_{CO_2}, H_2CO_3, HCO_3^-, and CO_3^{2-}. The four equilibrium constants connecting these species are K_1, K_2, K_H, and K_w. The values of these equilibrium constants vary with T, P, and S (Millero, 1995). To solve for the six unknowns we need to measure two of the four analytical parameters (Stumm and Morgan, 1996). Direct measurement of P_{CO_2} is the best approach, but if that is not possible then the most accurate and precise pair (Dickson, 1993) is Total CO_2 by the coulometric method (Johnson *et al.*, 1993) and pH by the colorimetric method (Clayton *et al.*, 1995).

The distributions of total CO_2 or DIC and alkalinity in the Atlantic, Indian and Pacific Oceans are compared with PO_4 and O_2 in Fig. 10-20 (Baes *et al.*, 1985). The main features are:

1. Uniform surface values.
2. General increase with depth.
3. Deep ocean values increase from the Atlantic to the Pacific.
4. DIC < Alk and Δ(DIC) > Δ(Alk).
5. pH profiles (not shown) are similar in shape to O_2.
6. P_{CO_2} profiles (not shown) mirror O_2.

Respiration of organic matter and dissolution of $CaCO_3$ are the main controls of the distribution of deep ocean Total CO_2 and alkalinity. These reactions (and their predicted effects on DIC and alkalinity) can be represented schematically as

$CH_2O + O_2 = CO_2 + H_2O$ Δ(DIC) = 1 Δ(Alk) = 0

$CaCO_3 = Ca^{2+} + CO_3^{2-}$ Δ(DIC) = 1 Δ(Alk) = 2

To a first approximation the deep ocean distributions shown in Fig. 10-20 can be reproduced if the particulate material dissolving in the deep sea has the ratio of 1 mol $CaCO_3$ to 4 mol organic carbon (Broecker and Peng, 1982).

10.4.3   Equilibrium Models of Seawater

Now that we have reviewed some basic aspects of the chemical composition of the ocean we can turn to a more fundamental question. What processes determine the composition of the ocean? Current evidence suggests that rivers are the most important contributors of dissolved substances to the ocean. Since there is geologic evidence that the concentration and composition of the ocean has been relatively constant over the last ~1.5 billion years, we must conclude that the river input must be balanced by removal.

10.4.3.1   Sillén's model

Sillén was a Swedish inorganic chemist who specialized in solution chemistry. In 1959, Sillén was asked as "an outsider" to give a lecture on the physical chemistry of seawater to the International Oceanographic Congress in New York (Sillén, 1961). Sillén proposed that the ionic composition of seawater might be controlled by equilibrium reactions between the dissolved ions and various minerals occurring in marine sediments. Goldschmidt (1937) had earlier proposed a schematic reaction for the geochemical balance:

igneous rock (0.6 kg) + volatiles (1.0 kg) →

seawater (1 L) + sediments (0.6 kg) + air (3 L)

This is a weathering reaction. Sillén argued that Goldschmidt's reaction could also go the other direction. The reverse reaction would be called reverse weathering.

The framework for constructing such multi-component equilibrium models is the Gibbs phase rule. This rule is valid for a system that has reached equilibrium and it states that

$$f = c + 2 - p$$

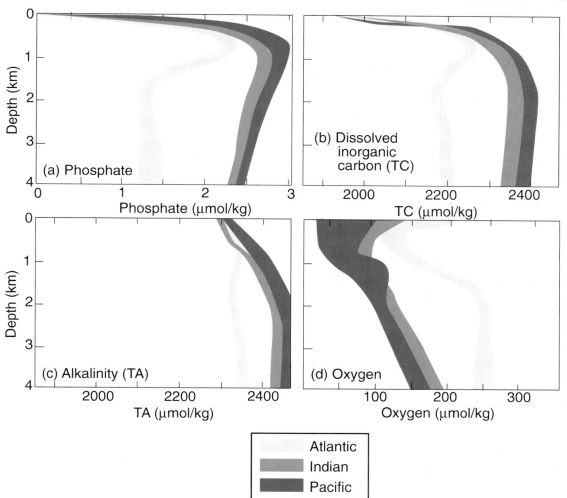

Fig. 10-20 Observed depth profiles of (a) phosphate, (b) dissolved inorganic carbon (TC), (c) alkalinity (TA), and (d) oxygen for the Atlantic, the Indian, and the Pacific Oceans as indicated. Data are from GEOSECS stations within 5° of the Equator in each ocean. (Modified from Baes *et al.* (1985).)

The number of degrees of freedom is represented by f. These are chosen from the list of all quantitatively related aspects of a system that can change. This includes T, P, and the concentrations of c components in each phase. c is the minimum number of components necessary to reproduce the system (ingredients), and p is the number of phases present at equilibrium. A phase is a domain with uniform composition and properties. Examples are a gas, a liquid solution, a solid solution, and solid phases.

In a mathematical sense f represents the difference between the number of independent variables (including T, P) and the number of constraints (equations). If the number of equations equals the number of unknown variables, we can solve for all the concentrations using equilibrium equations. (For more discussion of the phase rule see Stumm and Morgan, 1996). Sillén's approach was to mix components, pick a reasonable set of phases that might be present, and then see how many degrees of freedom there are to be fixed.

Sillén constructed his models in a stepwise fashion starting with a simplified ocean model of five components (HCl, H_2O, KOH, $Al(OH)_3$, and SiO_2) and five phases (gas, liquid, quartz, kaolinite, and potassium mica) (Sillén, 1967). His complete (almost) seawater model was composed of nine components. HCl, H_2O, and CO_2 are acids that correspond to the volatiles from the Earth. KOH, CaO, SiO_2, $NaOH$, MgO, and $Al(OH)_3$ correspond to the bases of the rocks. If there was an equilibrium assemblage of nine phases the system would have only two independent variables. Sillén argued that a plausible set could include a gas phase and a solution phase and the following seven solid phases:

- Calcite $CaCO_3$
- Quartz SiO_2
- Kaolinite $Al_2Si_2O_5(OH)_4$
- Illite $K_{0.59}(Al_{1.38}Fe_{0.73}Mg_{0.38})$
                   $(Si_{3.41}Al_{0.59})O_{10}(OH)_2$
- Chlorite $Mg_3(OH)_6Mg_3Si_4O_{10}$
                   $(OH)_{10}$
- Montmorillonite $Na_{0.33}Al_2(Si_{3.67}Al_{0.33})$
                   $O_{10}(OH)_2$
- Phillipsite (zeolite) $M_3Al_3Si_4O_{16}(H_2O)_6$,
                   where $M = Na + K + Ca + Mg$

If these phases all exist at equilibrium then $f = 2$. Sillén argued that we should fix T and $[Cl^-]$ (Cl^- does not enter any of the reactions and is thus conservative). If so, the composition of the aqueous and gas phases would be fixed. The implications are far reaching because these equilibria would fix P_{CO_2} of the atmosphere, the alkalinity of the ocean and thus the pH of the ocean!

10.4.3.2   MacKenzie and Garrels' chemical mass balance between rivers and oceans

Evaporation of river water will not make seawater. Instead, evaporation of the nearly neutral Na^+–Ca^{2+}–HCO_3^- river water produces a highly alkaline Na–HCO^-–CO_3^{2-} water such as found in the evaporitic lake beds of eastern California (Garrels and MacKenzie, 1967). In addition, on comparing the amount of material supplied to the ocean with the amount in the ocean, it may be seen that most of the elements could have been replaced many times (Table 10-12). Thus some chemical reactions must be occurring in the ocean to consume the river flux.

MacKenzie and Garrels (1966) approached this problem by constructing a model based on a river balance. They first calculated the mass of ions added to the ocean by rivers over 10^8 years. This time period was chosen because geologic evidence suggests that the chemical composition of seawater has remained constant over that period. They assumed that the river input is balanced only by sediment removal. The results of this balance are shown in Table 10-13.

In this balance SO_4 is removed by $CaSO_4$ and FeS_2 in proportion to their abundance in the sedimentary record (50/50). Ca is removed as $CaCO_3$ with enough Mg to correspond to the natural proportions. Chloride is removed as NaCl, and enough H_4SiO_4 is removed to make opal sediments. Some Na^+ is taken up and Ca^{2+}

Table 10-12   Number of times river constituents have passed through the ocean in 10 years assuming present annual worldwide river discharge, mean dissolved constituent concentration of rivers and ocean, and ocean volume of 1.37×10^{21} L (amounts in 10^{18} kg)

Constituent	Amount in ocean	Amount delivered by rivers to ocean in 10^8 years	Number of times constituents been "renewed" in 10^8 years
SiO_2	0.008	42.6	5300
HCO_3^-	0.19	190.2	1000
Ca^{++}	0.6	48.8	81
K^+	0.5	7.4	15
SO_4^{2-}	3.7	36.7	10
Mg^{++}	1.9	13.3	7
Na^+	14.4	20.7	1.4
Cl^-	26.1	25.4	1
H_2O	1370	3 333 000	2400

Table 10-13 Mass balance calculation for removal of river-derived constituents from the ocean (all units in 10^{21} mmol)

Reaction (balanced in terms of mmol of constituent)	Constituents									% of total products formed
	SO_4^{2-}	Ca^{2+}	Cl^-	Na^+	Mg^{2+}	K^+	SiO_2	HCO_3^-		
To be removed from ocean in 10^8 years	382	1220	715	900	554	189	710	3118		
Reaction 1	191	1220	715	900	554	189	710	3500	96 pyrite	3%
									287 kaolinite	8%
Reaction 2	0	1029	715	900	554	189	710	3500	191 "CaSO$_4$"	5%
Reaction 3	0	1029	715	900	502	189	710	3396	52 MgCO$_3$ in magnesium calcite	2%
Reaction 4	0	0	715	900	502	189	710	1338	1029 calcite or aragonite	29%
Reaction 5	0	0	0	185	502	189	710	1338	715 "NaCl"	20%
Reaction 6	0	0	0	185	502	189	639	1338	71 silica	2%
Reaction 7	0	24	0	139	502	189	639	1338	138 sodic montmorillonite	4%
Reaction 8	0	0	0	139	502	189	639	1290	24 calcite or aragonite	1%
Reaction 9	0	0	0	0	502	189	278	1151	417 sodic montmorillonite	12%
Reaction 10	0	0	0	0	0	189	218	147	100 chlorite	3%
Reaction 11	0	0	0	0	0	0	29	−42	378 illite	11%

1: $95.5FeAl_6Si_6O_{20}(OH)_4 + 191SO_4^{2-} + 47.8CO_2 + 55.7C_6H_{12}O_6 + 238.8H_2O \rightarrow 286.5Al_2Si_2O_5(OH)_4 + 95.5FeS_2 + 382HCO_3^-$.

2: $191Ca^{2+} + 191SO_4^{2-} \rightarrow 191CaSO_4$.

3: $52Mg^{2+} + 104HCO_3^- \rightarrow 52MgCO_3 + 52CO_2 + 52H_2O$.

4: $1029Ca^{2+} + 2058HCO_3^- \rightarrow 1029CaCO_3 + 1029CO_2 + 1029H_2O$.

5: $715Na^+ + 715Cl^- \rightarrow 715NaCl$.

6: $71H_4SiO_4 \rightarrow 71SiO_{2(s)} + 142H_2O$.

7: $138Ca_{0.17}Al_{2.33}Si_{3.67}O_{10}(OH)_2 + 46Na^+ \rightarrow 138Na_{0.33}Al_{2.33}Si_{3.67}O_{10}(OH)_2 + 23.5Ca^{2+}$.

8: $24Ca^{2+} + 48HCO_3^- \rightarrow 24CaCO_3 + 24CO_2 + 24H_2O$.

9: $486.5Al_2Si_{2.4}O_{5.8}(OH)_4 + 139Na^+ + 361.4SiO_2 + 139HCO_3^- \rightarrow 417Na_{0.33}Al_{2.33}Si_{3.67}O_{10}(OH)_2 + 139CO_2 + 625.5H_2O$.

10: $100.4Al_2Si_{2.4}O_{5.8}(OH)_4 + 502Mg^{2+} + 60.2SiO_2 + 1004HCO_3^- \rightarrow 100.4Mg_5Al_2Si_3O_{10}(OH)_8 + 1004CO_2 + 301.2H_2O$.

11: $472.5Al_2Si_{2.4}O_{5.8}(OH)_4 + 189K^+ + 189SiO_2 + 189HCO_3^- \rightarrow 378K_{0.5}Al_{2.5}Si_{3.5}O_{10}(OH)_2 + 189CO_2 + 661.5H_2O$.

released during ion exchange reactions in estuaries. At this point they still had to account for 15% of the initial Na, 90% of the Mg, 100% of the K, 90% of the SiO_2 and 43% of the HCO_3^-. To remove these excess ions MacKenzie and Garrels proposed reverse weathering type reactions of the general type:

X-ray amorphous clays + H_2SiO_4 + cations + HCO_3^-

$\rightarrow$ cation-rich aluminosilicates + CO_2 + H_2O

In their model they used a kaolinite-like clay for the degraded silicate and allowed Na, Mg, and K to react to form sodic montmorillonite, chlorite, and illite respectively. The balance is essentially complete with only small residuals for H_4SiO_4 and HCO_3^-. The newly formed clays would constitute about 7% of the total mass of sediments.

This last requirement has been the greatest stumbling block for accepting Sillén's and

MacKenzie and Garrels' equilibrium models. Most marine clays appear to be detrital and derived from the continents by river or atmospheric transport. Authigenic phases (formed in place) are found in marine sediments (e.g. Michalopoulos and Aller, 1995), however, they are nowhere near abundant enough to satisfy the requirements of the river balance. For example, Kastner (1974) calculated that less than 1% of the Na and 2% of the K transported by rivers is taken up by authigenic feldspars.

So while the equilibrium approach is attractive and certainly tells us the directions reactions tend to go, the experimental and empirical verification is lacking.

10.4.4   Kinetic Models of Seawater

The failure to identify the necessary authigenic silicate phases in sufficient quantities in marine sediments has led oceanographers to consider different approaches. The current models for seawater composition emphasize the dominant role played by the balance between the various inputs and outputs from the ocean. Mass balance calculations have become more important than solubility relationships in explaining oceanic chemistry. The difference between the equilibrium and mass balance points of view is not just a matter of mathematical and chemical formalism. In the equilibrium case, one would expect a very constant composition of the ocean and its sediments over geological time. In the other case, historical variations in the rates of input and removal should be reflected by changes in ocean composition and may be preserved in the sedimentary record. Models that emphasize the role of kinetic and material balance considerations are called kinetic models of seawater. This reasoning was pulled together by Broecker (1971) in a paper called "A kinetic model for the chemical composition of sea water."

10.4.4.1 ·Fast processes – internal cycling

The most obvious effects of nonequilibrium in the ocean are large variations in present-day composition. These arise mainly as the result of two processes.

1. *Biological cycling*. Organisms in surface seawater take up dissolved species during their growth. The remains of these organisms sink under the influence of gravity and gradually decompose by oxidation during respiration (release of C, N, P) and corrosion of hard parts (release of Ca, C, Si, trace elements like Ba, Cd, Zn, Cr, Ni, Se). This leads to a vertical segregation in the ocean of low concentrations in the surface and higher concentrations at depth.
2. *Oceanic circulation*. The process of ocean circulation described earlier yields an ocean circulation pattern that results in progressively older deep water as the water passes, in sequence from the Atlantic, Indian, to the Pacific Ocean. Surface water returns relatively quickly to the place of origin for the deep water.

The superposition of the vertical biological cycle on the horizontal ocean circulation pattern leads to three general features in the distribution of the elements involved in this cycle.

1. The warm surface ocean tends to have a constant composition.
2. The surface ocean is depleted relative to the deep ocean in those elements fixed by organisms.
3. Deep ocean concentrations increase progressively as the abyssal water flows (ages) from the North Atlantic, through the Indian Ocean to the North Pacific.

The characteristics are demonstrated in Fig. 10-21. As a result the elements influenced in this manner show various degrees of correlation. For element X, in terms of phosphorus:

$$[X] = a[P] + b$$

The moles X/moles P in average plankton is given by a, and b is the surface water concentration in phosphorus free water (water stripped of nutrients). In the case of P itself the surface ocean concentration is close to zero, while the deep Pacific has a concentration of 2.5 μM. For N, the N/P ratio of plankton is 16 and the surface water concentration is 0 μM. The predicted deep sea nitrate is 40 μM. The ratio of (deep)/(surface) is greater than 10. For calcium the Ca/P of

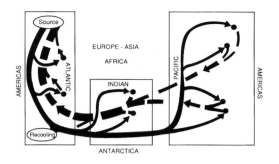

(a)

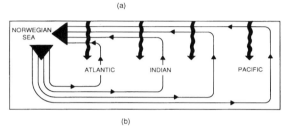

(b)

Fig. 10-21 (a) is an idealized map of the patterns of deep water flow (solid lines) and surface water flow (dashed lines). The large circles designate the sinking of NADW (North Atlantic Deep Water) in the Norwegian Sea and the recooling of water along the perimeter of the Antarctic Continent; the dark circles indicate the distributed upwelling which balances this deep water generation. (b) is an idealized vertical section running from the North Atlantic to the North Pacific showing the major advective flow pattern (thin lines) and the rain of particles (thick wavy lines). The combination of these two cycles leads to the observed distribution of nutrients. (Reproduced with permission from W. S. Broecker and T.-H. Peng (1982). "Tracers in the Sea," p. 34, Eldigio Press, Palisades, NY.)

plankton = 36 and the surface water content is $10\,000\,\mu M$. The predicted deep ocean concentration is 10 090 resulting in a deep ocean enrichment ratio of 1.01. These features are summarized for several elements in Table 10-14. Three elements (P, N, Si) show nearly complete depletion in surface water, reflecting the fact that Si limits diatom growth, while P and N limit the remaining organisms. Si shows the largest enrichment in the deep Pacific relative to the deep Atlantic suggesting that biological filtering is more efficient for Si than P. Hard parts of organisms undergo destruction at greater average depths than do the soft parts.

Three other elements (C, Ba, Ca) show a similar deep Pacific to surface distribution but of smaller magnitude (less than 10). Note that the present cycle requires less than 100% efficient surface removal, otherwise nutrients would be all in the Pacific after a one-way trip.

There are two important consequences of this superposition of biological cycling on the ocean circulation pattern that show up in the sediments.

1. It leads to lower diatom productivity in the Atlantic relative to the Pacific.
2. It leads to a tilting of the depth of $CaCO_3$ preservation in the sediments. The deep Pacific is more corrosive to $CaCO_3$ than the deep Atlantic (more CO_2 from respiration) and thus $CaCO_3$ is found in sediments 1500 m deeper in the Atlantic than in the Pacific.

10.4.4.2 Long-term processes – control of composition

In the previous section we considered only internal cycling. The questions we want to turn to now are:

1. What controls surface water concentrations?
2. What controls the P content of deep water and thus the deep-water content of other elements?

Broecker's (1971) approach was to form groups of elements that appear to be controlled by similar processes. We will follow that approach here while examining the important factors and time scales. The groups presented will differ from Broecker's in that we will include new information on hydrothermal processes not available at the time Broecker wrote his paper (Edmond *et al.*, 1979; McDuff and Morel, 1980). The groups used are kept as close as possible to Broecker's original list.

10.4.4.2.1 Group Ia (e.g., Cl). Elements in this group have long oceanic residence times. These elements are soluble and not reactive. The original source was degassing of the Earth's interior, which is either very slow now or complete. The main property of this group is geologic removal by formation of soluble salts in evaporite deposits.

Table 10.14 Concentration distributions in the sea for elements used in significant amounts by marine organisms[a]

Element	[S]	[DA]	[DP]	[DP]/[S]	([DP]-[S])/([DA]-[S])	Ref.
P[b]	<0.02	0.17	0.25	>10	1.7	d
N[b]	<0.2	2.1	3.3	>10	1.7	d
C[b]	205	227	248	1.25	1.9	d
Ca[b]	1000	1004	1009	1.01	~2	d
Si[c]	<100	1000	5000	>50	5	e
Ba[c]	9	12	27	3	6	e

[a] [S], [DA], [DP], represent respectively the concentrations of the given element in warm surface water, deep Atlantic water, and deep Pacific water.
[b] Amounts in 10^{-5} mol/L.
[c] Values are in g/L.
[d] Li *et al.* (1969).
[e] Wolgemuth and Broecker (1970).

The present sources to the ocean are the weathering of old evaporites (75% of river flux) and Cl^- carried by atmospherically cycled sea-salts (25% of river flux). Loss from the ocean occurs via aerosols (about 25%) and formation of new evaporites. This last process is sporadic and tectonically controlled by the closing of marginal seas where evaporation is greater than precipitation. The oceanic residence time is so long for Cl^- (~100 Myr) that an imbalance between input and removal rates will have little influence on oceanic concentrations over periods of less than tens of millions of years.

10.4.4.2.2 Group Ib (Mg, SO$_4$, probably K). The key property of this group is removal during seafloor hydrothermal circulation. This fits in with Broecker's original group I, tectonically controlled elements, but enlarged by two (Mg, K).

A simple model can be used to describe this control of the concentration. In this model the input is from rivers and the output is uptake by reactions in the ocean crust under hydrothermal systems. (An application of this model is given in Section 13.5). Thus

$$V_{riv}C_{riv} = V_{hydro}(C_{sw} - C_{exit\ fluid})$$

where V_{riv} is the river volume per unit time and V_{hydro} the hydrothermal circulation rate; C_{riv} is the river concentration and C_{sw} is the normal seawater concentration.

Hydrothermal vents have near-zero concentrations of Mg. Therefore

$$C_{sw} = (V_{riv}/V_{hydro})C_{riv}$$

The present-day best estimates are that V_{riv}/V_{hydro} is about 300. As V_{hydro} increases, e.g. faster spreading of ocean crust at ridges, C_{sw} responds. The dominant control is tectonics.

10.4.4.2.3 Group II (Ca, Na). This group includes the remaining cations with relatively long residence times. One important constraint is the charge balance of seawater, re-arranged in the following format:

$$2[Ca^{2+}] + [Na^+] - [HCO_3^-] = [Cl^-] + 2[SO_4^{2-}] - 2[Mg^{2+}] - [K^+]$$

Tectonic processes control the terms on the right-hand side, as already discussed. This also defines the sum of terms on the left-hand side. The control of the relative proportions of elements on the left-hand side is uncertain. The most plausible controls based on present data are:

1. Ca/Na by ion exchange in estuaries.
2. Ca/HCO$_3$ by calcium carbonate equilibria and control of carbon.

10.4.4.2.4 Group III (Si, P, C, N, trace elements). The common property of these elements is that they are biologically reactive and are deposited as thermodynamically unstable debris. For each of these elements kinetic controls can be hypothesized. A first step is to describe the present-day situation.

Let us define a two-box model for a steady-state ocean as shown in Fig. 10-22. The two well mixed reservoirs correspond to the surface ocean and deep oceans. We assume that rivers are the only source and sediments are the only sink. Elements are also removed from the surface box by biogenic particles (B). We also assume there is mixing between the two boxes that can be expressed as a velocity $V_{mix} = 2\,\text{m/yr}$ and that rivers input water to the surface box at a rate of $V_{riv} = 0.1\,\text{m/yr}$. The resulting ratio of V_{mix}/V_{riv} is 20.

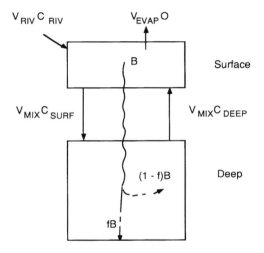

Fig. 10-22 Schematic of mass influx/outflux balance to surface/deep layers.

Water conservation can be expressed as

$$V_{riv} = V_{evap}$$
$$V_{down} = V_{up} = V_{mix}$$

The mass balance for biogenic elements in the surface box is

$$V_{riv}C_{riv} + V_{mix}C_{deep} = V_{mix}C_{surf} + B$$

For the lower box

$$V_{mix}C_{surf} + (1 - f)B = V_{mix}C_{deep}$$

where f is the fraction of the biogenic flux (B) buried and $(1 - f)$ is the fraction of B regenerated in the deep box. For the entire ocean the balance for biogenic elements is

$$V_{riv}C_{riv} = fB$$

Now that we have defined the biogenic model we can define two important properties. The first is the ratio of falling particles to the input to the surface. This property (g) is equivalent to the efficiency of bioremoval from surface waters:

$$g = \frac{B}{V_{riv}C_{riv} + V_{mix}C_{deep}}$$
$$= \frac{V_{riv}C_{riv} + V_{mix}C_{deep} - V_{mix}C_{surf}}{V_{riv}C_{riv} + V_{mix}C_{deep}}$$
$$= 1 - \frac{\left(\frac{V_{mix}}{V_{riv}}\right)\left(\frac{C_{surf}}{C_{riv}}\right)}{\left(\frac{V_{mix}}{V_{riv}}\right)\left(\frac{C_{deep}}{C_{riv}}\right) + 1}$$

As stated earlier, $V_{mix}/V_{riv} = 20$. The resulting values of g for several biogenic elements are given in Table 10-15. For Si, N, and P at least 95% of the elements brought to the surface are

Table 10-15 Parameters for element cycles within the sea

Element	[R]/[D]	[S]/[D]	g	f	fg
N	~0.20[a]	~0.05	0.95	0.01	0.01
P	~0.20[a]	~0.05	0.95	0.01	0.01
C	~0.10[b]	0.80	0.20	0.02	0.004
Si	~0.20	<0.05	~1.0	0.01	0.01
Ba	~2	0.30	0.75	0.12	0.09
Ca	0.04	0.99	0.01	0.12	0.001

[a]River value poorly known.
[b]Corrected for atmospheric recycling.

removed in particulate biogenic form. Only 20% of the C is removed in this form.

The second property is the fraction of particles sinking that are preserved in the sediments:

$$f = fB/B = V_{riv}C_{riv}/B$$
$$= V_{riv}C_{riv}/(V_{riv}C_{riv} + V_{mix}C_{deep} + V_{mix}C_{surf})$$
$$= 1/(1 + (V_{mix}/V_{riv})(C_{deep}/C_{riv} - C_{surf}/C_{riv}))$$

The values of f (Table 10-15) for N, P, C, and Si are 2% or less and are about 12% for Ba and Ca.

The product fg gives us the fraction of the elements that are permanently removed for each visit to the surface ocean. Conversely $1/f(fg)$ gives the number of times an element is recycled before it is permanently removed. For example, for a total ocean residence time of 1600 years, P goes through 105 cycles of 15 years each before being permanently removed.

We can finish this discussion of the biogenic elements by considering the implications of this kind of cycle on the long-term stability of the concentration. Phosphorus will be used as an example. For elements like P, that have a high value of g, the particulate flux responds directly to changes in the input. For P, the main input to the surface layer is upwelling and thus the biogenic particle flux $B = V_{mix}P_{deep}$. The system is now operating at close to 100% efficiency. P removal to the sediments depends on the O_2 level in the deep ocean. When there is less O_2 there is less efficient regeneration of P and thus f increases. The O_2 content of the deep sea depends on the balance between the input from above ($V_{mix}O_{2\ surf}$) and O_2 consumption during regeneration (e.g., respiration) of the flux B. Suppose the rate of ocean circulation increased. In this simple model we would parametrize this as saying V_{mix} increases. Because upwelling increases, productivity increases and B increases. Now we have a situation where both fluxes, B and ($V_{mix}O_{2\ surf}$), to the deep ocean increase. The net effect is to stabilize the concentration of P in the deep sea. Suppose there was an increase in river input ($V_{riv}C_{riv}$) with no change in mixing. Productivity and B increase without a coincident change in the O_2 flux ($V_{mix}O_{2\ surf}$ = constant). The increased B will result in more O_2 consumption (and PO_4 release) and thus lower O_2 levels in the deep sea. As a result, there will be more efficient P removal to the sediments. Thus f and fB increase, matching the increased river input. Again the net effect is to stabilize the deep P concentration. As f for P increases, the P content of the deep sea will tend to decrease, balancing the tendency to increase that was initially caused by the increased B. The general result is that the P (or N, Si, C) content of the deep sea seeks that level where upwelling of deep water brings PO_4 to the surface at a rate such that P in organisms resistant to decomposition just balances the amount of new PO_4 entering the sea. This negative feedback model tends to drive the PO_4 content of the ocean towards a regulated value where PO_4 loss equals PO_4 input.

10.4.5   Controls over Long Time Scales

Over multi-million year time scales, the concentration of CO_2 in the Earth's atmosphere is largely controlled by the geochemical carbon cycle, which consists of the exchange of carbon between the rock reservoir and the surficial reservoir (atmosphere, oceans, biosphere and soils) (Berner et al., 1983). This exchange occurs via the weathering of continental silicate rocks and organic matter, transport of and burial of weathering-derived carbonates and organic matter, geothermal breakdown (metamorphosis) of the weathering products at depth and subsequent degassing of CO_2 back to the surficial reservoir (Berner, 1990; 1991). The simplified representative equations for these processes are as follows, with the left-to-right reaction representing the surficial-to-rock transfer (Urey, 1952; Berner, 1991):

$$CO_2 + CaSiO_3 \underset{\text{metamorphism, magmatism}}{\overset{\text{weathering}}{\rightleftarrows}} CaCO_3 + SiO_2$$

$$CO_2 + MgSiO_3 \underset{\text{metamorphism, magmatism}}{\overset{\text{weathering}}{\rightleftarrows}} MgCO_3 + SiO_2$$

$$CO_2 + H_2O \underset{\text{weathering, metamorphism}}{\overset{\text{organic burial}}{\rightleftarrows}} CH_2O + O_2$$

The first two reactions, which constitute the major controls on atmospheric CO_2 on geologic time scales, are sometimes referred to as the Urey reactions. The third reaction denotes net photosynthesis and georespiration. The results of these models suggest that there has been significant variation in atmospheric CO_2 over the past 570 Myr with high levels during the Mesozoic and early Paleozoic and low levels during the Permo-Carboniferous and late Cenozoic.

Questions

10-1 There is some debate about what controls the magnesium concentration in seawater. The main input is rivers. The main removal is by hydrothermal processes (the concentration of Mg in hot vent solutions is essentially zero). First, calculate the residence time of water in the ocean due to (1) river input and (2) hydrothermal circulation. Second, calculate the residence time of magnesium in seawater with respect to these two processes. Third, draw a sketch to show this box model calculation schematically. You can assume that uncertainties in river input and hydrothermal circulation are 5% and 10%, respectively. What does this tell you about controls on the magnesium concentration? Do these calculations support the input/removal balance proposed above? Do any questions come to mind? Volume of ocean = 1.4×10^{21} L; River input = 3.2×10^{16} L/yr; Hydrothermal circulation = 1.0×10^{14} L/yr; Mg concentration in river water = 1.7×10^{-4} M; Mg concentration in seawater = 0.053 M.

10-2 The flux of oxygen can be in or out of the ocean. The oxygen partial pressure in the atmosphere is 0.20 atm and the Henry's Law constant is 1.26×10^{-3} M/atm. (a) As a result of photosynthesis, the nitrate concentration in seawater originally in equilibrium with the atmosphere has decreased by 20 μM. What is the new (nonequilibrium) oxygen concentration? (b) What is the flux of oxygen due to gas exchange? Use a diffusion coefficient of 2.0×10^{-5} cm^2/s and a film thickness of 50 μm.

10-3 Hydrothermal vents have been sampled at 21° along the East Pacific Rise. The pure end member hydrothermal solutions have a temperature of 350°C and the following major ion composition (von Damm *et al.* (1985). *Geochim. Cosmochim. Acta* **49**, 2197–2220). All concentrations are in mM and the pH is 3.4. Discuss the ways in which the vent solution composition and speciation differ from average seawater.

	Vent	Seawater		Vent	Seawater
Na^+	432	468	Cl^-	489	546
K^+	23	10.2	HCO_3^-	0	2
Mg^{2+}	0	5.32	CO_3^{2-}	0	0.2
Ca^{2+}	15	10.2	SO_4^{2-}	0.5	28
Mn^{2+}	0.96	0.001	H_2S	7.3	0
Fe^{2+}	1.7	10^{-6}			

10-4 A proponent of "reverse weathering" suggested that gibbsite, kaolinite, and quartz exist in equilibrium according to the following equation. In equilibrium expressions for these reactions, water will appear as the *activity*, rather than concentration. The activity can be approximated by the mole fraction of water. What is the activity of water if this equilibrium is maintained? Could this equilibrium exist in seawater, where the mole fraction of water is about 0.98? ΔG^0 values (kJ/mol): gibbsite -2320.4; kaolinite -3700.7; quartz -805.0; water -228.4.

$$Al_2Si_2O_5(OH)_4 \text{ (kaolinite)} + H_2O \rightleftarrows$$
$$Al_2O_3(H_2O)_3 \text{ (gibbsite)} + 2SiO_2 \text{ (quartz)}$$

Answers can be found on p. 509.

References

Alley, R. B., Mayewski, P. A., Sowers, T. *et al.* (1997). Holocene climatic instability; a prominent, widespread event 8200 years ago. *Geology* **25**(6), 483–486.

Alley, R. B., Meese, D. A., Shuman, C. A. *et al.* (1993). Abrupt increase in Greenland snow accumulation at the end of the Younger Dryas event. *Nature* **362**, 527–529.

Anderson, L. A. and Sarmiento, J. L. (1994). Redfield ratios of remineralization by nutrient data analysis. *Glob. Biogeochem. Cycles* **8**, 65–80.

Archer, D., Peltzer, E. T. and Kirchman, D. (1997). A timescale for dissolved organic carbon production in equatorial Pacific surface waters. *Glob. Biogeochem. Cycles* **11**, 435–452.

Baes, C. F., Björkström, A. and Mulholland, P. J. (1985). Uptake of carbon dioxide by the oceans. *In* "Atmospheric Carbon Dioxide and the Global Carbon Cycle" (J. R. Trabalka, ed.). *Report DOE/ER-0239*, US Department of Energy, Office of Energy Research, Washington, DC.

Barth, T. W. (1952). "Theoretical Petrology." Wiley, New York.

Behl, R. J. and Kennett, J. P. (1996). Brief interstadial event in the Santa Barbara Basin, NE Pacific, during the last 60 kyr. *Nature* **379**, 243–246.

Berner, R. A. (1990). Atmospheric carbon dioxide levels over Phanerozoic time. *Science*, **249**, 1382–1386.

Berner, R. A. (1991). A model for atmospheric CO_2 over Phanerozoic time. *Am. J. Sci.* **291**, 339–376.

Berner, E. K. and Berner, R. A. (1987). "The Global Water Cycle." Prentice-Hall, Englewood Cliffs, New Jersey.

Berner R. A., Lasaga, A. C., and Garrels, R. A. (1983). The carbonate–silicate geochemical cycle and its effect on atmospheric carbon dioxide over the past 100 million years. *Am. J. Sci.* **283**, 641–683.

Bond G. *et al.* (1997). A pervasive millennial-scale cycle in North Atlantic Holocene and glacial climates. *Science* **278**, 1257–1266.

Bond G. C., Heinrich, H., Broecker, W. S. *et al.* (1992). Evidence for massive discharges of icebergs into the North Atlantic Ocean during the last glacial period. *Nature* **360**, 245–249.

Boyle, E., and Edmond, J. M. (1975). Copper in surface waters south of New Zealand. *Nature* **253**, 107–109.

Brass, G. W. and Turekian. K. K. (1974). Strontium distribution in GEOSECS oceanic profiles. *Earth Planet. Sci. Lett.* **23**, 141–148.

Brewer, P. G. (1978). Direct observation of the oceanic CO_2 increase. *Geophys. Res. Lett.* **5**, 997–1000.

Broecker, W. S. (1971). A kinetic model for the chemical composition of seawater. *Quatern. Res.* **1**, 188–207.

Broecker, W. S. (1991). The great ocean conveyor. *Oceanography* **4**, 79–89.

Broecker, W. S. (1994). Massive iceberg discharges as triggers for global climate change. *Nature* **372**, 421–424.

Broecker, W. S. (1997). Thermohaline circulation, the Achilles Heel of our climate system: Will man-made CO_2 upset the current balance? *Science* **278**, 1582–1588.

Broecker, W. S. and Peng, T.-H. (1974). Gas exchange rates between air and sea. *Tellus* **26**, 21–35.

Broecker, W. S. and Peng, T.-H. (1982). "Tracers in the Sea." Eldigio Press, New York.

Buesseler, K. O. (1998). The decoupling of production and particulate export in the surface ocean. *Glob. Biogeochem. Cycles* **12**, 297–310.

Burton, K. W., Ling, H.-F. and O'Nions, R. K. (1997). Closure of the Central American Isthmus and its effect on deep-water formation in the North Atlantic. *Nature* **386**, 382–385.

Chavez, F. P. and Toggweiler, J. R. (1995). Physical estimates of global new production: The upwelling contribution. *In* "Upwelling in the Ocean: Modern Processes and Ancient Records" (C. P. Summerhayes, K.-C. Emeis, M. V. Angel, R. L. Smith and B. Zeitzschel, eds), pp. 313–320. John Wiley.

Chisholm S. W. and Morel, F. M. M. (eds) (1991). What controls phytoplankton production in nutrient-rich areas of the open sea? *Limnol. Oceanogr.* **36**, 1507–1965.

Christensen J. P., Murray, J. W., Devol, A. H. and Codispoti, L. A. (1987). Denitrification in continental shelf sediments has major impact on oceanic nitrogen cycle. *Glob. Biogeochem. Cycles* **1**, 97–116.

Clayton, T. D., Byrne, R. H., Breland, J. A. *et al.* (1995). The role of pH measurements in modern oceanic CO_2-system characteristics: precision and thermodynamic consistency. *Deep-Sea Res. II* **42**, 411–430.

Coale, K. H. and Bruland, K. W. (1988). Copper complexation in the Northeast Pacific. *Limnol. Oceanogr.* **33**, 1084–1101.

Coale, K. H., Johnson, K. S., Fitzwater, S. E. *et al.* (1996). A massive phytoplankton bloom induced by an ecosystem-scale iron fertilization experiment in the equatorial Pacific Ocean. *Nature* **383**, 495–501.

Cullen, J. J., Lewis, M. R., Davis, C. O. and Barber, R. T. (1992). Photosynthetic characteristics and estimated growth rates indicate grazing is the proximate control of primary production in the equatorial Pacific. *J. Geophys. Res.* **97**, 639–654.

Denton, G. H. and Hendy, C. H. (1994). Documentation of a Younger Dryas glacial advance in the New Zealand Alps. *Science* **264**, 1434–1437.

de Villiers, S. and Nelson, B. K. (1999). Detection of low-temperature hydrothermal fluxes by seawater Mg and Ca anomalies. *Science* **285**, 721–723.

Dickson, A. G. (1993). The measurement of seawater pH. *Mar. Chem.* **44**, 131–142.

Dortch, Q. (1990). The interactions between ammonium and nitrate uptake in phytoplankton. *Mar. Ecol. Prog. Ser.* **61**, 183–201.

Duce, R. A. and Tindale, N. W. (1991). Atmospheric transport of iron and its deposition in the ocean. *Limnol. Oceanogr.* **36**, 1715–1726.

Ducklow, H. W. and Harris, R. P. (1993). JGOFS: The North Atlantic Bloom Experiment. *Deep-Sea Res. II* **40**, 1–641.

Dugdale, R. C. and Goering, J. J. (1967). Uptake of new and regenerated forms of nitrogen in primary productivity. *Limnol. Oceanogr.* **12**, 196–206.

Dugdale, R. C. and Wilkerson, F. P. (1998). Silicate regulation of new production in the equatorial Pacific upwelling. *Nature* **391**, 270–273.

Dunne, J. P., Murray, J. W., Aufdenkampe, A., Blain, S. and Rodier, M. (1999). Silicon-nitrogen coupling in the equatorial Pacific upwelling zone. *Glob. Biogeochem. Cycles* **131**, 715–726.

Edmond, J. M., Measures, C., McDuff, R. E. *et al.* (1979). Ridge crest hydrothermal activity and the balance of the major and minor elements in the ocean: the Galapagos data. *Earth Planet. Sci. Lett.* **46**, 1–18.

Emerson, S., Quay, P., Karl, D. *et al.* (1997). Experimental determination of the organic carbon flux from open-ocean surface waters. *Nature* **389**, 951–954.

Emerson, S., Quay, P. D., Stump, C. *et al.* (1991). O_2, Ar, N_2 and ^{222}Rn in surface waters of the subarctic Pacific Ocean. *Glob. Biogeochem. Cycles*, **5**, 49–69.

Emerson, S., Quay, P. D., Stump, C. *et al.* (1995). Chemical tracers of productivity and respiration in the subtropical pacific. *J. Geophys. Res.* **100**, 15873–15887.

Eppley, R. W. and Peterson, B. J. (1979). Particulate organic matter flux and planktonic new production in the deep ocean. *Nature* **282**, 677–680.

Fairbanks, R. G. (1989). A 17,000-year glacio-eustatic sea level record: influence of glacial melting rates on younger dryas event and deep-ocean circulation. *Nature* **342**, 637–642.

Falkowski, P. G., Barber, R. T. and Smetacek, V. (1998). Biogeochemical controls and feedbacks on ocean primary production. *Science* **281**, 200–206.

Falkowski, P. G., Greene, R. and Geider, R. (1992). Physiological limitations on phytoplankton productivity in the ocean. *Oceanography* **5**, 84–91.

Fanning, K. A. (1992). Nutrient provinces in the sea: Concentration ratios, reaction rate ratios and ideal covariation. *J. Geophys. Res.* **97**, 5693–5712.

Fasham, M. J. R., Ducklow, H. W. and McKelvie, S. M. (1990). A nitrogen-based model of plankton dynamics in the ocean mixed layer. *J. Mar. Res.* **48**, 591–639.

Feely, R. A., Wanninkhof, R., Goyet, C. *et al.* (1997). Variability of CO_2 distributions and sea-air fluxes in the central and eastern equatorial Pacific during the 1991–94 El Niño. *Deep-Sea Res. II* **44**, 1851–1868.

Field, C. B., Behrenfeld, M. J., Randerson, J. T. and Falkowski, P. (1998). Primary production of the biosphere: Integrating terrestrial and oceanic components. *Science* **281**, 237–240.

Fine, R. A., Peterson, H., Rooth, C. G. H. and Ostlund, H. G. (1983). Cross-equatorial tracer transport in the upper waters of the Pacific Ocean. *J. Geophys. Res.* **88**, 763–769.

Fitzwater, S. E., Coale, K. H., Gordon, R. M. *et al.* (1996). Iron deficiency and phytoplankton growth in the equatorial Pacific. *Deep-Sea Res II* **43**, 995–1015.

Frausto da Silva, J. J. R. and Williams, R. J. P. (1991). "The Biological Chemistry of the Elements." Clarendon Press, Oxford.

Frost, R. W. and Franzen, N. C. (1992). Grazing and iron limitation in the control of phytoplankton stock and nutrient concentration: A chemostat analogue of the Pacific equatorial upwelling zone. *Mar. Ecol. Prog. Ser.* **83**, 291–303.

Frost, R. W. (1987). Grazing control of phytoplankton stock in the open subarctic Pacific ocean: a model assessing the role of mesozooplankton, particularly the large calanoid copepods *Neocalanus* spp. *Mar. Ecol. Prog. Ser.* **39**, 49–68.

Gaillard, J.-F. and Treguer, P. (1997). Antares I: France JGOFS in the Indian Sector of the Southern Ocean: Benthic and water column processes. *Deep-Sea Res. II* **44**, 951–1176.

Gammon, R. H., Cline, J. and Wisegarver, D. (1982). Chlorofluoromethanes in the Northeast Pacific Ocean: measured vertical distributions and application as transient tracers of upper ocean mixing. *J. Geophys. Res.* **87**, 9441–9454.

Garrels, R. M. and Mackenzie, F. T. (1967). Origin of the chemical compositions of some springs and lakes. *In* "Equilibrium Concepts in Natural Water Systems" (W. Stumm, ed.). Advances in Chemistry Series 67, pp. 222–274. American Chemical Society, Washington.

Garrels, R. M. and Thompson, M. E. (1962). A chemical model for seawater at 25°C and one atmosphere total pressure. *Am. J. Sci.* **260**, 57–66.

Goldschmidt, V. M. (1937). The principles of distribution of chemical elements in minerals and rocks. *J. Chem. Soc.* **1937**, 655–674.

Gordon, A. L. (1985). Indian-Atlantic transfer of thermocline water at the Agulhas retroflection. *Science* **227**, 1030–1033.

Gruber, N. and Sarmiento, J. L. (1997). Global patterns of marine nitrogen fixation and denitrification. *Glob. Biogeochem. Cycles* **11**, 235–266.

Hansell, D. and Carlson, C. A. (1998). Deep-ocean gradients in the concentration of dissolved organic carbon. *Nature* **395**, 263–266.

Hansell, D. A. and Peltzer, E. T. (1998). Spatial and temporal variations of total organic carbon in the Arabian Sea. *Deep-Sea Res. II* (in press).

Hansell, D. A., Bates, N. R. and Carlson, C. A. (1997). Predominance of vertical loss of carbon from surface waters of the equatorial Pacific Ocean. *Nature* **386**, 59–61.

Hartmann, D. L. (1994). "Global Physical Climatology." Academic Press, San Diego.

Haug, G. H. and Tiedemann, R. (1998). Effect of the formation of the Isthmus of Panama on Atlantic Ocean thermohaline circulation. *Nature*, **393**, 673–676.

Hedges, J. I. and Lee, C. (1993). Measurement of dissolved organic carbon and nitrogen in natural waters. *Mar. Chem.* **41**, 290pp.

Jenkins, W. J. (1980). Tritium and ^{3}He in the Sargasso Sea. *J. Mar. Res.* **38**, 533–569.

Jenkins, W. J. and Goldman, J. C. (1985). Seasonal oxygen cycling and primary production in the Sargasso Sea. *J. Mar. Res.* **43**, 465–491.

Johnson, K. S., Coale, K. H. and Jannasch, H. W. (1992). Analytical chemistry in oceanography. *Anal. Chem.* **64**, 1065A-1075A.

Johnson, K. S., Gordon, R. M. and Coale, K. H. (1997). What controls dissolved iron concentration in the world ocean? *Mar. Chem.* **57**, 137–161.

Johnson, K. M., Wills, K. D., Butler, D. B. *et al.* (1993). Coulometric total carbon dioxide analysis for marine studies: maximizing the performance of an automated gas extraction system and coulometric detector. *Mar. Chem.* **44**, 167–187.

Joyce, T. M., Warren, B. A. and Talley, L. D. (1986). The geothermal heating of the abyssal subarctic Pacific Ocean. *Deep-Sea Res.* **33**, 1003–1015.

Karl, D. M. and Michaels, A. F. (eds) (1996). Ocean Time-Series: Results from the Hawaii and Bermuda Research Programs. *Deep-Sea Res. II* **43**, 127–685.

Karl, D. M., Letelier, R., Hebel, D. *et al.* (1995). Ecosystem changes in the North Pacific subtropical gyre attributed to the 1991–92 El Niño. *Nature* **373**, 230–234.

Karl, D., Letelier, R., Tupas, L. *et al.* (1997). The role of nitrogen fixation in biogeochemical cycling in the subtropical North Pacific Ocean. *Nature* **388**, 533–538.

Kastner, M. (1974). The contribution of authigenic feldspars to the geochemical balance of alkalic metals. *Geochim. Cosmochim. Acta* **38**, 650–653.

Keigwin, L. D. (1982). Isotope paleoceanography of the Caribbean and east Pacific: role of Panama uplift in late Neogene time. *Science* **217**, 350–353.

Kessler, W. S. and McPhaden, M. J. (1995). The 1991–1993 El Niño in the central Pacific. *Deep-Sea Res. II* **42**, 295–334.

Ku, T. L., Knauss, K. G. and Mathieu, G. G. (1977). Uranium in open ocean: concentration and isotopic concentration. *Deep-Sea Res.* **24**, 1005–1017.

Landry, M. R., Barber, R. T., Bidigare, R. R. *et al.* (1997). Iron and grazing constraints on primary production in the central equatorial Pacific: and EqPac synthesis. *Limnol. Oceanogr.* **42**, 405–418.

Lasaga, A. C. (1980). The kinetic treatment of geochemical cycles. *Geochim. Cosmochim. Acta* **44**, 815–828.

Lasaga, A. C. (1981). Dynamic treatment of geochemical cycles: global kinetics. *In* "Kinetics of Geochemical Processes" (A. C. Lasaga and R. J. Kirkpatrick, eds), pp. 69–110. Mineral. Soc. Amer., Washington, DC.

Li, Y.-H. (1977). Confusion of the mathematical notation for defining the residence time. *Geochim. Cosmochim. Acta* **44**, 555–556.

Li, Y.-H. (1981). Ultimate removal mechanisms of elements from the ocean. *Geochim. Cosmochim. Acta* **45**, 1659–1664.

Li, T.-H., Takahasi, T. and Broecker, W. S. (1969). The degree of saturation of $CaCO_3$ in the oceans. *J. Geophys. Res.* **74**, 5507–5525.

Liss, P. S. and Merlivat, L. (1986). Air-sea gas exchange rates: introduction and synthesis. *In* "The Role of Air-Sea Exchange on Geochemical Cycling" (P. Buat-Menard, ed.), pp. 113–127. Reidel, Norwell, MA.

Liss, P. S. and Slater, P. G. (1974). Flux of gases across the air-sea interface. *Nature* **247**, 181–184.

Loukos, H., Frost, B., Harrison, D. E. and Murray, J. W. (1997). An ecosystem model with iron limitation of primary production in the equatorial Pacific at 140°W. *Deep-Sea Res. II* **44**, 2221–2249.

Mackenzie, F. T. and Garrels, R. M. (1966). Chemical mass balance between rivers and oceans. *Am. J. Sci.* **264**, 507–525.

Maier-Reimer, E., Mikolajewicz, U. and Crowley, T. J. (1990). Ocean general circulation model sensitivity experiment with an open American Isthmus. *Paleoceanography*, **5**, 349–366.

Manabe, S. and Stouffer, R. J. (1995). Simulation of abrupt climate change induced by freshwater input to the North Atlantic Ocean. *Nature* **378**, 165–167.

Mantyla, A. W. (1975). On the potential temperature in the abyssal Pacific Ocean. *J. Mar. Res.* **33**, 341–354.

Mantyla, A. W. and Reid, J. L. (1983). Abyssal characteristics of the world ocean waters. *Deep Sea Res.* **30**, 805–833.

Martin, J. H. (1990). Glacial–interglacial CO_2 change: The iron hypothesis. *Paleoceanography* **5**, 1–13.

Martin, J. H. and Fitzwater, S. E. (1988). Iron deficiency limits phytoplankton growth in the northeast Pacific subarctic. *Nature* **331**, 341–343.

Martin, J. H., Gordon, R. M., Fitzwater, S. and Broerkow, W. W. (1989). VERTEX: phytoplankton/iron studies in the Gulf of Alaska. *Deep-Sea Res.* **36**, 649–680.

Martin, J. H., Gordon, R. M. and Fitzwater, S. E. (1990). Iron in Antarctic waters. *Nature* **345**, 156–158.

Martin, J. H. *et al.* (1994). Testing the iron hypothesis in ecosystems of the equatorial Pacific Ocean. *Nature* **371**, 123–129.

McCarthy, M., Pratum, T., Hedges, J. and Benner, R. (1997). Chemical composition of dissolved organic nitrogen in the ocean. *Nature* **390**, 150–154.

McCarthy, J. J., Garside, C., Nevins, J. L. and Barber, R. T. (1996). New production along 140°W in the equatorial Pacific during and following the 1992 El Niño event. *Deep-Sea Res. II* **43**, 1065–1094.

McDuff, R. E. and Morel, F. M. M. (1980). The geochemical control of seawater (Sillen revisited). *Environ. Sci. Tech.* **14**, 1182–1186.

McPhaden, M. J. (1993). TOGA-TAO and the 1991–93 El Niño-Southern Oscillation event. *Oceanography* **6**, 36–44.

McPhaden, M. J., Busalacchi, A. J., Cheney, R. *et al.* (1998). The tropical ocean-global atmosphere observing system: A decade of progress. *J. Geophys. Res.* **103**, 14169–14240.

Menzel, D. W. (1974). Primary productivity, dissolved and particulate organic matter and the sites of oxidation of organic matter. *In* "The Sea," Vol. 5 (E. D. Goldberg, ed.). Wiley, New York.

Michaels, A. F., Bates, N. R., Buesseler, K. O. *et al.* (1994). Carbon-cycle imbalances in the Sargasso Sea. *Nature* **372**, 537–540.

Michalopoulos, P. and Aller, R. C. (1995). Rapid clay mineral formation in Amazon Delta sediments: reverse weathering and oceanic elemental cycles. *Science* **270**, 614–617.

Miller, C. B., Frost, B. W., Wheeler, P. A. *et al.* (1991). Ecological dynamics in the subarctic Pacific, a possible iron-limited ecosystem. *Limnol. Oceanogr.* **36**, 1600–1615.

Millero, F. J. (1993). What is PSU? *Oceanography* **6**, 67.

Millero, F. J. (1995). Thermodynamics of the carbon dioxide system in the ocean. *Geochim. Cosmochim. Acta* **59**, 661–677.

Morel, F. M. M., Reinfelder, J. R., Roberts, S. B. *et al.* (1994). Zinc and carbon co-limitation of marine phytoplankton. *Nature* **369**, 740–742.

Murray, J. W. (ed.) (1995). A U.S. JGOFS Process Study in the equatorial Pacific. *Deep-Sea Res. II* **42**, 275–903.

Murray, J. W. (ed.) (1996). A U.S. JGOFS Process Study in the equatorial Pacific. Part 2 *Deep-Sea Res. II* **43**, 687–1435.

Murray, J. W., Barber, R. T., Roman, M. R. *et al.* (1994). Physical and biological controls on carbon cycling in the equatorial Pacific. *Science* **266**, 58–65.

Murray, J. W., LeBorgne, R. and Dandonneau, Y. (eds) (1997). A JGOFS Process Study in the Equatorial Pacific. *Deep-Sea Res. II* **44**, 1759–2317.

Murray J. W., Ozsoy, E. and Top, Z. (1991). Temperature and salinity distributions in the Black Sea. *Deep-Sea Res.* **38**, S663-S689.

Murray, J. W., Young, J., Newton, J. *et al.* (1996). Export flux of particulate organic carbon from the central equatorial Pacific determined using a combined drifting trap-[234]Th approach. *Deep-Sea Res. II* **45**, 1095–1132.

Ostlund, G. G. and Fine, R. A. (1979). Oceanic distribution and transport of tritium. *IAEA-SM-232162*, pp. 303–314. Intl. Atom. Energy Agency, Vienna.

Ostlund, H. G., Dorsey, H. G. and Rooth, C. G. (1974). GEOSECS North Atlantic radiocarbon and tritium results. *Earth Planet. Sci. Lett.* **23**, 69–86.

Parsons, T. R., Takahashi, M. and Hargrave, B. (1977). "Biological Oceanographic Processes," 2nd edn. Pergamon, New York.

Peltzer, E. T. and Hayward, N. A. (1996). Spatial and temporal variability of total organic carbon along 140°W in the equatorial Pacific Ocean in 1992. *Deep-Sea Res. II* **43**, 1155–1180.

Peng, T.-H., Broecker, W. S., Mathieu, G. G. and Li, Y.-H. (1979). Radon evasion rates in the Atlantic and Pacific oceans as determined during the GEOSECS program. *J. Geophys. Res.* **84**, 2471–2486.

Perry, M. J. (1976). Phosphate utilization by an oceanic diatom in phosphorus-limited chemostat culture and in the oligotrophic waters of the central North Pacific. *Limnol. Oceanogr.* **21**, 88–107.

Philander, S. G. (1990). "El Niño, La Niña and the Southern Oscillation." Academic Press, San Diego.

Pickard, G. L. and Emery, W. J. (1982). "Descriptive Physical Oceanography," 4th edn. Pergamon Press, Oxford.

Platt, T. and Harrison, W. G. (1985). Biogenic fluxes of carbon and oxygen in the ocean. *Nature* **318**, 55–58.

Platt, T. and Sathyendranth, S. (1988). Oceanic primary production: Estimation by remote sensing at local and regional scales. *Science* **241**, 1613–1620.

Price, N. M. and Morel, F. M. M. (1990). Cadmium and cobalt substitution for zinc in a marine diatom. *Nature* **344**, 658–660.

Price, N. M. and Morel, F. M. M. (1991). Colimitation of phytoplankton growth by nickel and nitrogen. *Limnol. Oceanogr.* **36**, 1071–1077.

Quinby-Hunt, M. S. and Turekian, K. K. (1983). Distribution of elements in sea water. *EOS* **64**, 130–131.

Raven, J. A. and Johnston, A. M. (1991). Mechanisms of inorganic-carbon acquisition in marine phytoplankton and their implications for the use of other resources. *Limnol. Oceanogr.* **36**, 1701–1714.

Redfield, A. C., Ketchum, B. H. and Richards, F. A. (1963). The influence of organisms on the composition of seawater. *In* "The Sea," Vol. 2 (M. N. Hill, ed.), pp. 26–77. Wiley-Interscience, New York.

Rooth, C. G. and Ostlund, H. G. (1972). Penetration of tritium into the Atlantic thermocline. *Deep Sea Res.* **19**, 481–492.

Rue E. L. and Bruland, K. W. (1995). Complexation of iron (III) by natural organic ligands in the Central North Pacific as determined by a new competitive ligand equilibration/adsorptive cathodic stripping voltammetric method. *Mar. Chem.* **50**, 117–138.

Ryther, J. H. (1969). Photosynthesis and fish production in the sea. *Science* **166**, 72–76.

Sarmiento, J. L., Rooth, C. G. H. and Roether, W.

(1982). The North Atlantic tritium distribution in 1972. *J. Geophys. Res.* **87**, 8047–8056.

Siegenthaler, U. and Sarmiento, J. L. (1993). Atmospheric carbon dioxide and the ocean. *Nature* **365**, 119–125.

Sillén, L. G. (1961). The physical chemistry of seawater. *In* "Oceanography" (M. Sears, ed.), International Oceanographic Congress (New York, 1959), pp. 549–581. Publication 67, American Association for the Advancement of Science, Washington DC.

Sillén, L. G. (1967). The ocean as a chemical system. *Science* **156**, 1189–1196.

Smetacek, V., de Baar, H. J. W., Bathmann, U. V. *et al.* (eds) (1997). Ecology and biogeochemistry of the Antarctic Circumpolar current during austral spring: Southern Ocean JGOFS Cruise ANT X/6 of R.V. Polarstern. *Deep-Sea Res. II* **44**, 1–519.

Smith, S. (ed.) (1998). U.S. JGOFS Arabian Sea Process Study. *Deep-Sea Res. II* **45**.

Smith, W. H. F. and Sandwell, D. T. (1997). Global seafloor topography from satellite altimetry and ship depth soundings. *Science* **277**, 1956–1962.

Stommel, H. (1958). The abyssal circulation. *Deep-Sea Res.* **5**, 80–82.

Stommel, H. and Arons, A. (1960). On the circulation of the world ocean, 1 and 2. *Deep-Sea Res.* **6**, 140–154; 217–233.

Stuiver, M., Quay, P. D. and Ostlund, H. G. (1983). Abyssal water carbon-14 distribution and the age of the world oceans. *Science* **219**, 849–851.

Stumm, W. and Brauner, P. A. (1975). Chemical speciation. *In* "Chemical Oceanography," Vol. 1, 2nd edn. (J. P. Riley and G. Skirrow, eds), pp. 173–240. Academic Press, London.

Stumm, W. and Morgan, J. J. (1996). "Aquatic Chemistry," 3rd edn. Wiley, New York.

Suess, E. (1980). Particulate organic carbon flux in the oceans-surface productivity and oxygen utilization. *Nature* **288**, 260–263.

Sunda, W. and Guillard, R. R. L. (1976). The relationship between cupric ion activity and the toxicity of copper to phytoplankton. *J. Mar. Res.* **34**, 511–529.

Sunda, W. G., Swift, D. G. and Huntsman, S. A. (1991). Low iron requirements for growth in oceanic phytoplankton. *Nature* **351**, 55–57.

Tai, C.-K. and Wunsch, C. (1983). Absolute measurement by satellite altimetry of dynamic topography of the Pacific Ocean. *Nature* **301**, 408–410.

Takahashi, T., Broecker, W. S. and Langer, S. (1985). Redfield ratio based on chemical data from isopycnal surfaces. *J. Geophys. Res.* **90**, 6907–6924.

Turner D., Owens, N. and Priddle, J. (eds) (1995).

Southern Ocean JGOFS: The UK "Sterna" Study in the Bellingshausen Sea. *Deep-Sea Res. II* **42**, 907–1335.

UNESCO (1981). Background papers and supporting data on the Practical Salinity Scale 1978. *UNESCO Technical Papers in Marine Science No. 37.* UNESCO.

Urey, H. C. (1952). "The Planets, Their Origin and Development." New Haven, Yale Univ. Press.

Van Weering, T. C. E., Helder, W. and Schalk, P. (eds) (1997). Netherlands Indian ocean Program 1992–1993: First Results. *Deep-Sea Res. II* **44**, 1177–1480.

Wallace, J. M., Rasmusson, E. M., Mitchell, T. P. *et al.* (1998). On the structure and evolution of ENSO-related climate variability in the tropical Pacific: Lessons from TOGA. *J. Geophys. Res.* **102**, 14,241–14,259.

Wanninkhof, R. (1992). Relationship between gas exchange and wind speed over the ocean. *J. Geophys Res.* **97**, 7373–7381.

Warren, B. A. (1983). Why is no deepwater found in the North Pacific? *J. Mar. Res.* **41**, 327–347.

Waser, J. (1966). "Basic Chemical Thermodynamics." W. A. Benjamin, New York.

Whitfield, M. (1979). The mean oceanic residence time (MORT) concept – a rationalization. *Mar. Chem.* **8**, 101–123.

Whitfield, M. and Turner, D. R. (1979). Water-rock partition coefficients and the composition of seawater and river water. *Nature* **278**, 132–137.

Williams, P. M. (1971). The distribution and cycling of organic matter in the ocean. *In* "Organic Compounds in Aquatic Environments" (S. D. Faust and J. W. Hunter, eds). Dekker, New York.

Williams, P. J. LeB. (1975). Biological and chemical aspects of dissolved organic matter in seawater. *In* "Chemical Oceanography," Vol. 2 (J. P. Riley and G. Skirrow, eds). Academic Press, London.

Wilson, T. R. S. (1975). Salinity and the major elements of seawater. *In* "Chemical Oceanography," 2nd edn, Vol. I. (J. P. Riley and G. Skirrow, eds). Academic Press, London.

Wolgemuth, K. and Broecker, W. S. (1970). Barium in seawater. *Earth Planet. Sci. Lett.* **8**, 372–388.

Worthington, L. V. (1970). The Norwegian Sea as a Mediterranean basin. *Deep Sea Res.* **17**, 77–84.

Worthington, L. V. (1981). The water masses of the world ocean: Some results of fine-scale census. *In* "Evolution of Physical Oceanography" (B. A. Warren and C. Wunsch, eds). MIT Press, Cambridge, MA.

Zauker, F. and Broecker, W. S. (1992). Influence of interocean fresh water transports on ocean thermohaline circulation. *J. Geophys. Res.* **97**, 2765–2773.

Part Three

Biogeochemical Cycles

Biogeochemical cycles are the backbone of Earth system science and are the major focus of this book. In this part, we will use the background provided in Part One, along with the information on the major reservoirs provided in Part Two to tell a cohesive story about five biogeochemical cycles: those of carbon (Chapter 11), nitrogen (Chapter 12), sulfur (Chapter 13), phosphorus (Chapter 14), and the trace metals (Chapter 15). The goal of each one of these chapters is to provide an overview of the chemistry and isotopes of the element, describe the major reservoirs where the element resides and in what forms, provide estimates of the fluxes in and out of these reservoirs and the chemical transformations that accompany them, and finally to explain how the cycle has been altered by human activities. There is another element which is absolutely necessary for life on Earth that is not covered directly in this part: oxygen. As will be seen in the chapters that follow, oxygen is a part of other elemental cycles and often plays the part of an acid-maker and oxidizing agent. We have devoted an entire chapter to these processes, Chapter 16, which appears in the final part due to its integrative nature. What follows is a brief overview of the importance of each cycle, and a presentation of the highlights of each of these five chapters.

Carbon

The significance of the carbon cycle has relatively recently become appreciated by scientists and non-scientists alike, due to the buildup of carbon dioxide in the atmosphere. Out of all of the cycles we present here, it is perhaps the easiest one to use as an example of how humans have caused a change in an Earth system. Since fossil fuels are predominantly carbon, and are released as carbon dioxide when burned, we have essentially removed carbon from one reservoir (the lithosphere) and placed it directly into another (the atmosphere). Measurements of the concentration of CO_2 have shown an increase in CO_2 by about 15% since the measurements started in the 1950s and 30% since pre-industrial times. These findings are significant because CO_2 is radiatively active in the atmosphere (it is a greenhouse gas). The more permanent uptake of CO_2 is by the oceans. Since carbon is the major element in living things, there is a biospheric uptake by vegetation on the continents when they biochemically fix CO_2 into organic matter, but this carbon returns to the atmosphere during plant respiration. There are other reduced forms of carbon in the atmosphere, such as methane (CH_4) which also behave as greenhouse gases. A model of the carbon cycle is presented in this chapter. The data on the various fluxes and transformations were compiled using the ^{13}C and ^{14}C isotopes as tracers.

Nitrogen

Although carbon is the principal element in living organisms, the availability of nitrogen is often the limiting factor in plant growth. Thus, the cycle of nitrogen is intimately connected with the biosphere. Most of the biological interaction involves *nitrogen fixation*, which is any removal of N_2 from the atmosphere to form any other nitrogen compound. In plants these compounds are usually ammonia or amino acids. Since the human diet depends so heavily on

Earth System Science
ISBN 0-12-379370-X

agriculture, we have added nitrogen to soils in the form of fertilizer in order to increase crop yields. Another way humans have altered the nitrogen cycle is by burning fossil fuels, which releases nitrogen oxides into the atmosphere. Atmospheric nitrogen in the form of oxides are a source of pollution, and a significant form of photochemical smog. These compounds play roles in both the formation and destruction of ozone in the troposphere and stratosphere, and can produce nitric acid after reaction with water. The acid rain that is produced can cause acidification of soils and bodies of water if the underlying bedrock lacks alkaline species. Changes in the nitrogen cycle therefore result in chemical conditions that directly affect public health and environmental welfare.

Sulfur

A great deal of attention has been paid to sulfur in the last decade or so because of its ubiquity in atmospheric particles which reflect sunlight back to space and affect the optical properties of clouds. These particles arise when $SO_2(g)$ reacts with (a) water, to form sulfuric acid, H_2SO_4, or (b) ammonia, NH_3, to form ammonium sulfate, $(NH_4)_2SO_4$, or ammonium bisulfate, NH_4HSO_4. The anthropogenic source of SO_2 is generally from fossil fuel burning, whereas the major natural source is dimethyl sulfide (DMS), $(CH_3)_2S$, released from the oceans due to the activity of phytoplankton. The DMS oxidizes to SO_2 once it enters the atmosphere and eventually forms sulfuric acid, ammonium sulfate, or ammonium bisulfate, just as in the anthropogenic case. Worldwide, sulfates are the dominant constituent in fine (<1.0 μm diameter) particles, and are thought to exert a radiative effect on Earth that is similar in size to that of the CO_2 greenhouse effect, although in the opposite direction (cooling rather than warming). Changes in the sulfur cycle therefore affect climate. The sulfuric acid that is produced in high concentrations in areas with large amounts of coal burning can acidify soils and lakes in the same way that nitric acid does. Sulfate is rarely a limiting nutrient and is

very abundant in seawater. It is a very important oxidizing agent in anaerobic systems.

Phosphorus

Phosphorus is unusual in its chemistry compared to the other elements discussed here in that it exists in the environment almost completely in the P(V) oxidation state of phosphate, PO_4^{3-}. Therefore, except in very small amounts occurring in atmospheric particles, the atmosphere is essentially not a reservoir for phosphorus and does not participate in its biogeochemical cycling. However, phosphorus is an extremely important constituent in all living things, forming the backbone of DNA, deoxyribonucleic acid. It is also the main source of cellular energy, in the form of ATP, adenosine triphosphate. The natural cycling of phosphorus, therefore, is completely intertwined with biospheric interaction. In the ocean, phosphorus in surface waters is rapidly taken up by biota which sink to lower depths and decompose. On land, phosphorus enters the soil through the decay of dead organic material. As mentioned in Chapter 8, the soil is eventually transported by rivers, and eventually lays down in a marine sediment. The phosphorus present in solution can participate as a bionutrient anywhere along this path. There has been a human disruption of the phosphorus cycle through the use of detergents and fertilizers, which cause overgrowth of organisms in natural waters, thus altering the ecosystem. There are theories that phosphorus, through its action as a nutrient, can effect the cycling of other elements, especially carbon. This could have important climate ramifications over geologic time periods.

Trace Metals

In contrast with phosphorus, most metals can exist in a variety of oxidation states and physical forms, which makes them participants in all of the geospheres. However, because metals are generally trace elements for biota, most of the metal cycles are not significantly altered by biological interaction, but rather may affect the growth of organisms by acting as nutrients or poisons. Like the other cycles, we can identify the major ways that metals are mobilized into the environment. The major reservoir for all metals is in the lithosphere, so rock weathering and volcanic action are the main sources of natural metal cycling. From this starting point, metal ions enter the hydrosphere or the atmosphere, and are transformed depending on the conditions. The anthropogenic input of metals into the environment has received a great deal of attention in the last several decades because of the toxicity of the heavy metals. Mining, and subsequent use and disposal of metals into the atmosphere (e.g., from fly ash due to coal burning), the pedosphere (e.g., by burying metal ions), and the hydrosphere (e.g., by leaching of metal ions out of soils and sediments), have caused metals to enter the food chain. Probably the most well-known source of increased human consumption of metals is through fish and seafood, because some species accumulate the metals dissolved in water into their tissues.

11

The Global Carbon Cycle

Kim Holmén

11.1 Introduction

Although many elements are essential to living matter, carbon is the key element of life on Earth. The carbon atom's ability to form long covalent chains and rings is the foundation of organic chemistry and biochemistry. The biogeochemical cycle of carbon is necessarily very complex, since it includes all life forms on Earth as well as the inorganic carbon reservoirs and the links between them. Despite being a complicated elemental cycle, it is extensively studied and to date, probably the best understood elemental biogeochemical cycle. The possibility of global climatic change brought about by the enhanced greenhouse effect of fossil fuel CO_2 in the atmosphere has also prompted much carbon-related research.

There exists a multitude of review articles and books about the carbon cycle available with varied degrees of detail and points of emphasis: e.g., Bolin (1970a,b); Keeling (1973); Woodwell and Pecan (1973); Woodwell (1978); Bolin et al. (1979); Revelle (1982); Bolin and Cook (1983); Degens et al. (1984); Warneck (1988); Watson et al. (1990); IPCC (1992); Siegenthaler and Sarmiento (1993); Sundqvist (1993); Schimel et al. (1995); and Heimann (1997) to name a few.

This treatment of the carbon cycle is intended to give an account of the fundamental aspects of the carbon cycle from a global perspective. After a presentation of the main characteristics of carbon on Earth (Section 11.2), four sections follow: 11.3, about the carbon reservoirs within the atmosphere, the hydrosphere, the biosphere

and the lithosphere; 11.4, which covers some important fluxes between the reservoirs; 11.5, which gives brief accounts of selected models of the carbon cycle; and finally 11.6, describing natural and human-induced fluctuations in the carbon cycle. A recurring theme in this chapter will be to explore how mechanisms with different time scales in the carbon cycle influence the atmospheric cycle of CO_2. These time scales on which various components of the global carbon cycle interact with the atmosphere are indicated in Fig. 11-1. The reader should take note that the question of atmospheric CO_2 concentration, despite being of profound importance in today's debate about human induced climate change, is only one detail of the global carbon cycle. Carbon is present everywhere and indeed most material motions and transformations are linked in one way or another to the global carbon cycle.

The relevant time scales vary over many orders of magnitude, from millions of years (for processes controlled by the movement of the Earth's crust) to days and even seconds for processes related to air–sea exchange and photosynthesis. Depending on the problem studied, models are usually constructed only to include processes that work on time scales judged to be relevant in the particular study. For example, most models used to study mankind's perturbation of atmospheric CO_2 exclude the geologic processes working on time scales longer than 5000 years and only include those processes that actively respond to atmospheric P_{CO_2} changes on the decadal time scale. Although the CO_2 content in the atmosphere is modulated by changes

Earth System Science
ISBN 0-12-379370-X

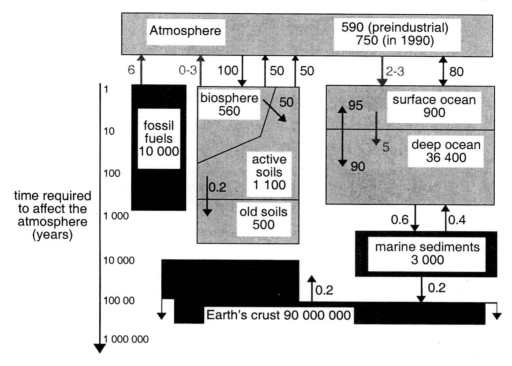

Fig. 11-1 Major reservoirs and fluxes of the global carbon cycle, including time scales. Numbers given are Pg C (1 Pg C = 10^{15} g C) Pg C/yr, respectively. (After Sundquist, 1993.)

in the exchange rates at the atmosphere–ocean and atmosphere–biosphere interface; the level of the CO_2 concentration in the atmosphere is ultimately determined by geologic processes occurring on very long time scales. The P_{CO_2}-controlled erosion rate, together with volcanism, releases carbon from the lithosphere into the ocean–atmosphere–biosphere system. This is counteracted by the sedimentation rate of carbon in the deep oceans. The balance between these two processes determines the long-term CO_2 level in the atmosphere.

There are more than a million known carbon compounds, of which thousands are vital to life processes. The carbon atom's unique and characteristic ability to form long stable chains makes carbon-based life possible. Elemental carbon is found free in nature in three allotropic forms: amorphous carbon, graphite, and diamond. Graphite is a very soft material, whereas diamond is well known for its hardness. Curiosities in nature, the amounts of elemental carbon on Earth are insignificant in a treatment of the

carbon cycle. Carbon atoms have oxidation states ranging from +IV to −IV. The most common state is +IV in CO_2 and the familiar carbonate forms. Carbonate exists in two reservoirs, in the oceans as dissolved carbon in the forms of $H_2CO_3(aq)$, HCO_3^- (aq), and CO_3^{2-}(aq), and in the lithosphere as solid carbonate minerals: $CaCO_{3(s)}$, $CaMg(CO_3)_2$, $FeCO_3$. Carbon monoxide, CO, is a trace gas present in the atmosphere with carbon in oxidation state +II. Assimilation of carbon by photosynthesis creates the reduced carbon pools of the Earth. Reduced carbon is present with variable oxidation states that will be discussed further below. Methane, CH_4, is the most reduced form of carbon with an oxidation state of −IV.

11.2 The Isotopes of Carbon

There are seven isotopes of carbon (^{10}C, ^{11}C, ^{12}C, ^{13}C, ^{14}C, ^{15}C, ^{16}C) of which two are stable (^{12}C and ^{13}C). The rest are radioactive with half-lives between 0.74 s (^{16}C) and 5726 years (^{14}C). Only the stable isotopes and ^{14}C (often referred to as "radiocarbon") are included in this treatment of the carbon cycle.

The most abundant isotope is ^{12}C, which constitutes almost 99% of the carbon in nature. About 1% of the carbon atoms are ^{13}C. There are, however, small but significant differences in the relative abundance of the carbon isotopes in different carbon reservoirs. The differences in isotopic composition have proven to be an important tool when estimating exchange rates between the reservoirs. Isotopic variations are caused by fractionation processes (discussed below) and, for ^{14}C, radioactive decay. Formation of ^{14}C takes place only in the upper atmosphere where neutrons generated by cosmic radiation react with nitrogen:

$$^{14}N + {}^1n \rightarrow {}^{14}C + {}^1p$$

The ^{14}C content of the material in a carbon reservoir is a measure of that reservoir's direct or indirect exchange rate with the atmosphere, although variations in solar also create variations in atmospheric ^{14}C content activity (Stuiver and Quay, 1980, 1981). Geologically important reservoirs (i.e., carbonate rocks and fossil carbon) contain no radiocarbon because the turnover times of these reservoirs are much longer than the isotope's half-life. The distribution of ^{14}C is used in studies of ocean circulation, soil sciences, and studies of the terrestrial biosphere.

Fractionation is another major process responsible for creating inhomogeneities in the isotope distribution. Physical, chemical, and biological processes may be sensitive to the molecular weights of the molecules (atoms) involved. Thus exchanges between reservoirs can discriminate between different isotopes; for example when plants take up CO_2 they preferentially take up ^{12}C, this makes organic carbon on average lighter than atmospheric carbon. The definition of δ used to describe variations in isotope composition was introduced in Chapter 5. For ^{13}C, δ^{13}C, in parts per thousand, ‰, is defined by

$$\delta^{13}C = \left[\frac{^{13}R_S}{^{13}R_0} - 1 \right] 1000 \quad (1)$$

Here $^{13}R_S$ is the ^{13}C/^{12}C ratio in the sample and $^{13}R_0$ is the ^{13}C/^{12}C ratio the accepted standard PDB (PeeDee belemnite, after a Cretaceous belemnite rock from the PeeDee formation in North Carolina). Craig (1957a) has determined R_0 to be 0.0112372.

The ^{14}C content of a sample is described in a similar manner. The basis for $^{14}R_0$ is an oxalic acid standard of the US National Bureau of Standards normalized for ^{13}C fractionation and corrected for radioactive decay since a reference date January 1, 1950 (Stuiver and Polach, 1977). The absolute value of $^{14}R_0$ is $1.176 \cdot 10^{-12}$ (Stuiver et al., 1981).

^{14}C content is usually reported in Δ^{14}C units. The Δ^{14}C scale was originally defined by Broecker and Olson (1959). The reasoning behind introducing the scale is that all variations in ^{14}C due to fractionation should be eliminated by correcting for the sample's observed ^{13}C/^{12}C ratio relative to that of postulated average terrestrial wood. The wood is assumed to have a δ^{13}C value of $-25‰$ and fractionation for the ^{14}C isotope is assumed to occur as the square of that for ^{13}C for all processes. The approximate expression for Δ^{14}C proposed by Broecker and Olsson (1959) is

$$\Delta^{14}C = \delta^{14}C_S - 2(\delta^{13}C_S + 0.025)(1 + \delta^{14}C_S) \quad (2)$$

Δ^{14}C is used frequently in modeling, since no corrections for fractionation are necessary when modeling fluxes between reservoirs.

11.3 The Major Reservoirs of Carbon

11.3.1 The Atmosphere

Carbon is present in the atmosphere mainly as CO_2, with minor amounts present as CH_4, CO and other gases. The CO_2 content of the atmosphere is one of the best known quantities of the global carbon cycle. Accurate measurements were begun in 1957 (Keeling et al., 1976a,b; Bacastow and Keeling, 1981) with other groups following in the 1960s (Bischof, 1981) and 1970s (Pearman, 1981). One of the latest results of this research is shown in Fig. 11-2, which is a global picture of the variations in CO_2 as function of latitude. The seasonal variations have a 6-month

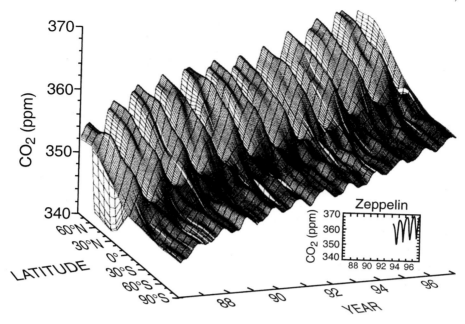

Fig. 11-2 Three-dimensional representation of the latitudinal distribution of atmospheric carbon dioxide in the marine boundary layer. Data from the NOAA CMDL cooperative air sampling network were used. The surface represents data smoothed in time and latitude. The Norwegian and Swedish flask sampling effort at Zeppelin Station is shown in the inset as flask monthly means. (Figure kindly provided by Dr Pieter Tans and Dr Thomas Conway of NOAA (CMDL).)

phase shift between the two hemispheres. The amplitude of the seasonal variations varies with latitude. The largest variations (10–15 ppmv) are seen at high latitudes, north of 50°N northern hemisphere. In southerly high latitudes the amplitude is only about 1 ppmv and along the equator there are small seasonal variations. The greater amplitude in the northern hemisphere is consistent with the occurrence of photosynthesis by extensive seasonal forests in that hemisphere that consume CO_2 in the spring and summer and respirate it in the fall and winter. Another very obvious fact in the CO_2 record is the increasing concentration caused by mankind's perturbation of the carbon cycle. This is mainly due to the combustion of fossil fuel, but carbon mobilized from carbon pools on land, mainly oxidation of phytomass and soil organic carbon, is also significant. The atmospheric CO_2 concentration at the end of 1997 was 365 ppmv at Mauna Loa. Estimates of the pre-industrial CO_2 content from recent ice-core results (Etheridge *et al.*, 1996) converge towards values close to

280 ppmv. Fossil fuel emissions (Marland and Boden, 1991; Marland and Rotty, 1984; Marland *et al.*, 1989; Rotty, 1981) are well known: the trade in coal and oil has considerable economic value and is therefore well documented. Assuming that all fossil fuel produced is oxidized within a few years from its removal from the ground, a good estimate of the total emissions can be made. During the past decades there has been an average observed airborne fraction of about 0.55. (The observed airborne fraction is defined as the observed CO_2 increase divided by the amount produced by fossil fuel combustion.) Since this quantity does not take any biospheric influences into account, it has limited value, although its use is widespread.

A recent development that has substantially raised our knowledge about the carbon cycle is the measurement of the O_2/N_2 atmospheric ratio (Heimann, 1997). Figure 11-3 shows the combined curves of CO_2 for Mauna Loa and South Pole as well as the recent development of data regarding the oxygen to nitrogen ratio in

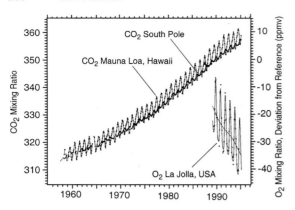

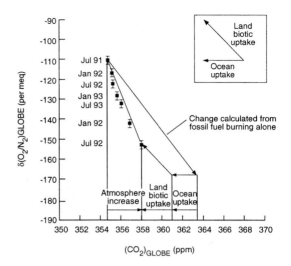

Fig. 11-3 Direct atmospheric measurements of the CO_2 concentration (left-hand scale) at Mauna Loa (Hawaii) and the South Pole station (Keeling *et al.*, 1995) together with the concurrently observed decrease in atmospheric oxygen content (right-hand scale) at La Jolla, CA after 1989. (Taken from Heimann (1997) with permission from the Royal Swedish Academy of Sciences.)

Figure 11-4 Globally and annually averaged oxygen versus CO_2 concentration from 1991 to 1994. The oxygen concentration is displayed as the measured O_2/N_2 ratio and expressed in "per meq" which denote the pm deviation from a standard ratio. The inset shows the directions of the state vector expected for terrestrial and oceanic uptake. The long arrow shows the expected atmospheric trend from fossil fuel burning if there were no oceanic and terrestrial exchanges. (Used with permission from Keeling *et al.* (1996). *Nature* **381**: 218–221, Macmillan Magazines.)

the air. We can note that there is a clear decline of the oxygen ration in the atmosphere consistent with the consumption of oxygen through combustion of fossil fuel. In Fig. 11-2 we saw that the seasonal variability of CO_2 is strongly dampened in the southern hemisphere; this is not true for atmospheric oxygen. To understand this one has to consider the carbonate chemistry in the oceans (see Section 10.4.2.5 of the previous chapter for an explanation of how dissolved inorganic carbon is dominated by carbonate and bicarbonate ions). Oxygen does not have the corresponding dissociation chemistry and will exchange freely between the surface waters and the atmosphere such that the seasonal cycle of oxygen in the surface waters is transferred into the atmosphere. The *increase* in CO_2 simultaneous with a *decrease* in oxygen is consistent with a carbon dioxide source from oxidation of highly reduced (organic) carbon (fossil fuels, biomass, or soil carbon). In Fig. 11-4 the combined usage of CO_2 and O_2 is utilized to calculate what reservoirs must have been changing in size to explain the trends of both gases during the past years. There has apparently been a substantial terrestrial sink active during the early 1990s. This is discussed more later in the chapter.

There are a number of ways to estimate the pre-industrial atmospheric CO_2 content. The total emissions (from fossil fuel combustion) during the period 1850 to 1982 are estimated at 173 Pg C with an uncertainty of less than 10 Pg (1 Pg = 10^{15} g). Assuming a constant airborne fraction of 0.54, a pre-industrial atmospheric CO_2 content of 614 Pg (290 ppmv) is calculated (Bolin *et al.*, 1981). This calculation does not take into account any biospheric emissions, nor is the assumption of a constant airborne fraction perfectly sound. Estimates of the pre-industrial CO_2 content based on the carbonate chemistry of "old" ocean (water that has yet to be contaminated by anthropogenic carbon emissions) give values ranging between 250 (Chen and Millero, 1979) and 275 ± 20 ppmv (Brewer, 1978). In a critical examination of CO_2 measurements performed in the 1880s, Wigley (1983) arrives at a CO_2 level around 260–270 ppmv. Finally, measurements of CO_2 content in air bubbles occluded in glacial ice from Antarctica and

Greenland (Neftel *et al.*, 1982; Barnola *et al.*, 1983; Etheridge *et al.*, 1996) give the least criticized data and indicate a value of 280 ± 10 ppmv (see Fig. 11-5).

Approximately 1% of the atmospheric carbon budget is maintained by methane (Ehhalt, 1974). Current global background levels of CH_4 are estimated at 1.7 ppmv, which corresponds to 3 Pg C (Blake and Rowland, 1988). Sources of methane are both natural and anthropogenic today. Important natural sources include fluxes from wetlands and enteric fermentation in wild animals (elephants and bison). Anthropogenic sources include rice paddies (to the extent they are placed in areas not previously part of natural wetlands), cattle (which have increased much more since the second world war than the decline in elephants and bison), coal mines, and leakage from gas fields (Crutzen, 1991; IPCC, 1990, 1992). Biomass burning is another substantial source of atmospheric methane. During the initial stages of burning there is usually an open flame in which essentially all carbon compounds are oxidized to CO_2. After the open flames have subsided there is, however, a long period of time when the remaining coals are fuming, releasing a multitude of less oxidized compounds (see Fig. 11-6) (Crutzen and Andreae, 1990; Hao *et al.*, 1996; Levine, 1994). Leakages from gasfields in the Former Soviet Union (FSU) were notorious. During the 1990s

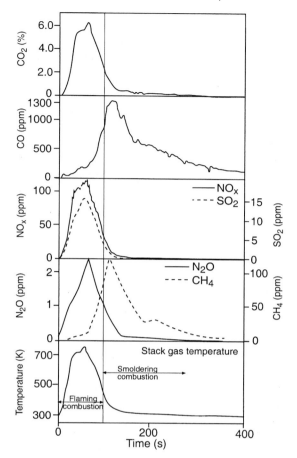

Fig. 11-6 Concentrations of gases in the smoke from an experimental fire of Trachypogon grass from Venezuela as a function of time and the stack gas temperature. The dotted line separates the flaming phase from the smoldering phase. Concentrations are in percent by volume for CO_2, in volume mixing ratios (ppm) for the other species (1% = 10 000 ppm). (Used with permission from Crutzen and Andreae (1990). *Science* **250**: 1669–1678, AAAS.)

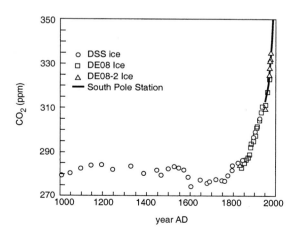

Fig. 11-5 CO_2 content of air bubbles trapped in glacial ice from Greenland and Antarctica, showing a pre-industrial concentration of ca. 280 ppmv.

the increase rate of atmospheric methane has decreased substantially probably to an extent due to infrastructure development in Russia combined with a general decrease in the gas extraction in FSU. The main sink for methane in the atmosphere is oxidation by the hydroxyl radical (·OH).

Oxidation of methane is one of the sources of atmospheric CO. Another internal source of importance is the oxidation of terpenes and isoprenes emitted by forests (Crutzen, 1983). Important are also biomass burning activities (Crutzen and Andreae, 1990; Hao et al., 1996; Levine, 1994). The carbon monoxide concentration in the atmosphere ranges from 0.05 to 0.20 ppmv in the remote troposphere (with considerable differences between the northern and southern hemispheres) which means that about 0.2 Pg of carbon is present as CO in the atmosphere.

Apart from CO_2, CH_4, and CO there are many gases containing carbon present in the atmosphere, terpenes, isoprenes, various compounds of petrochemical origin and others. We will not discuss them further, although some, like dimethylsulfide (DMS, $(CH_3)_2S$), are of great importance in the biogeochemical cycles of other elements. The total amount of atmospheric carbon in forms other than the three discussed is estimated at 0.05 Pg C (Freyer, 1979).

11.3.2   The Hydrosphere

Oceanic carbon is mainly present in four forms: dissolved inorganic carbon (DIC), dissolved organic carbon (DOC), particulate organic carbon (POC), and the marine biota itself. The marine biota, although it is a small carbon pool with a standing crop of about 3 Pg C (De Vooys, 1979) has a profound influence on the distribution of many elements in the sea (Broecker and Peng, 1982). Primary production in the photic zone is the major input of organic carbon in the oceans (Mopper and Degens, 1979). Labile (reactive) organic compounds are efficiently reoxidized in the mixed layer, whereas less than 10% of the primary production is distributed into the reservoirs of POC and DOC. Williams (1975) has used ^{14}C techniques to determine the average age of deep water DOC to be 3400 years. The DOC is thus clearly older than the turnover time of water in the deep oceans (100–1000 years) indicating the persistent nature of the dissolved organic compounds in the seas.

A detailed characterization of DOC is difficult to make; a large number of compounds have been detected but only a small portion of the total DOC has been identified. Identified species include amino acids, fatty acids, carbohydrates, phenols and sterols. The amount of carbon in the oceans as DOC is estimated to 1000 Pg and the amount present as POC is about 30 Pg (Mopper and Degens, 1979).

DIC concentrations have been studied extensively since the appearance of a precise analytical technique (Dyrssen and Sillén, 1967; Edmond, 1970). The aquatic chemistry of CO_2 has been treated extensively; reviews can be found in Skirrow (1975), Takahashi et al. (1980), and Stumm and Morgan (1981). When CO_2 dissolves in water it may hydrate to form $H_2CO_3(aq)$ which in turn dissociates to HCO_3^- and CO_3^{2-}. The conjugate pairs responsible for most of the pH buffer capacity in seawater are HCO_3^-/CO_3^{2-} and $B(OH)_3/B(OH)_4^-$ (with some minor contributions from silicate and phosphate). Although the predominance of HCO_3^- at the oceanic pH of 8.2 actually places the carbonate system close to a pH buffer minimum, its importance is maintained by the high DIC concentration (≈ 2 mM). Ocean water in contact with the atmosphere will, if the air–sea gas exchange rate is short compared to the mixing time with deeper waters, reach equilibrium according to Henry's Law.

Two further reactions to be considered are the ionization of water and the borate equilibrium:

$$H_2O + B(OH)_3(aq) \leftrightarrow B(OH)_4^-(aq) + H^+(aq);$$

$$K_B = \frac{[H^+][B(OH)_4^-]}{[B(OH)_3]} \quad (3)$$

In order to be able to solve for hydrogen ion concentration we define total borate (ΣB) and total carbon ($\Sigma C \equiv DIC$) as

$$\Sigma B = [B(OH)_3] + [B(OH)_4^-]$$

$$\Sigma C = [H_2CO_3^*] + [HCO_3^-] + [CO_3^{2-}]$$

$[H_2CO_3^*]$ represents the sum of $CO_2(aq)$ and H_2CO_3. Alkalinity, a capacity factor, representing the acid neutralizing capacity of the aqueous solution, is given by the following equation (ignoring influences from some minor components like phosphate and silicate) (see also Chapter 5):

$$Alk = [OH^-] - [H^+] + [B(OH)_4^-] \\ + [HCO_3^-] + 2[CO_3^{2-}] \quad (4)$$

Given any two of the four quantities ΣC, Alk, pH, P_{CO_2}, the other two can always be calculated provided appropriate equilibrium constants are available (the equilibrium constants depend on temperature, salinity and pressure). Hydrogen ion concentration, for example, be calculated from Alk and ΣC with the equation

$$\text{Alk} = \frac{[H^+]}{K_w} - [H^+] + \frac{\Sigma B \cdot K_B}{[H^+] + K_B} + \frac{K_1[H^+] + 2K_1K_2}{[H^+]^2 + [H^+]K_1 + K_1K_2} \Sigma C \quad (5)$$

K_1 and K_2 are the dissociation constants for H_2CO_3. Alkalinity and ΣC are the analyzed values and ΣB is calculated from salinity. P_{CO_2} can then be calculated by

$$P_{CO_2} = K_H \left(1 + \frac{K_1}{[H^+]} + \frac{K_1K_2}{[H^+]^2}\right)^{-1} \Sigma C \quad (6)$$

where K_H is the Henry's Law constant for CO_2 described in Chapter 5. At the pH of ocean water (about 8) most of the DIC is in the form of HCO_3^- and CO_3^{2-} (Fig. 11-7) with a very small proportion being $[H_2CO_3^*]$. Although $[H_2CO_3^*]$ changes in proportion to $CO_2(g)$, the ionic forms changes little as a result of the various acid–base equilibria. This fact is responsible for the "buffer" factor (buffer here refers to buffering of CO_2 exchange) also known as the "Revelle" factor (Revelle and Suess, 1957). See Chapter 4 for an application of this factor. The buffer factor is defined by

$$\beta = \frac{(\Delta P_{CO_2}/P_{CO_2})}{(\Delta \Sigma C/\Sigma C)} \quad (7)$$

Figure 11-8 shows how P_{CO_2}, pH, and β are dependent on ΣC and Alk. Note that in Fig. 11-8 that organic carbon formation (or the opposite process, respiration) is moving parallel to the alkalinity in the diagram (apart from a small alkalinity change due to nitrate and phosphate uptake) axis whereas calcium carbonate formation represents a vector that decreases DIC by one unit for every two units of alkalinity (moles and equivalents respectively). For current atmospheric P_{CO_2}, prevailing temperature and salinity, β is about 14 in polar regions and about 10 in equatorial waters. The significance of $\beta \sim 10$ is that for a 10% increase in atmospheric P_{CO_2} only a 1% increase in ΣC is necessary to reach a new equilibrium. The buffer factor's large value is important since it greatly constrains the ocean's ability to take up increases in atmospheric CO_2. Average DIC and Alk concentrations for the world oceans can be seen in Fig. 11-9. With an average DIC of 2.35 mmol/kg seawater and a world oceanic volume of 1370×10^6 km^3 the DIC carbon reservoir is estimated to be 37 900 Pg C (Takahashi *et al.*, 1981). The surface waters of the ocean contain a minor part of the DIC 700 Pg C. Nevertheless, the surface waters play an important role as a means of communication between the atmosphere and the deep oceans. Although DIC is a large carbon reservoir with lively exchange with the atmosphere, its importance as a sink for anthropogenic CO_2 emission is restricted by several factors. The static uptake capacity of the seawater (solubility and "buffer" factor), the slowness of attainment of equilibrium between the ocean water and atmosphere, the ventilation of the deep ocean, and the oceanic sedimentation rate all impose constraints on the role of the oceans as a sink for atmospheric CO_2. The sluggish deep water formation rates essentially dictated by the fact that the oceans are heated from above which creates a very stable stratification creates a central bottleneck in the carbon

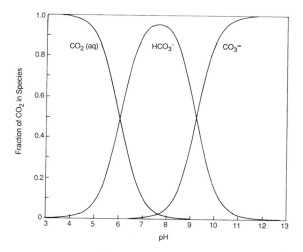

Fig. 11-7   Distribution of dissolved carbon species in seawater as a function of pH. Average oceanic pH is about 8.2. The distribution is calculated for a temperature of 15°C and a salinity of 35‰. The equilibrium constants are from Mehrbach *et al.* (1973).

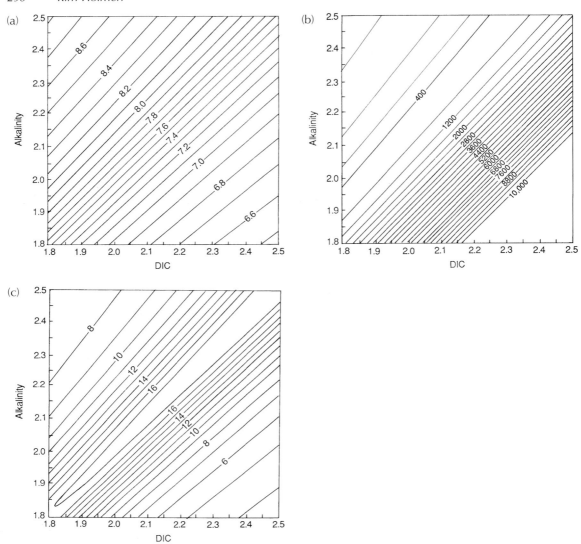

Fig. 11-8   Isolines of pH (a), P_{CO_2} (b), and the Buffer factor (c) plotted as functions of DIC and alkalinity. The lines have been calculated for a temperature of 15°C and a salinity of 35‰. The equilibrium constants for K_1 and K_2 are from Mehrbach *et al.* (1973), K_0 from Hansson (1973) and B calculated from salinity according to the formula given by Culkin (1965).

cycle. Despite a very rapid exchange of carbon between the surface waters and the atmosphere the carbon is not transported away from the atmosphere. The turnover time of carbon dioxide in the atmosphere is short (a few years) but the ability of the carbon cycle to transport excess carbon (like anthropogenic releases of fossil fuel) away from the atmosphere is much longer (decades to hundreds of years). A signifi-cant consequence of this is that releases of CO_2 into the atmosphere will create a perturbation of the atmospheric concentration of CO_2 that will require hundreds of years to fully equilibrate.

Oceanic surface water is everywhere super-saturated with respect to the two solid calcium carbonate species calcite and aragonite. Never-theless carbonate precipitation is exclusively controlled by biological processes, specifically

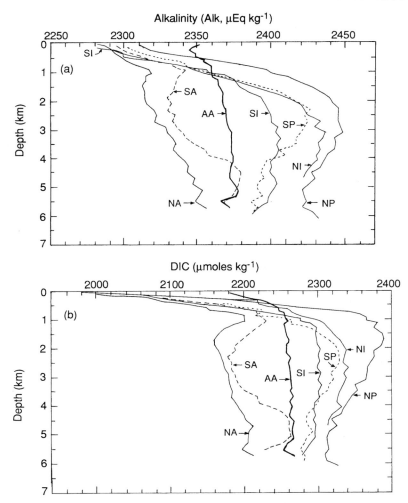

Fig. 11-9 (a) The vertical distributions of alkalinity (Alk) and dissolved inorganic carbon (DIC) in the world oceans. Ocean regions shown are the North Atlantic (NA), South Atlantic (SA), Antarctic (AA), South Indian (SI), North Indian (NI), South Pacific (SP), and North Pacific (NP) oceans. (Modified with permission from T. Takahashi *et al.*, The alkalinity and total carbon dioxide concentration in the world oceans, in B. Bolin (1981). "Carbon Cycle Modelling," pp. 276–277, John Wiley, Chichester.)

the formation of hard parts (i.e., shells, skeletal parts etc.). The very few existing accounts of spontaneous inorganic precipitation of $CaCO_3(s)$ (so-called "whitings") come from the Bahamas region of the Caribbean (Morse *et al.*, 1984).

The detrital rain of carbon-containing particles can be divided into two groups: the hard parts comprising calcite and aragonite and the soft tissue containing organic carbon. The composition of the soft tissue shows surprising uniformity, the average composition being $(CH_2O)_{106} (NH_3)_{16}PO_4$ (see Chapter 10, Section 10.3.1). The average composition of the particulate matter (here a composite of organic and inorganic particles) settling through the water column and subsequently being dissolved in the deep ocean is given by P:N:C:Ca:S = 1:15:131:26:50 (Broecker and Peng, 1982) with a $CaCO_3$-C/Org-C ratio of 1:4. Calculating an average composition of the carbon that actually is deposited in sediments is more difficult since the areas of deposition are different for organic and inorganic carbon. More than 90% of the deposition of organic material takes place on the continental shelves; soft tissues falling into the deep oceans are consumed by heterotrophic organisms before isolation from the water column within the sediments.

The solubility of calcite and aragonite increases with increasing pressure and decreasing temperature in such a way that deep waters are undersaturated with respect to calcium carbonate, while surface waters are supersaturated. The level at which the effects of dissolution are first seen on carbonate shells in the sediments is termed the lysocline and coincides fairly well with the depth of the carbonate saturation horizon. The lysocline commonly lies between 3 and 4 km depth in today's oceans. Below the lysocline is the level where no carbonate remains in the sediment; this level is termed the carbonate compensation depth.

The variations in ^{14}C seen in the deep oceans of the world (Fig. 11-10) show features created by radioactive decay. The radiocarbon distribution is an important tool for determining the replacement times of the deep oceans. Great care has to be taken when interpreting the ^{14}C distribution to take into account mixing between waters of different origin. This is especially true in the Atlantic, since the degree of isotopic equilibrium reached between the air and surface waters is different in the two source areas of Atlantic deep water, the Arctic and Antarctic surface waters (Broecker, 1979). This complication makes the apparent ^{14}C age in seawater not simply a measure of the time elapsed since isolation from the atmosphere but a complex blending of the effects of water-mass mixing and uneven degrees of isotopic equilibrium in the ocean.

Care must also be taken to not confuse the ^{14}C perturbation, e.g. from nuclear weapons testing with the ΔCO_2 perturbation lifetime. The former is largely due to fast isotopic exchange while the latter is controlled by a slower mass flux.

11.3.3 *The Terrestrial Biosphere*

Large amounts of carbon are found in the terrestrial ecosystems and there is a rapid exchange of carbon between the atmosphere, terrestrial biota, and soils. The complexity of the terrestrial ecosystems makes any description of their role in the carbon cycle a crude simplification and we shall only review some of the most important aspects of organic carbon on land. Inventories of the total biomass of terrestrial ecosystems have been made by several researchers, a survey of these is given by Ajtay *et al.* (1979).

Primary production maintains the main carbon flux from the atmosphere to the biota. In the process of photosynthesis, CO_2 from the atmosphere is reduced by autotrophic organisms to a wide range of organic substances. The complex biochemistry involved can be represented by the formula

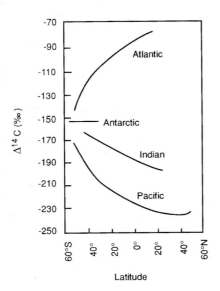

Fig. 11-10 The Δ^{14}C values of the cores of the North Atlantic, Pacific, and Indian Oceans deep waters. The oldest waters are encountered near 40°N in the Pacific Ocean. (Modified with permission from M. Stuiver *et al.* (1983). Abyssal water carbon-14 distribution and the age of the world oceans, *Science* **219**, 849–851, AAAS.)

$$CO_2 + H_2O \; \underset{\text{respiration}}{\overset{\text{Assimilation}}{\rightleftharpoons}} \; (CH_2O)_n + O_2 \quad (8)$$

Gross primary production (GPP) is the total rate of photosynthesis including organic matter consumed by respiration during the measurement period, while net primary production (NPP) is the rate of storage of organic matter in excess of respiration. There are two main routes taken to estimate the world NPP and standing phytomass. The first method is to classify the biosphere into ecosystems in which, from measurements of estimates, values for the primary productivity and phytomass are assigned. The alternative method is to use estimates made by prognostic models simulating the effects of environmental factors on productivity and phytomass.

The possible effects of increased atmospheric CO_2 on photosynthesis are reviewed by Goudriaan and Ajtay (1979) and Rosenberg (1981). Increasing CO_2 in a controlled environment (i.e., greenhouse) increases the assimilation rate of some plants, however, the anthropogenic fertilization of the atmosphere with CO_2 is probably unable to induce much of this effect since most plants in natural ecosystems are growth limited by other environmental factors, notably light, temperature, water, and nutrients.

Estimates of terrestrial biomass vary considerably, ranging from 480 Pg C (Garrels *et al.*, 1973) to 1080 Pg C (Bazilevich *et al.*, 1970). Bazilevich *et al.* attempted to estimate the magnitude of the biomass before mankind's perturbation of the ecosystems. The latest work that undoubtedly had the most data available estimates the total terrestrial biomass, valid as of 1970, as 560 Pg C (Olson *et al.*, 1983).

Terrestrial biomass is divided into a number of subreservoirs with different turnover times. Forests contain approximately 90% of all carbon in living matter on land but their NPP is only 60% of the total. About half of the primary production in forests yields twigs, leaves, shrubs, and herbs that only make up 10% of the biomass. Carbon in wood has a turnover time of the order of 50 years, whereas turnover times of carbon in leaves, flowers, fruits, and rootlets are less than a few years. When plant material becomes detached from the living, plant carbon is moved from the phytomass reservoir to litter. "Litter" can either refer to a layer of dead plant material on the soil or all plant materials not attached to a living plant. A litter layer can be a continuous zone without sharp boundaries between the obvious plant structures and a soil layer containing amorphous organic carbon. Decomposing roots are a kind of litter that seldom receives a separate treatment due to difficulties in distinguishing between living and dead roots. Average turnover time for carbon in litter is thus about 1.5 years, although caution should be observed when using this figure. For tropical ecosystems with mean temperatures above 30°C the litter decomposition rate is greater than the supply rate so storage is impossible. For colder climates NPP exceeds the rate of decomposition in the soil. The average temperature at which there is balance between production and decomposition is about 25°C. The presence of peat, often treated as a separate carbon reservoir, exemplifies the difficulty in defining litter. The total amount of peat is estimated at 165 Pg C (Ajtay *et al.*, 1979). Figure 11-11 illustrates this very strongly. The tropics have an extremely high NPP but very little carbon in the soil; whereas all higher latitude areas have the opposite relationship. The dynamics of the carbon reservoirs is very different on either side of the balance isotherm. Also thought provoking is the fact that a very large proportion of the areas that are covered today with carbon-rich soils have appeared in areas covered by ice shields during ice ages; much of the carbon in these soils today has probably been deposited since the last glaciation. A climatic change that moves the balance isotherm polewards would most likely give rise to a net flux of carbon to the atmosphere from regions that today are close to balance or in carbon accumulation zones. The zones of soil carbon accumulation are also the zones most likely to experience growth limitation due to lack of nutrients since the continuous deposition of carbon will always retain some nutrients (e.g., N and P). Another observation regarding Fig. 11-11 is that land-use changes occurring today in the tropics mobilize carbon to the atmosphere by the decrease in standing biomass; the companion flux of soil carbon oxidized upon plowing of virgin land is much smaller than for opening new agricultural land in temperate regions. This is something that occurred in Europe and North America during the nineteenth century. Many of

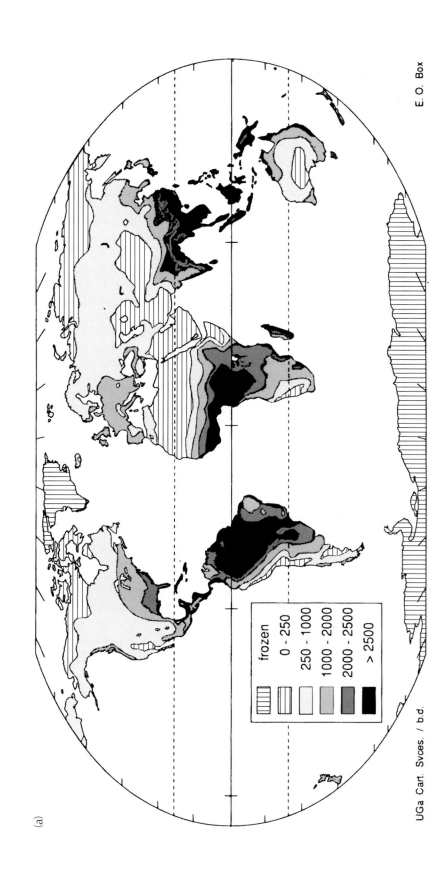

frozen

0 - 250

250 - 1000

1000 - 2000

2000 - 2500

> 2500

UGa Cart. Svces. / b.d.

E. O. Box

(a)

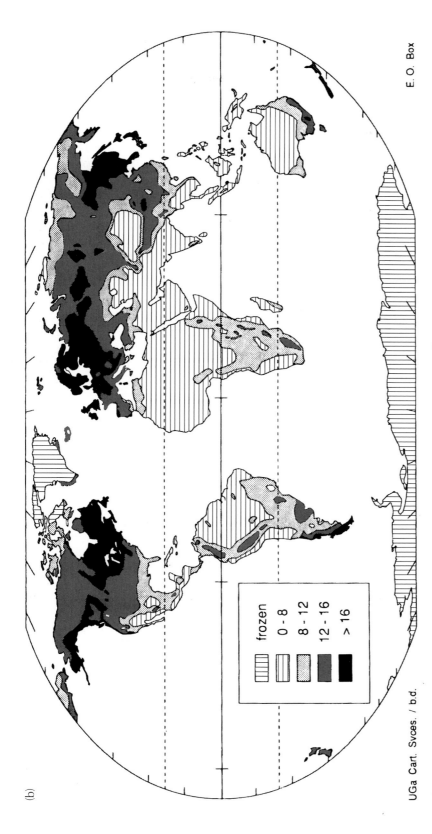

(b)

UGa Cart. Svces. / b.d.

E. O. Box

legend:
- frozen
- 0 - 8
- 8 - 12
- 12 - 16
- > 16

Fig. 11-11 (a) Global distribution of carbon produced annually, in grams of dry matter/m² per year. (b) Global distribution of carbon preserved in soils, in kg/m². (Both figures from Box and Meentemeyer (1991), used with permission from Elsevier Publishers.)

these lands are today being abandoned because of changes in agricultural practice; this gives rise to a net flux of carbon from the atmosphere to the standing biomass and soils (Sedjo, 1992). In the perspective of national carbon balances there are several complications; the net uptake today in temperate regions is only possible due to an early release of carbon due to land use changes. This illustrates two aspects of carbon exchanges and the terrestrial biosphere, the time scales of exchange can be long and simultaneously human behavior can alter the reservoirs rapidly upon change of human habits (see Fan *et al.*, 1998).

There is a group of organic compounds in terrestrial ecosystems that are not readily decomposed and therefore make up a carbon reservoir with a long turnover time. There are also significant structural differences between the marine and terrestrial substances (Stuermer and Payne, 1976). The soil organic matter of humus is often separated into three groups similar in structural characteristics but with differing solubility behavior in water solutions. Humic acids, fulvic acids, and humin are discussed in Chapter 8. Schlesinger (1977) presented an assessment of the various carbon pools for temperate grassland soil (Fig. 11-12). The undecomposed litter (4% of the soil carbon) has a turnover time measured in tens of years, the 22% of the soil carbon in the form of fulvic acids is intermediate with turnover times of hundreds of years. The largest part (74%) of the soil organic carbon (humins and humic acids) also has the longest turnover times (thousands of years).

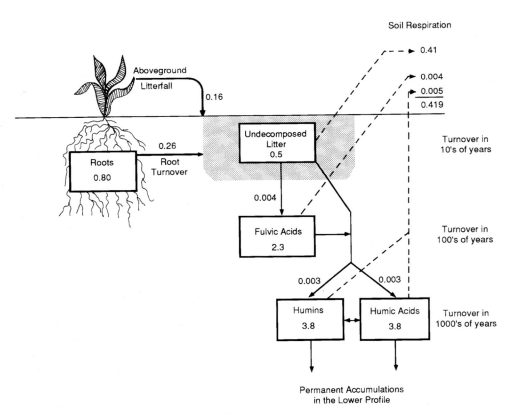

Fig. 11-12   Detrital carbon dynamics for the 0–20 cm layer of chernozem grassland soil. Carbon pools (kg C/m^2) and annual transfers (kg C/m^2 per year) are indicated. Total profile content down to 20 cm is 10.4 kg C/m^2. (Reproduced with permission from W. H. Schlesinger (1977). Carbon balance in terrestrial detritus, *Ann. Rev. Ecol. Syst.* **8**, 51–81, Annual Reviews, Inc.)

11.3.4 The Lithosphere

Although the largest reservoirs of carbon are found in the lithosphere, the fluxes between it and the atmosphere, hydrosphere, and biosphere are small. It follows that the turnover time of carbon in the lithosphere is many orders of magnitude longer than the turnover times in any of the other reservoirs. Many of the current modeling efforts studying the partitioning of fossil fuel carbon between different reservoirs only include the three "fast" spheres; the lithosphere's role in the carbon cycle has received less attention.

Fossil fuel burning is an example of mankind's ability to significantly alter fluxes between reservoirs. The burning of fossil fuel transfers carbon from the vast pool of reduced carbon in the lithosphere to the atmosphere, and hence to the biosphere, hydrosphere, soils and sediments. The elemental carbon reservoir is estimated from average carbon contents in different types of rocks, ranging from 0.9% elemental carbon in shales to 0.1% in igneous and metamorphic rocks (Kempe, 1979b) and the relative abundance of the rock types. The resulting estimate is 2×10^7 Pg C (Hunt, 1972), a single reservoir several orders of magnitude larger than the sum of all reservoirs discussed so far. Of the 2×10^6 Pg of recycled elemental carbon (recycled carbon has traveled at least once through the lithospheric cycle) in the lithosphere only 10^4 Pg make up the economically extractable reserves of oil and coal. Most of the reduced carbon species in the Earth's crust are highly dispersed and probably never will be used as fuels. The carbonate minerals distributed in sedimentary rocks represent a carbon reservoir that is even larger than the elemental carbon reservoir. About three-fourths of the carbon in the Earth's crust is present as carbonates. Several forms exist; the dominant biogenic forms are calcite and aragonite. Both are stoichiometrically $CaCO_3$ but calcite has six-coordinated Ca atoms and is capable of substituting several percent Mg into its lattice. Aragonite has nine-coordinated Ca atoms and several percent Sr can be incorporated into its lattice. Both forms can precipitate depending on the Ca/Mg ratio in the solution; for the present

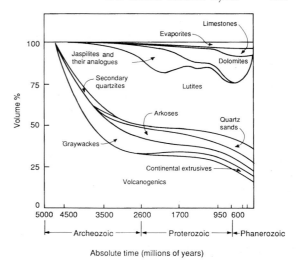

Fig. 11-13 Volume percent of sedimentary rocks as a function of age. (Modified with permission from A. B. Ronov (1964). On the post-Cambrian geochemical history of the atmosphere and hydrosphere, *Geochemistry* **5**, 493–506, American Geological Institute.)

ocean Mg-calcite or aragonite are precipitated. Dolomite $(CaMg(CO_3)_2)$ is a carbonate mineral of wide importance formed by diagenetic disintegration of Mg-rich calcites. Formation of dolomite is slow today and largely confined to evaporitic settings (Holland, 1978). The invasion of land by plants 600 million years ago, at the beginning of the Phanerozoic era increased the availability of CO_2 in the soil. There was a marked decrease of dolomitic sediments and increase in limestone sediments coherent with the appearance of terrestrial vegetation (Fig. 11-13).

11.4 Fluxes of Carbon between Reservoirs

Carbon is released from the lithosphere by erosion and resides in the oceans ca. 10^5 years before being deposited again in some form of oceanic sediment. It remains in the lithosphere on the average 10^8 years before again being released by erosion (Broecker, 1973). The amount of carbon in the ocean–atmosphere–biosphere system is maintained in a steady state by geologic processes; the role of biological processes is, however, of profound importance for the partitioning of carbon between the "fast" reservoirs.

Chemical weathering of crustal material can both add and withdraw carbon from the atmosphere. This has been discussed in Chapter 8. The oxidation of reduced carbon releases CO_2 to the atmosphere,

$$C^0(s) + O_2 \rightarrow CO_2$$

whereas dissolution of carbonates is associated with uptake of CO_2:

$$CaCO_3(s) + CO_2(g) + H_2O \rightarrow Ca^{2+}(aq) + 2HCO_3^-(aq)$$

Silicates also lead to uptake of CO_2. Weathering of a non-aluminum silicate like Mg-olivine may be written

$$Mg_2SiO_4(s) + 4CO_2 + 4H_2O \rightarrow 2Mg^{2+} + 4HCO_3^- + H_4SiO_4(aq)$$

An example of aluminosilicate weathering is the reaction of the feldspar albite to a montmorillonite-type mineral

$$2NaAlSi_3O_8(s) + 2CO_2 + 2H_2O \rightarrow Al_2Si_4O_{10}(OH)_2(s) + 2Na^+ + 2HCO_3^- + 2SiO_2(s)$$

In the weathering of carbonates, one mole of rock CO_2 is mobilized for each mole of atmospheric CO_2 consumed. The reverse reaction (sedimentation of carbonate in the oceans) will again release one mole equivalent of $CO_2(g)$. For the weathering of silicates there is a 1:1 relationship between $CO_2(g)$ consumed and HCO_3^- produced, in contrast to the 1:2 relationship for carbonate weathering. Estimates of global erosion rates are based either on average river data (Livingstone, 1963; Kempe, 1979b) or on global material balance calculations performed by extrapolating the material balance from a well documented area to the world (Kempe, 1979b).

The freshwater cycle is an important link in the carbon cycle as an agent of erosion and as a necessary condition for terrestrial life. Although the amount of carbon stored in freshwater systems is insignificant as a carbon reservoir (De Vooys, 1979; Kempe, 1979a), about 90% of the material transported from land to oceans is carried by streams and rivers.

Pure water in equilibrium with atmospheric CO_2 has a pH of about 5.6 at ambient temperatures. Although rainwater pH is affected by other airborne species with acid–base characteristics (Charlson and Rodhe, 1982), calculating the flux of carbon carried from the atmosphere to the surface by pH 5.6 rainwater is a good approximation, if a bit large. According to the weathering reaction, half the bicarbonate in stream water originates in the atmosphere. Rainwater clearly is unimportant as source for this carbon. Air within soils is the major source of CO_2 taking part in weathering reactions with carbonates and silicates. Bacterial decomposition of organic material together with root respiration maintain high partial pressures of CO_2 in soil air (Chapter 7 and Kempe, 1979a). Approximately 0.4 Pg C/yr are withdrawn from the atmosphere by weathering reactions, 0.1 Pg C/yr is released by oxidation of elemental carbon, yielding a net flux of 0.3 Pg C/yr from atmosphere to lithosphere (Holland, 1978). The importance of the presence of liquid water on the face of the Earth, despite the relatively small numbers indicated above, is of profound importance for the state of the planet. The process of precipitating carbonate minerals would probably not proceed without water; a consequence would then be that most of the carbon presently deposited in the lithosphere would be present as CO_2 in the atmosphere much like the atmospheres on Venus and Mars.

Garrels and Mackenzie (1971) calculated global river loads based on Livingstone's (1963) data. From these figures Kempe (1979a) deduced a total flux of 0.84 Pg/yr with rivers where about half is in the form of DIC and the rest being approximately evenly distributed between PIC, DOC, and POC.

Kempe (1979b) also estimates the flux of carbon from the lithosphere to oceans from glacial erosion (0.033 Pg/yr), global dust production (0.06 Pg/yr) and marine erosion (0.0045 Pg/yr) to give a combined flux of 0.1 Pg/yr. Although there is a general consensus that geologic processes control the amount of carbon in the ocean–atmosphere system, the mechanisms are debated. Walker (1977) assumes a steady-state model where the weathering rate is dependent on the CO_2 partial pressures in the atmosphere. An increase in P_{CO_2}

will give a higher weathering rate, which results in an increase of the cation content of the oceans. The increased cation concentration yields a higher precipitation rate of carbonates which removes carbon, thereby restoring the original balance. The long-term atmospheric carbon dioxide budget is thus governed by a volcanic source and consumption by the weathering of silicates. Volcanism does not depend on atmospheric P_{CO_2}, but weathering does. The equilibrium level of CO_2 is therefore determined by the demand that the weathering sink must just balance the volcanic source. Holland (1978) presents somewhat different arguments and deduces, from balance calculations, indications of a juvenile carbon flux from the Earth's interior of 0.08 Pg C/yr.

The exchange of carbon between the terrestrial biosphere and atmosphere goes through two channels. CO_2 is the major route with CH_4 making up about 1% of the exchange. Methane is mainly produced by enteric fermentation by animals, and anaerobic production in paddy fields, freshwater lakes, swamps, and marshes. Ehhalt (1974) estimates an upper limit for the herbivore production of methane from an assumption that 10% of the dry plant matter produced on land is consumed by herbivores and the food to methane conversion ratio for cattle is valid for all herbivores (cattle have the highest measured ratio). The present upper limit is 0.17 Pg C/yr with a range down to 0.08 Pg C/yr. Interestingly, the increase in world cattle population since the early 1940s accounts for about 25% of this figure. Methane emissions to the atmosphere from freshwater lakes, swamps and marshes are in the range of 0.15–0.22 Pg C/yr. The total flux of methane to the atmosphere is thus 0.5 Pg C/yr.

Carbon monoxide emissions from the terrestrial biosphere are small, but forest fires produce 0.02 Pg C/yr. Degradation of chlorophyll is dying plant material seems to be the largest CO-producing mechanism at 0.04–0.2 Pg C/yr (Freyer, 1979).

The exchange of CO_2 between the atmosphere and terrestrial biota is one of the prime links in the global carbon cycle. This is seen by studying the variations of ^{13}C in the atmosphere. Figure 11-14 presents atmospheric $\Delta^{13}C$ for the years

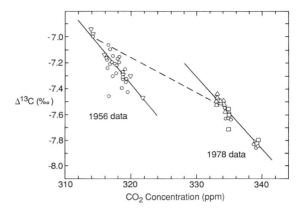

Fig. 11-14 Change in the relation between $\delta^{13}C$ and concentration of atmospheric CO_2 over 22 years. Mean change is shown as a dashed line. Solid lines show mixing relations for 1956 and 1978. __ denotes inferred northern hemispheric means for 1956 and 1978. (Modified with permission from C. D. Keeling *et al.* (1979). Recent trends in the $^{13}C/^{12}C$ of atmsopheric carbon dioxide, *Nature* **277**, 121–123, Macmillan Magazines.)

1956 and 1978. The lines are consistent with addition or subtraction of CO_2 with a $\Delta^{13}C$ of about $-27‰$, which could be derived from either fossil fuel or plant CO_2. It cannot be oceanic because surface water DIC has a $\Delta^{13}C$ of about $+2‰$ (Kroopnick, 1980). This confirms that the annual P_{CO_2} variations are primarily due to exchange with the terrestrial biosphere, and not caused by seasonal exchange with the oceans. This result has now been further confirmed by the already mentioned O_2/N_2 measurements (Heimann, 1997 and Fig. 11-3).

Many estimates of total terrestrial net primary production are available, ranging between 45.5 Pg C/yr (Lieth, 1972) and 78 Pg/yr (Bazilevich *et al.*, 1970). Ajtay *et al.* (1979) have revised the various estimates and methods involved, they also reassess the classifications of ecosystem types and the extent of the ecosystem surface area using new data and arriving at a total NPP of 60 Pg C/yr. Gross primary production is estimated to be twice net primary production, i.e., 120 Pg C/yr. This implies that about 60 Pg C/yr are returned to the atmosphere during the respiratory phase of photosynthesis. It is well known that carbon dioxide uptake by plants follows daily cycles; most plants take up CO_2

during the day and emit it at night. These diurnal patterns give rise to large variations in P_{CO_2} close to the vegetation sites (Fig. 11-15). The diurnal cycles that produce local changes are superimposed on the yearly cycles that give the hemispheric P_{CO_2} variations seen in Fig. 11-2.

The subsequent fate of the assimilated carbon depends on which biomass constituent the atom enters. Leaves, twigs, and the like enter litterfall, and decompose and recycle the carbon to the atmosphere within a few years, whereas carbon in stemwood has a turnover time counted in decades. In a steady-state ecosystem the net primary production is balanced by the total heterotrophic respiration plus other outputs. Non-respiratory outputs to be considered are fires and transport of organic material to the oceans. Fires mobilize about 5 Pg C/yr (Baes *et al.*, 1976; Crutzen and Andreae, 1990), most of which is converted to CO_2. Since bacterial heterotrophs are unable to oxidize elemental carbon, the production rate of pyroligneous graphite, a product of incomplete combustion (like forest fires), is an interesting quantity to assess. The inability of the biota to degrade elemental carbon puts carbon into a reservoir that is effectively isolated from the atmosphere and oceans. Seiler and Crutzen (1980) estimate the production rate of graphite to be 1 Pg C/yr.

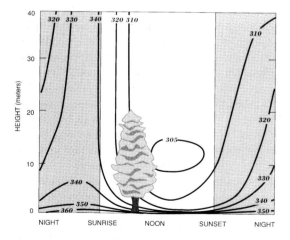

Fig. 11-15 Variation in the vertical distribution of carbon dioxide in the air around a forest with time of day. (Reprinted with permission from B. Bolin (1970). The carbon cycle. In "The Biosphere," p. 51, W. H. Freeman, NY.)

River transport of organic carbon, estimated earlier as 0.1 Pg C/yr, brings the sum of non-respiratory outputs to 7 Pg C/yr. Total respiration should therefore be around 50 Pg C/yr. This figure is in agreement with estimates of soil respiration rates determined from compilations of ecosystem types and their measured soil respiration rates (Ajtay *et al.*, 1979).

The exchange of carbon dioxide between ocean and atmosphere has been studied extensively, since the prevailing view is that fossil fuel derived CO_2 not remaining in the atmosphere has entered the oceans (Siegenthaler and Sarmiento, 1993). To appraise the ocean–atmosphere exchange we make use of the radiocarbon distribution in the oceans. All ^{14}C is produced in the atmosphere, hence all radiocarbon in the oceans must have entered through the air–sea interface. Under a steady-state assumption the net influx of ^{14}C must be balanced by the total decay within the oceans. Using our knowledge of the ^{14}C distribution in the oceans (Fig. 11-10) and the radiocarbon decay constant for ^{14}C we can calculate the flux of carbon by the following simple relations:

$$F_{ma} = F_{am}$$

$$F_{am}R_a = F_{ma}R_m + kR_0V_0C_0$$

where F_{ma} is the flux from the ocean mixed layer to atmosphere, F_{am} is the flux from atmosphere to mixed layer, k is the ^{14}C decay rate, R_a, R_m, and R_0 are the ^{14}C ratios in atmosphere, mixed layer, and average ocean respectively, V_0 is the ocean volume, and C_0 an average DIC concentration of the oceans. Solving for F_{am} we obtain the following:

$$F_{am} = (kR_0V_0C_0/(R_a - R_m) \approx 6.5 \times 10^{15} \text{ mol/yr} \tag{9}$$

The gross flux of carbon from atmosphere to ocean is thus ca. 80 Pg C/yr. There are several complications with the above calculation. The isotopic ratios must be steady-state values, which are unavailable due to the changes resulting from atmospheric atom bomb testing. The few available pre-bomb measurements from the late 1950s (Broecker *et al.*, 1960) together with $\Delta^{14}C$ determinations in corals (Druffel and Linick, 1978) are invaluable tools for determining a steady state R_m value. Nevertheless, we must be aware of the great sensitivity of the flux estimate to the R_m value since F_{am} is dependent on the reciprocal of $R_a - R_m$, which is a small number. Equation (9) can be reformulated to include the variations of surface water P_{CO_2} and the variations of $\Delta^{14}C$ in surface waters. Figure 11-16 shows latitudinal distributions of P_{CO_2} in the Atlantic and Pacific Oceans (Tans *et al.*, 1990). Surface water $\Delta^{14}C$ exhibits considerable variations with very low values in the Antarctic surface waters. This flux has also been constrained with ^{13}C studies (Quay *et al.*, 1992).

The two prime mechanisms of carbon transport within the ocean are downward biogenic detrital rain from the photic zone to the deeper oceans and advection by ocean currents of dissolved carbon species. The detrital rain creates inhomogeneities of nutrients illustrated by the characteristic alkalinity profiles (Fig. 11-9). The amount of carbon leaving the photic zone as sinking particles should not be interpreted as the net primary production of the surface oceans since most of the organic carbon is recycled

within the photic zone; only about 10% settles as detritus (Bolin *et al.*, 1979). There is considerable patchiness in the rate of oceanic primary production (Fig. 11-17) with high values in areas of intense upwelling, while areas of slow sinking motions (the subtropical gyres) show photosynthesis rates only one-tenth as large. De Vooys (1979) gives a thorough account of the many methods employed and uncertainties involved in estimating the net primary production of the aquatic environments. We adopt an estimate of total primary production of 50 Pg C/ yr but note that the range of estimates is huge (15–126 Pg C/yr).

If 10% of the total primary production settles as detrital material, about 5 Pg C leaves the photic zone annually. The $CaCO_3$ to organic carbon ratio in the detritus is usually taken as 1:4 (Broecker and Peng, 1982) which signifies a carbonate flux around 1 Pg C/yr. The inorganic/organic ratio can actually vary; Chen (1978) obtained ratios between 1:10 and 1:3 in a water column study. The detrital rain is balanced by a small river input of carbon and upwelling of deep water enriched in carbon from the decomposition of the detritus at depth. To balance the carbon budget of the photic zone, oceanic circulation must provide a net transport of about 5 Pg C/yr from deeper layers to the surface. Comparing Fig. 11-17 with a map of upwelling areas clearly shows that upwelling and primary production are coupled processes. We can also study Fig. 11-16 where the high P_{CO_2} values in equatorial regions are caused by the upwelling of carbon-rich and cold, deep water. This water is supersaturated with respect to atmospheric CO_2 when it upwells. This state enhanced further upon warming. The North Atlantic has low P_{CO_2} values caused by biological depletion of carbon and cooling of waters flowing northward (i.e., the Gulf Stream). The yearly circulation of carbon through the atmosphere from equatorial upwelling regions to high latitudes is around 0.01 Pg C/yr (Bolin and Keeling, 1963).

In a steady-state ocean the sediment deposition rate of a nutrient like phosphorus ought to be balanced by riverborne influx to the oceans; 1.5–4.0 Tg P are transported to the oceans by rivers (Richey, 1983). Assuming a C/P molar

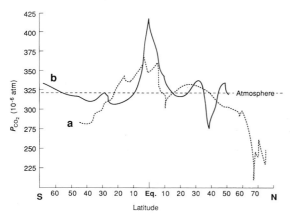

Fig. 11-16 Partial pressure of CO_2 in surface ocean water along the GEOSECS tracks: (a) the Atlantic western basin data obtained between August 1972 and January 1973; (b) the central Pacific data along the 180° meridian from October 1973 to February 1974. The dashed line shows atmospheric CO_2 for comparison. The equatorial areas of both oceans release CO_2 to the atmosphere, whereas the northern North Atlantic is a strong sink for CO_2. (Modified with permission from W. S. Broecker *et al.* (1979). Fate of fossil fuel carbon dioxide and the global carbon budget, *Science* **206**, 409–418, AAAS.)

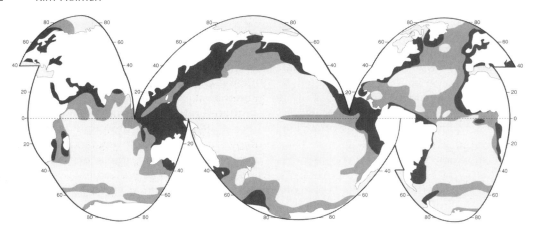

Fig. 11-17   Distribution of primary production in the world oceans (Degens and Mopper, 1976).

ratio of 106:1 in the sedimentary organic material, the corresponding carbon flux is in the range of 0.06 to 0.16 Pg C/yr. The sedimentation rate of organic material is also estimated from the total annual sedimentation rate of 6.1×10^{15} g/yr by applying an average organic carbon content of 0.5% (Kempe, 1979b), resulting in 0.03 Pg C/yr. Kempe (1979b) estimates inorganic sedimentation at 0.09–0.22 Pg C/yr based on an oceanic calcium balance calculation and the carbonate content in dated deep-sea cores.

11.5   Models of the Carbon Cycle

The descriptive account of the carbon cycle presented above is a first-order model. A variety of numerical models have been used to study the dynamics and response of the carbon cycle to different transients. This subject is an extensive field because most scientists modeling the carbon cycle develop a model tailored for their particular problem.

Box models have a long tradition (Craig, 1957b; Revelle and Suess, 1957; Bolin and Eriksson, 1959) and still receive a lot of attention. Most work is concerned with the atmospheric CO_2 increase, with the main goal of predicting global CO_2 levels during the next hundred years. This is accomplished with models that reproduce carbon fluxes between the atmosphere and other reservoirs on time

scales of 10–100 years, but does not include deep ocean circulation or sedimentary phenomena in detail. To study changes over thousands of years, the whole oceans, terrestrial biota, and soils must be considered. Extension to even longer time scales must deal with all geologic processes. On the other hand, many processes simulated in the fossil fuel studies can then be omitted or treated as instantaneous.

Simple three-box models with the atmosphere assumed to be one well-mixed reservoir and the oceans described by a surface layer and a deep-sea reservoir have been used extensively. Keeling (1973) has discussed this type of model in detail. The two-box ocean model is refined by including a second surface box, simulating an "outcropping" (deep-water forming) polar sea (e.g., Keeling and Bolin, 1967, 1968), and to include a better resolution of the main thermocline (e.g., Björkström, 1979). The terrestrial biota are included in a simple manner (e.g., Bolin and Eriksson, 1959) in some studies; Fig. 11-18 shows a model used by Machta (1972) where the role of biota is simulated by one reservoir connected to the atmosphere with a time lag of 20 years.

The inadequacy of the two-box model of the ocean led to the box-diffusion model (Oeschger et al., 1975). Instead of simulating the role of the deep sea with a well-mixed reservoir in exchange with the surface layer by first-order exchange processes, the transfer into the deep sea is maintained by vertical eddy diffusion. In

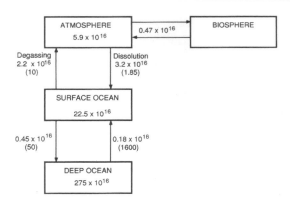

Fig. 11-18 A four-box model of the global carbon cycle. Reservoir inventories are given in moles and fluxes in mol/yr. The turnover time of CO_2 in each reservoir with respect to the outgoing flux is shown in brackets. (Reprinted with permission from L. Machta, The role of the oceans and biosphere in the carbon dioxide cycle, in D. Dryssen and D. Jagner (1972). "The Changing Chemistry of the Oceans," pp. 121–146, John Wiley.)

its original formulation, the box-diffusion model assumed that the eddy diffusivity remained constant with depth. Siegenthaler (1983) further developed the diffusion model to include polar outcropping areas. The box-diffusion model is in widespread use: for example an ambitious attempt (Peng *et al.*, 1983) to stimulate the changes in total carbon, ^{13}C and ^{14}C during the last hundred years uses a version of the Oeschger model combined with four boxes simulating the terrestrial system.

Box models and box-diffusion models have few degrees of freedom and they must describe physical, chemical, and biological processes very crudely. They are based on empirical relations rather than on first principles. Nevertheless, the simple models have been useful for obtaining some general features of the carbon cycle and retain some important roles in carbon cycle research (Craig and Holmén, 1995; Craig *et al.*, 1997; Siegenthaler and Joos, 1992).

There has been a tremendous development of various types of prognostic models of the carbon cycle during the past decades with increased refinement of both oceanic processes (see Siegenthaler and Sarmiento, 1993; Sarmiento *et al.*, 1992, 1998), terrestrial processes (Bonan,

1995a; Bunnell *et al.*, 1977; Cao and Woodward, 1998; Collatz *et al.*, 1992; Denning *et al.*, 1996a, b; Dickinson *et al.*, 1986; Dorman and Sellers, 1989; Foley *et al.*, 1996; Knorr and Heimann, 1995; Law *et al.*, 1996; Law and Simmonds, 1996; Nemry *et al.*, 1996; Potter *et al.*, 1993; Sellers *et al.*, 1996a, b) and atmospheric transport calculations (Erickson *et al.*, 1996; Fung *et al.*, 1983; Heimann and Keeling, 1989; Heimann *et al.*, 1989; Hunt *et al.*, 1996; Six *et al.*, 1996). Description of these models is beyond the scope of this volume, to describe but define the forefront of the modeling research today.

11.6 Trends in the Carbon Cycle

Throughout this chapter many of the arguments are based on an assumption of steady state. Before the agricultural and industrial revolutions, the carbon cycle presumably was in a quasi-balanced state. Natural variations still occur in this unperturbed environment; the Little Ice Age, 300–400 years ago, may have influenced the carbon cycle. The production rate of ^{14}C varies on time scales of decades and centuries (Stuiver and Quay, 1980, 1981), implying that the pre-industrial radiocarbon distribution may not have been in steady state.

Measurements of CO_2 concentrations in air bubbles trapped in glacial ice (Berner *et al.*, 1980; Delmas *et al.*, 1980; Jouzel *et al.*, 1993; Raynaud *et al.*, 1993) show that atmospheric P_{CO_2} was about 200 ppmv toward the end of the last glaciation 20 000 years ago (Fig. 11-19).

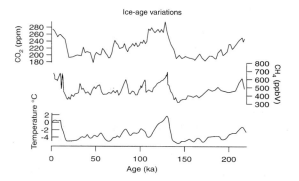

Fig. 11-19 Paleorecord of carbonaceous gases together with temperature. (After Jouzel *et al.*, 1993.)

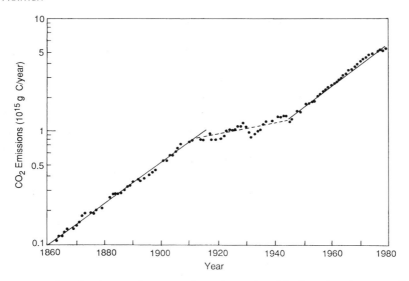

Fig. 11-20   Rate of transfer of carbon to the atmosphere due to fossil fuel combustion according to Rotty (1981).

Today, fossil fuel combustion undoubtedly accounts for a significant portion of the anthropogenic emissions although it is rivaled by carbon mobilized by deforestation and land-use changes (Woodwell *et al.*, 1983). The industrial revolution marked the onset of large-scale fossil fuel combustion in the early part of the 19th century. Around 1860 the emissions had reached 0.1 Pg C/yr (Fig. 11-20). The increase has been steady since that time although the rate of increase has changed during the past century. In 1860 essentially only coal was used; the use of oil began at the end of the 19th century followed by gas in the early decades of this century. Total fossil fuel emissions increased by 4% per year between 1860 and the beginning of the First World War, when fossil fuel consumption had reached 0.9 Pg C/yr (90% of which was coal). For the next 30 years (1914–1945) the yearly increase was about 1%, but after the Second World War the growth rate returned to 4%. Figure 11-21 shows the marked changes in fossil fuel use since 1973. In this period the annual rate of increase diminished to less than 2%, and striking changes in fuel mix occurred. Oil and gas use increased more rapidly than coal from the turn of the century to 1973 so that emissions from oil surpassed those from coal in the late 1960s.

The projection of future emissions of fossil fuel CO_2 is subject to a number of uncertainties. The growth rate has already shown great variations and the reserves of fossil fuels are not

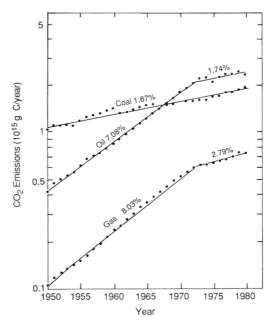

Fig. 11-21   Annual CO_2 emissions from each type of fossil fuel with growth rates (Rotty, 1981).

accurately known. Estimates of reserves range between 5000 and 10 000 Pg C, much of which would be costly to exploit with present techniques.

The atmospheric CO_2 content increased by about 1 ppmv per year during the period 1959–1978 (Bacastow and Keeling, 1981) with the South Pole P_{CO_2} increase lagging somewhat behind the Mauna Loa (19.5°N,155.6°W) data. This difference is consistent with our knowledge of interhemispheric mixing times and the fact that most fossil fuel emissions occur in the northern hemisphere (see also Conway *et al.*, 1994a).

In Fig. 11-22 (Heimann, 1997) the variability of the carbon cycle on interannual scales is shown. These processes still remain to be elucidated in detail but regional climate variability alters regional fluxes significantly. One can easily find indications of strong El Niño events in the record. El Niño has two main influences: one on the Pacific Ocean that is capped by warm water and thus a weaker source of atmospheric CO_2 from the upwelling waters, and the other being alterations of the temperature and precipitation fields in large areas. The latter proves to create a large flux of carbon to the atmosphere, mainly from drought stricken tropical regions

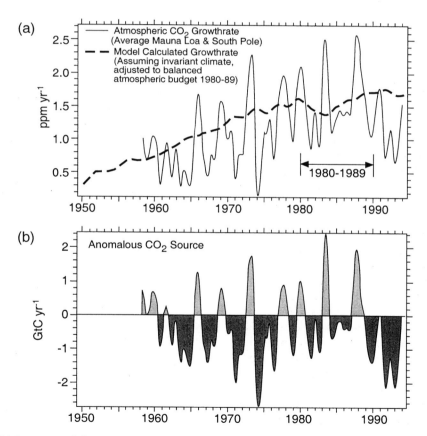

Fig. 11-22 (a) Interannual fluctuations of the growth rate of atmospheric CO_2 determined from the average of the seasonally adjusted records of the Mauna Loa and South Pole stations (see Fig. 11-3). (Data from Keeling *et al.*, 1995.) The dashed line is the growth rate that would result from an atmospheric balance which takes into account the documented CO_2 inputs from fossil fuel and changes in land use together with the uptake rates computed by an ocean and a terrestrial model. (b) Anomalous, presumably climate driven CO_2 source as implied by the difference between the solid and dashed line shown in part (a). (Both parts taken from Heimann (1997), with permission from the Royal Swedish Academy of Sciences.)

(notably northeast Brazil). The effect on the terrestrial components typically overshadows the decreased oceanic source during El Niño events (Gaudry *et al.*, 1987; Craig, 1998). Inter-annual variability in atmospheric CO_2 can also be induced by trends in global climate on other time scales; there are indications of changes in temperature and growing season during the latest decades (Bacastow *et al.*, 1985; Conway *et al.*, 1994b; Dai and Fung, 1993; Goulden *et al.*, 1996, Keeling *et al.*, 1995, 1996; Kindermann *et al.*, 1996; Myneni *et al.*, 1998). There are obviously many exciting questions regarding this issue that remain.

Fossil fuel emissions alter the isotopic composition of atmospheric carbon, since they contain no ^{14}C and are depleted in ^{13}C. Releasing radiocarbon-free CO_2 to the atmosphere dilutes the atmospheric ^{14}C content, yielding lower $^{14}C/C$ ratios ("the Suess effect"). From 1850 to 1954 the $^{14}C/C$ ratio in the atmosphere decreased by 2.0 to 2.5% (Fig. 11-23) (Suess, 1965; Stuiver and Quay, 1981). Then, this downward trend in ^{14}C was disrupted by a series of atmospheric nuclear tests. Many large fission explosions set off by the United States with high emission of neutrons took place in 1958 in the atmosphere and the Soviet Union held extensive tests during

1960–1963. Figure 11-24 shows the atmospheric $\Delta^{14}C$ trend during recent years. The two super-powers have ceased performing atmospheric bomb tests, which resulted in the spike-like injections of ^{14}C in the early 1960s. The further fate of these spikes in the global biogeochemical cycle has proven to be an unintentional but valuable tool when deducing carbon fluxes between reservoirs.

Agriculture has the explicit goal of harvesting organic matter and no accumulation of carbon in agricultural ecosystems is to be expected. It is important to recognize that an increased carbon fixation rate is not equivalent to an increased storage of carbon. Carbon accumulation in ecosystems is determined by the balance between net primary production and heterotrophic respiration, changes in both must be considered in the search for the missing carbon.

Eutrophication by the release of nitrogen and phosphorus in various forms could contribute significantly to the biotic storage of carbon. Anthropogenic releases of N and P correspond to the fixation of 9.6 and 17.6 Pg C/yr, respectively, if all N and P released was stored as trees in forests (Houghton and Woodwell, 1983). Most of the released nutrients are not available to be stored as wood, consequently the increased storage is probably much smaller than the potential value. Houghton and Woodwell (1983) conclude after considering eutrophication and other alterations of the environment that there is little evidence for an increased storage of carbon in the ecosystem of the Earth. This picture has later been modified; at least regionally there are likely increases of growth due to release of nutrients (McGuire *et al.*, 1992, 1997; Schimel *et al.*, 1996).

The terrestrial biota seem unable to take up much of the excess CO_2. In fact, a careful assessment of the impact of deforestation and land-use changes indicate that the terrestrial biota has been a considerable source of CO_2 during the past century (Bolin, 1977; Woodwell *et al.*, 1983). A complex effort to deduce mankind's impact on terrestrial biota using a bookkeeping model based on historical records on land use in all parts of the world (Moore *et al.*, 1981; Houghton *et al.*, 1983; Woodwell *et al.*, 1983) gives the curves in Fig. 11-25. Woodwell *et al.* (1983)

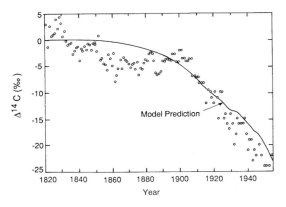

Fig. 11-23 Comparison between the Peng *et al.* (1983) model-derived Suess effect curve (solid line) and the observed $^{14}C/^{12}C$ trend (points) for atmospheric CO_2 as reconstructed by Stuiver and Quay (1981) from measurements of tree rings. (Reproduced with permission from W. Broecker *et al.* (1983). A deconvolution of tree ring based $\delta^{13}C$ record, *J. Geophys. Res.* **88**, 3609–3620, American Geophysical Union.)

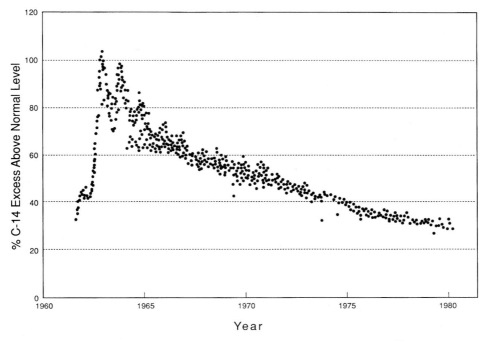

Fig. 11-24 Carbon-14 in the troposphere and the ocean surface water 1962–1981. ^{14}C values for ocean surface water during this period range from 0–15% with no trend over time. (Modified with permission from R. Nydal and K. Lövseth (1983). Tracing bomb ^{14}C in the atmosphere, *J. Geophys. Res.* **88**, 3621–3642, American Geophysical Union.)

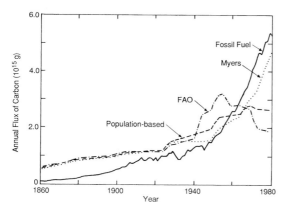

Fig. 11-25 Release of carbon from the biota and soils globally according to various estimates. The fossil fuel flux is from data of Rotty. (Modified with permission from G. M. Woodwell *et al.* (1983). Global deforestation: Contribution to atmospheric carbon dioxide, *Science* **222**, 1081–1086, AAAS.)

arrive at a carbon release in 1980 of 1.8 to 4.7 Pg C/yr from deforestation, which is comparable to the 5 Pg released from fossil fuels. Freyer and Belacy (1983) constructed a ^{13}C/^{12}C record from tree-ring studies (Fig. 11-26). Using a global carbon cycle model with box-diffusion model for the oceans and a simple four-box model for soil and land-biota, Peng *et al.* (1983) make use of this record of δ^{13}C changes in the atmosphere to deduce the CO_2 contribution from forest and soils. The CO_2 emissions arrived at for 1980 are about 1.5 Pg C/yr, similar to the minimum values of Woodwell *et al.* (1983), but the trends during the past century differ significantly (Fig. 11-27). There are numerous indications that the terrestrial biosphere has significant interannual variability that creates variability in the atmospheric CO_2 concentration.

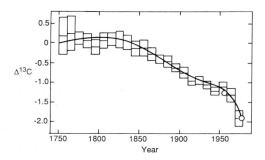

Fig. 11-26 Decade-averaged $\Delta^{13}C$ data of Northern hemisphere tree ring records from 1750–1979 and 7th-degree polynomial fit of the data. The vertical extension of blocks represents 95% confidence limits of the mean. The open circles give the ^{13}C change of -0.65% in atmospheric CO_2 observed from 1956 to 1978 by Keeling *et al.* (1979). (Adapted from Peng *et al.*, 1983.)

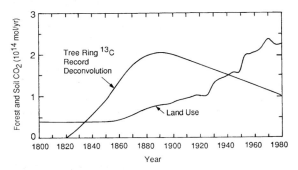

Fig. 11-27 Contrast between the $^{13}C/^{12}C$ derived forest-soil CO_2 scenario obtained from Peng *et al.* (1983)with that based on land use and stored carbon response functions obtained by Houghton *et al.* (1983). (Reproduced with permission from W. Broecker *et al.* (1983). A deconvolution of tree ring based $\delta^{13}C$ record, *J. Geophys. Res.* **88**, 3609–3620, American Geophysical Union.)

The role of carbon dioxide in the Earth's radiation budget merits this interest in atmospheric CO_2. There are, however, other changes of importance. The atmospheric methane concentration is increasing, probably as a result of increasing cattle populations, rice production, and biomass burning (Crutzen, 1983). Increasing methane concentrations are important because of the role it plays in stratospheric and tropospheric and because it is important to the radiation budget of the world.

As we have seen, civilization is altering the global carbon cycle in several ways. Even though our knowledge of the most relevant processes in the carbon cycle has increased considerably, it remains limited. Figure 11-1 summarizes the carbon cycle as described in this chapter. There are a number of central observations to point out regarding this figure. The amount of fossil fuel still remaining is sufficient to increase the atmospheric CO_2 content many-fold. The amount of carbon stored in soils and standing biomass is small compared to the fossil fuel reserves; it is sometimes argued that the planting of trees could decrease the impact of our fossil fuel releases. Even if we double the standing biomass (an unrealistic vision since most land is needed for food production) the amount of carbon stored will be small compared to the total amount of fossil fuel available. Most of the terrestrial biosphere and soil carbon reservoirs are easily changed on short time scales; even if there can be net storage during a decade (what seems to be the case during the 1990s) this carbon can very rapidly return to the atmosphere. Because the transport of carbon into the deep oceans is a slow process, release of CO_2 to the atmosphere will perturb the atmospheric concentration for several centuries even if emissions cease. We must also note that the airborne fraction observed during the past decades might very well change as a response to climatic change; for example if the ocean surface becomes warmer there will be an increased stability of the water column yielding a slower transport of water the abyss as well as a decreased solubility of carbon dioxide in the water. Both these effects thus make the oceans less prone to taking up the excess CO_2. Revisiting Fig. 11-19 we can also note that the observed perturbations during the past 200 years are extreme compared to the observed variations that have occurred during the past 200 000 years; CO_2 has increased from a pre-industrial global-mean value of about 280 ppmv to today's (December 1998) value of 366 ppmv. The climate system will undoubtedly respond to these dramatic changes in radiative forcing although the details of the response remain uncertain. The use of fossil fuels thus passes a potential climate change problem onwards to coming generations; if the climatic impacts of the enhanced greenhouse effect prove to create severe difficulties for mankind or nature centuries will be required to repair the atmosphere. Despite the uncertainties regarding the climatic impacts (IPCC, 1990, 1992; Kattenberg *et al.*, 1996) of the enhanced greenhouse effect this feature of the carbon cycle remains as a fundamental argument to avoid using fossil fuels whenever possible and actively seeking alternatives for the future.

References

For the interested reader, this reference list includes an extensive set of citations of the general and research literature on the carbon cycle.

Aber, J. D. and Driscoll, C. T. (1997). Effects of land use, climate variation, and N deposition on N cycling and C storage in northern hardwood forests, *Global Biogeochem. Cycles* **11**, 639–648.

Ajtay, G. L., Ketner, P. and Duvigneaud, P. (1979). Terrestrial primary production and phytomass. *In* "The Global Carbon Cycle" (B. Bolin, E. T. Degens, S. Kempe, and P. Ketner, eds), pp. 129–181. Wiley, New York.

Anklin, M., Barnola, J. M., Schwander, J., Stauffer, B. and Raynaud, D. (1995). Processes affecting the CO_2 concentrations measured in Greenland ice. *Tellus* **47B**, 461–470.

Arrhenius, S. (1896). On the influence of the carbonic acid in the air upon the temperature of the ground. *Philos. Mag.* **41**, 237–276.

Bacastow, R. B. (1976). Modulation of atmospheric carbon dioxide by the Southern Oscillation, *Nature* **261**, 116–118.

Bacastow, R. B. (1979). Dip in the atmospheric CO_2 level during the mid-1960s, *J. Geophys. Res.* **84**, 3108–3114.

Bacastow, R. B. and Björkström, A. (1981). Comparison of ocean models for the carbon cycle. *In* "Carbon Cycle Modeling" (B. Bolin, ed.), pp. 29–79. Wiley, New York.

Bacastow, R. B. and Keeling, C. D. (1981). Atmospheric carbon dioxide concentration and the observed airborne fraction. *In* "Carbon Cycle Modeling" (B. Bolin, ed.), pp. 103–112. Wiley, New York.

Bacastow, R. B., Adams, J. A., Keeling, C. D., Moss, D. J., Whorf, T. P. and Wong, C. S. (1980). Atmospheric carbon dioxide, the Southern Oscillation, and the weak 1975 El Niño, *Science* **210**, 66–68.

Bacastow, R. B., Keeling, C. D. and Whorf, T. P. (1985). Seasonal amplitude increase in atmospheric CO_2 concentration at Mauna Loa, Hawaii, 1959–1982, *J. Geophys. Res.* **90**, 10529–10540.

Baes, C. F., Goeller, H. E., Olson, J. S. and Rothy, R. M. (1976). "The Global Carbon Dioxide Problem," ORNL-5194, pp. 1–72. Oak Ridge National Laboratory, Oak Ridge, TN.

Bakwin, P. S., Tans, P. P., Zhao, C., Ussler III, W. and Quesnell, E. (1995). Measurements of carbon dioxide on a very tall tower, *Tellus, Ser. B*, **47**, 535–549.

Barnola, J. M., Raynaud, D., Neftel, A. and Oeschger, H. (1983). Comparison of CO_2 measurements by two laboratories on air from bubbles in polar ice. *Nature* **303**, 410–413.

Barnola, J. M., Anklin, M., Porcheron, J., Raynaud, D., Schwander, J. and Stauffer, B. (1995). CO_2 evolution during the last millennium as recorded by Antarctic and Greenland ice. *Tellus* **47B**, 264–272.

Barrie, L. A. (1986). Arctic air pollution: an overview of current knowledge, *Atmos. Environ.* **20**, 643–663.

Bazilevich, N. I., Rodin, L., Ye, L. and Roznov, N. N. (1970). Geographical Aspects of Biological Productivity. *Pap. V Congr. USSR Geogr. Soc.* USSR, Leningrad.

Berger A. (1988). Milankovitch theory and climate, *Reviews of Geophysics* **26** (4), 624–657.

Berner, W., Oeschger, H. and Stauffer, B. (1980). Information on the CO_2 cycle from ice core studies. *Radiocarbon* **22**, 227–235.

Bischof, W. (1981). The CO_2 content of the upper polar troposphere between 1963–1979. *In* "Carbon Cycle Modeling" (B. Bolin, ed.), pp. 113–116. Wiley, New York.

Björkström, A. (1979). A model of CO_2 interaction between atmosphere, oceans, and land biota. *In* "The Global Carbon Cycle" (B. Bolin, E. T. Degens, S. Kempe, and P. Ketner, eds), pp. 403–457. Wiley, New York.

Blake, D. R. and Rowland, F. S. (1988). Continuing worldwide increase in tropospheric methane 1978 to 1987. *Science* **239**, 1129–1131.

Bolin, B. (1970a). Changes of land biota and their importance for the carbon cycle. *Science* **196**, 613–615.

Bolin, B. (1970b). The carbon cycle. *Scient. Am.* **223** (3), 124–132.

Bolin, B. (ed.) (1981). "Carbon Cycle Modeling." Wiley, New York.

Bolin, B. and Bischof, W. (1970). Variations of the carbon dioxide content of the atmosphere in the northern hemisphere, *Tellus* **22**, 431–442.

Bolin, B. and Cook, R. (eds.) (1983). "The Major Biogeochemical Cycles and Their Interactions," *SCOPE* **21**. Wiley, New York.

Bolin, B. and Eriksson, E. (1959). Changes of the carbon dioxide content of the atmosphere and sea due to fossil fuel combustion. *In* "Atmosphere and Sea in Motion" (B. Bolin, ed.), pp. 130–142. The Rockefeller Institute Press.

Bolin, B. and Keeling, C. D. (1963). Large-scale atmospheric mixing as deduced from the seasonal and meridional variations of carbon dioxide, *J. Geophys. Res.* **68**, 3899–3920.

Bolin, B., Degens, E. T., Kempe, S. and Ketner, P. (eds.) (1979). "The Global Carbon Cycle," *SCOPE* **13**. Wiley, New York.

Bolin, B., Björkström, A., Keeling, C. D., Bacastow, R. and Siegenthaler, U. (1981). Carbon cycle modeling. *In* "Carbon Cycle Modeling" (B. Bolin, ed.), pp. 1–28. Wiley, New York.

Bonan, G. B. (1991). Atmosphere-biosphere exchange of carbon dioxide in boreal forests, *J. Geophys. Res.* **96**, 7301–7312.

Bonan, G. B. (1995a). Land-atmosphere CO_2 exchange simulated by a land surface process model coupled to an atmospheric general circulation model, *J. Geophys. Res.* **100**, 2817–2831.

Bonan, G. B. (1995b). Sensitivity of a GCM simulation to inclusion of inland water surfaces, *J. Clim.* **8**, 2691–2704.

Bonan, G. B. (1996a). A land surface model (LSM version 1.0) for ecological, hydrological, and atmospheric studies: Technical description and user's guide, *NCAR Tech. Note NCAR/TN-417+STR*, Natl. Cent. for Atmos. Res., Boulder, CO.

Bonan, G. B. (1996b). The NCAR land surface model (LSM version 1.0) coupled to the NCAR community climate model, *NCAR Tech. Note NCAR/TN-429+STR*, NCAR, Boulder, CO.

Bonan, G. B. (1996c). Sensitivity of a GCM simulation to subgrid infiltration and surface runoff, *Clim. Dyn.* **12**, 279–285.

Bonan, G. B. (1997). Effects of land use on the climate of the United States, *Clim. Change* **37**, 449–486.

Bonan, G. B. (1998). The land surface climatology of the NCAR Land Surface Model (LSM 1.0) coupled to the NCAR Community Climate Model (CCM3), *J. Clim.* in press.

Box, E. O. (1988). Estimating the seasonal carbon source-sink geography of a natural, steady-state terrestrial biosphere, *J. Appl. Meteorol.* **27**, 1109–1124.

Box, E.O. and Meentemeyer, V. (1991). Geographic modeling and modern ecology. *In* "Modern Ecology" (G. Esser and D. Overdieck, eds), pp. 773–804. Elsevier, Amsterdam.

Braswell, B. H., Schimel, D. S., Linder, E. and Moore III, B. (1997). The response of global terrestrial ecosystems to interannual temperature variability, *Science* **278**, 870–872.

Brewer, P. G. (1978). Direct observation of the oceanic CO_2 increase. *Geophys. Res. Lett.* **5** (12), 997–1000.

Broecker, W. S. (1973). Factors controlling CO_2 content in the oceans and atmosphere. *In* "Carbon and the Biosphere" (G. M. Woodwell and E. V. Pecan, eds), pp. 32–50. United States Atomic Energy Commission.

Broecker, W. S. (1979). A revised estimate for the radiocarbon age of North Atlantic Deep Water, *J. Geophys. Res.* **84**, C6, 3218–3226.

Broecker, W. S. and Olson, E. A. (1959). Lamont radiocarbon measurements VI. *Am. J. Sci.* Radiocarbon Suppl. 1, 111–132.

Broecker, W. S. and Peng, T.-H. (1982). "Tracers in the Sea." Eldigio Press, Palisades, New York.

Broecker, W. S., Gerard, R., Ewing, M. and Heezen, B. C. (1960). Natural radiocarbon in the Atlantic Ocean, *J. Geophys. Res.* **65**, 9, 2903–2931.

Broecker, W. S., Takahashi, T., Simpson, H. J. and Peng, T.-H. (1979). Fate of fossil fuel carbon dioxide and the global carbon budget. *Science* **206**, 409–418.

Bunnell, F. L., Tait, D. E. N., Flanagan, P. W. and Van Cleve, K. (1977). Microbial respiration and substrate weight loss, I. A general model of the influences of abiotic variables. *Soil Biol. Biochem.* **9**, 33–40.

Cao, M. and Woodward, F. I. (1998). Dynamic responses of terrestrial ecosystem carbon cycling to global climate change. *Nature* **393**, 249–252.

CDIAC (1990). "Carbon Dioxide Information Analysis Center Communications," Spring, 1990. Oak Ridge National Laboratory, Oak Ridge, TN.

Charlson, R. J. and Rodhe, H. (1982). Factors controlling the acidity of natural rainwater. *Nature* **295**, 683–685.

Chen, C.-T. and Millero, F. J. (1979). Gradual increase of oceanic CO_2. *Nature* **277**, 205–206.

Chen, C.-T.A. (1978). Decomposition of calcium carbonate and organic carbon in the deep oceans. *Science* **201**, 735–736.

Ciais, P., Tans, P. P., Trolier, M., White, J. W. C. and Francey, R. J. (1995). A large northern hemisphere terrestrial CO_2 sink indicated by the $^{13}C/^{12}C$ ratio of atmospheric CO_2, *Science* **269**, 1098–1102.

Coe, M. T. and Bonan, G. B. (1997). Feedbacks between climate and surface water in northern Africa during the middle Holocene, *J. Geophys. Res.* **102**, 11087–11101.

Collatz, G. J., Ball, J. T., Grivet, C. and Berry, J. A. (1991). Physiological and environmental regulation of stomatal conductance, photosynthesis, and transpiration: a model that includes a laminar boundary layer, *Agric. For. Meteorol.* **54**, 107–136.

Collatz, G. J., Ribas-Carbo, M. and Berry, J. A. (1992). Coupled photosynthesis-stomatal conductance model for leaves of C_4 plants, *Aust. J. Plant Physiol.* **19**, 519–538.

Conway, T. J., Tans, P., Waterman, L. S., Thoning, K. W., Masarie, K. A. and Gammon, R. H. (1988). Atmospheric carbon dioxide measurements in the remote global troposphere, 1981–1984. *Tellus* **40B**, 81–115.

Conway, T. J., Tans, P. P. and Waterman, L. S. (1994a). Atmospheric CO_2 records from sites in the NOAA/CMDL air sampling network. *In* "Trends '93: A Compendium of Data on Global Change," *Rep. ORNL/CDIAC-65* (T. A. Boden *et al.* eds), pp. 41–119. Oak Ridge Natl. Lab., Oak Ridge, Tenn.

Conway, T. J., Tans, P. P., Waterman, L. S., Thoning, K. W., Kitzis, D. R., Masarie, K. A. and Zhang, N. (1994b). Evidence for interannual variability of the carbon cycle from the National Oceanic and Atmospheric Administration/Climate Monitoring and Diagnostics Laboratory Global Air Sampling Network, *J. Geophys. Res.* **99**, 22831–22855.

Craig, H. (1957a). Isotopic standards for carbon and correction factors for mass-spectrometric analysis of carbon dioxide. *Geochim. Cosmochim. Acta* **12**, 133–149.

Craig, H. (1957b). The natural distribution of radiocarbon and the exchange time of carbon dioxide between atmosphere and sea. *Tellus* **9**, 1–17.

Craig, H., Horibe, Y. and Sowers, T. 1988. Gravitational separation of gases and isotopes in polar ice caps. *Science* **242**, 1675–1678.

Craig, S. G. (1998). The response of terrestrial carbon exchange and atmospheric CO_2 concentrations to El Niño SST forcing. *Report CM-94*, Int. Meteorol. Inst. in Stockholm, Dept. of Meteorol., Stockholm Univ.

Craig, S. G. and Holmén, K. J. (1995). Uncertainties in future CO_2 projections, *Global Biogeochem. Cycles* **9**, 139–152.

Craig, S. G. and Holmén, K. J. (1998). Atmospheric CO_2 simulated by the National Center for Atmospheric Research Community Climate Model. 2.

Interannual variability, *J. Geophys. Res.* submitted, 1998.

Craig, S., Holmén, K. and Björkström, A. (1997). Net terrestrial carbon exchange from mass balance calculations: an uncertainty estimate, *Tellus, Ser. B*, **49**, 136–148.

Craig, S. G., Holmén, K. J., Bonan, G. B and Rasch, P. J. (1998). Atmospheric CO_2 simulated by the National Center for Atmospheric Research Community Climate Model. 1. Mean fields and seasonal cycles, *J. Geophys. Res.* 13213–13235.

Crutzen, P. J. (1983). Atmospheric interactions–homogeneous gas reactions of C, N, and S containing compounds. *In* "The Major Biogeochemical Cycles and Their Interactions" (B. Bolin, ed.), pp. 67–112. Wiley, New York.

Crutzen, P. J. (1991). Methane's sinks and sources, *Nature* **350**, 380–381.

Crutzen, P. J. and Andreae, M. O. (1990). Biomass burning in the Tropics: impact on atmospheric chemistry and biogeochemical cycles, *Science* **250**, 1669–1678.

Crutzen, P. J. and Zimmerman, P. H. (1991). The changing photochemistry of the troposphere, *Tellus* **43**, 135–151.

Culkin, F. (1965). The major constituents of sea water. *In* "Chemical Oceanography" (J. P. Riley and G. Skirrow, eds), Chapter 4, pp. 121–161. Academic Press, London.

Dai, A. and Fung, I. Y. (1993). Can climate variability contribute to the "missing" CO_2 sink? *Global Biogeochem. Cycles* **7**, 599–609.

De Vooys, C. G. N. (1979). Primary production in aquatic environments. *In* "The Global Carbon Cycle" (B. Bolin, E. T. Degens, S. Kempe and P. Ketner, eds), pp. 259–292. Wiley, New York.

Degens, E. T. and Mopper, K. (1976). Factors controlling the distribution and early diagenesis of organic material in marine sediments. *In* "Chemical Oceanography" (J. P. Riley, ed.), Vol. 6, pp. 59–113. Academic Press, New York.

Degens, E. T., Kempe, S. and Spitzy, A. (1984). Carbon dioxide: a biogeochemical portrait. *In* "The Handbook of Environmental Chemistry" (O. Hutzinger, ed.), Vol. 1 (Part C). Springer-Verlag, New York.

Delmas, R. J. (1993). A natural artefact in Greenland ice-core CO_2 measurements. *Tellus* **45B**, 391–396.

Delmas, R. J., Ascencio, J.-M. and Legrand, M. (1980). Polar ice evidence that atmospheric CO_2 20 000 yr BP was 50% of present. *Nature* **284**, 155–157.

Denning, A. S., Collatz, G. J., Zhang, C., Randall, D. A., Berry, J. A., Sellers, P. J., Colello, G. D. and

Dazlich, D. A. (1996a). Simulations of terrestrial carbon metabolism and atmospheric CO_2 in a general circulation model. Part 1: Surface carbon fluxes, *Tellus, Ser. B*, **48**, 521–542.

Denning, A. S., Randall, D. A., Collatz, G. J. and Sellers, P. J. (1996b). Simulations of terrestrial carbon metabolism and atmospheric CO_2 in a general circulation model. Part 2: Simulated CO_2 concentrations, *Tellus, Ser. B*, **48**, 543–567.

Denning, A. S. (1995). Investigations of the transport, sources, and sinks of atmospheric CO_2 using a general circulation model, *Atmos. Sci. Pap.* **564**, Colo. State Univ., Fort Collins.

Denning, A. S., Fung, I. Y. and Randall, D. A. (1995). Latitudinal gradient of atmospheric CO_2 due to seasonal exchange with land biota, *Nature* **376**, 240–243.

Dickinson, R. E., Henderson-Sellers, A., Kennedy, P. J. and Wilson, M. F. (1986). Biosphere-Atmosphere Transfer Scheme (BATS) for the NCAR Community Climate Model, *NCAR Tech. Note TN-275+STR*, Nat. Cent. for Atmos. Res., Boulder, CO.

Dorman, J. L. and Sellers, P. J. (1989). A global climatology of albedo, roughness length and stomatal resistance for atmospheric general circulation models as represented by the simple biosphere model (SiB), *J. Appl. Meteorol.* **28**, 833–855.

Dörr, H. and Münnich, K. O. (1987). Annual variation in soil respiration in selected areas of the temperate zone, *Tellus, Ser. B*, **39**, 114–121.

Dougherty, R. L., Bradford, J. A., Coyne, P. I. and Sims, P. L. (1994). Applying an empirical model of stomatal conductance to three C-4 grasses, *Agric. For. Meteorol.* **67**, 269–290.

Druffel, E. M. and Linick, T. W. (1978). Radiocarbon in annual coral rings of Florida. *Geophys. Res. Lett.* **5**, 913–916.

Dyrssen, D. and Sillén, L. G. (1967). Alkalinity and total carbonate in sea water. A plea for P-T-independent data. *Tellus* **19**, 1, 113–120.

Edmond, J. M. (1970). High precision determination of titration alkalinity and total carbon dioxide concentration of sea water by potentiometric titration. *Deep-Sea Res.* **17**, 737–750.

Ehhalt, D. H. (1974). The atmospheric cycle of methane. *Tellus* **26**, 58–70.

Enting, I. G. (1985). A lattice statistics model for the age distribution of air bubbles in polar ice. *Nature* **315**, 654–655.

Enting, I. G. (1992). The incompatibility of ice-core CO_2 data with reconstructions of biotic CO_2 sources, II. The influence of CO_2-fertilised growth. *Tellus* **44B**, 23–32.

Enting, I. G., Wigley, T. M. L. and Heimann, M. (1994). Future emissions and concentration of carbon dioxide: Key ocean/atmosphere/land analyses. *Tech. Pap. 31*, Div. of Atmos. Res., Comm. Sci. and Ind. Res. Org., Melbourne, Victoria, Australia.

Enting, I. G., Trudinger, C. M. and Francey, R. J. (1995). A synthesis inversion of the concentration and $d^{13}C$ of atmospheric CO_2, *Tellus, Ser. B*, **47**, 35–52.

Enting, I. G., Trudinger, C. M., Francey, R. J. and Granek, H. Synthesis inversion of atmospheric CO_2 using the GISS tracer transport model, *Tech. Pap. 29*, Div. of Atmos. Res., Comm. Sci. and Ind. Res. Org., Melbourne, Victoria, Australia.

Erickson, D. J., Rasch, P. J., Tans, P. P., Friedlingstein, P., Ciais, P., Maier-Reimer, E., Six, K., Fischer, C. A. and Walters, S. (1996). The seasonal cycle of atmospheric CO_2: A study based on the NCAR Community Climate Model (CCM2), *J. Geophys. Res.* **101**, 15079–15097.

Etheridge, D. M., Steele, L. P., Langenfelds, R. L. and Francey, R. J. (1996). Natural and anthropogenic changes in atmospheric CO_2 over the last 1000 years from air in Antarctic ice and firn, *J. Geophys. Res.* **101**, 4115–4128.

Fan, S., Gloor, M., Mahlman, J., Pacala, S., Sarmiento, J., Takahashi, T. and Tans, P. (1998). A large terrestrial carbon sink in North America implied by atmospheric and oceanic carbon dioxide data and models. *Science* **282**, 442–446.

Farquhar, G. D. and Sharkey, T. D. (1982). Stomatal conductance and photosynthesis, *Ann. Rev. Plant Physiol.* **33**, 317–345.

Farquhar, G. D., von Caemmerer, S. and Berry, J. A. (1980). A biochemical model of photosynthetic CO_2 assimilation in leaves of C_3 species, *Planta* **149**, 78–90.

Feely, R. A., Gammon, R. H., Taft, B. A., Pullen, P. E., Waterman, L. S., Conway, T. J., Gendron, J. F. and Wisegarver, D. P. (1987). Distribution of chemical tracers in the eastern equatorial Pacific during and after the 1982–1983 El Niño/Southern Oscillation event, *J. Geophys. Res.* **92**, 6545–6558.

Feux, A. N. and Baker, D. R. (1973). Stable carbon isotopes in selected granitic, mafic, and ultramafic igneous rocks. *Geochim. Cosmochim. Acta* **37**, 2509–2521.

Field, C. B., Randerson, J. T. and Malmström, C. M. (1995). Ecosystem net primary production: Combining ecology and remote sensing, *Remote Sens. Environ.* **51**, 74–88.

Flowers, E. C., McCormick, R. A. and Kurfis, K. R. (1969). Atmospheric turbidity over the United States, 1961–1966, *J. Appl. Meteorol.* **8**, 955–962.

Foley, J. A., Prentice, I. C., Ramankutty, N., Levis, S.,

Pollard, D., Sitch, S. and Haxeltine, A. (1996). An integrated biosphere model of land surface processes, terrestrial carbon balance, and vegetation dynamics, *Global Biogeochem. Cycles* **10**, 603–628.

Francey, R. J., Tans, P. P., Allison, C. E., Enting, I. G., White, J. W. C. and Trolier, M. (1995). Changes in oceanic and terrestrial carbon uptake since 1982, *Nature* **373**, 326–330.

Freyer, H.-D. (1979). Atmospheric cycles of trace gases containing carbon. *In* "The Global Carbon Cycle" (B. Bolin, E. T. Degens, S. Kempe and P. Ketner, eds), pp. 101–128. Wiley, New York.

Freyer, H. D. and Belacy, N. (1983). $^{13}C/^{12}C$ records in Northern Hemispheric trees during the past 500 years – anthropogenic impact and climatic superpositions, *J. Geophys. Res.* **88**, 6844–6852.

Friedli, H., Lötscher, H., Oeschger, H., Siegenthaler, U. and Stauffer, B. (1986). Ice core record of the $^{13}C/^{12}C$ ratio of atmospheric CO_2 in the past two centuries. *Nature* **324**, 237–238.

Friedlingstein, P., Fung, I., Holland, E., John, J., Brasseur, G., Erickson, D. and Schimel, D. On the contribution of CO_2 fertilization to the missing biospheric sink. *Global Biogeochem. Cycles* **9**, 541–556.

Fung, I. Y. (1986). Analysis of the seasonal and geographical patterns of atmospheric CO_2 distributions with a three-dimensional tracer model. *In* "The Changing Carbon Cycle: A Global Analysis" (J. R. Trabalka and D. E. Reichle, eds), pp. 459–473. Springer-Verlag, New York.

Fung, I. Y., Prentice, K. C., Matthews, E., Lerner, J. and Russell, G. (1983). Three-dimensional tracer model study of atmospheric CO_2: Response to seasonal exchanges with the terrestrial biosphere, *J. Geophys. Res.* **88**, 1281–1294.

Fung, I. Y., Tucker, C. J. and Prentice, K. C. (1987). Application of Advanced Very High Resolution Radiometer vegetation index to study atmosphere-biosphere exchange of CO_2, *J. Geophys. Res.* **92**, 2999–3015.

Garrels, R. M. and Mackenzie. F. T. (1971). "Evolution of Sedimentary Rocks." Norton, New York.

Garrels, R. M., Mackenzie, F. T. and Hunt, C. (1973). "Chemical Cycles and the Global Environment." W. Kaufmann, Los Altos, California.

Gaudry, A., Monfray, P., Polian, G. and Lambert, G. (1987). The 1982–1983 El Niño: a 6 billion ton CO_2 release. *Tellus* **39B**, 209–213.

Gillette, D. A. and Box, E. O. (1986). Modeling seasonal changes of atmospheric carbon dioxide and carbon 13, *J. Geophys. Res.* **91**, 5287–5304.

Goudriaan, J. and Ajtay, G. L. (1979). The possible effects of increased CO_2 on photosynthesis. *In* "The

Global Carbon Cycle" (B. Bolin, E. T. Degens, S. Kempe and P. Ketner, eds), pp. 237–249. Wiley, New York.

Goulden, M. L., Munger, J. W., Fan, S.-M., Daube, B. C. and Wofsy, S. C. (1996). Exchange of carbon dioxide by a deciduous forest: Response to interannual climate variability, *Science* **271**, 1576–1578.

Grace, J., Lloyd, J., McIntyre, J., Miranda, A. C., Meir, P., Miranda, H. S., Nobre, C., Moncrieff, J., Massheder, J., Malhi, Y., Wright, I. and Gash, J. (1995). Carbon dioxide uptake by an undisturbed tropical rain forest in southwest Amazonia, 1992 to 1993, *Science* **270**, 778–780.

Haas-Laursen, D. E., Hartley, D. E. and Conway, T. J. (1997). Consistent sampling methods for comparing models to CO_2 flask data, *J. Geophys. Res.* **102**, 19059–19071.

Hack, J. J. (1994). Parameterization of moist convection in the National Center for Atmospheric Research community climate model (CCM2), *J. Geophys. Res.* **99**, 5551–5568.

Hack, J. J., Boville, B. A., Briegleb, B. P., Kiehl, J. T., Rasch, P. J. and Williamson, D. L. (1993). Description of the NCAR Community Climate Model (CCM2), *NCAR Tech. Note NCAR/TN-382+STR*, Nat. Cent. for Atmos. Res., Boulder, CO.

Hack, J. J., Boville, B. A., Kiehl, J. T., Rasch, P. J. and Williamson, D. L. (1994). Climate statistics from the National Center for Atmospheric Research community climate model (CCM2), *J. Geophys. Res.* **99**, 20785–20813.

Hansen, J., Ruedy, R., Sato, M. and Reynolds, R. (1996). Global surface air temperature in 1995: Return to pre-Pinatubo level, *Geophys. Res. Lett.* **23**, 1665–1668.

Hansson, I. (1973). A new set of acidity constants for carbonic acid and boric acid in sea water. *Deep-Sea Research* **20**, 461–478.

Hao, W., Ward, D. E., Olbu, G. and Baker, S. P. (1996). Emissions of CO_2, CO and hydrocarbons from fires in diverse African savanna ecosystems, *J. Geophys. Res.* **101**, 23577–23584.

Heimann, M. (1997). A review of the Contemporary Global Carbon Cycle and as Seen a Century Ago by Arrhenius and Högbom, *Ambio*, **26**, 17–24.

Heimann, M. and Keeling, C. D. (1989). A three-dimensional model of atmospheric CO_2 transport based on observed winds: 2. Model description and simulated tracer experiments. *In* "Aspects of Climate Variability in the Pacific and Western Americas," *Geophys. Monogr. Ser.* vol. 55 (D. H. Peterson, ed.), pp. 237–275, AGU, Washington, DC.

Heimann, M., Keeling, C. D. and Tucker, C. J. (1989). A three-dimensional model of atmospheric CO_2

transport based on observed winds: 3. Seasonal cycle and synoptic time scale variations. *In* "Aspects of Climate Variability in the Pacific and Western Americas," *Geophys. Monogr. Ser.* vol. 55 (D. H. Peterson, ed.), pp. 277–303, AGU, Washington, DC.

Holland, H. D. (1978). "The Chemistry of the Atmosphere and Oceans." Wiley-Interscience, New York.

Holtslag, A. A. M. and Boville, B. A. (1993). Local versus nonlocal boundary-layer diffusion in a global climate model, *J. Clim.* **6**, 1825–1842.

Houghton, R. A. (1993). Emissions of carbon from land-use change, paper presented at the 1993 Global Change Institute on the Carbon Cycle, Off. Interdisciplinary Earth Stud. Univ. Corp. Atmos. Res., Snowmass, Colorado, July 18–30.

Houghton, R. A. and Woodwell. G. M. (1983). Effect of increased C, N, P, and S on the global storage of C. *In* "The Major Biogeochemical Cycles and Their Interactions" (B. Bolin and R. B. Cook, eds), pp. 327–343. Wiley, New York.

Houghton, R. A. and Woodwell, G. M. (1989). Global climatic change, *Scient. Am.* **260**, 36–47.

Houghton, R. A., Hobbie, J. E., Melillo, J. M., Moore, B., Peterson, B. J., Shaver, G. R. and Woodwell, G. M. (1983). Changes in the carbon content of terrestrial biota and soils between 1860 and 1980: a net release of CO_2 to the atmosphere. *Ecol. Monogr.* **53**, 235–262.

Hunt, E. R. Jr., Piper, S. C., Nemani, R., Keeling, C. D., Otto, R. D. and Running, S. W. (1996). Global net carbon exchange and intra-annual atmospheric CO_2 concentrations predicted by an ecosystem process model and three-dimensional atmospheric transport model, *Global Biogeochem. Cycles* **10**, 431–456.

Hunt, J. M. (1972). Distribution of carbon in crust of Earth. *Bull. Am. Assoc. Pet. Geol.* **56**, 2273–2277.

IPCC (1990). Climate change. *In* "The IPCC Scientific Assessment" (J. T. Houghton, G. J. Jenkins and J. J. Ephraums, eds). Cambridge University Press, Cambridge.

IPCC (1992). Climate change. *In* "The Supplementary Report to The IPCC Scientific Assessment" (J. T. Houghton, B. A. Callander and S. K. Varney, eds). Cambridge University Press, Cambridge.

IPCC (1995). Climate change 1994. *In* "Radiative Forcing of Climate Change and An Evaluation of the IPCC IS92 Emission Scenarios" (J. T. Houghton, L. G. Meira Filho, J. Bruce, Hoesung Lee, B. A. Callander, E. Haites, N. Harris and K. Maskell, eds). Cambridge University Press, Cambridge.

Jansen, E. (1992). Deglaciation, impact on ocean circulation. *In* "Encyclopedia of Earth System Science" Vol. 2. Academic Press.

Jones, P. D. (1988). The influence of ENSO on global temperatures, *Clim. Monitor* **17**, 80–89.

Jones, P. D. (1994). Hemispheric surface air temperature variations: a reanalysis and an update to 1993, *J. Climate* **7**, 1794–1802.

Jones, P. D., Wigley, T. M. L. and Wright, P. B. (1986a). Global temperature variations between 1861 and 1984, *Nature* **322**, 430–434.

Jones, P. D., Raper, S. C. B., Bradley, R. S., Diaz, H. F., Kelly, P. M. and Wigley, T. M. L. (1986b). Northern Hemisphere surface air temperature variations: 1851–1984, *J. Clim. Appl. Meteor.* **25**, 161–179.

Jones, P. D., Raper, S. C. B. and Wigley, T. M. L. (1986c). Southern Hemisphere surface air temperature variations: 1851–1984, *J. Clim. Appl. Meteor.* **25**, 1213–1230.

Jouzel, J., Barkov, N. I., Barnola, J. M., Bender, M., Chappelaz, J., Genthon, C., Korlyakov, V. M., Lipenkov, V., Lorius, C., Petit, J. R., Raynaud, D., Raisbeck, G., Ritz, C., Sowers, T., Steivenard, M., Yiou, F. and Yiou, P. (1993). Extending the Vostok ice-core record of palaeoclimate to the penultimate glacial period, *Nature* **364**, 407–412.

Kattenberg, A., Giorgi, F., Grassl, H., Meehl, G. A., Mitchell, J. F. B., Stouffer, R. J., Tokioka, T., Weaver, A. J. and Wigley, T. M. L. (1996). Climate models–projections of future climate. *In* "Climate Change 1995: The Science of Climate Change" (J. T. Houghton *et al.*, eds), pp. 285–357. Cambridge University Press, Cambridge, U.K.

Keeling, C. D. (1973a). The carbon dioxide cycle. Reservoir models to depict the exchange of atmospheric carbon dioxide with the oceans and land plants. *In* "Chemistry of the Lower Atmosphere" (S. Rasool, ed.), pp. 251–329. Plenum Press, New York.

Keeling, C. D. (1973b). Industrial production of carbon dioxide from fossil fuels and limestone. *Tellus* **25**, 174–198.

Keeling, C. D. (1981). Standardization of notations and procedures. *In* "Carbon Cycle Modeling" (B. Bolin, ed.), pp. 81–101. Wiley, New York.

Keeling, C. D. (1991). CO_2 emissions – Historical record, global. *In* "Trends 91: A Compendium of Data on Global Change," *Rep. ORNL/CDIAC-46* (T. A. Boden, R. J. Sepanski and F. W. Stoss, eds), pp. 382–385. Oak Ridge Natl. Lab., Oak Ridge, TN.

Keeling, C. D. (1993). Surface ocean CO_2. *In* "The

Global Carbon Cycle" (M. Heimann, ed.), pp. 413–429. Springer-Verlag, Heidelberg.

Keeling, C. D. and Bacastow, R. B. (1977). Impact of industrial gases on climate. In "Energy and Climate," pp. 72–95. National Academy of Sciences, Washington, DC.

Keeling, C. D. and Bolin, B. (1967). The simultaneous use of chemical tracers in oceanic studies I. General theory of reservoir models. *Tellus* **19**, 566–581.

Keeling, C. D. and Bolin, B. (1968). The simultaneous use of chemical tracers in oceanic studies II. A three-reservoir model of the North and South Pacific Oceans. *Tellus* **20**, 17–54.

Keeling, C. D. and Whorf, T. P. (1991). Atmospheric CO_2 – Modern record, Mauna Loa. In "Trends 91: A Compendium of Data on Global Change," *Rep. ORNL/CDIAC-46* (T. A. Boden, R. J. Sepanski and F. W. Stoss, eds), pp. 12–15. Oak Ridge Natl. Lab., Oak Ridge, TN.

Keeling, C. D., Bacastow, R. B., Bainbridge, A. E., Ekdahl, C. A., Guenther, P. R., Waterman, L. S. and Chin, J. F. S. (1976a). Atmospheric carbon dioxide variations at Mauna Loa Observatory, Hawaii. *Tellus* **28**, 6, 538–551.

Keeling, C. D., Adams, J. A., Ekdahl, C. A. and Guenther, P. R. (1976b). Atmospheric carbon dioxide variations at the South Pole. *Tellus* **28**, 6, 552–564.

Keeling, C. D., Mook, W. G. and Tans, P. P. (1979). Recent trends in the $^{13}C/^{12}C$ ratio of atmospheric carbon dioxide. *Nature* **277**, 121–123.

Keeling, C. D., Bacastow, R. B., Carter, A. F., Piper, S. C., Whorf, T. P., Heimann, M., Mook, W. G. and Roeloffzen, H. (1989). A three-dimensional model of atmospheric CO_2 transport based on observed winds: 1. Analysis of observational data. In "Aspects of Climate Variability in the Pacific and Western Americas," *Geophys. Monogr. Ser.* vol. 55 (D. H. Peterson, ed.), pp 165–236. AGU, Washington, DC.

Keeling, C. D., Piper, S. C. and Heimann, M. (1989b). A three-dimensional model of atmospheric CO_2 transport based on observed winds: 4. Mean annual gradients and interannual variations. In "Aspects of Climate Variability in the Pacific and Western Americas," *Geophys. Monogr. Ser.* vol. 55 (D. H. Peterson, ed.), pp. 305–363. AGU, Washington, DC.

Keeling, C. D., Whorf, T. P., Wahlen, M. and van der Plicht, J. (1995). Interannual extremes in the rate of rise of atmospheric carbon dioxide since 1980, *Nature* **375**, 666–670.

Keeling, C. D., Chin, J. F. S. and Whorf, T. P. (1996).

Increased activity of northern vegetation inferred from atmospheric CO_2 measurements, *Nature* **382**, 146–149.

Keeling, R. F., Piper, S. C. and Heimann, M. (1996). Global and hemispheric CO_2 sinks deduced from changes in atmospheric O_2 concentration. *Nature* **381**, 218–221.

Kempe, S. (1979a). Carbon in the freshwater cycle. In "The Global Carbon Cycle" (B. Bolin, E. T. Degens, S. Kempe and P. Ketner, eds), pp. 317–342. Wiley, New York.

Kempe, S. (1979b). Carbon in the rock cycle. In "The Global Carbon Cycle" (B. Bolin, E. T. Degens, S. Kempe and P. Ketner, eds), pp. 343–377. Wiley, New York.

Kiehl, J. T. (1994). Sensitivity of a GCM climate simulation to differences in continental versus maritime cloud drop size, *J. Geophys. Res.* **99**, 23107–23115.

Kiehl, J. T., Hack, J. J., Bonan, G. B., Boville, B. A., Briegleb, B. P., Williamson, D. L. and Rasch, P. J. (1996). Description of the NCAR Community Climate Model (CCM3), *NCAR Tech. Note NCAR/TN-420+STR*, Natl. Cent. for Atmos. Res., Boulder, CO.

Kiehl, J. T., Hack, J. J., Bonan, G. B., Boville, B. B., Williamson, D. L. and Rasch, P. J. (1998). The National Center for Atmospheric Research Community Climate Model: CCM3. *J. Clim.* **11**, 1131–1150.

Kindermann, J., Würth, G., Kohlmaier, G. H. and Badeck, F.-W. (1996). Interannual variations of carbon exchange fluxes in terrestrial ecosystems, *Global Biogeochem. Cycles* **10**, 737–755.

Knorr, W. and Heimann, M. (1995). Impact of drought stress and other factors on seasonal land biosphere CO_2 exchange studied through an atmospheric tracer transport model, *Tellus, Ser. B*, **47**, 471–489.

Komhyr, W. D., Gammon, R. H., Harris, T. B., Waterman, L. S., Conway, T. J., Taylor, W. R. and Thoning, K. W. (1985). Global atmospheric CO_2 distribution and variations from 1968–1982 NOAA/GMCC CO_2 flask sample data, *J. Geophys. Res.* **90**, 5567–5596.

Koyama, T. (1963). Gaseous metabolism in lake sediments and paddy soils and the production of atmospheric methane and hydrogen, *J. Geophys. Res.* **68**, 3971–3973.

Kroopnick, P. (1980). The distribution of ^{13}C in the Atlantic Ocean. *Earth Planet. Sci. Letters* **49**, 469–484.

Kumar, A. and Hoerling, M. P. (1997). Interpretation and implications of the observed inter-El Niño variability. *J. Clim.* **10**, 83–91.

Kumar, M. and Monteith, J. L. (1981). Remote sensing of crop growth. *In* "Plants and the Daylight Spectrum" (H. Smith, ed.), pp. 133–144. Academic Press, New York.

Kutzbach, J., Bonan, G., Foley, J. and Harrison, S. P. (1996). Vegetation and soil feedbacks on the response of the African monsoon to orbital forcing in the early to middle Holocene, *Nature* **384**, 623–626.

Law, R. and Simmonds, I. (1996). The sensitivity of deduced CO_2 sources and sinks to variations in transport and imposed surface concentrations, *Tellus, Ser. B,* **48**, 613–625.

Law, R. M., Rayner, P. R., Denning, A. S., Erickson, D., Fung, I. Y., Heimann, M., Piper, S. C., Ramonet, M., Taguchi, S., Taylor, J. A., Trudinger, C. M. and Watterson, I. G. (1996). Variations in modeled atmospheric transport of carbon dioxide and the consequences for CO_2 inversions, *Global Biogeochem. Cycles* **10**, 783–796.

Legates, D. R. and Willmott, C. J. (1990a). Mean seasonal and spatial variability in global surface air temperature, *Theor. Appl. Climatol.* **41**, 11–21.

Legates, D. R. and Willmott, C. J. (1990b). Mean seasonal and spatial variability in gauge-corrected, global precipitation, *Int. J. Climatol.* **10**, 111–127.

Levine, J. S. (1994). Biomass burning and the production of greenhouse gases. *In* "Climate Biosphere Interaction: Biogenic Emissions and Environmental Effects of Climate Change" (R. G. Zepp, ed.). John Wiley and Sons.

Lieth, H. (1963). The role of vegetation in the carbon dioxide content of the atmosphere, *J. Geophys. Res.* **68**, 3887–3898.

Lieth, H. (1972). Über die primarproduktion der erde. *Z. Angew. Bot.* **46**, 1–37.

Lieth, H. (1975). Modeling the primary productivity of the world. *In* "Primary Productivity of the Biosphere" (H. Lieth and R. H. Whittaker, eds), pp. 237–263. Springer-Verlag, New York.

Likens, G. E. (ed.) (1981). "Some Perspectives of the Major Biogeochemical Cycles." Wiley, New York.

Liss, P. S. (1983). Gas transfer: experiments and geochemical implications. *In* "Air–Sea Exchange of Gases and Particles" (P. S. Liss and W. G. N. Slinn, eds), pp. 241–298. NATO ASI Series, Reidel, Holland.

Livingstone, D. A. (1963). Chemical composition of rivers and lakes. *In* "Data of Geochemistry," 6th edn (M. Fleischer, ed.), pp. 1–61. *US Geol. Surv. Prof. Pap. 440G.*

Machta, L. (1972). The role of the oceans and biosphere in the carbon dioxide cycle. *In* "The Changing Chemistry of the Oceans" (D. Dryssen and D. Jagner, eds). Wiley Interscience, London.

Machta, L. (1974). Global scale atmospheric mixing, *Advan. Geophys.* **18B**, 33–56.

Maisongrande, P., Ruimy, A., Dedieu, G. and Saugier, B. (1995). Monitoring seasonal and interannual variations of gross primary productivity, net primary productivity and net ecosystem productivity using a diagnostic model and remotely-sensed data, *Tellus, Ser. B,* **47**, 178–190.

Malmström, C. M., Thompson, M. V., Juday, G. P., Los, S.O., Randerson, J. T. and Field, C. B. (1997). Interannual variation in global-scale net primary production: Testing model estimates, *Global Biogeochem. Cycles* **11**, 367–392.

Marland, G. and Boden, T. A. (1991). CO_2 emissions – Modern record, global. In: Trends 91: A Compendium of Data on Global Change, *Rep. ORNL/ CDIAC-46* (T. A. Boden, R. J. Sepanski and F. W. Stoss, eds), pp. 386–389. Oak Ridge Natl. Lab., Oak Ridge, TN.

Marland, G. and Rotty, R. M. (1984). Carbon dioxide emissions from fossil fuels: A procedure for estimation and results for 1950–1982. *Tellus* **36B**, 232–261.

Marland, G., Boden, T. A., Griffin, R. C., Huang, S. F., Kanciruk, P. and Nelson, T. R. (1989). Estimates of CO_2 emissions from fossil fuel burning and cement manufacturing, based on the US Bureau of Mines cement manufacturing data, *Rep. ORNL/CDIAC-25, NDP-030*, Carbon Dioxide Information Analysis Center, Oak Ridge Natl. Lab., Oak Ridge, TN.

McGuire, A. D., Melillo, J. M., Joyce, L. A., Kicklighter, D. W., Grace, A. L., Moore III, B. and Vorosmarty, C. J. (1992). Interactions between carbon and nitrogen dynamics in estimating net primary productivity for potential vegetation in North America, *Global Biogeochem. Cycles* **6**, 101–124.

McGuire, A. D., Melillo, J. M., Kicklighter, D. W., Pan, Y., Xiao, X., Helfrich, J., Moore III, B., Vorosmarty, C. J. and Schloss, A. L. (1997). Equilibrium responses of global net primary production and carbon storage to doubled atmospheric carbon dioxide: Sensitivity to changes in vegetation nitrogen concentration, *Global Biogeochem. Cycles* **11**, 173–189.

Mehrbach, C., Culberson, C. H., Hawley, S. E. and Pytkowicz, R. M. (1973). Measurement of the apparent dissociation constants of carbonic acid in seawater at atmospheric pressure. *Limnol. Oceanog.* **18**, 6, 897–907.

Melillo, J. M., McGuire, A. D., Kicklighter, D. W., Moore III, B., Vorosmarty, C. J. and Schloss, A. L. (1993). Global climate change and terrestrial net primary production, *Nature* **363**, 234–240.

Melillo, J. M., Prentice, I. C., Farquhar, G. D., Schulze,

E.-D. and Sala, O. E. (1996). Terrestrial biotic responses to environmental change and feedbacks to climate. *In* "Climate Change 1995: The Science of Climate Change" (J. T. Houghton *et al.*, eds), pp. 445–481. Cambridge University Press, Cambridge.

Moore, B., Boone, R. D., Hobbie, J. E., Houghton, R. A., Melillo, J. M., Peterson, B. J., Shaver, G. R., Vorosmarty, C. J. and Woodwell, G. M. (1981). A simple model for analysis of the role of terrestrial ecosystems in the global carbon budget. *In* "Carbon Cycle Modeling" (B. Bolin, ed.), pp. 365–385. Wiley, New York.

Mopper, K. and Degens, E. T. (1979). Organic carbon in the ocean: nature and cycling. *In* "The Global Carbon Cycle" (B. Bolin, E. T. Degens, S. Kempe and P. Ketner, eds), pp. 293–316. Wiley, New York.

Morse, J. W., Millero, F. J., Thurmond, V., Brown, E. and Ostlund, H. G. (1984). The carbonate chemistry of Grand Bahama Bank Waters: after 18 years another look, *J. Geophys. Res.* **89**, C3, 3604–3614.

Myneni, R. B., Los, S. O. and Asrar, G. (1995). Potential gross primary productivity of terrestrial vegetation from 1982–1990, *Geophys. Res. Lett.* **22**, 2617–2620.

Myneni, R. B., Los, S. O. and Tucker, C. J. (1996). Satellite-based identification of linked vegetation index and sea surface temperature anomaly areas from 1982–1990 for Africa, Australia and South America, *Geophys. Res. Lett.* **23**, 729–732.

Myneni, R. B., Keeling, C. D., Tucker, C. J., Asrar, G. and Nemani, R. R. (1997). Increased plant growth in the northern high latitudes from 1981 to 1991, *Nature* **386**, 698–702.

Myneni, R. B., Tucker, C. J., Asrar, G. and Keeling, C. D. (1998). Interannual variations in satellite-sensed vegetation index data from 1981 to 1991, *J. Geophys. Res.* **103**, 6145–6160.

NAG. 1991. *The NAG Fortran Library Manual, Mark 15.* NAG Ltd, Oxford, U.K.

Nakazawa, T., Murayama, S., Miyashita, K., Aoki, S. and Tanaka, M. (1992). Longitudinally different variations of lower tropospheric carbon dioxide concentrations over the North Pacific Ocean, *Tellus, Ser. B*, **44**, 161–172.

Nakazawa, T., Morimoto, S., Aoki, S. and Tanaka, M. (1993). Time and space variations of the carbon isotopic ratio of tropospheric carbon dioxide over Japan, *Tellus, Ser. B*, **45**, 258–274.

Neftel, A., Oeschger, H., Schwander, J., Stauffer, B. and Zumbrunn, R. (1982). Ice core sample measurements give atmospheric CO_2 content during the past 40,000 yr. *Nature* **295**, 220–223.

Neftel, A., Moor, E., Oeschger, H. and Stauffer, B. (1985). Evidence from polar ice cores for the increase in atmospheric CO_2 in the past two centuries. *Nature* **315**, 45–47.

Neftel, A., Friedli, H., Moor, E., Lötscher, H., Oeschger, H., Siegenthaler, U. and Stauffer, B. (1994). Historical record from the Siple Station ice core. *In* "Trends 93: A Compendium of Data on Global Change," *Rep. ORNL/CDIAC-65* (T. A. Boden, D. P. Kaiser, R. J. Sepanski and F. W. Stoss, eds), pp. 11–14. Oak Ridge Natl. Lab., Oak Ridge, TN.

Nemry, B., François, L., Warnant, P., Robinet, R. and Gérard, J.-C. (1996). The seasonality of the CO_2 exchange between the atmosphere and the land biosphere: A study with a global mechanistic vegetation model, *J. Geophys. Res.* **101**, 7111–7125.

Nepstad, D. C., de Carvalho, C. R., Davidson, E. A., Jipp, P. H., Lefebvre, P. A., Negreiros, G. H., da Silva, E. D., Stone, T. A., Trumbore, S. E. and Vieira, S. (1994). The role of deep roots in the hydrological and carbon cycles of Amazonian forests and pastures, *Nature* **372**, 666–669.

Nisbet, E. G. (1992). Sources of atmospheric CH_4 in early postglacial time, *J. Geophys. Res.*, **97**, D12, 12859–12867.

Nordström, H. (1990). "Gräs." Natur och Kultur, Stockholm.

Nydal, R. and Lövseth, K. (1983). Tracing bomb ^{14}C in the atmosphere 1962–1980, *J. Geophys. Res.* **88**, C6, 3621–3642.

Oeschger, H., Siegenthaler, U., Schotterer, U. and Gugelmann, A. (1975). A box-diffusion model to study the carbon dioxide exchange in nature, *Tellus* **27**, 168–192.

Olson, J. S., Watts, J. A. and Allison, L. J. (1983). Carbon in live vegetation of major world ecosystems. United States Department of Energy, TR004.

Pearman, G. (1981). The CSIRO (Australia) Atmospheric CO_2 Monitoring Program. *In* "Carbon Cycle Modeling" (B. Bolin, ed.), pp. 117–120. Wiley, New York.

Pearman, G. I. and Beardsmore, D. J. (1984). Atmospheric carbon dioxide measurements in the Australian region: Ten years of aircraft data, *Tellus, Ser. B*, **36**, 1–24.

Pearman, G. I. and Hyson, P. (1980). Activities of the global biosphere as reflected in atmospheric CO_2 records, *J. Geophys. Res.* **85**, 4468–4474.

Pearman, G. I. and Hyson, P. (1981). The annual variation of atmospheric CO_2 concentration observed in the Northern Hemisphere, *J. Geophys. Res.* **86**, 9839–9843.

Pearman, G. I. and Hyson, P. (1986). Global transport and inter-reservoir exchange of carbon dioxide with particular reference to stable isotopic distributions, *J. Atm. Chem.* **4**, 81–124.

Pearman, G. I., Hyson, P. and Fraser, P. J. (1983). The global distribution of atmospheric carbon dioxide: 1. Aspects of observations and modeling, *J. Geophys. Res.* **88**, 3581–3590.

Peng, T.-H., Broecker, W. S., Mathieu, G. G. and Li, Y.-H. (1979). Radon evasion rates in the Atlantic and Pacific Oceans as determined during the GEOSECS program, *J. Geophys. Res.* **84**, 2471–2486.

Peng, T.-H., Broecker, W. S., Freyer, H. D. and Trumbore, S. (1983). A deconvolution of the tree ring based $\delta^{13}C$ record, *J. Geophys. Res.* **88**, C6, 3609–3620.

Post, W. M., King, A. W. and Wullschleger, S. D. (1997). Historical variations in terrestrial biospheric carbon storage, *Global Biogeochem. Cycles* **11**, 99–109.

Potter, C. S., Randerson, J. T., Field, C. B., Matson, P. A., Vitousek, P. M., Mooney, H. A. and Klooster, S. A. (1993). Terrestrial ecosystem production: A process model based on global satellite and surface data, *Global Biogeochem. Cycles* **7**, 811–841.

Quay, P. D., Tillbrook, B. and Wong, C. S. (1992). Oceanic uptake of fossil fuel CO_2: Carbon-13 evidence, *Science* **256**, 74–79.

Raich, J. W. and Schlesinger, W. H. (1992). The global carbon dioxide flux in soil respiration and its relationship to vegetation and climate, *Tellus, Ser. B*, **44**, 81–99.

Randall, D. A., Dazlich, D. A., Zhang, C., Denning, A. S., Sellers, P. J., Tucker, C. J., Bounoua, L., Berry, J. A., Collatz, G. J., Field, C. B., Los, S. O., Justice, C. O. and Fung, I. Y. (1996). A revised land surface parameterization (SiB2) for GCMs. Part III: The greening of the Colorado State University General Circulation Model, *J. Clim.* **9**, 738–763.

Randerson, J. T., Thompson, M. V., Malmström, C. M., Field, C. B. and Fung, I. Y. (1996). Substrate limitations for heterotrophs: Implications for models that estimate the seasonal cycle of atmospheric CO_2, *Global Biogeochem. Cycles* **10**, 585–602.

Randerson, J. T., Thompson, M. V., Conway, T. J., Fung, I. Y. and Field, C. B. (1997). The contribution of terrestrial sources and sinks to trends in the seasonal cycle of atmospheric carbon dioxide, *Global Biogeochem. Cycles* **11**, 535–560.

Raynaud, D., Jouzel, J., Barnola, J. M., Chappellaz, J., Delmas, R. J. and Lorius, C. (1993). The ice record of greenhouse gases. *Science* **259**, 926–934.

Rayner, P. J. and Law, R. M. (1995). A comparison of modelled responses to prescribed CO_2 sources, *Tech. Pap. 36*, Div. of Atmos. Res., Comm. Sci. and Ind. Res. Org., Melbourne, Victoria.

Redfield, A. C. (1934). On the proportions of organic derivatives in sea water and their relation to the composition of plankton. *In* "James Johnson Memorial Volume," Liverpool, pp. 176–192.

Redfield, A. C., Ketchum, B. H. and Richards, F. A. (1963). The influence of organisms on the composition of sea water. *In* "The Sea" (M. N. Hill, ed.), Vol. 2, pp. 26–77. Wiley-Interscience, New York.

Revelle, R. (1982). Carbon dioxide and world climate. *Scient. Am.* **247**, 2, 33–41.

Revelle, R. and Suess, H. E. (1957). Carbon dioxide exchange between atmosphere and ocean, and the question of an increase of atmospheric CO_2 during the past decades, *Tellus* **9**, 18–27.

Revelle, R. and Munk, W. (1977). The carbon dioxide cycle and the biosphere. *In* "Energy and Climate," pp. 140–158. National Academy of Sciences, Washington, DC.

Richey, J. E. (1983). C, N, P, and S cycles: major reservoirs and fluxes; the phosphorus cycle. *In* "The Major Biogeochemical Cycles and Their Interactions" (B. Bolin and R. B. Cook, eds), pp. 51–56. Wiley, New York.

Ronov, A. B. (1964). Common tendancies in the chemical evolution of the Earth's crust, ocean, and atmosphere, *Geochem.* **8**, 715–743.

Ropelewski, C. F. and Halpert, M. S. (1987). Global and regional scale precipitation patterns associated with the El Niño/Southern Oscillation. *Mon. Wea. Rev.* **115**, 1606–1626.

Rosenberg, N. J. (1981). The increasing CO_2 concentration in the atmosphere and its implications on agricultural productivity. I. Effects on photosynthesis, transpiration and water use efficiency. *Climat. Change* **3**, 265–279.

Rotty, R. M. (1981). Data for global CO_2 production from fossil fuels and cement. *In* "Carbon Cycle Modeling" (B. Bolin, ed.), pp. 121–125. Wiley, New York.

Sarmiento, J. L. (1993). Atmospheric CO_2 stalled, *Nature*, **365**, 697–698.

Sarmiento, J. L., Orr, J. C. and Siegenthaler, U. (1992). A perturbation simulation of CO_2 uptake in an ocean general circulation model, *J. Geophys. Res.* **97**, 3621–3645.

Sarmiento, J. L., Hughes, T. M. C., Stouffer, R. J. and Manabe, S. (1998). Simulated response of the ocean carbon cycle to anthropogenic climate warming. *Nature* **393**, 245–249.

Schimel, D. S., Braswell, B. H., Holland, E. A., McKeown, R., Ojima, D. S., Painter, T. H., Parton, W. J. and Townsend, A. R. (1994). Climatic, edaphic, and biotic controls over storage and turnover of carbon in soils, *Global Biogeochem. Cycles* **8**, 279–293.

Schimel, D. S., Enting, I. G., Heimann, M., Wigley, T. M. L., Raynaud, D., Alves, D. and Siegenthaler, U. (1995). CO_2 and the carbon cycle. *In* "Climate Change 1994: Radiative Forcing of Climate Change and an Evaluation of the IPCC IS92 Emission Scenarios" (J. T. Houghton *et al.* eds), pp 35–71. Cambridge University Press, Cambridge.

Schimel, D. S., Braswell, B. H., McKeown, R., Ojima, D. S., Parton, W. J. and Pulliam, W. (1996). Climate and nitrogen controls on the geography and timescales of terrestrial biogeochemical cycling, *Global Biogeochem. Cycles* **10**, 677–692.

Schlesinger, W. H. (1977). Carbon balance in terrestrial detritus. *Ann. Ecol. Syst.* **8**, 51–81.

Schnell, R. C., Odh, S.-Å and Njau, L. N. (1981). Carbon dioxide measurements in tropical East African biomes, *J. Geophys. Res.* **86**, 5364–5372.

Schwander, J. and Stauffer, B. (1984). Age difference between polar ice and the air trapped in its bubbles. *Nature* **311**, 45–47.

Schwander, J., Barnola, J. M., Andrié, C., Leuenberger, M., Ludin, A., Raynaud, D. and Stauffer, B. (1993). The age of the air in the firn and the ice at Summit, Greenland, *J. Geophys. Res.* **98**, 2831–2838.

Sedjo, R. A. (1992). Temperate forest ecosystems in the global carbon cycle, *Ambio*, **21**, 274–277.

Seiler, W. and Crutzen, P. J. (1980). Estimates of gross and net fluxes of carbon between the biosphere and the atmosphere from biomass burning. *Climat. Change* **2**, 226–247.

Sellers, P. J. (1985). Canopy reflectance, photosynthesis and transpiration, *Int. J. Remote Sens.* **6**, 1335–1372.

Sellers, P. J., Mintz, Y., Sud, Y. C. and Dalcher, A. (1986). A simple biosphere model (SiB) for use within general circulation models. *J. Atmos. Sci.* **43**, 505–531.

Sellers, P. J., Berry, J. A., Collatz, G. J., Field, C. B. and Hall, F. G. (1992). Canopy reflectance, photosynthesis, and transpiration. III. A reanalysis using improved leaf models and a new canopy integration scheme, *Remote Sens. Environ.* **42**, 187–216.

Sellers, P. J., Randall, D. A., Collatz, G. J., Berry, J. A., Field, C. B., Dazlich, D. A., Zhang, C., Collello, G. D. and Bounoua, L. (1996a). A revised land surface parameterization (SiB2) for atmospheric GCMs. Part I: Model formulation, *J. Clim.* **9**, 676–705.

Sellers, P. J., Los, S. O., Tucker, C. J., Justice, C. O., Dazlich, D. A., Collatz, G. J. and Randall, D. A. (1996b). A revised land surface parameterization (SiB2) for atmospheric GCMs. Part II: The generation of global fields of terrestrial biophysical parameters from satellite data, *J. Clim.* **9**, 706–737.

Shea, D. J., Trenberth, K. E. and Reynolds, R. W. (1990). A global monthly sea surface temperature climatology, *NCAR Tech. Note NCAR/TN-345+STR*, 167 pp, Natl. Cent. for Atmos. Res., Boulder, CO.

Siegenthaler, U. (1983). Uptake of excess CO_2 by an outcrop-diffusion model of the ocean, *J. Geophys. Res.* **88**, C6, 3599–3608.

Siegenthaler, U. (1990). El Niño and atmospheric CO_2, *Nature* **345**, 295–296.

Siegenthaler, U. and Oeschger, H. (1987). Biospheric CO_2 emissions during the past 200 years reconstructed by deconvolution of ice core data, *Tellus* **39B**.

Siegenthaler, U. and Joos, F. (1992). Use of a simple model for studying oceanic tracer distributions and the global carbon cycle, *Tellus* **44B**, 186–207.

Siegenthaler, U. and Sarmiento, J. L. (1993). Atmospheric carbon dioxide and the ocean. *Nature* **365**, 119–125.

Simmons, A. J. and Strüfing, R. (1983). Numerical forecasts of stratospheric warming events using a model with a hybrid vertical coordinate, *Q. J. R. Meteorol. Soc.* **109**, 81–111.

Six, C., Fischer, A. and Walters, S. (1996). The seasonal cycle of atmospheric CO_2: A study based on the NCAR Community Climate Model (CCM2), *J. Geophys. Res.* **101**, 15079–15097.

Skirrow, J. (1975). The dissolved gases – carbon dioxide. *In* "Chemical Oceanography," 2nd edn (J. P. Riley and G. Skirrow, eds), Vol. 2, pp. 1–192. Academic Press, London.

Smith, T. M. and Shugart, H. H. (1993). The transient response of terrestrial carbon storage to a perturbed climate, *Nature* **361**, 523–526.

Stuermer, D. H. and Payne, J. R. (1976). Investigation of seawater and terrestrial humic substances with carbon-13 and proton nuclear magnetic resonance. *Geochim. Cosmochim. Acta* **40**, 1109–1114.

Stuiver, M. (1980). ^{14}C distribution in the Atlantic Ocean, *J. Geophys. Res.* **85**, 2711–2718.

Stuiver, M. and Polach, H. A. (1977). Discussion reporting of ^{14}C data. *Radiocarbon* **19**, 355–363.

Stuiver, M. and Quay, P. D. (1980). Changes in atmospheric carbon-14 attributed to a variable sun. *Science* **207**, 11–19.

Stuiver, M. and Quay, P. D. (1981). Atmospheric ^{14}C changes resulting from fossil fuel CO_2 release and cosmic ray flux variability. *Earth Planet. Sci. Lett.* **53**, 349–362.

Stuiver, M., Ostlund, H. G. and McConnaughey, T. A. (1981). GEOSECS Atlantic and Pacific ^{14}C distribution. *In* "Carbon Cycle Modeling" (B. Bolin, ed.), pp. 201–221. Wiley, New York.

Stuiver, M., Quay, P. D. and Ostlund, H. G. (1983).

Abyssal water carbon-14 distribution and the age of the world oceans. *Science* **219**, 849–851.

Stumm, W. and Morgan, J. J. (1981). "Aquatic Chemistry," 2nd edn. Wiley, New York.

Suess, H. E. (1965). Secular variations of the cosmic-ray-produced carbon-14 in the atmosphere and their interpretation, *J. Geophys. Res.* **70**, 5937–5952.

Sundquist, E. T. (1993). The global carbon dioxide budget. *Science* **259**, 934–941.

Taguchi, S. (1996). A three-dimensional model of atmospheric CO_2 transport based on analyzed winds: Model description and simulation results for TRANSCOM, *J. Geophys. Res.* **101**, 15 099–15 109.

Takahashi, T., Broecker, W. S., Werner, S. R. and Bainbridge, A. E. (1980). Carbonate chemistry of the surface waters of the world oceans. *In* "Isotope Marine Chemistry" (E. P. Goldberg, Y. Horibe and K. Sarubashi, eds), pp. 291–326. Uchida Rokakuho, Tokyo.

Takahashi, T., Broecker, W. S. and Bainbridge, A. E. (1981). The alkalinity and total carbon dioxide concentration in the world oceans. *In* "Carbon Dioxide Modeling" (B. Bolin, ed.), pp. 271–286. Wiley, New York.

Tanaka, M., Nakazawa, T. and Aoki, S. (1987). Time and space variations of tropospheric carbon dioxide over Japan, *Tellus, Ser. B*, **39**, 3–12.

Tans, P. P., Conway, T. J. and Nakazawa, T. (1989). Latitudinal distribution of the sources and sinks of atmospheric carbon dioxide derived from surface observations and an atmospheric transport model, *J. Geophys. Res.* **94**, 5151–5172.

Tans, P. P., Fung, I. Y. and Takahashi, T. (1990). Observational constraints on the global atmospheric CO_2 budget, *Science* **247**, 1431–1438.

Taylor, J. A. (1989). A stochastic Lagrangian atmospheric transport model to determine global CO_2 sources and sinks – a preliminary discussion, *Tellus, Ser. B*, **41**, 272–285.

Thompson, M. L., Enting, I. G., Pearman, G. I. and Hyson, P. (1986). Interannual variation of atmospheric CO_2 concentration, *J. Atm. Chem.* **4**, 125–155.

Thompson, M. V., Randerson, J. T., Malmström, C. M. and Field, C. B. (1996). Change in net primary production and heterotrophic respiration: How much is necessary to sustain the terrestrial carbon sink?, *Global Biogeochem. Cycles* **10**, 711–726.

Walker, J. C. G. (1977). "Evolution of the Atmosphere." Macmillan, New York.

Warnant, P., François, L., Strivay, D. and Gérard, J.-C. (1994). CARAIB: A global model of terrestrial biological productivity, *Global Biogeochem. Cycles* **8**, 255–270.

Warneck, P. (1988). "Chemistry of the Natural Atmosphere." International Geophysics Series, Vol 41. Academic Press, San Diego.

Watson, R. T., Rodhe, H., Oeschger, H. and Siegenthaler, U. (1990). Greenhouse gases and aerosols. *In* "Climate Change, the IPCC Scientific Assessment" (J. T. Houghton, G. J. Jenkins and J. J. Ephraums, eds), pp. 1–40. Cambridge University Press, New York.

Weiss, R. F. (1974). Carbon dioxide in water and seawater: the solubility of a non-ideal gas. *Marine Chem.* **2**, 203–215.

Wigley, T. M. L. (1983). The pre-industrial carbon dioxide level. *Climat. Change* **5**, 315–320.

Wigley, T. M. L. (1991). A simple inverse carbon cycle model. *Global Biogeochem. Cycles* **5**, 373–382.

Wigley, T. M. L. (1993). Balancing the carbon budget: Implications for projections of future carbon dioxide concentration changes, *Tellus* **45B**, 409–425.

Wigley, T. M. L. (1994). How important are carbon cycle uncertainties? In: "Climate Change and the Agenda for Research" (T. Hanisch, ed.), pp. 169–191. Westview, Boulder, CO.

Williams, P. J. le B. (1975). Biological and chemical aspects of dissolved organic material in sea water. *In* "Chemical Oceanography," 2nd edn (J. P. Riley and G. Skirrow, eds), Vol. II, pp. 301–363. Academic Press, London.

Williams, R. T. (1982). "Transient Tracers in the Ocean Preliminary Hydrographic Report," Vols 1–4. Scripps Institution of Oceanography, La Jolla, CA.

Williamson, D. L. and Rasch, P. J. (1994). Water vapor transport in the NCAR CCM2, *Tellus, Ser. A*, **46**, 34–51.

Wofsy, S. C., Goulden, M. L., Munger, J. W., Fan, S.-M., Bakwin, P. S., Daube, B. C., Bassow, S. L. and Bazzaz, F. A. (1993). Net exchange of CO_2 in a mid-latitude forest, *Science* **260**, 1314–1316.

Wolter, K. (1997). Trimming problems and remedies in COADS. *J. Clim.* **10**, 1980–1997.

Wong, C. S., Chan, Y.-H., Page, J. S., Smith, G. E. and Bellegay, R. D. (1993). Changes in the equatorial CO_2 flux and new production estimated from CO_2 and nutrient levels in Pacific surface waters during the 1986/87 El Niño, *Tellus, Ser. B*, **45**, 64–79.

Woodwell, G. M. (1978). The carbon dioxide question. *Scient. Am.* **238**, 38–43.

Woodwell, G. M. and Pecan, E. V. (eds) (1973). "Carbon and the Biosphere." United States Atomic Energy Commission, Washington, DC.

Woodwell, G. M., Hobbie, J. E., Houghton, R. A., Melillo, J. M., Moore, B., Peterson, B. J. and Shaver, G. R. (1983). Global deforestation: contribution to atmospheric carbon dioxide. *Science* **222**, 1081–1086.

Zhang, G. J. and McFarlane, N. A. (1995). Sensitivity of climate simulations to the parameterization of cumulus convection in the Canadian Climate Centre General Circulation Model, *Atmos.-Ocean* **33**, 407–446.

Zinsmeister, A. R. and Redman, T. C. (1980). A time series analysis of aerosol composition measurements, *Atmos. Environ.* **14**, 201–215.

12

The Nitrogen Cycle

Daniel A. Jaffe

12.1 Introduction

In most natural systems available or "fixed" nitrogen is usually the limiting factor in plant growth. This realization led to the invention and massive use of nitrogen fertilizers during the 20th century and ever increasing crop yields per acre of farmed land. Without this use of nitrogenous fertilizers, the Earth could not support its current population of six billion people (Smil, 1997).

At the same time, the widespread use of fossil fuels releases not only carbon dioxide, but nitrogen oxides as well. These nitrogen oxides contribute to urban photochemical smog and acid precipitation. The combined effect of these two anthropogenic processes, agriculture and fossil fuel combustion, is similar in magnitude to natural nitrogen fixation. These substantial modifications to the global nitrogen cycle have important implications in a number of areas including photochemical smog, climate, stratospheric ozone, regional eutrophication, and ecosystem diversity, which will be discussed in this chapter.

12.2 Chemistry

Nitrogen has five valence electrons and can take on oxidation states between +5 and −3. Most of the nitrogen compounds we will discuss either have nitrogen bonded to carbon and hydrogen, in which case the oxidation state of the nitrogen is negative (N is more electronegative than either C or H); or have nitrogen bonded to O, in which case the nitrogen has a positive oxidation state.

Table 12-1 lists the most common nitrogen compounds that exist in the natural world, by oxidation state. In addition it also lists the boiling point for each compound as well as its heat of formation ($\Delta H^0(f)$) and free energy of formation ($\Delta G^0(f)$). For comparison, the data on H_2O are also included.

To fully understand some of the major players in the nitrogen cycle, we should also consider some of the industrial and social implications of these compounds.

1. HNO_3. Nitric acid is a very strong acid; about 6.8 million metric tons per year are manufactured for industrial purposes in the US. Most of it is produced from ammonia by the catalytic oxidation to NO, which is then further oxidized to NO_2. Addition of water forms HNO_3. Most of the nitric acid produced is used in the manufacture of fertilizers, and a lesser amount is used to make explosives.

In the troposphere, nitrogen oxides react to also produce HNO_3. The oxidants are free radicals produced photochemically, such as HO_2, RO_2, and OH. The HNO_3 produced in this manner is an important contributor to "acid rain."

In its pure form, nitric acid is a liquid with a high vapor pressure (47.6 torr at 20°C), so that in the lower atmosphere HNO_3 exists as a gas, in an aerosol or in a cloud droplet. When nitric acid reacts with a base a nitrate salt is produced. If

Earth System Science
ISBN 0-12-379370-X

Table 12-1   Chemical data on important nitrogen compounds

Oxidation state	Compound	b.p. (°C)	$\Delta H^0(f)$	$\Delta G^0(f)$
			(kJ/mol, 298 K)	
+5	$N_2O_5(g)$	11	115	
	$HNO_3(g)$	83	−135	−75
	$Ca(NO_3)_2(s)$		−900	−720
	$HNO_3(aq)$		−200	−108
+4	$NO_2(g)$	21	33	51
	N_2O_4		9	98
+3	$HNO_2(g)$		−80	−46
	$HNO_2(aq)$		−120	−55
+2	$NO(g)$	−152	90	87
+1	$N_2O(g)$	−89	82	104
0	$N_2(g)$	−196	0	0
−3	$NH_3(g)$	−33	−46	−16.5
	$NH_4^+(aq)$		−72	−79
	$NH_4Cl(s)$		−201	−203
	$CH_3NH_2(g)$		−28	28
	$H_2O(g)$	100	−242	−229

the atmospheric base is ammonia, NH_4NO_3 is the result:

$$NH_3(g) + HNO_3(g \text{ or } aq) \rightarrow NH_4NO_3(s \text{ or } aq)$$

If the reaction is between two gas-phase species, then this reaction could be a source of cloud condensation nuclei, or simply a means to neutralize an acidic aerosol. Although there are some questions concerning the measurement of atmospheric HNO_3, (Lawson, 1988) most measurements indicate that gaseous HNO_3 concentrations predominate over particle NO_3^-.

In addition, there are numerous other nitrate salts. These are all high-melting, colorless solids, and are very soluble in water. Many of these, such as KNO_3 and $NaNO_3$, are mined for use in fertilizers and explosives. Prior to the industrial production of HNO_3 from ammonia, the mining of these salts was the major means for producing explosives. During World War I, Germany's supply of $NaNO_3$ from Chile was cut off. The Haber process ($N_2 + 3H_2 \rightarrow 2NH_3$), developed by Fritz Haber only a few years earlier, allowed Germany to produce ammonia, and therefore nitric acid, to make nitrate salts for explosives.

This allowed Germany to continue its war effort even without its Chilean source of $NaNO_3$.

2. *NO_2*. Nitrogen dioxide is a brown/yellow gas at room temperature due to its light absorption at wavelengths shorter than 680 nm. NO_2 dimerizes into the colorless N_2O_4 (and indeed there is a very small amount of N_2O_4 found in urban atmospheres). Nitrogen dioxide has a very irritating odor, and is quite toxic. It is produced by the oxidation of NO, so that the concentrations of these two gases are coupled in the atmosphere. Since NO is a by-product of virtually all combustion processes, NO and NO_2 are generally found in much higher concentrations in urban areas than in the natural background, and this is a significant source of photochemical smog.

3. *NO*. Nitric oxide, or nitrogen monoxide, is a colorless gas at room temperature. As we have already seen, it is industrially produced by the oxidation of ammonia. However, with respect to the urban environment, a more significant process is the high temperature reaction of N_2 with O_2 (as in a car engine) to produce NO.

4. N_2O. Nitrous oxide is also a colorless gas at room temperature. Its principal uses are as "laughing gas" and as an aerosol propellant. Industrially, it is not produced in large quantities. In many respects N_2O is analogous to CO_2. It has the same linear structure, the same number of electrons (isoelectronic), and a similar (low) reactivity. However, CO_2 is more soluble in water as a result of the acid–base reaction of CO_2 and water. It is the low reactivity of N_2O that results in a long tropospheric lifetime, and therefore its eventual transport to the stratosphere, where it is believed to be a primary control on the concentration of ozone in the stratosphere. Concentrations of N_2O in the troposphere are increasing and this is significant since it is a "greenhouse" gas and plays a significant role in stratospheric ozone chemistry.

5. N_2. Nitrogen is a colorless gas at room temperature. It is generally considered to be a very stable molecule; however, it is not its thermodynamic stability, but rather kinetics that accounts for its low reactivity. This is shown by the values of the thermodynamic parameters given in Table 12-1. In the presence of oxygen, N_2 is thermodynamically unstable with respect to aqueous NO_3^-, but the high activation energy necessary to break the N_2 triple bond results in its chemical inertness. If we lived in a world dominated by equilibrium chemistry, most atmospheric N_2, and all of the O_2 would be consumed, yielding an ocean containing approximately 0.1 M HNO_3 (Lovelock, 1979). As a result of its non-polar nature N_2 has a low solubility in water, but with its high partial pressure in the atmosphere it is the most prevalent nitrogen species in the ocean. Nitrogen constitutes some 78% by volume of the atmosphere, and is industrially separated by the liquefaction and distillation of air.

6. NH_3. Ammonia is a colorless gas. It is a strong base, forms hydrogen bonds, is soluble in water, and is a fairly reactive molecule. Each year 12.4 million metric tons are manufactured by the Haber process ($N_2 + 3H_2 \rightarrow 2NH_3$ at 400°C and 250 atm), principally for nitric acid production, which is then used to make fertilizers and explosives. As a fertilizer, ammonia can be utilized in three ways: first by direct injection

of the boiling ($-33°C$) liquid. This method works only because most soils are moist and acidic, so that the NH_3 dissolves in the wet soil before it evaporates. The second method is to use ammonia in ammonium salts, such as $NH_4Cl(s)$ or $NH_4NO_3(s)$. The third method is to oxidize the ammonia to HNO_3, and use it in a nitrate salt. In the atmosphere, ammonia is the primary gaseous base, so that it will react with acids either in the gas or aqueous phase to produce an ammonium salt, as in these reactions:

$$NH_3(g) + HCl(g) \rightarrow NH_4Cl(s)$$
$$NH_3(g) + H_2SO4(l) \rightarrow (NH_4)_2SO_4(s)$$

A significant proportion of the total reduced nitrogen in the atmosphere exists as aqueous or aerosol ammonium ion, with lesser amounts of gaseous ammonia (Quinn *et al.*, 1988). Ammonia has a very irritating odor, and is toxic at low concentrations. Ammonium salts contain the tetrahedral NH_4^+ ion. Some ammonium salts, like NH_4NO_3 and $(NH_4)_2SO_4$, are manufactured on a large scale, 6.0 and 1.8 million metric tons per year respectively. Ammonium sulfate is used principally as a fertilizer, whereas ammonium nitrate is also used in explosives. Since plants can utilize nitrogen in both the -3 and $+5$ oxidation states, ammonium nitrate is particularly well suited for use as a fertilizer; however, it must be handled with caution. In April 1947, a ship loaded with ammonium nitrate in Texas City, Texas, exploded, killing a total of 576 people both onboard and in the surrounding city.

There are also many significant nitrogen compounds that are a necessary part of all organisms. Most of these have nitrogen in a -3 oxidation state.

7. *Amines.* Amines ($R—NH_2$) are an organic derivative of ammonia, where an alkyl group ($—R$) replaces one or more of the hydrogens. The simplest amine is methylamine, $CH_3—NH_2$. An amine can have one, two, or three alkyl groups, as in trimethylamine, $(CH_3)_3—N$. Like ammonia, amines are fairly basic and participate in hydrogen bonding; low molecular weight amines are quite water soluble. As the alkyl groups get larger the amine begins to show

more properties of an organic molecule than of ammonia, and water solubility decreases. The amines are generally odorous compounds with relatively high boiling points, due to hydrogen bonding. Methylamine, and other amines, are often found in the flesh of rotting fish, and this could represent a pathway into the atmosphere for these compounds. Amines are bases, and their reaction with acids in the atmosphere is probably their principal removal mechanism (as for ammonia):

$$CH_3–NH_2(g) + H_3O^+(aq)$$
$$\rightarrow NH_3–NH_3^+(aq) + H_2O$$

The alkyl ammonium ion is quite soluble in water. Like ammonia, amines may also be oxidized. As would be expected from their structure, amines are fairly polar. Amines (including aromatic amines) are quite common in the biological world. For example many vitamins (such as thiamine and niacinamide) and all alkaloids (e.g. caffeine, cocaine, nicotine, and lysergic acid) contain an amine functional group. Many of these groups are found in ring systems, with pyridine rings (C_5H_5N) being quite common.

8. *Amides.* Amides occur quite frequently in nature, most significantly in urea and proteins. Urea, $NH_2—CO—NH_2$, is an important carrier of nitrogen between animals and plants. Animals metabolize proteins and amino acids, and excrete large amounts of urea. Plants break urea down into ammonia, which they can utilize. For this reason, urea is also an excellent fertilizer, and five million metric tons are manufactured each year through the high-temperature reaction of CO_2 and NH_3. Animal excrement and urea fertilizers are thought to be significant sources of atmospheric ammonia (Freney *et al.*, 1983).

Proteins are also important nitrogen compounds. They constitute much of the cell materials, and are present in every type of organism known. In humans, muscle tissue, skin, and hair is mostly protein, about half of the dry weight of our bodies. From a chemical point of view, proteins are polymers of amino acids, alpha amine derivatives of carboxylic acids. Only about 20 different amino acids are actually found in proteins. It is the large number of variations in the protein chain, using only these 20 amino acids that gives rise to the great diversity of proteins. Numerous other organic nitrogen compounds are found in natural systems in small amounts, some of which are very toxic or carcinogenic. These include certain types of nitro compounds, cyano compounds, or nitrosamines, for example.

Having considered many of the different type of compounds we will be discussing, we should briefly consider some of the uses of the data presented in Table 12-1. The thermodynamic relationships can be used to calculate equilibrium concentrations for such gases as NO and NO_2 in air. This only requires the relationship between ΔG^0 and the equilibrium constant. A more complex analysis would include the contribution of photolytically driven reactions, to derive steady-state concentrations for many trace species. Holland (1978) has presented such an analysis for an abiotic world, including the input energy from sunlight. His results show that virtually all nitrogen compounds are currently found at much higher levels than the calculated steady-state value, indicating that other important processes have been ignored. This can only happen if the chemical reactions are slow compared to the rates of other processes. Lovelock (1979) has stated that on Earth, the disagreement between abiotic steady-state calculated concentrations and measured atmospheric concentrations indicates the presence of life, and suggests that atmospheric chemical probes can be used as a simple means to detect life on any planetary system. If chemical measurements of a planet's atmosphere show that it is close to thermodynamic equilibrium, substantial numbers of living organisms are probably not present. In natural systems, some thermodynamic equilibria do exist, but in many cases, environmental systems are in some type of dynamic non-equilibrium steady state. On Earth, the presence of life significantly alters the atmospheric concentrations of many trace species.

12.3 Biological Transformations of Nitrogen

Our next task is to consider the various ways that nitrogen is processed by the biosphere.

These mechanisms are the primary mover in the terrestrial and oceanic nitrogen cycles. It is important to remember that even though much of our discussion of the nitrogen cycle revolves around transfer of nitrogen between the major global reservoirs (atmosphere, aquatic, and biosphere), these fluxes represent only a small portion of the nitrogen transferred within the biosphere–soil and biosphere–aquatic systems. Rosswall (1976) estimates that on a global basis, the "internal" *biological* nitrogen cycle accounts for 95% of all nitrogen fluxes. The important processes are indicated schematically in Fig. 12-1. The various terms are easier to identify if one remembers that they are defined from the perspective of the organism.

1. *Nitrogen fixation* is any process in which N_2 in the atmosphere reacts to form *any* nitrogen compound. Biological nitrogen fixation is the enzyme-catalyzed reduction of N_2 to NH_3, NH_4^+, or any organic nitrogen compound.
2. *Ammonia assimilation* is the process by which NH_3 or NH_4^+ is taken up by an organism to become part of its biomass in the form of organic nitrogen compounds.
3. *Nitrification* is the oxidation of NH_3 or NH_4^+ to NO_2^- or NO_3^- by an organism, as a means of producing energy.
4. *Assimilatory nitrate reduction* is the reduction of NO_3^-, followed by uptake of the nitrogen by the organism as biomass.
5. *Ammonification* is the breaking down of organic nitrogen compounds into NH_3 or NH_4^+.

6. *Denitrification* is the reduction of NO_3^- to any gaseous nitrogen species, generally N_2 or N_2O.

All of the above processes are mediated by various types of microorganisms. Some of these processes are energy producing, and some of these occur in symbiotic relationships with other organisms. It is appropriate to start discussion of the processes with nitrogen fixation, since this is the only means by which nitrogen can be brought into natural systems (in the absence of artificial fertilization). Similarly, it is principally denitrification that removes nitrogen from the biosphere.

Biological nitrogen fixation is the ultimate source of nitrogen in all living organisms, in the absence of industrial fertilizers. It can be done by a variety of bacteria and algae, both symbiotic and free living, although, in general, the symbiotic organisms are thought to be quantitatively more significant. There are two major limitations to biological nitrogen fixation. The first is that the process takes a large amount of input energy to overcome the high activation energy of the nitrogen triple bond. Despite the fact that ΔG^0 for the production of NH_3 from N_2 and H_2 is negative at 25°C, only those organisms with highly developed catalytic systems are able to fix nitrogen. The second limitation is that nitrogen fixation is a reductive process, highly sensitive to the presence of O_2, and so only those organisms that live in anaerobic environments, or can provide an anaerobic environment, will fix nitrogen. Burns and Hardy (1975) present a good review of biological nitrogen fixation.

In terrestrial systems the symbiotic bacteria, particularly strains of genus *Rhizobium*, are a significant source of nitrogen fixation. These bacteria are found on the roots of many leguminous plants (clover, soybeans, chickpeas, etc.), and have been used agriculturally as a means of replenishing soil nitrogen ("green manures") (Smil, 1997). Many of these organisms are anaerobes, or have developed mechanisms to maintain an anaerobic environment (Granhall, 1981). Other symbiotic diazotrophs (nitrogen-fixing organisms) exist, but the *Rhizobium* have been the most extensively studied (e.g., Postgate, 1982).

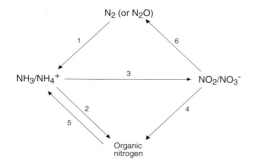

Fig. 12-1 Biological transformations of nitrogen compounds. The numbers refer to processes described in the text.

Most diazotrophic (nitrogen-fixing) organisms utilize the nitrogenase enzyme. This enzyme has been the focus of intensive research in recent years due to the possibilities of utilizing it to improve industrial nitrogen fixation (Postgate, 1982). It has been isolated from 20 to 30 different prokaryotic organisms and appears to have very similar properties regardless of the source. Nitrogenase consists of two metalloproteins; one, a Mo–Fe protein, which serves to bind the N_2 to the enzyme, probably at the metal site, the other an Fe protein, which is the source of electrons for the reduction. Both of these metalloproteins are very sensitive to O_2. It is interesting to note that whereas microorganisms can fix N_2 at a partial pressure of 0.8 atm and 20°C, the industrial fixation requires 250 atm and 400°C!

Once nitrogen has been fixed in the soil or aquatic system as NH_3 or NH_4^+, there are two major pathways it can follow. It can be oxidized to NO_2^-/NO_3^-, or assimilated by an organism to become part of its biomass. The latter process is termed ammonia assimilation, and for those organisms that can directly utilize ammonia, this represents a significant nitrogen source. Since many plants obtain most of their nitrogen from nitrate, via reductive assimilation, direct ammonia assimilation yields a significant energy savings, and therefore gives those organisms a competitive advantage. Free NH_4^+ ion does not exist for long in aerobic soils before nitrification occurs, and so NO_3^- is the prevalent form of nitrogen in aerobic soil and aquatic environments (Delwiche, 1981). Ammonia assimilation does not involve nitrogen transfer to other reservoirs, and is quantitatively less significant than nitrification.

Nitrification consists of two energy yielding steps: the oxidation of ammonium to nitrite, and the oxidation of nitrite to nitrate. These equations are generally represented as follows:

$$NH_4^+ + \tfrac{3}{2}O_2 \rightarrow NO_2^- + H_2O + 2H^+$$
$$\Delta G^0 = -290 \text{ kJ/mol}$$

$$NO_2^- + \tfrac{1}{2}O_2 \rightarrow NO_3^- \qquad \Delta G^0 = -82 \text{ kJ/mol}$$

There are several organisms that utilize nitrification as an energy source (Delwiche, 1981). The first step in the process is principally done by bacteria of genus *Nitrosamonas*, and the second by *Nitrobacter*, both autotrophic organisms. These organisms utilize CO_2 as a carbon source and obtain their energy from the oxidation of NH_4^+. Heterotrophic bacteria, which utilize organic compounds rather than CO_2, can also perform nitrification; however, these are thought to be quantitatively much less significant than the autotrophs (Bremner and Blackmer, 1981). In the oxidation of NH_4^+ to NO_3^-, hydoxylamine (NH_2OH) and other less stable compounds are likely intermediates, and NO and N_2O are almost certainly intermediates, probably as enzyme bound complexes. These intermediates have brought attention to nitrification as a possible source of atmospheric constituents, particularly N_2O.

Once in a soil or aquatic system, nitrate has two major pathways, it can serve as a terminal electron acceptor under anaerobic conditions (denitrification), or it can be simultaneously reduced and assimilated into an organism's biomass. The latter process is termed assimilatory nitrate reduction, and is likely to be dominant when reduced nitrogen is in low supply, as during aerobic conditions. This represents a primary input of nitrogen for most plants and many microorganisms. Most plants can assimilate both NH_4^+ and NO_3^-, even though there is an energy cost in first reducing the NO_3^-. After the NO_3^- has been taken up by the root system of a plant and reduced, it then follows the same pathway by which NH_4^+ is incorporated into its biomass (Kikby, 1981).

Besides nitrogen fixation, the only other major source of reduced nitrogen is the decomposition of soil or aquatic organic matter. This process is called ammonification. Heterotrophic bacteria are principally responsible for this. These organisms utilize organic compounds from dead plant or animal matter as a carbon source, and leave behind NH_3 and NH_4^+, which can then be recycled by the biosphere. In some instances heterotrophic bacteria may incorporate a complete organic molecule into their own biomass. The majority of the NH_3 produced in this way stays within the biosphere; however, a small portion of it will be volatilized. In addition to this source, the breakdown of animal excreta also contributes to atmospheric ammonia. It is generally believed that volatilization from animal excreta is more significant than ammonification, as a source of atmospheric ammonia. Freney *et al.* (1983) present a good review on the overall ammonia volatilization process.

Denitrification is the only process in which the major end-product is removed from the biological nitrogen cycle. It is the principal means of balancing the input flux from nitrogen fixation. Generally, N_2 is the end-product of denitrification; however, NO, and particularly N_2O, are also common. Microorganisms use NO_3^- as a terminal electron sink (oxidant) in the absence of O_2, as in waterlogged anaerobic soils. The overall process of oxidizing an organic compound, and reducing the NO_3^-, is an energy producing process for the microorganisms. There are approximately 17 genera of facultative anaerobic bacteria that can utilize NO_3^- as an oxidizing agent, and these are thought to be widespread. Denitrification occurs via a well-known series of intermediates, including NO_2^-, NO, and N_2O. The ratio of N_2:N_2O production during denitrification is an interesting topic. Generally the major product is N_2, accounting for 80–100% of the nitrogen released (Delwiche, 1981). Söderlund and Svensson (1976) used a global terrestrial average of 16:1 for N_2:N_2O in calculating the N_2O flux to the atmosphere. Under certain environmental conditions, nitrous oxide can become a major product. Generally, conditions that increase the amount of N_2O production also decrease the overall rate of denitrification. For example, at lower pH values, and higher O_2 concentrations, the proportion of N_2O increases, but the overall rate of denitrification decreases.

From a biogeochemical cycle perspective, the biological nitrogen cycle is not a simple system. It is clear that that the largest nitrogen input to the biosphere is biological fixation, and the primary loss from the biosphere is denitrification. However, from a biogeochemical perspective, it is not the exchange of N_2 that is the most important process, but rather the exchange of various trace gases, such as NO, NO_2, N_2O, and NH_3. Trace gases are released into the atmosphere at various stages in the biological nitrogen cycle. To understand the behavior of these trace gases, it is necessary to consider a wide range of biological processes.

12.4   Anthropogenic Nitrogen Fixation

Human activities result in the fixation of huge amounts of nitrogen on a daily basis. This occurs as a result of three different processes:

1. Direct *intentional* production of NH_3 and HNO_3 (mostly for fertilizer)
2. *Unintentional* production of NO during combustion (fossil fuels and biomass)
3. Biological nitrogen fixation as a result of agricultural practices (e.g., planting clover so as to replenish nitrogen on farmlands).

As mentioned in the introduction, the rapid growth in fertilizer use over the past 50 years has allowed us to feed the six billion inhabitants on the planet today (e.g., Smil, 1997). Over the same time period, fossil fuel usage has shot up dramatically and is now a major source of "fixed" nitrogen, both in developed *and* developing countries around the world. As a result of the rapid growth in these activities, the total anthropogenic N fixation (the sum of these three processes) is now similar to the global natural N fixation and this is thought to be an issue of some concern (e.g., Galloway *et al.*, 1995; Vitousek *et al.*, 1997a,b).

Production and use of industrial nitrogen fertilizers is a late 20th century phenomenon. Industrially produced nitrogen fertilizers come in many forms, including NH_3, $(NH_4)SO_4$, NH_4NO_3, and urea. Starting around 1950, when production was less than 5 Tg N/yr, it has increased to a current value of around 80 Tg N/yr. Smil (1997) shows that fertilizer use is closely linked to rapid population growth over this same time period. Estimates by Galloway *et al.* (1995) suggest that this increase will continue into the 21st century, with most of the growth in fertilizer use coming in the developing world. This is also the region where most of Earth's population growth is predicted to occur. Although industrially produced fertilizers have substantially increased agricultural yields, this process is not without its concerns, and these will be described in Section 12.7 below.

Unintentional N fixation occurs every time we burn fossil fuels to produce nitric oxide, NO. This occurs as a result of the high-temperature combination of N_2 and O_2 (described previously). In urban regions the NO is the primary cause for photochemical smog. Eventually most of the NO is converted into HNO_3, which is a major contributor to acid rain. Galloway *et al.* (1995) estimate that NO produced from fossil fuel combustion has doubled over the past 30 years (from ~ 10 to 20 Tg N/yr) and will double again over the next 30, with most of the increase in developing regions.

The use of plants, such as clover and soybeans, by humans to increase available nitrogen is a much older practice and not changing as rapidly as the other nitrogen fixing processes. The flux is also much less certain. Galloway *et al.* (1995) estimate that the current value for nitrogen fixation by legumes and other planted vegetation is ~ 40 Tg N/yr, with only a modest (20%) increase over the past 20 years.

12.5   Atmospheric Chemistry

The atmospheric chemistry of nitrogen is quite complex and involves literally hundreds or thousands of chemical reactions. Although the fluxes are much smaller than the biological fluxes, these processes are important for a variety of reasons, including impacts on climate, stratospheric ozone, and photochemical smog. In this section we present an overview of the most important processes.

12.5.1   *Homogeneous Gas Phase Reactions*

Photochemistry plays a significant role in nitrogen's atmospheric chemistry by producing reactive species (such as OH radicals). These radicals are primarily responsible for all atmospheric oxidations. However, since the photochemistry of the atmosphere is quite complex, it will not be dealt with in detail here. For an in-depth review on tropospheric photochemistry, the reader is referred to Logan *et al.* (1981), Finlayson-Pitts and Pitts (1986), Crutzen and Gidel (1983) or Crutzen (1988).

In most cases, the direct reaction of N_2 with O_2 is slow under ambient conditions. It is the presence of numerous odd electron species (for example, OH, HO_2, and RO_2 radicals) that are photochemically produced and responsible for most of the oxidizing reactions of nitrogen species in the atmosphere. Some of the important reactions are shown below:

$$O_3 + h\nu \rightarrow O_2 + O(^1D)$$

$O(^1D)$ is an electronically excited oxygen atom. It can decay back to a ground state oxygen atom (^3P) (which will regenerate an ozone molecule), or else it can react with water to produce two OH radicals:

$$O(^1D) + H_2O \rightarrow 2OH$$

The OH radical is a primary oxidizer in the atmosphere, oxidizing CO to CO_2 and CH_4 and higher hydrocarbons to CH_2O, CO, and eventually CO_2. OH and other radical intermediates can oxidize CH_4 and NO in the following sequence of reactions:

$$OH + CH_4 \rightarrow H_2O + CH_3$$
$$M + CH_3 + O_2 \rightarrow CH_3O_2 + M$$
$$CH_3O_2 + NO \rightarrow NO_2 + CH_3O$$
$$CH_3O + O_2 \rightarrow HO_2 + CH_2O$$
$$HO_2 + NO \rightarrow NO_2 + OH$$

Net reaction:
$$CH_4 + 2O_2 + 2NO \rightarrow CH_2O + 2NO_2 + H_2O$$

Hydroxyl, OH, acts as a catalyst for the oxidation of NO to NO_2. NO_2 molecules can react with the OH radical to produce HNO_3, which may be removed in precipitation. This is how most tropospheric NO_x eventually gets removed, either in wet or dry deposition.

$$NO_2 + OH + M \rightarrow HNO_3 + M$$

NO_2 may also photolyze and produce a ground state O atom, which will go on to produce ozone:

$$NO_2 + h\nu \rightarrow NO + O \;(\lambda < 410 \text{ nm})$$

$$O_2 + O + M \rightarrow O_3 + M$$

These reactions are important in a cycle that oxidizes CO and hydrocarbons and produces ozone, in the presence of NO_x ($NO + NO_2$). In photochemical smog, ozone can build up to unhealthy levels of several hundred parts per billion (ppb) as a result of these reactions. There are many other reactions that occur, some of which may be significant at various times, including the destruction of O_3 by NO, production and loss of HONO (nitrous acid) and peroxyacetyl nitrate (PAN). These reactions, and many more, represent a complex set of chemical interactions. For our purposes here, it is only necessary to note the major features.

1. The oxidation of CO and all hydrocarbons to CO_2 is indirectly driven by ozone and sunlight via the OH radical.
2. In the presence of sufficient NO_x (roughly 30 parts per trillion) this oxidation produces ozone.
3. Most NO and NO_2 eventually gets removed as HNO_3.
4. The lifetime of gaseous NO_x in the troposphere is on the order of 1–30 days (Söderlund and Svensson, 1976; Garrels, 1982; Crutzen, 1988).
5. Conversion of NO_x to PAN can result in significantly greater disbursement (e.g., Honrath and Jaffe, 1992).

Nitrogen oxides also play a significant role in regulating the chemistry of the stratosphere. In the stratosphere, ozone is formed by the same reaction as in the troposphere, the reaction of O_2 with an oxygen atom. However, since the concentration of O atoms in the stratosphere is much higher (O is produced from photolysis of O_2 at wavelengths less than 242 nm), the concentration of O_3 in the stratosphere is much higher.

$$O_2 + hv \rightarrow 2O$$

$$O_2 + O + M \rightarrow O_3 + M$$

Stratospheric ozone production is balanced by various catalytic destruction sequences:

$$O_3 + X \rightarrow XO + O_2$$
$$XO + O \rightarrow X + O_2$$

Net: $O_3 + O \rightarrow 2O_2$

where X can be any of the radical species NO, H, or Cl. For example substituting NO for X yields

$$O_3 + NO \rightarrow NO_2 + O_2$$
$$NO_2 + O \rightarrow NO + O_2$$

Net: $O_3 + O \rightarrow 2O_2$

Considering natural stratospheric ozone production/destruction as a balanced cycle, the NO_x reaction sequence is responsible for approximately half of the loss in the upper stratosphere, but much less in the lower stratosphere (Wennberg et al., 1994). Since this is a natural steady-state process, this is not the same as a long term O_3 loss. The principal source of NO to the stratosphere is the slow upward diffusion of tropospheric N_2O, and its subsequent reaction with O atoms, or photolysis (McElroy et al., 1976).

$$N_2O + hv \rightarrow N_2 + O$$
$$N_2O + O(^1D) \rightarrow 2NO$$

The first of these reactions results in the generation of a single ozone atom, whereas the second reaction produces two NO molecules which leads to catalytic ozone destruction as shown above. The relative rates of these two reactions is in an approximate ratio 9:1, favoring the first. Since NO is a catalyst for O_2 destruction (a single NO will destroy many ozone molecules before being removed), N_2O is believed to exert a significant control on stratospheric O_3 concentrations, although the effect is quite complex (e.g., Wennberg et al., 1994).

A ground level source and stratospheric sink for N_2O is consistent with the observed vertical concentration gradient (Weiss, 1981). The chemistry of stratospheric ozone is complex and closely tied to nitrogen species. Data from Antarctica suggest that anthropogenically produced chlorofluorocarbons fragment and result in substantial ozone depletion in the stratosphere (Farman et al., 1985). This process is accelerated in the extremely cold vortex that forms over Antarctica each winter. Measurements suggest that the nitrogen oxides, which would generally remove Cl fragments, are tied up in ice particles in polar stratospheric clouds, which form only at very low temperatures. Once the nitrogen oxides are "frozen out," the Cl fragments can go on to catalytically destroy ozone. Although the chemistry is complex, it is clear that N_2O and NO_x in the stratosphere play a critical role in the chemistry of ozone (Solomon and Schoeberl, 1988; Toon and Turco, 1991).

12.5.2   Heterogeneous Atmospheric Reactions

Heterogeneous processes play a role in several ways including: gas–particle conversions, gas uptake by cloudwater and precipitation, exchange of gases into or from the oceans, and exchange of gases into or from soil.

In the atmosphere, precipitation is the most important means to remove the NH_4^+ and NO_3^- ions. These ions are produced by reactions such as the following:

$$NH_3(g) + HNO_3(g) \rightarrow NH_4NO_3(s)$$
$$NH_3(g) + HNO_3(aq) \rightarrow NH_4^+(aq) + NO_3^-(aq)$$
$$NH_3(g) + H_2O(l) \rightarrow NH_3(aq)$$
$$NH_3(g) + H_2SO_4(l) \rightarrow NH_4(HSO_4)(s)$$
$$NH_3(g) + NH_4(HSO_4)(s) \rightarrow (NH_4)_2SO_4(s)$$

All of these species are very soluble in a rain or cloud drop and are an important source of atmospheric aerosols. For ammonia and ammonium, the condensed phases (l and s) represent approximately two-thirds of the total atmospheric burden, whereas for nitric acid and nitrates, about two-thirds is in the gas phase (Söderlund and Svensson, 1976).

The oceans represent a large potential reservoir for any gaseous compound. The solubility of a particular compound is governed by its chemical structure, temperature, pH, and other chemical properties of the solution, as well as its atmospheric concentration. For some atmospheric gases, such as NH_3 and NO_x, the oceans are probably a net sink (through the mechanism of precipitation), since virtually all measurements of oceanic air show significantly lower concentrations than is found over continental regions. This seems reasonable in light of the fact that most sources for NH_3 and NO_x are in continental areas. However, recent measurements of gaseous and aerosol ammonia in seawater and air indicate that the oceans are supersaturated with respect to ammonia in some regions. Thus they are a likely source, locally, of gaseous ammonia, while at the same time being a *net* global sink for it (Quinn *et al.*, 1988).

Loss of nitrogen compounds from soils is also a major pathway into the atmosphere for some compounds (e.g., N_2O, NO, and NH_3). As in the aquatic systems, parameters that play an important role in this process include: the nature of the compound; soil temperature, water content, pH, aeration of the soil; and a concentration gradient of the gas in question.

12.6   The Global Nitrogen Cycle

The global nitrogen cycle is often referred to as the nitrogen cycles, since we can view the overall process as the result of the interactions of various biological and abiotic processes. Each of these processes, to a first approximation, can be considered as a self-contained cycle. We have already considered the biological cycle from this perspective (Fig. 12-1), and now we will look at the other processes, the ammonia cycle, the NO_x cycle, and the fixation/denitrification cycle.

12.6.1   Nitrogen Inventories

We will consider the inventories of nitrogen in the following compartments: terrestrial, oceanic, and atmospheric. In general, there is more agreement among researchers over the values for the nitrogen burdens, than for the fluxes. In considering these inventories it is significant to recall that 99.96% of the non-crustal nitrogen exists as uncombined atmospheric N_2 and it is this fact that causes nitrogen to often be a limiting nutrient in the condensed phase.

The principal form of nitrogen in terrestrial systems is as dead soil organic matter, with biomass accounting for only about 4%, and inorganic nitrogen about 6.5%, on a global average (Söderlund and Svensson, 1976). There is, however, a large difference in the distribution of nitrogen in the tropics and the polar regions, with tropical regions having a larger proportion of nitrogen contained as biomass. Most of the reservoirs for organic nitrogen and biomass have been estimated by knowledge of the carbon content of soils and biomass and an appropriate ratio of carbon to nitrogen. The inventories of inorganic forms of nitrogen have a higher relative uncertainty.

Dissolved N_2 is the principal form of nitrogen in the oceans, accounting for 95% of the total oceanic nitrogen. The remainder of the oceanic nitrogen is principally NO_3^- and dead organic matter. The oceans hold about 0.5% of the total non-crustal nitrogen (as N_2). In contrast to CO_2, where the oceans are a significant reservoir, the oceans contain only about 15% of the total N_2O, due to its lower solubility. As in the terrestrial compartment, organic nitrogen compounds are estimated from knowledge of the carbon content and an appropriate C/N ratio, and the inorganic nitrogen inventories have a higher relative uncertainty.

In the atmosphere, N_2 is the principal nitrogen component and over 99% of the remaining nitrogen is found as N_2O. The other trace nitrogen species all have reactivities and removal mechanisms that result in residence times of less than a year and low atmospheric concentrations. Gaseous ammonia, for example, decreases significantly with height due to its removal by acidic gases, acidic aerosols, and liquid water (Hoell *et al.*, 1980). The short residence times for these species results in concentration fields, which are highly inhomogeneous, and thus difficult to quantify on a global basis. The inventory accuracy of many of the trace species, particularly NH_3, are limited by scant data, especially in the southern hemisphere and remote regions of the globe. For NO_x, a large number of recent field campaigns have significantly increased our understanding of this important species (e.g., Emmons *et al.*, 1997). Based on the use of a global three-dimensional model, Jaffe *et al.* (1997) estimate that the troposphere contains 0.6 Tg N of reactive nitrogen oxides (primarily NO_x, PAN, and HNO_3), with another 1–1.5 Tg N present in the stratosphere (mostly as HNO_3). Figures 12-2, 12-3, and 12-4 show the distribution of various forms of nitrogen in the atmospheric, oceanic, and terrestrial reservoirs.

12.6.2   Fluxes of Nitrogen

Figure 12-5 shows a schematic diagram of the NH_3/NH_4^+ cycle. The largest source of atmos-

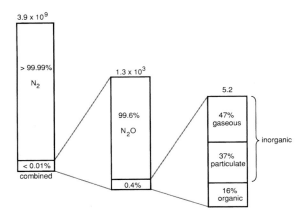

Fig. 12-2   Partitioning of the various forms of nitrogen in the atmosphere. Units are Tg N. (Reprinted with permission from R. Söderlund and T. Rosswall, The nitrogen cycles. In O. Huntizger (1982). "The Natural Environment and the Biogeochemical Cycles," p. 70, Springer-Verlag, Heidelberg.)

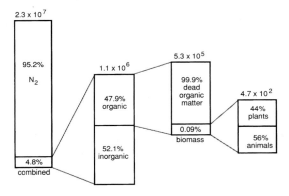

Fig. 12-3   Partitioning of the global inventories of nitrogen in the aquatic system. Units are Tg N. (Reprinted with permission from R. Söderlund and T. Rosswall, The nitrogen cycles. In O. Huntizger (1982). "The Natural Environment and the Biogeochemical Cycles," p. 71, Springer-Verlag, Heidelberg.)

pheric ammonia is ammonification and volatilization from animal excreta (Freney *et al.*, 1983). This includes a substantial fraction, probably more than half of the total, due to domestic animals. Direct anthropogenic emissions, including combustion and fertilizers, are much smaller. The majority of this $NH_3(g)$ is returned as NH_4^+ in precipitation or as $NH_3(g)$ via dry deposition. The ammonia is then available again to the biosphere, and the cycle is repeated.

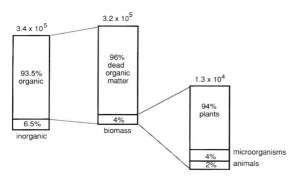

Fig. 12-4 Partitioning of the global inventories of nitrogen in the terrestrial system. Units are Tg N. (Reprinted with permission from R. Söderlund and T. Rosswall, The nitrogen cycles. In O. Huntizger (1982). "The Natural Environment and the Biogeochemical Cycles," p. 72, Springer-Verlag, Heidelberg.)

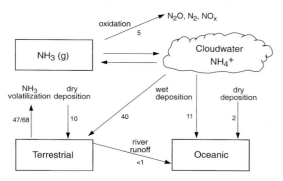

Fig. 12-5 The ammonia–ammonium cycle. Each arrow represents one flux. The magnitude of the flux is given in Tg N/yr^{-1}. Where two numbers are given, the top value is the anthropogenic contribution and the lower is the total flux (natural + anthropogenic).

In the NO_x cycle (Fig. 12-6), gaseous emissions of NO, and much smaller emissions of NO_2, are balanced by dry deposition of NO_2 and HNO_3 and wet deposition of NO_3^-. The principal sources of NO_x are anthropogenic combustion (both fossil fuels and biomass). Microbial processes in soils, lightning, and natural forest fires are much smaller NO_x sources (Galloway *et al.*, 1995). In the atmosphere, NO_x is converted to HNO_3 via photochemical oxidation, and therefore has a short residence time, on the order of a few days. Wet deposition occurs mainly as NO_3^- in precipitation, and since the anthropogenic emissions of NO_x occur mainly in urban areas, HNO_3 is a significant contributor to acid precipitation in adjacent regions (within a few thousand km). Anthropogenic acid deposition results from both sulfur and nitrogen oxides; however, since emissions of NO_x are increasing more rapidly than SO_2 in many regions, NO_3^- is becoming an increasingly important contributor to acid deposition (Mayewski *et al.*, 1990; Sirois, 1993). Once nitrate is deposited back to the terrestrial or oceanic systems it can be incorporated into biomass, enter the fixation–denitrification cycle, or accumulate in the ocean.

As mentioned previously, the fixation–denitrification cycle (Fig. 12-7) is the most heavily perturbed by humans. This is due to both the increasing use of nitrogenous fertilizers and the planting of nitrogen-fixing plants. One of the

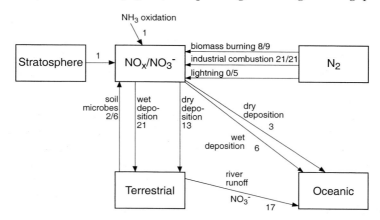

Fig. 12-6 NO_x–NO_3^- cycle. Each arrow represents one flux. The magnitude of the flux is given in Tg N/yr. Where two numbers are given, the top values is the anthropogenic contribution and the lower number is the total flux (natural + anthropogenic).

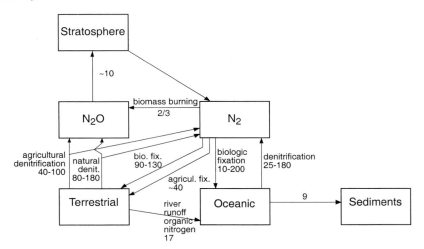

Fig. 12-7. Fixation–denitrification cycle. Each arrow represents one flux. The magnitude of the flux is given in Tg N/yr. Where two numbers are given, the top value is the anthropogenic contribution and the lower number is the total flux (natural + anthropogenic).

most important results of this is the increasing emissions and concentration of atmospheric N_2O.

A summary of the major N fluxes is presented in Table 12-2 and Fig. 12-8 shows a summary of the N fluxes for atmospheric NH_3, N_2O, and NO_x.

12.7   Human Impacts

Vitousek *et al.* (1997a,b,c) describe a range of concerns associated with the large anthropogenic perturbations to the nitrogen cycle. This includes groundwater contamination from NO_3^- (agricultural runoff), eutrophication (agricultural runoff and atmospheric NO_3^- deposition), radiative forcing of climate (N_2O and tropospheric O_3), stratospheric chemistry, photochemical ozone "smog," acid precipitation (HNO_3 deposition), changes in species diversity (due to N deposition), and N fertilization of the global carbon cycle. We provide a brief overview of each of these issues.

12.7.1   Groundwater/Eutrophication

As described previously, agricultural fertilizer contains large amounts of fixed nitrogen, mostly

in the form of NH_3, NH_4^+, or NO_3^-. Some of this N is utilized by the growing crops, but a significant amount is not taken up by the biomass and instead washes off the farm and into either ground or surface waters. This can cause two different problems, NO_3^- toxicity and eutrophication.

Although NO_3^- is not usually thought of as a "toxic" chemical, it does cause several health problems including methemoglobinemia in infants (blue-baby syndrome) and may also be linked to stomach cancer. Agricultural runoff can lead to significant, potentially harmful, concentrations of NO_3^- in ground or surface water.

Eutrophication can occur when NO_3^- (or other nutrients such as PO_4^{3-}) accumulate in lakes, ponds, or estuaries from agricultural runoff, sewage, or phosphate detergents. These nutrients will accelerate plant growth, often leading to algal blooms, oxygen depletion and sometimes mass fish death. Eutrophication can also lead to substantial impacts on the overall aquatic ecosystem. Impacts due to eutrophication are well known for the Baltic Sea, Black Sea, Chesapeake Bay, and other regions (Vitousek *et al.*, 1997b). One example of the human influence on aquatic NO_3^- is the results reported by Turner and Rabalais (1991). These authors have shown that NO_3^- concentrations in the Mississippi River have more than doubled (since 1965)

Table 12-2 Fluxes in the global nitrogen cycle (units are Tg N/yr)

	Stedman and Shetter (1983)	Jaffe (1992)	Galloway *et al.* (1995)	Range [b]
Terrestrial-atmospheric				
1 Natural biological fixation (pre-agriculture)	110		90–130	90–170
2 Fixation due to planting of nitrogen fixing plants			43	
3 Total biological fixation (1+2 above)		150		
4 Industrial fixation (fertilizer production)		40	78	30–78
5 Natural denitrification (pre-agriculture)	124.5	124	80–180	80–243
6 Additional denitrification due to agriculture		23	50–110	
7 Microbial NO_x production	10	8	4	4–89
8 Microbial N_2O production (natural)	38	7	8	12–69
9 Ammonia volatilization	82[a]	122	68	16–244
10 Biomass burning N_2O production			2	
11 Biomass burning NO_x production	5	12	8	5–15
12 Anthropogenic N_2O production (all sources)	11	5	3.4	3–11
Oceanic–atmospheric				
13 Biologic fixation	40	40	40–200	10–200
14 Denitrification	30.5	30	150–180	25–180
Atmospheric–atmospheric				
15 NO_x production by lightning	3	5	3	0.5–10
16 NO_x production by industrial combustion	20	20	21	15–40
Terrestrial–oceanic				
17 River run-off		34	34	14–40

[a] NH only.
[b] For additional flux estimates not reported here refer to Jaffe (1992).

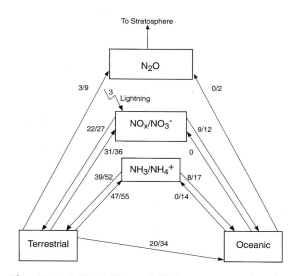

Fig. 12-8 NH_3, NO_x and N_2O sources and sinks. Where two numbers are given, the top value is the anthropogenic contribution and the lower number is the total flux (natural + anthropogenic).

due to a variety of human causes. Similar results are known for other regions of the globe (Vitousek *et al.*, 1997b).

12.7.2 Climate: N_2O and Tropospheric O_3

Due to the imbalance of sources and sinks, atmospheric N_2O is increasing by 3 Tg N/yr or 0.2%/yr. Figure 12-9 shows average N_2O mixing ratios from four stations in the NOAA–CMDL network, Barrow, Mauna Loa, Samoa, and the South Pole (data are from the NOAA–CMDL and can be obtained from www.cmdl.noaa.gov). The most recent IPCC estimate gives a total N_2O source of 16 Tg N, 7 Tg of which are a result of human activities (IPCC, 1997). The largest contribution to the anthropogenic N_2O sources is 3 Tg N from

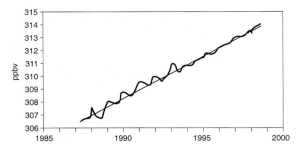

Fig. 12-9 Global N_2O concentrations based on NOAA–CMDL observations at BRW, MLO, SMO, and SPO. Data are from the NOAA–CMDL and can be obtained from www.cmdl.noaa.gov.

agricultural soils, mostly lost after applying fertilizer.

With respect to nitrogen fertilizers, the subject of gaseous losses due to denitrification has been extensively studied. Rolston (1981) presents a good review of this topic, and gives data suggesting that anywhere from 10–75% of fertilizer nitrogen may be lost by this process (typically in the range of 20%). Various crop management practices have been developed to counter this problem. For example, having large amounts of NO_3^-, organic carbon and water will increase denitrification, whereas limiting one of these factors will decrease it.

Matson *et al.* (1998) describe an experiment on fertilizer usage and gaseous emissions on a wheat farm in Mexico. In this experiment, the authors used both a traditional/high fertilizer approach and a reduced fertilizer method, but where the fertilizer was applied in a manner that was more efficiently used by the growing plants. The results showed that when using a reduced fertilizer strategy crop yields per hectare were similar to the high fertilizer case, but gaseous N emissions (NO and N_2O) were about 10% of the base case. Since this strategy used much less fertilizer, this protocol also gave the best results with respect to the farmer's profits. This is important in that it shows that it is possible to reduce agricultural emissions of NO and N_2O and improve the farmer's bottom line.

Tropospheric ozone is also radiatively active and there is good evidence that it is now about twice its pre-industrial concentration, at least in the northern hemisphere. This increase is due to increasing emissions of NO_x from fossil fuel combustion, followed by photochemical ozone production (Crutzen, 1988; Logan, 1985; Volz and Kley, 1988). This human caused change contributes significantly to radiative forcing of climate (Marenco *et al.*, 1994; IPCC, 1995). Logan (1994) conducted a detailed evaluation of free tropospheric ozone sonde data. In general, this analysis shows ozone trends of about 1–2% per year over the last two decades over the US, Europe, and Japan, with the higher trends observed over Japan and Europe. For the US and Europe the O_3 concentrations appear to have approximately leveled out, whereas the Japanese stations continue to show increases. This is consistent with the fact that NO_x emissions in the US and Europe are not increasing, whereas they are increasing rapidly (5%/year) in East Asia (Kato and Akimoto, 1992).

The increases in both N_2O and tropospheric O_3 are leading to increased radiative forcing. Based on the concentration changes between pre-industrial times, 1850 to 1992, CO_2, N_2O, and tropospheric O_3 contribute 1.56, 0.14, and $0.4 W/m^2$ to global average radiative forcing (IPCC, 1995). However, due to its shorter lifetime (2–4 weeks), tropospheric ozone is quite inhomogeneous in the troposphere and so the uncertainty in calculating its contribution to radiative forcing is significantly greater than for the more long-lived species. Based on projected future trends of energy consumption and agricultural emissions, all of these are expected to continue to increase through the 21st century.

12.7.3 Stratospheric Chemistry

As mentioned previously, N_2O plays an important role in stratospheric chemistry by providing the dominant source of NO_x in the stratosphere (see Section 12.5). What is more difficult to predict is how stratospheric chemistry will change as a result of continued increases in the concentration of atmospheric N_2O. Early research suggested that increased N_2O would lead to significant reductions in stratospheric O_3. However, more current reports suggest that stratospheric NO_x plays a key role in "protecting" stratospheric O_3 from more significant losses from CFC-produced Cl radicals (e.g., Solomon and Schoeberl, 1988; Toon and Turco, 1991). Exactly how future increases in N_2O will impact stratospheric O_3 is something of an open question at present.

An additional area of concern with respect to stratospheric ozone is possible direct emissions of NO_x into the stratosphere by high-flying supersonic aircraft. This issue has come up repeatedly over the past 20 years, as air travel and pressure from commercial airlines has increased. However, despite substantial research effort to understand stratospheric chemistry, the question is complicated by the changing levels of stratospheric chlorine, first due to a rapid accumulation of tropospheric CFCs, followed by a rapid decline in CFC emissions due to the Montreal Protocol. To quote from the from the 1994 WMO/UN Scientific assessment of ozone depletion, executive summary (WMO 1995):

> Atmospheric effects of supersonic aircraft depend on the number of aircraft, the altitude of operation, the exhaust emissions, and the background chlorine and aerosol loadings. Projected fleets of supersonic transports would lead to significant changes in trace-species concentrations, especially in the North-Atlantic flight corridor. Two-dimensional model calculations of the impact of a projected fleet (500 aircraft, each emitting 15 grams of NO_x per kilogram of fuel burned at Mach 2.4) in a stratosphere with a chlorine loading of 3.7 ppb, imply additional (i.e., beyond those from halocarbon losses) annual-average ozone column decreases of 0.3–1.8% for the Northern hemisphere. There are, however, important uncertainties in these model results, especially in the stratosphere below 25 km. The same models fail to reproduce the observed ozone trends in the stratosphere below 25 km between 1980 and 1990. Thus, these models may not be properly including mechanisms that are important in this crucial altitude range.

12.7.4   Photochemical Smog

Unhealthy concentrations of ozone due to photochemical production from nitrogen oxides are a daily occurrence for millions of people who live in large urban centers. This is especially true for inhabitants of large cities in the warmer climates, such as Los Angeles, Mexico City, Athens, and Beijing. For example, in the 1970s Los Angeles exceeded the US EPA O_3 standard around 175–200 days per year (Lents and Kelly, 1993), however Los Angeles is now making progress. In the early 1990s, Los Angeles exceeded the O_3 health standard on "only" 100–150 days per year and the peak concentrations also declined considerably. This change is a result of tightened vehicle emission standards, improved engine reliability and increased controls on non-vehicular sources, despite having more people driving nearly twice as many vehicle miles as in 1970. Nonetheless, there are serious concerns about whether cities such as Los Angeles can ever meet the existing ozone standard.

In 1997, the US Environmental Protection Agency tightened the O_3 standard, changing it from 120 ppbv as a 1 h average, to 80 ppbv as an 8 h standard. This was done due to substantial new evidence that O_3 health effects occur at this lower level. Since Los Angeles can not meet the current standard, it seems highly unlikely they will be able to meet this new standard. While for a long time Los Angeles could reasonably be called "the ozone capital of the world," it is now probably being exceeded by rapidly developing cities in other countries. For example, both Mexico City and Beijing have serious O_3 smog problems that probably exceed the problems in Los Angeles.

Ozone also causes significant damage to vegetation. In some regions where intensive industry and agriculture coexist, there is the possibility for substantial impacts on food production due to ozone. This is because ozone damage to crops can occur at mixing ratios as low as 60 ppbv and also due to the fact that the application of nitrogen fertilizers will increase local NO emissions (as described above). Thus to some extent there is a "self-limiting" effect from adding additional fertilizer in that the increased NO emissions will result in decreased crop yields due to ozone damage. Based on the photochemical model of Chameides *et al.* (1994), crop yields in regions of China are already being impacted by a few percent. This impact will continue and worsen as China continues to industrialize.

While the subject of photochemical ozone production has been extensively studied, there are still some remaining uncertainties. The essential reactions have already been presented in Section 12.5, and will only be briefly discussed here. In all high-temperature combustion processes, particularly power plants and automobiles, NO is produced by the direct reaction of $N_2 + O_2$, and from nitrogen-containing fuels. This NO can then be oxidized by a variety of mechanisms to NO_2. In the presence of NO_x and sunlight, the oxidation of CO, CH_4, and other hydrocarbons results in ozone production. In an urban environment, the diurnal cycle of these trace species will generally exhibit a characteristic pattern of concentration maxima first in NO, then NO_2, followed by O_3 around midday (National Academy of Sciences, 1977). Evidence indicates that natural hydrocarbons along with anthropogenically produced NO_x are important precursors to urban and rural ozone (Liu *et al.*, 1987, 1988).

As seen in Table 12-2, global NO_x production is dominated by anthropogenic sources. In an urban environment, virtually all NO_x is from fossil fuel combustion.

12.7.5 Acid Rain

Acid precipitation, or acid rain, can causes significant impacts on freshwater, coastal, and forested ecosystems (e.g., Likens *et al.*, 1996). Both NO_3^-, from NO_x emissions, and SO_4^{2-} from SO_2 emissions contribute significantly to acid rain. The relative ratio of SO_4^{2-}/NO_3^- in precipitation will be substantially determined by the regional emissions of SO_2/NO_3. In developed countries, uncontrolled combustion of coal and high-sulfur fuel oil led to significant emissions of SO_2, relative to NO_x. Due to strict control of smokestack SO_2 emissions in some regions and increasing NO_x emissions from automobiles, the relative contribution of NO_3^- is expected to increase (Sirois, 1993; Mayewski *et al.*, 1990).

In remote ice cores, SO_4^{2-} and NO_3^- concentrations have increased due to anthropogenic emissions (Mayewski *et al.*, 1986, 1990). This is due to the fact the precursor compounds (e.g.,

NO_x) are exported from the source regions. For example, Honrath and Jaffe (1992) found elevated concentrations of nitrogen oxides at Barrow, Alaska during spring, as compared to summer. This is largely due to decreased removal processes during winter for some nitrogen oxides, such as peroxyacetyl nitrate.

12.7.6 Species Diversity

Every day the planet loses forever a large number of plant and animal species. This current rate of extinction exceeds anything in the past history of life on Earth. Most of this loss is due to loss in habitat, especially in the tropics. However, some species are lost due to changes in the nutrient balance and resulting changes in the ecosystem structure. Vitousek *et al.* (1997b,c) documents losses of plant diversity in several regions due to wet and dry deposition of anthropogenic NO_3^-. This occurs because in regions with relatively high nitrogen deposition, a smaller number of species will flourish. According to Vitousek *et al.* (1997b) this reduced diversity makes the ecosystem less stable, as for example, during times of drought.

12.7.7 N Fertilization and the Global Carbon Cycle

Due to the fact that nitrogen is a limiting nutrient in many ecosystems, additions of fixed nitrogen can significantly increase plant growth. This is termed "nitrogen fertilization." In regions where anthropogenic nitrogen is deposited two results are possible: reduced vegetative growth due to acid precipitation (see above) or increased growth due to nitrogen fertilization. Typically, the initial nitrogen deposition will stimulate growth; however, later, a plant can become "nitrogen saturated" and no longer respond to additional nitrogen inputs (Mellilo *et al.*, 1989; Vitousek, 1997b,c). To the extent that nitrogen fertilization causes increased plant growth, this will result in increased uptake of atmospheric CO_2 and increased global biomass.

One of the problems with understanding

perturbations to the global carbon cycle has been the problem of the "missing sinks." This describes the fact that the relatively well-known anthropogenic sources of CO_2 significantly exceeded the annual atmospheric increase of CO_2. Some of the anthropogenic emissions are being taken up by the oceans, but based on a number of quantitative models, it is unlikely that the oceans are adsorbing all of the missing sink (IPCC, 1995).

Based on a variety of evidence, a number of researchers have now concluded that terrestrial biomass must be taking up a significant fraction of the annual anthropogenic carbon emissions. For example Tans *et al.* (1990) used CO_2 observations and a global model to calculate regional sources and sinks and concluded that a large carbon sink must be operative in the northern hemisphere. Schindler and Bayley (1993) used a biogeochemical approach to conclude that northern forests are storing an additional 1.0–2.3 Tg C/yr due to deposition of anthropogenic nitrogen. Hudson *et al.* (1994), using a global three-dimensional ocean–atmosphere–biosphere carbon model, reached a similar conclusion. Thus, it would appear that increased carbon uptake due to anthropogenic nitrogen fertilization is responsible for sequestering a large fraction of the 6 Tg of carbon emitted each year by human activities.

On the surface it would seem that global nitrogen fertilization is beneficial in that it reduces the concentration of CO_2 in the atmosphere and thus its radiative forcing. However, it raises the question of how long global biomass can continue to respond in this way. Should the northern forests switch over from nitrogen fertilization to nitrogen saturation, or if other nutrients become limiting in these ecosystems, then the rate of rise of atmospheric CO_2 would increase (assuming emissions remained the same). Overall, this uncertainty is an important limitation on our ability to predict future concentrations of atmospheric CO_2.

12.7.8   The Future

Global population, fertilizer use, and fossil fuel combustion are all expected to continue to grow. Galloway *et al.* (1995) provided estimated N fluxes due to fertilizer production and fossil fuel combustion, by regions, for the present and the year 2020 (see Table 12-3). Based on these estimates, global fertilizer production will increase by more than 70% and fossil fuel emissions of NO_x will increase by 115%! As can be seen from Table 12-3, most of this increase is predicted to occur in the developing world as their large populations attempt to reach the living standard and lifestyles of the developed world.

Questions

12-1 How would the nitrogen cycle change if life on Earth were suddenly absent? What would be the time scale for these changes?

12-2 If, as a result of anthropogenic activities, nitrogen is being removed from the atmospheric reservoir (as N_2) to the oceanic reservoir (as NO_3^-), how long would it take to detect this change? Is this a thermodynamically favorable process?

Table 12-3   Current and estimated (for 2020) impacts on the global N cycle (after Galloway *et al.*, 1995)

	Emissions of NO_x due to fossil fuel combustion emissions (Tg N/yr)		Fertilizer production and usage (Tg N/yr)	
	Present	2020	Present	2020
Developed world	14	16	29	30
Less developed world	6	29	50	104
Asia (excluding Japan)	4	13	36	85
Global	21	46	78	134

12-3 How have agriculture and deforestation changed the global rates of nitrogen fixation and denitrification? How can increased agricultural productivity be sustained without using industrially produced fertilizers?

12-4 Discuss the importance of atmospheric N_2O. Why is it important to know something about its natural and anthropogenic sinks? What role might atmospheric N_2O play in the control of planetary climate? (See e.g., Lovelock, 1979.)

12-5 Describe the trends in the ozone concentrations in the troposphere and stratosphere, and the total ozone column. What roles do nitrogen oxides play in these changes?

12-6 What are the key reactions that result in the formation of photochemical smog? How do *increases* in NO_x emissions lead to *lower* peak ozone concentrations in some areas? Would you advocate the lowering of NO_x emission standards in some areas? What strategies would you suggest for cities in developing countries to avoid becoming like Los Angeles?

12-7 Knowing that the average precipitation on Earth is approximately 1 m per year, calculate the global mean concentration of NO_3^- in rainwater assuming that 50% of all NO_x is removed in wet deposition as HNO_3. Assuming this were the only source of acidity, what would the pH be for this rainwater? Now redo this calculation using the NO_x emissions for 2020.

12-8 From the data in Fig. 12-4 and Table 12-3, calculate the lifetime for atmospheric N_2O. Would you expect the atmospheric N_2O growth rate (Fig. 12-9) to remain about the same, greater or slower in the year 2020 as compared to today? Explain.

12-9 Assuming the current emissions and sinks remain about the same, estimate the global atmospheric CO_2 mixing ratio in the year 2050. Now repeat this calculation, but this time assume that the terrestrial biosphere no longer continues to sequester some of this anthropogenic carbon.

References

Bremner, J. M. and Blackmer, A. M. (1981). Terrestrial nitrification as a source of atmospheric nitrous oxide. *In* "Denitrification, Nitrification, and Atmospheric Nitrous Oxide" (C. C. Delwiche, ed.). Wiley, New York.

Burns, R. C. and Hardy, R. W. F. (1975). "Nitrogen Fixation in Bacteria and Higher Plants." Springer-Verlag, New York.

Chameides, W. L., Kasibhatla, P. S., Yienger, J. and Levy II., H. (1994). Growth of Continental-Scale Metro-Agro-Plexes, Regional Ozone Pollution, and World Food Production. *Science* **264**, 74–77.

Crutzen, P. J. (1988). Tropospheric ozone: an overview. *In* "Tropospheric Ozone-Regional and Global Scale Interactions" (I. S. A. Isaksen, ed.). NATO ASI Series C, Vol. 227. D. Reidel Publ. Co., Boston, MA.

Crutzen, P. J. and Gidel, L. T. (1983). A 2-dimensional photochemical model of the atmosphere 2. the tropospheric budgets of the anthropogenic chlorocarbons CO, CH_4, CH_3Cl and the effects of various NO_x sources on tropospheric O_3. *J. Geophys. Res.* **88**, 6641–6661.

Delwiche, C. C. (1981). The nitrogen cycle and nitrous oxide. *In* "Denitrification, Nitrification, and Atmospheric Nitrous Oxide" (C. C. Delwiche, ed.). Wiley, New York.

Emmons, L. K. *et al.* (1997). Climatologies of NO_x and NO_y: A comparison of data and models. *Atmos. Env.* **31**, 1851–1904.

Farman, J. C., Gardiner, B. G. and Shanklin, J. D. (1985). Large losses of total ozone in Antarctica reveal seasonal ClO_x/NO_x interaction. *Nature* **315**, 207–210.

Finlayson-Pitts, B. and Pitts, J. (1986). "Atmospheric Chemistry – Fundamentals and Experimental Techniques." Wiley, New York.

Freney, J. R., Simpson, J. R. and Denmead, O. T. (1983). Volatilization of ammonia. *In* "Gaseous Loss of Nitrogen from Plant-Soil Systems" (J. R. Freney and J. R. Simpson, eds), Martinus Nijhoff, Dr. W. Junk Publishers, Boston.

Galloway J. N. *et al.* (1995). Nitrogen fixation: Anthropogenic enhancement – environmental response. *Global Biogeochem. Cycles* **9**, 235–252.

Garrels, R. M. (1982). Introduction: chemistry of the troposphere – some problems and their temporal frameworks. *In* "Atmospheric Chemistry" (E. D. Goldberg, ed.), Dahlem Konferenzen, pp. 3–16. Springer-Verlag, New York.

Granhall, U. (1981). Biological nitrogen fixation in relation to environmental factors and functioning of natural ecosystems. *In* "Terrestrial Nitrogen Cycles" (F. E. Clark and T. Rosswall, eds), *Ecological Bulletin* **33**, 131–145. Swedish Natural Science Research Council, Stockholm.

Hoell, J. M., Harward, C. N. and Williams, B. S. (1980). Remote infrared heterodyne radiometer measurements of atmospheric ammonia profiles. *Geophys. Res. Lett.* **7**, 325–328.

Holland, H. D. (1978). "The Chemistry of the Atmosphere and Oceans." Wiley, New York.

Honrath, R. E. and Jaffe, D. A. (1992). The seasonal cycle of nitrogen oxides in the arctic troposphere at Barrow, Alaska, *J. Geophys. Res.* **97**, 20 615–20 630.

Hudson R. J. M. *et al.* (1994). Modeling the global carbon cycle: Nitrogen fertilization of the terrestrial biosphere and the "missing" CO_2 sink. *Global Biogeochem. Cycles* **8**, 307–333.

Intergovernmental Panel on Climate Change (IPCC) (1996). "Climate Change 1995: The Science of Climate Change" (J. T. Houghton, L. G. Meira Filho, B. A. Callender, N. Harris, A. Kattenberg and K. Maskell, eds). Cambridge University Press, Cambridge.

Intergovernmental Panel on Climate Change (IPCC) (1997). "IPCC Guidelines for National Greenhouse Gas Inventories," Chapter 4. OECD, Paris, France.

Jaffe, D. A. (1992). The nitrogen cycle in global biogeochemical cycles. *In* "Global Biogeochemical Cycles" (S. S. Butcher, R. J. Charlson, G. H. Orians, G. V. Wolfe, eds). Academic Press, New York.

Jaffe, D. A., Berntsen, T. and Isaksen, I. S. A. (1997). A global 3D chemical transport model; 2. Nitrogen oxides and non methane hydrocarbon results. *J. Geophys. Res.* **102**, 21 281–21 296.

Kato, N. and Akimoto, H. (1992). Anthropogenic emissions of SO_2 and NO_x in Asia: Emission inventories, *Atmos. Environ.* **26A**, 2997–3017.

Kikby, E. A. (1981). Plant growth in relation to nitrogen supply. *In* "Terrestrial Nitrogen Cycles" (F. E. Clark and T. Rosswall, eds), *Ecological Bulletin* **33**, 249–271. Swedish Natural Science Research Council, Stockholm.

Lawson, D. R. (1988). The nitrogen species methods comparison study: An overview. *Atmos. Environ.* **22**, 1517.

Lents, M. and Kelly, W. J. (1993). Clearing the air in Los Angeles. *Scient. Am.* Oct., 32–39.

Likens, G.E., Driscoll, C. T. and Buso, D. C. (1996). Long-term effects of acid rain: response and recovery of a forest ecosystem. *Science* **272**, 244–246.

Liu, S. C., Trainer, M., Fehsenfeld, F. C., Parrish, D. D., Williams, E. J., Fahey, D. W., Hubler, G. and Murphy, P. C. (1987). Ozone production in the rural troposphere and the implications for regional and global ozone distributions. *J. Geophys. Res.* **92**, 4191–4207.

Liu, X., Trainer, M. and Liu, S. C. (1988). On the nonlinearity of the tropospheric ozone production. *J. Geophys. Res.* **93**, 15 879–15 888.

Logan, J. A. (1994). Trends in the vertical distribution of ozone: An analysis of ozonesonde data, *J. Geophys. Res.* **99**, 25 553–25 585.

Logan, J. A. (1985). Tropospheric ozone: seasonal behavior, trends, and anthropogenic influences. *J. Geophys. Res.* **90**, 10 463–10 482.

Logan, J., Prather, M. J., Wofsy, S. C. and McElroy, M. B. (1981). Tropospheric chemistry: a global perspective. *J. Geophys. Res.* **86**, 7210–7254.

Lovelock, J. (1979). "Gaia: a New Look at Life on Earth." Oxford University Press, New York.

Marenco, A., Gouget H., Nedelec P., Pages, J.-P. and Karcher, F. (1994). Evidence of a long-term trend in tropospheric ozone from Pic du Midi data series: consequences: positive radiative forcing. *J. Geophys. Res.* **99**, 16 617–16 632.

Matson, P. A. *et al.* (1998). Integration of environmental, agronomic and economic aspects of fertilizer management. *Science* **280**, 112–114.

Mayewski, P. A. *et al.* (1990). An ice-core record of atmospheric response to anthropogenic sulfate and nitrate. *Nature* **346**, 554–556.

Mayewski, P. A., Lyons, W. B., Spencer, M. J., Twickler, M., Dansgaard, W., Koci, B., Davidson, C. I. and Honrath, R. E. (1986). Sulfate and nitrate concentrations from a South Greenland ice core. *Science* **232**, 975–977.

McElroy, M. B., Elkins, J. W., Wofsy, S. C. and Yung, Y. L. (1976). Sources and sinks for atmospheric N_2O. *Rev. Geo. Space Phys.* **14** (2), 143–150.

Mellilo, J. M., Steudler, P. A., Aber, J. D. and Bowden, R. D. (1989). Atmospheric deposition and nutrient cycling. *In* "Dahlem Workshop on Exchange of Trace Gases Between Terrestrial Ecosystems and the Atmosphere" (M. O. Andreae and D. S. Schimel, eds). Wiley Interscience Publishers, New York.

National Academy of Sciences (1977). "Nitrogen Oxides." National Academy of Sciences, Washington, DC.

NOAA, US Dept. of Commerce (1987). "Geophysical Monitoring for Climatic Change," No. 15, Summary Report 1986, pp. 85–90. Boulder, CO.

Postgate, J. R. (1982). Biological nitrogen fixation: fundamentals. *In* "The Nitrogen Cycle" (W. D. P. Stewart and T. Rosswall, eds), pp. 73–83. The Royal Society, London.

Quinn, P. K., Charlson, R. J. and Bates, T. S. (1988). Simultaneous observations of ammonia in the atmosphere and ocean. *Nature* **335**, 336–338.

Rolston, D. E. (1981). Nitrous oxide and nitrogen gas production in fertilizer loss. *In* "Denitrification, Nitrification, and Atmospheric Nitrous Oxide" (C. C. Delwiche, ed.), pp. 127–149. Wiley, New York.

Rosswall, T. (1976). The internal nitrogen cycle between microorganisms, vegetation, and soil. *In* "Nitrogen, Phosphorus and Sulfur – Global Cycles" (B. H. Svensson and R. Söderlund, eds),

Ecol. Bull. No. 22, pp. 157–167, SCOPE. Swedish Natural Science Research Council, Stockholm.

Schindler, D. W. and Bayley, S. E. (1993). The biosphere as an increasing sink for atmospheric carbon: Estimates from increased nitrogen deposition. *Global Biogeochem. Cycles* **7**, 717–733.

Sirois, A. (1993). Temporal variations in sulphate and nitrate concentrations in precipitation in eastern North America: 1979–1990. *Atmos. Env.* **27A**, 945–963.

Smil, V. (1997). Global Population and the Nitrogen Cycle. *Scient. Am.* July, 76–81.

Söderlund, R. and Svensson, B. H. (1976). The global nitrogen cycle. *In* "Nitrogen, Phosphorus and Sulfur – Global Cycles" (B. H. Svensson and R. Söderlund, eds), *Ecol. Bull. No. 22*, pp. 23–73, SCOPE. Swedish Natural Science Research Council, Stockholm.

Solomon, S. and Schoeberl, M. R. (1988). Overview of the polar ozone issue. *Geophys. Res. Lett.* **15**, 845–846.

Stedman, D. H. and Shetter, R. E. (1983). The global budget of atmospheric nitrogen species. *In* "Trace Atmospheric Constituents" (S. E. Schwartz, ed.). Wiley, New York.

Tans, P. P., Fung, I. Y. and Takahashi, T. (1990). Observational constraints on the global atmospheric CO_2 budget. *Science* **247**, 1431–1438.

Toon, O. and Turco, R. (1991). Polar stratospheric clouds and ozone depletion. *Scient. Am.* **264**, 68.

Turner, R. E. and Rabalais, N. N. (1991). Changes in Mississippi River water quality this century. *BioScience* **41**, 140–147.

Vitousek, P. M. *et al.* (1997a). Human domination of the earth's ecosystems. *Science* **277**, 494–499.

Vitousek, P. M. *et al.* (1997b). Human alteration of the global nitrogen cycle: Sources and consequences. *Ecol. Apps.* **7**, 737–750.

Vitousek, P. M. *et al.* (1997c). Human alteration of the global nitrogen cycle: Causes and consequences. *In* "Issues in Ecology" Vol. 1, Ecol. Soc. America, Washington, DC. Also available on the web at http://www.sdsc.edu/~ESA/.

Volz, A. and Kley, D. (1988). Evaluation of the Montsouris series of ozone measurements made in the nineteenth century. *Nature* **332**, 240–242.

Weiss, R. F. (1981). The temporal and spatial distribution of tropospheric nitrous oxide. *J. Geophys. Res.* **86**, 7185–7195.

Wennberg, P. O. *et al.* (1994). Removal of stratospheric O_3 by radicals: In-situ measurements of OH, HO_2, NO, NO_2, ClO and BrO. *Science* **266**, 398–404.

World Meteorological Organization (WMO) (1995). "Scientific Assessment of Ozone Depletion: 1994." WMO Global Ozone Research and Monitoring Project, Report No. 37, Geneva.

13

The Sulfur Cycle

R. J. Charlson, T. L. Anderson, and R. E. McDuff

13.1 Introduction

Sulfur, the fourteenth most abundant element in the Earth's crust, plays a variety of important roles in the chemical functioning of the Earth. In its reduced oxidation state, sulfur is a key nutrient to life, providing, for example, structural integrity to protein-containing tissues. In its fully oxidized state, sulfur exists as sulfate, SO_4^{2-}, the second most abundant anion in rivers (after bicarbonate, HCO_3^-) and in seawater (after chloride, Cl^-) and is the major cause of acidity in both natural and polluted rainwater. This link to acidity makes sulfur a key player in natural weathering of rocks and such environmental problems as "acid rain." Sulfate in the atmosphere has been identified as the dominant component of cloud condensation nuclei in both remote and polluted settings (Bigg *et al.*, 1984). Thus, it has important interactions with clouds (and perhaps the global radiative energy balance which is sensitive to clouds) and with the hydrologic cycle. Finally, of the major elemental cycles (i.e., C, N, O, P, S), the sulfur cycle is one of the most heavily perturbed by human activity. It is estimated that anthropogenic emissions of sulfur into the atmosphere (largely from coal combustion) are currently about double natural sulfur emissions (see Table 13-2).

With the exception of ionic sulfides formed from highly electropositive elements (i.e., Na, K, Ca, Mg), sulfur bonding in natural environments is covalent. When fully oxidized, however, the covalently bonded sulfur atom exists within the sulfate ion which forms either sulfuric acid (a gas or liquid) or ionic compounds such as $CaSO_4 \cdot 2H_2O$ (gypsum) and $(NH_4)_2SO_4$ (ammonium sulfate). Covalent bonding occurs in organosulfur compounds. Hence, the chemistry of sulfur involves chemical complexity not found in elements at the edges of the periodic table.

Sulfur exists naturally in several oxidation states, and its participation in oxidation/reduction reactions has important geochemical consequences. For example, when an extremely insoluble material, FeS_2, is precipitated from seawater under conditions of bacterial reduction, Fe and S may be sequestered in sediments for periods of hundreds of millions of years. Sulfur can be liberated biologically or volcanically with the release of H_2S or SO_2 as gases.

There are nine known isotopes of sulfur of which four are stable:

Isotope	Average crustal abundance (%)
^{32}S	95.0
^{33}S	0.76
^{34}S	4.22
^{36}S	0.014

The prevalence of sulfur's second most abundant isotope, ^{34}S, along with the fractionation known to occur in many biogeochemical processes, make isotopic studies of sulfur a potentially fruitful method of unraveling its sources and sinks within a given reservoir.

Earth System Science
ISBN 0-12-379370-X

13.2   Oxidation States of Sulfur

Table 13-1 includes many of the key naturally occurring molecular species of sulfur, subdivided by oxidation state and reservoir. The most reduced forms, S($-$II), are seen to exist in all except the aerosol form, in spite of presence of free O_2 in the atmosphere, ocean and surface waters. With the exception of H_2S in oxygenated water, these species are oxidized very slowly by O_2. The exception is due to the dissociation in water of H_2S into $H^+ + HS^-$. Since HS^- reacts quickly with O_2, aerobic waters may contain, and be a source to the atmosphere of, RSH, RSR etc. but not of H_2S itself. Anaerobic waters, as in swamps or intertidal mudflats, can contain H_2S and can, therefore, be sources of H_2S to the air.

S($-$II and $-$I) also are found in minerals, notably in metal and metaloid sulfides. As many as 95 sulfide minerals appear in standard lists, where sulfur is bound to a wide variety of other elements: Ag, Fe, Cd, Hg, Mn, Ca, Te, Se, As, Sn, Cu, Pb, Pt, Sb, Co, Ni, Mo, Rn, W. Of these, FeS_2 (pyrite) is the most abundant.

Highly oxidized forms of organic sulfur exist in folic acid and sulfolipids, but the major form of sulfur in organisms is in amino acids. There are two sulfur-containing amino acids: cysteine and methionine. Although sulfur bonds assume a variety of oxidation states in living organisms, we list the amino acid sulfur as oxidation state $-$II because the sulfur is generally bonded to carbon or hydrogen. Thus, methionine has a —SCH_3 bond like dimethyl sulfide, and cysteine is a mercaptan which can convert to smaller gaseous molecules like CH_3SH. A key biological function of sulfur is to provide disulfide (—SS—) linkages between amino acids within protein molecules, thus giving three-dimensional structure to proteins and strength or mechanical structure to tissues composed of proteins. Sulfur in methionine is crucial to the methylation reaction, a biosynthetic process. The amount of sulfur in organisms varies but is typically of the same order of magnitude as phosphorus, e.g. about 0.25% (by dry weight). Even though sulfur is an essential element for biota, its abundance, 29 mmol/kg in seawater, is often so large that other elements (e.g., N and P) provide limits to growth, while S does not.

The most oxidized form of sulfur, S($+$VI), is predominantly sulfate, SO_4^{2-}. Sulfate particles ranging in composition from pure sulfuric acid (H_2SO_4) to fully neutralized ammonium sulfate ((NH_4)$_2SO_4$) are ubiquitous constituents of the atmosphere (see Chapter 7).

In the ocean, sulfate exists as a free ion, SO_4^{2-}. In sedimentary rock, sulfate is found in evaporite minerals (i.e., minerals produced by the evaporation of seawater), with gypsum, $CaSO_4 \cdot 2H_2O$, being the most common. By contrast, the intermediate oxidation state, S($+$IV) has only a transitory existence in the atmosphere and in some volcanic and industrial emissions to the atmosphere. Gaseous SO_2 is soluble in water, leading to the presence of HSO_3^- and SO_3^{2-} ions which are unstable under aerobic conditions, producing SO_4^{2-} as the stable end-product (Chapter 5). Oxidation of SO_2 to sulfate occurs in the gas phase as well due to the presence of the strong oxidizing agent, OH•.

Elemental sulfur also occurs naturally, with production either by biological or inorganic processes. In either case, it appears that a higher oxidation state of sulfur may react with the $-$II (sulfide) state to yield a zero oxidation state product, or that S(0) is an intermediate in the oxidation of S($-$II). There are several genera of sulfate-reducing bacteria, two of which are widely recognized: *Desulfovibrio* and *Desulfotomaculum*. These organisms utilize sulfur in the sulfate ion as an electron acceptor and produce H_2S. This H_2S is then available to react with iron and form the insoluble precipitate, FeS_2 (pyrite). Other organisms can utilize H_2S and produce elemental sulfur, S(0). Elemental sulfur also is produced inorganically, for example by the following reaction in volcanoes:

$$2H_2S + SO_2 \rightarrow 2H_2O + 3S$$

For completeness, we mention the existence of compounds of mixed oxidation states. Here, two or more atoms of sulfur exist in the molecule or ion, each having a different oxidation state. Numerous examples of these species are known but details of their natural existence are obscure. It is suggested (Grinenko and Ivanov, 1983) that

Table 13-1 Some naturally occurring sulfur compounds

Oxidation state	Gas	Aerosol	Aqueous	Soil	Mineral	Biological
$-II$	H_2S, RSH RSR OCS CS_2	—	H_2, HS^-, S^{2-} RS^-	S^{2-}, HS^- MS	S^{2-} HgS CuS_2 etc	Methionine $CH_3S(CH_2)_2CHNH_2COOH$ Cysteine $HSCH_2CHNH_2COOH$ Dicysteine
$-I$	$RSSR$	—	$RSSR$	SS^{2-}	FeS_2	—
0	$CH_3SOCH_3^+$	—		S_8	—	—
II			$S_2O_3^{2-}$	—	—	—
IV	SO_2	$SO_2 \cdot H_2O$ HSO_3^-	$SO_2 \cdot H_2O$ HSO_3^- SO_3^{2-} $HCHO \cdot SO_2$	SO_3^{2-}		
VI	SO_3	H_2SO_4, HSO_4^- SO_4^{2-} $(NH_4)_2SO_4$ etc. Na_2SO_4 CH_3SO_3H	SO_4^{2-} HSO_4^-, SO_4^{2-} $CH_3SO_3^-$	$CaSO_4$ $ROSO_3$	$CaSO_4 \cdot H_2O$ $MgSO_4$	

thiosulfate, $S_2O_3^{2-}$, ion can be produced by bacteria in waterlogged soils, paddy fields and the like. The formal oxidation state for the two sulfur atoms is II, but the two sulfur atoms are chemically different. Such species may play important roles as intermediates between the major species such as S^{2-} and SO_4^{2-} (Jorgensen, 1990).

Returning to Table 13-1, we see that sulfur is found in gaseous, aerosol, aqueous, soil, mineral, and biological forms. The gaseous forms, which are found in the atmosphere, are generally of the lower oxidation states $-$II or $+$IV, while most of the aqueous form is S($+$VI) as SO_4^{2-}. The vapor pressures of non-ionic S($-$II) compounds are large enough that transfer of these species occurs from bodies of water to the atmosphere. Sulfate ion in solution is relatively inert, and much of its processing is due to transport by water or evaporation of the water to form solid evaporites. The sulfate minerals (evaporites like $CaSO_4$) are water soluble, while the sulfide minerals (such as FeS_2) tend to be highly insoluble.

13.3   Sulfur Reservoirs

The preceding discussion of oxidation states demonstrates the existence of sulfur in solid, liquid, and gaseous forms and in living organisms. It is thus convenient to organize the remainder of the discussion along these same lines. Figure 13-1 is a box diagram showing the key reservoirs and the approximate burdens of sulfur in each (Freney *et al.*, 1983).

The vast majority of sulfur at any given time is in the lithosphere. The atmosphere, hydrosphere, and biosphere, on the other hand, are where most transfer of sulfur takes place. The role of the biosphere often involves reactions that result in the movement of sulfur from one reservoir to another. The burning of coal by humans (which oxidizes fossilized sulfur to SO_2 gas) and the reduction of seawater sulfate by phytoplankton which can lead to the creation of another gas, dimethyl sulfide (CH_3SCH_3), are examples of such processes.

The remainder of this chapter, which discusses the cycling of sulfur, is divided into an

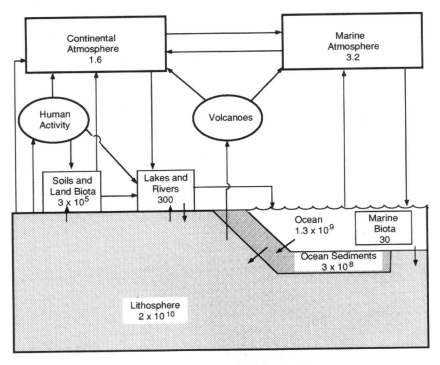

Fig. 13-1   Major reservoirs and burdens of sulfur. Units are Tg (10^{12} g) S.

atmospheric part and an oceanic/solid earth part. The amount of sulfur in the atmosphere at any given time is small, and the lifetime of most sulfur compounds in air is relatively short (e.g., days) because the fluxes through the atmosphere are large. Sulfur in the ocean as SO_4^{2-} is cycled much more slowly, and the primary interactions in that cycle are with the solid earth.

13.4 The Atmospheric Cycle of Sulfur

13.4.1 Transformations of Sulfur in the Atmosphere

Figure 13-2 summarizes the chemical and physical transformations of sulfur compounds that occur in the atmosphere. Most of the chemical transformations involve the oxidation of sulfur. The key oxidizing agents are thought to be the OH· radical, for the gas phase, and H_2O_2, O_3, and OH· in the aqueous phase. Many of these transformations, however, can only be identified as "multistep" processes – that is, the detailed chemistry is not currently understood. The rates at which most of the transformations occur are also poorly understood and have been estimated only semiquantitatively. The amounts of sulfur within the various reservoirs (Fig. 13-1) are better known because they can be directly measured, although these data are greatly complicated by the patchy and episodic nature of the distribution of atmospheric sulfur species. Among the fluxes (see Fig. 13-6 and Table 13-2), the best data are available for anthropogenic SO_2 emissions and SO_4^{2-} deposition to the surface in rainwater. Recent improvements have also been made in quantifying the natural emissions of reduced sulfur gases, especially from the world oceans. In sum, the qualitative picture of the atmospheric sulfur cycle now appears to be in good focus, although many quantitative details remain to be filled in.

The most important pathway of sulfur through the atmosphere involves injection as a low-oxidation-state gas and removal as oxidation-state VI sulfate in rainwater (Fig. 13-2, paths 1, 4, 5, 6, 7, 8, 9, 10, 12, and 13.) Since this pathway involves a change in chemical oxidation state and physical phase, the lifetime of

sulfur in the atmosphere is governed by both the kinetics of the oxidation reactions and the frequency of clouds and rain. We will argue below that the overall process is fast – on the order of days – meaning that the atmospheric sulfur cycle is a regional phenomenon and that the distribution of nearly all sulfur species in the atmosphere is necessarily "patchy" over the globe.

13.4.2 Sources and Distribution of Atmospheric Sulfur

Biological processes result in the production of a variety of reduced sulfur-containing gases. The six most important of these are H_2S (hydrogen sulfide), CS_2 (carbon disulfide), OCS (carbonyl sulfide), CH_3SH (methyl mercaptan), CH_3SCH_3 (dimethyl sulfide or DMS), and CH_3SSCH_3 (dimethyl disulfide or DMDS). Of these, DMS has been shown to dominate sulfur emissions from the open ocean (Andreae, 1985, 1986; Andreae and Raemdonck, 1983) and may, therefore, modulate the sulfur cycle over a large portion of the globe. A varying mixture of all six of these gases is found over terrestrial ecosystems (Adams *et al.*, 1981). Estimates of biogenic sulfur emissions have been revised substantially downward over the last 15 years, according to a review by Bates *et al.* (1992). Referring to the final column in Table 13-2, we see that annual terrestrial biogenic emissions (F4) are currently estimated at 0.4 Tg S and marine biogenic emissions (F14) at 15 Tg S, each with large uncertainties due to measurement difficulties and the sporadic nature of the emissions. Other natural sources of low-oxidation state sulfur to the atmosphere are biomass burning and volcanoes (9 Tg S/yr), each of which emit sulfur mostly as SO_2 gas. Volcanoes are a sporadic source, of secondary importance to the tropospheric sulfur cycle but capable of causing huge fluctuations locally and in the stratospheric reservoir. Biomass burning has both anthropogenic and natural components (often hard to separate). The sulfur output from this source is highly uncertain; it is not included in the fluxes in Table 13-2.

Emissions of sulfur to the atmosphere by

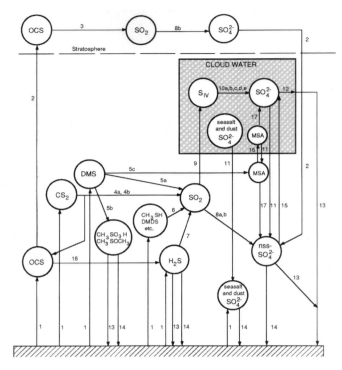

Fig. 13-2 The chemical and physical transformations of sulfur in the atmospheric cycle. Circles are chemical species, the box represents cloud–liquid phase. DMS = CH_3SCH_3, DMDS = CH_3SSCH_3, S_{IV} = $(SO_2)_{aq}$ + HSO_3^- + SO_3^{2-} + $CH_2OHSO_3^-$, and MSA (methane sulfonic acid) = CH_3SO_3H. The chemical transformations are as follows:

1. Surface emissions
2. Tropospheric/stratospheric exchange
3. $OCS + hv \rightarrow S + S + CO$
   $S + O_2 \rightarrow SO + O$
   $SO + O_2 \rightarrow SO_2 + O$
4a. $CS_2 + OH \rightarrow CS_2OH$
   $CS_2OH \rightarrow$ multistep $\rightarrow COS + SO_2$
4b. $CS_2 + hv \rightarrow CS_2^*$
   $CS_2^* + O_2 \rightarrow CS + SO_2$
   $CS + O_2 \rightarrow COS + O$
   $CS + O_3 \rightarrow COS + O_2$
5a. $CH_3SCH_3 + OH \rightarrow$ multistep $\rightarrow SO_2$
5b. $CH_3SCH_3 + OH \rightarrow$ multistep $\rightarrow CH_3SOCH_3$
5c. $CH_3SCH_3 + OH \rightarrow$ multistep $\rightarrow CH_3SO_3H$
6. $CH_3SH + OH \rightarrow$ multistep $\rightarrow SO_2$
7. $H_2S + OH \rightarrow$ multistep $\rightarrow SO_2$
8a. $SO_2 + OH \rightarrow HSO_3$
   $HSO_3 + O_3 \rightarrow HO_2 + SO_3$
   $SO_3 + H_2O \rightarrow H_2SO_4$
8b. $SO_2 \rightarrow SO_4^{2-}$ (Heterogeneous reaction)

9. $(SO_2)_g \leftrightarrow (SO_2)_{aq}$
   $(SO_2)_{aq} + H_2O \rightarrow HSO_3^- + H^+$
   $(HSO_3^- \leftrightarrow H^+ + SO_3)$
   $CH_2(OH)_2 + HSO_3^- \leftrightarrow H_2O + CH_2OHSO_3^-$
10a. $(H_2O_2)_g \leftrightarrow (H_2O_2)_{aq}$
   $HSO_3^- + (H_2O_2)_{aq} \rightarrow$ multistep $\rightarrow H^+ + SO_4^{2-}$
10b. $(O_3)_g \leftrightarrow (O_3)_{aq}$
   $HSO_3^- + (O_3)_{aq} \rightarrow$ multistep $\rightarrow H^+ + SO_4^{2-}$
10c. $(HO_2)_g \leftrightarrow (HO_2)_{aq}$
   $(HO_2)_{aq} \rightarrow H^+ + O_2^-$
   $(HO_2)_{aq} + O_2 \xrightarrow{H_2O} (H_2O_2)_{aq} + OH^-$
   $HSO_3^- + (H_2O_2)_{aq} \rightarrow$ multistep $\rightarrow 2H^+ + SO_4^{2-}$
10d. $HSO_3^- + O_2 \rightarrow (OH)_{aq}$
   $HSO_3^- + (OH)_{aq} \rightarrow$ multistep $\rightarrow H^+ + SO_4^{2-}$
10e. $HSO_3^- + O_2 \rightarrow$ multistep $\rightarrow H^+ + SO_4^{2-}$
11. Evaporation
12. SO_4^{2-} in cloudwater $\rightarrow SO_4^{2-}$ in rainwater
13. Washout, rainout
14. Dry deposition
15. Cloud nucleation
16. $COS + OH \rightarrow$ multistep $\rightarrow H_2S$
17. $MSA \rightarrow SO_4^{2-}$ by some mechanism

(Modified with permission from R. J. Charlson, W. L. Chameides, and D. Klay (1985). The transformations of sulfur and nitrogen in the remote atmosphere. *In* "The Biogeochemical Cycling of Sulfur and Nitrogen in the Remote Atmosphere" (J. N. Galloway, R. J. Charlson, M. O. Andreae, and H. Rodhe, eds), pp. 67–80, D. Reidel Publishing Company, Dordrecht.)

Table 13-2   Major sulfur fluxes (Tg S/yr)

Flux	Description	Kellogg et al. (1972)	Friend (1973)	Granat et al. (1976)	Möller (1984a,b)	Ivanov (1983) Natural	Ivanov (1983) Anthro.	This work
1. Continental part of the cycle								
F1a	Emission to atmosphere from fossil fuel burning and metal smelting	50	65	65	75	—	113	77[a]
F1b	Effluents from chemical industry and mining	—	—	—	—	—	29	29
F1c	Soils to rivers (anthro. portion is pollution of rivers from fertilizers)	—	26	—	—	—	28	8[b]
F2	Aeolian emission (dust)	—	—	0.2	—	20	—	20
F3a	Volcanic emissions to continental atmosphere	1	1	1.5	1	14	—	3[a,c]
F4	Biogenic gases (land)	—	58	5	35	18	—	0.4[a]
F5	Gravitational settling of large (aeolian) particles to land	—	—	—	—	12	—	18
F6	Washout and dry deposition of gases and fine particles to land	111	121	71	—	25	47	59.3
F7	Transport to oceanic atmosphere	5	8	18	—	35	66	42.1
F8	Transport from oceanic atmosphere	4	4	17	—	20	—	19
F9	Weathering to soil	—	42[d]	66[d]	—	114[d]	—	26
F10	Weathering to rivers	—			—		—	93
F11	Burial of sulfur in sediments from continental water bodies	—	—	—	—	—	—	35
F12	River runoff to oceans	—	136	122	—	104	104	104
2. Oceanic part of the cycle (see also F7, F8, and F12 above)								
F3b	Volcanic emissions to marine atmosphere	1	1	1.5	1	14	—	6[a,c]
F13	Aeolian emissions (seasalt)	47	44	44	175	140	—	140
F14	Biogenic gases (marine)	—	48	27	35	23	—	15[a]
F15	Washout and dry deposition of gases and particles to oceans	72	96	73	—	258	—	184.1
F16	Burial of sulfur in oceanic sediments	—	—	—	—	139	—	69
F17	Deposition of marine sulfate via thermal vent reactions at mid-ocean ridges	—	—	—	—	—	—	43[e]
F18	Lithification of marine sediments	—	—	—	—	—	—	69[f]

[a] Bates *et al.* (1992).
[b] Deduced to balance budget for "soils and land biota" reservoir.
[c] Bates *et al.* (1992) global total for volcanic emissions reapportioned as 1/3 continental and 2/3 marine.
[d] Number shown represents the sum of fluxes F9 and F10.
[e] Deduced to balance budget for "ocean" reservoir.
[f] Deduced to balance budget for "ocean sediments" reservoir.

humans are almost entirely in the form of SO_2. The main sources are coal burning and sulfide ore smelting. The total anthropogenic flux is estimated to be about 80 Tg S/yr (Ivanov, 1983; Bates *et al.*, 1992) and is thus substantially greater than the natural flux of low-oxidation-state sulfur to the atmosphere. Clearly, the atmospheric sulfur cycle is intensely perturbed by human activity. To estimate the spatial extent of this perturbation, we will need some idea of the residence time of sulfur in the atmosphere.

The definition of turnover time is total burden within a reservoir divided by the flux out of that reservoir – in symbols, $\tau = M/S$ (see Chapter 4). A typical value for the flux of non-seasalt sulfate (nss-SO_4^{2-}) to the ocean surface via rain is 0.11 g S/m^2 per year (Galloway, 1985). Using this value, we may consider the residence time of nss-SO_4^{2-} itself and of total non-seasalt sulfur over the world oceans. Appropriate vertical column burdens (derived from the data review of Toon *et al.*, 1987) are 460 μg S/m^2 for nss-SO_4^{2-} and 1700 μg S/m^2 for the sum of DMS, SO_2, and nss-SO_4^{2-}. These numbers yield residence times of about 1.5 days for nss-SO_4^{2-} and 5.6 days for total non-seasalt sulfur. We might infer that the oxidation process is frequently

slower than the rain removal process of particulate sulfate, although within the accuracy of these estimates it is safer simply to note that the two processes seem to be of similar duration. Direct estimates of the residence time of DMS with respect to oxidation by the OH radical are in the range of 1.5 to 2 days (Andreae, 1985 and references therein).

As shown in Figs 13-3 and 13-4, the flux of sulfate in rainwater over polluted industrial regions is of order 1 g S/m^2 per year or about 10 times the remote marine flux. Since natural sources of sulfur are probably much weaker over continents, this indicates a massive human perturbation of the sulfur cycle within industrialized areas and a potentially strong perturbation over much larger areas depending on how far atmospheric sulfur is transported on average before being deposited. Figures 13-3 and 13-4 give the impression of a regional phenomenon with a horizontal scale of about 1000 km. This scale is confirmed by sulfur budget studies in polluted regions. For example, Ottar (1978) estimated that 80% of European sulfur emissions are deposited over Europe. For mid-latitude weather, horizontal transport over a spatial scale of order 1000 km corresponds to a

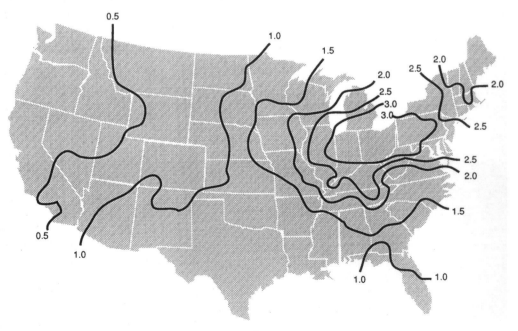

Fig. 13-3   Precipitation-weighted concentration of SO_4^{2-} (mg/L) over North America for 1987.

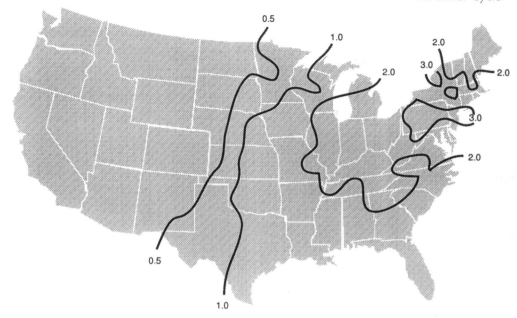

Fig. 13-4   Wet sulfate ion deposition ($g/m^2/yr$) for 1987.

time scale of a few days. Studies of the rate of oxidation of SO_2 to sulfate in polluted regions indicate that a conversion rate of about 1% per hour is a good average. This corresponds to a residence time for SO_2 gas of about 2 days. Once converted to sulfate, washout by rain should occur on a time scale of a few days.

For both polluted and remote conditions, therefore, the cycling of sulfur from low oxidation state gas to sulfate particles and then back to the surface in rain takes place on a time scale of a few days.

Some sulfur gases are directly absorbed at the surface, and particulate sulfur (aerosol) has some flux to the surface due to Brownian motion and gravitational settling. Together, these processes are referred to as "dry deposition" (to distinguish them from the rain removal mechanism). Dry deposition may be an important removal process for SO_2 in polluted areas and is probably fairly minor for SO_2 and the reduced sulfur gases over the oceans (Galloway, 1985). The dry deposition of sulfate particles (other than seasalt and dust sulfate) is also minor. In any case, these additional sinks for atmospheric sulfur can only have the effect of shortening its residence time.

An interesting exception to the patchiness of atmospheric sulfur compounds is carbonyl sulfide (OCS). This compound, which may be emitted directly or produced by the oxidation of CS_2, is highly stable against further oxidation (until it reaches the stratosphere) and so is unavailable for rapid wet removal. As a result, OCS has a long residence time (~ 1 year) as well as a large and globally uniform concentration (~ 500 pptv) in the troposphere. Being inert in the troposphere, this compound, like the chlorofluorocarbons, is available for gradual mixing into the stratosphere. Here it can be broken apart by ultraviolet radiation and oxidized to SO_2 and sulfate (see top of Fig. 13-2). Through this mechanism, OCS is thought to be the major source of sulfate particles in the stratosphere during volcanically quiescent periods (Crutzen, 1976). In the troposphere, while chemically unimportant, it is actually the largest reservoir of sulfur. This situation is analogous to that of nitrogen, where the dominant species is N_2, which, being so inert, is usually ignored in discussions of atmospheric nitrogen chemistry.

Finally, there is a major flux of sulfur through the atmosphere in both seasalt particles (~ 140 Tg S/yr) and terrestrial dust (~ 20 Tg S/yr). In each case, the form of sulfur is sulfate, originating mostly as the mineral gypsum in the case of dust and as sulfate ion in seawater in the case of seasalt. Already in its fully oxidized, stable state, this component of atmospheric sulfur does not participate in atmospheric redox reactions, nor does it contribute to the acidity of cloud or rainwater. Instead, it is simply returned to the surface, via dry or wet removal, in the same chemical form in which it was emitted. Sulfate in both dust and seasalt particles is best viewed as an inert, secondary constituent, merely "along for the ride," with little geochemical consequence. (It is very important to measure this component, however, so that it may be subtracted from total sulfate measurements in order to derive the non-seasalt sulfate quantity of chemical interest.) Since it does not need to be oxidized prior to wet removal and has, in addition, a large sink via dry deposition, the residence time of seasalt and dust sulfate is significantly shorter than that of other sulfur species. Because both seasalt and dust particles are relatively large, they are subject to significant removal by gravitational settling.

We have seen that, except for OCS, sulfur species in the atmosphere have residence times that are short (days) such that their geographical distribution is patchy. This perspective on the atmospheric sulfur cycle has important implications. While human emissions certainly constitute an overwhelming perturbation within heavily industrialized regions, there may be even larger areas of the globe in which the human influence on the sulfur cycle is relatively unimportant, e.g., much of the southern hemisphere (see discussion below on the remote marine sulfur cycle). In addition, we see that the sulfur cycle can only sensibly be studied on a regional basis and that the calculation of global budgets will necessarily be a painstaking process of making myriad measurements over a wide range of regions and seasons to allow accurate averaging.

13.4.3   The Remote Marine Atmosphere

Let us turn now to a detailed, box-model investigation of a regional sulfur cycle. The discussion so far suggests that the sulfur cycle over much of the ocean should be largely unin-

fluenced by human or other continental input. The absence of complex, polluted-air chemistry, along with a high degree of horizontal spatial homogeneity, should provide the simplest possible system for studying the transformations introduced in Fig. 13-2. We begin by assuming that the atmosphere is in a steady state (fluxes into and out of each box are equal) and that the remote marine environment is a closed system with respect to sulfur (no fluxes from or to the continents). We ignore seasalt sulfate. Finally, we assume that DMS is the only significant reduced sulfur species emitted from the ocean surface.

Figure 13-5 is the box model of the remote marine sulfur cycle that results from these assumptions. Many different data sets are displayed (and compared) as follows. Each box shows a measured concentration and an estimated residence time for a particular species. Fluxes adjoining a box are calculated from these two pieces of information using the simple formula, $S = M/\tau$. The flux of DMS out of the ocean surface and of nss-SO_4^{2-} back to the ocean surface are also quantities estimated from measurements. These are converted from surface to volume fluxes (i.e., from μg S/(m^2 h) to ng S/(m^3 h)) by assuming the effective scale height of the atmosphere is 2.5 km (which corresponds to a reasonable thickness of the marine planetary boundary layer, within which most precipitation and sulfur cycling should take place). Finally, other data are used to estimate the factors for partitioning oxidized DMS between the MSA and SO$_2$ boxes, for SO$_2$ between dry deposition and oxidation to sulfate, and for nss-SO_4^{2-} between wet and dry deposition.

We begin our analysis by comparing the surface fluxes. According to the indicated partitioning factors, 74% of the 11 μg DMS-S/m^2/h emitted from the ocean surface should be returned as nss-SO_4^{2-} in rain. This leads to a predicted wet deposition flux of nss-SO_4^{2-} of 8.1 μg S/(m^2/h), which is 37% lower than the measured flux of 13 μg S/(m^2/h). Since the estimated accuracy of the DMS emission flux is $\pm$50% (Andreae, 1986), this is about as good agreement as can be expected. It indicates that our "closed system" assumption is at least a reasonable first approximation. (A more sophisticated treatment would consider sulfur oxida-

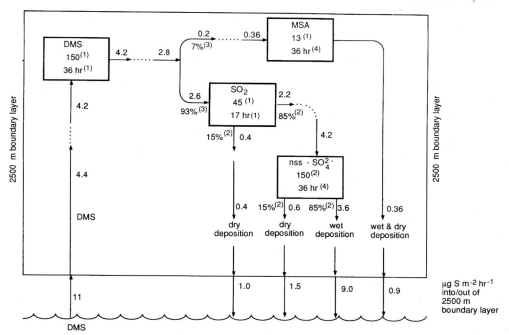

Fig. 13-5 The sulfur cycle in the remote marine boundary layer (adapted from Andreae, 1986). Within the 2500 m boundary layer, burden units are ng S/m^3 and flux units are ng S/(m^3 h). Fluxes within atmospheric layer are calculated from the burden and the residence time. Ellipses indicate that calculations based on independent measurements are being compared. The measured total deposition of 13 ± 7 mg S/(m^2 h)[2] agrees well with the calculated sum of 13.4 mg S/(m^2 h), suggesting that a consistent model of the remote marine sulfur cycle within the planetary boundary layer can be constructed based on biogenic DMS inputs alone. Data: (1) Andreae (1986); (2) Galloway (1985); (3) Saltzman *et al.* (1983); (4) sulfate aerosol lifetime calculated earlier in this chapter based on marine rainwater pH; the same lifetime is applied to MSA aerosol. (Modified with permission from P. J. Crutzen, D. M. Whelpdale, D. Kley, and L. A. Barrie (1985). The cycling of sulfur and nitrogen in the remote atmosphere. *In* "The Biogeochemical Cycling of Sulfur and Nitrogen in the Remote Atmosphere" (J. N. Galloway, R. J. Charlson, M. O. Andreae, and H. Rodhe, eds), pp. 203–212. D. Reidel Publishing Company, Dordrecht.)

tion on hydrated seasalt particles (Sievering *et al.*, 1992; Chameides and Stelson, 1992) and sulfur emissions from ships (Capaldo *et al.*, 1999.))

The two estimates of the flux into the DMS box are in excellent agreement, tending to support our 2.5 km assumed boundary layer height. However, the flux out of the DMS box is about 40% larger than necessary to support the fluxes through the MSA and SO$_2$ boxes. This might suggest a missing sink for DMS. Could DMSO, another known oxidation product of DMS whose concentrations and lifetimes have not been carefully studied, fill this gap? An even larger descrepancy exists between the two estimates of the flux from SO$_2$ to nss-SO$_4^{2-}$, which differ by almost a factor of two. The fact that the

flux out of the DMS box (multiplied by the appropriate partitioning factors) provides better agreement with the flux into the nss-SO$_4^{2-}$ box suggests that the error may lie in the measured SO$_2$ concentration (too low) or the estimated SO$_2$ lifetime (too high). In any case, the various data sets are in reasonable agreement and all of the above comparisons suggest further measurements which would help to refine our understanding of these atmospheric processes.

13.4.4 The Global Atmospheric Sulfur Budget

Figure 13-5 is an example of direct application of modelling concepts from Chapter 4 to the tropospheric portion of the sulfur cycle. As such, it illustrates two important aspects of the cycle approach:

1. This approach allows independent data sets to be compared.
2. It suggests which measurements would be most helpful to determine more precisely the main features of the system.

Another way in which the cycle approach is often used is in the development of global atmospheric budgets. Here, the only assumption made is that the atmosphere, taken as a whole, is in a steady state. Thus, if the total burden of S compounds in the air is not increasing or decreasing, the global fluxes into the atmosphere must be equal to the total of fluxes out of the atmosphere. A comparison of the flux out (from rainwater analyses) with known fluxes from industrial production led Eriksson (1959, 1960) and many others to suggest that a "missing source" was needed to balance the budget. Indeed, this notion has provided substantial impetus over the past 20 years to study natural sources, including emission of reduced sulfur from swamps (Adams *et al.*, 1981), emission from volcanoes (Warneck, 1988), and biological production in and emission from the oceans (Andreae and Raemdonck, 1983).

Figure 13-6a (Ivanov, 1983) is a depiction of the natural global sulfur budget. Figure 13-6b depicts the budget with natural and anthropogenic sources. Table 13-2 serves to explain Fig. 13-6 and includes the wide range of estimates of various fluxes, and demonstrates the degree of uncertainty inherent in such approaches.

Comparison of Figs 13-6a and 13-6b clearly demonstrates the degree to which human activity has modified the cycle of sulfur, largely via an atmospheric pathway. The influence of this perturbation can be inferred, and in some cases measured, in reservoirs that are very distant from industrial activity. Ivanov (1983) estimates that the flux of sulfur down the Earth's rivers to the ocean has roughly doubled due to human activity. Included in Table 13-2 and Fig. 13-6 are fluxes to the hydrosphere and lithosphere, which leads us to these other important parts of the sulfur cycle.

13.5 Hydrospheric Part of the Cycle of Sulfur

The ocean plays a central role in the hydrospheric cycling of sulfur since the major reservoirs of sulfur on the Earth's surface are related to various oceanic depositional processes. In this section we consider the reservoirs and the fluxes focusing on the cycling of sulfur through this oceanic node.

There are three major sulfur reservoirs at the Earth's surface (Table 13-3): as S($-$II) in sedimentary shales, as S($+$VI) in evaporite deposits and as S($+$VI) in seawater. The rate of cycling through these reservoirs is closely related to the fluxes of sulfur to and from the ocean. However, sulfur is unusual among the *major* constituents of natural waters in that its fluxes have been significantly modified by human activities. Thus, a useful starting point is the pre-industrial cycle of Fig. 13-6a. The main features are (1) a tightly closed loop through which sulfate is carried by seaspray into the atmosphere, but quickly returned to the ocean, and (2) a larger, slower loop in which sulfate is derived by weathering, carried by rivers to the ocean, and returned to the continents through the cycling of rocks (Chapter 9). Over geologic periods of time, the ocean composition is thought to be relatively constant, with river inputs to the ocean balanced by depositional processes. What processes maintain this balance?

13.5.1 *Oceanic Outputs*

There are both uncertainties and controversies concerning the removal of sulfur from the ocean. FeS_2 (pyrite) in sedimentary shales is most often formed in reducing marine sediments, particularly when these sediments underlie waters of high biological productivity. In these areas, the flux of (dead) organic carbon to the sediments greatly exceeds the rates at which oxygen can diffuse into the sediments (Chapter 8). Iron (III) hydroxides and then sulfate become the terminal electron acceptors for the oxidation of organic matter, thereby providing a source of Fe^{+2} and HS^- for pyrite formation.

Various workers have estimated the rate of pyrite formation. Berner (1972) summed the

(a)

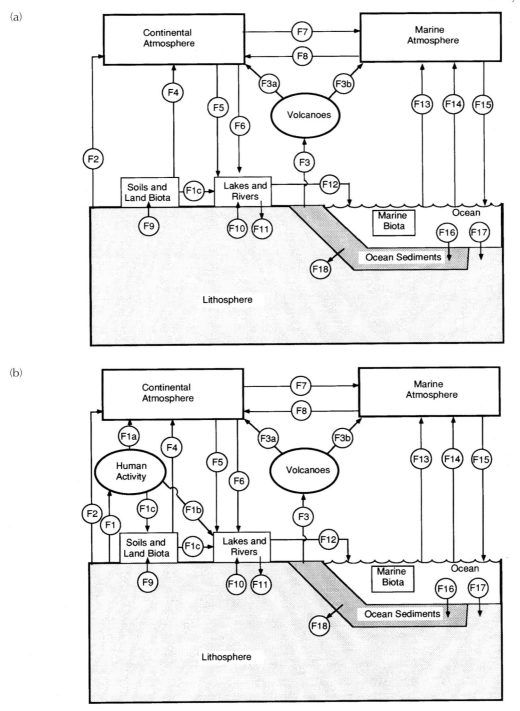

(b)

Fig. 13-6   Major fluxes of the global biogeochemical sulfur cycle excluding (a) and including (b) human activity (modified from Ivanov, 1983). Numbers in circles designate fluxes described in Table 13-2.

Table 13-3 Major sulfur reservoirs (Tg S)

Reservoir	Major form(s)	Burden
Continental atmosphere[a]	OCS, SO_4^{2-}, SO_2, DMS	1.6
Marine atmosphere[a]	OCS, SO_4^{2-}, SO_2, DMS	3.2
Soils and land biota[a]	Reduced	3×10^5
Lakes and rivers[b]	SO_4^{2-}	300
Marine biota[c]	Reduced	30
Seawater[a,d]	SO_4^{2-}	1.3×10^9
Ocean sediments[a,e]	Gypsum ($CaSO_4$)	3×10^8
	Pyrite (FeS_2)	
Rest of lithosphere[a,e]	Pyrite (FeS_2)	2.4×10^{10}
	Gypsum ($CaSO_4$)	

[a] Freney *et al.* (1983).
[b] Calculated from mass of surface freshwater (1.3×10^{20} g H_2O) and average sulfate concentration (2.5×10^{-6} g S/g H_2O).
[c] Calculated from carbon in ocean biota (3000 Tg) and approximate mass ratio for sulfur to carbon of 1:100.
[d] Volkov and Rozanov (1983).
[e] Migdisov *et al.* (1983).

sulfur accumulation rates of various sediment types in proportion to their areal coverage and found a flux of about 10% of the river flux. Li (1981) carried out a similar calculation and finds 30% of the river flux, probably indicative of the uncertainty of the approach. Toth and Lerman (1977) established that the decrease of sulfate with depth in sediment pore waters is a function of sedimentation rate. This information was used to estimate the diffusive flux of sulfur into sediments driven by pyrite formation, again a value about 10% of the river flux. Apparently, pyrite formation, while measureable, is not the dominant removal process.

Evaporite deposition is a much more episodic process and thus difficult to quantify. Because seawater is significantly undersaturated with respect to common evaporitic minerals, like gypsum and halite, evaporites are only formed when restricted circulation develops in an ocean basin in which evaporation exceeds precipitation. A geologically recent example is the Mediterranean Sea of 5–6 Myr ago. At this time excess evaporation exceeded the supply of ocean water through shallow inlet(s) from the Atlantic Ocean. As salinity increased, first $CaSO_4$, then NaCl precipitated. Over time, salt deposits 2–3 km thick formed. This thickness represents about 40 desiccations of the entire volume of the Mediterranean. How rapidly would the Mediterranean have to evaporate to remove all the sulfate introduced by rivers to the ocean? The time scale τ is given by $\tau = V_{Med}C_{oc}/F_r$, where F_r is the river flux of sulfate, V_{Med} is the volume of the Mediterranean, and C_{oc} is the concentration of sulfate in the ocean. This time is about 25 000 yr, which could be readily achieved (2000 m of water depth in 25 000 yr is only 8 cm/yr of excess evaporation). A second point to consider is that the basin will eventually fill. One meter of seawater evaporates to a layer of solid just over a centimeter thick. Evaporite formation could only be sustained in the Mediterranean for 2–3 Myr, removing only about one-third of the ocean's sulfate in this time. Evaporite formation can match the river flux, but it must be taking place on a large scale.

A solution, still controversial, has been recently proposed. This is the loss of sulfate from seawater during hydrothermal circulation through mid-ocean ridges (Edmond *et al.*, 1979). The flow of water through these systems is estimated to be about 1.4×10^{14} L/yr, about 0.4% of the flow of rivers. However, sulfate is quantitatively removed, yielding a flux of 125 Tg S/yr, capable of balancing the river flux. The controversy is whether the chemistry involved in removing sulfate is the formation of anhydrite ($CaSO_4$) or reduction and subsequent precipitation as metal sulfides. Anhydrite forms on simple heating of seawater. As the ocean crust ages and cools it would redissolve, creating an equal and opposite flux. In contrast, metal sulfides are much more resistant to subsequent interaction with seawater.

13.5.2   Oceanic Inputs

Referring again to Fig. 13-6a, the materials that constitute the oceanic sinks for sulfur are recycled over time by exposure to weathering on the continents. The rates at which these processes occur help to regulate the flux of sulfur into rivers.

The evaporite source is characterized by co-variation of sulfate (from gypsum) and chloride (from halite). That elements can be recycled from the ocean to land by movement of salt-bearing aerosols (so-called "cyclic salts") has confused the interpretation of river flux data somewhat. While this cycling generally follows the ratio of salts in the sea, the S/Cl ratio is an exception. Taking the S/Cl ratio of the cyclic component to be 2 (based on compositional data for marine rains) and assuming that all chloride in rivers is cyclic, an upper limit for the cyclic influence can be calculated.

$$F_{r,Cl}(SO_4^{2-}/Cl)_{seawater}$$
$$= (310 \text{ Tg Cl/yr})(0.046 \text{ g S/g Cl})$$
$$= 29 \text{ Tg S/yr}$$

where $F_{r,Cl}$ is the riverine flux of Cl to seawater.

However, not all the chloride is cyclic, a fact first appreciated in recent years. An example comes from a detailed study of river geochemistry conducted in the Amazon Basin. In the inland regions, rains typically have a chloride content of 10 μM, while major inland tributaries have chloride contents of 20–100 μM. These data suggest that only 25% of the Cl is cyclic, whereas 75% is derived by weathering of evaporites. Indeed, 90% of this 75% can be shown to have its origin in the Andean headwaters, derived from evaporites that make up only 2% of the area of the Amazon Basin (Stallard and Edmond, 1981). As the ratio of sulfate to chloride in evaporite deposits is generally much higher

than seawater (gypsum ($CaSO_4 \cdot 2H_2O$) precipitates first as seawater is evaporated), dissolution of evaporites represents one of the principal sources of riverine sulfate.

The other principal source is weathering of sedimentary or igneous sulfides, mainly pyrite, by the oxidation:

$$FeS_2 + \tfrac{15}{4}O_2 + \tfrac{7}{2}H_2O \rightarrow 2SO_4^{2-} + Fe(OH)_3(s) + 4H^+$$

What are the relative contributions of these two sources? Two approaches have been taken. One is to establish the geology and hydrology of a basin in great detail. This has been carried out for the Amazon (Stallard and Edmond, 1981) with the result that evaporites contribute about twice as much sulfate as sulfide oxidation. The other approach is to apply sulfur isotope geochemistry. As mentioned earlier, there are two relatively abundant stable isotopes of S, ^{32}S, and ^{34}S. The mean 34/32 ratio is 0.0442. However, different source rocks have different ratios, which arise from slight differences in the reactivities of the isotopes. These deviations are expressed as a difference from a standard, in the case of sulfur the standard being a meteorite found at Canyon Diablo, Arizona.

Evaporitic sulfur has a range of sulfur isotopic composition from +10‰ to +30‰, while sedimentary sulfur is depleted in the heavy isotope and has a range of isotopic composition of about −40‰ to +10‰. Most of this variation reflects systematic changes with geological age. The source fractions of a river water can be estimated from an isotopic mass balance:

$$\delta^{34}S_{evap}X_{evap} + \delta^{34}S_{pyrite}X_{pyrite} = \delta^{34}S_{mean \text{ river}}$$

where X_i is the fraction from that source. This approach has been applied to the Volga River, which has a $\delta^{34}S$ of +6‰, suggesting again about 2/3 from evaporites and 1/3 from sulfides (calculated taking $\delta^{34}S_{evap}$ as +15‰ and $\delta^{34}S_{pyrite}$ as −15‰).

Accepting these relative proportions from evaporites (2/3) and sulfides (1/3), the characteristic times, τ_i, of cycling of the evaporite sulfur and sulfide sulfur reservoirs can be estimated from the reservoir sizes (R_i) in Table 13-3, and the river flux of sulfur. For evaporites:

$$\tau_{evaporite} = (93 \times 10^6 \text{ Tg})/(2 \times 104/3 \text{ Tg/yr})$$
$$= 140 \text{ Myr}$$

and for sulfides:

$$\tau_{pyrite} = (48 \times 10^6 \text{ Tg})/(104/3 \text{ Tg/yr}) = 140 \text{ Myr}$$

The characteristic times of sulfur cycling through the two reservoirs are nearly identical. Sulfur cycling through evaporites is on a time scale similar to the cycling of the evaporites themselves, (about 200 Myr, Garrels and Mackenzie, 1971) suggesting that physical factors limit the rate of weathering of this very soluble component. Sulfur cycling through shales is considerably faster than the cycling of the overall shale reservoir (about 600 Myr).

Questions

13-1 What would be the approximate sulfate concentration of rainwater globally for the following cases (assume that rainfall is uniformly 75 cm/yr):

(a) The only source of sulfur were DMS and it was uniform over the globe.
(b) In addition to (a) consider the sulfur from industry uniformly distributed over the globe.
(c) Same as (b), but assume that all industry is in the northern hemisphere and consider the hemispheres separately.
(d) Compare these concentrations to the data for eastern North America (Fig. 13-4).

13-2 Sulfur and oxygen are in the same column of the periodic table. List their chemical similarities and differences and consider the biogeochemical consequences of each.

13-3 Estimate the total amount of oxidation (mol/yr) caused by the reduction of SO_4^{2-} to $S(-II)$. Compare this to the total amount of oxidation by O_2.

13-4 After considering the Redfield–Ketchum–Richards ratio (see Chapter 10) for C:N:P consider the analog for sulfur in land and marine biota. What key biochemical species are the major determinants of the C:S relationship in biota?

13-5 Hypothetical problem for chemists: consider the global cycle of selenium which has many chemical similarities to sulfur. Construct a box diagram for the global selenium cycle based on known similarities and differences of Se and S.

13-6 This problem is a first-order attempt to quantify the possible anthropogenic perturbation of the northern hemisphere (NH) marine sulfur cycle. First, assume that present day anthropogenic sulfur emissions result in 20 Tg S/yr being transported from North America to the atmosphere over the NH Atlantic and 10 Tg S/yr being transported from Asia to the atmosphere over the NH Pacific. Assume a uniform concentration in the N–S direction, average westerly wind speeds of 10 m/s, that both ocean regions are 4000 km in N–S extent, that the NH Atlantic is 5500 km from east to west and that the NH Pacific is 8500 km from east to west. Next, assume that biogenic emissions of DMS from the ocean to the atmosphere are uniformly 0.1 g S/m² per year (the average value from Andreae and Raemdonck, 1983). Now if we use a 2-day average residence time (or e-folding time – see Chapter 4) for sulfur in the atmosphere, what percentage of total atmospheric sulfur would be anthropogenic sulfur at the middle and western edge of each ocean?

References

Adams, D. F., Farwell, S. O., Robinson, E., Park, M. R., and Bamsberger, W. L. (1981). Biogenic sulfur source strengths. *Environ. Sci. Technol.* **15**, 1493–1498.

Andreae, M. O. (1985). The emission of sulfur to the remote atmosphere. *In* "The Biogeochemical Cycling of Sulfur and Nitrogen in the Remote Atmosphere" (J. N. Galloway, R. J. Charlson, M. O. Andreae, and H. Rodhe, eds). Reidel, Dordrecht.

Andreae, M. O. (1986). The ocean as a source of atmospheric sulfur compounds. *In* "The Role of Air–Sea Exchange in Geochemical Cycling" (P. Buat-Menard, ed.). Reidel, Dordrecht.

Andreae, M. O. and Raemdonck. H. (1983). Dimethyl sulfide in the surface ocean and the marine atmosphere: a global view. *Science* **221**, 744–747.

Bates, T. S., Lamb, B. K., Guenther, A., Dignon, J., and Stoiber, R. E. (1992). Sulfur emissions to the atmosphere from natural sources. *J. Atmos. Chem.* **14**, 315–337.

Berner, R. A. (1972). Sulfate reduction, pyrite formation and the oceanic sulfur budget. *In* "The changing chemistry of the oceans" (D. Dyrssen and D. Jagner, eds). Wiley-Interscience, Stockholm.

Bigg, E. K., Gras, J. L., and Evans, C. (1984). Origin of Aitken particles in remote regions of the Southern Hemisphere. *J. Atmos. Chem.* **1**, 203–214.

Capaldo, K., Corbett, J. J., Kasibhatla, P., Fischbeck, P., and Pandis, S. N. (1999). Effects of ship emissions on sulphur cycling and radiative climate forcing over the ocean. *Nature*, **400**, 743–746.

Chameides, W. L. and Stelson, A. W. (1992). Aqueous-phase chemical processes in deliquescent sea-salt aerosols: a mechanism that couples the atmospheric cycles of S and sea salt. *J. Geophys. Res.* **97**, 20565–20580.

Crutzen, P. J. (1976). The possible importance of CSO for the sulphate layer of the stratosphere. *Geophys. Res. Lett.* **3**, 73–76.

Edmond, J. M., Measures, C., McDuff, R. E., Chan, L. H., Collier, R. W., Grant, B., Gordon, L. I., and Corliss, J. B. (1979). Ridge crest hydrothermal activity and the balances of the major and minor elements in the ocean: the Galapagos data. *Earth Planet. Sci. Lett.* **46**, 1–18.

Eriksson, E. (1959). The yearly circulation of chloride and sulfur in nature: meteorological, geochemical and pedological implications, Part I. *Tellus* **11**, 375–603.

Eriksson, E. (1960). The yearly circulation of chloride and sulfur in nature: meteorological, geochemical and pedological implications, Part II. *Tellus* **12**, 63–109.

Freney, J. R., Ivanov, M. V., and Rodhe, H. (1983). The sulphur cycle. *In* "The Major Biogeochemical Cycles and Their Interactions, SCOPE 21" (B. Bolin and R. B. Cook, eds). Wiley, Chichester.

Friend, J. P. (1973). The global sulfur cycle. *In* "Chemistry of the Lower Atmosphere" (S. I. Rasool, ed.). Plenum Press, New York.

Galloway, J. N. (1985). The deposition of sulfur and nitrogen from the remote atmosphere. *In* "The Biogeochemical Cycling of Sulfur and Nitrogen in the Remote Atmosphere" (J. N. Galloway, R. J. Charlson, M. O. Andreae, and H. Rodhe, eds). Reidel, Dordrecht.

Garrels, R. M. and Mackenzie, F. T. (1971). "Evolution of sedimentary rocks." W. W. Norton, New York.

Granat, L., Rodhe, H., and Hallberg, R. O. (1976). The global sulphur cycle. *In* "Nitrogen, Phosphorus and Sulphur – Global Cycles" (B. H. Svensson and R. Söderlund, eds) pp. 89–134. SCOPE Report 7, Ecol. Bull., Stockholm.

Grinenko, V. A. and Ivanov, M. V. (1983). Principal reactions of the global biogeochemical cycle of sulphur. *In* "The Global Biogeochemical Sulfur Cycle, SCOPE 19" (M. V. Ivanov and J. R. Freney, eds). Wiley, Chichester.

Ivanov, M. V. (1983). Major fluxes of the global biogeochemical cycle of sulphur. *In* "The Global Biogeochemical Sulphur Cycle, SCOPE 19" (M. V. Ivanov and J. R. Freney, eds). Wiley, Chichester.

Jorgensen, B. B. (1990). A thiosulfate shunt in the sulfur cycle of marine sediments. *Science* **249**, 152–154.

Kellogg, W. W., Cadle, R. D., Allen, E. R., Lazrus, A. L., and Martell, E. A. (1972). The sulfur cycle. *Science* **175**, 587–596.

Li, Y. H. (1981). Geochemical cycle of elements and human perturbation. *Geochim. Cosmochim. Acta*, **45**, 2037–2084.

Migdisov, A. A., Ronov, A. B., and Grinenko, V. A. (1983). The sulfur cycle in the lithosphere, Part I: Reservoirs. *In* "The Global Biogeochemical Sulfur Cycle, SCOPE 19" (M. V. Ivanov and J. R. Freney, eds). Wiley, Chichester.

Möller, D. (1984a). Estimation of the global man-made sulphur emission. *Atmos. Environ.* **18**, 19–27.

Möller, D. (1984b). On the global natural sulphur emission. *Atmos. Environ.* **18**, 29–39.

Ottar, B. (1978). An assessment of the OECD study on long range transport of air pollutants (LRTAP), *Atmos. Environ.* **12**, 445–454.

Saltzman, E. S., Savoie, D. L., Zika, R. G., and Prospero, J. M. (1983). Methane sulfonic acid in the marine atmosphere. *J. Geophys. Res.* **88**, 10897–10902.

Sievering, H., Boatman, J., Gorman, E., Kim, Y., Anderson, L., Ennis, G., Luria, M., and Pandis, S. (1992). Removal of sulphur from the marine boundary layer by ozone oxidation in sea-salt aerosols. *Nature*, **360**, 571–573.

Stallard, R. F. and Edmond, J. M. (1981). Geochemistry of the Amazon 1. Precipitation chemistry and the marine contribution to the dissolved load at the time of peak discharge. *J. Geophys. Res.* **86**, 9844–9858.

Toon, O. B., Kasting, J. F., Turco, R. P., and Liu, M. S. (1987). The sulfur cycle in the marine atmosphere. *J. Geophys. Res.* **92**, 943–963.

Toth, D. J. and Lerman, A. (1977). Organic matter reactivity and sedimentation rates in the ocean. *Am. J. Sci.* **277**, 265–285.

Volkov, I. I. and Rozanov, A. G. (1983). The sulfur supply in oceans, Part I: Reservoirs and fluxes. *In* "The Global Biogeochemical Sulphur Cycle, SCOPE 19" (M. V. Ivanov and J. R. Freney, eds). Wiley, Chichester.

Warneck, P. (1988). "Chemistry of the Natural Atmosphere." Academic Press, New York.

14

The Phosphorus Cycle

Richard A. Jahnke

Phosphorus is one of the most important elements on Earth. It participates in, influences, or controls many of the biogeochemical processes occurring in the biosphere. Feedbacks between the P and other global chemical cycles have been suggested to control many basic characteristics of the biosphere such as the oxygen content of the atmosphere. To understand the interaction between P and other biogeochemical processes and elemental distributions, it is necessary to understand the distribution of P on the Earth's surface and the processes that control its distribution. The strategy of this chapter, therefore, is to (1) discuss the chemical forms in which P is present in the environment; (2) describe the processes that control its distribution in terrestrial, aquatic, and oceanic systems; and (3) define the major P reservoirs on the Earth's surface and the rate at which P is exchanged between these reservoirs. Because all of these subjects must be addressed within this single chapter, the discussion is somewhat superficial and intended to expose the reader to the individual topics rather than to provide a thorough discussion of each. References in each section provide more detailed presentations of the individual topics.

14.1 Occurrence of Phosphorus

The global occurrence of P differs from that of the other major biogeochemical elements, C, N, S, O, and H in several very important aspects. First, while gaseous forms of P can be produced in the laboratory, none have ever been found in significant quantities in the natural environment. Thus, although some P is transported within the atmosphere on dust particles and dissolved in rain and cloud droplets, the atmosphere generally plays a minor role in the global P cycle. It should be noted, however, that at certain locations this small atmospheric source of P could be important. An example is the surface waters in the central gyres of the oceans where extremely low standing stocks of P are observed and the transport of P from other potential sources is very slow.

The second significant difference between P and the other major biogeochemical elements is that oxidation–reduction reactions play a very minor role in controlling the reactivity and distribution of P in the natural environment. While several oxidation states for P are chemically possible, these forms are generally restricted to controlled laboratory settings. In natural systems, therefore, P is almost exclusively present in the +V oxidation state where it is found as phosphate (PO_4^{3-}), a tetrahedral oxy-anion. Nearly all dissolved and particulate forms of P are combined, complexed or slightly modified forms of this ion. In general, the biogeochemical cycle of P is synonymous with that of phosphate.

Finally, P also differs from other elements in that it is overwhelmingly dominated by a single, stable isotopic form containing 15 protons and 16 neutrons. There are only two naturally occurring radioactive forms of P: ^{32}P and ^{33}P, which are produced in the atmosphere by nuclear reactions with argon. A small amount of ^{32}P is

Earth System Science
ISBN 0-12-379370-X

also contributed by ^{32}Si decay. Because these isotopes have extremely short half-lives (^{32}P half-life, 14.3 days; ^{33}P half-life, 25.3 days), their activities in the environment are always very low and account for a minute portion of the total P in the Earth. Nevertheless, modern analytical capabilities have made the study of these isotopes possible, providing new insights in aquatic biogenic processes (Waser *et al.*, 1996; Lal and Lee, 1988; Benitez-Nelson and Buessler, 1999).

14.1.1 Dissolved Inorganic Forms of Phosphorus

Phosphate, PO_4^{3-}, is the fully dissociated anion of triprotic phosphoric acid, H_3PO_4:

$$H_3PO_4 \leftrightarrow H^+ + H_2PO_4^- \leftrightarrow 2H^+ + HPO_4^{2-}$$
$$\leftrightarrow 3H^+ + PO_4^{3-}$$

The dissociation constants for these equilibria for freshwater and seawater are listed in Table 14-1. The proportion of the individual protonated species in distilled water and seawater over the pH range of 2 to 10 is shown in Fig. 14-1. At the pH of freshwater systems (very roughly 6–7), $H_2PO_4^-$ is the dominant phosphate species. The high ionic strength of seawater and the presence of cations such as Ca^{2+}, Mg^{2+}, and Na^+, which form ion pairs with the PO_4^{3-} species, significantly alter the dissociation of phosphoric acid. In seawater at a pH of 8, HPO_4^{2-} dominates. The importance of ion pairs on the PO_4^{3-} activity in seawater is further demonstrated in Fig. 14-2. It is clear that ion pairs with dissolved cations play a central role in controlling the aqueous PO_4^{3-} speciation and

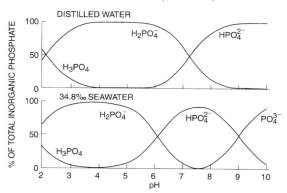

Fig. 14-1 Extent of dissolution of phosphoric acid species as a function of pH in distilled and sea waters (Atlas, 1975).

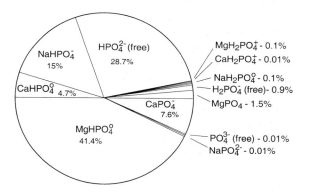

Fig. 14-2 Calculated speciation of PO_4^{3-} in seawater of 34.8 parts per thousand salinity at 20°C and a pH of 8.0 (Atlas, 1975).

that free PO_4^{3-} ions constitute a very small proportion of the total present (0.01% at standard seawater conditions). The chemical reactivity of PO_4^{3-} in aqueous systems is therefore highly dependent on the composition and pH of the solution. Acid–base and complexation reactions are not only important in seawater systems, but also influence the reactivity of PO_4^{3-} in groundwater and freshwater systems.

Another important class of inorganic PO_4^{3-} compounds are the condensed or polyphosphates. In these compounds, two or more phosphate groups bond together via P–O–P bonds to form chains or in some cases cyclic compounds. Although polyphosphates generally account for only a small portion of the total P present in

Table 14-1 Dissociation constants of phosphoric acid at 25°C

	Distilled water[a] (pK)	Seawater[b] (pK)
$H_3PO_4 \leftrightarrow H^+ + H_2PO_4^-$	2.2	1.6
$H_2PO_4^- \leftrightarrow H^+ + HPO_4^{2-}$	7.2	6.1
$HPO_4^{2-} \leftrightarrow H^+ + PO_4^{3-}$	12.3	8.6

[a] Stumm and Morgan (1981).
[b] Atlas (1975).

natural waters, they are extremely reactive compounds and are routinely used in industrial and commercial applications. Condensed phosphates form soluble complexes with many metal cations and are used, therefore, as water softeners.

14.1.2 Particulate Inorganic Forms of Phosphorus

Phosphorus is the tenth most abundant element on Earth with an average crustal abundance of 0.1% and may be found in a wide variety of mineral phases. There are approximately 300 naturally occurring minerals in which PO_4^{3-} is a required structural component. Phosphate may also be present as a trace component in many minerals either by the substitution of small quantities of PO_4^{3-} into the crystal structure or by the adsorption of PO_4^{3-} onto the mineral surface (Nriagu and Moore, 1984; Slansky, 1986).

By far the most abundant phosphate mineral is apatite, which accounts for more than 95% of all P in the Earth's crust. The basic composition of apatite is listed in Table 14-2. Apatite exhibits a hexagonal crystal structure with long open channels parallel to the *c*-axis. In its pure form, F^-, OH^-, or Cl^- occupies sites along this axis to form fluorapatite, hydroxyapatite, or chlorapatite, respectively. However, because of the "open" nature of the apatite crystal lattice, many minor substitutions are possible and "pure" forms of apatite as depicted by the general formula in Table 14-2 are rarely found.

Table 14-2 The chemical formula of apatites

General formula: $Ca_{10}(PO_4)_6X_2$

  $X = F^-$ Fluorapatite
  OH^- Hydroxyapatite
  Cl^- Chlorapatite

Possible substitutes for Ca^{2+}:
  Na^+, K^+, Ag^+, Sr^{2+}, Mn^{2+}, Mg^{2+}, Zn^{2+}, Cd^{2+}, Ba^{2+}, Sc^{3+}, Y^{3+}, rare earth elements, Bi^{3+}, U^{4+}

Possible substitutes for PO_4^{3-}:
  CO_3^{2-}, SO_4^{2-}, CrO_4^{3-}, AsO_4^{3-}, VO_4^{3-}, $F \cdot CO_3^{3-}$, $OH \cdot CO_3^{3-}$, SiO_4^{4-}

Of the possible substituting ions, CO_3^{2-} ion is by far the most important followed by Na^+, SO_4^{2-} and Mg^{2+}. The most common form of natural apatite in sedimentary rocks is francolite, a substituted form of carbonate fluorapatite deposited in marine systems. The substitution of CO_3^{2-} ions into the mineral lattice has a substantial effect on apatite solubility (Jahnke, 1984). More studies are required, however, before the effects of all substituting ions are understood and an accurate assessment of the solubility of complex, natural apatites can be made.

The importance of this mineral is perhaps best demonstrated by the diversity of its sources. Assuming that nearly all of the P present is in the form of apatite, igneous rocks contain between 0.02% and 1.2% apatite. Apatite is also produced by organisms (including man) as structural body parts such as teeth, bones, and scales. After an organism dies, these components tend to accumulate in sediments and soils. In some locations, these constituents are reworked and concentrated by physical processes to form economically important deposits. By far the largest accumulations of P on the Earth's surface are massive sedimentary apatite deposits (phosphorites). The mining of these deposits provides 82% of the total world PO_4^{3-} production and 95% of the total remaining reserves (Howard, 1979). Phosphate rock may also form by the accumulation of bird or bat droppings (guano) and subsequent diagenetic alteration and crystallization. The mining of guano can be important locally, although this constitutes a negligible fraction of the world's phosphate rock production.

In general, the major phosphorite deposits are of marine origin and occur as sedimentary beds ranging from a few centimeters to tens of meters in thickness. There is still debate concerning the exact mechanism or mechanisms by which phosphorites form (Sheldon, 1981). The relative roles of direct inorganic precipitation, solid-phase replacement of CO_3^{2-} by PO_4^{3-} in biogenic calcium carbonates, biologically mediated precipitation and concentration of dispersed apatite grains through physical reworking processes may vary considerably among different deposits (Froelich *et al.*, 1988). Ultimately, however, the

PO_4^{3-} that is incorporated into the authigenic apatite is thought to be supplied primarily via the decomposition of organic materials at the sea floor. A variety of additional processes, such as cycling at redox boundaries or incorporation by microbial communities, may act to elevate pore water PO_4^{3-} concentrations, promoting precipitation (Schuffert *et al.*, 1994; 1998; Follmi, 1996). Because of the ultimate connection to organic matter input, modern phosphorite formation tends to be located along continental margins which exhibit strong coastal upwelling and high biological primary production (Follmi, 1996). While these deposits are most obvious, finely dispersed apatite appears to be forming at many continental margin locations (Ruttenberg and Berner, 1993).

The occurrence of other phosphate minerals is also important in certain locations. In sediments, the formation of vivianite, $Fe_3(PO_4)_2 \cdot H_2O$, and struvite, $NH_4MgPO_4 \cdot 6H_2O$, have been reported (Emerson and Widmer, 1978; Murray *et al.*, 1978). Minerals that may form as secondary weathering products in PO_4^{3-} deposits include crandallite, $CaAl_3(PO_4)_2(OH)_5 \cdot 8H_2O$, brushite, $CaHPO_4 \cdot 2H_2O$, whitlockite, $Ca_3(PO_4)_2$, vivianite, and others. Nriagu (1976) has described with thermodynamic arguments the weathering of PO_4^{3-} minerals in terrestrial systems. His calculations suggest that in neutral to acidic soils, calcareous PO_4^{3-} minerals may be progressively weathered to forms richer in aluminum, passing first through crandallite, to eventually wavellite, $Al_3(OH)_3(PO_4)_2 \cdot 5H_2O$.

Phosphate is also ubiquitous as a minor component within the crystal lattices of other minerals or adsorbed onto the surface of particles such as clays, calcium carbonate, or ferric oxyhydroxides (Ruttenberg, 1992). Therefore, in general, transport of these other particulate phases represents an important transport pathway of P as well.

14.1.3 Organic Forms of Phosphorus

Many of the most fundamental biochemicals required for life contain P generally linked to long, complicated organic molecules by phosphate ester bonds. Among the impressive list of compounds in which phosphate is a necessary constituent are the nucleic acids, DNA and RNA, discussed in Chapter 3. In these compounds (Fig. 14-3a), phosphates covalently link the mononucleotide units forming long polymers which, depending on the composition of the attached base, encode all necessary genetic information.

Phosphate also plays a central role in the transmission and control of chemical energy within the cells primarily via the hydrolysis of the terminal phosphate ester bond of the adenosine triphosphate (ATP) molecule (Fig. 14-3b). In addition, phosphate is a necessary constituent of phospholipids, which are important components in cell membranes, and as mentioned before, of apatite, which forms structural body parts such as teeth and bones. It is not surprising, therefore, that the cycling of P is closely linked with biological processes. This connection is, in fact, inseparable as organisms cannot exist without P, and their existence controls, to a large extent, the natural distribution of P.

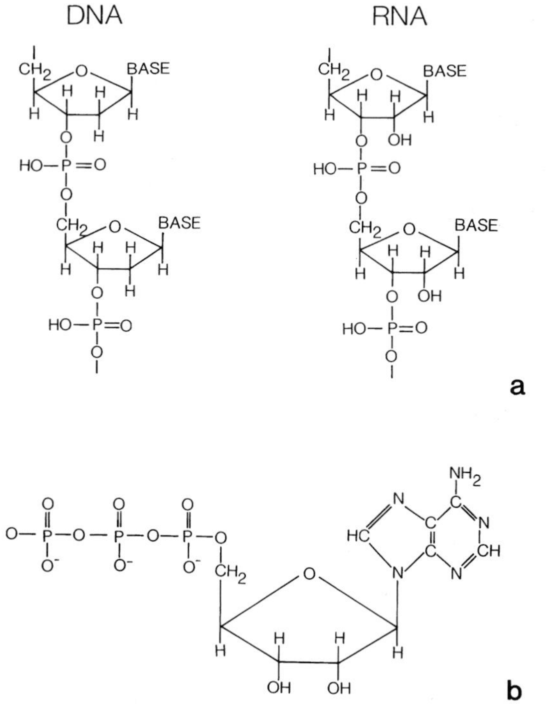

Fig. 14-3 Structure of DNA and RNA (a) and ATP (b).

Because these P-containing compounds are abundant in all organisms, P is one of the major components of all organisms. In general, marine microorganisms contain 105 to 125 carbon atoms for every P atom (Redfield *et al.*, 1963; Peng and Broecker, 1987; Anderson and Sarmiento, 1994). Because of the increased abundance of structural parts not involving phosphorus, the average C:P ratio of benthic macroalgae and sea grasses and terrestrial plants is much higher, approximately 550:1 and 800:1, respectively (Atkinson and Smith, 1983; Deevey, 1970).

Dissolved organic compounds that contain P also constitute an important pool of reactive P, particularly in the euphotic zone where it may exceed dissolved inorganic phosphate concentrations (Karl and Yanagi, 1997; Smith *et al.*, 1986). The chemical form of the dissolved organic P is not well known but is reported to be dominated by phosphate esters and phosphonates and differs significantly from fresh organic materials, suggesting considerable remineralization and transformation (Clark *et al.*, 1998; Nanny and Minear, 1997). The dissolved organic P pool appears to be readily available to organisms (Bjorkman and Karl, 1994) and may provide a "buffer," providing utilizable phosphate to phytoplankton between episodic inputs from other sources (Jackson and Williams, 1985).

14.2 Sub-Global Phosphorus Transfers

A global representation of the P cycle, by necessity, will be general. It will combine a wide variety of P-containing components into relatively few reservoirs and will parameterize intricate processes and feedback mechanisms into simple first-order transfers. To appreciate the rationale behind the construction of such a model and to understand its limitations, the transfers of P within a hypothetical terrestrial ecosystem and in a generalized ocean system will be discussed first.

14.2.1 *Freshwater Terrestrial Ecosystems*

The dominant processes controlling the movements of P through terrestrial ecosystems are schematically presented in Fig. 14-4. In a general way, the overall movement of P on the continents may be envisioned as the constant erosion of P from continental rocks and transport in both dissolved and particulate form by rivers to the ocean, stopping occasionally along this pathway to interact with biological and mineralogical systems.

Physical and chemical erosion of continental rocks (1) introduces particulate and dissolved P to the soil system. Approximately 5% of the mobilized P is present in dissolved form and is

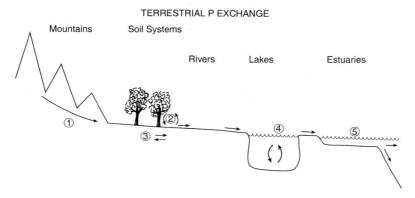

Fig. 14-4 Schematic representation of the transport of P through the terrestrial system. The dominant processes indicated are: (1) mechanical and chemical weathering of rocks, (2) incorporation of P into terrestrial biomass and its return to the soil system through decomposition, (3) exchange reactions between soil interstitial waters and soil particles, (4) cycling in freshwater lakes, and (5) transport through the estuaries to the oceans of both particulate and dissolved P.

readily available to enter biological systems (2) and to react with soil particles (3). The majority (approximately 95%) of the P in rivers remains in particulate form, with approximately 40% in organic phases and the remainder either trapped in the mineral lattices of the particulate matter or sorbed onto particle surfaces (such as FeOOH coating mineral grains or as colloids). Much of this P will be transported with the suspended material or bedload downstream until it eventually reaches the estuaries and the oceans, never having entered the biological cycles. However, even a small release from the particulate phase would significantly add to the reactive dissolved pool. Release from particles appears to occur in estuaries in response to increasing salinity. Developing a quantitative understanding of this release remains an important research topic (Kaul and Froelich, 1984; Conley *et al.*, 1995; Chambers *et al.*, 1995).

As reactive P is transported through the terrestrial system, it is assimilated into plants and subsequently into the rest of the biosphere (2). Although many elements are required for plant life, in many ecosystems P is the least available and, therefore, limits overall primary production (Schindler, 1977; Smith *et al.*, 1986). Thus, in many instances the availability of P influences or controls the cycling of other bioactive elements. When organisms die, the organic P compounds decompose and the P is released back into the soil–water system. This cycle of uptake and release may be repeated numerous times as P makes its way to the oceans.

Inorganic reactions in the soil interstitial waters also influence dissolved P concentrations. These reactions include the dissolution or precipitation of P-containing minerals or the adsorption and desorption of P onto and from mineral surfaces. As discussed above, the inorganic reactivity of phosphate is strongly dependent on pH. In alkaline systems, apatite solubility should limit groundwater phosphate whereas in acidic soils, aluminum phosphates should dominate. Adsorption of phosphate onto mineral surfaces, such as iron or aluminum oxyhydroxides and clays, is favored by low solution pH and may influence soil interstitial water concentrations. Phosphorus will be exchanged between organic materials, soil interstitial waters, and mineral phases many times on its way toward the ocean.

Lakes (4) also constitute an important component of the terrestrial P system. Because much of mankind's activities occur on or adjacent to lakes and because P availability has such a major influence on the biological community, P cycling in lakes has been extensively studied. The hypothetical P and temperature profiles for a temperate lake in summer are displayed in Fig. 14-5. Briefly, in summer, warming of the surface layers produces strong stratification which restricts exchange between the lighter, warm surface water and the colder, denser deep water. During photosynthesis, the dissolved P in the photic zone is incorporated into plants and is eventually transported below the thermocline on sinking particles. Most of this P is released back to the lake's deep water during the degradation of the organic matter but due to the stratification, this P is transported back to the surface photic zone very slowly. The constant stripping of P from the surface layers by

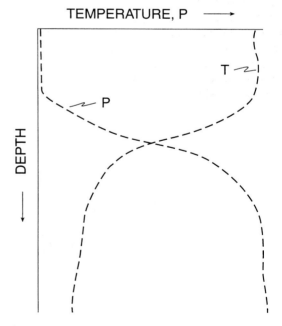

Fig. 14-5 Typical distribution of P and temperature in a temperate lake in summer. Thermal stratification restricts exchange between surface and deep waters. Phosphorus is depleted in the surface waters by the sinking of biologically produced particles.

production and subsequent sinking of particles results in extremely low levels of P in the surface waters which often limit overall biological productivity.

In this context, it is easy to envision the potential influence of increased P input to the surface layers via anthropogenic sources. If P is continually added to the photic zone, productivity will not be limited but will simply continue unchecked. This results in large amounts of organic particles sinking below the thermocline. As this material decomposes, it consumes oxygen. Since exchange with the surface layer is restricted, the deep waters will become depleted in oxygen (cf. Lehman, 1988). If enough organic materials sink into the deep waters, the oxygen will be thoroughly consumed via decomposition. In extreme conditions, this can result in the formation of anoxic deep waters and fish kills.

As cooling occurs in the late fall and early winter, the thermal stratification breaks down, permitting mixing of the deep and surface layers. This allows the surface layers to be replenished with P. During the winter months, biological productivity in a temperate lake is limited by the availability of light rather than nutrients.

In the hypothetical terrestrial system depicted in Fig. 14-4, the P eroded from the land is eventually transported to the estuaries. As in lakes, soils, and rivers, many chemical and biological processes act to control the transport of P within and from the estuary (Lucotte and d'Anglejan, 1988; Jonge and Villerius, 1989). Dissolved P may be removed from solution onto the particulate phase and deposited in the sediments. On the other hand, the change in the solution composition may cause P to be released from the particulate load. The P that is transported from the estuaries to the ocean bonded to particles will rapidly settle to the sea floor and be incorporated into the sediments. The reactive forms of P, primarily dissolved but including particulate forms that can be easily solubilized, will enter the surface ocean and participate in the biological cycles. Determining what proportion of the transported P is reactive is a critical step in the elucidation of the marine P budget (Froelich *et al.*, 1982).

14.2.2   The Oceanic System

Over much of the ocean (exclusive of upwelling regions, high-latitude areas and specific high-nutrient, low-chlorophyll regions) the vertical distribution of dissolved PO_4^{3-} is represented by the shape of the profile displayed in Fig. 14-6, which is similar to the shape observed for the

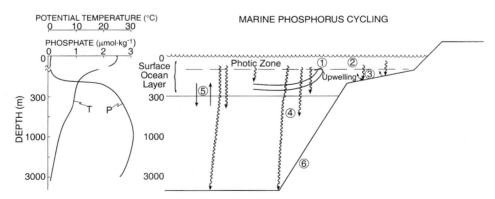

Fig. 14-6   Profiles of potential temperature and phosphate at 21° 29'N, 122° 15'W in the Pacific Ocean and a schematic representation of the oceanic processes controlling the P distribution. The dominant processes shown are: (1) upwelling of nutrient-rich waters, (2) biological productivity and the sinking of biogenic particles, (3) regeneration of P by the decomposition of organic matter within the water column and surface sediments, (4) decomposition of particles below the main thermocline, (5) slow exchange between surface and deep waters, and (6) incorporation of P into the bottom sediments.

temperate lake in summer. Also included in the figure are the major processes responsible for controlling this shape. In general, dissolved P is nearly undetectable in the euphotic zone (generally the upper 20–100 m) and increases to maximum concentrations of 1–3 μM at approximately 1000 m. The distribution can best be envisioned as the balance between the incorporation of P into organisms with the eventual sinking of some fraction of this P from the surface waters and the constant slow rate of return of P to the surface layer by physical processes. The majority of ocean deep water is formed in the North Atlantic and slowly spreads sequentially to the South Atlantic, Indian, and finally to the Pacific Oceans. Because of the continuous rain of particulate P into the deep waters from the surface layers, the deepwater PO_4^{3-} concentration increases progressively from the North Atlantic to the Pacific.

Unlike the temperate lake, stratification in the ocean does not completely break down in the winter. Except for specific high-latitude regions of the ocean, processes other than deep convective overturn are responsible for returning the P stored in the deep waters to the surface. A relatively slow exchange occurs between water layers everywhere in the ocean and this supplies some P to the surface ocean (process 5 in Fig. 14-6) from the intermediate layers below. More important sources of P to the photic zone are the major upwelling regions generally located adjacent to the western, sub-tropical continental margins (1) and in equatorial divergence zones. In the western margins, the prevailing winds tend to transport surface water offshore. This water is replaced by nutrient-rich water from below. At these locations P input (along with other required nutrients) is not limited by slow diffusive transport processes but is enhanced many-fold by the upward advection of water. For this reason, upwelling areas are capable of supporting extremely high rates of biological primary production and abundant populations of higher organisms. Thus, the major fisheries of the world are concentrated in upwelling regions such as off Peru. A significant amount of P is also returned to the surface ocean in cold, high-latitude regions where decreased stratification results in greater vertical mixing than in the temperate and equatorial regions.

Once in the photic zone, P is readily incorporated into biogenic particles (2) via the photosynthetic activities of plants and some fraction of the biogenic materials subsequently sinks. Increasingly, improved technologies permit the dynamics of this system to be followed by studying ^{32}P and ^{33}P distributions. The majority of the particles decompose in the surface layer or in shallow sediments and the P is recycled directly back into the photic zone (3) to be reincorporated into biological particles. A small portion of the particles produced in the surface layers, however, does escape the surface layers and sinks into the deep ocean. Most of these particles eventually decompose (4) and the cycle is repeated. A very small fraction of these particles, however, escapes decomposition and is incorporated into the sediment (6). P appears to be buried in sediments primarily as organic P, apatitic P (including both authigenic apatite and fish debris), P associated with other mineral phases (primarily $CaCO_3$ and $FeOOH$) and P loosely sorbed on to other solid phases (Ruttenberg, 1993; Follmi, 1996; Anschutz *et al.*, 1998).

14.3   The Global Phosphorus Cycle

The main reservoirs and exchange pathways of the global P cycle are schematically presented in Fig. 14-7. This representation is primarily taken from Lerman *et al.* (1975) and modified to include atmospheric transfers. The mass of P in each reservoir and rates of exchange are taken from Mackenzie *et al.* (1993) and Follmi (1996).

In choosing these reservoirs to describe the P cycle, compromises were made to maintain a general focus and global scale and yet avoid being too general and hence lose information about important transfers and reservoirs. The following is a brief discussion of the rationale behind the choice of the reservoir definitions and their estimates. For the purpose of discussion, the reservoirs have been numbered as presented in Lerman *et al.* (1975) with the addition of the atmosphere (reservoir 8). The total P content of each reservoir and comments concerning the estimate are provided in Table 14-3.

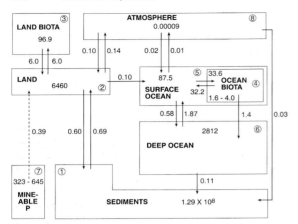

Fig. 14-7 The global phosphorus cycle. Values shown are Tmol and Tmol/yr for reservoirs and fluxes, respectively. ($T = 10^{12}$).

The reservoir representing the land (2) is defined as the amount of P contained in the upper 60 cm of the soil. This rather narrow definition of the land reservoir is made because it is through the upper portions of the soil system that the major interactions with the other P reservoirs occur. Specifically, most plants receive their nutritive P needs from the upper soil horizons and the return of P to the soil system by the decomposition of plant matter is also concentrated in this upper soil zone. Similarly, the major interactions with the atmosphere, ground waters, and rivers occur near the

soil surface. Finally, phosphate in the form of fertilizer is applied directly to the soil surface. Thus, in attempting to represent the land and its interaction with other reservoirs, the surface soil horizon most directly interacts with all components and best represents the dynamic nature of this reservoir. Phosphorus in soils deeper than 60 cm and in crustal rocks is included in the sediment reservoir (1). This reservoir accounts for all of the particulate P that exchanges with the other reservoirs only on longer time scales.

The land biota reservoir (3) represents the phosphorus contained within all living terrestrial organisms. The dominant contributors are forest ecosystems with aquatic systems contributing only a minor amount. Phosphorus contained in dead and decaying organic materials is not included in this reservoir. It is important to note that although society most directly influences and interacts with the P in lakes and rivers, these reservoirs contain little P relative to soil and land biota and are not included in this representation of the global cycle.

The ocean system is separated into three major reservoirs that best represent the dominant pools and pathways of P transport within the ocean. The surface ocean reservoir (5) is defined as the upper 300 m of the oceanic water column. As discussed in an earlier section and displayed in Fig. 14-6, the surface layer roughly corresponds to the surface mixed layer where all

Table 14-3   The mass of P in the major reservoirs (mol P $\times$ 10^{-12})

Reservoir	P content	Comments
1. Sediments	1.29×10^8	Van Wazer (1961)
2. Land	6460	Computed from land area of 133×10^6 km^2, soil thickness of 60 cm, density of 2.5 g/cm^3 and mean P content of 0.1% (Taylor, 1964)
3. Land biota	96.9	Computed from an estimate of the N in land biota (12×10^4 tons N; Delwiche, 1970); and a mean P:N atomic ratio in land plants (1.8:16; Deevey, 1970)
4. Oceanic biota	1.6–4.0	Mackenzie *et al.* (1993)
5. Surface Ocean	87.5	Computed from assumed mean concentration of 25 mg/m^3 of dissolved P, 300 m thick surface layer and area of 3.61×10^8 km^2
6. Deep ocean	2812	Computed from assumed mean concentration of 80 mg/m^3 of dissolved P, a deep water thickness of 3000 m, and same surface area as above.
7. Mineable P	323–645	Mean values reported by Stumm (1973), Ronov and Korzina (1960) and Van Wazer (1961) and the value reported by Mackenzie *et al.* (1993)
8. Atmospheric	0.00009	Graham (1977)

the photosynthetic uptake of P and the majority of the decomposition and release of P from sinking organic matter occur. Therefore, the most active exchange of P between solution and ocean biota occurs in this zone. Also, the 300 m depth roughly corresponds to the top of the main thermocline, which restricts exchange between surface and deep waters, thus representing a natural boundary. Additionally, dissolved PO_4^{3-} brought to the ocean via rivers is introduced directly to the surface layer.

The oceanic biota reservoir (4) is also within the surface layers. Although organisms reside at all depths within the ocean, the overwhelming majority reside within the photic zone where phytoplankton dominate. The oceanic biota reservoir only contains roughly 1/30 as much P as the land biota reservoir. This is primarily because oceanic biomass is composed of relatively short-lived organisms, while land biomass is dominated by massive long-lived forests.

The deep ocean (6) is the portion of the water column from 300 m to 3300 m and is the largest ocean reservoir of dissolved P. However, since the deep ocean is devoid of light, this P is not significantly incorporated into ocean biota. Mostly, it is stored in the deep waters until it is eventually transported back into the photic zone via upwelling or eddy diffusive mixing.

The sediment reservoir (1) represents all phosphorus in particulate form on the Earth's crust that is (1) not in the upper 60 cm of the soil and (2) not mineable. This includes unconsolidated marine and fresh water sediments and all sedimentary, metamorphic and volcanic rocks. The reason for this choice of compartmentalization has already been discussed. In particulate form, P is not readily available for utilization by plants. The upper 60 cm of the soil system represents the portion of the particulate P that can be transported relatively quickly to other reservoirs or solubilized by biological uptake. The sediment reservoir, on the other hand, represents the particulate P that is transported primarily on geologic time scales.

The atmospheric reservoir (8) represents P contained on dust particles. Because the mean residence time of dust in the air is very short, the standing stock of P in the atmosphere is relatively small. Mineable P (7) is simply an estimate of the total amount of P stored in economic deposits.

14.3.1 Fluxes between Reservoirs

A summary of the estimated fluxes between reservoirs and the methods of calculation is presented in Table 14-4. The fluxes between reservoirs are chosen to represent the principal pathways by which P is transported between reservoirs. The notation used here is the same as that presented by Lerman et al. (1975) with the first number representing the reservoir from which the P originates and the second number representing the receiving reservoir. It is important for the reader to understand that the evaluation of the fluxes is extremely difficult. This is, in part, caused by the introduction of P from anthropogenic sources obscuring natural levels. Also, as stated earlier, most P is present in particulate form and is not biologically active. Thus, the evaluation of the P that is actively transferred between the inorganic reservoirs and the biota or as the dissolved component in rivers must always be made in the presence of a large background. Small exchanges between the relatively non-reactive particulate P and reactive (primarily dissolved) P could significantly alter the flux estimate.

The transfer of P from land to terrestrial biota (F_{23}) represents the sum of terrestrial biological productivity. There is no significant gaseous form of P, nor is there a major transfer of living organisms between the freshwater–terrestrial system and the oceans. The terrestrial biota system is, therefore, essentially a closed system where the flux of P to the biota (F_{23}) is balanced by the return of P to the land from the biota (F_{32}) due to the decay of dead organic materials.

The transfer of P from the continents to the ocean is separated into two distinct pathways. The flux of reactive P (F_{25}) is estimated via measurements of dissolved organic and inorganic P in rivers. A small correction (33%; after Kaul and Froelich, 1984) is added to the measured values to account for P released from particles within the estuaries. This P is transported directly to the surface ocean and is

Table 14-4   Summary of the flux of P between reservoirs (mol P $\times 10^{12}$)

Transfer	Flux	Comments
F_{12}	0.69	Computed from combined rates of mechanical and chemical denudation of continents (2×10^9 tons/yr; Garrels and Mackenzie, 1971) and a mean P content of crustal material of 0.1% (Taylor, 1964)
F_{23}, F_{32}	6.00	Computed from total C fixed annually (560 Pg C/yr; Sundquist, 1993) and a mean P:C atomic ratio of land biota of 1:510 (Delwiche and Likens, 1977)
F_{25}	0.10	Garrels et al. (1973) and adjusting upward 33% to account for release from particles (after Kaul and Froelich, 1984)
F_{54}	33.6	Computed from rate of N fixation of oceanic biota of 7.5×10^9 tons N/yr (Vaccaro, 1965) and a mean P:N atomic ratio in ocean biota of 1:16 (Redfield et al., 1963)
F_{45}	32.2	Computed assuming that 96% of oceanic biota recycled within upper 300 m
F_{46}	1.40	Difference between fluxes F_{54} and F_{45}
F_{56}	0.58	Computed from the P content of surface layer given in Table 14-3 and a water exchange rate between surface and deep ocean of 2 m/yr (Broecker, 1971)
F_{65}	1.87	Computed as F_{56} using the P content of the deep ocean given in Table 14-3
F_{61}	0.11	Calculated that the ocean is in steady state
F_{72}	0.39	Stumm (1973); Mackenzie et al. (1993)
F_{21}	0.60	Mackenzie et al. (1993)
F_{28}	0.14	Graham (1977)
F_{82}	0.10	Graham (1977)
F_{58}	0.01	Graham (1977)
F_{85}	0.02	Graham (1977)
F_{81}	0.03	Graham (1977)

available for biological uptake. The other pathway by which P is transported to the oceans is that associated with particulate materials, either suspended in river waters or simply transported to the ocean as bedload. The P in these materials is considered locked in the solid structure and not available for biological uptake. In general, these particles rapidly settle to the ocean bottom and are incorporated into the sediments. This removal is rapid enough that this flux is represented as the direct transport of P from the land reservoirs to the sediments (F_{21}).

Separating "reactive" and "non-reactive" P in the estuarine system is extremely difficult. Since the majority of P is associated with the particles, a small exchange between the particulate and the dissolved fraction would alter the "reactive" flux (F_{25}) significantly. Kaul and Froelich (1984) have investigated the fate of dissolved and particulate P in a small estuary. They determined that although P is readily incorporated into plants in the estuary, all of the dissolved P brought to the estuary by the river is eventually transported to the ocean. In addition, they found

that one-third of the "reactive" P that leaves the estuary is derived from fluvial particulate matter. Thus, estimates of the flux of "reactive" P from the land to the surface ocean based on dissolved P concentration data may be 50% too low. While uncertainty exists, the value of F_{25} has been increased by 33% as a rough attempt to account for this release.

A small flux is shown between the land and atmosphere. This represents the transport of dust particles to the atmosphere (F_{28}) and the deposition of these particles back on land either as dry deposition or associated with atmospheric precipitation (F_{82}). Similarly, fluxes that represent the transport of seasalt from the surface ocean to the atmosphere (F_{58}) and the deposition of soluble (F_{85}) and insoluble (F_{81}) atmospheric forms are also shown. As already discussed for the river fluxes, the insoluble particulate flux is represented as a direct transport of P to the sediment reservoir.

The natural circulation of the oceans also exchanges waters between the deep and surface ocean reservoirs. Because biological uptake constantly strips P from the surface layers, the P concentration is much less in the surface reservoir than the deep reservoir. Thus, although the continuity of water demands equal volumes of water to be exchanged between the reservoirs, far more P is carried by the upwelling water (F_{65}) than by the downwelling waters (F_{56}).

By far, the largest exchange of P between reservoirs occurs between the surface ocean and ocean biota. These large numbers attest to the tremendous role of primary production in controlling P dynamics in the ocean. However, the relatively small standing stock also attests to the short life-span and rapid turnover of these organisms. Most of the recycling occurs in the surface waters and, hence, the flux of P into the biota (F_{54}) is nearly equaled by the release of P from the organisms (F_{45}). However, a small fraction (4%) of the P incorporated into the biota sinks out of the surface layers before being solubilized by decomposition. Although this particulate flux to the deep ocean reservoir (F_{46}) is a small fraction of the total amount incorporated into oceanic biota, ocean productivity is so great that this flux is quantitatively quite large.

Nearly all of the detrital particles sinking into the deep ocean decompose and release the associated P. A small percentage (approximately 8%), however, do survive and accumulate on the sea floor. This P is then buried in the sediments (F_{61}) and represents the ultimate removal of P from the ocean.

The basic cycle of P transport from continents to oceans to sediments is then completed by the slow geologic processes that eventually return marine sediments to the continents (F_{12}). One additional flux is considered which represents the P mined and placed on agricultural fields in the form of fertilizers (F_{72}). Notice that there are no fluxes into the reservoir marked mineable P. This is undoubtedly incorrect. However, at time scales important to human beings, the formation of mineable P deposits is assumed to be too slow to be included. Notice also that because of this, the representation of the P cycle is not in a steady state.

14.3.2 The Steady-State Cycle

The total burden, sum of inputs or exports, and average residence times for the reservoirs are listed in Table 14-5. As discussed in Chapter 4, the residence time of an element within a reservoir reflects the reactivity and exchange of that element with other reservoirs. A short residence time suggests that removal processes are rapid and significant over short time scales compared to the amount in the reservoir.

From Table 14-5 it is obvious that the residence time of P in the atmosphere is extremely short. This does not represent chemical reaction and removal of P from the atmosphere but rather the rapid removal of most phosphorus-containing particles that enter the atmosphere.

More informative is the comparison between the residence times of P in the land and ocean biota. Although there is 5.6 times more biological incorporation of P in the oceans, the standing stock is only 1.7–4.1% of that on land. The residence time of a P atom incorporated into

Table 14-5 Summary of reservoir amounts, total fluxes, and residence time

Reservoir	A (mol × 10^{-12})	Σ fluxes (mol a^{-1} × 10^{-12})	Residence time (years)
Atmosphere	0.00009	0.15	0.0006 (5.3 h)
Land biota	96.9	6.0	16.2
Land	6460	9.81	949
Surface ocean	87.5	34.2	2.56
Ocean biota	1.6–4.0	33.6	0.048–0.19 (18–69 d)
Deep ocean	2812	1.98	1420
Sediments	1.29×10^8	0.71	1.82×10^8
Total ocean system	2902	0.12	24 180

oceanic biota is relatively short (2–7 days) compared to the 16-year-residence time of P in land biota. This disparity represents a fundamental difference in the types of organisms in the two reservoirs. Whereas oceanic biota are dominated by single celled, short-lived planktonic plants and microbes, terrestrial biomass is dominated by forests.

The residence time of P in the deep ocean is 1400 years and is dependent primarily on the rate at which P is transported to the surface waters via upwelling. The ocean system also demonstrates an important characteristic of global reservoir models. The residence times of P in the oceanic biota, surface ocean, and deep ocean are 2–7 days, 2.6 years, and 1400 years, respectively. These relatively short times suggest that P is recycled rapidly throughout the ocean. However, since this cycling is almost exclusively among oceanic reservoirs and not to outside reservoirs, the residence time of P in the entire ocean system is relatively long, 24 180 years. This value is well within the range 16 000–38 000 years estimated by Ruttenberg (1993) but much shorter than the 80 000 years estimated earlier by Froelich *et al.* (1982). Increased estimates on input and burial rates account for this decrease in the estimated residence time. Based on these values, the average P atom is cycled approximately 17 times between the deep water and surface waters before being removed to the sediments. Also, each time a P atom reaches the surface waters, it is cycled between the oceanic biota and the dissolved inorganic pool 14–52 times before being transported to the deep water. Thus, the average P atom is incorporated into the ocean biota a total of 238–884 times during its stay in the ocean.

14.3.3   Perturbations

One of the main goals in establishing a global understanding of P cycling is to evaluate the influence of changing conditions on the P distribution. Perhaps the most obvious and visible mechanism by which the P cycle may be altered is through anthropogenic input. This includes P applied to the soil in the form of fertilizers as well as P used in detergents and various industrial applications. The amount of P added to the environment in this manner may be estimated from the amount of P mined each year. We have all witnessed or have read about the detrimental effects of large P inputs on some freshwater systems. The influence of this added P on the global cycle is less obvious.

Lerman *et al.* (1975) considered several cases in which mankind's activities perturbed the natural cycle. If we assume that all mined P is supplied to the land as fertilizer and that all of this P is incorporated into land biota, the mass of the land biota will increase by 20%. This amount is small relative to the P stored in the land reservoir. Since P incorporated into land biota must first decompose and be returned to the land reservoir before being transported further, there is essentially no change in the other reservoirs. Thus, although such inputs would significantly alter the freshwater–terrestrial ecosystem locally where the P release is concentrated, the global cycle would be essentially unaffected.

A greater perturbation results if one assumes that the rate at which P is mined doubles every 10 years and that the dissolved river-borne flux to the ocean increases in proportion to the mining rate. At the time of Lerman *et al.*'s calculation, known reserves implied that mineable P would be exhausted in 60 years under these conditions. Since then, additional deposits have been identified. Nevertheless, this calculation still effectively demonstrates the sensitivity of the P reservoirs to changes of this magnitude. At the end of the 60-year period, the P contained in the surface ocean has increased by 38% and the P present in ocean biota is 30% greater than present (assuming no other limiting factors for biological production). The other reservoirs will not change significantly. After 60 years, increased P input will then cease and after approximately 150 years, the system will then return to present-day levels. Additional discussion of the impacts of perturbations to the P cycle due to increased land weathering rates and decreased terrestrial productivity are discussed in Mackenzie *et al.* (1993).

There may also be natural fluctuations within this cycle that occur over time scales ranging from thousands of years (glacial–interglacial) to millions of years (Follmi, 1996). Overall, the global P burial rate is estimated to have varied by one order of magnitude or less over the last 160 million years. Maxima in P burial are estimated for the Callovian–Oxfordian boundary (approx. 157 Myr ago), Valanginian–Hauterivian boundary (approx. 135 Myr ago), late Albian (approx. 98 Myr ago), Campanian (approx. 82 Myr ago), and Paleocene (approx. 60 Myr ago). The changes in total P burial rates appear to be correlated with changes in the burial of biogenic P. These variations suggest that there have been relative variations in continental weathering rates, proportion of mobilized P that is dissolved, and overall sediment accumulation.

Because P is such an important element in biological systems, the P cycle may influence other biogeochemical cycles and processes. One example is the potential link between the P content of deep ocean water and atmospheric CO_2. Atmospheric CO_2 levels are increasing rapidly presently due to the burning of fossil fuels and deforestation. Because CO_2 influences the efficiency with which solar radiation is absorbed in the atmosphere and, hence, the temperature at the Earth's surface, there is great concern as to how this increase will influence the global climate. However, analyses of gas pockets trapped in ice cores taken from Greenland and Antarctica suggest that there have been natural fluctuations in the concentration of atmospheric CO_2 and that during the last glacial period, atmospheric CO_2 was 80 ppmv less than the modern pre-industrial value (Neftel *et al.*, 1982). Thus, to predict the consequences of the build-up of CO_2 in the atmosphere, one must understand the natural variations. Broecker (1982) has suggested that one mechanism by which atmospheric CO_2 may be altered is that of changing the PO_4^{3-} concentration in the deep waters of the ocean. His argument is as follows.

The present average PO_4^{3-} concentration of deep ocean water is 2.2 μmol/kg. When a parcel of deep water is transported to the photic zone, this PO_4^{3-} is completely incorporated into plants. Note that this assumes that net primary productivity is not limited by the availability of other micronutrients. In short-term laboratory studies, this assumption is clearly not true in that it has been demonstrated

that the availability of fixed N or iron can also limit production. Because certain organisms are capable of fixing N from the large N_2 gas pool whereas there is no alternative source or substitute for PO_4^{3-}, it is likely that on longer time scales, PO_4^{3-} limits productivity. Because the overall chemical composition of marine organisms is relatively constant, the complete utilization of the upwelled PO_4^{3-} also determines the amount of dissolved inorganic carbon and alkalinity that is removed from the surface waters and transported downward on sinking particles. This, in turn, alters the solution chemistry and partial pressure of CO_2 in the surface waters and, in the long term, the partial pressure of CO_2 in the atmosphere as well. If the PO_4^{3-} concentration of the water entering the photic zone were to change, the resulting partial pressure of CO_2 would also be affected. In fact, Broecker (1982) speculates that if the PO_4^{3-} concentration in the deep water during the last glacial period averaged 3.2 μmol/kg, the resulting atmospheric CO_2 would be 80 ppmv less than the modern pre-industrial value.

Recently, it has also been suggested that the burial rates of organic carbon and phosphorus in marine sediments are inversely correlated (Ingall and Jahnke, 1994, 1997; Van Cappellen and Ingall, 1994). This provides a mechanism for stabilizing atmospheric oxygen through geological time by providing a controlling feedback mechanism between organic carbon burial and photosynthetic oxygen production (Van Cappellen and Ingall, 1996; Coleman and Holland, in press). Understanding such potential linkages are fundamental to understanding Earth systems and to predicting how they may change in the future. Thus, natural variations in the transfer rates of P between reservoirs may have profound effects on other geochemical cycles and climate. The elucidation of such interactions between cycles is the focus of many exciting research programs currently underway.

Questions

14-1 Due to anthropogenic inputs, the mean pH of a lake has decreased from 6.5 to 6.0. Assuming no significant changes in the chemical composition and ionic strength of the lake waters and a total dissolved inorganic P concentration of 1 μM, by what factor will the free PO_4^{3-} change?

14-2 A farmer in the Imperial Valley has been irrigating and fertilizing fields for many years. Because of the generally hot conditions, much of this irrigation water is lost to evaporation, increasing the salt content of the soil and groundwater. Assuming that the composition of the salt is roughly similar to that of seawater, how might this salt buildup influence the availability of P to the plants? Will the farmer need to increase the amount of fertilizer used?

14-3 Because of a primitive sewage system and the use of fertilizers, the citizens of a small community have added a significant amount of P to a nearby lake. Recognizing that this will stimulate biological production and may cause anoxia in the deep waters when the lake is stratified, the community decides to install a pumping system to continually exchange the deep water with the surface water. What will this do to the productivity of the lake? Will this prevent the deep water from becoming anoxic?

14-4 In recent years, many of the world's forests have been cut down and replaced with short-lived crops. What effect, if any, might this have on the: (1) P stored in the land biota reservoir; (2) exchange rate of P between the land biota and the land reservoirs; (3) exchange rate between the land reservoir and the surface ocean?

14-5 It has been suggested by Broecker (1982) that glacial–interglacial $p CO_2$ differences could be explained if the glacial ocean contained 1.5 times more dissolved P than at present. If the exchange fluxes with the non-oceanic reservoirs remained the same, what would be the residence time of P in the ocean during the last glacial period? How does this compare to glacial–interglacial time scales and what does it suggest about the exchange fluxes?

References

Anderson, L. A. and Sarmiento, J. L. (1994). Redfield ratios of remineralization determined by nutrient data analysis. *Global Biogeochem. Cycles* **8**, 65–80.

Anschutz, P., Zhong, S., Sundby, B., Mucci, A., and Gobeil, C. (1998). Burial efficiency of phosphorus and the geochemistry of iron in continental margin sediments. *Limnol. Oceanogr.* **43**, 53–64.

Atkinson, M. J. and Smith, S. V. (1983). C:N:P ratios of benthic marine plants. *Limnol. Oceanogr.* **28**, 568–574.

Atlas, E. L. (1975). Phosphate equilibria in seawater and interstitial waters. Ph.D. Thesis, Oregon State University.

Benitez-Nelson, C. R. and Buesseler, K. O. (1999). Variability of inorganic and organic phosphorus turnover rates in the coastal ocean. *Nature* **398**, 502–505.

Bjorkman, K. and Karl, D. M. (1994). Bioavailability of inorganic and organic phosphorus compounds to natural assemblages of microorganisms in Hawaiian coastal waters. *Mar. Ecol. Prog. Ser.* **111**, 265–273.

Broecker, W. S. (1971). A kinetic model for the chemical composition of seawater. *Quatern. Res.* **1**, 188–207.

Broecker, W. S. (1982). Ocean chemistry during glacial time. *Geochim. Cosmochim. Acta* **46**, 1689–1706.

Chambers, R. M., Fourqurean, J. W., Hollibaugh, J. T., and Vink, S. M. (1995). Importance of terrestrially-derived particulate phosphorus to phosphorus dynamics in a west coast estuary. *Estuaries* **18**, 518–526.

Clark, L. L., Ingall, E. D., and Benner, R. (1998). Marine phosphorus is selectively remineralized. *Nature* **393**, 426–428.

Coleman, A. S. and Holland, H. D. (in press, January 2000). The global diagenetic flux of phosphorus from marine sediments to the oceans: redox sensitivity and the control of atmospheric oxygen levels. *In* "Marine Authigenesis: from Microbial to Global" (C. R. Glenn, L. Prevot-Lucas and J. Lucas, eds), SEPM Publication No. 66.

Conley, D. J., Smith, W. M., Cornwell, J. C., and Fisher, T. R. (1995). Transformation of particle-bound phosphorus at the land-sea interface. *Estuar. Coast. Shelf Sci.* **40**, 161–176.

Deevey, E. S., Jr. (1970). Mineral cycles. *Scient. Am.* **223**, 149–158.

Delwiche, C. C. (1970). The nitrogen cycle. *Scient. Am.* **223**, 137–146.

Delwiche, C. C. and Likens, G. E. (1977). Biological response to fossil fuel combustion products. *In* "Global Chemical Cycles and Their Alterations by Man" (W. Stumm, ed.), pp. 73–88. Dahlem Konferenzen, Berlin.

Emerson, S. and Widmer, G. (1978). Early diagenesis in anaerobic lake sediments II. Equilibrium and kinetic factors controlling the formation of iron phosphate. *Geochim. Cosmochim. Acta* **42**, 1307–1316.

Follmi K. B. (1996). The phosphorus cycle, phosphogenesis and marine phosphate-rich deposits. *Earth-Sci. Rev.* **40**, 55–124.

Froelich, P. N., Bender, M. L., Luedtke, N. A., Heath, G. R., and DeVries, T. (1982). The marine phosphorus cycle. *Am. J. Sci.* **282**, 474–511.

Froelich, P. N., Arthur, M. A., Burnett, W. C., Deakin, M., Hensley, V., Jahnke, R., Kaul, L., Kim, K.-H., Roe, K., Soutar, A., and Vathakanon, C. (1988). Early diagenesis of organic matter in Peru continental margin sediments: phosphate precipitation. *Mar. Geol.* **80**, 309–343.

Garrels, R. M. and MacKenzie, F. T. (1971). "Evolution of Sedimentary Rocks." W. W. Norton & Co., Inc., New York.

Garrels, R. M., MacKenzie, F. T., and Hunt, C. A. (1973). "Chemical Cycles and the Global Environment." W. Kaufmann, Los Altos, CA.

Graham, W. F. (1977). Atmospheric pathways of the phosphorus cycle. Ph.D. Thesis, University of Rhode Island.

Howard, P. F. (1979). Phosphate. *Econ. Geol.* **74**, 192–194.

Ingall, E. D. and Jahnke, R. A. (1994). Evidence for enhanced phosphorus regeneration from marine sediments overlain by oxygen depleted waters. *Geochim. Cosmochim. Acta* **58**, 2571–2575.

Ingall, E. and Jahnke, R. (1997). Influence of water-column anoxia on the elemental fractionation of carbon and phosphorus during sediment diagenesis. *Mar. Geol.* **139**, 219–229.

Jackson, G. A. and Williams, P. M. (1985). Importance of dissolved organic nitrogen and phosphorus in biological nutrient cycling. *Deep Sea Res.* **32**, 223–235.

Jahnke, R. A. (1984). The synthesis and solubility of carbonate fluorapatite. *Am. J. Sci.* **284**, 58–78.

Jonge, V. N. de and Villerius, L. A. (1989). Possible role of carbonate dissolution in estuarine phosphate dynamics. *Limnol. Oceanogr.* **34**, 332–340.

Karl, D. M. and Yanagi, K. (1997). Partial characterization of the dissolved organic phosphorus pool in the oligotrophic North Pacific Ocean. *Limnol. Oceanogr.* **42**, 1398–1405.

Kaul, L. W. and Froelich, P. N. Jr. (1984). Modeling estuarine nutrient geochemistry in a simple system. *Geochim. Cosmochim. Acta* **48**, 1417–1434.

Lal, D. and Lee, T. (1988). Cosmogenic ^{32}P and ^{33}P used as tracers to study phosphorus recycling in the upper ocean. *Nature* **333**, 752–754.

Lehman, J. T. (1988). Hypolimnetic metabolism in Lake Washington: Relative effects of nutrient load and food web structure on lake productivity. *Limnol. Oceanogr.* **33**, 1334–1347.

Lerman, A., MacKenzie, F. T., and Garrels, R. M. (1975). Modeling of geochemical cycles: Phosphorus as an example. *Geo. Soc. Am. Mem.* **142**, 205–218.

Lucotte, M. and d'Anglejan, B. (1988). Seasonal changes in the phosphorus-iron geochemistry of the St. Lawrence estuary. *J. Coastal Res.* **4**, 339–349.

Mackenzie, F. T., Ver, L. M., Sabine, C., Lane, M., and Lerman, A. (1993). C, N, P, S global biogeochemical cycles and modeling of global change. In "Interactions of C, N, P, and S Biogeochemical Cycles and Global Change" (R. Wollast, F. T. Mackenzie, and L. Chou, eds), pp. 1-61. Springer-Verlag, New York.

Murray, J. W., Grundmanis, V., and Smethie, W. M. Jr. (1978). Interstitial water in the sediments of Saanich Inlet. *Geochim. Cosmochim. Acta* **42**, 1011–1026.

Nanny, M. A. and Minear, R. A. (1997). Characterization of soluble unreactive phosphorus using ^{31}P nuclear magnetic resonance spectroscopy. *Mar. Geol.* **139**, 77–94.

Neftel, A., Oeschger, H., Swander, J., Stauffer, B., and Zumbrunn, R. (1982). New measurements on ice core samples to determine the CO_2 content of the atmosphere during the last 40 000 years. *Nature* **295**, 220–223.

Nriagu, J. O. (1976). Phosphate-clay mineral relations in soils and sediments. *Can. J. Earth Sci.* **13**, 717–736.

Nriagu, J. O. and Moore, P. B. (1984). "Phosphate Minerals." Springer-Verlag, New York.

Peng, T. H. and Broecker, W. S. (1987). C:P ratios in marine detritus. *Global Biogeochem. Cycles* **1**, 155–162.

Redfield, A. C., Ketchum, B. H., and Richards, F. A. (1963). The influence of organisms on the composition of seawater. In "The Sea," Vol. 2 (M. N. Hill, ed.), pp. 26–77. Interscience Publishers, New York.

Ronov, A. B. and Korzina, G. A. (1960). Phosphorus in sedimentary rocks. *Geochemistry* **8**, 805–829.

Ruttenberg, K. C. (1992). Development of a sequential extraction method for different forms of phosphorus in marine sediments. *Limnol. Oceanogr.* **37**, 1460–1482.

Ruttenberg, K. C. (1993). Reassessment of the oceanic residence time of phosphorus. *Chem. Geol.* **107**, 405–409.

Ruttenberg, K. C. and Berner, R. A. (1993). Authigenic apatite formation and burial in sediments from non-upwelling, continental margin environments. *Geochim. Cosmochim. Acta* **57**, 991–1007.

Schindler, D. W. (1977). Evolution of phosphorus limitation in lakes. *Science* **195**, 260–262.

Schuffert, J. D., Jahnke, R. A., Kastner, M., Leather, J., Sturtz, A., and Wing, M. R. (1994). Rates of formation of modern phosphorites off western Mexico. *Geochim. Cosmochim. Acta* **58**, 5001–5010.

Schuffert, J. D., Kastner, M., and Jahnke, R. A. (1998). Carbon and phosphorus burial associated with modern phosphorite formation. *Mar. Geol.* **146**, 21–31.

Sheldon, R. P. (1981). Ancient marine phosphorites. *Ann. Rev. Earth Planet. Sci.* **9**, 251–284.

Slansky, M. (1986). "Geology of Sedimentary Phosphates." Elsevier, New York.

Smith, S. V., Kimmerer, W. J., and Walsh, T. W. (1986). Vertical flux and biogeochemical turnover regulate nutrient limitation of net organic production in the North Pacific Gyre. *Limnol. Oceanogr.* **31**, 161–167.

Stumm, W. (1973). The acceleration of the hydrogeochemical cycling of phosphorus. *Water Res.* **7**, 131–144.

Stumm, W. and Morgan, J. J. (1981). "Aquatic Chemistry." Wiley-Interscience, New York.

Sundquist, E. T. (1993). The global carbon dioxide budget. *Science* **259**, 934–941.

Taylor, S. R. (1964). Abundance of chemical elements in the continental crust: A new table. *Geochim. Cosmochim. Acta* **28**, 1273–1285.

Vaccaro, R. F. (1965). Inorganic nitrogen in seawater. *In* "Chemical Oceanography," Vol. 1 (J. P. Riley and G. Skirrow, eds), pp. 365–408. Academic Press, New York.

Van Cappellen, P. and Ingall, E. D. (1994). Benthic phosphorus regeneration, net primary production, and ocean anoxia: A model of the coupled marine biogeochemical cycles of carbon and phosphorus. *Paleoceanography* **9**, 677–692.

Van Cappellen, P. and Ingall, E. D. (1996). Redox stabilization of the atmosphere and oceans by phosphorus-limited marine productivity. *Science* **271**, 493–496.

Van Wazer, F. (ed.) (1961). "Phosphorus and Its Compounds," Vol. 2. Interscience Publishers, New York.

Waser, N. A. D., Bacon, M. P., and Michaels, A. F. (1996). Natural activities of ^{32}P and ^{33}P and the $^{32}P/^{33}P$ ratio in suspended particulate matter and plankton in the Sargasso Sea. *Deep-Sea Res.* **43**, 421–436.

15

Trace Metals

Mark M. Benjamin and Bruce D. Honeyman

15.1 Introduction

Industrialized society is built upon the use of metals. Unlike many of the synthetic organic compounds used in industry, medicine, and agriculture, metals are part of natural biogeochemical cycles. Human activity influences their cycling in two interrelated ways: by altering the rate at which metals are transported among different reservoirs and by altering the form of the metals from that in which they were originally deposited.

Metals and other elements of economic interest are deposited when geochemical conditions reduce their mobility. Deposits range in quality from nearly pure elements, such as native copper, to highly disseminated deposits of marginal economic value. In addition, there are natural background levels of nearly all of the elements in what constitutes the average crustal rocks – the shales, sandstones, and igneous and metamorphic rocks that make up the continents and ocean floors. In the absence of human activities, elements are released to terrestrial and aquatic environments at rates corresponding to natural chemical and mechanical erosion times. Mining, construction, and large-scale changes to the natural environment alter the rate of release of elements to that part of the biogeochemical environment we call the ecosphere. These alterations, in turn, have a cascading effect on the rate at which metals are exchanged among various reservoirs in the ecosphere. In this way, the release of metals from, for example, the crustal reservoir, can affect biota in aquatic systems far from the original deposit site.

One of the characteristics of the cycle of metal mobilization and deposition is that the form of the metal is changed. This change in speciation of a metal has a profound effect on its fate. The link between metal speciation and fate is the central theme of this chapter.

This chapter differs from previous ones in that it describes the cycling of several chemical elements. These elements have many similar properties and can be considered as a group, but each also has properties that make it unique. One of the most important properties that distinguishes them from other elements is their tendency to bond reversibly with a very large number of compounds. The availability and nature of such compounds in a system can control the transport and fate of metals. In view of the importance of these reactions, this chapter starts with a broad overview of global metal cycling, which is then interpreted in the context of a discussion of the nature of metals and their chemical reactions. Finally, these concepts are applied in some detail to the natural and perturbed biogeochemical cycling of two specific metals.

15.2 Metals and Geochemistry

15.2.1 Metal Abundance and Availability

The average composition of the Earth's crust is essentially the composition of igneous rocks,

Earth System Science
ISBN 0-12-379370-X

since metamorphic and sedimentary rocks constitute a relatively insignificant portion of the total crustal mass. Eight elements – O, Si, Al, Fe, Ca, Na, K, and Mg – make up nearly 99% of the total elemental mass; the remaining elements are differentiated throughout the crust according to their particular chemical properties. Geochemical differentiation based upon chemical affinities was first described by Goldschmidt (1954) in his proposal for a general geochemical classification scheme. In his framework, elements are considered to be siderophiles, chalcophiles, lithophiles, or atmophiles depending on their relative affinities for minerals containing iron, sulfide, or silicate, or for the atmosphere, respectively. Such a classification scheme was a significant advance in our understanding of the distribution of elements, making it possible to relate the general geochemical character of elements to their position in the Periodic Table and to fundamental chemical properties such as electronegativity and ion size.

With regard to geochemical cycling (as well as for economic considerations) it is important to distinguish between the abundance of an element and its availability. The availability of an element is related not only to its relative abundance on Earth but also to the stability of minerals of which it is a major constituent. Thus, a number of elements (e.g., copper, mercury, tin, and arsenic) which are scarce in terms of their average crustal abundance are easily isolated due to their ability to form mineral deposits. The most unavailable elements are those that form no major minerals of their own. Many of the rarer elements are available for economic use only to the extent that they are obtained as by-products of the extraction of more abundant elements. Tellurium, for example, is produced during the electrolytic refining of copper.

15.2.2 Metal Mobilization

The availability of a metal describes one aspect of its potential to cycle among biogeochemical reservoirs. The initiation of the cycling process is called mobilization. Metals may be mobilized, that is, made available for transport away from their region of deposition, when the geochemi-

cal character of the depositional environment changes. These changes may be due to either natural or anthropogenic causes.

Natural mobilization includes chemical, mechanical, and biological weathering and volcanic activity. In chemical weathering, the elements are altered to forms that are more easily transported. For example, when basic rocks are neutralized by acidic fluids (such as rainwater acidified by absorption of CO_2), the minerals contained in the rocks can dissolve, releasing metals to aqueous solution. Several examples are listed below of chemical reactions that involve atmospheric gases and that lead to the mobilization of metals:

$$Al(OH)_3 + 3H_2CO_3 \rightarrow Al^{3+} + 3HCO_3^- + 3H_2O$$

$$Fe(OH)_3 + 3H_2CO_3 \rightarrow Fe^{3+} + 3HCO_3^- + 3H_2O$$

$$ZnS + 2O_2 \rightarrow Zn^{2+} + SO_4^{2-}$$

$$PbCO_3 + H_2CO_3 \rightarrow Pb^{2+} + 2HCO_3^{2-}$$

Biological and volcanic activities also have roles in the natural mobilization of elements. Plants can play multiple roles in this process. Root growth breaks down rocks mechanically to expose new surfaces to chemical weathering, while chemical interactions between plants and the soil solution affect solution pH and the concentration of salts, in turn affecting the solution–mineral interactions. Plants also aid in decreasing the rate of mechanical erosion by increasing land stability. These factors are discussed more fully in Chapters 6 and 7.

Volcanic activity has a significant effect on the mobilization of metals, particularly the more volatile ones, e.g., Pb, Cd, As, and Hg. Effects of volcanism are qualitatively different from those of the weathering and other near-surface mobilization processes mentioned above, in that volcanism transports materials from much deeper in the crust and may inject elements into the atmospheric reservoir.

15.2.3 Human Activities as Geochemical Processes

That humans have significantly altered the biogeochemical cycles of many metals is no longer an arguable point. What is uncertain is the magnitude of the effects from these alterations, particularly in the long term. Given that the flux of energy and materials through the biosphere is self-regulating, at least within certain limits, the issue becomes one of evaluating the ability of the biosphere to assimilate anthropogenic metal inputs and predicting the rate and types of changes that will occur as a new steady-state condition is approached. These sorts of predictions are hampered not only by the state of our understanding of the complex interactions that occur between metals and the environment, but also because the determination of background metal concentrations in uncontaminated environments is so difficult. In view of the worldwide dissemination of anthropogenic materials that has already occurred, locating an uncontaminated site is extremely difficult. For example, concentrations of mercury, selenium, and sulfur above background levels have been found in arctic ice sheets. These elemental "signals" correlate with the beginning of worldwide industrialization (Weiss *et al.*, 1971a, b). In addition, there are problems associated with preparing and analyzing samples containing extremely low concentrations of metals without contaminating them.

Despite the difficulties, there have been many efforts in recent years to evaluate trace metal concentrations in natural systems and to compare trace metal release and transport rates from natural and anthropogenic sources. There is no single parameter that can summarize such comparisons. Frequently, a comparison is made between the composition of atmospheric particles and that of average crustal material to indicate whether certain elements are enriched in the atmospheric particulates. If so, some explanation is sought for the enrichment. Usually, the contribution of seaspray to the enrichment is estimated, and any enrichment unaccounted for is attributed to other natural inputs (volcanoes, low-temperature volatilization processes, etc.) or anthropogenic sources.

A second approach is to compare total mining production of a metal to an estimate of its total natural flux, making the implicit assumption that all mined materials will be released to the environment in the near future (a reasonable assumption when comparing with geologic processes).

Finally, some authors have computed metal loading to the environment from specific human activities, such as discharges of wastewater, and compared these with natural release rates. While the details of the computations and conclusions vary, the general observation for many metals is that anthropogenic contributions to metal ion transport rates and environmental burdens are approaching and in many cases have already exceeded natural contributions. A few such comparisons are provided in Tables 15-1–15-4.

The amount of metals released as the by-product of a single activity, the burning of coal, illustrates the potential importance of anthropogenic sources. Inorganic, non-combustible materials present in coal constitute the ash that remains after combustion. Fly ash (the ash that leaves the furnace and is collected by flue-scrubbing devices) is particularly enriched in metals, and in 1975 approximately 33 million tonnes of fly ash were produced in the United States (Theis and Wirth, 1977). This represents an average of 270 000 tonnes of fly ash per year for a typical 1000 MW power plant. A trace element with a concentration of 1 mg/kg (ppm) in the fly ash would be produced as waste at an average rate of 270 kg per 1000 MW plant per year. The composition of a few fly ash samples is shown in Table 15-5. As, Cd, Co, Cr, Cu, Hg, Pb, Se, V and Zn are present in concentrations ranging from 1 to 1000 ppm(mass). Furthermore, several trace metals including As, Pb, Cd, Se, Cr, and Zn are concentrated on the surfaces of the smallest particles, which have the greatest likelihood to escape the plant and be transported significant distances in the atmosphere. When these ashes are ponded or exposed to rain, the pH of the water can decrease to less than 4 or increase to greater than 13, depending on the chemistry of the fly-ash matrix. The fraction of the total metal solubilized under these conditions is a sensitive function of pH and, although this fraction is usually 10% or less, it can exceed 50%.

Once the anthropogenic release rates of metals are established, the next critical step is to evaluate their fate upon discharge to receiving

Table 15-1   Natural and anthropogenic sources of atmospheric emissions[a]

Element	Natural rate (10^2 tonnes/yr)	Anthropogenic rate (10^2 tonnes/yr)	Anthropogenic/natural ratio
Al	48 900	7 200	0.15
Ti	3 500	520	0.15
Sm	4.1	1.2	0.29
Fe	27 800	10 700	0.39
Mn	605	316	0.52
Co	7	4.4	0.63
Cr	58	94	1.6
V	65	210	3.2
Ni	28	98	3.5
Sn	5.2	43	8.2
Cu	19	263	13.6
Cd	0.3	5.5	19.0
Zn	36	840	23.5
As	2.8	78	27.9
Se	0.4	14	33.9
Sb	1	38	38.0
Mo	1.1	51	44.7
Ag	0.06	5	83.3
Hg	0.04	11	27.5
Pb	5.9	2 030	34.6

[a] Lantzy and Mackenzie (1979).

Table 15-2 Comparison between artificial and natural rates (10^3 tonnes/yr) of global metal injection into the oceans and atmosphere[a]

	Input from industrial world's municipal waste water	Input from combustion[b]	Natural weathering[c]
Cd	3	—	36
Cr	55	1.5	50
Cu	42	2.1	250
Fe	440	1400	24 000
Pb	15	3.6	110
Mn	7.4	7.0	250
Ni	17	3.7	11
Ag	2.3	0.07	11
Zn	100	7	720

[a] Galloway (1979).
[b] Bertine and Goldberg (1971).
[c] Turekian (1971).

Table 15-3 Calculated present-day fluxes ($\mu g/(cm^2 yr)$) of heavy metals into the sediments of Lake Erie[a]

Element	Anthropogenic	Natural
Cd		
Stn 1	0.36	0.16
Stn 3	0.02	0.02
Stn 7	0.54	0.09
Cu		
Stn 1	12.0	7.8
Stn 3	0.15	0.47
Stn 7	8.8	4.8
Pb		
Stn 1	11.8	4.3
Stn 3	0.33	0.44
Stn 7	10.9	2.4
Zn		
Stn 1	36.2	14.8
Stn 3	1.0	0.68
Stn 7	30.6	5.9

[a] Nriagu (1979).

waters. Sediment analyses are often useful in this regard because changes in metal concentration as a function of depth in the sediment can indicate historical trends. Also, the concentrations of metals are typically much greater in sediments than in the water column and are therefore easier to analyze and evaluate. Bruland *et al.* (1974) used metal/aluminum ratios in sediment to determine the magnitude of the anthropogenic component of the heavy metal transport rate to a Southern California basin. Assuming that aluminum has had a uniform rate of transport to the sediments over the past century from crustal rock sources, they concluded that Pb, Cr, Cd, Zn, Cu, Ag, V, and Mo are now accumulating at higher rates than a century or more ago. For all of these metals, the anthropogenic component represented at least one-third of the natural emission rate, and for Pb, Ag, and Mo, the anthropogenic rate exceeded the natural rate. A qualitatively similar conclusion applies to trace metal transport into the northern portion of Chesapeake Bay (Helz, 1976).

In a subsequent section, trace metal movement through some other aquatic systems will be reviewed and analyzed in the context of specific chemical reactions. First, though, the most important chemical reactions of metal ions are reviewed.

15.3 An Overview of Metal Ion Chemistry

15.3.1 Introduction

Metals, like many of the elements discussed in previous chapters, can exist in nature in several different oxidation states. When bonded to other

Table 15-4 Inventory of sources and sinks of heavy metals in Lake Erie[a]

Source	Flux (tonnes/yr)			
	Cadmium	Copper	Lead	Zinc
Detroit River (import from Upper Lakes)	—	1640	630	5220
Tributaries, USA	—	100	52	271
Tributaries, Ontario	—	31	19	140
Sewage discharges	5.5	448	283	759
Dredged spoils	4.2	42	56	175
Atmospheric inputs	39	206	645	903
Shoreline erosion	7.9	190	221	308
Total, all sources	—	2477	1906	7776
Export, Niagara River and Welland Canal	—	1320	660	4440
Retained in sediments	—	1157	1246	3376

[a] Nriagu (1979).

Table 15-5 Comparison of elemental concentrations in size-classified fly-ash fraction[a]

Element	Concentration ($\mu g/g$)			
	Fraction 1 (18.5^{b} μm)	Fraction 2 (6.0 μm)	Fraction 3 (3.7 μm)	Fraction 4 (2.4 μm)
Cr	28	53	64	68
Ni	25	37	43	40
Zn	68	189	301	590
Cu	56	89	107	137
Cd	0.4	1.6	2.8	4.6
Pb	73	169	226	278
As	13.7	56	87	132
Se	19	59	78	198
V	86	178	244	327

[a] Coles *et al.* (1979).
[b] Mass median diameters determined by centrifugal sedimentation.

elements, metal ions are almost always assigned a positive oxidation number and are somewhat electrophilic. Because of this they are stabilized by association with electron-rich atoms. In particular, atoms that have a free electron pair can "donate" some of their electron density to the metal to form a bond. Such atoms are Lewis bases. The most common and environmentally important donor atoms are oxygen, nitrogen, and sulfur. The bonds they form with metal ions range in strength from relatively weak associations such as those between a dissolved metal ion and water to very strong covalent bonds. These types of bonds are significant in both aqueous phase reactions and in the formation of insoluble compounds.

The characteristic affinity of metals for electron-rich donor atoms leads to an important distinction between the geochemical behavior of metal ions and that of the elements discussed earlier in this text. Specifically, metal ions in a single oxidation state can bind to donor atoms that are part of a variety of cationic, anionic, or neutral molecules. Thus, the metals can be found in and can move through the environment as parts of molecules spanning the complete range of charge, molecular weight, bioavailability, and other chemical characteristics. In addition, since metals constitute an extremely small fraction of biological organisms,

the connection between metal transport through the environment and biological activity is generally less direct and of less quantitative importance than for, say, oxygen or phosphorus. (However, recent evidence is pointing to a more significant role for biota in controlling metal concentrations and transport in the ocean.) In a relative sense, then, purely chemical reactions are of more importance for metals than for the elements discussed earlier. The strength of the chemical bonding between the donor atoms O, N, and S and metals is the overriding factor controlling the geochemical cycling of metals. Once these reactions are understood, the behavior of metals, with respect to both transport and biological impacts, can be placed in a more logical framework.

15.3.2 *Oxidation–Reduction Reactions*

Metals exist in nature primarily in positive oxidation states, and many form stable compounds in more than one oxidation state. The formal oxidation number of the most common form can range from +1 to +6. The stable form in a given environment depends on the oxidation potential and chemical composition of that environment. Often the stable form at the Earth's surface in the presence of molecular oxygen is different from that which is stable in anoxic sediments or waters.

Thermodynamically, virtually all metals in the elemental form are unstable with respect to redox reactions in environments where they are exposed to air and water, i.e., virtually all environments where they are used. Those metals least likely to oxidize (corrode) were long ago given the distinguished title "noble metals." Efforts to prevent metals from corroding, and the cost of repairing and replacing metal structures that have done so, runs into the billions of dollars annually. Thus, one characteristic feature of the society's use of metals is that the metals are continuously, albeit slowly, "degrading" to a less useful form from the moment they are put into use.

The different oxidation states of a metal can have dramatically different chemical properties, which in turn affect their biogeochemical forms and significance. For example, almost $4\,g/L$ ferrous iron, Fe(II), can dissolve in distilled water maintained at pH 7.0. However, if the water is exposed to air and the iron is oxidized to Fe(III) essentially all the iron will precipitate, reducing the soluble Fe concentration by more than eight orders of magnitude. Oxidation state can also affect a metal ion's toxicity. For instance, the toxicity of As(III) results from its ability to inactivate enzymes, while As(V) interferes with ATP synthesis. The former is considerably more toxic to both aquatic organisms and humans.

The thermodynamically stable oxidation state of a metal in a given environment is a function of the prevailing oxidation potential. The value of the potential is given by the Nernst equation, which is described in Chapter 5 for the generic reduction half-cell:

$$\text{Reactants} + ne^- \rightarrow \text{Products}$$

The electrochemical potential that would cause the reactants and products to be equilibrated with each other is

$$E = E^0 - \left(\frac{2.303RT}{nF}\right) \log \frac{[\text{Products}]}{[\text{Reactants}]}$$

If the equilibrium half-cell potentials for two redox reactions are different, electrons will be transferred from the reduced species in the reaction with the less positive potential to the oxidized form with the more positive potential. The process is repeated until all exchangeable electrons have the same equilibrium potential. Water chemistry texts describe rapid graphical or computerized approaches to solve for the concentrations of all species once this equilibrium condition has been attained.

While these calculations provide information about the ultimate equilibrium conditions, redox reactions are often slow on human time scales, and sometimes even on geological time scales. Furthermore, the reactions in natural systems are complex and may be catalyzed or inhibited by the solids or trace constituents present. There is a dearth of information on the kinetics of redox reactions in such systems, but it is clear that many chemical species commonly found in environmental samples would not be present if equilibrium were attained. Furthermore, the conditions at equilibrium depend on the concentration of other species in the system, many of which are difficult or impossible to determine analytically. Morgan and Stone (1985) reviewed the kinetics of many environmentally important reactions and pointed out that determination of whether an equilibrium model is appropriate in a given situation depends on the relative time constants of the chemical reactions of interest and the physical processes governing the movement of material through the system. This point is discussed in some detail in Section 15.3.8. In the absence of detailed information with which to evaluate these time constants, chemical analysis for metals in each of their oxidation states, rather than equilibrium calculations, must be conducted to evaluate the current state of a system and the biological or geochemical importance of the metals it contains.

To summarize, an evaluation of the oxidation state of metals in an environment is central to determining their probable fate and biological significance. Redox reactions can lead to orders of magnitude changes in the concentration of metals in various phases, and hence in their mode and rate of transport. While equilibrium calculations are a valuable tool for understanding the direction in which changes are likely to occur, field measurements of the concentrations of metals in their various oxidation states are always needed to evaluate metal speciation, since equilibrium has often not yet been attained.

15.3.3   Volatilization

15.3.3.1   Volatilization from the solid state

The extent to which any chemical species is volatilized is governed by its vapor pressure, which is sensitive to temperature. Most metals and their compounds have very low vapor pressures at normal temperatures, low enough that their tendency to vaporize can be ignored. The major exceptions are metallic Hg and organometallic compounds. Nevertheless, in some environments significant quantities of metals can be volatilized either as elements or inorganic compounds such as oxides or carbonates. The most obvious such environments are high-temperature furnaces such as in smelters or fossil-fuel-burning power plants and in regions of geothermal activity or vulcanism. While the oxides, sulfates, carbonates, and sulfides of a metal all have somewhat different volatilities, the most volatile metals, regardless of the anion with which they are associated, are Hg, As, Cd, Pb, and Zn. Metallic Hg and organometallic compounds may be transported significant distances as gases and eventually be removed from the atmosphere by dissolution in rain droplets. However, most volatilized metals condense rapidly as they cool and fall to the surface associated with particulate matter, either as dry deposition or scavenged by precipitation. Some of the condensed particles are light enough to carry long distances, and those that fall to Earth may be resuspended by wind action or washed into a water body by surface runoff. Particles produced by high–temperature combustion processes are mostly in the $<2 \mu m$ size range and typically have atmospheric residence times of 7 to 14 days, while those generated by soil erosion are larger ($>5 \mu m$) (Hardy *et al.*, 1985). Anthropogenic particles are typically enriched in trace metals (normalized to the concentration of Al) by a factor of 100 to 10 000 compared with atmospheric particulates gener-

ated by natural erosion and wind action. As an indication of the importance of volatilization of metals to their overall biogeochemical budgets, Galloway (1979) estimated that volatilization of As, Hg, and Se overwhelms total dust and volcanic emanation rates by factors of 7.5, 625, and 7.3, respectively.

15.3.3.2   Volatilization from solution

The equilibrium volatility of a species dissolved in water is characterized by its equilibrium constant for the reaction $X(g) \leftrightarrow X(aq)$, i.e., its Henry's Law constant. These equilibrium constants are related, but are not directly comparable to vapor pressures of the corresponding dry solids because of the effect of the solvent, water. (The most appropriate direct comparison involves the calculation of fugacities. The fugacity, or escaping tendency of a chemical species from the environment in which it exists, depends upon both the concentration of the chemical species of interest and the strength of its interactions with the surrounding (solvent) molecules. Details of the calculation are provided in chemical thermodynamics textbooks and in a number of articles by Mackay and co-workers (e.g. Mackay, 1979).) Suffice it to say that when volatile metal species, such as methylmercury, are present in water, there virtually always exists a driving force to strip them out of the aqueous phase, since their partial pressures in air are essentially zero. Thus, equilibrium between gaseous and aqueous phases is rarely a limiting factor. Rather, the factor limiting volatilization of these species is usually the rate at which they move through the water column and across the water–air interface.

15.3.3.3   Volatilization of mercury and lead

The ratio of anthropogenic emissions to total natural emissions is highest for the atmophilic elements Sn, Cu, Cd, Zn, As, Se, Mo, Hg, and Pb (Lantzy and Mackenzie, 1979). In the case of lead, atmospheric concentrations are primarily the consequence of the combustion of leaded gasoline. For many years, lead was used as a gasoline additive, in the form of an organometal compound, tetraethyl lead. When the fuel was burned, most of the lead was converted to inorganic forms and released in the exhaust. The widespread use of this compound as an anti-knock agent in gasoline engines led to its dispersion everywhere automobiles travel, and from there to very remote locations. As with metals near smelters, lead concentrations in the soil near major roadways decrease rapidly with distance from the source, but significant amounts of the metal are transported by wind, either directly after being emitted or by resuspension after a period of deposition.

Atmospheric fluxes of lead in the United States rose steadily from the first decades of this century, reaching a maximum in the early 1970s (see Eisenrich *et al.*, 1986 and references therein). Passage of the Clean Air Act of 1972 and its subsequent amendments resulted in dramatic reductions in atmospheric lead concentrations, although lead fluxes worldwide still remain 10–1000 times background levels (Settle *et al.*, 1982; Settle and Patterson, 1982).

Volatilization is also a dominant transport mode for mercury, which is the most volatile metal in its elemental state. As with lead, a key reaction that can increase the volatility of mercury is formation of an organometallic compound. In this case, the reactions take place in water and are primarily biological, being mediated by bacteria commonly found in the upper levels of sediments. These reactions and their importance in the global mercury cycle are discussed in some detail later in the chapter.

To summarize, metals can be transferred into the gas phase in high-temperature processes either in their elemental form or as inorganic compounds, and these compounds can then be transported long distances as gases or in other physical–chemical forms. Many organometallic compounds are also volatile. In a few cases, natural organometallic compounds may be formed that are volatile. These compounds are formed by microorganisms in mildly reducing aquatic environments and are then transported to the surface and across the air–water interface to enter the gas phase. Methylmercury compounds are probably the most important and certainly have been the most widely studied of these because of their central role in the bioaccumulation of mercury; others, such as methylar-

sine and organotin compounds, are environmentally important as well. Anthropogenic release to the atmosphere overwhelms natural sources for Hg, Cd, Cu, Ag, Zn, Pb, As, and Se, and possibly other metals.

15.3.4   *Complexation Reactions*

The stability of liquid water is due in large part to the ability of water molecules to form hydrogen bonds with one another. Such bonds tend to stabilize the molecules in a pattern where the hydrogens of one water molecule are adjacent to oxygens of other water molecules. When chemical species dissolve, they must insert themselves into this matrix, and in the process break some of the bonds that exist between the water molecules. If a substance can form strong bonds with water, its dissolution will be thermodynamically favored, i.e., it will be highly soluble. Similarly, dissolution of a molecule that breaks water-to-water bonds and replaces these with weaker water-to-solute bonds will be energetically unfavorable, i.e., it will be relatively insoluble. These principles are presented schematically in Fig. 15-1.

Cationic metal ions form strong bonds with water molecules. By orienting themselves in such a way that the metal "faces" an oxygen atom, the negative charge on the oxygen is partially distributed onto the metal, forming the analog of a strong hydrogen bond. Similarly, some of the charge on the metal ion is neutralized. (Recall the electrophilic nature of metal ions.) These bonds are strong enough that, in most aqueous environments, most metal ions are surrounded by an "inner hydration sphere" of four to eight strongly bound water molecules, as well as a loosely attached "outer hydration sphere" of variable size. Although chemical convention represents dissolved metal ions as Me^{n+}, a more accurate designation is $Me(H_2O)_x^{n+}$. The strength of the metal-to-water bond increases with decreasing size and increasing charge of the metal ion, and also depends on the distribution of electrons around it.

The water molecules in the inner hydration sphere can undergo dissociation reactions just as water molecules far from a dissolved metal ion

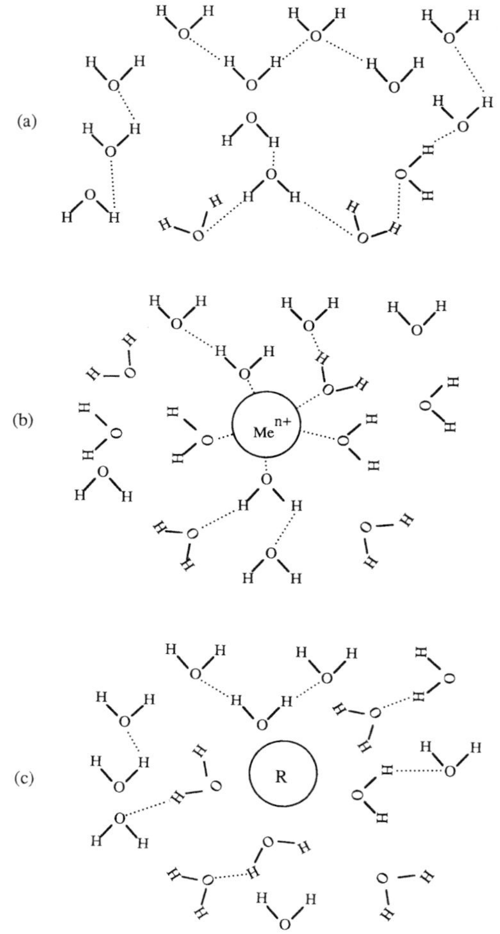

Fig. 15-1 Schematic representation of the change in water structure (water molecule orientation) due to the presence of a charged (hydrophilic) solute. (a) Pure water. (b) A solute forming strong bonds with water (dissolution favorable). (c) a solute forming weak bonds with water (dissolution unfavorable).

do, but the presence of the metal ion changes the equilibrium constant for this reaction. The magnitude of this alteration is directly related to the strength of the metal-to-water bond. If the metal attracts a large portion of the electron density from the oxygen, it weakens the oxygen-to-hydrogen bonds and enhances the tendency for the water molecule to dissociate. This leads to dissolved species that are typically designated hydrolyzed metal ions, $Me(OH)_y^{n-y}$, but which are more accurately portrayed as $Me(H_2O)_x(OH)_y^{n-y}$. As an extreme example, we

can think of a metal like molybdenum bonding so strongly to four oxygen atoms that they lose their ability to bond to any hydrogen at all, forming MoO_4^{3-}. In such a situation, the oxygens are not considered part of the hydration sphere at all, but as covalently bonded to the molybdenum. Nevertheless, in a qualitative sense this reaction is no different from the hydrolysis reactions described above. Comparison of a metal hydrolysis reaction with hydrolysis of pure water indicates that if the equilibrium constant for metal hydrolysis is larger than 10^{-14} (the dissociation constant of pure water) the metal acts as an acid, i.e., it has a greater tendency to release hydrogen ions to solution than pure water (Fig. 15-2). Furthermore, since metal ions are surrounded by several waters of hydration, they can act as multi-protic acids, releasing four, five, or even six hydrogen ions to solution and acquiring a net negative charge in the process.

The comparisons between pure water and a solution containing dissolved metal ions can be extended by considering the behavior of other dissolved ions such as Cl^-, SO_4^{2-}, HCO_3, S^{2-}, and dissolved organic molecules. Just as metal ions do, these ions tend to orient themselves in a way that maximizes the bond strength between them and other constituents of the solution. For instance, the water molecules surrounding a chloride ion are somewhat structured by the negative charge of the chloride, so that the positive (hydrogen) ends of the water molecule are closer to the ion than is the oxygen. Typically, this tendency to attract and orient a hydration sphere is much weaker for anions than for metals. If a metal ion is dissolved in the same solution as the chloride, then an additional possibility presents itself. The chloride ion could replace one of the waters of hydration surrounding the metal, thereby exchanging a water–metal bond and a water–chloride bond for a metal–chloride bond and a water–water bond (Fig. 15-2c). Equilibrium constants for such reactions are called stability constants, and are tabulated in a number of sources. Values for these constants can range considerably. Very large values imply that metals and ligands are virtually certain to form complexes, even if both ions are present in very low concentrations.

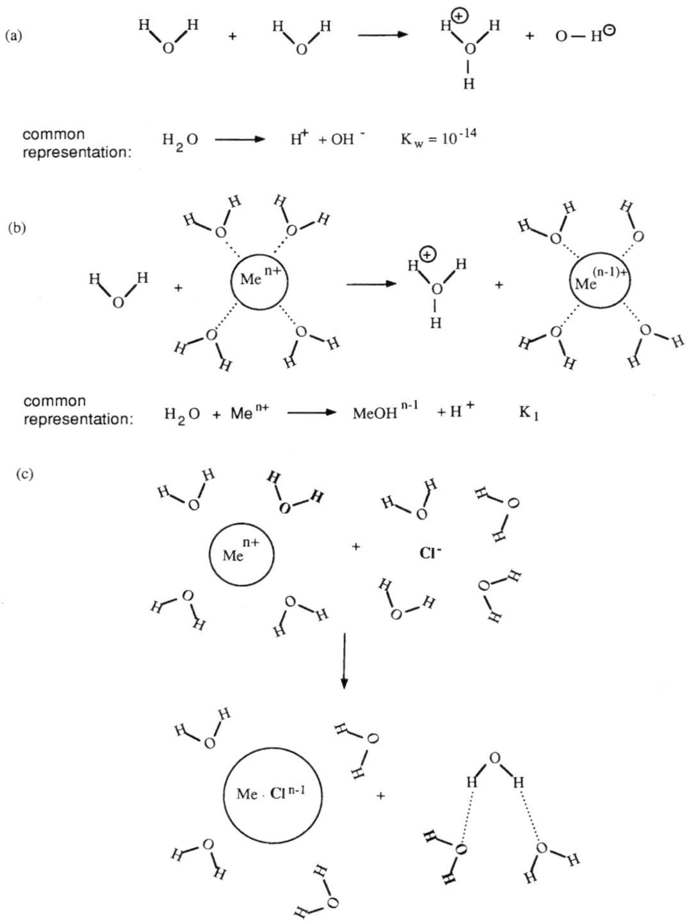

Fig. 15-2 Comparison of water dissociation in bulk solution (a) and in the hydration sphere of a metal ion (b). Exchange of water of hydration for a chloride ion (c) forms the Me–Cl complex (from Manahan, 1979).

The concept of exchange of water molecules in the hydration sphere of a metal ion for other dissolved species can be extended to include "mixed ligand complexes," i.e., those in which water molecules have been replaced by two or more different types of ligands, and "multi-dentate" complexes, those in which a single molecule can bind to the metal through more than one atom, and therefore cause more than one water of hydration to be released for each ligand molecule binding (Fig. 15-3). As might be expected, the binding strength typically decreases as ligands are added, and the bond strength of a multi-dentate ligand is typically greater than that of mono-dentate ligands, as a result of the multiple bonds formed. When a

ligand forms a strong multi-dentate complex it is referred to as a chelating agent, and the complex is called a chelate.

It is important to recognize that although many ligands are anions and hence have an electrostatic attraction for the metal ion, this is not a requirement. For instance, the neutral ammonia molecule (NH_3) is a strong complexing agent for many metals. Furthermore, if electrostatic attraction were critical to formation of these types of bonds, then complexation would cease once the positive charge on the metal was neutralized. Complexation can continue until several of the molecules in the hydration sphere are replaced, yielding a chemical species consisting of a positively charged central

(a)

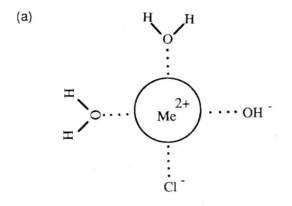

(b)

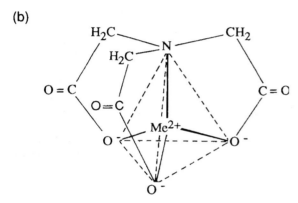

Fig. 15-3 Mixed ligand and multi-dentate complexes: (a) a hypothetical Me(OH)Cl^0 complex; (b) nitrilotriacetate chelate of a divalent metal ion in a tetrahedral configuration.

metal ion surrounded by as many as six negatively charged ligands, and carrying a net negative charge of -3 or -4.

Complexation reactions are of crucial importance in biogeochemical cycling of metals because a large fraction of the total dissolved metal may be complexed. Biological effects can be sensitive to the types of complexes present. In many cases free metal ions (fully hydrated metal ions) are more toxic than complexed ones, especially those complexed by multi-dentate organic ligands. There is also evidence that some organisms secrete chelating agents in response to high metal concentrations in their environment, presumably as a defense against metal ion toxicity.

On the other hand, organisms can also produce chelating agents to acquire metals that are necessary for certain metabolic functions. These chelating agents are often extremely specific for a given metal and are used to "collect" metals from solution or maintain a desired concentration of metals inside the cell.

In addition to those complexing agents produced by organisms to regulate their environment, numerous complexing and chelating agents are present in aquatic systems as a result of the normal exchange of metabolites between organisms and their environment, cell death and decay or as a result of anthropogenic activities. The high-molecular-weight compounds known as fulvic and humic acids are particularly important. These acidic polymers are primarily the residue from organic decay processes and do not have a well-defined chemical structure. However, like synthetic chelating agents, they contain high concentrations of electron-rich donor atoms, especially oxygen in carboxylic and phenolic groups, which can complex or chelate metal ions. Sulfur- and nitrogen-containing portions of the molecules may also be important in complexing such metals. These molecules are especially effective chelators since they are large and flexible, and can have several complexing sites on each molecule. At least in theory, they may be able to arrange themselves in such a way that several donor groups are at the optimal positions to form very strong bonds.

The speciation (ignoring organic complexes) of selected metals in various aquatic environments is described in Table 15-6. The relative strengths of the complexes of a given metal, and differences among metals in the strength of binding to a given ligand, are apparent. Also apparent is the fact that the dominant form of a given metal often changes when the local environment changes pH or ionic strength (as in an estuary). Important changes also occur when the environment changes from aerobic to anaerobic (for instance in the bottom of a seasonally anoxic lake), or when the concentration of dissolved organic matter changes, such as near a sewage outfall.

Summarizing this section, metal ions may be present in aqueous solution surrounded by and

Table 15-6 Model results for metal speciation in natural waters[a,b]

| | Freshwater | | | Seawater | |
| | Inorganic | | Inorganic and organic pH 7 | Inorganic pH 8.2 | Inorganic and organic pH 8.2 |
	pH 6	pH 9			
Ag^+	72, Cl	65, Cl, CO_3	65, Cl	<1, Cl	<1, Cl
Al^{3+}	<1, OH, F	<1, OH		<1, OH	
Cd^{2+}	96, Cl, SO_4	47, CO_3, OH	87, org, SO_4	3, Cl	1, Cl
Co^{2+}	98, SO_4	20, CO_3, OH		58, Cl, CO_3, SO_4	63, Cl, SO_4
Cr^{3+}	<1, OH	<1, OH		<1, OH	
Cu^{2+}	93, CO_3, SO_4	<1, CO_3, OH	<1, org	9, CO_3, OH, Cl	<1, org, CO_3
Fe^{2+}	99	27, CO_3, OH		69, Cl, CO_3, SO_4	
Fe^{3+}	<1, OH	<1, OH	<1, org, OH	<1, OH	<1, OH, org
Hg^{2+}	<1, Cl, OH	<1, OH^-		<1, Cl	
Mn^{2+}	98, SO_4	62, CO_3	91, SO_4	58, Cl, SO_4	25, Cl, SO_4
Ni^{2+}	98, SO_4	9, CO_3		47, Cl, CO_3, SO_4	50, org, Cl, SO_4
Pb^{2+}	86, CO_3, SO_4	<1, CO_3, OH	9, CO_3, org	3, Cl, CO_3, OH	2, CO_3, OH
Zn^{2+}	98, SO_4	6, OH, CO_3	95, SO_4, org	46, Cl, OH, SO_4	25, OH, Cl, org

[a] Data for inorganic freshwater and inorganic seawater from Turner *et al.* (1981). Data for systems with inorganics and organics from Stumm and Morgan (1981). Six organic ligands included corresponding to 2.3 mg/L total soluble organic carbon. Stability constants and inorganic composition of model water were not identical in the two studies, so comparisons are qualitatively valid but may have minor quantitative inconsistencies.

[b] Each entry has the % of total metal present as the free hydrated ion, then the ligands forming complexes, in decreasing order of expected concentration. For instance, in inorganic freshwater at pH 9, Ag is present as the free aquo ion (65%), chloro-complexes (25%), and carbonato-complexes (9%).

bonded to water molecules or a wide variety of inorganic and organic complexing agents. The bonds formed by complexation are often reversible, and their strength varies widely depending on the chemical and physical properties of the metal and ligand. As a result of complexation reactions, concentrations of total dissolved metal in an aquatic system can be orders of magnitude higher than those of free metal ions. Thus, if the concentration of free metal ion in a system is limited by virtue of low solubility, complexes can still cause large amounts of that metal to dissolve and be transported among the various geochemical reservoirs.

15.3.5 Precipitation Reactions

If metal concentrations in solution become large, soluble complexes that contain more than one metal ion can form. Eventually these can grow so large that a three-dimensional network is created in which most of the ions in the interior are not in contact with the bulk solution, and a separate phase is formed. In systems where large soluble polymeric metal complexes are stable, the exact point at which the new phase forms is open to question and is somewhat a matter of definition. In most practical situations, however, these large polymers have very limited stability and the system undergoes a dramatic change from a completely soluble state containing monomeric and relatively small polymeric species to one containing an identifiable second phase. The activation energy required to form this new phase is large, and solutions must be highly supersaturated for precipitation to be initiated in particle-free systems. On the other hand, the activation energy for growth of solids once a solid–liquid boundary has been established is much smaller, so once precipitation has started it can often proceed rapidly. In natural aquatic systems, the problem of high activation energies for formation of new phases is usually circumvented by new solids forming on the surfaces of existing solids with a compatible crystal structure.

Solubility equilibria are described quantitatively by the equilibrium constant for solid dissolution, K_{SP} (the solubility product). Formally, this equilibrium constant should be written as the activity of the products divided by that of the reactants, including the solid. However, since the activity of any pure solid is defined as 1.0, the solid is commonly left out of the equilibrium constant expression. The activity of the solid is important in natural systems where the solids are frequently not pure, but are mixtures. In such a case, the activity of a solid component that forms part of an "ideal" solid solution is defined as its mole fraction in the solid phase. Empirically, it appears that most solid solutions are far from ideal, with the dilute component having an activity considerably greater than its mole fraction. Nevertheless, the point remains that not all solid components found in an aquatic system have unit activity, and thus their solubility will be less than that defined by the solubility constant in its conventional form.

Metal precipitates of biogeochemical significance are primarily oxides, hydroxides, carbonates, and sulfides. Not surprisingly, these are also the ligands with which many metals form strong complexes. However, the correspondence between the strength of bonding in a solid phase and that in a soluble complex is far from exact, because of the effects of the bonds with water in the latter cases. All of these anions have concentrations that are strongly dependent on pH. Because of this, a change in pH is one of the most important driving forces for precipitation or dissolution reactions. The other dominant factor is the oxidation–reduction potential of the solution. Metal sulfides tend to be extremely insoluble, but they can only exist in environments where the redox potential is sufficiently reducing for sulfur to exist in the $-II$ oxidation state (H_2S, HS^- or S^{2-}). Thus, some metals are soluble in the oxidized layers of sediments but are precipitated in lower anoxic layers, where sulfate (SO_4^{2-}) is reduced to sulfide.

Model calculations that illustrate these points have been performed by Morel *et al.* (1975) for the speciation of a sewage–seawater mixture. They concluded that sulfides of Cu, Cd, Pb, Zn, Ag, Hg, and Co would be the predominant form of those metals in the sewage, along with oxidized precipitates $Cr(OH)_3$ and Fe_2O_3, cyanocomplexes of Ni^{2+}, and free Mn^{2+}. However, as dilution and oxidation of the sewage by seawater proceed and the $S(-II)$ concentration decreases, the sulfides dissolve and all the metals except Fe eventually go into solution. Their calculations show that $CoCO_3(s)$, $ZnCO_3(s)$, and $CuO(s)$ may form as intermediates between the initial (sulfidic solid) and final (soluble) endstates.

In addition to effects on the concentration of anions, the redox potential can affect the oxidation state and solubility of the metal ion directly. The most important examples of this are the dissolution of iron and manganese under reducing conditions. The oxidized forms of these elements (Fe(III) and Mn(IV)) form very insoluble oxides and hydroxides, while the reduced forms (Fe(II) and Mn(II)) are orders of magnitude more soluble (in the absence of $S(-II)$). The oxidation or reduction of the metals, which can occur fairly rapidly at oxic–anoxic interfaces, has an important "domino" effect on the distribution of many other metals in the system due to the importance of iron and manganese oxides in adsorption reactions. In an interesting example of this, it has been suggested that arsenate accumulates in the upper, oxidized layers of some sediments by diffusion of As(III), Fe(II), and Mn(II) from the deeper, reduced zones. In the aerobic zone, the cations are oxidized by oxygen, and precipitate. The solids can then oxidize, as As(III) to As(V), which is subsequently immobilized by sorption onto other Fe or Mn oxyhydroxide particles (Takamatsu *et al.*, 1985).

While the solubility constants for various potential solids can indicate which solid is thermodynamically stable under a given set of conditions, reactions involving precipitation or dissolution of a solid are typically more subject to kinetic limitations than are reactions that take place strictly in solution. As a result, a thermodynamically unstable solid (often referred to as "metastable") may form and remain in place for geological time periods. Such is frequently the case when an insoluble but relatively less stable solid-phase precipitates, and the thermodynamically stable phase has a different crystal structure. Often the only way for the more stable solid to form is by dissolution of ions from the first solid and reprecipitation of the second. Since the first solid is quite insoluble itself, dissolution occurs very slowly. Since this causes the solution to be only very slightly supersaturated with respect to the second solid, the driving force for its formation is also small. Thus, for example, goethite (α-FeOOH) is more stable than lepidocrocite (γ-FeOOH), but lepidocrocite forms more rapidly than goethite when iron sulfide is oxidized by oxygen, and lepidocrocite can be found in geologic formations long after the oxidation has taken place.

An important result of the concepts discussed in this section and the preceding one is that precipitation and complexation reactions exert joint control over metal ion solubility and transport. Whereas precipitation can limit the dissolved concentration of a specific species (Me^{n+}), complexation reactions can allow the total dissolved concentration of that metal to be much higher. The balance between these two competing processes, taking into account kinetic and equilibrium effects, often determines how much metal is transported in solution between two sites.

15.3.6   Adsorption

The importance of one other type of reaction that metal ions undergo has been recognized and studied extensively in the past 40 years. This reaction is adsorption, in which metal ions bind to the surface of particulate matter and are thereby transported as part of a solid phase even though they do not form an identifiable precipitate. Conceptually, these reactions can be thought of as hybrids between complexation and precipitation reactions. Most studies of these reactions have used metal oxides or hydroxides as the solid (adsorbent) phase, and the

reactions will be discussed in that context here, although sorption onto other solids may be important as well.

At a solid–solution interface, atoms are fixed in place by their attachment to the solid, but are freer to move and react with solution components than atoms in the solid. Put another way, these atoms are half in solution and half out. To the extent that their bonding requirements are not completely met by the bond to the bulk solid, they are energetically driven to react with solution components. In the case of a ferric oxide, for instance, an oxygen atom may be attached to an iron atom on the solid side, and may bind to a hydrogen ion on the solution side, to form the analog of an FeOH complex. If we treat the FeO$^-$ surface group as a reactant, we can write an equilibrium reaction describing this interaction. Furthermore, as shown in Fig. 15-4a we can postulate that the surface group might react with a second hydrogen ion to form the analog of a hydrated Fe ion. The result is

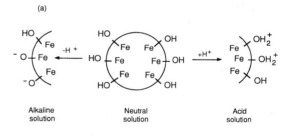

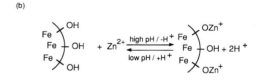

Fig. 15-4 Analogy between dissolved ligands and adsorbents (surface-bound ligands): (a) surface acid–base reactions; (b) surface complexation of free metals; (c) formation of "mixed-ligand" surface complexes.

that each surface Fe atom can behave as a diprotic acid, binding or releasing hydrogen ions in response to changes in solution pH. A second result is that the surface acquires a net charge, which may be positive or negative depending on the pH and the affinity of the surface for hydrogen ions. This charge is neutralized by ions of opposite charge accumulating in solution near the surface. This structure of the oxide–solution interface – a net surface charge balanced by a swarm of ions in the solution near the surface – is called the electrical double layer. This effect is also discussed in Chapter 8.

Consider the situation of a metal ion, say Zn^{2+}, dissolved in a solution containing FeOOH. Just as in the cases described earlier, the zinc ion might be present as a free aquo ion, or as a hydroxo- or other complex. However, if an iron oxide surface is available, another possibility is that the Zn might bind to a surface oxide site. The species that forms in such a case is analogous to a metal–ligand complex, with the surface serving as the ligand. The empirical evidence is that in many cases the most stable arrangement is for the Zn or other dissolved metal ion to form such complexes (Fig. 15-4b). As would be expected, since the reaction is effectively a competition between the metal and hydrogen ions to bind to the oxide, metal ion adsorption is enhanced at high pH and diminished at low pH (Fig. 15-5a). This conceptual model for adsorption is also consistent with the observation that the adsorptive bond strength of most metal ions onto oxide surfaces is strongly correlated with their ability to form hydroxo complexes in solution (Fig. 15-5b). That is, those metals that bind strongly to the oxygen atom of water molecules in solution also bond strongly to surface bound oxygen atoms.

As with the complexation reactions described earlier involving only dissolved species, the electrostatic attraction between a negatively charged surface and a positively charged metal ion can enhance the attraction of these reactants for one another, but much of the driving force for the reaction is provided by the formation of a chemical bond between the surface and the metal. Because of this, many metals can bind even to a positively charged surface, and in other cases can cause the surface charge to change sign from negative to positive as a result of their adsorption.

Dissolved metal complexes such as $MeCl_x^{n-x}$ or $Me(HCO_3)_y^{n-y}$ are also potential adsorbing species and can form the analog of mixed-ligand complexes at the surface (Fig. 15-4c). In most cases these simple inorganic complexes tend to adsorb somewhat less strongly than aquo or hydroxo complexes. This may be partially accounted for simply by statistical factors: when one of the metal's waters of hydration is replaced by another ligand, there are fewer waters of hydration that can be exchanged for the surface oxide ligand. Also, the bond strength of the complex-to-surface bond may be less than the aquo metal-to-surface bond because the ligand satisfies some of the bonding requirements of the metal ion. In any case, complexation by simple inorganic ligands generally acts to retain metal ions in solution at the expense of metal ion adsorption. Formation of Cd–Cl complexes, for instance, has been cited to explain the desorption of Cd from suspended matter as it passes through an estuary (Paulson *et al.*, 1984).

The effects on metal ion adsorption of ligands that can themselves adsorb strongly can be quite different from that described above. Many multi-atomic ligands can bond to oxide surfaces through atoms different from those they use to

---

Fig. 15-5 Comparative adsorption of several metals onto amorphous iron oxyhydroxide systems containing 10^{-3} M Fe_T and 0.1 M $NaNO_3$. (a) Effect of solution pH on sorption of uncomplexed metals. (b) Comparison of binding constants for formation of soluble Me-OH complexes and formation of surface Me–O–Si complexes; i.e. sorption onto SiO_2 particles. (c) Effect of solution pH on sorption of oxyanionic metals. (Figures (a), (c) reprinted with permission from Manzione, M. A. and Merrill, D. T. (1989). "Trace Metal Removal by Iron Coprecipitation: Field Evaluation," EPRI report GS-6438, Electric Power Research Institute, California. Figure (b) reprinted with permission from Balistrieri, L. *et al.* (1981). Scavenging residence times of trace metals and surface chemistry of sinking particles in the deep ocean, *Deep-Sea Res.* **28A**: 101–121, Pergamon Press.)

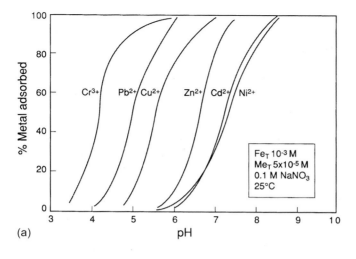

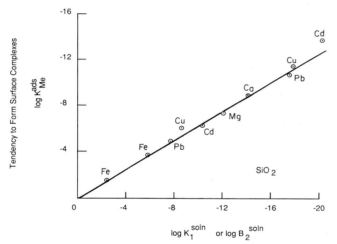

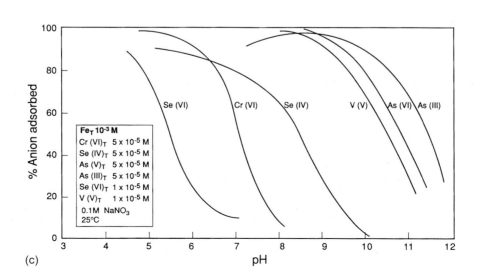

bind to metal ions. For example, the sulfidic sulfur of thiosulfate ($—S—S—O_3^{2-}$) is thought to bind to polarizable metal ions such as Ag^+ or Cd^{2+}, while the sulfate group is more likely to bond to an oxide surface site. In such a case, the ligand may simultaneously bind to both the surface and the metal, thereby acting as a bridge between them. To explain the interactions in such a system, it is necessary to discuss ligand adsorption briefly.

The adsorption of anions, including metalloids such as SeO_4^{2-}, MoO_4^{2-}, and CrO_4^{2-}, anionic ligands such as PO_4^{3-} and SO_4^{2-}, and fulvic acids, is similar to that of cationic metals except that it usually has the inverse dependence on solution pH. That is, the sorption reaction involves competition with OH^- ions for the surface, and ligand adsorption is therefore strong at low pH, where competition is weak (Fig. 15-5c). The complete set of equilibria describing the interactions in a system containing dissolved metal and ligand molecules and a solid that can adsorb either of these is complicated, but conceptually it involves a balance among all the relatively simple individual reactions discussed thus far. These interactions are summarized schematically in Fig. 15-5c. Low pH favors adsorption of the free ligand and ligand-bridged metals. High pH favors adsorption of free metal ions and metal-bridged ligands. The complexation of the metal and ligand in solution may or may not be a function of pH. The net interaction of all these factors can lead to solutions where the adsorption of metal ions is enhanced or diminished compared to that in ligand-free systems. Interesting examples of some of these interactions have been provided by Davis and co-workers, using both synthetic ligands and natural organic matter as the complexing agents (Fig. 15-6).

In addition to the interactions discussed above, which all depend in part on the ionizability, or at least polarizability, of the surface and the adsorbates, hydrophobic parts of ligands may bind to corresponding parts of surfaces. Thus, if a metal ion is complexed or irreversibly bonded to a hydrophobic molecule, the metal may be incorporated into the bulk or surface of a particle via hydrophobic interaction between the molecule and the solid phase. Such interactions may be quantitatively significant in systems with high concentrations of dissolved and particulate organic matter.

The fraction of the total metal bound to particulate matter in an aquatic system can be large, at times dominating the fraction in bulk solution. Davies-Colley *et al.* (1984) investigated the distribution of Cd among various model sediment components and concluded that for "typical" oxidized estuarine sediments, most of the trace metal will be bound to hydrous iron oxides and organic matter, with the organic-bound fraction of Cu being greater than that of Cd. Manganese oxides and clays contribute negligibly to the total binding capacity for both metals, according to their work.

Another approach to assess the partitioning of metals among the phases comprising natural particulate matter is to sequentially and selectively extract or dissolve portions of natural particulate matter. Based on the release of trace metals accompanying each step, associations between the trace metal and the extracted phase are inferred. Both of the above approaches have drawbacks, and at this time it is impossible to predict in advance how and to what extent metals and particulate matter will bond to one another in a natural system. Despite the uncertainties, empirical results can often be interpreted using the framework provided here,

Fig. 15-6 Adsorbing, complexing ligands can enhance sorption of metals at low pH and interfere with it at high pH. This is shown (a) in a strictly inorganic solution (line shows adsorption curve with ligands absent) and (b) in a solution with natural organic matter as the complexing agent. (c) Metals that do not form strong complexes with the ligand are unaffected by its presence. (Figure (a) modified with permission from Davis, J. A. and Leckie, J. O. (1978). Effect of adsorbed complexing ligands on trace metal uptake by hydrous oxides, *Environ. Sci. Technol.* **12**, 1309–1315, American Chemical Society. Figures (b), (c) reprinted with permission from Davis, J. A. (1984). Complexation of trace metals by adsorbed natural organic matter, *Geochim. Cosmochim. Acta* **48**, 605–615, Pergamon Press.)

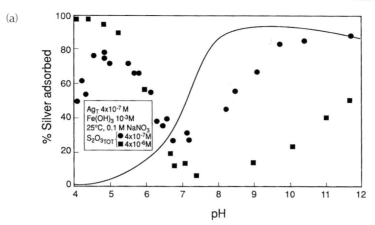

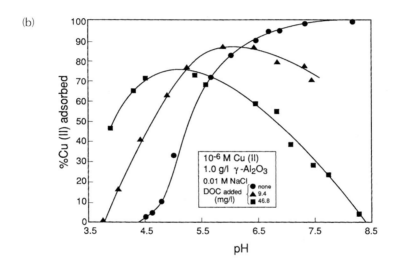

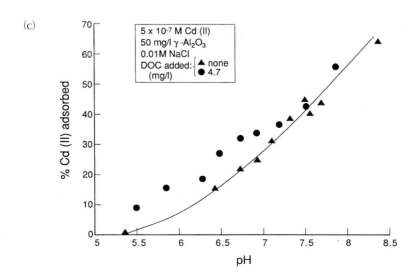

offering insights into the behavior of metals in complex systems and ultimately leading to improved predictive capability. Experimental measurements of the overall fractionation of metals between particulate and soluble phases in a few aquatic systems are presented in Fig. 15-7.

A related phenomenon is the adsorption of organic matter, metals, and particulate matter at another interface – that between the bulk water and air. At the air–water boundary, a microlayer of approximately 50 μm thickness is established with physical–chemical properties different from those of the bulk water. These properties cause metals, particles, microorganisms, and dissolved organic matter to accumulate in concentrations 5 to 1000 times greater than in bulk solution (Hardy *et al.*, 1985). While this layer does not contain a significant fraction of the total metal in the aquatic system, it comprises much of the material that is transferred to the atmosphere as spray and represents an important region through which substances must pass to enter the bulk water from the atmosphere. During the several hours that particles are thought to spend in this enriched environment, important dissolution, complexation, photo-

chemical, and biological reactions may take place.

15.3.7   Reactions Involving Organisms

Metals play an essential role in many enzyme systems, and virtually all metals can be toxic at concentrations that exceed the levels at which they are required or are normally found in the environment. Microorganisms play a central role in converting inorganically and organically bound metals to other chemical forms and transporting metals among various compartments of aquatic ecosystems as adsorbed or absorbed species. The effects of metals on biological systems is an established field of study in itself, while the importance of biological reactions in local or global metal cycling has only recently begun to become clear. An overview of the current understanding in these areas is provided in this section, focusing once again on the importance of metal speciation to the understanding of the system.

15.3.7.1   Effects of organisms on metal speciation and cycling

The biological contribution to metal cycling has been mostly studied in metal systems for which biological reactions contribute significantly to the total global flux, or for which biological transformations have significant implications in humans. Mercury and arsenic fall in this category. The biological cycles of these metals were first described clearly by Wood (1974). He emphasized that they, and perhaps many other metals, have natural biological cycles even though the metals themselves are not biologically essential. Biological conversions of metals may be evolutionary responses that microorganisms have developed as detoxification mechanisms. However, at times these "detoxification" reactions in one group of organisms actually intensify the toxicity of the metal to higher organisms. This is the case with of mercury and arsenic, which are both methylated by lower organisms.

In the case of mercury (Fig. 15-8), Wood suggested that reduction of Hg^{2+} to Hg^0 and

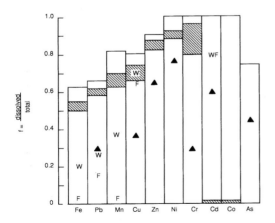

Fig. 15-7   Relationship between dissolved and total heavy metal concentrations in several rivers. Cross-hatched bands represent range of values from the Ruhr (Imhoff *et al.*, 1980); W and F represent winter and fall values at a selected station in the Mississippi (Eisenreich *et al.*, 1980); Triangles represent values from the Rhine river (Davis, 1984).

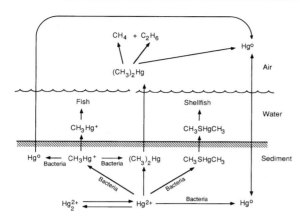

Fig. 15-8 The mercury cycle, demonstrating the bioaccumulation of mercury in fish and shellfish. Reprinted with permission from "An Assessment of Mercury in the Environment" (1978) by the National Academy of Sciences, National Academy Press, Washington, DC.

ing its oxidation state or forming organometallic compounds, they still may play an important role in the cycling of many metals. For instance, Wood pointed out that bacteria can facilitate the mobilization of mercury from mineral deposits by oxidizing sulfide and thereby allowing mercury which had been sequestered in the extremely insoluble solid cinnabar (HgS) to dissolve. The same mechanism can be important in release of many other metal sulfides. Additionally, recent evidence has shown that the cycling of several metals in the open ocean is closely tied to the cycling of particulate organic matter, i.e., microorganisms (Jones and Murray, 1984). Figure 15-9 shows correlations of cadmium and phosphate in the ocean off the coast of California. Like phosphate, Cd is depleted in the surface waters due to biological uptake and released to the aqueous phase in deeper water as the sinking organic solids decay. It is not yet

alkylation to form methyl- or dimethylmercury can both be viewed as detoxification reactions, because all the products are volatile and can be lost from the aqueous phase. Organisms can also convert the methylated forms to Hg^0, which is more volatile and less toxic. However, both the methylated and the reduced forms are more toxic to humans and other mammals than is Hg^{2+}.

The arsenic cycle in ocean waters and sediments also has important biological steps (Andreae, 1979). Arsenate (As(V)) can be biologically converted into arsenite (As(III)) and at least eight different organo-arsenic compounds, all presumably representing detoxification processes mediated by bacteria in reduced sediments or by algae in the water column. On the other hand, biologically catalyzed demethylation and perhaps oxidation of arsenite serve to return arsenate to the water. In the bulk water below 400 m, one study found the ratio of arsenite to arsenate to be at least 12 orders of magnitude greater than that expected from equilibrium calculations, indicating the importance of biological reactions, since the kinetics of the abiological reactions controlling this equilibrium must be slow.

Even if microorganisms do not mediate a reaction with the metal directly, e.g., by chang-

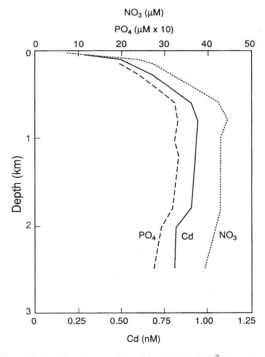

Fig. 15-9 Depth profiles for NO_3^-, PO_4^{3-}, and Cd observed at station 64 off the coast of California in April 1977. (Reprinted with permission from Bruland, K. W. *et al.* (1978). Cadmium in Northeast Pacific waters, *Limnol. Oceanogr.* **23**, 618–625, Society for Limnology and Oceanography.)

known whether the metal uptake is active and carried out only by living organisms, or is passive and involves adsorption on dead cells as well. Similar correlations have been reported for Ni, Zn, and Cu.

The range of processes that must be considered in the cycle of metals is described in Fig. 15-10 (Nelson *et al.*, 1977). Both the complexity of metal cycle analysis in a real system and the importance of speciation are well-stated by Andreae (1979) in his overview of the arsenic cycle in seawater:

> The biological cycle of arsenic in the surface ocean involves the uptake of arsenate by plankton, the conversion of arsenate to a number of as yet unidentified organic compounds, and the release of arsenite and methylated species into the seawater. Biological demethylation of the methylarsenicals and the oxidation of arsenite by as yet

unknown mechanisms serve to regenerate arsenate. The concentrations of the arsenic species are then controlled primarily by the relative rates of biologically mediated reactions, superimposed on processes of physical transport and mixing. The presence of arsenite in the deep ocean at concentrations far from thermodynamic equilibrium further emphasizes the importance of kinetic restraints on redox equilibration in the ocean.

15.3.7.2   Effects of metals on organisms: Toxicity and bioavailability

Recognition of the importance of metal speciation has had an enormous impact in the area of toxicity studies. Before this recognition, data were often difficult to interpret even from a single study, let alone from a range of investiga-

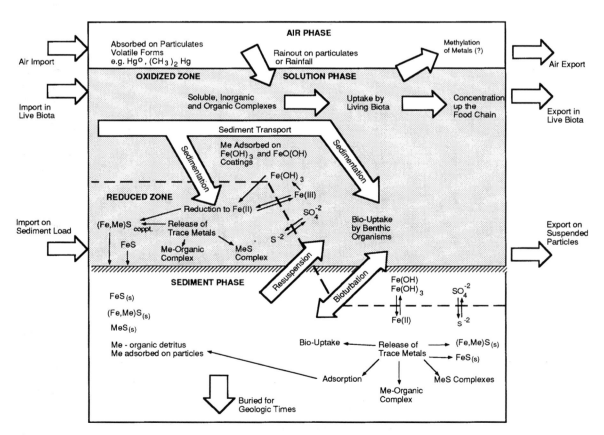

Fig. 15-10   Summary of reactions and processes important in metal biogeochemical cycling (after Nelson *et al.*, 1977).

tions each with slightly different conditions. Recent analysis of speciation has led to a remarkably consistent conclusion: for all metals that have been studied (Cu, Cd, and Zn have been investigated the most) the toxicity of metals to algae, various aquatic invertebrates, and fish is strongly related to the activity of the free metal ion, regardless of the concentration of total metal in solution. This means, for example, that a solution containing 1.0 mg/L total Cu and enough organic matter so that 99% of the copper is strongly complexed, will be less toxic than an otherwise comparable solution with only 0.05 mg/L total dissolved Cu and little or no complexing organic matter. Other investigations have reached the same conclusion with respect to metal ion availability as an essential nutrient.

Strongly complexed metals, especially those complexed by chelating agents such as EDTA or by natural humic material, appear to be completely unavailable and non-toxic. A typical experiment showing such a result is summarized in Fig. 15-11. Weaker complexes formed with monodentate ligands such as chloride or carbonate also provide some protection against toxicity, although it is not yet clear whether these complexes are as innocuous as the chelates. Of course, the overall effect of potentially toxic metals depends not only on metal speciation but on all aspects of solution chemistry and on the identity and previous exposure history of the test organism. For example, one of the unresolved issues in this field has to do with the effects of water hardness (conferred primarily by dissolved Ca and Mg ions) and alkalinity on metal toxicity. In some studies these variables are negatively correlated with toxicity, and in others they seem to have no effect. Undoubtedly the results depend partially on the ability of the hardness ions to compete with the toxic ones for specific binding sites and on other responses of the organism to changes in the concentrations of these ions, quite apart from the response to the toxin. Regardless of the ultimate resolution of the remaining issues, it is clear that consideration of speciation has been the key step in rationalizing studies of toxicity and bioavailability in recent years and will be a crucial part of any such future studies.

(a)

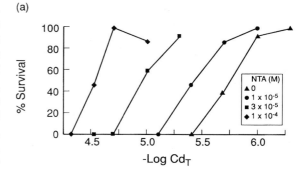

(b)

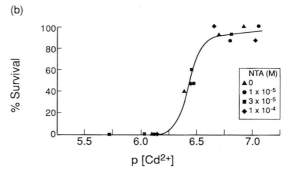

Fig. 15-11 Effects of strong complexation on metal ion toxicity. (a) Increasing concentration of NTA, a strong multi-dentate complexing agent, decreases the toxicity of Cd to grass shrimp. All systems have equal concentrations of total Cd. (b) When the results are replotted showing survival as a function of Cd^{2+} concentration, the data for all concentrations of NTA collapse to a single curve. (Reprinted with permission from W. G. Sunda *et al.* (1978). Effect of chemical speciation on toxicity of cadmium to grass shrimp, *Palaemonetes pugio*: importance of free cadmium ions, *Environ. Sci. Technol.* **12**, 409–413, American Chemical Society.)

15.3.7.3 Toxic sediments

The widespread use of many metals such as silver, cadmium, copper, mercury, nickel, lead, and zinc has resulted in their accumulation in the environment. Sediments are often the repositories of toxic metals (e.g., Table 15-2). For example, copper is used as an anti-biofouling agent in marine paints and many harbor sediments contain markedly elevated levels of copper.

A considerable number of studies has demonstrated that the adverse biological effects of dissolved metals is largely regulated by the *activity* (i.e., "effective" concentration) of free metal ions (see Section 15.3.7.2), although some reports indicate that other metal species or variables may, at times, be important (e.g., Erickson *et al.*, 1996). One challenge facing environmental scientists is the development of methodologies for assessing the toxicity of sediments contaminated with toxic metals.

As is the case with assessments of the toxicity of dissolved trace metals, the development of sediment quality criteria (SQC) must be based on the fraction of sediment-associated metal that is bioavailable. Bulk sediments consist of a variety of phases including sediment solids in the silt and clay size fractions, and sediment pore water. Swartz *et al.* (1985) demonstrated that the bioavailable fraction of cadmium in sediments is correlated with interstitial water cadmium concentrations. More recent work (e.g., Di Toro *et al.*, 1990; Allen *et al.*, 1993; Hansen *et al.*, 1996; Ankley *et al.*, 1996, and references therein) has demonstrated that the interstitial water concentrations of a suite of trace metals is regulated by an extractable fraction of iron sulfides.

The role of iron sulfides in regulating sediment interstitial water concentrations is shown by the following reactions (Di Toro *et al.*, 1990):

$$FeS(s) \leftrightarrow Fe^{2+} + S^{2-}$$

$$Cd^{2+} + S^{2-} \leftrightarrow CdS(s)$$

where '(s)' indicates that the substance is a solid. Anoxic sediments contain a large reservoir of (amorphous) iron sulfides. The first reaction represents the dissociation of iron sulfide; cadmium (the second reaction) associates with sulfide to form the insoluble cadmium sulfide. The overall reaction is

$$Cd^{2+} + FeS(s) \leftrightarrow Fe^{2+} + CdS(s)$$

which is strongly distributed to the products' side (the right-hand side) of the equation. The molar ratio of Cd^{2+} to FeS is 1:1; consequently, Cd will be essentially unavailable in the free aquo form (i.e., Cd^{2+}) until the FeS has been exhausted. Only if the ratio of Cd to FeS (the

acid volatile sulfide) is greater than 1 will the Cd^{2+} be potentially bioavailable in substantial amounts.

Figure 15-12 is a schematic illustration of a technique known as acid volatile sulfides/simultaneously extracted metals analysis (AVS/SEM). Briefly, a strong acid is added to a sediment sample to release the sediment-associated sulfides, acid volatile sulfides, which are analyzed by a cold-acid purge-and-trap technique (e.g., Allen *et al.*, 1993). The assumption shown in Fig. 15-12 is that the sulfides are present in the sediments in the form of either FeS or MeS (a metal sulfide). In a parallel analysis, metals *simultaneously* released with the sulfides (the simultaneously extracted metals) are also quantified, for example, by graphite furnace atomic absorption spectrometry. Metals released during the acid attack are considered to be associated with the phases operationally defined as "exchangeable," "carbonate," "Fe and Mn oxides," "FeS," and "MeS."

Sediment toxicity is evaluated by equilibrating 200 ml of the target sediment with 600 ml of water (e.g., saltwater for estuarine sites). Lethality tests are run for a specific period of time (e.g., 10 days) after which the indicator organism is counted for mortality. Figure 15-13 shows generalized results for a hypothetical suite of sediments. Percent mortality of the indicator organism is plotted as a function of the SEM/AVS ratio. Significant organism mortality ($\geq 24\%$) appears at SEM/AVS molar ratios greater than 1 as anticipated from the discussion on metal sulfides given above.

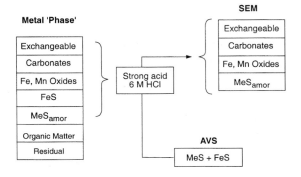

Fig. 15-12 Schematic illustration of the acid volatile sulfides (AVS)/simultaneously extractable metal (SEM) analysis. Refer to the text for details.

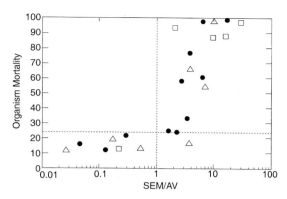

Fig. 15-13 Organism mortality as a function of SEM/AVS ratio for a hypothetical sediment. This figure is generalized from results typical of sediment toxicity tests (e.g., Hansen *et al.*, 1996). Organisms evaluated in such tests include amphipods and polychaetes. The symbols represent different sediments. The vertical line at 10^0 is positioned at an SEM/AVS ratio of 1.0; the horizontal line at 24% represents the limit of toxicity, that is, mortality $\leqslant 24\%$ is defined as not the consequence of toxicity.

15.3.8 Geochemical Kinetics

To this point, we have emphasized that the cycle of mobilization, transport, and redeposition involves changes in the physical state and chemical form of the elements, and that the ultimate distribution of an element among different chemical species can be described by thermochemical equilibrium data. Equilibrium calculations describe the potential for change between two end states, and only in certain cases can they provide information about rates (Hoffman, 1981). In analyzing and modeling a geochemical system, a decision must be made as to whether an equilibrium or non-equilibrium model is appropriate. The choice depends on the time scales involved, and specifically on the ratio of the rate of the relevant chemical transition to the rate of the dominant physical process within the physical–chemical system.

Comparisons of time scales for various physical processes have been discussed by Lerman (1979) and Schwartzenbach and Imboden (1983). In general, physical processes in lakes have characteristic times in the range of 10^{-4} to 10^2 years. Oceanic processes span times from days to thousands of years (Broecker, 1974)

and geologic events may occur catastrophically (e.g., earthquakes or landslides exposing new weathering surfaces) or very slowly, as in the case of continental subduction. If the characteristic chemical reaction times (τ_{chem}) are short compared to the characteristic times of the dominant physical processes (τ_{phys}) then it is probably appropriate to consider the chemical transition in terms of equilibrium concepts. If τ_{chem} is large compared to τ_{phys} then reactions will proceed only slightly, if at all, from their initial conditions. When τ_{chem} is of the same order as τ_{phys} then quasi-equilibrium or kinetic descriptions may be employed (Morgan and Stone, 1985; Morel, 1983; Keck, 1978). Thus, a description of the chemical transitions occurring in each reservoir of a biogeochemical system must consider the relative physical and chemical time scales characteristic of that reservoir. For example, the hydrolysis of Fe(III) has a characteristic time (τ_{chem}) of 3.2×10^{-7} seconds (Hemmes *et al.*, 1971). Since most natural physical processes are much slower than this rate, the Fe(III) hydrolysis reaction can be considered to be at equilibrium in those systems. In contrast, the dissolution of silica proceeds at a much slower rate, with a characteristic time of approximately 8 years. The oxygenation of Mn(II) is more ambiguous. It has characteristic times ranging from weeks to tens of years depending on whether the reaction is homogeneous, mediated by bacteria, or catalyzed by the presence of metal-oxide surfaces (Morgan and Stone, 1985).

The suite of chemical reactions taking place in a geochemical compartment, such as the atmosphere or a lake, are often quite complex, involving higher-order reaction kinetics and multiple, linked reactions (e.g., Pankow and Morgan, 1981). Nevertheless, the principle of comparing the relative time scales of physical versus chemical process is still valid. What is needed is an understanding of the rate controlling reaction(s) and the dominant physical processes. A variety of chemical reactions and their characteristic times are shown in Fig. 15-14. Also shown on the figure are the major physical processes extant in lakes and the range of their characteristic times. As described in Chapter 4, the characteristic time is the time required for the

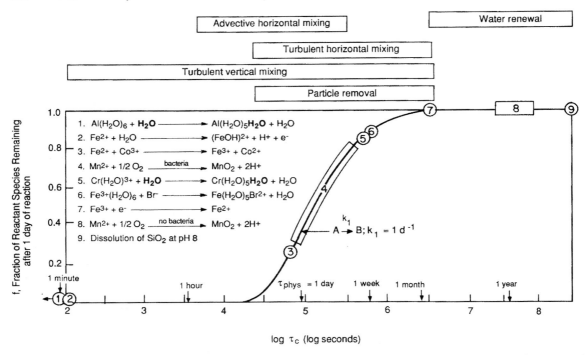

Fig. 15-14 Examples of characteristic times for lake transport processes and reactions, and extent of reaction of several environmentally significant reactions in a system with $\tau_{phys} = 1$ day.

transformation to proceed to within the fraction $1/e$ of completion. This is also known as the system response time (Lasaga, 1980). The characteristic times of the transformations shown in Fig. 15-14 range from approximately 1 h to half a year. Of the chemical reactions shown, a significant number have characteristic times of the order of the predominant physical processes, suggesting that non-equilibrium approaches to the geochemical modeling need to be considered.

15.4   Observations on Metals in Natural Systems

15.4.1   *Combining Physical and Chemical Information*

All the factors mentioned in the previous sections play a role in the movement of metals through their overall biogeochemical cycle: injection into the atmosphere, deposition onto land or water surfaces, transport via rivers and

estuaries to the oceans, and sedimentation and ultimate burial in the sediments. The physical–chemical form in which each metal is transported in the aquatic phases of the cycle will depend on the specific metal and its interactions with other dissolved and suspended constituents in accord with the principles discussed above. However, it is important to keep in mind that the physical processes of fluid flow and sediment transport must be combined with the chemical reactions to gain insight into the functioning of the complete system. This point has been well-stated by Turekian (1977) in a discussion of metal concentrations in the ocean:

> Why are the oceans so depleted in these trace metals? Certainly it is not for the lack of availability from rock weathering or because of constraints imposed by the solubility of any unique compound of these elements. The reason must lie in the dynamics of the system of delivery of the metals to the oceans and their subsequent behavior in an ocean that cannot be simulated by simple in vitro experiments involving homogeneous reaction kinetics.

In this section, an overview of the net transport of metals through rivers, estuaries, and the oceans provides an example of how these chemical and physical forces interact.

Recent studies have reported that from 10% to more than 90% of the trace metal load of streams is carried in the particulate fraction, with the fraction of particle-bound Fe, Pb, Cu, As, and Zn typically being greater than that of Cd, Co, and Ni. As noted earlier, the exact balance will depend on the concentration of inorganic and organic complexing ligands, the type and quantity of particulate matter available, and the concentrations of other ions competing for the binding sites. In interpreting these data one must bear in mind that the fraction defined as being "particulate" does not represent an absolute measure; rather, it is defined by the separation technique employed. Any material that will not pass a 0.45 μm filter is often defined as particulate. However, a significant fraction of the "soluble" trace metals, Fe and Mn, and organic material that does pass the filter can subsequently be filtered out using filters with smaller pores. Furthermore, much of the 0.45 μm filterable material is converted to non-filterable matter (still using 0.45 μm filters) when the concentrations of Ca^{2+}, Mg^{2+}, and Na^+ increase to levels they attain in estuaries. This process and its implications merit some special attention.

15.4.2   Particle and Metal Interactions in Estuaries

Estuaries exhibit physical and chemical characteristics that are distinct from oceans or lakes. In estuaries, water renewal times are rapid (10^{-3} to 10^{-4} years compared to 1 to 10 years for lakes and 10^4 years for oceans), redox and salinity gradients are often transient, and diurnal variations in nutrient concentrations can be significant. The biological productivity of estuaries is high and this, coupled with accumulation of organic debris within estuary boundaries, often produces anoxic conditions at the sediment–water interface. Thus, in contrast to the relatively constant chemical composition of the oceans, the chemical environment in an estuary varies dramatically in time and space.

An estuary can be defined as a system in which ocean waters have been diluted by freshwater from land drainage. Many ions are more abundant in coastal or oceanic waters than river water, e.g., Na, K, Mg, Ca, Cl, SO_4^{2-}, and HCO_3^-. However, nutrients (N, P, and Si) and many transition series metals, including Fe, are more concentrated in river water. Dissolved and particulate organic matter (DOM and POM) are one to two orders of magnitude more concentrated in rivers than in ocean water. As end-member waters mix, dissolved constituents originally in each of the end-member waters may remain in solution during mixing (i.e., behave conservatively), interact with sinking particulate matter, or precipitate out of solution. The wide variety of mixing regimes and end-member waters makes it difficult to generalize about the estuarine behavior of many metals of interest to environmental geochemists. Generally, however, the major components of seawater – Na, Ca, K, and SO_4 – behave conservatively. Some elements, such as zinc, may behave conservatively in some estuaries but be rapidly removed in others. Iron, by contrast, is removed rapidly and efficiently in nearly all estuarine environments, i.e., its behavior is highly non-conservative.

As river water (Fig. 15-15a) high in dissolved and particulate iron and other inorganics (PIM) and organic matter enters an estuarine mixing zone (Fig. 15-15b), rapid flocculation of the DOM and coagulation of particulate matter occurs (Fig. 15-15c). The extent to which this occurs is a function of the salinity of the water mixture: flocculation and coagulation generally increase with increasing salinity up to a salinity of about 15‰, with greater salinities having little added effect. While systems containing only colloidal iron oxides have been shown to coagulate as salinity increases, the presence of organic matter increases the coagulation rate and leads to the formation of Fe oxide–organic aggregates. The efficiency of metal removal by the flocculation process generally follows the relative order of trace metal–DOM complex stabilities: $Fe^{3+} > Al^{3+} > Cu^{2+} \sim Ni^{2+}$, etc. Reactions that may initiate the iron–organic removal process

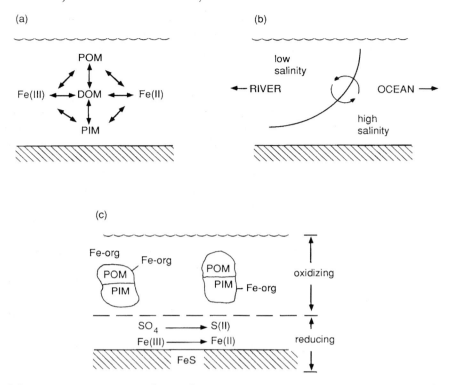

Fig. 15-15 Schematic representation of Fe and organic matter interactions in an estuary. POM = particulate organic matter; DOM = dissolved organic matter; PIM = particulate inorganic matter.

include coagulation of pre-existing colloidal particles or precipitation of previously soluble humic acids, and adsorption of trace metals onto particles before or after either of the above processes.

The efficient removal of iron in estuaries allows very little of the initial river-borne iron to escape the estuary to the coastal waters. "Soluble" iron concentrations are extremely low in river water and in estuaries. Nearly all of the iron entering the estuary is transported in particulate form to sediments on the estuary floor where, due to the high accumulation of organic matter, reducing conditions are often met. Fe(III) is reduced in anaerobic sediments and, since S is often also accumulating in the sediments, Fe(II) may be precipitated as iron sulfides. Resuspension of sediments may reintroduce iron to the oxygenated layer of the estuary where it can be rapidly oxidized, forming new colloidal particles, which can sorb other metals, associate with suspended organic matter

and return to the sediments in repeated episodes of internal cycling.

The ultimate result is that not only Fe, but most metals that interact strongly with organic matter or oxide adsorbents, are likely to settle out of the water column in estuaries. Not only does this process reduce the metal flux reaching the ocean, but 50% of the time the current at the bottom of an estuary actually facilitates the movement of metals upstream.

While Fe- and Mn-oxides certainly are important in the binding and transport of the trace metals in the estuarine system, organic matter appears to play a dominant role in the coagulation step. For instance, adjustment of the solution pH to values where humics normally precipitate (pH = 1 to 3) causes all the components (Fe and Mn, organics, and trace metals) to coagulate into filterable particles. By contrast, adjustment of the pH upward, which would normally favor precipitation of Fe- and Mn-oxides, has no effect. In view of this it is not surprising that the tendency of trace metals to be removed from estuarine waters corresponds roughly with their tendency to bind to organics, with Cu being the most strongly affected. In one study, for instance, approximately 40% of the dissolved Cu, 15% of the dissolved Cd and Ni, and less than 5% of the dissolved Co and Mn were filterable after coagulation was induced by addition of Ca and Mg to river water (Sholkovitz and Copland, 1981). As noted earlier, desorption of metals owing to the formation of soluble Me–Cl complexes may partially or completely counteract this process for some metals.

To summarize, even though some Fe- and Mn-oxides may dissolve in the anoxic sediments of estuaries and thereby release trace metals, re-oxidation of the Fe and Mn in the upper layers of the sediments, coagulation of Fe–Mn–humic–trace metal mixtures in the water column, and circulation patterns within the estuary all combine to return trace metals to the upstream particulate fraction. Turekian has suggested that while these processes allow significant internal cycling of trace metals in the estuary, they prevent any more than a small fraction of the total trace metal load from reaching the ocean. In describing the role that soils and rocks play in attenuating the movement of metals from the land to water bodies and the scavenging of metals from the water column by particles in the ocean in combination with the estuarine processes, Turekian dubbed the overall process "the great particle conspiracy," working to maintain soluble trace metal concentrations at extremely low and remarkably consistent levels throughout the world's open oceans.

15.4.3   Application of Chemical Principles to Evaluate Field Data

Examples were given earlier of a few systems for which the theoretical speciation of metals was calculated, and a few others in which it was analyzed experimentally, at least into some subgroups if not specific species. These approaches have rarely been combined to critically evaluate data or hypotheses regarding metal transport through a real system. One study, already noted earlier in this chapter, can serve as a model for combining theoretical considerations with field data. In that study, Morel *et al.* (1975) used data for the partitioning of metals between the soluble and particulate fractions to estimate the redox potential of primary-treated sewage. Knowing the composition of the Southern California coastal waters into which the sewage was discharged, they were then able to interpret the relative changes in the concentrations of several metals as a function of dilution and oxidation. Combining chemical analyses with data for sedimentation rates of metals and organic carbon, they suggested that most metals are not mobilized in the nearfield deposition zone, and they offered plausible explanations for the deviation of Pb and Cd from this pattern. As the authors noted, extension of their results to other systems is not warranted, but extension of the methodology, i.e., combination of experimental data with chemical speciation models to test hypotheses and suggest profitable research directions, certainly is.

15.4.4   Acid Mine Drainage (AMD)

From the beginnings of ecology as a discipline, the mining industry has been at the center of the battle over preservation versus exploitation. As discussed in the introduction to this chapter, human activities such as mining, power production from fossil fuels and discharges of industrial and municipal wastes not only increase the rate at which metals enter the biosphere but may also drastically alter the speciation of metals from what it would be in the undisturbed geologic cycle.

A major consequence of the activities associated with the exploitation of mineral deposits (i.e., exploration, the development of mines and processing facilities, the extraction and concentration, which is also called beneficiation, of ores containing the desired minerals, and the decommissioning or abandonment of mine facilities) is the production of extremely large volumes of unwanted materials. Waste volumes vary from ca. 30% of the mass of the ore in the case of gypsum and other non-metals, to about 50% for base metals to more than 80% for strip-mined coal (see Ripley *et al.*, 1996, and references therein).

In a number of milling and mining operations (e.g., metal mines containing sulfide minerals either in the ore or surrounding rock, or coal or lignite deposits) the oxidation of sulfides, particularly iron pyrite, is of primary concern for the contamination of water. The oxidation process, which generates acid drainage from the deposit, occurs under natural conditions but is greatly accelerated as the consequence of mining activities. In this case it is then referred to as acid mine drainage (AMD).

The existence of acid drainage has been recognized for some time:

> The spring water issuing through fissures in the hills, which are only masses of coal, is so impregnated with bituminous and sulphurous particles as to be frequently nauseous to the taste and prejudicial to the health. (T. M. Morris, 1803, in MacKenthum, 1969)

However, the seriousness of the problem, and its underlying causes, have only relatively recently been addressed by the mining industry (Ripley *et al.*, 1996). As a consequence, many mining operations conducted before the 1980s were not managed to control the deleterious effects of acid drainage.

Some of the basic processes that contribute to the generation of acid drainage are discussed in earlier portions of this section (e.g., the oxidation of Fe(II) to Fe(III)). Although, as Langmuir (1997) points out, the presence of iron pyrite in coal or metal deposits does not "predetermine" that mining operations will generate significant amounts of AMD, the amount and nature of the pyrite present is an important factor in the generation of AMD.

Basically, the oxidation of iron pyrite, FeS_2, results in the production of iron(III) sulfate and sulfuric acid, H_2SO_4. However, two overall reaction stoichiometries are possible and each will yield a different acid generation capacity (e.g., Langmuir, 1997; Baird, 1995):

Oxidation by O_2

$$FeS_2 + \tfrac{7}{2}O_2 + H_2O \rightarrow Fe^{2+} + 2SO_4^{2-} + 2H^+$$

Oxidation by Fe^{3+}

$$FeS_2 + 14Fe^{3+} + 8H_2O \rightarrow 15Fe^{2+} + 2SO_4^{2-} + 16H^+$$

In both reactions, Fe^{2+} is produced as a consequence of the oxidation of sulfur from the -1 oxidation state of pyritic sulfur to the $+6$ state of sulfate.

A second oxidation step is the oxidation of Fe^{2+} to Fe^3:

pH < 3

$$Fe^{2+} + \tfrac{1}{4}O_2 + H^+ \rightarrow Fe^{3+} + \tfrac{1}{2}H_2O$$

pH > 3

$$Fe^{2+} + \tfrac{1}{4}O_2 + \tfrac{5}{2}H_2O \rightarrow Fe(OH)_3 + 2H^+$$

The abiotic rate of the first oxidation reaction is slow; the rate of the second reaction increases with increasing pH. The second iron oxidation reaction produces $Fe(OH)_3(s)$, ferric hydroxide. "Yellow boy," a limonitic precipitate, is produced when the ferric hydroxide mixes with ferric sulfates; when formed, "Yellow boy" gives receiving waters an unappealing yellow tint.

The rates of oxidation are significantly increased in the presence iron- or sulfur-oxidizing bacteria. Iron-oxidizing bacteria, which are aerobic, include *Gallionella* and *Sphaerotilus*, which are autotrophs (i.e., obtaining carbon for synthesis of new biomolecules from carbonate species) and heterotrophs (require organic carbon), respectively. Sulfur-oxidizing bacteria include *Thiobacillus thiooxidans* and *Ferrobacillus ferrooxidans*; all such bacteria are aerobic and autotrophic.

15.5   Examples of Global Metal Cycling

In this final section, the global cycles of two metals, mercury and copper, are reviewed. These metals were chosen because their geochemical cycles have been studied extensively, and their chemical reactions exemplify the full gamut of reactions described earlier. In addition, the chemical forms of the two metals are sufficiently different from one another that they behave differently with respect to dominant transport modes and pose different risks to organisms.

15.5.1 *Mercury*

Mercury occupies a unique (and infamous) place in environmental history because it was the first chemical for which a direct connection was proven between relatively low concentrations in a natural water system, bioaccumulation up the food chain, and a serious health impact on a human population at the top of the food chain. Within a relatively few years, such epidemics were documented at two sites in Japan, and somewhat less definitive evidence suggesting mercury poisoning was reported in Canada. In addition, consumption of animals that had been fed grain seed contaminated by methylmercury fungicide led to acute poisoning of a family in the United States, and, in one of the worst cases of acute human exposure to an environmental hazard, over 6000 Iraqis were hospitalized with symptoms of mercury poisoning after consuming home-made bread made from seed wheat treated with this same type of fungicide. In the latter case, over 500 deaths were reported by hospitals, and it is likely that many other affected individuals did not report to the hospital.

15.5.1.1 *Global mercury cycling*

Unlike most heavy metals, the natural and anthropogenic cycles of mercury are dominated by atmospheric transport. Metallic mercury has the highest vapor pressure of any heavy metal, and it is released in geochemically significant quantities by volcanic activity, volatilization from land and ocean surfaces, and in high-temperature industrial processes such as smelting of minerals and burning of fossil fuels. Figure 15-16a,b provides an estimate of the natural and anthropogenic mercury cycles (Mason *et al.*, 1994). The two parts demonstrate many salient points, among which are the following:

1. Exchange rates of mercury between the atmosphere and the land and between the atmosphere and the ocean are much greater than transport from the land directly into the ocean via riverine discharge.

2. While the natural exchange of mercury between the land and atmosphere and the atmosphere and oceans is balanced, human activity has tipped this balance. There is now about three times more mercury in the atmosphere and fluxes of more than four times to and from the atmosphere.

3. The transport rate of mercury flowing from the land to the oceans in rivers has been increased by a factor of about three by human activity. While the increased rate is still relatively less important than the total transport of Hg through the atmosphere, it can represent a significant stress on the exposed organisms, particularly since the increased flux is unevenly distributed. That is, human activity has created local environments where the transport of mercury or its concentration in a river or estuary is many tens of times higher than background levels.

4. The average residence times for mercury in the atmosphere, terrestrial soils, oceans, and oceanic sediments are approximately 1 yr, 1000 yr, 3200 yr, and 2.5×10^8 yr, respectively. (See Bergan *et al.* (1999) for more details on atmospheric residence times.)

15.5.1.2 *Environmentally important forms of mercury*

There are several environmentally significant mercury species. In the lithosphere, mercury is present primarily in the +II oxidation state as the very insoluble mineral cinnabar (HgS), as a minor constituent in other sulfide ores, bound to the surfaces of other minerals such as oxides, or bound to organic matter. In soil, biological reduction apparently is primarily responsible for the formation of mercury metal, which can then be volatilized. Metallic mercury is also thought to be the primary form emitted in high-temperature industrial processes. The insolubility of cinnabar probably limits the direct mobilization of mercury where this mineral occurs, but oxidation of the sulfide in oxygenated water can allow mercury to become available and participate in other reactions, including bacterial transformations.

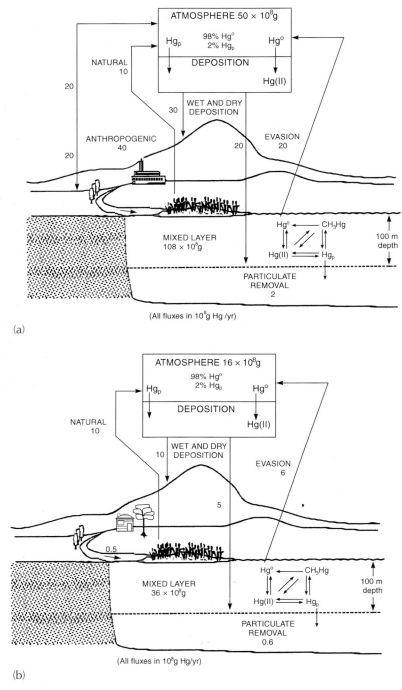

Fig. 15-16 The (a) present day and (b) pre-industrial global cycles for mercury. Units are 10^8 g Hg (burdens) and 10^8 g Hg/yr (fluxes). Hg_p refers to mercury in particles. Redrawn from Mason *et al.*, 1994.

In aqueous solution, mercury forms strong complexes with both organic and inorganic ligands. In particular, mercury is strongly complexed by compounds containing reduced sulfur atoms or carboxyl groups, by hydroxide, and by chloride. As a result, much of the dissolved mercury in freshwater is associated with dissolved organic matter, most likely humic and fulvic acids. Even in extremely "pure" river water, containing few dissolved minerals or organic compounds, mercury exists not as the divalent free aquo metal ion, but as a neutral or anionic hydroxo complex, depending on the solution pH. In seawater, dissolved mercury exists primarily as chloro and possibly organic complexes and mixed ligand complexes of chloride, bromide, and hydroxide.

Dissolved mercury also reacts to form strong complexes with inorganic and probably organic particulate matter. Figure 15-17 shows the distribution of mercury between dissolved and adsorbed phases in an idealized system consisting of water, an artificially prepared oxide and various concentrations of chloride. The mercury is strongly sorbed in the absence of the chloride, but as more of this ligand is added to the system, the mercury remains in solution at pH values where it had previously sorbed. Chemically, the effect of the chloride is to convert the mercury from Hg–OH complexes to Hg–Cl or Hg–Cl–OH complexes. These complexes apparently do not sorb as strongly as the Cl-free mercury species. While the conditions in any natural water are more complex, it is clear that this type of interaction may have a significant effect on the form and fate of mercury under changing environmental conditions, for instance in water moving through an estuary.

Finally, in addition to being an unusual metal because of its high volatility, mercury is unusual because of the importance of biological reactions in altering its form. It has been pointed out that biological activity is critical in converting oxidized (+2) mercury to metallic (0) mercury in soils, allowing subsequent volatilization. Mercury also undergoes biomethylation, which is of overriding significance with respect to its toxicity. In this reaction, which takes place in the upper layers of fresh- and saltwater sediments, divalent mercury forms covalent bonds with organic methyl groups. The organo-mercury compounds are more volatile and more easily bioaccumulated than their inorganic precursors. They may sorb to organic matter in the sediment, but once released they are rapidly taken up by organisms and are concentrated up the food chain, eventually becoming available for human consumption. This is the path that led to the first recognized epidemic of mercury poisoning affecting humans, at Minamata Bay, Japan. The rate of biomethylation of mercury is proportional to the concentration of inorganic mercury in the system and the concentration of organisms capable of carrying out the reaction. Thus, it has been suggested that limiting the nutrient supply discharged to receiving waters may be

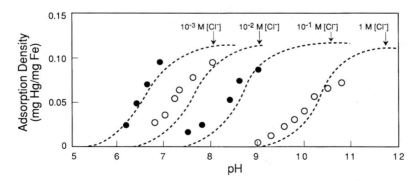

Fig. 15-17 The effect of chloride on adsorption of mercury by hydrous iron oxide at constant total mercury concentration of 3.4×10^{-5} M. The lines represent the predicted adsorption assuming that Hg–Cl complexes do not sorb at all. (Reprinted with permission from P. V. Avotins, "Adsorption and coprecipitation studies of mercury on hydrous iron oxides," 1975, Ph.D. dissertation, Stanford University, Stanford, CA.)

an effective method of limiting biomethylation in contaminated areas. Biodemethylation has also been observed, and may be a microorganismal response to mercury intoxication.

15.5.1.3 *Mercury movement in aquatic systems: two case studies*

Although there have been numerous efforts to describe the global mercury flux, studies in which fluxes of the individual species in a well-defined natural system were evaluated are rare. Two studies that have been reported – one for a highly contaminated saltwater system, the other for a relatively less contaminated stretch of a river – provide an interesting synthesis of the above information. In the former case, the movement of mercury out of Minamata Bay over a period of about 25 years was estimated (Kudo and Miyahara, 1983). The results support a remarkably strong affinity between the discharged mercury and bottom sediments, so strong in fact that the majority of the mercury discharged to the Bay remained there, attached to bottom sediments, 25 years after all Hg discharges had ceased. This result is all the more surprising considering the rapid flushing rate of the Bay: the residence time of water in the Bay is only 2.5 days due to vigorous tidal action. Even so, it is estimated that between 1960 and 1975 the rate of mercury elimination from the Bay to the sea was only 0.4% per year, which could easily be accounted for by sediment transport. The rate of movement of mercury from the Bay to the sea increased by approximately a factor of 10 in the period 1975–78, which the authors of the report hypothesize was caused by increased sediment movement generated by an increase in the traffic of large ships in the Bay during this period. As supporting evidence for the speed and tenacity with which mercury binds to sediments in the Bay, the investigators point out that the sea sediments just outside of Minamata Bay were not contaminated by the initial mercury discharges, indicating that sorption or other processes immobilizing mercury in the Bay must have occurred very quickly, on a time scale of hours. Thus, whatever the immobilization process is, it appears to be rapid, efficient, and long-lasting.

A detailed study of mercury in a 3-mile (4.9 km) section of the Ottawa River (Kudo, 1983), represents one of the few comprehensive studies in which the transport of mercury in several physical and chemical forms was evaluated simultaneously. This stretch of river had received mercury as an industrial discharge for many years, a practice that was stopped 3 years prior to the start of the study. In addition some mercury is derived from natural weathering reactions in the river's headwaters region. The vast majority of the total mercury in the test section ($>96\%$) was associated with the bed sediments, and in conformity with the study of Minamata Bay described above, this mercury was not easily released to the solution phase. Thus, despite the relatively large reservoir of mercury in the sediments, slow bed sediment velocities limited the significance of sediment-mediated mercury transport. Although the water contained only 13 ng/L total dissolved Hg, it accounted for 58% of the total mercury flux through the system, while 41% could be attributed to suspended soil transport. Of the total mercury in the system, less than 6% was methylmercury. This mercury species was apparently less strongly sorbed to suspended sediments than was inorganic mercury, and as a result the aqueous phase accounted for 80% of its transport, with the remaining 20% moving with the suspended sediments. While the exact percentages would undoubtedly vary with such parameters as suspended sediment load, average water velocity, and sediment composition, these results are useful as a reference point with which other systems can be compared. This study is also representative of the type of analysis necessary for an evaluation of heavy metal movements in water systems and the risks they pose to aquatic and human populations.

Mercury provides an excellent example of the importance of metal speciation in understanding biogeochemical cycling and the impact of human activities on these cycles. Mercury exists in solid, aqueous, and gaseous phases, and is transported among reservoirs in all these forms. It undergoes precipitation–dissolution, volatilization, complexation, sorption, and biological reactions, all of which alter its mobility and its effect on exposed populations. The effect of all these reactions on the environmental behavior of mercury has been indicated in this section. The importance of analyzing speciation is perhaps most evident from the fact that the most toxic form of mercury and the only form that has been identified as having affected large populations of humans, methylmercury, is relatively insignificant in the global mercury balance. In the next section, a similar summary will be presented for copper. Despite significant differences from mercury with respect to certain chemical properties and environmental transport modes, it will once again be shown that an understanding of chemical speciation is critical to an appreciation of the metal's environmental behavior.

15.5.2   Copper

The biogeochemical cycling of copper has been extensively studied. While some questions remain regarding the gross inventories of copper in various environmental reservoirs, studies dating from as early as the turn of the century provide considerable insight into the flux rates among the reservoirs. The speciation of copper in natural waters has also been well-studied, and in recent years a rough consensus seems to have been reached regarding the dominant copper species in "typical" environments, although the exact values of the stability constants of some important copper complexes are still somewhat uncertain.

Unlike mercury, copper is known to be a metabolically essential element for virtually all organisms. It displays a well-established property of many of the heavy metals, being essential to growth at low concentrations and toxic at high concentrations. The requirement for Cu stems from its inclusion in several proteins, in which it is always coordinated to N, S, or O ligands (Moore and Ramamoorthy, 1984). At the other extreme, copper's toxic properties have been exploited to reduce growth of unwanted organisms such as algae and fungus in water bodies and in soils. Copper biogeochemistry has been discussed and reviewed frequently in recent years. In this section, only an outline of the literature will be provided, with emphasis on how the speciation of copper controls its behavior.

15.5.2.1   Global cycling

The global cycling of copper has been reviewed by Nriagu (1979) and is described schematically in Fig. 15-18. Like most heavy metals and in contrast to mercury, the flux of metal from terrestrial to oceanic reservoirs is dominated by transport in rivers. Copper reaching the oceans by atmospheric transport is of the same order of magnitude as that by three strictly anthropogenic sources: direct discharge of wastes and sludge into the oceans, discharge of domestic and industrial wastes into rivers which eventually discharge into the ocean, and the use of copper-containing anti-fouling paints. Each of these sources is several hundred times less than the copper burden carried to the oceans by riverine runoff as part of the natural biogeochemical cycle. The transport that does occur through the atmosphere is essentially all via particulate matter, not transport of gaseous copper species, and roughly 90% of this particulate matter is injected into the atmosphere as a result of smelting, fossil fuel burning and other human activities. Despite the relative insignificance of human inputs to the global cycle of copper, atmospheric transport is thought to be responsible for most of the transport to some inland water systems, e.g. the Great Lakes, and there is no question that in localized ecosystems anthropogenic inputs can dominate over natural ones.

15.5.2.2   Copper in rocks, minerals, and atmospheric dust

Copper exists in crustal rocks at concentrations ranging from about 10 to a few hundred ppm, with 70 ppm being about average. In addition, at least 20 copper minerals have been identified, containing copper in the 0, +I, or +II oxidation state. These are primarily sulfides, hydroxides, and carbonates, of which chalcopyrite ($CuFeS_2$), is most common. Copper is also found in relatively high concentrations in deep-sea ferromanganese nodules, in many cases at concentrations

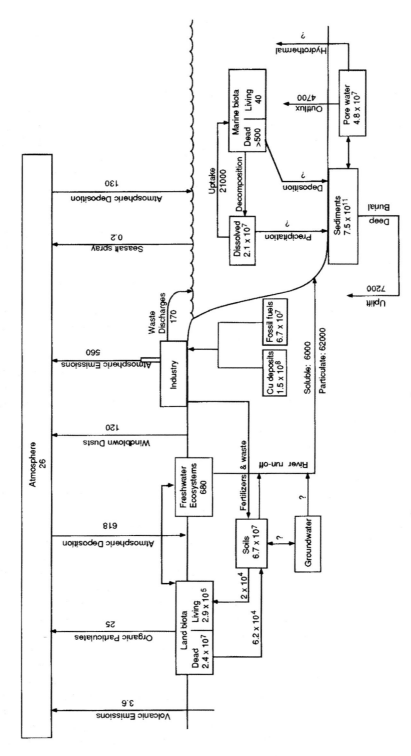

Fig. 15-18 The global copper cycle. Units are 10^8 g Cu (burdens) and 10^8 g Cu/yr (fluxes). (Reprinted with permission from J. O. Nriagu (1979). "Copper in the Environment, Part I: Ecological Cycling", Wiley-Interscience, NY.)

greater than 0.5% and occasionally greater than 1.0%.

Solids containing oxidized anions (carbonates, sulfates, hydroxides, and oxides) are the dominant forms of Cu in airborne particulate matter. In the few studies that have addressed the reactions of these particles in atmospheric washout, about 50% of the copper has been found to be soluble. Since the solubility is strongly dependent on pH, acid precipitation and acidification of receiving waters may have a significant effect on the form and fate of airborne copper.

15.5.2.3 Reactions of copper in aquatic systems

15.5.2.3.1 Redox reactions and complexation.
Copper in solution, as in solids, can carry either a +1 or +2 charge. The divalent form is the stable one in oxygenated water and is the prevalent form in natural waters. The distinction between the two forms is particularly significant to copper transport because they tend to form complexes with different ligands. Cuprous (Cu^+) ions form strong bonds with Cl, which probably controls its speciation in marine systems, but these have been little studied since Cu(I) is rarely detected in such systems other than in anoxic sediments. By contrast, cupric ions (Cu^{2+}) form strong complexes with hydroxide, carbonate, phosphate, and ammonia, and its complexes with organic molecules are typically the second strongest (after mercury) of any of the heavy metals. Even in waters with low concentrations of dissolved organic carbon, dissolved copper is associated primarily with organic matter, i.e., humic or fulvic acids. Turner *et al.* (1981) computed that in seawater with a total humic acid concentration of 10^{-6} M as carbon, organic complexes could account for 47% of the total dissolved copper, 2% of the lead, and less than 0.1% of the Mn, Co, Ni, Zn, Cd, or Hg. Metals other than Cu are prevented from complexing with the organics by a combination of competition for the ligand by Ca^{2+} and Mg^{2+} and complexation of the metals themselves by inorganic ligands.

$$Me^{n+} + (Cl^- \text{ or } CO_3^{2-}) \leftrightarrow Me/anion \text{ complex}$$

$$(Ca^{2+} \text{ or } Mg^{2+}) + Hum \leftrightarrow (CaHum \text{ or } MgHum)$$

In seawater most of the non-organically bound Cu^{2+} is complexed with inorganic carbon. In freshwater, because both types of competition shown above are less significant than in seawater, and because humic and fulvic acid concentrations are greater than in seawater, all metals, including Cu^{2+}, are more likely to exist as organic complexes. Thus, frequently >90% of the dissolved copper in rivers is reported to be organically bound. (It should be remembered that this may include significant quantities of suspended Cu–Fe–humate colloids.) The tendency to form such complexes would be even greater in waters contaminated by organic discharges.

Equilibrium complexation constants for Cu reactions with natural organic matter and the details of Cu speciation are bound to remain somewhat uncertain, since the composition of the complexing molecules varies from site to site. What is not in dispute is that the fraction of dissolved copper present as free aquo Cu^{2+} is probably very small in any natural water. In extremely pristine waters, hydroxide and carbonate complexes may dominate, but organic complexes usually dominate in waters containing more than a few tenths of a mg/L organic carbon.

15.5.2.3.2 Adsorption.
Considering the similarities between adsorption and complexation reactions, it is not surprising that Cu^{2+} is among the most strongly sorbing of the heavy metals, and that it sorbs onto both inorganic and organic solids. Sorption of copper onto oxides and clays has been investigated extensively, both because of its importance to biogeochemical cycling and as part of studies of the availability of soil-bound copper to plants. Typical results showing the relative adsorptive strength of copper and several other metals in inorganic systems were presented in Fig. 15-5.

In a completely inorganic system, binding of Cu^{2+} to suspended and bottom sediments would remove much of the Cu from the dissolved phase. The formation of Cu–organic complexes and the presence of organic solids complicates the analysis, however. In essence, free Cu^{2+} has a strong tendency to react with many components of the system, and when these components are tested one at a time, each is able to sequester most of the Cu in the system. In a real system these reactants compete and interact with one another, leading to complex speciation patterns that depend on the relative concentrations of each component and solution pH. The general picture that emerges from studies of the speciation of Cu in real and simulated natural waters is that natural organic complexing agents can bind Cu^{2+} so strongly that direct copper-to-surface bonding is effectively inhibited. Rather, the copper remains bonded to the organic matter under almost all conditions realistically expected in natural waters, and is adsorbed only under conditions where the organics sorb. Copper bound to organic particulates, or to organic matter adsorbed onto inorganic particulates, is thus the most likely form of Cu to reach bottom sediments, and speciation studies have consistently found Cu to be primarily associated with organics in aerobic sediments. The role of coagulation of colloidal matter in facilitating transport of Cu to estuarine sediments has already been discussed. In anaerobic sediments, reduction of SO_4^{2-} to S^{2-} leads to precipitation of extremely insoluble CuS or mixed metal sulfides containing Cu.

15.5.2.4 Biological effects

The strength of the copper–organic bond and the well-established dual potential of copper as a biological stimulant or inhibitor has led to many studies of the effect of copper speciation on growth. A few of these studies were mentioned earlier, showing that strong organic and inorganic complexes of most metals, including Cu, are non-toxic. The general conclusion that hydrated Cu^{2+} appears to be more biologically active than chelated or certain organically or inorganically complexed forms, bears repeating. This fact becomes more significant when put in the context of the very small ratio of Cu^{2+} to total Cu in most natural waters. More than for any other metal, it has been demonstrated for copper that toxicity does not correlate well with total copper or even total soluble copper in a system. Some uncertainty remains regarding the toxicity of certain inorganic complexes (par-

ticularly the hydroxo- and carbonato-complexes) and the roles of alkalinity, hardness, and other heavy metals in modifying copper toxicity. Nevertheless, studies of Cu toxicity interpreted in terms of speciation have been extremely successful and have helped establish the value of such an approach. Analysis of speciation has shifted from the state-of-the-art to the mainstream in toxicity studies in the past decade and will undoubtedly continue to shed light on the biological effects of all heavy metals.

15.6 Summary

This chapter provides an overview of the important processes and reactions controlling biogeochemical cycling of metals. While each metal has some unique properties, several generalizations apply. The most important of these have to do with metal speciation, that is, the physical–chemical form in which the metal exists in a given environment. Because metals tend to undergo a greater number and variety of relatively rapid, reversible reactions than the elements discussed in previous chapters, they are found in the environment in solid, aqueous, or gaseous phases, associated with literally thousands of different compounds. These reactions often reflect the affinity of metal ions for other atoms with free electron pairs, in particular O, N, or S.

The critical processes controlling global metal cycling are volatilization (exchange between aqueous and gaseous phases), adsorption–precipitation–dissolution (exchange between aqueous and solid phases), and complexation (conversion among various dissolved forms of the metal). For most metals, transport as a gaseous species is of little quantitative importance except in very high-temperature environments. A few metals, most notably mercury, can exist as gases at ambient temperatures, and several metals form volatile organometallic compounds that may dominate transport of the metal in local environments. Transport of particles suspended in the air is an important process for distributing many metals to regions far from their sources.

Transport in solution or aqueous suspension is the major mechanism for metal movement from the land to the oceans and ultimately to burial in ocean sediments. In solution, the hydrated metal ion and inorganic and organic complexes can all account for major portions of the total metal load. Relatively pure metal ores exist in many places, and metals from these ores may enter an aquatic system as a result of weathering. For most metals a more common sequence is for a small amount of the ore to dissolve, for the metal ions to adsorb onto other particulate matter suspended in flowing water, and for the metal to be carried as part of the particulate load of a stream in this fashion. The very insoluble oxides of Fe, Si, and Al (including clays), and particulate organic matter, are the most important solid adsorbents on which metals are "carried."

The distribution of metals between dissolved and particulate phases in aquatic systems is governed by a competition between precipitation and adsorption (and transport as particles) versus dissolution and formation of soluble complexes (and transport in the solution phase). A great deal is known about the thermodynamics of these reactions, and in many cases it is possible to explain or predict semi-quantitatively the equilibrium speciation of a metal in an environmental system. Predictions of complete speciation of the metal are often limited by inadequate information on chemical composition, equilibrium constants, and reaction rates.

Metals act upon and are acted upon by biota in important ways. Through their effect on the chemical environment in soils, sediments, or open bodies of water, organisms can help dissolve, complex, or precipitate metals. Additionally, as noted above, suspended organic particulates can be important adsorbents for metals, carrying them through an aquatic system to its outlet or through a water column to the bottom sediments. Organisms can also directly mediate reactions involving metals, such as by formation of organometal compounds.

Metals, in turn, can either stimulate or inhibit biological activity. In this regard, the different effects of dissolved metal species are especially important. For most metals, the hydrated metal ion and simple inorganic complexes have a much greater effect on organisms (in both the stimulatory and inhibitory ranges) than large organic complexes or adsorbed metal ions.

The full appreciation of the overriding importance of metal speciation in evaluating the transport and effects of metals in an environment is a relatively recent event. As more information is gathered on the forms in which metals exist and are transported through various environmental compartments, it will become possible to predict more accurately the response of the biological communities exposed to the metals and hopefully avert or mitigate the adverse effects.

Questions

15-1 Imagine that an industrial spill allows an acidic solution containing chromate ion (CrO_4^{2-}) to enter a turbid, slow-moving stream containing organic-rich runoff. The streamwater pH is lowered to about 5 by the spill. As the water moves downstream, some of the particles settle, eventually being buried in the sediment. Other particles remain suspended, and as additional tributaries mix in, the pH of the water returns to a normal level of around 7.5. Describe the various fates that chromate ions might experience in this system.

15-2 In the treatment of water for public consumption, iron chloride is frequently added to the water. After pH adjustment, the iron forms a precipitate of $Fe(OH)_3$, which is thought to enhance coagulation and subsequent settling of suspended colloids. Discuss what you think would happen to dissolved copper in the water when this iron solid precipitates. Consider two "types" of water supply: a relatively pristine supply generated by snowmelt from a remote area, and a supply high in natural organic matter (humic matter) generated by percolation of rainwater through organic-rich vegetated zones.

15-3 Using the equilibrium constants below, calculate the concentrations of free (uncomplexed) cadmium ion in a freshwater with a chloride concentration of 15 mg/L, and in seawater containing 17 000 mg/L chloride. Ignore complexation with other ions.

$$Cd^{2+} + Cl^- \leftrightarrow CdCl^- \qquad \log K_1 = 2.0$$
$$CdCl^- + Cl^- \leftrightarrow CdCl_2 \qquad \log K_2 = 0.7$$
$$CdCl_2 + Cl^- \leftrightarrow CdCl_3^- \qquad \log K_3 = 0.0$$

15-4 Calculate the specific surface site concentration (moles of surface sites per gram) for quartz (SiO$_2$) particles 1 μm in diameter. Assume a site density, n_s, of 5 sites/nm^2 (1 nm = 10^{-9} m). What will be the concentration of silica surface sites if the suspended quartz is present at 2 ppm (mg/L)?

References

Allen, H. E., Fu, G. and Deng, B. (1993). Analysis of acid-volatile sulfide (AVS) and simultaneously extracted metals (SEM) for the estimation of potential toxicity in aquatic sediments. *Environ. Toxicol. Chem.* **12**, 1441–1453.

Andreae, M. O. (1979). Arsenic speciation in seawater and interstitial waters: the role of biological-chemical interactions on the chemistry of a trace element. *Limnol. Oceanog.* **24**, 440–452.

Ankley, G. T., Di Toro, D. M., Hansen, D. M. and Berry, W. J. (1996). Technical basis and proposal for deriving sediment quality criteria for metals. *Environ. Toxicol. Chem.* **15**, 2056–2066.

Avotins, P. V. (1975). Adsorption and coprecipitation studies of mercury on hydrous iron oxides. Ph.D. Dissertation, Stanford University, Stanford, CA.

Baird, C. (1995). "Environmental Chemistry." W. H. Freeman and Co., New York.

Balistrieri, L., Brewer, P. G. and Murray, J. W. (1981). Scavenging residence times of trace metals and surface chemistry of sinking particles in the deep ocean. *Deep-Sea Res.* **28A**, 101–121.

Bergan, T., Gallardo, L. and Rodhe, H. (1999). Mercury in the global troposphere: A three-dimensional study. *Atmos. Environ.* **33**, 1575–1585.

Bertine, K. K. and Goldberg, E. D. (1971). Fossil fuel combustion and the major sedimentary cycle. *Science* **173**, 233–235.

Broecker, W. S. (1974). "Chemical Oceanography." Harcourt Brace Jovanovich, New York.

Bruland, K. W., Bertine, K., Koide, M. and Goldberg, E. D. (1974). History of metal pollution in Southern California coastal zone. *Environ. Sci. Technol.* **8**, 425–432.

Bruland, K. W., Knauer, G. A. and Martin, J. H. (1978). Cadmium in Northeast Pacific waters. *Limnol. Oceanog.* **23**, 618–625.

Coles, D. G., Ragaini, R. C., Ondor, J. M., Fisher, G. L., Silberman, D. and Prentice, B. A. (1979). Chemical studies of stack fly ash from a coal-fired power plant. *Environ. Sci. Technol.* **13**, 455–459.

Davies-Colley, R. J., Nelson, P. O. and Williamson, K. J. (1984). Copper and cadmium uptake by estuarine sedimentary phases. *Environ. Sci. Technol.* **18**, 491–499.

Davis, J. A. (1984). Complexation of trace metals by adsorbed natural organic matter. *Geochim. Acta* **48**, 679–691.

Davis, J. S. and Leckie, J. O. (1978). Effect of adsorbed complexing ligands on trace metal uptake by hydrous oxides. *Environ. Sci. Technol.* **12**, 1309–1315.

Di Toro, D. M., Mahony, J. D., Hansen, D. J., Scott, K. J., Hinks, M. B., Mayhr, S. M. and Redmond, M. S. (1990). Toxicity of cadmium in sediments: the role of acid volatile sulfide. *Environ. Toxicol. Chem.* **9**, 1487–1502.

Eisenreich, S. J., Hoffman, M. R., Rastetter, D., Yost, E. and Maier, W. J. (1980). Metal transport phases in the upper Mississippi River. *In* "Particulates in Water" (M. Kavanaugh and J. O. Leckie, eds), Adv. in Chem. Series #189. American Chem. Soc., Washington, DC.

Eisenrich, S. J., Metzer, N. A. and Urban, N. R. (1986). Response of atmospheric lead to decreased use of lead in gasoline. *Environ. Sci. Technol.* **20**, 171–174.

Erickson, R. J., Benoit, D. A., Mattson, V. R., Nelson, H. P. and Leonard, E. N. (1996). The effects of water chemistry on the toxicity of copper to flathead minnows. *Environ. Toxicol. Chem.* **15**, 181–193.

Galloway, J. N. (1979). Alteration of trace metal geochemical cycles due to the marine discharge of wastewater. *Geochim. Cosmochim. Acta* **43**, 207–218.

Goldschmidt, V. M. (1954). "Geochemistry." Oxford University Press, Fairlawn, NJ.

Hansen, D. J., Berry, W. J., Mahony, J. D., Boothman, W. S., Di Toro, D. M., Robson, D. L., Ankley, G. T., Ma, D., Tan, Q. and Pesch, C. E. (1996). Predicting the toxicity of metal contaminated field sediments using interstitial concentration of metals and acid-volatile sulfide normalizations. *Environ. Toxicol. Chem.* **12**, 2080–2094.

Hardy, J. T., Apts, C. W., Crecelius, E. A. and Fellingham, G. W. (1985). The sea-surface microlayer: fate and residence times of atmospheric metals. *Limnol. Oceanog.* **30**, 93–101.

Helz, G. R. (1976). Trace element inventory for the Northern Chesapeake Bay with emphasis on the influence of man. *Geochim. Cosmochim. Acta* **40**, 573–580.

Hemmes, P., Rich, L. D., Cole, D. L. and Eyring, E. M. (1971). Kinetics of hydrolysis of ferric ion in dilute aqueous solution. *J. Phys. Chem.* **75**, 929–932.

Hoffman, M. R. (1981). Thermodynamic, kinetic and extra-thermodynamic considerations in the development of equilibrium models for aquatic systems. *Environ. Sci. Technol.* **15**, 345–353.

Imhoff, K. R., Koppe, P. and Dietz, F. (1980). Heavy metals in the Ruhr River and their budget in the catchment area. *Progr. Water Technol.* **12**, 735–749.

Jones, C. J. and Murray, J. W. (1984). Nickel, cadmium and copper in the northeast Pacific off the coast of Washington. *Limnol. Oceanog.* **29**, 711–720.

Keck, J. C. (1978). Rate-controlled constrained equilibrium method for treating reactions in complex systems. *In* "Maximum Entropy Formalism" (R. D. Levine and M. Tribus, eds). M.I.T. Press, Cambridge, MA.

Kudo, A. (1983). Physical/chemical/biological removal mechanisms of mercury in a receiving stream. *In* "Toxic Materials – Methods for Control" (N. E. Armstrong and A. Kudo, eds). The Center for Research in Water Resources, University of Texas at Austin, Austin, TX.

Kudo, A. and Miyahara, S. (1983). Migration of mercury from Minamata Bay. *In* "Toxic Materials – Methods for Control" (N. E. Armstrong and A. Kudo, eds). The Center for Research in Water Resources, University of Texas at Austin, Austin, TX.

Langmuir, D. (1997). "Aqueous Environmental Geochemistry." Prentice-Hall, New York.

Lantzy, R. J. and Mackenzie, F. T. (1979). Atmospheric trace metals: global cycles and assessment of man's impact. *Geochim. Cosmochim. Acta* **43**, 511–525.

Lasaga, A. (1980). The kinetic treatment of geochemical cycles. *Geochim. Cosmochim. Acta.* **44**, 815–828.

Leckie, J. O., Appleton, A. R., Ball, N. B., Hayes, K. F. and Honeyman, B. D. (1986). Adsorptive removal of trace elements from fly-ash pond effluents onto iron oxyhydroxide. Final Report EPRI-RP-910-1, Electric Power Research Institute, Palo Alto, CA.

Lerman, A. (1979). "Geochemical Processes: Water and Sediment Environments." Wiley-Interscience, New York.

Mackay, D. (1979). Finding fugacity feasible. *Environ. Sci. Technol.* **13**, 1218–1223.

MacKenthum, K. M. (1969). "The Practice of Water Pollution Biology." US Dept of Interior, Federal Water Pollution Control Administration.

Manahan, S. E. (1979). "Environmental Chemistry," 3rd edn. Willard Grant Press, Boston, MA.

Mason, R. P., Fitzgerald, W. F. and Morel, F. M. M. (1994). The biogeochemical cycling of elemental mercury: Anthropogenic influences. *Geochim. Cosmochim. Acta* **58**, 3191–3198.

Moore, J. W. and Ramamoorthy, W. (1984). "Heavy Metals in Natural Waters: Applied Monitoring and Impact Assessment." Springer-Verlag, New York.

Morel, F. M. M. (1983). "Principles of Aquatic Chemistry." Wiley-Interscience, New York.

Morel, F. M. M., Westall, J. C., O'Melia, C. R. and Morgan, J. J. (1975). Fate of trace metals in Los Angeles County wastewater discharge. *Environ. Sci. Technol.* **9**, 756–761.

Morgan, J. J. and Stone, A. T. (1985). Kinetics of chemical processes of importance in lacustrine environments. *In* "Chemical Processes in Lakes" (W. Stumm, ed.). Wiley, New York.

NAS (1978). "An Assessment of Mercury in the Environment," National Academy of Sciences, Washington, DC.

Nelson, M. B., Davis, J. A., Benjamin, M. M. and Leckie, J. O. (1977). "The Role of Iron Sulfides in Controlling Trace Heavy Metals in Anaerobic Sediments: Oxidative Dissolution of Ferrous Monosulfides and the Behavior of Associated Trace Metals." Air Force Weapons Laboratory, Technical Report 425.

Nriagu, J. O. (1979). "Copper in the Environment, Part I: Ecological Cycling." Wiley-Interscience, New York.

Pankow, J. F. and J. J. Morgan (1981). Kinetics for the aquatic environment. *Environ. Sci. Technol.* **15**, 1155–1164.

Paulson, A. J., Feely, R. A., Curl, H. C. and Gendron, J. F. (1984). Behavior of Fe, Mn, Cu, and Cd in the Duwamish River estuary downstream of a sewage treatment plant. *Water Res.* **18**, 633–641.

Ripley, E. A., Redman, R. E. and Crowder, A. A. (1996). "Environmental Effects of Mining." St. Lucie Press, Delray Beach, FL.

Schwartzenbach, R. P. and Imboden, P. M. (1983). Modelling concepts for hydrophobic pollutants in lakes. *Ecol. Modelling* **22**, 171.

Settle, D. M. and Patterson, C. C. (1982). Magnetites and sources of precipitation and dry deposition fluxes of industrial and natural leads to the North Pacific at Enewetak. *J. Geophys. Res.* **87**, 8857–8869.

Settle, D. M., Patterson, C. C., Turekian, K. K. and Cochran, J. K. (1982). Lead precipitation fluxes at tropical ocean sites determined from ^{210}Pb measurements. *J. Geophys. Res.* **87**, 1239–1245.

Sholkovitz, E. R. and Copland, D. (1981). The coagulation, solubility, and adsorption properties of Fe, Mn, Cu, Ni, Cd, Co, and humic acids in a river water. *Geochim. Cosmochim. Acta* **45**, 181–189.

Stumm, W. and Morgan, J. J. (1981). "Aquatic Chemistry," 2nd edn. Wiley-Interscience, New York.

Sunda, W. G., Engel, D. W. and Thuotte, R. M. (1978). Effect of chemical speciation on toxicity of cadmium to grass shrimp, *Palaemonetes pugio*: importance of free cadmium ion. *Environ. Sci. Technol.* **12**, 409–413.

Swartz, R. C. Ditzworth, G. R., Schultz, D. W. and Lamberson, J. O. (1985). Sediment toxicity to a marine infaunal amphipod: cadmium and its interaction with sewage sludge. *Mar. Environ. Res.* **18**, 133–153.

Takamatsu, T., Kawashima, M. and Koyama, M. (1985). The role of Mn-rich hydrous manganese oxide in the accumulation of arsenic in lake sediments. *Water Res.* **19**, 1029–1032.

Theis, T. L. and Wirth, J. L. (1977). Sorptive behavior of trace metals on fly ash in aqueous systems. *Environ. Sci. Technol.* **11**, 1096–1100.

Turekian, K. K. (1971). Rivers, tributaries, and estuaries. *In* "Impingement of Man on the Ocean" (D. W. Wood, ed.). Wiley, New York.

Turekian, K. K. (1977). The fate of metals in the oceans. *Geochim. Cosmochim. Acta* **41**, 1139–1144.

Turner, D. R., Whitfield, M. and Dickson, A. G. (1981). The equilibrium speciation of dissolved components in freshwater and seawater at 25°C and 1 atm pressure. *Geochim. Cosmochim. Acta* **45**, 855–881.

Weiss, H. V., Koide, M. and Goldberg, E. D. (1971a). Selenium and sulfur in a Greenland ice sheet: relation to fossil fuel combustion. *Science* **172**, 261–263.

Weiss, H. V., Koide, M. and Goldberg, E. D. (1971b). Mercury in a Greenland ice sheet: evidence of recent input by man. *Science* **174**, 692–694.

Wood, J. (1974). Biological cycles for toxic metals in the environment. *Science* **183**, 1049–1052.

Part Four

Integration

Earth system science consists of both reductionist and integrated components. Parts of this transdisciplinary field are clearly disciplinary, exemplified by the chapters in Part Two. Other parts are more integrative in the sense that the disciplines of the Earth Sciences plus chemistry, physics, and biology are all needed in the study of the elemental cycles as found in Part Three. Yet other levels of integration emerge when specific aspects of the Earth system are examined. We include in Part Four examples ranging from an integrated view of the acid–base and redox balances of Earth, to the coupling of biogeochemical cycles and climate, to a closely related view of the record of biogeochemistry as found in ice sheets and glaciers, to the many ways that humans modify the biogeochemistry of Earth. Undoubtedly, there are many other examples of integration that could be developed; however, incomplete though these may be, they do convey a variety of ways in which Earth systems can be viewed. As we cautioned in Chapter 1, the whole of Earth system science as an integrative discipline is a new and emerging branch of science. It is far from complete; yet, it is clear from the examples in Part Four that many aspects of science that are important to the life and even the existence of humans require explicit integration in order to be applied. We repeat that this integration does not happen on its own and that it requires a global systems view that is often or usually missing in reductionist science (see Section 1.7 of Chapter 1).

Chapter 16 considers a broad but closely related set of chemical processes that control the acid–base and oxidation–reduction (or redox) balances of the planet. It will be seen that both sets of processes depend greatly on the chemical properties and the biogeochemical cycling of the same key elements (C, N, O, S) and of course, water. Further, it will become apparent that all of the reservoirs of the planet from Part Two are involved. While some of the acid–base and redox processes can be viewed usefully from a single sphere or single discipline perspective, they all eventually have connections of one or another sort to the Earth system and the broader integrative picture.

Climate and its connections to biogeochemistry are the focus of Chapter 17. The fundamental nature of the inherently variable climate of Earth presents numerous puzzles, especially the currently inexplicable stability of climate for the last ca. 10^4 years. It seems possible that this stability is somehow connected to biogeochemistry, again through the key elements C, N, O, and S. In order to explore these connections, it is necessary to define the key processes that contribute to climate. It also is necessary to clearly define and delineate the concepts of *forcings*, *feedbacks*, and *responses*. This delineation allows examination of the roles of biogeochemistry, and most especially of feedbacks. The latter can be viewed as a hierarchy that range from the simplest physical ones to the most complex biogeochemical ones.

Gaining a historical perspective on how the Earth functions in a climatic and biogeochemical sense is made possible largely through studies of the chemistry and isotopic character of ice in the large ice sheets (mainly antarctic and Greenland) and in glaciers. Chapter 18 presents an integrated review of the record of the Earth's biogeochemistry that is obtained by chemical and physical analysis of old ice, some of which has been sequestered for hundreds of thousands

Earth System Science
ISBN 0-12-379370-X

of years. Importantly, all of this old ice originated in the atmosphere; materials trapped in it were subject to atmospheric transport and processing, requiring an understanding of atmospheric chemistry for interpretation of the record. While some evidence, especially for times greater than are represented in old ice (e.g., more than a few hundred thousand years) is provided by sedimentary and even metamorphic rocks, the old ice constitutes the most detailed and complete record – by far. Perhaps the most important record is that of gases (especially CO_2 and CH_4) trapped in the ice, literally providing paleosamples of the atmosphere. Isotopes, especially ^2H (deuterium, D) and ^{18}O in the frozen water itself yield proxy information of temperature.

Finally, Chapter 19 addresses the increasingly apparent modification of the biogeochemical processes of Earth by humans. Major changes such as the ca. 30% increase in atmospheric CO_2, increases in methane and N_2O concentrations, and rainwater acidification are among the more obvious examples, while changes in the biosphere such as de- and re-forestation and changes in species diversity are evident. Again, it will be seen that the key elements of C, N, O, and S come into play. Isotopes of H, C, and O play important roles, especially for inference of climatic changes.

16

The Acid–Base and Oxidation–Reduction Balances of the Earth

Robert J. Charlson and Steven Emerson

16.1 Introduction

Biogeochemical cycles interact with each other in many complex ways. Among these interactions are two related and important chemical features of the planet: acid–base balances and oxidation–reduction systems. The acid–base balances have a wide range of consequences in a variety of aqueous phase systems, including control of the weathering and solubility of many minerals, biological influences (including toxicity), and control of numerous aqueous phase and heterogeneous reaction rates. While it is often taken for granted that one or another acid–base balance is a fixed parameter, these systems are in fact extremely sensitive to small changes in the controlling biogeochemical factors and also to anthropogenic influences such as "acidic precipitation."

It is also often taken for granted that many of the Earth's subsystems are exposed to free oxygen (O_2), leading to a range of one-way reactions of reduced materials (such as organic carbon or metal sulfides) to an oxidized form. As pointed out many times in earlier chapters, the oxidation–reduction status of the planet is the consequence of the dynamic interactions of biogeochemical cycles. As is the case with the acid–base balances, there is considerable sensitivity to perturbations of "redox" conditions, sometimes dramatically as in the case of bodies of water that suddenly become anaerobic because of eutrophication. Another extreme example is the loss of ozone in the so-called "ozone hole."

16.1.1 Similarities and Differences of the Fundamental Concepts of Acid–Base and Oxidation–Reduction Balances

Referring to the discussion of the fundamental concepts regarding half cells and the Nernst equation in Chapter 5 (Section 5.3.1) it is possible to briefly summarize the similarities and differences of these two sets of systems. It is important to recognize the ways in which they are different when considering the behavior of complex multivariate systems such as the oceans and clouds, or a lake–river system.

First, the simple thermodynamic description of $p\varepsilon$ (or E_h) and pH are both most directly applicable to the liquid aqueous phase. Redox reactions can and do occur in the gas phase, but the rates of such processes are described by chemical kinetics and not by equilibrium concepts of thermodynamics. For example, the acid–base reaction

$$NH_3\,(g) + HCl\,(g) \rightarrow NH_4Cl\,(gas\ or\ solid)$$

and the redox reaction

$$CO + OH\cdot \rightarrow CO_2H\cdot$$

are known or supposed to occur in the atmosphere. However, they are not normally considered along with the acid–base and redox

Earth System Science
ISBN 0-12-379370-X

systems in the aqueous phase. One of the most important reasons for separating the aqueous and gas phases is that the transfer between them is often governed by physical transport rates and the systems are usually not in thermodynamic equilibrium. Hence, we find that the gas phase and aqueous phase systems are treated separately, although it might at first seem logical to couple them.

Second, the concepts of pH and $p\varepsilon$ (or E_h) for aqueous phase systems can be written down as mathematical analogs, as cited by Stumm and Morgan (1970, 1981):

$$pH = -\log_{10}[H^+]$$

$$p\varepsilon = -\log_{10}[e^-]$$

where $[e^-]$ is the activity of free electrons in the system. Just as a large value of $[H^+]$ implies a low pH and high relative acidity, a large value of $[e^-]$ implies low $p\varepsilon$ and strong reducing conditions, and vice versa. However mathematically convenient this may be, it fails to adequately recognize the true nature of the fundamental differences of acid–base and redox reactions.

Acid–base reactions are strictly involved with the exchange of protons (H^+ ions) for example, within the aqueous phase. Redox reactions, on the other hand, involve a change in *oxidation state* of the reactants, requiring the exchange of electrons from one chemical species to another. Indeed, while charge balance is maintained in both redox and acid–base reactions, the former involves changes in oxidation state while the latter does not. The change in the oxidation state or electron density around an element has a more pronounced influence on ultimate molecular form and thus other important properties (solubility, reaction pathway, toxicity, etc.) than does the loss or gain of hydrogen ions.

To summarize, understanding the acid–base and redox systems of the Earth requires careful, if separate, consideration of the aqueous and gas phases and their interplay with solid phases. It also requires recognition of the fundamental similarities and differences of these two related systems.

16.1.2   *Reservoirs in which Acid–Base and/or Redox Reactions and Balances Occur*

Foregoing any discussion of the very slow processes within the lithosphere, the immediate focal points are the atmosphere, the hydrosphere, and the interfaces between them, and the solid phases (sediment, the pedosphere, and lithosphere). For the aqueous phase the reservoirs are:

- Condensed water in clouds.
- Rainwater/snow water.
- Freshwater on the continents/land:
  - surface runoff
  - groundwater
  - streams and rivers
  - lakes.
- Saline waters:
  - saline lakes and inland seas
  - oceans.

In addition to these relatively simple liquid phase aqueous systems, it is necessary to identify situations in which any of these aqueous phase reservoirs come into physical and chemical contact with solid surfaces (e.g., rocks, biomass, sediments, soils, magma etc.). In general, the presence of two or more phases (liquid plus one or more solid phase) provides important constraints on the chemical reactions that may occur within the system as a whole.

This hierarchical list also conveys the notion of a range of system response times and sensitivities, with the atmospheric cases being fast and sensitive to perturbations and the oceans being slowest and least subject to short-term changes.

We will first consider acid–base balances, then redox systems. Finally, we will illustrate in conclusion that both the ultimate H^+ ion concentration (pH) and electron concentration ($p\varepsilon$) result from interactions of biogeochemical cycles.

16.2   A Hierarchy of Acid–Base Balances

The simplest acid–base relationships should involve the smallest number of molecular species *and* the smallest number of phases. Adding species and involving more phases produces a hierarchy of increasing chemical complexity and, consequently, increasing algebraic difficulty in describing equilibrium conditions. The simplest case in our list of reservoirs is liquid water in clouds, even though this putative simplest system actually requires consideration of ten or more variables simultaneously.

Before proceeding through a hierarchy of examples, a word about the term *equilibrium* is in order, particularly as it applies to the dynamically changing components of the Earth system. It is a fact that any particular chemical system itself will rarely be in true equilibrium, just as the physical systems of Earth are not ever really in a perfect steady state. The equilibrium conditions are extremely relevant because they describe the tendency of the system to which termodynamically favorable reactions tend. That is, no matter what the condition is, all systems are moving toward equilibrium.

16.2.1 Key Acids and Bases

Not surprisingly, the acid–base balances within the Earth system almost all involve elements of high abundance, i.e., elements that have low atomic number. In many cases, the acidic molecule is an oxygen-containing oxidation product of an element. Table 16-1 lists the main acids and bases in the global environment. The sources of these acids are chemical reactions of reduced forms of the element involved. Both gas and aqueous phase reactions exist for production of acids.

- Gas phase examples:

$$OH + SO_2 \rightarrow HOSO_2$$

$$HOSO_2 + O_2 \rightarrow SO_3 + HO_2$$

$$SO_3 + H_2O \rightarrow H_2SO_4 \, (g)$$

- Aqueous phase example:

$$\tfrac{1}{2}H_2O + HS^- + \tfrac{7}{4}O_2 \rightarrow SO_4^{2-} + 2H^+$$

Table 16-1   Dominant acids and bases in the Earth system

A. Acids
1.   Carbon based:
     H_2CO_3 (carbonic acid) aqueous phase
     $HC_2H_3O_2$ (acetic acid) aqueous, gas phases
     $HCHO_2$ (formic acid) aqueous, gas phases
     $H_2C_2O_4$ (oxalic acid) aerosol particles; solid phase
     RCOOH (many carboxylic acids)
     Various: macromolecular acids (fulvic, humic)
2.   Nitrogen based:
     HNO_3 (nitric acid) gas, aqueous, solid phases
     HNO_2 (nitrous acid) gas, aqueous phases
3.   Sulfur based:
     H_2SO_4 (sulfuric acid) aerosol, aqueous phases
     H_2SO_3 (sulfurous acid) aqueous phase
     CH_3SO_3H (from oxidation of $(CH_3)_2S$, methane sulfonic acid) aerosol, aqueous phases

B. Bases
1.   Nitrogen based:
     NH_3, RNH_2, R_2NH, R_3N (ammonia and amines) gas and aqueous phases
2.   Igneous mineral based:
     Al, Mg silicates (weathering reaction in Chapters 8 and 9)
3.   Sedimentary mineral based (examples):
     $CaCO_3$
     $Al(OH)_3$
     (Chapters 8 and 9)

16.2.2 The Simplest Case of Liquid Water in Cloudy Air

Referring back to Chapter 7 (Section 7.7), we see that clouds form at low supersaturations because of vapor-pressure depression by solutes from cloud condensation nuclei (CCN). This solute plus any added from the dissolution of *component* gas phase species (e.g., CO_2, SO_2, NH_3, and HNO_3) are then the starting *constituents* that enter into the interactions that lead toward equilibrium. Given the tiny sizes of the aerosol particles that act as CCN (0.01 to perhaps 0.1 μm) and the small amount of aerosol substances in air (0.1–10 μg/m^3), it might seem possible to simply neglect the solute of aerosol origin and consider only the solutes coming in from the gas phase. Indeed, many textbooks assume that, in the absence of pollution, the dominant solute in cloud and rainwater is dissolved CO_2. After all, 300+ ppm by volume of CO_2 translates to about 0.5 g/m^3 of possible solute, so how could microgram levels matter? As happens to be the case, CO_2 is not very water soluble, and its aqueous form H_2CO_3 is a very weak acid; hence, it will be seen that CO_2 is seldom of any importance to the acid base balance of the cloud/rainwater even though it produces the dominant solute in rivers (HCO_3^-).

We will illustrate the necessity of including solute from CCN by a simple calculation, recalling that pH = 5.6 is the supposed equilibrium value for water in contact with 300 ppm of CO_2. (That calculation will appear later.) In clean, marine air, the concentration of submicrometer aerosol particles (by far the most numerous) is small, say 0.25 μg m^{-3}. It is known from measurements that the molecular form is often NH_4HSO_4, and we assume it is all dissolved in 0.125 g/m^3 of liquid water in a cloud – which is typical for fair-weather marine clouds. Thus the *average* concentration of sulfate ion [SO_4^{2-}], mol/L, is

$$[SO_4^{2-}] = \frac{2.5 \times 10^{-7}\,\text{g/m}^3}{(1.25 \times 10^{-1}\,\text{g/m}^3)(96\,\text{g/mol})}$$

$$\approx 2 \times 10^{-8}\,\frac{\text{mol}}{\text{g H}_2\text{O}} \approx 2 \times 10^{-5}\,\text{mol/L}$$

Because the second dissociation constant of H_2SO_4 is large (ca. 10^{-2}), and because about half the cation in the system is H^+ and half NH_4^+, we then have

$$[NH_4^+] = [H^+] = [SO_4^{2-}] \approx 2 \times 10^{-5}\,\text{mol/L}$$

and pH = 4.7, a value often observed in unpolluted marine settings. Note that pH = 4.7 is almost a factor of 10 more acidic (i.e., a whole unit of pH) than would be the case for CO_2 alone. By this simple calculation, it is easy to see that high levels of H_2SO_4 in polluted air (5–10 μg m^{-3}) can indeed cause low pH rainwater.

Table 16-2 presents what might be termed the minimum set of constituents that must be considered in the case of cloud/rainwater. If we consider the amount of water, L, to be fixed by atmospheric physical processes, the minimum number of input components that can vary are: SO_2, NH_3, CO_2, and whatever solute is present from the CCN, often one or another sulfate compound between H_2SO_4 and $(NH_4)_2SO_4$. Occasionally, salt particles from the ocean surface may be sufficiently abundant to provide enough solute to influence the pH via the inherent alkalinity of seawater, and we will consider that as a second, somewhat more complicated possibility.

Table 16-2 also provides a list of 15 equations that can be solved simultaneously to yield the equilibrium condition (Taylor *et al.*, 1983). Furthermore, if the concentration of each species is calculated as a function of pH (the so-called master-variable diagram or Sillén diagram, named after Sillén (1967) who popularized the method, it is possible to examine various sensitivities in the system, e.g., to the addition of more solute (see explanatory box on Sillén diagrams).

Figure 16-1 is a master-variable diagram corresponding approximately to the previous clean marine case, illustrating that HCO_3^- derived from CO_2 is only important at pH > 7, and that at equilibrium H^+, NH_4^+, and SO_4^{2-} are the dominant species. Figure 16-2 extends this approach to the small population of droplets without any SO_4^{2-} in them that are nucleated on particles of seasalt that is present. In this case, pH = 6.7 and the dominant cation is seawater "alkalinity" or Ak$^+$ (alkalinity in seasalt is the sum of cation concentration due to dissolved

Table 16-2   List of input components for the simplest case of the acid–base balance of unpolluted marine clouds. Also shown are the mass conservation statements, chemical equilibrium expressions and constants, and the requirement for charge balance

Equilibrium	Equilibrium constant	Assumed value at 5°C
(1) $H_2O \leftrightarrow H^+ + OH^-$	$K_w = [H^+][OH^-]$	1.82×10^{-15} mol^2/L^2
(2) SO_2 (g) $+ H_2O \leftrightarrow SO_2 \cdot H_2O$	$K_{HS} = \dfrac{P_{SO_2}}{[SO_2 \cdot H_2O]}$	0.379 atm L/mol
(3) $SO_2 \cdot H_2O \leftrightarrow H^+ + HSO_3^-$	$K_{1S} = \dfrac{[H^+][HSO_3^-]}{[SO_2 \cdot H_2O]}$	0.0206 mol/L
(4) $HSO_3^- \leftrightarrow H^+ + SO_3^{2-}$	$K_{2S} = \dfrac{[H^+][SO_3^{2-}]}{[HSO_3^-]}$	8.88×10^{-8}/L
(5) NH_3 (g) $+ H_2O \leftrightarrow NH_3 \cdot H_2O$	$K_{HN} = \dfrac{P_{NH_3}}{[NH_3] \cdot H_2O]}$	7.11×10^{-3} atm L/mol
(6) $NH_3 \cdot H_2O \leftrightarrow NH_4^+ + OH^-$	$K_b = \dfrac{[NH_4^+][OH^-]}{[NH_3 \cdot H_2O]}$	1.5×10^{-5} mol/L
(7) CO_2 (g) $+ H_2O \leftrightarrow CO_2 \cdot H_2O$	$K_{HC} = \dfrac{P_{CO_2}}{[CO_2 \cdot H_2O]}$	16.6 atm L/mol
(8) $CO_2 \cdot H_2O \leftrightarrow H^+ + HCO_3^-$	$K_{1C} = \dfrac{[H^+][HCO_3^-]}{[CO_2 \cdot H_2O]}$	2.94×10^{-7} mol/L
(9) $HCO_3^- \leftrightarrow H^+ + CO_3^{2-}$	$K_{2C} = \dfrac{[H^+][CO_3^{2-}]}{[CO_2 \cdot H_2O]}$	2.74×10^{-11} mol/L
(10) $\Sigma[CO_2] = C_1$ (mol/m^3)		
(11) $\Sigma[S(IV)] = C_2$ (mol/m^3)		
(12) $\Sigma[N\,(-III)] = C_3$ (mol/m^3)		
(13) $\Sigma[$Strong acid anions$] = C_4$ (mol/m^3)		
(14) Liquid water content $= $ (L/m^3)		
(15) $\Sigma[+] = \Sigma[-]$		

After Charlson and Rodhe (1982).

carbonates such as $CaCO_3$, $MgCO_3$ etc.), while the dominant anions are HSO_3^- from SO_2 and HCO_3^- from CO_2. Interestingly, these two categories of droplets can coexist in the same cloud if the submicrometer sulfate and supermicrometer seasalt particles are not chemically mixed with each other.

16.2.3   Anthropogenic Modifications of the Acid–Base Balance of Rainwater: Alkalinity in Cloud Water "Acid Rain"

It is easy to see from the simple, clean marine case in Fig. 16-1 that the constituents controlling the pH are H^+, NH_4^+ as cations and SO_4^{2-} as the sole important anion. Any ionic solute that is added to this mixture, e.g. H_2SO_4 from the oxidation of anthropogenic SO_2, will simply add more moles of solute to the system. In the real case of acid rain in industrial regions, many solutes are added; including some that are alkaline, trace metals, NH_3, and significant amounts of the strong acids H_2SO_4 and HNO_3. Sometimes, organic acids are also added, particularly, formic, acetic, and oxalic acids. If the acid that is added is a strong acid, the increased anion concentration is essentially pH

Sillén or "Master Variable" Diagrams

Examining the equilibrium expressions in Table 16-2, the equations can be put into a logarithmic form, e.g., for the dissociation of a simple acid, HA,

$$\log [A^-] - \log [HA] = pH + \log K_a \quad K_a = \frac{[H^+][A^-]}{[HA]}$$

becomes

$$\log K_a = \log [H^+] + \log [A^-] - \log [HA]$$

or

$$\log [A^-] - \log [HA] = pH + \log K_a$$

Because this is in the simple algebraic form:

$$y = mx + b$$

where x = pH (the independent variable), y = the dependent variables (log $[A^-]$ − log $[HA]$), m = the slope (plus one), and b is a constant (intercept), it lends itself to a simple graphical presentation. If more equations are needed (as in Table 16-2), each can be put into a form where pH is the independent of "master" variable and each of the species concentrations represents a dependent variable. For many species in very dilute systems $[A^-] \gg [HA]$ so that the plot of log $[A^-]$ vs. pH is a straight line on log–log paper. The same holds for systems with a fixed gas phase, e.g., CO_2. Subsequently the concentrations of the dependent variables are plotted vs. pH, as in Fig. 16-1.

This graphical presentation contains all of the information that is normally used in numerical or algebraic calculations. The advantage of using the graphical method is "in giving in a single glance a clear picture of the situation," as stated by Sillén (1967).

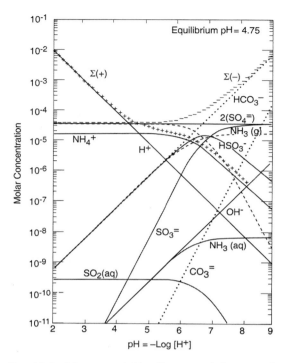

Fig. 16-1 Master-variable diagram of clean marine cloud at a model altitude of 875 m. Equilibrium occurs where $\Sigma[+] = \Sigma[-]$, i.e. charge balance. Input conditions are 0.2 $\mu g/m^3$ of SO_4^{2-} aerosol, roughly half neutralized by NH_4^+ in the amount 0.039 $\mu g/m^3$. 0.1 ppb by volume SO_2 and 0.125 g/m^3 of liquid water, 340 ppm CO_2, temperature is 278 K. (After Twohy *et al.*, 1989.)

independent and the horizontal line for $2[SO_4^{2-}]$ (i.e., the equivalent concentration of sulfate) is subsequently elevated as in Fig. 16-3. Water that lacks material to react with the added acid is said to be unbuffered, as is indeed the usual case for rainwater. The equilibrium pH decreases as the amount of H_2SO_4 added increases from cases c through a.

Another way to illustrate the sensitivity of the acid–base balance to perturbations appears in the form of titration curves, as in Fig. 16-4. While this indicates a significant temperature dependence (due to the dependence of the equilibrium constants and gas solubilities on temperature), the main feature is the strong falloff in pH when the concentration of strong acid anion exceeds that of the weak base, in this case ammonia. Except in certain agricultural areas with high NH_3 source strengths (e.g., dairy-intensive areas like the Netherlands), the total ammonia available in the atmosphere is low, yielding curves like those lower in the figure. This is the essential nature of the problem of acidic precipitation – a large decrease of pH with the addition of small amounts of acid.

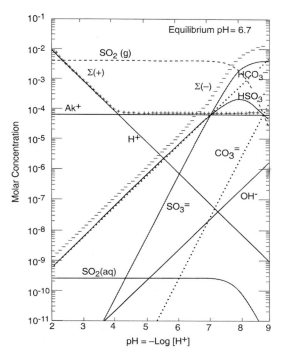

Fig. 16-2 Same as Fig. 16-1, except for the ca. 0.001 g/m³ of water nucleated on alkaline seasalt particles in the same cloud. 1 μg/m³ of seasalt.

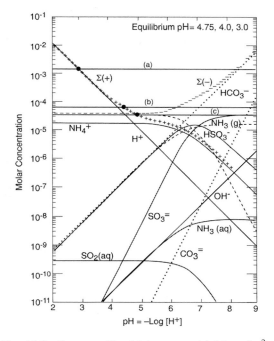

Fig. 16-3 Same as Fig. 16-1, except (a) 10 μg/m³ of H₂SO₄ aerosol; (b) 0.4 μg/m³ and (c) 0.2 μg/m³.

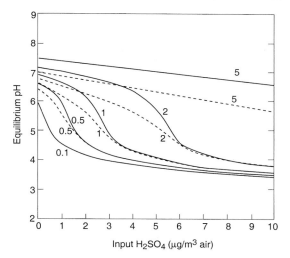

Input H₂SO₄ (μg/m³ air)

Fig. 16-4 pH sensitivity to SO_4^{2-} and NH_4^+. Model calculations of expected pH of cloud water or rainwater for cloud liquid water content of 0.5 g/m. 100 pptv SO_2, 330 ppmv CO_2, and NO_3. The abscissa shows the assumed input of aerosol sulfate in $\mu g/m^3$ and the ordinate shows the calculated equilibrium pH. Each line corresponds to the indicated amount of total $NH_3 + NH_4^+$ in units of $\mu g/m^3$ of cloudy air. Solid lines are at 278 K, dashed ones are at 298 K. The familiar shape of titration curves is evident, with a steep drop in pH as the anion concentration increases due to increased input of H_2SO_4. (From Charlson, R. J., C. H. Twohy and P. K. Quinn, "Physical Influences of Altitude on the Chemical Properties of Clouds and of Water Deposited from the Atmosphere." NATO Advanced Research Workshop Acid Deposition Processes at High Elevation Sites, Sept. 1986. Edinburgh, Scotland.)

16.2.4 Acid–Base Balances in Freshwater Systems

Freshwaters, ranging from surface runoff water to small streams, lakes, and rivers, exhibit a very wide range of solute compositions and concentrations that are dictated by the total history of processes that have influenced the composition, starting with the formation of cloud water. Figure 16-5 shows a Sillén diagram for "average" river water, using data from Stumm and Morgan (1981). Also shown in a dashed line in Fig. 16-5 is the estimate from Fig. 16-1 for the pH in the pristine rainwater case, showing the

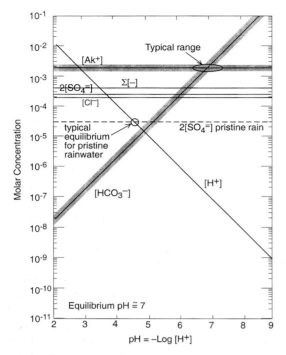

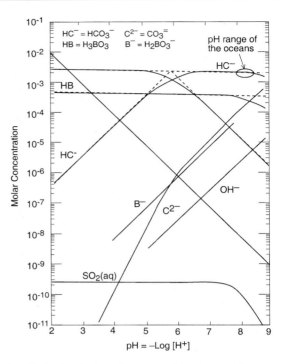

Fig. 16-5 Sillén (master variable) diagram for "average" river water, using data from Stumm and Morgan (1970).

Fig. 16-6 Sillén (master variable) diagram for seawater. Replotted from Sillén's own graph (Sillén, 1967).

radical changes that occur between rainwater and streamwater. Starting with dissolution of minerals at the continental (rock or soil) surface, water is further processed by ion exchange reactions in soils and sediments, adding alkali and alkaline-earth metal ions that dominate the cation load. Anions are added as well, particularly Cl^- and SO_4^{2-} from minerals and weathering reactions and, most importantly, HCO_3^- from dissolved, biologically produced CO_2 in soil water.

The HCO_3^- in rivers is greatly enhanced over that in rain (the latter being due to atmospheric CO_2 alone), mainly because of high CO_2 partial pressure in soils from bacterial and fungal decay of biomass and from root respiration. This CO_2 is highly variable temporally and from place to place, so $[HCO_3^-]$ in soil water and river water is also highly variable. The main cations (Na^+, K^+, Mg^{2+}, and Ca^{2+}) are variable as well, causing a range of compositions as indicated in Fig. 16-5.

16.2.5   The Acid–Base Balance of the Ocean

Referring back to Chapter 10, the chemical composition of the ocean can be approximated by an equilibrium model (actually proposed by Sillén) and can be explained more completely by a dynamical model. Of the constituents in seawater, two emerge as being dominant in controlling the pH: HCO_3^-, and alkalinity (Ak^+) which is the excess of cations like Ca^{2+}, Mg^{2+} etc. above and beyond the primary solute anions Cl^- and SO_4^{2-}. Figure 16-6 is replotted from Sillén's own graph depicting the acid–base balance of seawater. The high pH of seawater, typically $8 < pH < 8.4$ (shown) is a balance of Ak^+ (not shown) and HCO_3^-.

16.3   Oxidation–Reduction Balances of the Earth System

Just as was the case with acid–base interactions, numerous elements are involved in the oxidation–reduction or *redox* processes that may also be thought of in terms that are global. Redox reactions are often broken down into half reactions in which electrons (e^-) are added or taken away from element or molecules (Chapter 5, Section 5.2.1) analogous to hydrogen ions in acid–base reactions. This approach makes possible a clear method for describing equilibrium thermodynamics of the redox reactions.

Oxidation–reduction reactions in water are dominated by the biological processes of photosynthesis and organic matter oxidation. A very different set of oxidation reactions occurs within the gas phase of the atmosphere, often a consequence of photochemical production and destruction of ozone (O_3). While such reactions are of great importance to chemistry of the atmosphere – e.g., they limit the lifetime in the atmosphere of species like CO and CH_4 – the global amount of these reactions is trivial compared to the global O_2 production and consumption by photosynthesis and respiration.

An intriguing aspect of environmental redox reactions is the interplay between thermodynamics and kinetics. Reaction rates vary from a second or less for some of the fast electron transfer reactions in the atmosphere to millions of years for some organic matter degradation reactions. Global-scale redox reactions are driven by photosynthesis and photochemistry, both of which use energy from the sun to create local centers of thermodynamically unstable compounds, like plants of all sizes or reducing compounds in sunlit surfaces of natural waters and the atmosphere. This process sets the stage for the cascade of thermodynamically favorable, energy-yielding reactions that oxidize reduced compounds and tend to decrease the free energy of the planet. In water, nearly all oxidation reactions are microbially (enzymatically) catalyzed, creating a vast range of rates toward thermodynamic equilibrium. While kinetically slow, degradation of organic matter by a variety of oxidation reactions is efficient enough to destroy all but a small amount of the compounds produced with the sun's energy before they are buried. That this small fraction escapes oxidation is extremely important to the global redox balance because it frees an equivalent amount of molecular oxygen to the atmosphere creating an environment suitable for higher forms of life. It also releases sufficient oxygen to produce the ozone layer, blocking out damaging solar UV radiation.

16.3.1   Redox Couples in the Environment

Only a few elements are dominant in the electron transfer reactions in the environment (Table 16-3). There are vastly more elements that undergo redox transformations at Earth temperature and pressures, but carbon, nitrogen, oxygen, sulfur, iron, and manganese are by far quantitatively the most important because of their abundance and different oxidation states. Redox reactions consist of two half reactions (Table 16-4) in which one element gains electrons (it is reduced) and in the other loses electrons (the element is oxidized). The most ubiquitous of these reactions is between oxygen and carbon in which carbon is reduced from CO_2 to organic matter during photosynthesis (Table 16-4a). We use the simple approximation of carbohydrate, CH_2O, to represent organic matter. The opposite reaction oxic respiration is the thermodynamically favored respiration reaction (Table 16-4b). Two other globally important oxidation reactions by O_2 are the oxidation of reduced iron and sulfur (Table 16-4c and d) in sedimentary rocks.

The reduced forms of sulfur and iron and other elements are created during anoxic respiration of organic carbon in the absence of oxygen (Table 16-4e,f). Note that any reaction in which an element loses electrons during transformation to a higher oxidation state is called an oxidation, whether or not the reaction involves oxygen. The energy released during the different oxidation reactions can be calculated from thermodynamic values for the free energies of formation in which all free energy values are related to the standard hydrogen electrode (see Chapter 5, Section 5.3.1). The free energy changes for the important half reactions are given in Table 16-5. When these reactions are coupled with organic matter oxidation, the largest amount of free energy is released in the reaction with oxygen, with oxidation by

Table 16-3 Oxidation states of key elements. The *oxidation state* of an element is its charge or valence state in various chemical species. Many elements have more than one oxidation state. In nature you can usually *determine the oxidation state* of an element in a compound by assuming $O(-II)$ and $H(I)$

Element	Oxidation state	Chemical species
Carbon	$C(-IV)$	CH_4 (methane)
	$C(0)$	C (graphite), CH_2O (glucose or other carbohydrate)
	$C(+IV)$	CO_2, HCO_3^-, CO_3^{2-} (carbonate system)
Hydrogen	$H(0)$	H_2 (hydrogen gas)
	$H(I)$	H^+ (hydrogen ion)
Nitrogen	$N(-III)$	NH_3, NH_4^+ (ammonia)
	$N(0)$	N_2 (nitrogen gas)
	$N(+V)$	NO_3^- (nitrate ion)
Oxygen	$O(-II)$	H_2O, CO_2, SO_4^{2-}
	$O(0)$	O_2 (oxygen gas)
Sulfur	$S(-II)$	H_2S, HS^- (hydrogen sulfide)
	$S(0)$	S (elemental sulfur)
	$S(+II)$	$S_2O_3^{2-}$ (thiosulfate ion)
	$S(+VI)$	SO_4^{2-} (sulfate ion)
Iron	$Fe(II)$	Fe^{2+}, $Fe(OH)_2$, $FeCO_3(s)$ (ferrous ion, ferrous hydroxide, ferrous carbonate solid)
	$Fe(III)$	Fe^{3+}, $Fe_2O_3(s)$ (ferric ion, ferric oxide solid)
Manganese	$Mn(II)$	Mn^{2+}, $MnCO_3(s)$ (manganous ion, manganous carbonate solid)
	$Mn(IV)$	$MnO_2(s)$ (manganese dioxide solid)

reduction of NO_3^-, Mn (IV), Fe (III), SO_4^{2-}, and production of CH_4 yielding successively less energy.

We shall see later in the chapter that the global burden of free O_2 is essentially all in the atmosphere. Since most reduced compounds are solids on the continental surface, in soils and sediments, oxidation by O_2 occurs primarily at the interface of the atmosphere with oxidizable surfaces. The actual sequence of organic matter oxidation reactions in nature is most clearly demonstrated in the chemistry of pore waters of sediments where organic matter is degraded and where the sediments are bathed by oxygen-containing bottom water. Studies of the concentration of electron acceptors (oxidants) in sediment pore waters reveal that the sequence of oxidant use follows the order in which the gain of free energy is the greatest. Figure 16-7 is a schematic compilation of many different sets of pore water measurements. The whole range is rarely observed in a single setting because the supply of organic matter to the sediments is

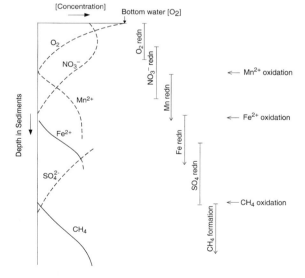

Fig. 16-7 Changes in concentration of metabolites in sediment. The reaction order follows the sequence dictated by thermodynamics (i.e., the most energy releasing reactions occur first). (Modified from Froelich *et al.*, 1979.)

Table 16-4   Examples of redox reactions, consisting of an oxidation and reduction half reaction

---

(a) *Photosynthetic production of reduced carbon in carbohydrate*

| Carbon reduced | $(CO_2 \rightarrow C)$ | $CO_2 + 4H^+ + 4e^- \Leftrightarrow H_2O + CH_2O$ |
| Oxygen in water oxidized | $(H_2O \rightarrow O_2)$ | $2H_2O \Leftrightarrow O_2 + 4H^+ + 4e^-$ |

Net

$$CO_2 + H_2O \Leftrightarrow O_2 + CH_2O$$

(b) *Organic matter oxidation*

| Carbon oxidized | $(C \rightarrow CO_2)$ | $CH_2O + H_2O \Leftrightarrow CO_2 + 4H^+ + 4e^-$ |
| Oxygen reduced | $(O_2 \rightarrow H_2O)$ | $O_2 + 4H^+ + 4e^- \Leftrightarrow 2H_2O$ |

Net

$$CH_2O + O_2 \Leftrightarrow CO_2 + H_2O$$

(c) *Iron oxidation*

| Iron oxidized | $(Fe^{2+} \rightarrow Fe^{3+})$ | $4Fe^{2+} \Leftrightarrow 4Fe^{3+} + 4e^-$ |
| Oxygen reduced | $(O_2 \rightarrow H_2O)$ | $O_2 + 4H^+ + 4e^- \Leftrightarrow 2H_2O$ |

Net

$$4Fe^{2+} + O_2 + 4H^+ \Leftrightarrow 4Fe^{3+} + 2H_2O$$

(d) *Sulfide oxidation*

| Sulfide oxidized | $(HS^- \rightarrow SO_4^{2-})$ | $HS^- + 4H_2O \Leftrightarrow SO_4^{2-} + 9H^+ + 8e^-$ |
| Oxygen reduced | $(O_2 \rightarrow H_2O)$ | $2O_2 + 8H^+ + 8e^- \Leftrightarrow 4H_2O$ |

Net

$$2O_2 + HS^- \Leftrightarrow SO_4^{2-} + H^+$$

(e) *Sulfate reduction*

| Carbon oxidized | $(C \rightarrow CO_2)$ | $2CH_2O + 2H_2O \Leftrightarrow 2CO_2 + 8H^+ + 8e^-$ |
| Sulfate reduced | $(SO_4^{2-} \rightarrow H_2S)$ | $SO_4^{2-} + 10H^+ + 8e^- \Leftrightarrow H_2S + 4H_2O$ |

Net

$$2CH_2O + SO_4^{2-} + 2H^+ \Leftrightarrow H_2S + 2CO_2 + 2H_2O$$

(f) *Iron reduction*

| Carbon oxidized | $(C \rightarrow CO_2)$ | $CH_2O + H_2O \Leftrightarrow CO_2 + 4H^+ + 4e^-$ |
| Iron reduced | $(Fe^{3+} \rightarrow Fe^{2+})$ | $4Fe^{3+} + 4e^- \Leftrightarrow 4Fe^{2+}$ |

Net

$$CH_2O + H_2O + 4Fe^{3+} \Leftrightarrow CO_2 + 4H^+ + 4Fe^{2+}$$

(g) *Nitrate reduction*

| Carbon oxidized | $(C \rightarrow CO_2)$ | $5CH_2O + 5H_2O \Leftrightarrow 5CO_2 + 20H^+ + 20e^-$ |
| Nitrate reduced | $(NO_3^- \rightarrow N_2)$ | $4NO_3^- + 24H^+ + 20e^- \Leftrightarrow 2N_2 + 12H_2O$ |

Net

$$4NO_3^- + 5CH_2O + 4H^+ \Leftrightarrow 5CO_2 + 2N_2 + 7H_2O$$

---

exhausted before methane formation occurs if the concentration of oxygen in the bottom waters is high. In environments where organic matter supply is high and oxygen concentration in bottom waters low, the reactions proceed from SO_4^{2-} reduction to CH_4 formation.

Normally in chemistry one does not expect reaction rates or sequences to follow thermodynamic energy gain. Rather, they are dictated by activation energy barriers (with the exception of linear free energy relationships). Since virtually all the organic matter oxidation reactions in nature are microbially mediated, the observation in Fig. 16-7 demonstrates that the microbial (enzyme) catalysis effectively reduces activation energy barriers so that the maximum amount of available free energy can be utilized by the opportunistic bacteria.

Table 16-5   The standard free energy of reaction, ΔG_r^0, for the main environmental redox reactions

Reaction	ΔG_r^0 (1/2 reaction, kJ/mol)	ΔG^0 (whole reaction, kJ/mol)
Oxidation		
$CH_2O^a + H_2O \rightarrow CO_2 + 4H^+ + 4e^-$	-27.4	
Reduction[b]		
$4e^- + 4H^+ + O_2 \rightarrow 2H_2O$	-491.0	-518.4
$4e^- + 4.8H^+ + 0.8NO_3^- \rightarrow 0.4N_2 + 2.4H_2O$	-480.2	-507.6
$4e^- + 8H^+ + 2MnO\,(s) \rightarrow 2Mn^{2+} + 4H_2O$	-474.5	-501.9
$4e^- + 12H^+ + 2Fe_2O_3\,(s) \rightarrow 4Fe^{2+} + 6H_2O$	-253.22	-280.6
$4e^- + 5H^+ + \frac{1}{2}SO_4^{2-} \rightarrow \frac{1}{2}H_2S + 2H_2O$	-116.0	-143.4
$4e^- + 4H^+ + CH_2O \rightarrow CH_4 + H_2O$	-7.0	-34.4

[a] CH_2O represents organic matter. $\Delta G_f = -129$ kJ/mol.
[b] The free energies of formation are (kJ/mol) (Stumm and Morgan, 1981): $H^+ = 0$; MnO_2 (pyrolusite) $= -465.1$; $Mn^{2+} = -228.0$; $H_2O = -237.0$; $NO_3^- = -111.3$; $N_2 = O$; $Fe^{2+} = -73.9$; Fe_2O_3 (hematite) $= -742.7$; $SO_4^{2-} = -744.6$; $H_2S = -27.9$; $CH_4 = -34.4$; $CO_2 = -394.4$; $O_2 = +16.3$.

16.3.2   Thermodynamic Disequilibrium and Microbial Catalysis of Oxidation Reactions

While the sequence of oxidation reactions in nature follow thermodynamic predictions for free energy gain, this does *not* mean the environment approaches equilibrium with respect to redox reactions. This is evident given the existence of organic life in our oxygen-containing environment, but it is also the case in more subtle forms in natural waters. A classic example of this is illustrated by the concentrations of redox couples (elements that exist in different oxidation states) in environments in which oxygen-containing and oxygen-depleted waters coexist. Concentrations of a variety of redox couples measured in Saanich Inlet, a fjord with restricted circulation on Vancouver Island, Canada, are presented in Fig. 16-8. The oxygen/hydrogen sulfide interface is at about 130 m water depth indicating a dramatic change in the redox potential at this horizon. Values of the $p\varepsilon$ (see Chapter 5, Section 5.3.2 regarding $p\varepsilon$) and $p\varepsilon_w$ ($p\varepsilon_w$ is the $p\varepsilon$ at the pH of the environment, pH = 7.4 in this case) of eight measured redox couples are compared with the values calculated from their measured activities, $p\varepsilon_C$, in Table 16-6. The calculated $p\varepsilon$ values, $p\varepsilon_C$, vary by 17 orders of magnitude in the region around the O_2/H_2S interface. Couples with vastly different $p\varepsilon$ values exist in the oxygen containing region where reduced species I^-, Mn^{2+}, and $Cr(OH)^{2+}$ coexist with oxidized forms resulting in $p\varepsilon_C$ values of between 6.6 and 12.8. The state of disequilibrium is less dramatic in the waters containing hydrogen sulfide where $p\varepsilon_C$ values range between -2.6 to -3.4 for the N_2/NH_4^+, $FeOOH/Fe^{2+}$ and SO_4^{2-}/HS^- couples.

These measurements indicate that it is not possible to identify a single value of $p\varepsilon$ surrounding the O_2/H_2S interface in the environment. Redox couples do not respond to the $p\varepsilon$ of the environment with the same lability as hydrogen ion donors and acceptors. There is no clear electron buffer capacity other than the most general states of "oxygen containing" or "H_2S containing." The reason for the vast differences in $p\varepsilon_C$ in the oxic waters is the slow oxidation kinetics of the reduced forms of the redox couples. The reduced species for which the kinetics of oxidation by O_2 has been most widely studied is Mn^{2+}. This oxidation reaction

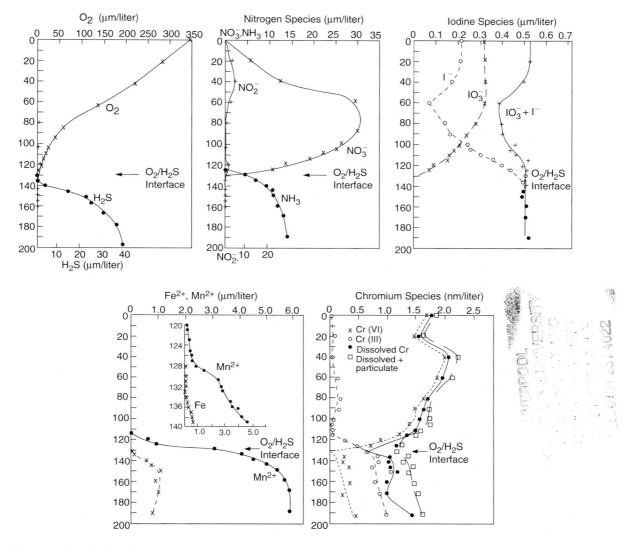

Fig. 16-8 Depth distribution of the concentration of redox couples in a partially anoxic fjord, Saanich Inlet. (Redrawn from Emerson *et al.*, 1979.)

is slow and known to be catalyzed by solid surfaces and by bacteria. Recent studies at natural seawater concentration levels in the absence of catalysis indicate that the residence time for homogeneous oxidation is on the order of 6–7 years at 25°C and pH = 8 (von Langen *et al.*, 1997). Results of studies of the catalysis of this oxidation reaction in laboratory experiments and on field samples taken from the O_2/H_2S interface of Saanich Inlet are presented in Table 16-7. The oxidation rate constant normalized for different

O_2, Mn, pH, and solid concentrations indicates that the character of the solid is important partly because some surfaces bind Mn^{2+} more strongly and partly because they facilitate the electron transfer differently. Catalysis by enzymes is clearly the most effective oxidation enhancing process as indicated by the laboratory studies with spores and material from the O_2/H_2S interface of Saanich Inlet. Microbial catalysis in this environment reduces the oxidation lifetime of Mn^{2+} to about one day. This example illustrates

Table 16-6   Redox half reactions, $p\varepsilon^0$, $p\varepsilon_w^0$, and calculated $p\varepsilon$ values from the redox species distributions $p\varepsilon_c$ for the zone 125–135 m in Saanich Inlet

Equation	$p\varepsilon^{0a}$	$p\varepsilon_w^{0\,b}$	$p\varepsilon_c^{\,c}$
$\frac{1}{4}O_2(g) + H^+ + e^- \rightarrow \frac{1}{2}H_2O$	+20.75	+13.35	+12.8
$\frac{1}{5}NO_3^- + \frac{6}{5}H^+ + e^- \rightarrow \frac{1}{10}N_2(g) + \frac{3}{5}H_2O$	+21.05	+12.17	+11.2
$\frac{1}{6}IO_3^- + H^+ + e^- \rightarrow \frac{1}{6}I^- + \frac{1}{2}H_2O$	+18.35	+10.95	+10.5
$\frac{1}{2}MnO_2(s) + 2H^+ + e^- \rightarrow \frac{1}{2}Mn^{2+} + H_2O$	+21.85	+7.05	+8.9
$\frac{1}{3}CrO_4^{2-} + 2H^+ + e^- \rightarrow \frac{1}{3}Cr(OH)_2^+ + \frac{2}{3}H_2O$	+22.03	+7.23	+6.6
$\frac{1}{6}N_2(g) + \frac{4}{3}H^+ + e^- \rightarrow \frac{1}{3}NH_4^+$	+4.65	−5.22	−2.6
$FeOOH(s) + 3H^+ + e^- \rightarrow Fe^{2+} + 2H_2O$	+8.25	−6.55	−2.7
$\frac{1}{8}SO_4^{2-} + \frac{9}{8}H^+ + e^- \rightarrow \frac{1}{8}HS^- + \frac{1}{2}H_2O$	+4.25	−4.08	−3.4

[a] $p\varepsilon^0 \equiv \log K$; $\log K$ estimated from free energy (ΔG_r^0) data. $p\varepsilon^0$ is the $p\varepsilon$ to expect at equilibrium if the ratio of the oxidized and reduced species are at unit activity:

$$\ln\left\{\frac{a_{red}}{a_{ox}}\right\} = \ln 1 = 0$$

[b] $p\varepsilon_w^0$ is the $p\varepsilon^0$ normalized to constant pH:

$$p\varepsilon_w^0 = p\varepsilon^0 + n\,(-7.4); \text{ pH here} = 7.4.$$

[c] $p\varepsilon_c$ is the calculated $p\varepsilon$ using the measured activities of the oxidized and reduced species:

$$p\varepsilon_c = p\varepsilon^0 - \log\left\{\frac{a_{red}}{a_{ox}a_{H^+}^n}\right\}$$

assuming: pressure $(N_2) = 0.76$, $[SO_4^{2-}] = 28 \times 10^{-3}$ mol/kg; pH = 7.4.

Table 16-7   The surface or enzyme-catalyzed reaction rate constant, k_{Mn}, for oxidation of Mn^{2+} normalized for oxygen concentration $[O_2]$, pH and particulate concentration $[X]$. $d[Mn^{2+}]/dt = k_{Mn}[Mn][O_2][OH]^2[X]$

Catalytic surface	Particulate concentration (mol/l)	Medium	pH	PO$_2$ (atm)	Mn(II)$_0$ (μmol/l)	k_{Mn} 10^{18}(mol/l)$^{-4}$/d	Reference
Silica	8×10^{-3}	0.2 M NaClO$_4$	9.1	1.0	50	0.003	a
α-FeOOH	10×10^{-3}	0.1 M NaClO$_4$	8.6	1.0	50	0.08	a
γ-FeOOH	1×10^{-3}	0.1 M NaClO$_4$	8.0	1.0	50	2.0	a
MnO$_x$	$<0.4 \times 10^{-3}$	1.6 mM DIC	9.0–9.5	1.0	450	0.1–0.3	b
Spores[e]	$\sim 1 \times 10^{-6}$	Seawater	7.8	0.2	3	3000–16 000	c
Saanich Inlet[e] Bacteria	$\sim 0.5 \times 10^{-6}$	Seawater	7.4	0.01	1	800,000	d

[a] Davies (1985).
[b] Morgan (1967).
[c] Hastings and Emerson (1986).
[d] Tebo and Emerson (1986).
[e] $[X]$ is taken to be the concentration of particulate Mn.

the generally empirical nature of our current understanding of the rate of electron transfer reactions in nature and why it is presently not possible to predict many of these rates even though they frequently control the distribution of redox couples in the environment.

16.3.3   Oxidation Processes in the Atmosphere

Owing to the high availability of O_2 in air, redox reactions there are extremely one-sided, with reduced compounds that may enter the atmosphere becoming oxidized at various rates. However, while O_2 is the source of the oxidizing agent, mechanistically most of the oxidation reactions proceed via shorter-lived species containing oxygen, often with an odd number of O atoms. Examining Fig. 7-11, we can group together several sets of highly reactive oxidizing agents:

- Odd oxygen: O, O_3.
- Odd hydrogen: OH, HO_2.
- Odd nitrogen: NO, NO_2, NO_3, N_2O_5, HNO_3, $ClONO_2$ etc.
- Odd halogen: ClO, HOCl.

"Odd" implies uneven numbers of the atom. With few exceptions, these and related oxidizing agents are free radicals; i.e., they are molecules with an odd number of electrons such that one of the electrons is unpaired. This leads to higher reaction rate constants, often higher by many orders of magnitude than for ordinary molecules. Ozone, of course, is not a free radical, and as might be guessed, it is not as reactive as free radicals like OH, the hydroxyl radical.

Interestingly, many of these free radicals are produced from photochemical reactions in the atmosphere of O_2 and O_3, for example

$$\text{Stratosphere:} \quad O_2 + h\nu \rightarrow 2O$$

$$O + O_2 + M \rightarrow O_3 + M$$

$$\text{Troposphere:} \quad NO_2 + h\nu \rightarrow NO + O$$

$$O + O_2 + M \rightarrow O_3 + M$$

$$\text{Both:} \quad O_3 + h\nu \rightarrow O(^1D) + O_2$$

$$O(^1D) + H_2O \rightarrow 2OH$$

So the ultimate source of oxidant is O_2, although the actual species involved in the oxidation is not O_2.

An important example of an atmospheric oxidation reaction is found in the main sink for carbon monoxide:

$$CO + OH \rightarrow CO_2$$

Another one is the primary atmospheric sink reaction for carbonyl sulfide:

$$OH + OCS \rightarrow CO_2 + SH$$

For a more complete discussion of these reac-

tions, consult Chapter 7. Beyond that, a thorough analysis of the redox state of the atmosphere can be found in textbooks on atmospheric chemistry, e.g., Wayne (1991).

16.3.4 Global Mass Balance of Oxygen and Carbon

Photosynthesis and oxic respiration dominate global redox processes as well as the cycles of oxygen and carbon. Because oxygen is a relatively insoluble gas it resides predominantly in the atmosphere (Fig. 16-9) whereas the largest "short-term" reservoir of carbon is dissolved inorganic carbon in the oceans (see Chapter 10 on oceanography and Chapter 11 on the carbon cycle). About half of the global photosynthesis occurs on land and half in the sea. This must be roughly the case because the $\delta^{18}O$ of molecular oxygen in the atmosphere is about +23‰ and the sources from land and ocean photosynthesis are roughly +26 and +20‰, respectively. Of the nearly $10\,000 \times 10^{12}$ moles of oxygen fixed annually by photosynthesis, nearly all of it is consumed again by respiration, with only a small fraction of the marine photosynthesis escaping oxidation via burial. Since about 1.5 mol of O_2 are produced for every mole of organic carbon fixed in the ocean (the value is more like 1.1 on land), removal of organic matter from the ocean environment by carbon burial represents a source of O_2 to the atmosphere. Only about 0.1% of the carbon fixed by photosynthesis is buried (13 $\times$ 10^{12} mol C/yr, which is equivalent to 20 $\times$ 10^{12} mol O_2/yr, Fig. 16-9). The residence time of atmospheric oxygen with respect to this net source is

$$\tau_{O_2} = \frac{37.5 \times 10^{18} \text{ mol } O_2}{20 \times 10^{12} \text{ mol } O_2/\text{yr}} = 1.9 \times 10^6 \text{ yr}$$

If the global oxygen production caused by the small fraction of carbon fixed by photosynthesis that is buried were not balanced by an oxygen consumption term, and other processes remained the same, the O_2 content of the atmosphere would roughly double in about 2 million years. This is a short time geologically, particularly because it is believed there has been

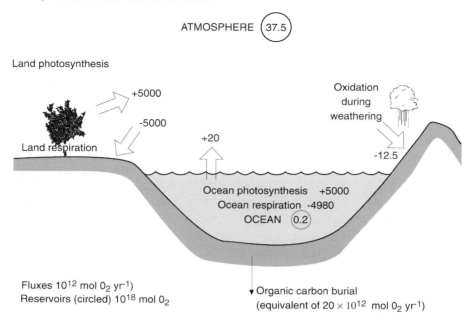

Fig. 16-9 The global oxygen balance. Ocean photosynthesis is estimated from satellite color (Antoine *et al.*, 1996). It is assumed that photosynthesis on land and in the ocean are about equal. The global carbon burial rate is from Hedges and Keil (1995). Organic C fluxes are converted to oxygen fluxes using a stoichiometry of $\Delta O_2/\Delta OC = 1.5$ in the marine environment and 1.1 on land. Oxygen consumption during weathering of C, S, and Fe in sedimentary rocks, in the ratio of 6:2:1, is from Holland (1978). At steady state organic carbon burial and O_2 consumption during weathering must be equal, which is certainly possible within the errors of these two estimates.

enough oxygen in the atmosphere to support animal life for at least 600 million years.

The primary oxygen sink that balances the burial of organic matter in the sea is oxidative weathering of reduced compounds in sedimentary rocks. Organic matter dominates the reduced redox elements being oxidized in these rocks, but sulfide as pyrite and ferrous iron also play an important role. According to Holland (1978) the present rate of oxidation of reducing compounds of C, S^{2-}, and Fe^{2+} in sedimentary rocks consumes 12.5×10^{12} mol of O_2 annually. Notably, both sulfide and reduced iron exist in sedimentary rocks because the oxidized forms of these compounds were reduced during anaerobic organic matter oxidation via equations (e) and (f) in Table 16-4 where the "proto" rocks were being formed. Thus, the reduced compounds in sedimentary rocks are either organic C or its proxy – elements reduced during the oxidation of organic carbon.

The oxygen budget in Fig. 16-9 is written using present estimates of oxygen produced via organic matter burial in the oceans and independently determined O_2 consumption rates by sedimentary rock weathering reactions. In order for the atmospheric oxygen concentration to be stable over millions of years, these two quantities must balance. Given the certainty of these estimates there is remarkable agreement today, but why should this be so? Why does global respiration oxidize all but a small fraction of the photosynthetically produced organic matter, and why is its electron donor capacity almost exactly equal to that of rocks that were formed on average several 100 million years ago? What process keeps the global redox system in whack? One is forced to assume that there are feedback mechanisms among the concentration of oxygen in the atmosphere, the oxidation of reduced organic carbon, sulfur, and iron in sedimentary rocks and the burial of organic matter in marine sediments. This can be achieved if the oxidation reactions are dependent in some way on the oxygen concentration in the atmosphere. If all of the reduced elements in sedimentary rocks were not oxidized, then the burial of these compounds, would increase along with the flux of O_2 to the atmosphere, and ultimately, the oxygen content of the atmosphere. The increased atmospheric oxygen concentration would then oxidize more of the reduced compounds bringing the system back to a steady state. By the same argument, a more efficient oxidation of all reducing compounds and the organic matter produced during photosynthesis, would decrease the net flux of oxygen to the atmosphere and ultimately the concentration of atmospheric O_2, with the feedback of enhanced organic carbon burial. This argument is illustrated in Fig. 16-10 which is adapted from Holland (1978). The figure illustrates that the atmospheric oxygen concentration is poised between the processes of oxygen production and oxygen consumption. However, the details of the mechanisms controlling this delicate balance between organic matter production and oxidation, remain undiscovered. They represent one of the fascinating mysteries of the global redox processes.

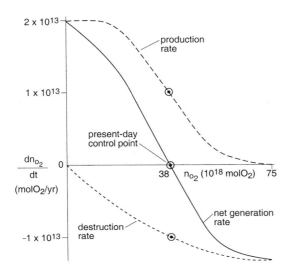

Fig. 16-10 A schematic representation of how the oxygen reduction and production rate, dn_{O_2}/dt combine to determine the size of the global oxygen reservoir. (Adapted from Holland, 1978.)

16.4 Conclusion

Having reviewed the acid–base and redox balances of Earth, it is easy to see that they are both the consequences of the interaction of several biogeochemical cycles. In the case of the acid–base balances, the cycles of C, N, and S stand out, along with the hydrologic cycle. For the redox balances, C, N, O, S, and Fe, and Mn are the dominant cycles. Importantly, *both* acid–base and redox systems are heavily mediated by biota. Beyond these simple generalizations lie a number of still open scientific questions including the intriguing challenge to find the feedbacks that are involved in both systems. The stability of atmospheric O_2 is an important case where feedbacks are likely, but many other examples can be presumed to exist.

References

Antoine, D., Andre, J.-M., and Morel, A. (1996). Oceanic primary production, 2. Estimation at global scale from satellite (coastal zone color scanner) chlorophyll. *Global Biogeochem. Cycles* **10**, 57–70.

Charlson, R. J. and Rodhe, H. (1982). Factors controlling the acidity of natural rainwater, *Nature* **295**, 683–685.

Davies, S. H. R. (1985). "Mn(II) oxidation in the presence of metal oxides." PhD dissertation, California Institute of Technology, Pasedena, California.

Davies, S. and Morgan, J. J. (1989). Manganese (II) oxidation kinetics on metal oxide surfaces, *J. Colloid Interface Sci.* **129**, 63–77.

Emerson, S., Cranston, R., and Liss, P. (1979). Redox species in a reducing fjord: Equilibrium and kinetic considerations, *Deep-Sea Res.* **26**, 859–878.

Froelich, P. M., Klinkhammer, G. P., Bender, M. L. *et al.* (1979). Early oxidation of organic matter in pelagic sediments of the eastern equatorial Atlantic; suboxic diagenesis, *Geochem. Cosmochim. Acta* **43**, 1075–1090.

Hastings, D. and Emerson, S. (1986). Oxidation of manganese by spores of a marine Bacillus: Kinetics and thermodynamic considerations. *Geochem. Cosmochim. Acta* **50**, 1819–1824.

Hedges, J. and Keil, R. (1995). Sedimentary organic matter preservation: an assessment and speculative synthesis, *Marine Chem.* **49**, 81–115.

Holland, H. (1978). "The Chemistry of the Atmosphere and Oceans." Wiley Interscience, New York.

Morgan, J. J. (1967). Chemical equilibria and kinetic properties of manganese in natural waters. *In* "Principles and Applications of Water Chemistry" (S. D. Faust and J. V. Hunter, eds), pp. 561–623. Wiley, New York.

Sillén, L. G. (1967). *In* "Equilibrium Concepts in Natural Water Systems," *Adv. Chem. Soc.* **67**, 45–56. American Chemical Society, Washington, DC.

Stumm, W. and Morgan, J. J. (1970). "Aquatic Chemistry," 1st edn. John Wiley and Sons, New York.

Stumm, W. and Morgan, J. J. (1981). "Aquatic Chemistry," 2nd edn. John Wiley and Sons, New York.

Taylor, G. S., Baker, M. B., and Charlson, R. J. (1983). Heterogeneous interactions of the C, N, and S Cycles in the atmosphere: The role of aerosols and clouds. *In* "The Major Biogeochemical Cycles and their Interactions" (B. Bolin and R. B. Cook, eds). John Wiley and Sons, Chichester.

Tebo, B. and Emerson, S. (1986). Microbial manganese (II) oxidation in the marine environment: A quantitative study. *Biogeochemistry* **2**, 149–161.

Twohy, C. H., Austin, P. H., and Charlson, R. J. (1989). Chemical consequences of the initial diffusional growth of cloud droplets: a marine case. *Tellus* **4** (B), 51–50.

von Langen, P. J., Johnson, K. S., Coale, K. H. *et al.* (1997). Oxidation kinetics of manganese (II) in seawater at nanomolar concentrations, *Geochem. Cosmochim. Acta* **61**, 4945–4954.

17

The Coupling of Biogeochemical Cycles and Climate: Forcings, Feedbacks, and Responses

R. J. Charlson

17.1 The Climate System

The word *climate* is used loosely to mean the aggregate of all components of weather averaged over a lengthy period – usually decades, centuries, or longer. As discussed earlier in Chapters 1 and 7, all processes that contribute to the current climate are driven by the flow of energy from the sun. On very long time scales, the Earth's internal energy plays a role via its influence on tectonics and the size and location of the continents. Because the Earth absorbs some of the solar energy that it receives, and because over time scales of years to decades the heat balance of the Earth is in a steady state, the Stefan–Boltzmann equation can be used to define the global-mean or effective temperature of the Earth:

$$\text{Energy absorbed} = \text{Energy emitted}$$
$$\pi R_e^2 (1 - A) S_0 = \varepsilon 4 \pi R_e^2 \sigma T_e^4 \qquad (1)$$

where R_e is the radius of the Earth, A is its albedo (reflectivity for sunlight), S_0 is the solar flux, σ is the Stefan–Boltzmann constant, and T_e is the temperature of the planet as seen radiatively from space. Emissivity, ε, is generally assumed to be unity (i.e., a black body), the surface of which is composed largely of the effective infrared emitter, H_2O. Since $A \approx 0.3$

(controlled mainly by clouds!) and $S_0 = 1370 \text{ W}/\text{m}^2$ we can write

$$T_e = \left[\frac{S_0 (1 - A)}{4\sigma} \right]^{1/4} = \left(\frac{s}{\sigma} \right)^{1/4} \qquad (2)$$

where s is the mean flux absorbed by the Earth ($s \approx 240 \text{ W}/\text{m}^2$). For this steady state to be met, $T_e = 255$ K, which is well below the freezing point of water and well below the current global mean surface temperature of ca. 288 K. This temperature really does exist at a level high in the atmosphere from which the emitted flux emerges, which is around 6 km. This ca. 6 km level is an effective average over the entire emission spectrum, and includes an emission height of zero (i.e., at the surface) for the 8–11 μm infrared window region and the height of the tropopause for other wavelengths. The primary reason that the Earth emits from such a high altitude in the atmosphere is that the atmosphere below that height strongly absorbs the infrared radiation characteristic of a low-temperature (250–300 K) black body such that emissions from lower altitudes are absorbed in the atmosphere and cannot escape directly to space. These concepts form the basis for the so-called *greenhouse effect*, already described qualitatively in Chapter 7. Figure 17-1 depicts the processes involved in the radiation balance.

Earth System Science
ISBN 0-12-379370-X

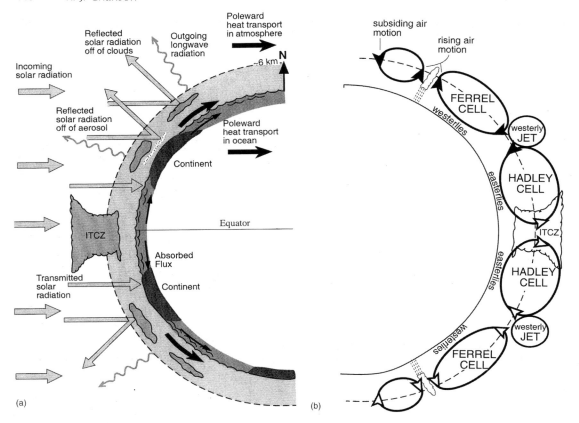

Fig. 17-1   The global climate system. (a) Energy fluxes, including incoming solar radiation, reflected radiation, emitted longwave radiation (from an effective altitude of ca. 6 km), and atmospheric and oceanic heat flux toward the polar regions. (b) The atmospheric circulation corresponding to part (a). Refer back to Fig. 7-4 and associated text for a discussion of the general circulation.

Two points now emerge:

1. The chemical composition of the atmosphere *controls* the absorptivity and emissivity for infrared or long-wave radiation. Hence, the presence in the air of H_2O, CO_2, CH_4, and other strong absorbers becomes a key factor in climate. And, since these gases are participants in biogeochemical cycles, one large aspect of the coupling becomes apparent; they are connected via the greenhouse effect.

2. Heat transfer processes besides pure radiative transfer are involved in control of the temperature of the air, especially below the effective emission height of ~6 km. Referring back to Chapter 7, we see that vertical motions of air in the troposphere are a main factor dictating that temperature decreases as altitude increases – air loses internal energy

as it gains altitude and potential energy. The global-mean lapse rate ($-dT/dz$) is 6.5 K/ km. Thus the average temperature of the Earth's surface is ~288 K given that the temperature at ca. 6 km is 255 K.

17.1.1   Recognizing the Fundamental Processes

These two points taken together illustrate that the temperature at the Earth's surface depends on both a radiative balance *and* all of the meteorologic processes that transport heat within the lower atmosphere *and* of course, all the oceanographic factors that transport heat in the ocean as well. So, at this juncture we must abandon the simple picture of a global-mean radiative heat balance and recognize that in addition, the climate of the Earth is not horizontally uniform, nor is it inherently stable. Any combination of cloud contribution to global albedo and meteorological processes that results in a radiative balance at some range of heights within the atmosphere can yield a steady-state heat balance. This implies that the surface temperature and its geographical distribution (dependent variables) are also not inherently stable.

Central to understanding climate, its variability and its geographical characteristics are the specific roles of the atmospheric and oceanic processes that distribute the energy absorbed from solar radiation. These processes generally move energy from low to high latitude and from the surface to higher altitude where radiation to space can occur. The main transports are (1) the vertical and horizontal atmospheric flux of sensible heat or enthalpy, (2) transport of heat by ocean currents, (3) the transfer of sensible heat to latent heat in the process of evaporation of water, (4) the release of latent heat as sensible heat by formation of clouds, and (5) very importantly, the formation of clouds that control the albedo of the planet *and* the distribution and amount of solar energy absorbed. A schematic of these processes is shown in Fig. 17-1.

17.1.2 Climate Models

The processes listed above that transfer energy reaching the Earth as solar radiation, convert it to heat, and send it back to space behave as a linked system. Figure 17-2 illustrates this web of interactions of the processes that are described in typical atmospheric models. An even more complex picture would emerge if the oceans were included, and of course many more details have to be added in order to include the biosphere. Refer back to Fig. 4-14 for a schematic representation of the components of a global climate model. This diagram shows the main sets of equations that are used and the main heat exchanges. Note especially that both Figs 17-2 and 4-13 derive from modeling approaches that emphasize a physical approach to climate modeling, i.e., they include no chemistry or biology. Figure 17-3a,b extends this picture to include the

biological and industrial exchanges with the atmosphere of a few key substances – notably greenhouse gases, aerosols, and their influences on clouds.

17.1.3 The Coupled System

Keeping in mind the entire set of components in the climate system as depicted in Figs 17-2, 4-13, and 17-3, we can now re-examine Fig. 1-2 to emphasize that biogeochemical cycles are coupled with the climate system. The temperature (as inferred from the record of the deuterium to hydrogen ratio in Antarctic ice) covaries with CO_2, CH_4 and other species that derive from biological processes. Two simple, if extreme, possibilities can be drawn:

1. Changes in physical climate cause changes in biological activity, but not vice versa; *or*
2. biological activities drive the climate.

The range of possibilities in between these two extremes would be:

3. The biosphere and its processes are an integral part of the physical climate system and are coupled to and have influenced changes in climate.

It is scenario 3 that is most consistent with the data depicted in Fig. 1-2. Given that the physical climate system is strongly influenced by gases in the atmosphere that absorb and emit infrared radiation (e.g., H_2O, CO_2, CH_4, etc.), and since the amounts of these species in the air depend to some extent (for some, a great extent) on the functioning of the biosphere, it is logical to view the climate of the Earth as a coupled physical, chemical, and biological entity.

Taken together, Figs 17-2, 4-13, 17-3, and 1-2 constitute a complex image of the Earth's climate system, including most of the factors that are known to be involved. However, such diagrams fail to adequately represent the dynamical nature of the totality of interactions of all of the parts. In order to explore these interactions, the natural variability of climate, and changes due to external perturbations, we must now introduce the key notions of *forcings, feedbacks, and responses*.

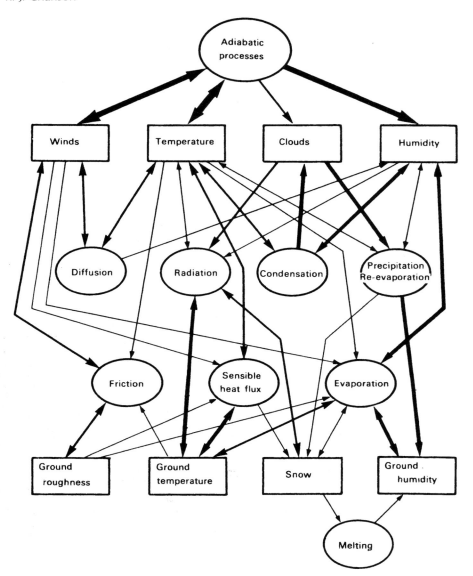

Fig. 17-2   The web of interactions in the atmospheric part of the global climate system. The strength of the interactions is qualitatively depicted by the thickness of the line. Bidirectional interactions have two arrowheads, unidirectional ones have only one. (From Houghton (1984), reprinted with permission from Cambridge University Press.)

17.2   The Dynamics of the Climate System: Forcings, Feedbacks, and Responses

The actual workings of the coupled biogeochemical and physical climate system, the ways that it responds to external perturbations, and the ways that it approaches or departs from a steady-state depend on myriad functional relationships between all the factors that are involved. The daunting complexity of the interwoven web of interactions might seem to preclude any understanding of the whole system, suggesting instead reductionist study of the pieces. However, looking again at Fig. 1-2 we

(a)

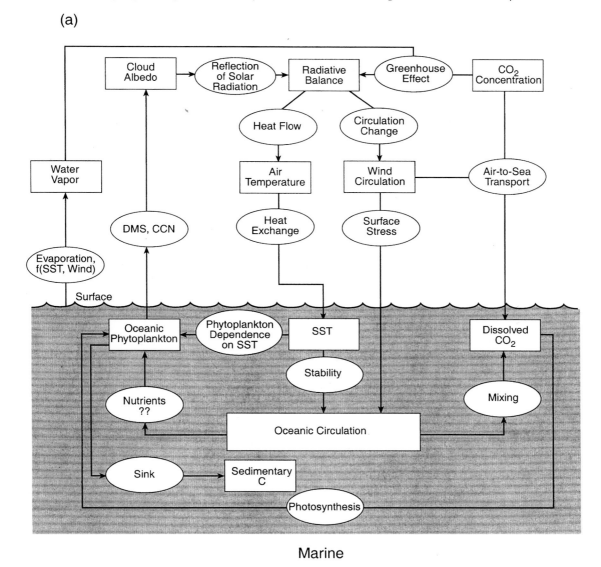

Marine

Fig. 17-3   (a) A simplified picture of the coupling of biogeochemistry and global radiative balance in marine systems. SST = Sea surface temperature.

see that the Earth has had stable, interglacial climates several times, as well as ice ages. The ice-core record indicates that for the duration of the record, the temperatures of the interglacial times were very similar, and the current interglacial time has been remarkably stable for about 10 000 years. The question then arises as to what allows this complex system to be stable? Why is it not subject to continuous, wild, chaotic swings among all the possible climatic states?

In an effort to answer such questions, we turn to another branch of natural sciences that has evolved from consideration of the overwhelming complexity of the functioning of living organisms. Physiology and its various branches (human physiology, plant physiology, etc.) view organisms as functioning systems, indeed systems that have stable steady states. As is well known to design engineers of complex systems like electronic devices or airplanes, such stability

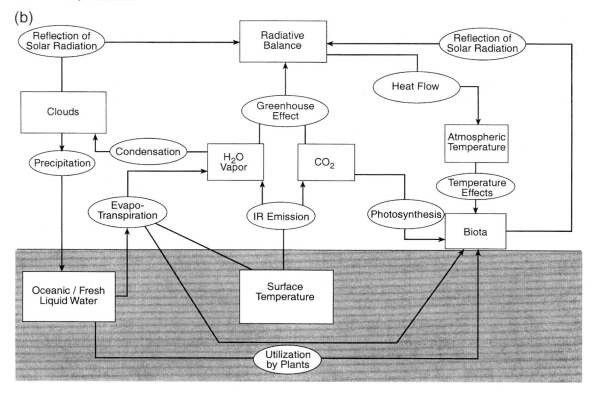

Terrestrial

Fig. 17-3   (b) Coupling of biogeochemistry and global radiative balance in terrestrial systems.

is achieved using feedbacks,. A prime example is the body temperatures of normal humans – it is nearly constant at 37°C thanks to biochemical feedbacks generated by the hypothalamus gland. Similarly, temperature inside a house is regulated by a thermostat that provides feedback – it turns off the furnace when the temperature rises above a set point. What, then, might provide thermostasis for a planet?

17.2.1   Definitions

17.2.1.1   Forcings

In order to understand the stable states of complex systems, it is useful to understand how the system responds to external perturbations. Externally imposed changes in the energy balance of the planet are referred to as *forcings*,

given in watts per square meter (W/m^2). A change in solar luminosity would be an example of a forcing, and indeed the solar flux is calculated to have increased by ca. 30% over the last 4.5×10^9 years. Small variations in the solar flux also occur because of systematic variations in the Earth's orbit and tilt on timescale of 10^4–10^5 years, often referred to as the *Milankovich effect*. Other forcings exist that also are external to the climate system such as the change in planetary albedo due to volcanic dust being injected into the stratosphere from large eruptions. Anthropogenic (human-induced) changes in heat balance are also considered to be external to the climate system and thus are also *forcings*.

17.2.1.2   Feedbacks and responses

Changes in the physical characteristics of the Earth that are *internal* to the climate system can participate in *feedback* processes. A simple example of a feedback will serve to illustrate their nature: if increased concentrations of CO_2 causes T to increase, the vapor pressure of water will increase where evaporation occurs and the water content of the atmosphere will also increase. The net result is a *further* increase of the total content of greenhouse gases (inasmuch as H_2O vapor is a powerful greenhouse gas), which *amplifies* the initial warming caused by CO_2 increase. This defines an *amplifying (or positive) feedback*.

The nature of such processes can be depicted as a feedback loop, as shown in Fig. 17-4. Using the nomenclature in this figure and continuing with enhanced evaporation of water vapor as our physical example of a feedback that is completely internal to the climate system, we have additional heating (ΔQ_F) caused by the additional greenhouse effect of the increased water vapor concentration in the atmosphere. The evaporation was caused by the original forcing, ΔQ. In this case, E, the physical effect, is the increased water vapor and F, the resultant feedback that causes ΔQ_F, is the greenhouse effect of the added water vapor. The reader desiring a more thorough mathematical treatment of feedbacks should consult an appropriate text, e.g., Peixoto and Oort (1992).

While most climate models consider feedbacks as being dependent on temperature (usually T_s), there are many other dependent variables in the climate system that could be involved, for example solar irradiance at the ground or rainfall. However, it is customary to describe these mathematically as functions of T_s,

(a) No feedback

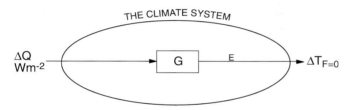

(b) With feedbacks that are internal to the climate system

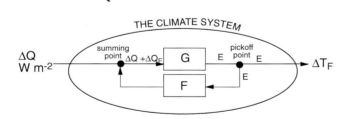

Where: E = the physical result of the gain operating on $\Delta Q + \Delta Q_F$
ΔQ = forcing, W m⁻²
ΔQ_F = effective additional forcing due to feedback (can be positive or negative)
G = gain. In this simple climate system G has units K (W m⁻²)⁻¹
F = feedback (W m⁻² K⁻¹)
ΔT_F = temperature response with feedback, K.
$\Delta T_{F=0}$ = temperature response with no feedback, K.

Fig. 17-4 Schematic of the climate system with and without feedbacks, that depend on temperature.

leaving surface temperature as the conventional single index of climate.

It is important to keep in mind the separation of forcings, feedbacks, and responses. Considerable progress in climate research has been made in recent years by making this separation, which was first suggested by Dickinson and Cicerone (1986), in recognition of the fact that forcings can be calculated with more confidence than responses. There is a fundamental dilemma, however, in defining forcings: should they be defined before the system has responded or afterwards? While there is some ambiguity extant in the practices of the climate modeling community, the easiest way to define forcings as being separate from feedbacks is to define forcings as partial derivatives, i.e. with all other parameters held constant. For example, for a forcing agent x (like ΔCO_2, ΔCH_4, etc.), forcing becomes the integral of the partial derivative of heat balance with respect to the change in x, assuming *all* other factors are held constant:

$$\text{Forcing} \equiv \int_{x_0}^{x_1} \frac{\partial Q}{\partial x} \, dx = Q(x_1) - Q(x_0) \quad (3)$$

Responses clearly include the changes in any and all dependent climate variables that occur due to the sum total effect of forcings and feedbacks. Again, just as is the case in defining feedbacks, ΔT_s is the conventional single index of response. Thus having defined forcings (ΔQ, in W/m^2), feedbacks, and responses (ΔT_s, in K) we can discuss the nature and magnitude of forcings and feedbacks in turn.

17.3 Forcings of Climate

17.3.1 Natural Forcings

The simplest forcing to define and describe – although by no means easy to quantify – is the change or variability of the so-called solar constant. The IPCC (1990, 1995, 1996) provides several estimates of this forcing, e.g.,

- Current variability due to orbital changes
  -0.035 W/m^2 per decade
- Change in solar output since 1850
  $+0.3 \pm 0.2$ W/m^2

Another class of natural forcings derives from the global (or near global) increase in albedo caused by dust and other aerosol substances lofted into the stratosphere by large volcanoes. Such events are extremely sporadic, and there is no indication of a secular trend over the last century. Individual volcanic events such as Krakatoa in 1883 or more recently, Pinatubo in 1991, caused measurable global cooling while the aerosol remained in the stratosphere (2–3 years total). Large-scale (hemispheric to global) decreases in surface temperature of ~ 0.5 K are well documented, although the forcings were measured for only one case, Pinatubo. The latter averaged ca. -1.5 W/m^2 for the period mid-1991–1993 with a peak of ca. -3 W/m^2, as observed by satellite radiometers. It is important to recognize that the correlation of cooling with measured negative forcings supports the notion that climate (i.e., T_s) responds to forcing; however, because the volcanic events are short in duration (1–3 years) a full response of the climate system cannot develop. For example, ocean circulations cannot fully respond in less than a decade or decades, so volcanic perturbation cannot provide a complete measure of climate sensitivity. Nonetheless, the *apparent* sensitivity of ca. 0.3 K/(W/m^2) does give a sense of the rough magnitude of responses to be expected from other forcings, assuming that the entire climatic state does not change.

17.3.2 Anthropogenic Forcings

17.3.2.1 Greenhouse gases

The activities of humans, most importantly the combustion of fossil fuels and biomass, produce quantities of gases that absorb and emit infrared radiation – the so-called *greenhouse gases* – as well as aerosol particles that reflect and absorb solar radiation, absorb some infrared radiation and also are involved in cloud microphysics. The greenhouse gases are by far the best understood, while *direct* aerosol effects (scattering and absorption of sunlight) are quantified but uncertain. *Indirect* aerosol effects on clouds are not yet quantified in any convincing way.

The effects of different greenhouse gases vary greatly depending upon their individual infrared absorption spectra, how those spectra overlap with the absorption of other species (especially H_2O) and how much absorption by each of those gases existed naturally before the anthropogenic increases began. Table 17-1 provides algebraic expressions that were determined through detailed studies of infrared spectra (shown in Fig. 17-5) for relating the forcing by each gaseous species to its change in concentration. It is critically important to recognize that the dependence of forcing on changed CO_2 levels is much weaker per molecule than for other species, e.g., CFC-11 or CFC-12. These manmade chemicals absorb infrared radiation in a part of the spectrum where water vapor and CO_2 do not already have strong bands. On the other hand, the manmade increase of CO_2 is so large (currently ca. 25% since the mid-1800s – see Chapter 11) that it is the largest anthropogenic input to the greenhouse effect (not counting feedbacks).

Table 17-2 summarizes the estimates of global-mean climate forcing by greenhouse gases in ca. 1995, and a "business as usual" forecast by IPCC (1995) for the year 2025. While these figures are useful for comparing the forcings by the different greenhouse gas species,

Table 17-1 Expressions used to derive radiative forcing for past trends and future scenarios of greenhouse concentrations (IPCC, 1990)

Trace gas	Radiative forcing approximation, ΔF in W/m^2
CO_2	$\Delta F = 6.3 \ln\left(\dfrac{[CO_2]}{[CO_2]_0}\right)$ where $[CO_2]$ is concentration in ppmv Valid for concentrations < 1000 ppmv
CH_4	$\Delta F = 0.036 \{[M]^{1/2} - [M]_0^{1/2}\} - \{f([M], [N]_0) - f([M]_0, [N]_0)\}$ where $[M]$ is concentration of CH_4 in ppbv and $[N]$ is concentration of N_2O in ppbv and $f([M], [N])$ is the methane–nitrous oxide overlap term[a] valid for $[M] < 5$ ppmv
N_2O	$\Delta F = 0.14 \{[N]^{1/2} - [N]_0^{1/2}\} - \{f([M]_0, [N]) - f([M]_0, [N]_0)\}$ with $[M]$ and $[N]$ as above valid for $[N] < 5$ ppmv
CFC-11	$\Delta F = 0.22 ([X] - [X]_0)$ where $[X]$ is concentration of CFC-11 in ppbv valid for $[X] < 2$ ppbv
CFC-12	$\Delta F = 0.28 ([Y] - [Y]_0)$ where $[Y]$ is concentration of CFC-12 in ppbv valid for $[Y] < 2$ ppbv
Stratospheric H_2O vapor	$\Delta F = 0.011 ([M]^{1/2} - [M]_0^{1/2})$ with $[M]$ as above
Tropospheric O_3	$\Delta F = 0.02 ([O] - [O]_0)$ where $[O]$ is concentration of ozone in ppbv

[a] $f([M],[N]) = 0.47 \ln [1 + 2.01 \times 10^{-5} ([M][N])^{0.75} + 5.31 \times 10^{-15} [M] ([M][N])^{1.52}$, with $[M]$ and $[N]$ in ppbv.

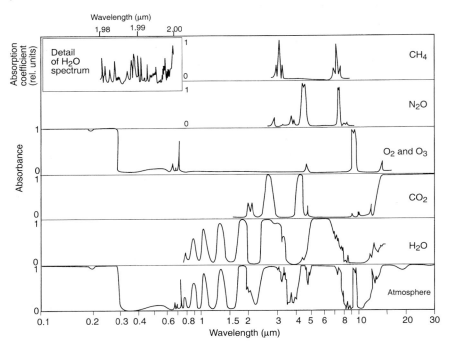

Fig. 17-5 Absorption spectra for H_2O, CO_2, O_2, N_2O, CH_4, and the absorption spectrum of the atmosphere. (Reprinted from Fleagle and Businger, 1963.)

Table 17-2 Current and projected global-mean anthropogenic greenhouse forcing, W/m^2

Greenhouse gas	1995	2025
CO_2	1.5	2.9
CH_4	0.4	0.7
N_2O	0.1	0.2
Chlorofluorocarbons	0.3	0.5
Total	2.3	4.6

it is somewhat misleading to utilize a single global-mean value because the forcings are not geographically uniform.

Figure 17-6 is a model-calculated map of the anthropogenic greenhouse forcing around 1990 (after Kiehl and Briegleb, 1993). This figure reveals an important feature of the global-change picture. Even though the gases themselves are almost perfectly well mixed throughout the atmosphere, their forcings are *not* uniform. The causes of this apparent paradox are twofold – first, the forcing by the key gases

(CO_2, CH_4, N_2O, etc.) depend on the water vapor content and the cloud amount since there are overlaps of the spectral features of the gases with those of H_2O vapor and liquid (see Fig. 17-5 for IR spectra of greenhouse gases.) Thus, where it is moist (e.g., in some parts of the tropics), changes in greenhouse gases are somewhat less effective than where it is dry (e.g., over subtropical deserts). Second, the magnitude of the greenhouse forcing is a strong function of the upwelling flux of infrared radiation from the surface, the integral of which is proportional to T_s^4. Because T_s in the tropics and subtropics is around 300–310 K and in a cold place like the Antarctic plateau T_s is only 220–230 K, the amount of the infrared flux that is absorbed is much greater where it is hot at the surface than where it is cold.

17.3.2.2   Aerosols

Aerosol particles – the visible haze in polluted air – reflect and absorb sunlight, *directly* influencing the heat balance over large ($\geq$1000 km)

How Much Light Do Aerosols Reflect Away?

Atmospheric sulfate aerosol scatters light in all directions. About 15 to 20% of the light is scattered back into space. The backscattering constitutes the direct effect of atmospheric aerosol on incoming radiation. The light-scattering efficiency of aerosol, represented by the Greek letter alpha (α), is high even at low humidity: each gram represents an area of about 5 m^2. Moisture increases the scattering by making the aerosol expand. At the global average relative humidity, the efficiency doubles, to almost 10 m^2/g. One can use this value to estimate the magnitude of the direct effect of anthropogenic sulfate.

The rate at which light is lost from the solar beam is defined by the scattering coefficient, represented by the Greek letter sigma (σ, expressed in units of per meter). This value is determined by the amount of aerosol mass, M (in grams per cubic meter), multiplied by the light-scattering efficiency: $\sigma = \alpha M$. When both sides of this equation are integrated over altitude, z, a dimensionless quantity called the aerosol optical depth and represented by the Greek letter delta (δ) results:

$$\int_0^\infty \sigma\,dz = \delta = \alpha \int_0^\infty M\,dz = \alpha B$$

Here B is the world average burden of anthropogenic sulfate aerosol in a column of air, in grams per square meter. The optical depth is then used in the Beer Law (which describes the transmission of light through the entire vertical column of the atmosphere). The law yields $I/I_0 = e^{-\delta}$, where I is the intensity of transmitted radiation, I_0 is the incident intensity outside the atmosphere and e is the base of natural logarithms. In the simplest case, where the optical depth is much less than 1, δ is the fraction of light lost from the solar beam because of

scattering. The question, then, is just how large δ is or, more properly, that part of it that results from manmade sulfate.

This global average burden of anthropogenic sulfate aerosol can be estimated by considering the entire atmospheric volume as a box. Because the lifetime of sulfate aerosol is short, the sum of all sulfate sources, Q, and its lifetime in the box, t, along with the area of the earth, determine B:

$$B = \frac{Qt}{\text{area of the Earth}}$$

About half the manmade emissions of sulfur dioxide become sulfate aerosol. That implies that currently 35 Tg per year of sulfur in sulfur dioxide is converted chemically to sulfate. Because the molecular weight of sulfate is three times that of elemental sulfur, Q is about 105 Tg per year. Studies of sulfate in acid rain have shown that sulfates persist in the air for about five days, or 0.014 year. The area of the Earth is 5.1×10^{14} m^2. Substituting these values into the equation for B yields about 2.8×10^{-3} g/m^{-2} for the burden.

This apparently meager amount of material produces a small but significant value for the aerosol optical depth. Using the value of scattering efficiency (α) of 5 m^2/g and a factor of two for the increase in scattering coefficient because of relative humidity, the estimated anthropogenic optical depth becomes $\delta = 0.028$. This value means that about 3% of the direct solar beam fails to reach the Earth's surface because of manmade sulfate. A smaller amount – perhaps 0.5% – is lost to space. This scattering operates over the noncloudy parts of the earth. About half the earth is cloudy at any given time, so that globally 0.2 to 0.3% is lost.

(From R. J. Charlson and Wigley, T. M. L. (1994). "Sulfate Aerosol and Climatic Change," *Scientific American*, Inc. All rights reserved.)

---

regions. They also *indirectly* influence the albedo and perhaps the lifetime of clouds through their action as cloud condensation nuclei. Many types of anthropogenic aerosols are implicated, especially sulfates from the

atmospheric oxidation of SO$_2$, condensed organic matter from biomass combustion, black carbon or soot from all kinds of combustion of carbon-based fuels, and dust from disturbed soils. Figure 17-7 shows the model

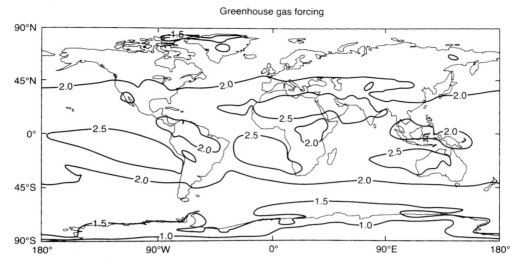

Fig. 17-6 Calculated geographical distribution of the climate forcing (W/m^2) by anthropogenic greenhouse gases alone, from pre-industrial periods to ca. 1990. From IPCC (1995), after Kiehl and Briegleb (1993). See text for interpretation. (Reprinted by permission from IPCC.)

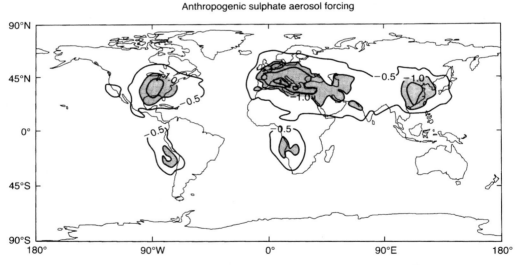

Fig. 17-7 Calculated geographical distribution of the direct climate forcing (W/m^2) by one anthropogenic aerosol component, sulfates, from pre-industrial periods to ca. 1990. (Reprinted with permission from IPCC, 1995.)

calculated geographical distribution of direct forcing by one type of aerosol – anthropogenic sulfates.

Figure 17-8 summarizes and compares the anthropogenic and natural forcings for the period 1800–1990, showing the complex nature of the issue of anthropogenic climate forcing.

17.4   Feedbacks

We have already posed the hypothesis based on Fig. 1-2 that climate is stabilized by negative feedbacks. The known and proposed feedbacks are listed below, starting with the purely physical examples, which are the simplest and best

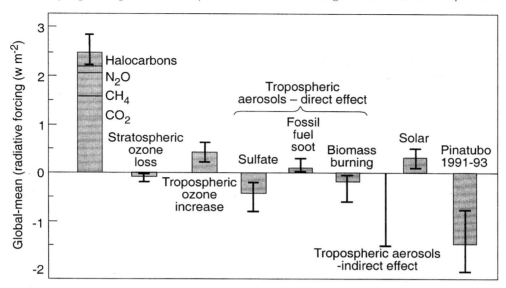

Fig. 17-8 A comparison of global-mean forcing by a variety of anthropogenic agents from pre-industrial periods to ca. 1990, compared to the measured forcing by the aerosol from the June 1991 eruption of Mt. Pinatubo. Shown are approximate averages for June 1991–1993. Peak forcing was ca. -2.7 ± 1 W/m^2 in August and September 1991 averaged over latitudes from $\pm 40°$. Caution is advised in comparing global means because of geographical non-uniformities (see Figs 17-6 and 17-7). (Adapted from data in IPCC, 1996.)

understood. The latter examples involve chemical and biological interaction and are much more complex.

17.4.1 Physical Feedbacks

17.4.1.1 Stefan–Boltzmann and water vapor feedbacks

As discussed earlier, it is customary to represent many feedbacks as functions of the surface temperature, T_s. The Stefan–Boltzmann feedback, is perhaps the simplest, most fundamental, and best understood of all such processes. As T_s increases, the emission of longwave (infrared) radiation going out into the atmosphere and to space also increases. This process helps cool the surface of the planet, and is thus a stabilizing (or negative) _feedback_. A different amplifying (positive) feedback that occurs is due to increase in evaporation of water vapor upon increase in T_s. This in turn increases the already large water-vapor greenhouse effect, leading to the so-called water-vapor feedback.

Although thermodynamically it is relatively simple to determine the amount of water vapor that enters the atmosphere using the Clausius–Clapeyron equation (see, e.g., Chapter 6, Equation (1)), its resultant atmospheric residence time and effect on clouds are both highly uncertain. Therefore this seemingly easily describable feedback is very difficult to quantify.

17.4.1.2 Albedo feedbacks of snow, ice, and clouds

Another family of feedbacks arises because the radical differences in the albedo (reflectivity) of ice, snow, and clouds compared to the rest of the planetary surface, which causes a loss of the absorption of solar radiation and thereby cools the planet. Indeed, the high albedo of snow and ice cover may be a factor that hastens the transition into ice ages once they have been initiated. Of course, the opposite holds due to decreasing albedo at the end of an ice age. As simple as this concept may appear to be, the cloud-albedo feedback is not easy to quantify because clouds reflect solar radiation (albedo effect) but absorb infrared radiation (greenhouse effect). The net effect of low clouds is to cool the planet, while thin high clouds (cirrus) warm it. Hence, even the sign of the total cloud feedback is not well established and depends in complex ways on the vertical temperature and humidity structure of the atmosphere.

17.4.1.3   Dynamical feedbacks

There are a number of dynamical (fluid-mechanical) feedbacks that relate to the rate at which energy is transported from low to high latitudes and to the vertical motions that produce the cooling needed to form clouds (Hartmann, 1994). These types of feedbacks cause changes in weather parameters such as average wind speed, frequency of occurrence of high- and low-pressure systems, tropical storms, etc. One of the consequences of this set of feedbacks is that ice-age temperature *change* near the equator was only a few degrees Celsius, while near the poles it was ca. -10 or $-20°C$, indicating less efficient transport of heat poleward during ice ages.

These physical feedbacks are currently included in global climate models, although with considerable uncertainty. These large, computer-based constructs generally allow such feedbacks to be "tuned" so that the model simulates the current climate. The existence of this practice emphasizes the high uncertainty even in these best-known feedbacks. Furthermore, models do not necessarily include all of the possible feedbacks. Lindzen (1990) points out that both positive (amplifying) and negative (stabilizing) feedbacks may exist but that most appearing in contemporary climate models are the ones that happen to amplify somewhat the predicted climatic response to increases in CO_2 concentration. It may well be the case that there are other, presently unknown dynamical feedbacks that would have the opposite effect.

17.4.2   Chemical Feedbacks

Chemical reactions can be a part of the feedback picture. Perhaps the best studied of the many possibilities is the reaction of CH_4 with the tropospheric free radical $·OH$, which is the primary sink for CH_4:

$$OH + CH_4 + O_2 \xrightarrow{net} CH_2O_2 + H_2O$$

If this is all that happens, it would seem likely that increased source strength of CH_4 would cause OH to decrease, leading to an increased lifetime of CH_4. A simple picture of a feedback emerges: increased CH_4 could cause increased CH_4 lifetime, thereby increasing the CH_4 content. However, other processes go on at the same time. For example, OH is both produced and consumed by other pollutant reactions, and there are chain reactions in which OH is first consumed and then reproduced later in the chain. To complicate this picture further, another sink for CH_4 is in the stratosphere where it is oxidized to form H_2O. Aside from the fact that water vapor happens to be a very effective greenhouse gas in the stratosphere, this increased H_2O increases the relative humidity in the stratosphere, possibly increasing the formation of high-altitude clouds. This is expected to happen in regions of low temperatures, such as near the poles. The type of clouds that form are *polar stratospheric clouds*, which have also been implicated in the destruction of stratospheric ozone (see Fig. 7-11).

Many other chemical feedback possibilities have been suggested (see, e.g., IPCC, 1995, 1996) and the complexities of them are recognized but unresolved in terms of both their mechanistic and quantitative aspects. Again referring back to Fig. 7-11, a web of reactions occur in both the troposphere and stratosphere, linking together the chemical processing of reduced carbon compounds, oxides of nitrogen, halogens, numerous free radicals (especially OH and HO_2), and ozone. These complexities make it difficult to quantify chemical feedbacks that depend on climate-active greenhouse gases. Hence, given that CH_4 has increased over time, it is possible that a feedback has occurred so that its lifetime has increased (i.e. the sink has decreased). On the other hand, sources could have increased. Either a decreased sink or an increased source (or both) have clearly caused sources to exceed sinks by ca. 7-8% over the period 1985-1995, the time that high-resolution data became available (see Fig. 19-3).

Physical climatic feedbacks involving CH_4 have also been suggested (see, e.g., IPCC, 1996). For example, a great deal of methane is trapped in permafrost in the form of *clathrates* (methane hydrates), which upon melting might release the gas into the atmosphere, causing an enhanced CH_4 greenhouse effect. However, there are many factors involved in both the production and consumption of CH_4. It can be released from soils at any latitude if they are highly anaerobic, or waterlogged. Warm wet soils are especially effective at releasing methane, for example in rice paddies. At this point, no quantitative estimate for that feedback can be given.

Possibilities exist for the involvement of halogenated species such as CCl_2F_2 (CFC-12) or CCl_3F (CFC-11) inasmuch as they can influence the column amounts of stratospheric O_3 which is both a strong absorber of solar ultraviolet radiation and an absorber and emitter of infrared radiation. (Refer back to Fig. 7-11 for a survey of the chemical reactions that are involved.)

17.4.3 Biogeochemical Feedbacks

The most complex of the feedback systems are those in which biota are directly involved. Indeed, one feedback of climate on CH_4 may well be an example; microorganisms in soils are a known CH_4 sink, and the rate at which they consume CH_4 is temperature dependent.

While there would seem to be a very large number of possible biogeochemical feedbacks, only a few have been identified and even these are not quantified. Two main classes of feedbacks can be defined; those in which biota influence albedo, and those that involve changes in the composition of the atmosphere.

Direct biological influences on the albedo that have been suggested include the darkening of land surfaces by vegetation, the likely increase of albedo with deciduous leaf drop and, in the ocean, the change in reflectivity of the sea due to organisms. Increases in albedo of the ocean have been observed via satellite imagery over large (100–1000 km) regions due to blooms of cocolithophores, which have calcareous (calcium-containing) hard parts. The almost milky appearance of the water causes more sunlight to be reflected, and diminishes the penetration of light into the ocean, thus decreasing the depth of the photic zone.

Influences of biota on the composition of the atmosphere are better understood, the best known being a slight global increase of the rate of photosynthesis caused by increased CO_2. The result is the production of more biomass and a decrease in the rate of accumulation of anthropogenic CO_2 in the atmosphere. This feedback may seem to be an indication that the biosphere will consume some of the extra CO_2, thus helping to solve the problem of anthropogenic greenhouse enhancement. But, it is very clear that the strength of the feedback is not anywhere near strong enough to do that since CO_2 concentration is continuing to increase. Details of this are discussed in Chapter 11.

Another family of feedbacks involving biota arise via the process of evapotranspiration in which the rate of water vapor is transferred from the land surface to the atmosphere is mediated by plants. Several consequences have been proposed that include influences of biota on the greenhouse effect of water vapor as well as relative humidity and clouds. Lovelock (1988) suggested that tropical forests might be kept cool by increasing cloud cover in response to higher relative humidity released through enhanced evapotranspiration (via the clouds' influences on albedo). Yet another connection arises because tree-covered land has different turbulence properties above it than bare soil, which also influences the cloud cover above.

A more complex feedback has been proposed that involves the production of dimethylsulfide by certain classes of marine phytoplankton. Four observations in the remote marine atmosphere formed the basis of this idea:

1. Sulfate compounds (e.g., NH_4HSO_4) form a major constituent of aerosol particles in remote, unpolluted marine air.
2. These particles can be a major fraction of the cloud condensation nuclei.
3. Cloud albedo is calculated to be sensitive to the droplet population. Twomey (1977) showed theoretically that albedo is enhanced by the addition of particles to the atmosphere.

4. The only continuous major natural source of the sulfur in these aerosol particles is dimethyl sulfide from marine phytoplankton (algae).

The putative feedback involves the influence of emissions of this aerosolgenic gas, $(CH_3)_2S$, that influences cloud albedo and hence either the temperature of the seawater in which the phytoplankton live or the amount of light available for their photosynthesis. Figure 17-9 represents the hypothetical feedback loop, and emphasizes that even the sign of the feedback is not known. Contradictory evidence has been developed

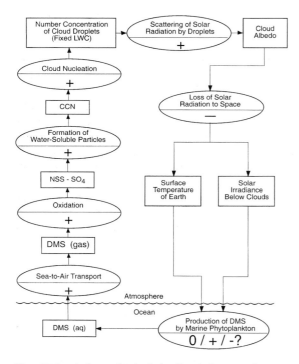

Fig. 17-9 A hypothetical feedback loop indicating dimethylsulfide (DMS) from marine phytoplankton. The rectangles are measurable quantities, and the ovals are processes linking the rectangles. The sign (+ or −) in the oval indicates the effect of a positive change of the quantity in the preceding rectangle on that in the succeeding rectangle. The most uncertain link in the loop is the effect of cloud albedo on DMS emission; its sign would have to be positive in order to regulate the climate. (From Charlson *et al.* (1987). Reprinted by permission from Nature, Copyright © (1987) Macmillan Magazines Ltd.)

regarding the sign. For example, sulfates and methane sulfonic acid were both elevated in Antarctic ice cores during the last ice age, suggesting that the feedback might have acted as an amplifier of climatic change (a minus sign in the $0/+/-$? oval in Fig. 17-9). On the other hand, a strong positive correlation was observed seasonally at Cape Grim, Tasmania, of higher levels of $(CH_3)_2S$, higher CCN and higher temperature in summer time; and methane sulfonic acid was low during ice ages in Greenland (both yielding a plus sign).

While this feedback may or may not be climatically relevant, it does serve to illustrate the nature of biogeochemical feedbacks. It seems likely that many such complex systems exist, and that they may indeed be factors that influence climate. To return to the introduction to this chapter, it is not possible to rule out biogeochemical feedbacks as factors that have stabilized climate over the past ca. 10^4 years.

One class of feedbacks has an exceedingly large potential for effectiveness: those that depend on precipitation, evaporation, and the ratio of the two. Changes in the timing of precipitation/evaporation are also important, indeed for the very existence of land biota. This is a complex set of feedbacks since the availability of liquid water in the right amounts at the right time must also coincide with the existence of appropriate amounts and types of nutrients in that same water.

Yet another level of complexity has been proposed in which the biogeochemical feedbacks actually influence the evolution of the organisms. Figure 17-10 advances an hypothesis that might explain how feedbacks could produce self-regulation (homeostasis) without teleology (or intent on the part of the biota). Lenton (1998) proposed a framework to support Gaia theory, which all depends on the notion that the physical climate of Earth is strongly influenced or even controlled by biota.

The final judgment is not in yet on whether biotic feedbacks are strong or weak; however, some level of connection of biota to climate has yet to be disproved. Indeed, the presence of O_2 and O_3 in the atmosphere are a direct consequence of photosynthesis, and their role and that of CO_2 in climate cannot be disputed.

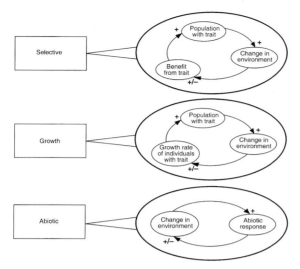

Fig. 17-10 A hierarchy of environmental feedbacks. Three levels are identified, abiotic (purely geochemical and geophysical) feedback, feedback on growth, and feedback on natural selection (see Chapter 3). "Trait" always refers to an environment-altering trait, and "growth" includes reproduction. At each level, both positive and negative feedbacks are possible, and these are illustrated in a general form. A plus symbol indicates a direct relationship. For example, an increase in the population with a particular environment-altering trait increases the resulting change in the environment. A plus/minus symbol indicates a relationship that can be either direct or inverse depending on specific conditions. For example, a change in the environment may increase or decrease the growth rate of individuals carrying the responsible trait, depending on the state of the environment and the direction in which it is being altered. When all the links of a complete feedback loop are positive, the feedback is positive; when one link is negative, the feedback is negative. Steps up the hierarchy are often additive. The activities of organisms can alter an underlying geochemical or geophysical feedback, while feedback on selection may be superimposed on underlying feedbacks on growth. (The spread of a trait is often subject to a direct positive feedback that is not shown – the larger a population, the larger its rate of growth.)

17.4.4 Missing Forcings and Feedbacks

The key to understanding the natural variability of climate and its response to natural or anthropogenic forcings is to understand fully both forcings and feedbacks. Forcings are relatively well understood; for example, the uncertainty in forcing by anthropogenic CO_2 and CH_4 is only 10–20%. Aerosol forcings have also at least been identified, although one of them – their influence on cloud albedo – is not currently quantifiable. However, the feedback problem, in all of its physical, chemical, and biogeochemical complexity, is still a barrier to understanding the system responses.

There are clearly some feedbacks that have not been quantified and, probably, some that have not yet been identified. One feedback that can be identified but not well quantified is the feedback of the cloud portion of the global albedo on surface temperature. The global-mean albedo is very sensitive to *both* cloud cover and cloud albedo. The global albedo also is strongly influenced (increased) by the extent of sea-ice and snow or glacial cover (the so-called ice-albedo feedback). Indeed, it has been pointed out recently that one of the stable climatic states of Earth is to be totally frozen – into a global ice age. This "snowball Earth" climate is suggested to have existed numerous times in the distant past, terminated each time by increases in the greenhouse effect due to high levels of atmospheric CO_2 from continued volcanic emissions (Hoffman *et al.*, 1998).

Going back to Equation (1), and writing T_{e1} for the radiative temperature at time 1, and T_{e2} for temperature at time 2 with constant S_0 and R_e,

$$\frac{T_{e1}}{T_{e2}} = \left(\frac{1 - A_1}{1 - A_2}\right)^{1/4}$$

If we let T_1 be the predictive temperature that we have now, 255 K, we can quickly see what happens if A changes from A_1, to a new value A_2. The current albedo of Earth is ca. 0.3, and current fractional cloud cover is ca. 0.5. Ice and snow cover are minimal and most of the Earth is oceans with $A_{ocean} \approx 0.1$. Forests have $A_{forests} \approx 0.1$ so most of the noncloudy Earth has $A \approx 0.1$. This gives an expression for average cloud albedo:

$$(0.5)(0.1) + (0.5)(A_{cloud}) \approx 0.3$$

so A_{cloud} currently is ca. 0.5. Now if the fractional area covered by cloud increases to 1 while

A_{cloud} stays at 0.5; A_{earth} increases to 0.5 (a trivial but important hypothetical case):

$$T_e \approx 234 \text{ K}$$

a drop of ca. 20 K. It is unlikely to have 100% cloud cover because that would require very different atmospheric dynamics, but it is an upper limit. Clearly, even a small change in either the area covered by cloud or the albedo of cloud (A_{cloud}) can cause enormous changes in the radiative temperature of Earth. Indeed a $\pm 1\%$ change in A_{cloud} would cause a ΔT_e of ca. ∓ 0.2 K. T_s would likely have larger change due to feedbacks. One message in this simple calculation is that the planetary temperature apparently has *not* varied by anywhere near 20 K over the past 10^4 years, thus clouds must have remained nearly constant in *both* their fractional coverage of Earth *and* their albedo. (Or, even more mysteriously, the two variables could have covaried perfectly to maintain thermostasis.) What feedbacks could keep the cloud contribution to albedo so constant? One partial explanation may be that as much air moves upward (causing clouds to form) as goes down, for continuity reasons.

Since feedbacks may have a large potential for control of albedo and therefore temperature, it seems necessary to highlight them as targets for study and research. Besides the simple example above of cloud area or cloud extent, there are others that can be identified. High-altitude ice clouds, for example, (cirrus) have *both* an albedo effect and a greenhouse effect. Their occurrence is very sensitive to the amount of water vapor in the upper troposphere and to the thermal structure of the atmosphere. There may also be missing feedbacks.

17.5 Climatic States and Responses

It is clear from the records of ice ages (see, e.g., Fig. 1-2) that Earth can have and has had climates that are different from our current state. Other, more extreme possibilities have been suggested, each of which could be stable for considerable periods of time. The frozen "snowball Earth" already mentioned is analogous to present-day Mars, where its CO_2 greenhouse effect amounts to only 3 K (compared with 33 K for present-day Earth, where water vapor is dominant). Another notable period – the middle Cretaceous, from ca. 120 to 90 million years ago – appears to have been extraordinarily warm. Fossil evidence suggests that plant habitats were up to $15°$ closer to the poles than their current latitudinal position. Studies of sediments show that coal was formed from peat in areas north of the Arctic Circle. Isotopic data indicate that deep ocean water was as much as $15–20°C$ warmer than it is now. This period also is noteworthy because of the large amounts of today's fossil fuels that were laid down, indicating that the carbon cycle was very different from today. Many features of Earth were different then, including the location and size of the continents as well as the ocean circulation, so no simple definitive explanation for the warmth of this period is possible. One factor can be suggested based on climate models and the likelihood of enhanced volcanism during that period: the greenhouse effect of CO_2 at levels of four to six times that of the present.

Another extreme climate would occur if large amounts of the water in the oceans were evaporated, yielding a positive feedback and a runaway greenhouse effect. A still warmer situation would arise if much or all of the present-day carbonate rock were dissolved and released as CO_2 to the atmosphere. These sorts of "runaway" greenhouse effects would occur in large part because the wavelengths emitted by a black-body decrease as temperature increases. On Earth at present the maximum emission occurs at ca. $10–15$ μm wavelength. If the temperature is such that this maximum moves (via the Wien effect) to the region of the $4–7$ μm absorption band of H_2O, the greenhouse effect of water vapor is greatly enhanced. A "runaway" greenhouse effect, which is roughly comparable to Venus today, would appear to be a condition from which there is no return.

Tying together forcings and feedbacks with responses is an enormous puzzle. For example, the decreased CO_2 during ice ages certainly contributed to the cooling, but by itself was not enough to explain the total cooling during the ice age. Similarly, the orbital changes (Milankovich effect) also caused forcings that are too small to explain the whole systematic temperature shift. As will be seen in the following chapter, the ice age records in glaciers and sediments offer an intriguing glimpse into the workings of the Earth system but are not yet fully explained.

Thus, the current pressing questions of how to understand and deal with human-induced increases of greenhouse gases cannot be answered with the current knowledge of the extant feedbacks and system response characteristics. Many attempts have been made to detect the global influence of increases in greenhouse gases on global and regional temperatures, and not surprisingly a range of results have been found. From the side of the skeptics, Lindzen (1990) states that the observed temperature changes are consistent with what is known about natural variability, while Hansen and Lebedeff (1988) and Santer *et al.* (1996) claim that the pattern of temperature changes is what is expected from the known anthropogenic forcings. The IPCC (1996) issued the famous if somewhat equivocal statement, "The balance of evidence suggests a discernible human influence on global climate," based substantially on the work of Santer *et al.* (1996). However, the correlations such as were used by IPCC cannot prove causality. What is needed is a more thorough understanding of the processes that control climate, most especially the feedbacks. It seems necessary to face the reality that the physical climate system is coupled to a wide variety of chemical and biogeochemical processes. Most importantly, it would appear that feedbacks involving the biosphere must be included as a key factor in the coupling.

At this point let us pose some important, thought-provoking questions: Will anthropogenic climate forcing be sufficient to cause a shift in the climate *state*? Do other climatic states exist besides those mentioned above? Will the response to the gradual increase of anthropogenic forcings also be gradual or will these be abrupt changes? These are all matters of current research and cannot be answered at present. To end this chapter, it seems appropriate to quote Roger Revelle and colleagues (1965): "Through his worldwide industrial civilization, Man is unwittingly conducting a vast geophysical experiment." We can add the obvious point that we do not know enough yet to be able to predict the results. Given that the Earth, Mars, and Venus evolved at about the same time at similar distances from the sun, it is likely that they accreted similar abundances of excess volatile compounds, notably H_2O and CO_2. Given that the climates of Earth's two sister planets are radically different because of the ways the greenhouse gases are dispersed and cannot support a complex biosphere, it is critically important to understand the factors that might cause shifts to radically different climatic states.

References

Charlson, R. J., Lovelock, J. E., Andreae, M. O., and Warren, S. G. (1987). "Oceanic phytoplankton, atmospheric sulfur, cloud albedo and climate." *Nature* **326**, 655–661.

Dickinson, R. E. and Cicerone. R. J. (1986). Future global warming from atmospheric trace gases. *Nature* **319**, 109–115.

Fleagle, R. G. and Businger, J. A. (1963). "An Introduction to Atmospheric Physics." Academic Press, New York and London.

Hansen, J. E. and Lebedeff, S. (1988). Global surface air temperatures: update through 1987. *Geophys. Res. Lett.* **19**, 323–326.

Hartmann, D. L. (1994). "Global Physical Climatology." Academic Press, San Diego.

Hoffman, P. F., Kaufman, A.J., Halverson, G. P., and Schrag, D.P. (1998). A neoproterozoic snowball earth. *Science* **281**, 1342–1346.

Houghton, J. T. (1984). "The Global Climate." Cambridge University Press, Cambridge.

Intergovernmental Panel on Climate Change (1990). "Climate Change: The IPCC Scientific Assessment" (J. T. Houghton, G. J. Jenkins, and J. J. Ephraums, eds). Cambridge University Press, Cambridge.

Intergovernmental Panel on Climate Change (1995). "Climate Change, 1994: Radiative Forcing of Climage Change and An Evaluation of the IPCC IS92 Emission Scenarios" (J. T. Houghton, L. G. M. Filho, J. Bruce, H. Lee, B. A. Callander, E. Haites, N. Harris, and K. Maskell, eds). Cambridge University Press, Cambridge.

Intergovernmental Panel on Climate Change (1996). "Climate Change 1995: The Science of Climate Change" (J. T. Houghton, L. G. M. Filho, B. A.

Callander, N. Harris, A. Kattenberg, and K. Maskell, eds). Cambridge University Press, Cambridge.

Kiehl, J. T. and Briegleb, B. P. (1993). The relative roles of sulfate aerosols and greenhouse gases in climate forcing. *Science* **260** (16 April), 311–314.

Lenton, T. M. (1998). Gaia and natural selection. *Nature* **394**, 439–447.

Lindzen, R. S. (1990). Some coolness concerning global warming. *Bull. Am. Met. Soc.* **71**(3), 288–299.

Lovelock, J. (1988). "The Ages of Gaia." W. W. Norton, New York and London.

Peixoto, J. P. and Oort, A. H. (1992). "Physics of Climate." American Institute of Physics, New York.

Revelle, R., Broecker, W., Craig, H., Keeling, C. D., and Smagorinsky, J. (1965). Atmospheric carbon dioxide. *In* "Restoring the Quality of Our Environment," Report of the Environmental Pollution Panel, President's Science Advisory Committee, The White House. Washington, DC, p. 126.

Santer, B. D., Taylor, K. E., Wigley, T. M. L., Johns, T. C., Jones, P. D., Karoly, D. J., Mitchell, J. F. B., Oort, A. H., Penner, J. E., Ramaswamy, V., Schwartzkopf, M. D., Stouffer, R. J., and Tett, S. (1996). A search for human influences on the thermal structure of the atmosphere. *Nature* **382**, 39–46.

Twomey, S. (1977). "Atmospheric Aerosols." Elsevier, Amsterdam.

18

Ice Sheets and the Ice-Core Record of Climate Change

Kurt M. Cuffey and Edward J. Brook

18.1 Introduction

18.1.1 Historical Context

In the mid-19th century, Louis Agassiz and colleagues recognized that many features of the landscapes of northern Europe and North America could be sensibly explained only as products of glaciation (Agassiz, 1840). These features include bouldery ridges (called moraines) crossing flat lands, enormous boulders found far from their sources, deep lake basins, and ancient channels of enormous rivers. The realization that much of the now densely populated regions of Europe and North America were rather recently buried by ice lead to a startling revelation: the Earth experiences environmental changes of spectacular magnitude and global significance. Subsequently, careful analyses of the glacial deposits revealed that there had been at least several major episodes of glaciation. In the 1950s and 60s new geochemical and paleontologic tools painted a much more complete picture of these ice ages. In particular, the oxygen isotopic composition of seawater, whose history can be learned by analyzing cores of ocean-floor deposits, is a measure of continental ice-sheet volume. These data revealed that there were 30 or so major periods of glaciation over the past two million years, and these occurred with considerable temporal regularity.

Such major environmental changes present a tremendous opportunity for studying Earth systems and biogeochemical cycling. These changes may be viewed as global-scale experiments that have already been conducted for us. By elucidating their nature (the histories of both forcings and environments), we can learn much about the behavior of coupled Earth systems.

Records of past environmental change are preserved in a broad range of Earth materials. Past environments are inferred from "proxy" records, meaning measurements of physical and chemical parameters of marine and terrestrial sediment, polar ice, and other materials that were in some way influenced by their environment during accumulation. Examples of proxy records are the distribution of glacial deposits, the isotopic composition of terrestrial and marine sediments and ice, the abundance and species composition of plant and animal fossils, and the width of tree rings.

18.1.2 This Chapter

This chapter is a brief introduction to Earth's historical environmental changes, with emphasis on the recent ice-age cycles. We chose this emphasis because preservation of these environmental records is much better than for earlier times, and because the ice ages constitute drastic changes in global environment which have

Earth System Science
ISBN 0-12-379370-X

occurred throughout the past two million years (the Quaternary) despite the constancy of the major boundary conditions controlling climate (solar luminosity, positions of the continents). Understanding the causes of these changes is crucial for assessing the modern environment's potential for change. Current understanding is far from complete.

We will further focus much of this chapter on paleoclimate records obtained by analyzing ice cores extracted from the polar ice sheets. There are many important types of paleoclimate records, but the ice-sheet records have special importance. The ice sheets are the only significant accumulations of atmospheric sediment and therefore contain unique information about atmospheric composition and processes. In particular, atmospheric gases are trapped in glacial ice, so ice-core analysis can rather directly reveal changes in greenhouse gas composition through time. Readers wishing to see a more comprehensive treatment of the material in this chapter are referred to Crowley and North (1991) for general paleoclimatology, and Hammer *et al.* (1997) for ice-core studies.

The goal of this chapter is twofold: first, to present some of the most important aspects of Earth's environmental history, and second, to communicate the power, complexity, and incompleteness of paleoclimatology through a presentation of one of its most important tools.

18.2 Quaternary Climate Change

18.2.1 Context

Over the last 60 to 70 million years the Earth has gradually cooled by 5 to 15°C and sea level has fallen by more than 200 m. Possible causes of this trend include changes in land distribution, ocean circulation, and atmospheric CO_2. The land area in northern high latitudes has increased, and this may have allowed the growth of large ice sheets, resulting in cooler climate due to the increase in Earth's albedo (Chapter 7) (Turekian, 1996). Models and proxy measurements (Cerling *et al.*, 1993; Berner, 1990) suggest that atmospheric CO_2 content decreased substantially during this time, from values of

1000–2000 ppmv in the Cretaceous to 200–280 ppmv by the late Quaternary. One proposed explanation for this CO_2 drawdown is continental weathering accompanying the uplift of high mountain ranges (Raymo and Ruddiman, 1992). Mountain uplift could also cool and dry the global climate by directly affecting large-scale atmospheric circulation and precipitation patterns (Ruddiman and Kutzbach, 1991). There is considerable debate, however, about whether increased mountain uplift was a cause or consequence of the climatic cooling of the past five million years which finally plunged Earth into the ice ages (Molnar and England, 1990). Nonetheless, in all these scenarios, the root cause of Earth's cooling from the balmy Cretaceous to the modern glacial ages is the slow tectonic motions of the Earth's lithospheric plates (see Chapter 9).

18.2.2 The Tempo of Quaternary Climate Change: Oxygen Isotopes and Orbital Forcing

In addition to being relatively cool, the climate of the last two million years has been quite variable. The early studies of glacial deposits on land established that Earth's climate was much colder than present in the very recent geologic past, only a little over 10 millennia ago. Further studies indicated that periods of glaciation were interrupted by warm climates. More recently, a detailed picture of Quaternary climate changes has emerged from studies of ocean sediments. The largest amount of paleoclimate information from these sediments has come from studies of variations in abundance and species composition of planktonic (surface-water dwelling) and benthic (sea-floor dwelling) microscopic plants and animals, and the oxygen and carbon isotopic composition of the calcium carbonate shells of these organisms.

In modern sediments particular assemblages of species are characteristic of particular environmental conditions. Therefore it is possible to use species assemblages in ancient sediments to infer past sea-surface temperatures and other variables, and these techniques provide a wealth of paleoclimate information. Complementary information is provided by the oxygen isotopic composition of marine plankton's calcium carbonate shells, which has varied significantly throughout the Quaternary. The $\delta^{18}O$ (see Box 1) of these shells is a function of both the $\delta^{18}O$ and the temperature of the water in which the shells grow. The measurements of $\delta^{18}O$ on sea-core sediments therefore reveal a combined history of ocean temperature and ocean $\delta^{18}O$. Emiliani (1955) initially interpreted the Quaternary marine $\delta^{18}O$ record as primarily an indicator of temperature change. Subsequent work (Shackleton, 1967) revealed that much of the signal instead resulted from changes in sea water $\delta^{18}O$, due to changes in volume of continental ice sheets and glaciers (see Section

18.3.2.1 below). The marine $\delta^{18}O$ record is now interpreted primarily as a record of changes in continental ice volume. Continental ice volume expands (and ocean $\delta^{18}O$ increases) during cooling climates as northern hemisphere ice sheets form and grow and the Antarctic ice sheet expands. Thus marine $\delta^{18}O$ is a proxy for global climatic temperature, with high marine $\delta^{18}O$ indicating cold climates. Because large-scale climate changes are required to grow large continental ice sheets, marine $\delta^{18}O$ is one of the single best proxies for global climatic changes.

Numerous studies of marine $\delta^{18}O$ histories (and the strongly correlated species assemblages) demonstrate that global climate variations have followed a characteristic quasi-cyclic pattern over the past 2 Myr (Fig. 18-1). Dominant in the last 700 kyr of these records is an approximately 100 kyr cycle of glacial and interglacial periods (prior to this time an ~40 kyr cycle dominated). This 100 kyr cycle has a characteristic asymmetrical "sawtooth" shape, with gradual descents into cold climate followed by abrupt terminations of glacial conditions and return to warm interglacial climate (Broecker and Denton, 1989). The warm interglacial climates are generally of brief duration relative to the whole cycle.

Much of the variation in these time series for the past 700 kyr can be described by a combination of a 100 kyr cycle plus additional cycles with periods of ~20 and ~40 kyr. This result immediately suggests that the ice-age cycles are caused by variations in the amount and seasonality of solar radiation reaching the Earth (insolation), because the ~20, ~40, and ~100 kyr periods of climate history match the periods of cyclic variations in Earth's orbit and axial tilt. The hypothesis that these factors control climate was proposed by Milutin Milankovitch in the early part of the 20th century and is widely known as "Milankovitch Theory." It is now generally accepted that the Milankovitch variations are the root cause of the important 20 and 40 kyr climate cycles. The 100 kyr cycle, however, proves to be a puzzle. The magnitude of the insolation variation at this periodicity is relatively trivial, but the 100 kyr cycle dominates the climate history of the last 700 kyr. Further,

---

Box 1: Definition of δ

The two most common isotopes of oxygen are ^{16}O and ^{18}O, and the two natural isotopes of hydrogen are ^{1}H and ^{2}H or D (deuterium). If a sample of carbonate contains C_{18} moles of ^{18}O and C_{16} moles of ^{16}O, then define the heavy:light ratio as $R_{18} = C_{18}/C_{16}$. Similarly, natural water molecules are mostly of three types: $H_2^{16}O$, $H_2^{18}O$, and $HD^{16}O$. For a given sample of water, call the number of moles of each W, W_{18}, and W_D, respectively. Then define the heavy:light ratios as $R_{18} = W_{18}/W$ and $R_D = W_D/W$. For either water or carbonate, the δ are defined as deviations of these ratios from standard values for these ratios (call the ratios of the standards S_{18} and S_D):

$$\delta^{18}O = \frac{R_{18}}{S_{18}} - 1 \qquad (A1)$$

$$\delta D = \frac{R_D}{S_D} - 1 \qquad (A2)$$

For water, the standard ratios are approximately the average ratios for modern seawater (these standards are called SMOW, or Standard Mean Ocean Water), so $\delta^{18}O$ and δD both equal zero for modern seawater (if a standard other than SMOW is used, this will not be true). δ values are typically very small numbers and so are usually multiplied by 1000, in which case the units are "per mil."

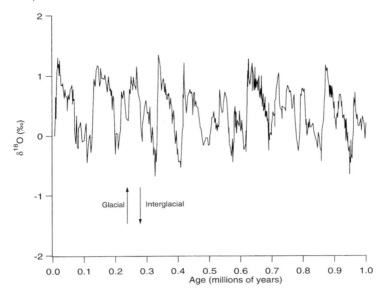

Fig. 18-1 Benthic foraminiferal oxygen isotope record from 3477 m water depth in the eastern tropical Pacific ocean from Ocean Drilling Program site 677 (Shackleton *et al.*, 1990). $^{18}O/^{16}O$ ratios are expressed in the δ notation relative to the SMOW standard. Note the strong 100 kyr periodicity characteristic of Quaternary climate records.

the insolation forcings do not directly predict the asymmetrical "sawtooth" form of the main cycles.

The three orbital parameters that control the amount and seasonality of incoming radiation are the tilt of the Earth's spin axis, the eccentricity of the Earth's orbit, and the precession of the equinoxes (Fig. 18-2). The tilt of the spin axis varies between 22° and 25°, with a period of 41 kyr. The eccentricity of the Earth's elliptical orbit varies from nearly circular to 6%, with periods of 100 and 400 kyr. Precession of the equinoxes refers to the continual change in the timing of the seasons relative to the position of the Earth along its eccentric orbit (which determines the Earth to sun distance). Two processes influence equinox precession. First, the Earth's spin axis does not have a fixed orientation in space, but rather wobbles in such a way that the geographic north pole describes a full circle in space every 26 000 years (Polaris is not always the North Star). Second, the ellipse of the orbit itself also rotates slowly, with a period of 22 000 years. The interaction of these two generates precession cycles with periods of 19 000 and 23 000 years.

What is the effect of these orbital variations on the input of solar radiation to the Earth? The precession cycles cause changes in the seasonal distribution of insolation. The seasons exist because the Earth's spin axis is tilted. Northern hemisphere summer occurs when the Earth's spin axis points toward the sun, and on the summer solstice the sun lies in the vertical plane containing the spin axis. Currently the summer solstice occurs at a position along the elliptical orbit for which the Earth–sun distance is near its maximum value (aphelion). Through time, the precession cycle shifts the position of the solstice along the orbit so that the Earth–sun distance on the day of summer solstice varies through time. Northern hemisphere summers are colder, and southern hemisphere summers warmer, if the summer solstice occurs when the Earth–sun distance is a maximum, as it is in the modern day. Eleven thousand years ago, when ice sheets were melting, the situation was reversed; northern hemisphere summer occurred while the Earth was closest to the sun (near perihelion), and summertime northern hemisphere insolation was greater than that today. Thus precession creates cyclical varia-

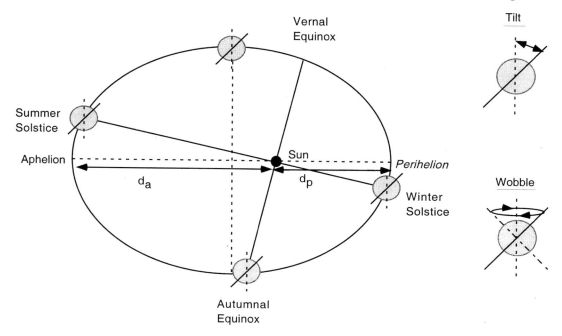

Fig. 18-2 Orbital parameters that control the seasonal and annual receipt of radiation at the Earth's surface. The Earth's rotational axis tilts and wobbles as shown. The eccentricity of the orbit is $(d_p - d_a)/(d_p + d_a)$ which is zero for a circle but for the Earth's orbit is sometimes as large as 0.06. (After Hartmann (1994, p. 303) and Turekian (1996, p. 80).)

tions in seasonal radiation at periods of 19 and 23 kyr, and these variations are opposite in the northern and southern hemispheres. Compared to other Milankovitch forcings, precession causes the largest seasonal variations at low latitudes. The net global radiation received over a year is not affected by precession.

Tilt variations also do not affect the annual total of solar energy received by the whole Earth, but do change the annual total for polar regions (simultaneously for both hemispheres). Tilt also affects the seasonal insolation at high latitudes, with greater tilt leading to warmer summers and cooler winters in both hemispheres.

Eccentricity variations change the average Earth–sun distance such that annual insolation changes by approximately 0.2%, or 0.5 W/m² at the Earth's surface (Crowley and North, 1991). This is small compared to the 10% seasonal changes associated with the other Milankovitch cycles. That a 100 kyr periodicity characterizes both the eccentricity and the glacial–interglacial

cycles may reflect an important role for the eccentricity cycle in causing exceptional highs of seasonal insolation (in concert with the other Milankovitch forcings) which may be necessary to terminate glaciations. How, and if, the eccentricity cycle influenced Quaternary climate is a major question in current research. An alternative, and not yet fully tested, explanation for the 100 kyr cycle is that there actually is a significant astronomical forcing of this period, related to tilt of the ecliptic with respect to a horizon of interstellar dust (Muller and MacDonald, 1997). Despite this uncertainty concerning the 100 kyr cycle, the marine oxygen isotope record provides convincing evidence that changes in the Earth's orbit and tilt are a root cause of some climate changes throughout the Quaternary (Hays *et al.*, 1976). Considerable effort has been devoted to understanding this connection. The match between periodicities of insolation and climate is robust, but many analyses have shown that the timing of insolation changes and global climate changes are correlated only if insolation is taken to be that for summer at mid–high latitudes in the northern hemisphere. This suggests the fundamental link between Milankovitch forcings and climate involves increased glacier melt rates during very warm summers, without opposing changes in very cold winters. In this way, greater seasonal contrast leads to the growth and decay of the great ice sheets on North America and Eurasia.

Though an important part of the Quaternary climate puzzle, this view of the ice-age cycles is by itself quite incomplete. If Milankovitch forcings alone govern climate change then:

1. Most climate variability will occur with 20 and 40 kyr periods.
2. Climate change will be globally asynchronous.
3. Global climate change will occur slowly and gradually.

We have already shown that 1 is not generally true. Later in this chapter we show that 2 and 3 are not generally true either. Thus there are additional important elements of the climate change engine.

18.2.3   A Closer Look at Glacial to Interglacial Climate Change

There is a much better record of the last major transition from glacial to interglacial climate than for earlier portions of the climate cycles. This transition therefore provides a useful starting point for examining the character of major climate changes, though each of these may have had unique elements.

18.2.3.1   The Ice-Age Earth

Approximately 20 kyr ago glaciation had attained its maximum extent (a time called the last glacial maximum or LGM). Much of Canada and the northern United States, as well as northern Europe and Asia, were covered by large ice sheets that in places exceeded 3 km in thickness (Fig. 18-3). Although the growth of these major ice sheets was the most dramatic change of the Earth's surface, smaller ice caps and glaciers also advanced during the last glacial maximum. In

Antarctica the West Antarctic ice sheet expanded significantly during the LGM (Denton *et al.*, 1989), while the East Antarctic ice sheet expanded modestly. Concurrent with the growth of major ice sheets was a drop in sea level, with a maximum sea level depression during the LGM of approximately 120 m (Shackleton, 1987).

The Earth was drier than present during the LGM, and the hydrologic cycle slower. Polar snow accumulation rates were significantly depressed. Delivery of continental dust to the ice sheets and the ocean was enhanced, suggesting overall higher levels of aridity. Many studies of fossil pollen, and geomorphic features like sand dunes, support this view. Some regional exceptions existed; wet areas during the LGM included the Great Basin of North America, and portions of Russia, the Middle East, southern South America, and southern Australia (COHMAP, 1988; Crowley and North, 1991). Studies of terrestrial fossils, and pollen in particular, reveal equatorward shifts of vegetation zones during colder climates, and suggest that local climate changes were highly variable. Viewed as a whole, the terrestrial and marine records consistently show that environmental changes in polar regions were much greater than those in equatorial regions.

18.2.3.2   Global Temperature Change

In the 1970s the CLIMAP project compiled the analyses of marine sediment fossil assemblages (CLIMAP members, 1976) to produce global maps of sea surface temperatures during the LGM. These suggested that sea surface temperatures in the mid to high latitudes were depressed by up to 6 to 10°C, that areas of seasonal sea ice were larger, and that polar fronts were shifted to lower latitudes. Pollen-based reconstructions of mid–high-latitude land temperatures suggest temperature depressions of 4 to 8°C in the mid latitudes, and larger changes adjacent to the ice sheets. CLIMAP reconstructions for many areas of the tropics indicate little or no change in sea surface temperature, in contrast to the mid–high-latitude data. This appears inconsistent with locations of

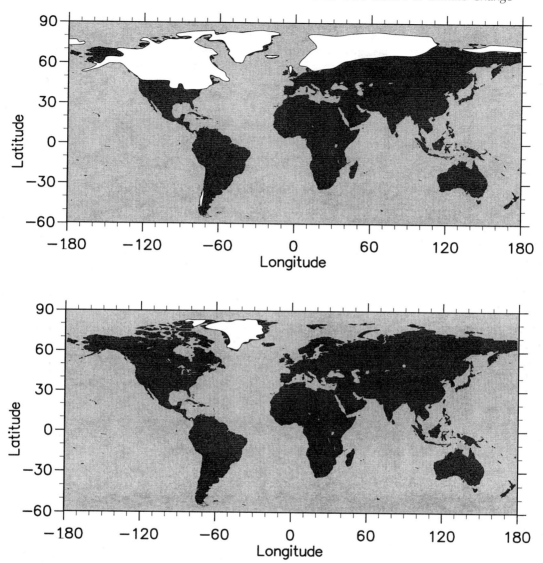

Fig. 18-3 Extent of ice sheet cover (shown as white) during the LGM (top panel) compared to modern (bottom panel), excluding Antarctica. Locations of ice sheets are approximate.

glacial deposits in New Guinea, Africa, South America, and Hawaii, all of which show that snowlines were depressed during the glacial period by up to 1000 m below modern levels. The latter suggests a much larger tropical temperature depression, approximately 5–6°C based on the 6°C/km decrease of temperature with increasing altitude in the troposphere. This conclusion is supported by newer paleotemperature proxies, including the noble gas composition of glacial age groundwater in tropical aquifers (Stute *et al.*, 1992, 1995), and the Sr/Ca ratio of coral skeletons (Guilderson *et al.*, 1994). However, it is possible that tropical temperature depression varied regionally. The magnitude and patterns of glacial–interglacial temperature change in the tropics remain subjects of active research.

18.2.3.3   Ocean Circulation

In the modern world, oceanic heat transport has a prominent role in determining global and regional climates. Colder ocean waters during the LGM suggest reduced LGM oceanic heat transport on average, but such changes cannot be inferred without knowledge of circulation changes too. Several geochemical proxies record ocean circulation changes. Most important are the carbon isotopic composition ($\delta^{13}C$) and the Cd/Ca ratio of planktonic and benthic foraminifera (Curry *et al.*, 1988; Boyle *et al.*, 1988). Cycling of both parameters depends on biological productivity in surface waters. Surface-water dwelling organisms preferentially remove ^{12}C and Cd from surface waters and the rain of their carcasses to the deep sea causes surface waters to be depleted in Cd and enriched in ^{13}C.

After transport to the deep sea (see Chapter 10), these surface-derived waters gradually enrich in Cd and ^{12}C. More vigorous transport (meaning more vigorous formation of deep water and vertical mixing of the global ocean) causes deep waters to have lower Cd/Ca and higher $\delta^{13}C$, and this signal is recorded by benthic foraminifera. Much attention has focused on the deep-water formation sites in the North Atlantic Ocean, because these are coupled to the tremendous northward transport of heat by the Gulf Stream. Studies of foraminifera Cd/Ca and $\delta^{13}C$ show that the formation of North Atlantic Deep Water was greatly reduced during the LGM, and replaced by formation of shallower intermediate water. This probably means that Gulf Stream heat transport was greatly reduced. Longer time-series indicate that variability in deep-water production was a general characteristic of glacial–interglacial cycles (Raymo *et al.*, 1989). Further, rapid variations in deep-water formation correlate with rapid climate changes during the last glacial period (see 18.3.2.4). Thus changes in ocean heat transport likely have a large role in the climate cycles (Broecker and Denton, 1989).

18.3   Ice Sheets as Paleoclimate Archives

We now explore in detail the methods and results of ice-core paleoclimatology. The reward for this effort is an astonishing expansion of our knowledge of past environments, remarkable both for its implications and its level of detail.

18.3.1   Ice-Core Basics

In the interiors of ice sheets, there are boundaries (usually coincident with topographic ridges on the ice-sheet surface) separating regions where ice flows in one direction from regions where ice flows in the opposite direction. These boundaries are analogous to drainage divides separating watersheds, and are therefore called ice divides. At the ice divide, the ice motion is predominately vertical downward (Fig. 18-4) and sluggish so that ice deposited as snow on the surface at this one location accumulates there for a long time. Due to the spreading of the ice flow lines, a block of ice deposited at the surface is horizontally stretched as it descends, which causes the block to thin vertically, because volume of the block is conserved. Thus, the age of ice increases exponentially with depth beneath the ice divide. By extracting a core of ice at, or near, an ice divide, one can obtain a continuous record of atmospheric sediment that contains more and more years per length of core near the bottom. Core extraction is usually done with a drill that mechanically cuts a cylinder of ice, typically 10 to 15 cm in diameter. The locations of many important core sites are shown in Fig. 18-5.

For an ice sheet of thickness H in equilibrium with a climate supplying accumulation at a rate $\dot{a}$ (thickness of ice per unit time), the vertical velocity near the ice-sheet surface is $\dot{a}$ and this velocity decreases to zero at the ice-sheet bed. A characteristic time constant for the ice core is $H/\dot{a}$. The longest histories are therefore obtained from the thick and dry interiors of the ice sheets (particularly central East Antarctica, where $H/\dot{a} = 2 \times 10^5$ yrs). Unfortunately, records from low $\dot{a}$ sites are also low resolution, so to obtain a high-resolution record a high $\dot{a}$ site must be used and duration sacrificed (examples are the Antarctic Peninsula ($H/\dot{a} = 10^3$) and southern Greenland ($H/\dot{a} = 5 \times 10^3$)).

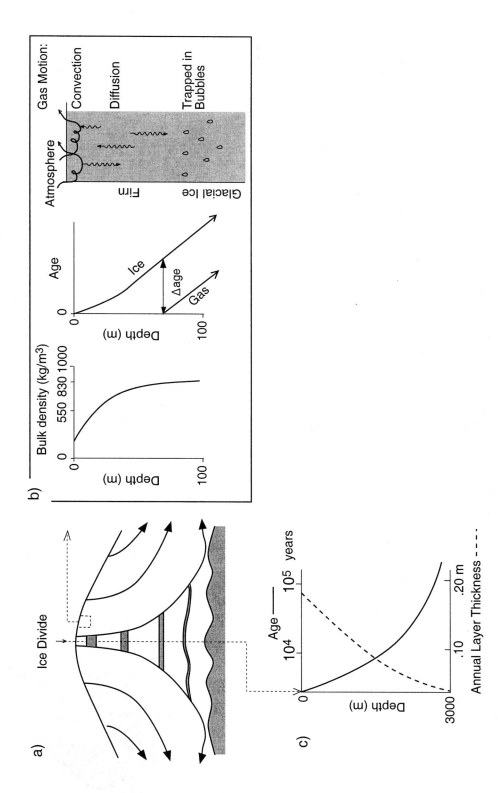

Fig. 18-4 (a) Cross-section through an ice divide showing flow lines and thinning of layers. (b) Closer look at the upper 100 m: characteristic density variations, age–depth relations for ice and gas, and mechanisms of gas transport. (c) Characteristic depth–age relation and annual layer thicknesses, with numbers chosen to represent central Greenland. Age axis is non-linear.

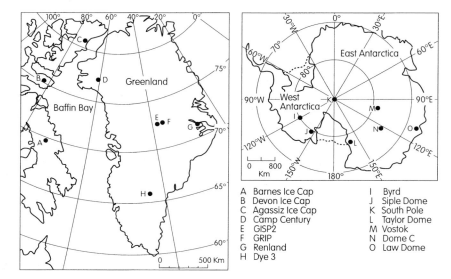

Fig. 18-5 Locations of important ice-core sites in Greenland and Antarctica (after Paterson, 1994). The two central Greenland sites, GISP2 and GRIP, are collectively known as Summit.

18.3.1.1 Depth–age scales

Obtaining an accurate and detailed depth–age relationship for an ice core is, of course, a necessary task for learning paleoclimate histories. Approximate time scales can be calculated using numerical models of ice and heat flow for the core site (Reeh, 1989), constrained by estimates of the modern accumulation rate and by measurements of ice thickness from radio-echo-sounding surveys.

Once a core is extracted, measurements of seasonally varying properties allow identification of annual layers (Hammer *et al.*, 1987). Counting these along the core, as one would count rings in a tree, yields a depth–age scale of high precision and very good accuracy. Useful seasonally varying properties include dust content, isotopic composition, concentrations of some chemical impurities, electrical conductivity (a proxy for core acidity), and physical stratigraphy (seen as variations in transmitted light seen through a sliced core). Annual layers must be preserved in the ice sheet for absolute age determination to be possible in this fashion. Annual layer preservation is good at high accumulation-rate sites in Greenland and

West Antarctica, but not at the very dry sites in East Antarctica. In addition to such continual counting methods, one can determine the absolute age of ice by identifying the fallout from volcanic eruptions of known age.

The rapid global mixing of gases in the Earth's atmosphere, in a few years' time, enables powerful techniques for learning the depth–age scale of an ice core by correlating the variation of gases along one core to the same variations along another core for which absolute ages are known, and to marine sediment cores. Of particular use are the $\delta^{18}O$ of O_2 gas (Sowers *et al.*, 1993; Bender *et al.*, 1994) and methane (Blunier *et al.*, 1998; Steig *et al.*, 1998).

A very important complication in interpreting ice core records, and in defining depth–age relations, is the fact that snow transforms to ice 50 to 100 m below the surfaces of most polar ice sheets. This means the gas trapped in ice is actually younger than the solid ice at the same depth, and that a variety of processes can transport and redistribute gases in this snowy upper layer (called the firn). To understand this, and to prepare for subsequent discussions, we must discuss how snow converts to ice near the ice sheet surface.

18.3.1.2   Firn and gas trapping

At the ice-sheet surface, motion of water vapor very rapidly converts faceted and angular snow to solid ice spherules. Subsequently, the weight of overlying accumulation forces these into a more compact configuration by slip at their boundaries and creep deformation. Thereby the density of a given layer steadily increases through time, from a value of approximately $300 \, kg/m^3$ at deposition to an ultimate value of approximately $920 \, kg/m^3$ for glacial ice. Thus there is a marked increase of density with depth below the surface, with most of the variation occurring in the upper 100 m. Correspondingly, the air-filled pore spaces between ice grains become smaller and more isolated as depth (or age) increases. These pores are finally isolated as distinct bubbles at a bulk density of around $830 \, kg/m^3$. Above this level, which is generally 40 to 100 m deep, air spaces are interconnected and open to the atmosphere above. This entire zone is called the firn. In the firn, gases are free to move to and from the atmosphere, by wind-driven convection in the upper 10 m and by diffusion below this (Fig. 18-4). Thus the gas trapped in bubbles is younger than the age of the adjacent ice. This gas-age–ice-age difference (often referred to as Δ_{age}) today ranges from 30 years at ice-core sites with very high accumulation rates (> 100 cm/yr) to greater than 2000 years at low-accumulation-rate sites (<2 cm/yr). During glacial times, when snow accumulation rates were lower, Δ_{age} was larger. Below the firn is "glacial ice," which retains visible bubbles until depths of 1500 m or so, where the gases finally are incorporated into the solid lattice as clathrates.

The interstitial air trapped during this process preserves a largely unaltered record of the composition of past atmospheres on time scales as short as decades and as long as several hundred thousand years. Such records have provided critical information about past variations in carbon dioxide (CO_2), methane (CH_4), nitrous oxide (N_2O), carbon monoxide (CO), and the isotopic composition of some of these trace species. In addition, studies of the major elements of air: nitrogen, oxygen, and argon, and their isotopic composition, have contributed greatly to understanding of past climates and the processes that trap gases in ice. Gas records are discussed in Sections 18.3.2.3 and 18.3.5.

18.3.2   Thermometry: Tools and Results

18.3.2.1   Stable isotopes in water

One of the most important measurements made on an ice core is the isotopic composition of the water molecules composing the ice (Dansgaard, 1964), specifically $\delta^{18}O$ and δD (see Box 1). Several complementary techniques that also reveal temperature histories are described further in Sections 18.3.2.2 and 18.3.2.3. Consider a schematic cross-section through the Earth from low latitudes to the poles (Fig. 18-6). Water evaporates from the oceans in the subtropics (some moisture, not shown here, is contributed by the continents and the polar oceans as well) and is transported generally poleward. Along this path, the air mass cools as it rises and radiates heat, and consequently water vapor condenses and precipitates, progressively depleting the air mass of water along this path. Eventually this air mass snows onto the ice sheet. At each stage of this process, when water vapor is converted to liquid or solid, the heavy isotope of oxygen is preferentially removed from the vapor. Thus the vapor becomes progressively lighter and lighter isotopically along this path (see Box 2), so that the snow accumulating on the ice sheet is substantially isotopically lighter than the original source. In fact, ice sheets are so much isotopically lighter than the ocean that the growth and disappearance of ice sheets through the glacial–interglacial cycles causes significant changes to the δ of the ocean, despite its tremendous volume. The ocean becomes heavier by 1.2‰ during extremes of glacial climate. This is the primary explanation for the $\delta^{18}O$ variations of deep-sea sediments (discussed in Section 18.2.2).

Further considering the system displayed in Fig. 18-6 the total isotopic "lightening" of the vapor is a direct function of the fraction of water mass removed from the air. This fraction in turn is a function of the net cooling of the air mass. This suggests that δ measured at the ice core site

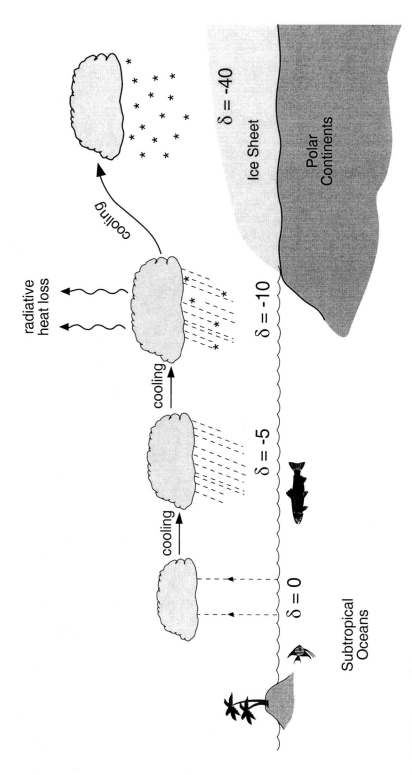

Fig. 18-6 Characteristic air-mass trajectory and corresponding per mil isotopic composition of precipitation, along a transect from the subtropics to a polar ice sheet. This is a highly schematic view of the true atmospheric system.

Box 2: Stable Isotope Tutorial

The global atmospheric circulation acts as an enormous filtration system, which depletes high-latitude precipitation of heavy isotope-bearing water molecules. Because of this system, measurements of the stable isotopic composition of the ice sheets and of ocean-floor sediments reveal very important paleo-environmental information (see Sections 18.2.2, 18.3.2, and 18.3.3). Here we examine this filtration system at a physical level. This system was first understood by a great Danish geochemist named Willi Dansgaard (Dansgaard, 1964).

Tutorial: Equilibrium Processes

Inside the air mass moving poleward in Fig. 18-6 there are n_{16}^v moles of $H_2^{16}O$ vapor, n_{18}^v moles of $H_2^{18}O$ vapor, and n_D^v moles of HDO vapor. Write n_j^v to signify either of the heavy-isotope-bearing vapors (n_D^v or n_{18}^v). Define the isotopic ratio of the vapor as $R_{vj} = n_j^v / n_{16}^v$. If the condensation of the vapor to form liquid water or solid ice is an equilibrium process, and this condensation is quickly removed from the air mass as precipitation, the corresponding isotopic ratio in the precipitation (R_{pj}) is proportional to that of the vapor

$$R_{pj} = \alpha R_{vj} \qquad (B1)$$

where the fractionation factor α is slightly greater than one, and is different for the deuterium and ^{18}O systems. Call these α_{18} and α_D respectively. At Earth surface temperatures, values for these range from 1.008 to 1.025 for α_{18} and 1.07 to 1.25 for α_D and these are temperature dependent, and different for liquid–vapor and ice–vapor equilibrium (Horita and Wesolowski, 1994; Majoube, 1971; Merlivat and Nief, 1967). The definition of δ for the precipitation is (see Box 1)

$$\delta_{pj} = \frac{R_{pj}}{R_{sj}} - 1 \qquad (B2)$$

where either R_s is a standard ratio. Note that the δ values for vapor and condensate (δ_v and δ_p) in equilibrium are related by

$$(1 + \delta_{pj}) = \alpha(1 + \delta_{vj}) \qquad (B3)$$

Differentiating both (B1) and (B2) shows that changes in the composition of precipitation and the composition of the vapor are related by

$$\frac{d\delta_{pj}}{1 + \delta_{pj}} = \frac{d\alpha_j}{\alpha_j} + \frac{dR_{vj}}{R_{vj}} \qquad (B4)$$

From the definition of R_v:

$$\frac{dR_{vj}}{R_{vj}} = \frac{dn_j^v}{n_j^v} - \frac{dn_{16}^v}{n_{16}^v} \qquad (B5)$$

Mass conservation and equilibrium fractionation together require that

$$\frac{dn_j^v}{dn_{16}^v} = \alpha_j \frac{n_j^v}{n_{16}^v} \qquad (B6)$$

Combining (B4), (B5), and (B6) shows that

$$\frac{d\delta_{pj}}{1 + \delta_{pj}} = \frac{d\alpha_j}{\alpha_j} + (\alpha_j - 1)\frac{dn_{16}^v}{n_{16}^v} \qquad (B7)$$

As water is almost entirely $H_2^{16}O$ we can replace n_{16}^v in (B7) with N_v where N_v is the total moles of water vapor in the air mass:

$$\frac{d\delta_{pj}}{1 + \delta_{pj}} = \frac{d\alpha_j}{\alpha_j} + (\alpha_j - 1)\frac{dN_v}{N_v} \qquad (B8)$$

which describes a Rayleigh distillation process (Merlivat and Jouzel, 1979). The quantity N_v in a specified amount of air is proportional to the ratio of vapor pressure to total pressure, in accordance with the gas law. When precipitating, the air is nearly saturated with water vapor, so N_v per mass of air is directly proportional to the saturation vapor pressure e_s and inversely proportional to the total pressure P

$$N_v \propto \frac{e_s}{P} \qquad (B9)$$

The saturation vapor pressure is strictly a function of temperature as indicated by the Clausius–Clapeyron equation

$$e_s \propto \exp\left[\frac{-\Lambda}{T}\right] \qquad (B10)$$

where the constant $\Lambda = 5435K$. It is this basic thermodynamic principle that makes δ a thermometer. We can now integrate Equation (B8) to see how the isotopic composition of precipitation δ_{pj} changes as an air mass cools and its mixing ratio decreases (recall the j refers either to the ^{18}O system or the deuterium system). It is instructive to first consider an approximation wherein α_j has a constant (average) value $\bar{\alpha}_j$. In this case Equation

(B8) is easily integrated between initial and subsequent δ_{pj} values (δ_{pj}^0 and δ_{pj}) and vapour contents (N_{v0} and N_v) to show

$$\delta_{pj} = (1 + \delta_{pj}^0)\left[\frac{N_v}{N_{v0}}\right]^{\bar{\alpha}_j - 1} - 1 \qquad (B11)$$

into which we substitute (B9) and (B10) to show that, in terms of initial and subsequent temperatures (T_o and T) and pressures (P_0 and P)

$$1 + \delta_j = (1 + \delta_{pj}^0)\left[\frac{P_0}{P}\right]^{\bar{\alpha}_j - 1} \exp\left[(\bar{\alpha}_j - 1)\Lambda \frac{T - T_0}{TT_0}\right] \qquad (B12)$$

The thermometric nature of δ is clearly seen, and in particular the dependence of δ on the temperature drop from initial condensation temperature to final condensation temperature. In addition, δ depends on its initial value (the source-region composition δ_j^0) and on the net pressure change. This pressure dependence is not important for temporal changes in δ at ice core sites in the interiors of Antarctica and Greenland, but is important for the spatial variation in δ across the ice sheets. It may also be important at sites closer to the ice sheet margins, where elevation may have changed considerably through glacial–interglacial cycles. An example of the predicted isotope-temperature distribution with constant α using (B12) is shown in Fig. 18-23. Now let us abandon our heuristic assumption of constant α. In reality, the fractionation is significantly enhanced at low temperatures; between $20°C$ and $-40°C$ ($\alpha_{18} - 1$) increases by a factor of 1.3 and ($\alpha_D - 1$) increases by a factor of 1.6. This includes a 30% and 10% increase, respectively, at the transition from liquid–vapor equilibrium to ice–vapor equilibrium. This temperature dependence is not the primary reason for the thermometric properties of δ but it does enhance the sensitivity of δ to T in the polar regions, and must be included in calculations. An example of integrating Equation (B11) to solve for δ with the proper temperature dependence is shown in Fig. 18-23. Achieving a good $\delta(T)$ prediction at low temperatures requires incorporating non-equilibrium fractionations (Jourel and Merlivat, 1984).

Meteoric Water Line and Deuterium Excess

Now consider the simultaneous evolution of $\delta^{18}O$ and δD. Suppose an air mass having initial compositions $\delta_j = 0$ cools and precipitates, always as an equilibrium process. According to the preceding analysis, measurements of both isotopic ratios in precipitation will plot on a curve with a slope approximately given by

$$\frac{d\delta D}{d\delta^{18}O} \approx \frac{\alpha_D - 1}{\alpha_{18} - 1} \qquad (B13)$$

which is very nearly a constant with a value between 7 and 9. This is the primary explanation for the Meteoric Water Line (see Section 18.3.3), which has a slope of 8 (Fig. 18-24). Deviations from a line of slope eight and zero intercept are called deuterium excess, $d = \delta D - 8\delta^{18}O$.

Conclusion

Equilibrium isotopic fractionations explain the gross behavior of $\delta^{18}O$ and δD in precipitation of mid to high latitudes.

---

should be useful as some sort of thermometer. Most directly, it is a measure of the total temperature drop from source to the precipitating clouds over the ice-core site (see Box 2). Indeed, if one visits many locations on an ice sheet and measures the mean annual δ and the mean annual temperature at these sites, a very strong correlation between δ and temperature is found (Fig. 18-7). This is true both for the Greenland ice sheet and the Antarctic ice sheet, and for high-latitude precipitation in general (Jouzel *et al.*, 1987). The strength of these correlations suggests that measurements of δ along an ice core may give an accurate, quantitative view of temperature history at the core site.

The utility of δ as a thermometer gains further support from direct measurements showing correlation of temperature and δ of precipitation through time over seasonal cycles (Shuman *et al.*, 1995). At longer temporal scales, temperature measurements in boreholes and gas composition measurements both provide temperature information which can be compared to δ. Results from such comparisons have so far

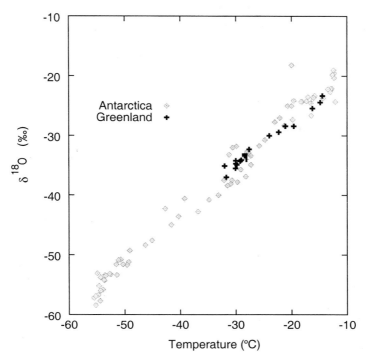

Fig. 18-7 Observed correlation of isotopic composition with temperature for near-surface snow in Greenland (Johnsen *et al.*, 1989), and Antarctica (Dahe *et al.*, 1994).

supported the use of δ as a qualitative thermometer, but not a quantitative one.

There are many reasons to suspect that δ may not be a faithful quantitative thermometer. First, δ of the initial source water(s) can change through time, as ocean composition changes due both to ice-sheet growth and to changes in water flux from the continents. Source-water δ also can change through time as the source-water location (or, more generally, the spatial distribution of sources) changes, as a result of atmospheric circulation changes. Second, as described above, δ is primarily sensitive to the temperature difference between source and site. Therefore, coincident variation of temperature at source and site can significantly dampen δ changes at the ice-core site. δ will work best as a thermometer for the core site if the source temperature changes not at all. Independent variations in source and site temperature could possibly result in a complicated δ record whose meaning is ambiguous.

Important complications occur at the ice-core site. The seasonal variation of δ is much larger than the δ change accompanying climate changes, even for the great glacial–interglacial transitions. Yet precipitation does not necessarily fall uniformly throughout the year. Relatively small changes in the seasonal timing of precipitation could therefore cause changes in δ that may be misinterpreted as significant changes in climatic temperature. Precipitation biasing can be a problem at shorter temporal scales too, in that precipitation over the ice sheets accompanies advection of relatively warm air masses, so there will generally be a warm-weather bias to the stable isotope records.

All these potential complications mean that independent temperature-history information is very useful, though no other temperature reconstruction techniques yet match the very high resolution of these isotopes.

18.3.2.2   Borehole temperatures

The modern distribution of temperature at

depth in an ice sheet also contains important paleoclimate information, and in particular provides a relatively direct measure of past climatic temperature at the ice-sheet surface. Near an ice divide, heat is transported downward in the ice via conduction and advection. Given a steady surface temperature (constant climate), the temperature at depth will evolve to have a relatively isothermal upper layer, due to downward transport of ice, and warmer ice at depth, due to heat flux from the Earth and to heat generation in the rapidly deforming basal layers (Fig. 18-8). If the climate warms abruptly, and maintains the new warmer temperature, the heat transfer processes will send a wave of warmth downward into the ice. Subsequent measurements of the temperature–depth distribution will reveal a cold spot, which is a direct remnant of the previous, cooler climate. By mathematically reversing this natural heat flow, we can reconstruct a history of climatic temperature variations through time. This mathematical reversal is accomplished by heat- and ice-flow modeling coupled to geophysical inverse techniques (Dahl-Jensen and Johnsen, 1986; Cuffey and Clow, 1997).

There are severe limits on the recoverable paleoclimate information by this method (MacAyeal *et al.*, 1993). Reconstructions are only possible for long-term average temperatures and are restricted to the last glacial period and more recent times.

The power of this technique is due to the fact that the temperature–depth profile is a direct remnant of paleotemperatures at the ice-sheet surface. It provides a quantitatively accurate measure of long-term average temperatures. This allows the stable isotope records to be calibrated for major climate events (Cuffey *et al.*, 1995).

18.3.2.3   Firn gas thermometry

The mobility of gases in the firn column leads to temperature-dependent changes in the composition of gas trapped in ice at the base of the firn (Severinghaus *et al.*, 1998). If there is a rapid change in climatic temperature at the ice-sheet surface, a steep temperature gradient will temporarily exist throughout the firn. This will temporarily cause thermal fractionation of gas-

phase isotopes, and consequently gases trapped into the ice during this time will have an unusual composition in terms of isotopic ratios such as $^{15}N/^{14}N$. Thus, measurements of gas isotopic ratios along an ice core will show spikes corresponding to abrupt temperature changes in the distant past. Such data yield paleotemperature information in two ways. First, the magnitudes of these spikes are themselves measures of the magnitude of the ancient temperature change. Second, the along-core separation of the gas-phase spike from the solid ice δ change is a direct measure of the gas–ice age difference, Δ_{age}. This in turn enables an estimate of temperature prior to the climate change, because Δ_{age} depends on temperature and accumulation rate (the density–depth profile being a function of both of these) (Herron and Langway, 1980).

18.3.2.4   Thermometry: Results

Ice-core isotopic records from both hemispheres (Fig. 18-9) clearly show the cycling of Earth's climate between cold glacial and warm interglacial phases (Jouzel *et al.*, 1996; Johnsen *et al.*, 1997). The interglacials have been relatively brief, approximately one quarter the duration of the glacials. The glacials generally become progressively colder throughout most of their duration, and then terminate rather abruptly. This long-term pattern of temperature variations is identical to that of global ice volume. The polar temperature histories have also been similar to the ice volume history in having significant variation at ~20 and ~40 kyr periodicities, in correlation with northern hemisphere insolation variations. And in both, the (unexplained) 100 kyr glacial–interglacial cycle dominates. The average ice-sheet surface temperature during the last glacial period was 15°C colder than at present in Greenland and possibly 8°C colder in central Antarctica. The shorter-duration cores from mid- and low-latitude mountain ice caps also show, qualitatively, a substantial temperature difference from glacial to interglacial (Thompson *et al.*, 1995).

There is a striking contrast between the millennial variability of polar interglacial and glacial temperature histories. Whereas climate changes were frequent and large throughout

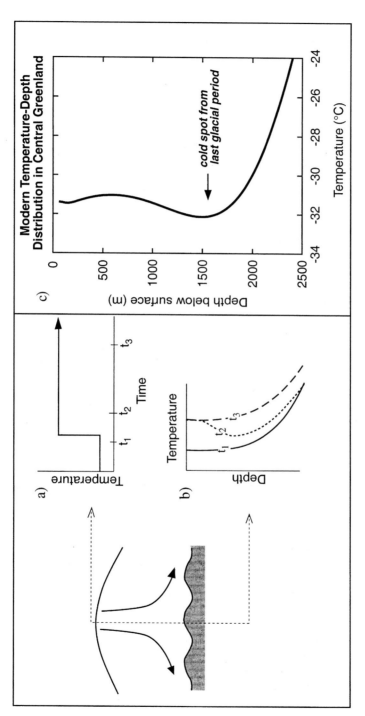

Fig. 18-8 Characteristic temperature–depth distributions at an ice divide. For a climatic temperature history as shown in (a) the temperature–depth distribution changes as shown in (b). Following the step increase in surface temperature, the initial steady temperature profile (t_1 in (b)) is altered by a warming wave (e.g., at time t_2) but eventually reaches a new steady profile by time t_3. (c) Temperature data from Greenland measured by Gary Clow of the US Geological Survey, showing wiggles due to climate variations (Cuffey *et al.*, 1995).

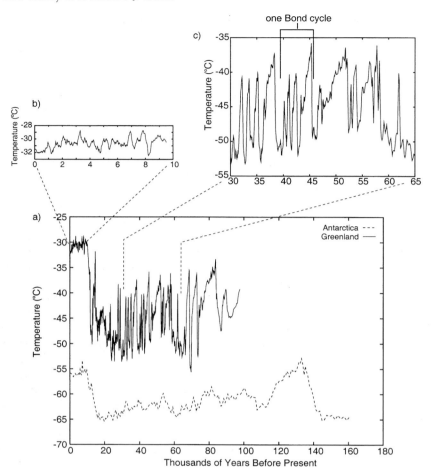

Fig. 18-9 (a) Temperature histories derived by borehole-temperature calibration of isotopic records, for Antarctica (Salamatin *et al.*, 1998; Jouzel *et al.*, 1993) and Greenland (Cuffey *et al.*, 1995; Grootes *et al.*, 1993). (b) A closer look at the Holocene. (c) A closer look at some of the rapid climate changes (D–O events) during the last glacial period. The magnitudes of these temperature swings are not yet calibrated, and may be only half as large as depicted here.

the glacial period in both polar regions, the polar climate of the current interglacial (the Holocene) has been relatively stable (Dansgaard *et al.*, 1993). The large, rapid millennial-scale climate change events during the glacial period are called Dansgaard–Oeschger (D–O) events in honor of their original discoverers. These many substantial climate changes within the last glacial period are definitively recorded as abrupt events in all ice cores from Greenland (Johnsen *et al.*, 1997), and also appear in Antarctic cores in muted form (Bender *et al.*, 1994). There are several lines of compelling evidence which

argue that these isotopic changes are in fact temperature changes, and not simply changes in one of the other controls on δ: including the presence of gas isotopic spikes of appropriate magnitude, and coincident changes in CH_4, accumulation rate, and atmospheric impurity deposition (see Sections 18.3.4, 18.3.5.2, and 18.3.6).

The abrupt climate oscillations seen in the Greenland cores (D–O events) occurred in groups that outline a sawtooth pattern of initial abrupt warming followed by gradual cooling, each group lasting about 7 kyr (Bond *et al.*,

1993). These are called Bond cycles (Figure 18-9). The D–O events within each Bond cycle show generally decreasing durations and maxima through time. Moreover, the Bond cycles are correlated to changes in sea-floor deposits, and in particular the cold extremes of the Bond cycles are coincident with massive iceberg discharge events from the North American ice sheet into the North Atlantic (Heinrich events; Broecker, 1994). This correlation, plus coincident changes in methane (see Section 18.3.5.2 below), plus the expression of these events in Antarctic cores and sea cores (e.g., Kennett and Ingram, 1995) show conclusively that these rapid environmental changes had global significance.

The most recent of these abrupt climate changes (11 500 years ago; Fig. 18-9) was the termination of the Younger Dryas event, a 1000-year return to glacial conditions during the longer-term warming from glacial to interglacial (see Section 18.3.4). The termination of this cold climate was accomplished in a geologic instant – a mere 10 years. In this time, the temperature jumped by probably 5 to 10°C (Severinghaus *et al.*, 1998) and the accumulation rate doubled. Earlier D–O events had many similarly abrupt transitions. Note, however, the lack of a prominent Younger Dryas event in the Antarctic.

The D–O events are far too rapid to be caused by insolation changes, and they most likely result from changes in ocean circulation. Their prominence and clarity in the Greenland cores relative to the Antarctic ones is due to the proximity of Greenland to the sites of deep-water formation in the North Atlantic and the tremendous amount of heat being delivered to them by the Gulf Stream (Broecker and Denton, 1989).

Though extremely stable compared to the glacial climate, the polar Holocene climate did vary. One notable rapid climate change event, a dip to lower temperature, occurred approximately 8 kyr ago (Alley *et al.*, 1997). On average, the early Holocene was modestly warmer than the present. The past several thousand years have been cooler, especially during the 16th–19th centuries, a time known loosely as the Little Ice Age (LIA). There are direct records of this cold period from many places around the globe, and glaciers were substantially larger in mountain ranges of both hemispheres.

18.3.3 Deuterium Excess

Measurements of both the $\delta^{18}O$ and the δD of precipitation over much of the globe reveal a very striking linear relationship between the two (an explanation is given in Box 2) called the Meteoric Water Line. The slope of this line is almost exactly $d\delta D/d\delta^{18}<O = 8$ (Craig, 1961). If all isotopic fractionations in the water cycle occur during equilibrium processes, the intercept of this line would be approximately zero. Instead, the measured intercept is 8.6. This offset results from isotopic separation during evaporation from the ocean surface (Merlivat and Jouzel, 1979). Because this isotopic separation depends on evaporation rate it also depends on temperature, relative humidity and wind speed, all of which are interesting climatic parameters. Furthermore, the slope value of 8 also depends on air mass temperature early in its trajectory. For both of these reasons, measurements of deuterium excess (defined as $d = \delta D - 8\delta^{18}O$) along an ice core yield paleoclimate information (Jouzel *et al.*, 1982; Johnsen *et al.*, 1989), not about climate over the ice sheets but about climate of low to mid latitudes.

Analyses of d in modern snow (both its average value and its seasonal cycle) in Greenland and Antarctica show that the vapor source regions for the high-altitude interiors of the polar ice caps are located in the subtropics. Ice deposited in Antarctica during the last glacial period has average d lower by 4‰ than the modern value, reflecting both colder ocean surface temperature and also higher relative humidity. In ice-age ice from Greenland, d varies from 4 to 8‰ in anti-correlation with the temperature changes of the D–O events. This indicates a substantial northward shift of moisture sources during the warm phases of climate oscillations within the glacial period.

18.3.4 Accumulation Rates: Ancient Precipitation Gages

Accumulation rate is the net rate of mass addition to the ice-sheet surface (generally reported as thickness of ice per time), which equals the snowfall rate minus rate of loss by wind scour, sublimation and (at warm sites) melt. Snowfall rate is a climatic parameter of great interest, as it depends on atmospheric circulation and water content. Snowfall rate is approximately equal to accumulation rate at all but the driest of ice core sites.

To infer accumulation rate history from an ice core, one needs to measure the thickness of annual layers (either directly if annual layers are resolvable, or by differentiating the depth–age scale determined by other means) and then correct for the thinning of these layers caused by the ice flow (the vertical strain: Fig. 18-4). Estimates of vertical strain can be very uncertain for the deep part of an ice core. But vertical strain will not change rapidly with depth. Thus, if annual layers are resolvable one can learn relative accumulation rate changes across climate transitions with great confidence.

An alternative method for inferring accumulation rate relies on assuming that the rain of cosmogenic nuclides such as ^{10}Be onto the ice sheet surface is known. Then high accumulation rate dilutes the cosmogenic nuclide so its concentration as measured in the ice core is inversely proportional to accumulation rate.

Accumulation rate histories from both Greenland and Antarctica all show substantially lower accumulation rates during glacial periods (Fig. 18-10), rates being as low as 25% of the modern value in central Greenland and 30% in East Antarctica. Much of this reduction is probably a consequence of thermodynamics; at saturation the content of water vapor in air is a strongly increasing function of temperature. During cold periods, the air masses over the ice sheets must be very dry. In central Greenland, accumulation rate inferences have annual resolution across the termination of the last glacial period, because annual layers are still readily distinguishable in ice of this age. At the end of the cold Younger Dryas period, accumulation rate doubled (Alley *et al.*, 1993). Remarkably, this happened within a single decade. The earlier doubling of accumulation rate at 14 600 yr B.P. (before present) was similarly abrupt (Fig. 18-10).

18.3.5   *Trace Gases*

18.3.5.1   *Ice-core records of CO_2*

As explained in Section 18.3.1, the polar ice sheets trap and preserve air. Analyses (Sowers *et al.*, 1997) of this air reveal the compositions of past atmospheres. One of the most important constituents that can be measured in ice samples is carbon dioxide (CO_2). Excepting water vapor, a direct record of which is not preserved in polar ice, CO_2 is the most important of Earth's greenhouse gases, and past CO_2 concentrations reveal significant variations in the global carbon cycle. The most reliable records of atmospheric CO_2 are from Antarctic ice cores. Records from Greenland ice are not completely reliable due to high concentrations of carbonate dust in glacial-age ice there.

Ice-core records spanning the industrial revolution clearly show a large increase in CO_2 levels over the last 200–250 years (Neftel *et al.*, 1982; Etheridge *et al.*, 1996). CO_2 concentrations were approximately 275 ppmv (parts per million by volume) prior to 1750, and then began a steady rise to current levels of ~360 ppmv. The most detailed ice-core record comes from Law Dome, a site in east Antarctica with a very high accumulation rate (~1.2 m/yr) and accurate chronology (Fig. 18-11). Where this record overlaps the modern direct measurements of atmospheric CO_2 agreement between the two records is excellent (Fig. 18-11).

Longer records show that atmospheric CO_2 concentrations follow global climate trends (Barnola *et al.*, 1987). Over the last two glacial–interglacial climate cycles, CO_2 levels were approximately 280 ppmv during interglacial periods and substantially lower, approximately 190–210 ppmv, during the cold glacial (Fig. 18-12). A detailed record from the Byrd ice core illustrates the CO_2 changes through the last glacial to interglacial transition (Fig. 18-13). CO_2 levels began to rise about 18–20 kyr B.P. and reached interglacial levels by ~10 000 kyr B.P.

The nature and causes of the atmospheric CO_2 concentration variations on different time scales are very interesting and only partially understood. The "anthropogenic transient" has been

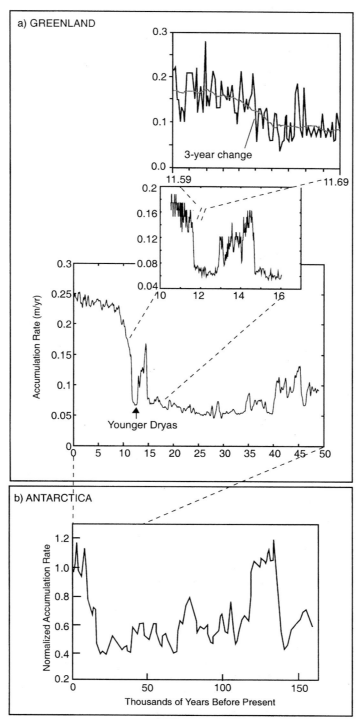

Fig. 18-10 (a) Accumulation rate history for Greenland (Alley *et al.*, 1993; Cuffey and Clow, 1997), with insets focusing on the Younger Dryas cold interval and the very rapid termination of the last glaciation, with substantial change that occurred in only 3 years noted (after Alley *et al.*, 1993). (b) Accumulation-rate history, normalized to the modern rate, for Antarctica (Raisbeck *et al.*, 1987), derived from [10]Be analysis.

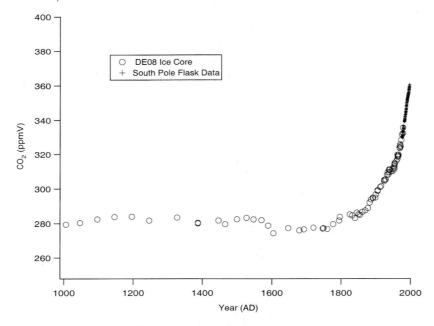

Fig. 18-11   Records of atmospheric CO_2 in Antarctica for the past 1000 years. Open circles are ice-core data from Law Dome, on the coast of east Antarctica (Etheridge *et al.*, 1996). Plus signs are direct measurements of CO_2 in air samples collected monthly at the South Pole (NOAA Climate Monitoring and Diagnostics Laboratory, Boulder, Colorado).

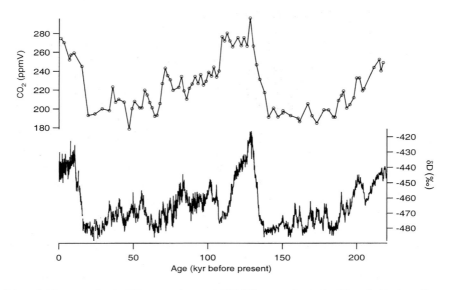

Fig. 18-12   Record of atmospheric CO_2 over the past 220 000 years from the Vostok ice core (Jouzel *et al.*, 1993; Barnola *et al.*, 1987), and the corresponding deuterium isotope record temperature proxy (Jouzel *et al.*, 1993) (see Section 18.3.2).

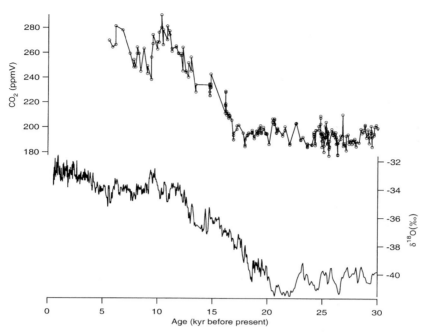

Fig. 18-13 High-resolution measurements of CO_2 over the last glacial–interglacial transition from the Byrd ice core in west Antarctica (Neftel *et al.*, 1982). Also plotted is the oxygen isotope record temperature proxy from the Byrd core (Johnsen *et al.*, 1972). The time scale for the records plotted here is from Sowers and Bender (1995).

studied in detail. The period A.D. 1550–1800 was characterized by CO_2 levels up to 6 ppm lower than the preceding 500 years (Fig. 18-11). The Little Ice Age (LIA) period of global cooling occurred during this time period, and cooler temperatures may have resulted in increased oceanic uptake of CO_2 (due to the temperature dependence of its solubility), and enhanced wind-driven mixing of surface waters at high latitudes. Global cooling may also have influenced carbon uptake by terrestrial ecosystems. The rise in CO_2 that began at about A.D. 1750 is believed to be due to a combination of land use changes (deforestation and other agricultural processes that release carbon to the atmosphere) and fossil fuel burning, with land use changes dominating prior to about A.D. 1900. Between A.D. 1000 and 1800, CO_2 levels were between 275–284 ppmv (Fig. 18-11), and a value of 280 ppmv is often cited as the "pre-industrial" CO_2 concentration. Records from the Byrd ice core show that during the entire Holocene period (0–10 000 years B.P.) CO_2 levels actually varied between 245–280 ppm (Neftel *et al.*, 1982)

with a prominent minimum of 245–255 ppm at about 8000 years B.P.

The glacial-interglacial change in CO_2 is about 80 ppm, and is not well understood. The shift occurred over about 8000 years during both the last and penultimate deglaciation. During the last deglaciation this rise in CO_2 preceded sea-level rise and therefore the melting of ice sheets by 4 to 8 millennia (Sowers *et al.*, 1993). During the penultimate deglaciation, the CO_2 rise probably preceded sea-level rise by a few millennia (Broecker and Henderson, 1998). Both observations suggest an important role for CO_2 in the warming during deglaciation and the CO_2 change may have forced as much as 50% of the total glacial–interglacial warming (see Section 18.4.6 below).

Proposed explanations for glacial–interglacial CO_2 change involve changes in ocean chemistry (see Chapter 10), either surface ocean total inorganic carbon (often abbreviated Σ_{CO_2}) levels or changes of the calcium carbonate budget of the ocean (Archer and Maier-Reimer, 1994; Broecker and Henderson, 1998). Much of the work on this subject has focused on the transition from low glacial CO_2 levels to high interglacial CO_2 levels, depicted in detail in Fig. 18-12.

Changes in Σ_{CO_2} could result from increases or decreases in the rate of organic carbon production in surface waters. Sinking organic material removes carbon from surface waters to the deep sea. Enhancement of this "biological pump" would decrease surface water P_{CO_2} and draw down atmospheric CO_2. If primary productivity (the rate at which plant organic matter is formed per unit area) during glacial times were higher than during interglacial times, atmospheric CO_2 levels would have been lower due to this effect. One proposed mechanism for stimulating this additional productivity is enhanced phytoplankton growth due to higher fluxes of iron (a micronutrient) to the ocean surface in the dustier glacial atmosphere (Martin, 1990).

Mechanisms involving changes in the ocean's calcium carbonate budget (Chapter 10) invoke imbalances in the production, dissolution, and burial of calcium carbonate (the details of which are beyond the scope of this chapter) that can cause a rise or fall in atmospheric CO_2. For example, if $CaCO_3$ accumulation increased at the beginning of the glacial–interglacial transition, CO_2 levels would rise, via the reaction $Ca^{2+} + 2HCO_3^- = CaCO_3 + CO_2 + H_2O$. Possible reasons for changes in $CaCO_3$ accumulation include growth or demise of coral reefs or shallow water carbonate sediments as sea-level rises or falls, changes in the ecology of marine plankton such that $CaCO_3$ precipitating organisms are more or less abundant than other types of plankton, or changes in alkalinity of river waters. Because no proposed mechanisms appear to be fully supported by the paleoclimate record (Broecker and Henderson, 1998) the problem of adequately explaining the glacial–interglacial CO_2 rise remains a major challenge.

18.3.5.2   Ice-core records of CH_4

Methane (CH_4), the Earth's third most important greenhouse gas (behind water vapor and CO_2) is produced almost exclusively by terrestrial sources (see Chapters 7 and 11). Natural methane sources include anaerobic microbial respiration in wetlands, termites, wild animals (ruminants), and a variety of other minor sources. Major anthropogenic sources are rice paddies, domestic cattle and other animals, and fossil fuel production. The primary methane sink is oxidation by OH in the troposphere, with a small fraction of the total methane production oxidized in soils and in the stratosphere (Chapter 7).

Ice-core records covering the past three centuries (Etheridge *et al.*, 1992; Blunier *et al.*, 1993) show that methane concentration rose at the beginning of the industrial revolution, largely in parallel with increasing CO_2 levels (Fig. 18-14). During the period from A.D. 900 to 1700, methane levels were between ~ 675 and 725 ppbv, and started to rise sharply after about A.D. 1750 (Blunier *et al.*, 1993). As this chapter is written, the global average value is 1730 ppbv (Dlugokencky *et al.*, 1998). Levels are slightly lower in the southern hemisphere due to dominance of northern hemisphere sources. The ice-core values from the youngest section of the record shown in Fig. 18-14 overlap, and agree well with, direct measurements at a site in Australia, supporting the fidelity of the ice core CH_4 record. The growth rate of atmospheric methane increased from about 1 ppbv/yr in

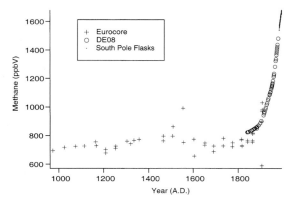

Fig. 18-14 Ice-core methane record for the past 1000 years. Plus signs are data from Eurocore in central Greenland (Blunier *et al.*, 1993), and open circles are data from DE08, an ice core in East Antarctica (Etheridge *et al.*, 1992). Dots are monthly atmospheric data from the South Pole (NOAA Climate Monitoring and Diagnostics Laboratory in Boulder, Colorado).

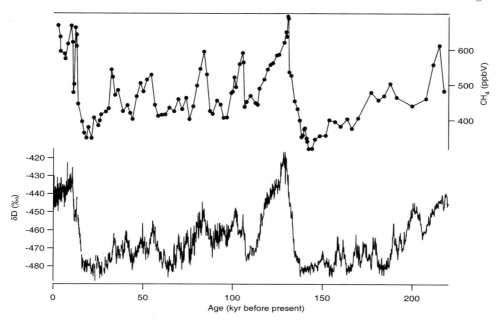

Fig. 18-15   Ice-core methane record from Vostok, Antarctica, for the past 220 000 years (Chappellaz *et al.*, 1990; Jouzel *et al.*, 1993).

1850 to 15 ppbv/yr in 1975 (Etheridge *et al.*, 1992), with the exception of a period from 1920 to 1945 when the growth rate was stable. The general increase in growth rate during the industrial revolution presumably reflects the expansion of anthropogenic methane sources.

Longer ice-core records show that methane concentrations have varied on a variety of time scales over the past 220 000 years (Fig. 18-15) (Jouzel *et al.*, 1993; Brook *et al.*, 1996). Wetlands in tropical (30° S to 30° N) and boreal (50° N to 70° N) regions are the dominant natural methane source. As a result, ice-core records for preanthropogenic times have been interpreted as records of changes in methane emissions from wetlands. Studies of modern wetlands indicate that methane emissions are positively correlated with temperature, precipitation, and net ecosystem productivity (Schlesinger, 1996).

Over the past 220 000 years methane concentrations ranged between 350 and 750 ppbv, compared to modern values in excess of 1700–1800 (Fig. 18-15). Over tens of millennia, methane variations appear to correspond to northern hemisphere insolation changes, correlate with Vostok paleotemperatures (Chappel-laz *et al.*, 1990) and have a strong 20 000 year periodicity, corresponding to the Earth's precession cycle, which dominantly influences low-latitude climate. Changes in tropical monsoon strength have been proposed as a possible link; a more active monsoon and warmer temperatures could cause both greater wetland area and greater methane production in the tropics (Chappellaz *et al.*, 1990). Recently, analysis of new 400 000+ year record from the Vostok site (Delmotte *et al.*, 1998) suggests that a 40 000 year periodicity is also evident.

High-resolution methane records from Greenland (which in some cases resolve sub-century variation) considerably expand this picture of methane variability (Chappellaz *et al.*, 1993; Brook *et al.*, 1996). These records confirm the long-term patterns but also show higher-frequency millennial-scale variability closely associated with rapid temperature changes inferred from isotopic records; all of the interstadial events in the central Greenland δ^{18}O record have correlated increases in atmospheric methane mixing ratio (Fig. 18-16). The methane response to rapid climate change during the last glacial period demonstrates that these rapid

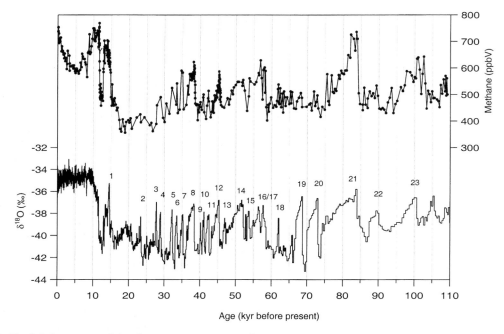

Fig. 18-16 Methane record for the past 110 000 years from the GISP2 ice core (Brook *et al.*, 1996; Brook *et al.*, in press). Numbers are labels for each of the warm "interstadial" periods within the cold glacial climate.

climate changes had widespread effects, because methane sources have a broad geographic distribution.

18.3.5.3   Ice-core records of N_2O and other gases

Nitrous oxide (N_2O) is a trace gas that contributes to the planetary greenhouse effect and to loss of stratospheric ozone (Chapters 7 and 12). Ice-core data for N_2O are considerably sparser than for CH_4 and CO_2. Machida *et al.* (1995) produced a high quality N_2O record from a shallow Antarctic ice core (Fig. 18-17a) for the period A.D. 1750–1960. Their results show preanthropogenic levels of 273–280 ppbv, with an N_2O minimum at about A.D. 1830, possibly coincident with the minimum in CO_2. After about 1850, N_2O levels rose, reaching values of 310 ppbv in the early 1990s. The preanthropogenic data from Machida *et al.* agree extremely well with similar measurements of N_2O from South Pole firn air samples (Battle *et al.*, 1996). The origin of the N_2O rise during the anthropogenic period is not well understood. Possible

sources include fertilizer, deforestation, combustion, and biomass burning.

Glacial–interglacial variations in N_2O have been described by Leuenberger and Siegenthaler (1992) who showed that N_2O levels during the glacial maximum at about 25 000 years B.P. were ∼ 190 ppbv (Fig. 18-17b). They found intermediate values of N_2O (200–250 ppbv) between 30 000 and 40 000 years B.P.

Other trace gases important in atmospheric chemistry and climate (for example carbonyl sulfide and carbon monoxide) may also be measured in polar ice, and development of these and other measurements is underway in a number of laboratories around the world.

18.3.6   Non-Gaseous Impurities: Chemicals and Particles

In addition to the trace gases, many impurities, both soluble chemicals and insoluble particles, transfer from the atmosphere to the ice sheets and have measurable concentrations in ice cores. Such measurements are a window onto the

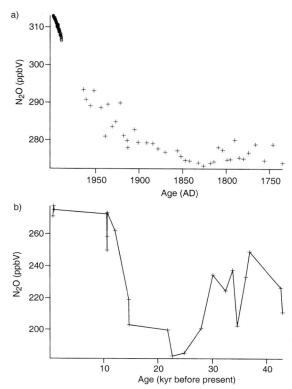

Fig. 18-17 Ice core records of N_2O. (a) Data of Machida *et al.* (1995) from the H15 ice core, east Antarctica, for the time period 1750–1950, and monthly atmospheric N_2O measurements at the South Pole from the NOAA Climate and Diagnostics Laboratory, Boulder, CO, for the period 1989–1998. (b) Data from Leuenberger and Siegenthaler (1992) from the Byrd ice core in West Antarctica.

history of atmospheric burdens and fluxes (from sources and to the ice sheets) for a great variety of aerosols (Delmas, 1992). These determine important radiative and chemical properties of the atmosphere, and are functions of atmospheric circulation intensity and the geography and strengths of sources and sinks. Inferring atmospheric loads and fluxes from concentrations in ice sheets is difficult and currently is plagued by important unsolved problems. A main difficulty is that we still do not have a quantitative understanding of the processes by which many species of atmospheric impurities are incorporated into ice.

Unlike the trace gases discussed in Section 18.3.5, most of the impurities discussed here

have very short atmospheric residence times and are therefore not globally mixed. The records from Antarctica and Greenland therefore give unique information about the polar regions of each hemisphere.

Impurities travel from atmosphere to ice sheet surface either attached to snowflakes or as independent aerosols. These two modes are called wet and dry deposition, respectively. The simplest plausible model for impurity deposition describes the net flux of impurity to ice sheet (which is directly calculated from ice cores as the product of impurity concentration in the ice, C_i, and accumulation rate, $\dot{a}$) as the sum of dry and wet deposition fluxes which are both linear functions of atmospheric impurity concentration C_a (Legrand, 1987):

$$C_i\dot{a} = k_d C_a + k_w \dot{a} C_a \qquad (1)$$

where k_d and k_w are coefficients pertaining to dry and wet deposition, respectively. If an impurity deposits primarily by dry deposition, snowfall will act as a dilutant, so the ice-core concentration, C_i, will be inversely correlated to accumulation rate through time. If wet deposition predominates, C_i will be independent of accumulation rate and directly proportional to atmospheric concentration. In a general case, a plot of $\dot{a}C_i$ versus $\dot{a}$ from ice-core data will, according to this simple model, yield a line whose slope and intercept are both proportional to ancient C_a and whose intercept indicates the importance of dry deposition. A change in C_a, across a climate transition is inferable from ratios of slopes and intercepts *if* the values k_d and k_w do not themselves change with climate. Herein lies the difficulty; this assumption is questionable and the model itself is a substantial simplification of reality for some species.

Most of the non-gaseous impurities in ice were once atmospheric aerosols. Atmospheric aerosols raining onto an ice sheet are of two types: primary aerosols, which are incorporated directly into the atmosphere as aerosols (these include continental dust and sea spray), and secondary aerosols which form in the atmosphere from gases. In addition to aerosol-derived impurities, some soluble gases in the atmosphere (HNO_3 HCl, H_2O_2, and NH_3) adsorb directly onto ice, and so are measured in a core as soluble ions rather than as constituents of trapped air bubbles. Some reactions occur during the aerosol phase which change the chemistry of primary aerosols. Primarily, secondary aerosol acids can react with salts in water droplets to separate the salts into metal ions and HCl gas. In modern Greenland and Antarctic snow, the secondary aerosol gas-derived acids dominate the chemistry, except very close to the coasts where sea salts can be important.

18.3.6.1   Sulfur cycle: SO_4^{2-} and MSA

One of the most important of the acids is H_2SO_4, which (see Chapter 13 on the sulfur cycle) primarily results from dissolution of SO_2 in water droplets. Both SO_2 and the strong acid MSA are products of oxidation of DMS gas, which is emitted by organisms in the near-surface ocean. MSA is incorporated unaltered in the ice, so MSA measurements provide a relatively direct measure of DMS production and possible related albedo effects (Chapter 13). (It does not provide a direct measure because H_2SO_4/MSA partitioning is not fixed.) The best measure of the relative contribution of biogenic gases to the total sulfur input is thought to be the MSA fraction (Whung *et al.*, 1994), defined as the ratio MSA/(MSA+ SO_4^{2-}).

Histories of MSA and SO_4^{2-} archived (Fig. 18-18) in the Greenland ice sheet (Saltzman *et al.*, 1997) show that both atmospheric MSA concentrations and MSA fraction were lower during the cold glacial period than during the Holocene. The sulfur cycle–climate feedback was therefore probably a negative one in the North Atlantic, with increases in aerosol production accompanying, and opposing, climate warming from glacial to Holocene. However, MSA concentrations were low during the time of most rapid melt of the ice sheets (when presumably the ocean surface had relatively low salinity), and rose slowly throughout most of the Holocene, not peaking until approximately 3500 years ago. This slow response contrasts starkly with the rapid fluctuations of most climate parameters during deglaciation, and implies that the feedback is a delayed one. The most likely explanation for the slow response of

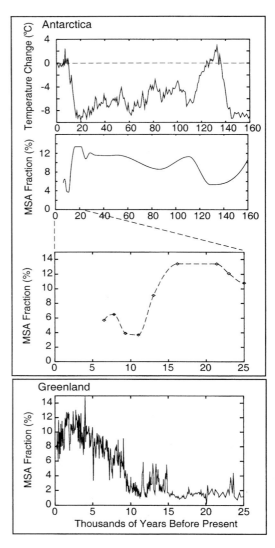

Fig. 18-18   Top two panels show the history of Antarctic MSA fraction (Legrand *et al.*, 1991), and coincident isotopic temperature history (Jouzel *et al.*, 1993). Bottom two panels present a closer look at the last 25 kyr, for both Antarctica and Greenland (Saltzman *et al.*, 1997).

the biogenic sulfur output, and for the glacial–interglacial contrast, is an ecological change, specifically a change in relative faunal abundances.

MSA and SO_4^{2-} histories from Antarctica (Legrand *et al.*, 1991) imply a very different response (Fig. 18-18) for the southern oceans than for those of the north. While SO_4^{2-} not derived from seasalts was only modestly higher during glacial times (by less than a factor of two after accumulation correction), both MSA and MSA fraction were substantially higher during the coldest part of the last glacial period, approximately five times higher than Holocene values. Thus, in contrast to the North Atlantic, the southern oceans apparently yielded a substantial increase in biogenic sulfur flux during the glacial climate (and the associated climate feedback was positive). One possible explanation for this is that increased nutrient flux to the oceans via primary aerosols (see Section 18.3.5.1) fueled more biological activity.

Two further aspects of SO_4^{2-} records deserve special mention. First, the recent and substantial anthropogenic production of SO_2 has a clear signature in increased SO_4^{2-} concentrations in Greenland accumulation, but not in northwestern Canada or the Antarctic (Mayewski *et al.*, 1993). Second, large or local volcanic eruptions can produce substantial SO_4^{2-} spikes.

18.3.6.2 Nitrogen cycle: NO_3^- and NH_4^+

The second important acid in polar snow is HNO_3, which, along with other NO_3^- as well as NH_3, can record nitrogen cycle perturbations. HNO_3 and NO_3^- are products of NO_x compound oxidation (see Chapter 12 on the nitrogen cycle), and NO_3^- is additionally a constituent of salts, such as $Ca(NO_3)_2$, which are primary aerosols. A major motivation for studying NO_3^- in ice cores is the role of its NO_x precursors in controlling the oxidative capacity of the atmosphere. Both NO_3^- and NH_4^+ are very interesting in that their histories do not correlate well with the vast majority of other ions and climate measures, especially in Greenland ice.

There are a wide variety of initial sources of NO_3^- for the ice sheets, including bacterial emissions, biomass burning, photochemical reactions, and lightning. These are generally low–mid-latitude continental sources. This very complicated mixed source renders interpretations of ice-core NO_3^- concentrations difficult. A further complication results from possible limitations on delivery of NO_3^- to ice-core sites by atmospheric circulation, due to the large distance from

source regions to ice sheets coupled with limited residence times.

As with SO_4^{2-}, anthropogenic activity has caused a modern increase in NO_3^- in Greenland firn (e.g., Mayewski *et al.*, 1990). This appears to be the only unambiguous result from the many NO_3^- measurements. Atmospheric nitrate concentrations were approximately 20% lower over Greenland during glacial times than at present (Fuhrer and Legrand, 1997), which may reflect decreased nitrogen gas emissions from soils during the cold climate, but which may also reflect greater difficulty of transport resulting from the large ice sheets. In Antarctic ice, glacial-age nitrate concentrations are higher than in Holocene ice, and correlate strongly with primary aerosols from continental sources. Legrand *et al.* (1988) infer that much of this NO_3^- is from continental salts, and that the remainder, representing the gas-derived acid, had a lower concentration during the glacial than during the Holocene, as it did over Greenland.

Ammonium sources are biogenic and are also largely mid- or low-latitude continental. Main sources are emissions from soils and macrobiota, and fires. As with NO_3, NH_4^+ concentration depends importantly both on source strength and efficiency of transport from source to ice sheet. Atmospheric NH_4^+ over Greenland was generally lower during the cold glacial climates, due to a combination of reduced source strength (resulting from colder and drier climate on the North American continent) and reduced transport of mid-latitude air into the Arctic (Meeker *et al.*, 1997). But most of the NH_4^+ variation is uncorrelated to the global climate, and in fact late Holocene NH_4^+ concentrations are similar to those of the last glacial maximum. Greenland NH_4^+ appears to be strongly correlated with summertime insolation in the northern hemisphere, except when the continental ice sheets were large, during which times NH_4^+ was not delivered to the ice sheet efficiently.

18.3.6.3 Primary aerosols: seasalts and continental dust

The concentration of primary aerosols (Fig. 18-19) was substantially higher in the atmosphere

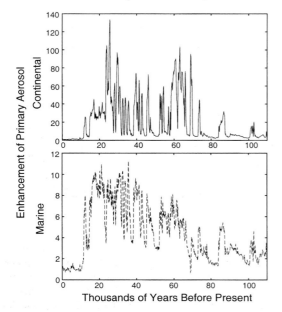

Fig. 18-19 Enhancement, relative to modern values, of marine and continental primary aerosols in Greenland through the last ice age (derived from Mayewski *et al.*, 1997).

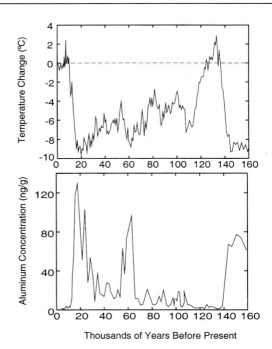

Fig. 18-20 Correlation of temperature history (Jouzel *et al.*, 1993; Salamatin *et al.*, 1998) and continent-derived dust reaching Antarctica (De Angelis *et al.*, 1987).

of cold glacial climates than in that of warm interglacials (De Angelis *et al.*, 1987; Mayewski *et al.*, 1997). Many primary aerosols derive from seasalts (Na, Mg, Cl, K, and some Ca, SO_4, NO_3), whose entrainment in the atmosphere and deposition onto ice sheets increases strongly with increasing ocean surface wind speed and, less importantly, with decreasing sea-ice aerial cover. Other important primary aerosols are dust from the continents, especially insoluble particles (mostly aluminosilicates, for which Al is used as an index), but also $CaCO_3$ and $CaSO_4$, which are soluble. Entrainment and deposition of primary continental aerosol increases as a function of increased wind speed, increased aridity, and increased source area.

The markedly higher seasalt concentrations during glacial times are due to a combination of greater ocean surface wind strength in both hemispheres as well as stronger meridional transport (Herron and Langway, 1985). Estimates of the surface wind strength increase are 3 to 8 m/s. The markedly higher continental dust flux probably results from both increased wind strength and expansion of arid areas.

Average dust grain size was larger during glacial times too, which indicates an increase in wind speed. The greater wind strength implied by these proxy records strongly suggests that the average equator-to-pole temperature gradients were larger during the cold climate. The isotopic signature of continental dust in both Greenland and Antarctica reveals the primary geographic source regions for ice age dust (Biscaye *et al.*, 1997), which are central Asia and Patagonia, respectively. These areas must have been particularly dry then.

The increase in dust in the interior of Antarctica strongly implies that the flux of chemicals from continents to the southern oceans was also much higher during glacial times (Fig. 18-20). Because biological activity in vast areas of the southern oceans is limited by lack of nutrients (including iron) it is quite possible that this increase of dust flux stimulated marine biological activity where none exists during interglacials (the iron-fertilization hypothesis; see Section 18.3.5.1), which could contribute to reduction of atmospheric CO_2. The temporal pattern of increases in continent-derived atmospheric dust is interesting in this context because of relative timing; it is clear from the ice-core records that dust flux increases occurred only well after temperature and greenhouse gas concentrations had begun to fall at the start of the last glacial period (De Angelis *et al.*, 1987).

Examples of inferred enhancements of atmospheric primary aerosol concentration in the glacial atmosphere relative to the modern are factors of 4 to 7 for insoluble particles from continents, and 3 for seasalts (Alley *et al.*, 1995), over Greenland.

18.4 Some Lessons in Environmental History

18.4.1 Summary of the Ice-Age Experiment

Over the past two million years, changes of global average radiative energy input to the Earth have been regular but minor, while changes of seasonal energy input have been significant but asynchronous between north and south hemispheres. These insolation changes cause substantial global environmental changes, encompassing changes in atmospheric and oceanic composition and circulation, changes in the physical surfaces of continents, changes in the biosphere, and changes in associated chemical cycles. The transformation of a minor or geographically localized forcing into a major and global response is a direct consequence of both the strong coupling between diverse Earth systems and the operation of feedbacks.

The strong coupling of Earth systems is spectacularly evident in the ice-age cycles as nearly synchronous changes in temperature, precipitation, wind strength, ocean current strength, terrestrial and marine biota, continental ice volume, and atmospheric concentrations of greenhouse gases and aerosols. Many of these changes, or their direct consequences, are elegantly recorded in the 250 kyr environmental history retrieved from Vostok in the heart of the Antarctic ice sheet (Fig. 18-21), and in the history of the last major climate transition retrieved from central Greenland (Fig. 18-22).

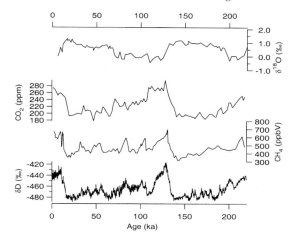

Fig. 18-21 The last 250 000 years of environmental history, recorded in the central Antarctic ice sheet. Bottom three panels are data from the Vostok ice core (Lorius *et al.*, 1990; Jouzel *et al.*, 1993). Top panel is marine data representing global ice volume (Shackleton *et al.*, 1990).

In the global ice-age experiment, important positive feedbacks include long-term changes in continental-ice cover, sea-ice cover, greenhouse gas concentrations, oceanic heat transport, atmospheric aerosol load, and vegetation. Strongly positive rapid feedbacks are also operative. These most likely involve atmospheric water vapor and clouds, and seasonal snow cover, which together amplify temperature changes by a factor of two to four (see 18.4.6). Negative feedbacks are harder to identify, but probably involve poleward atmospheric heat transport, biogenic sulfur emissions in the northern hemisphere, and some aerosols. Many feedbacks that are not yet understood probably exist in addition to these, especially involving controls on biogeochemical cycling.

The Earth systems also act as a frequency modulator in the ice-age cycles. The dominant forcings of 20 and 40 kyr periodicity are directly expressed in environmental changes, but these are considerably over-shadowed by the 100 kyr periodicity of glacial–interglacial cycling, for which there is no significant forcing. This suggests that key elements of the Earth's environmental systems have a long response time. Possible candidates are changes in the dynamic characteristics of the ice sheets themselves and

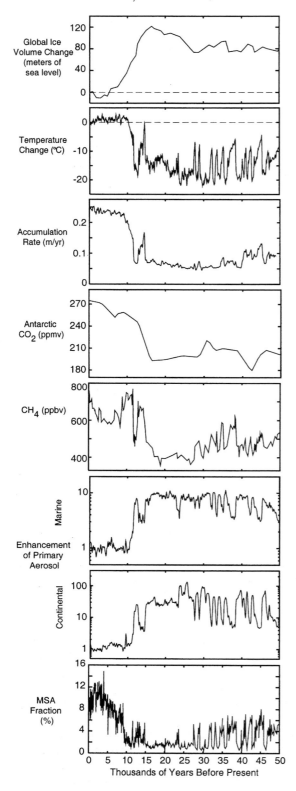

the dependence of ice-sheet melt and calving rates on elevation changes caused by isostatic response of the Earth's crust, which is rate-limited by creeping flow of the mantle (Peltier, 1987).

Another feature of the coupled Earth system exhibited in the ice-age experiment is that the global environment has modes, or relatively stable states that it maintains despite continual changes in external forcings. A manifestation is that extremes of climate have generally consistent values; the temperature minima attained during each of the past four glacial cycles are all very similar, as are each of the interglacial temperature maxima, despite the changing character of the insolation forcing. Climate modes likely reflect, in part, multiple stable configurations for ocean circulation. The consistency of extrema despite variable forcings, and the constancy of some climate periods despite continual changes in forcings (as during the Holocene) both indicate internal regulation of the Earth's environmental systems. Self-regulation generally results from feedbacks (see Chapter 17).

18.4.2   Some Large and Abrupt Climate Changes

Despite gradual changes in forcings, some very large climate changes have occurred in very little time – on time scales that are short even compared to human lives and to cultural changes. The clearest example is the abrupt warming and increase in precipitation at the final termination of the last glacial period (the end of the Younger Dryas), which is recorded at high resolution in Greenland ice and shown by coincident methane changes and by southern hemisphere geologic records (Denton and

Fig. 18-22   The last 50 000 years of environmental history, recorded in central Greenland (GISP2 and GRIP ice cores), plus the CO_2 record from Vostok, Antarctica, and global ice volume measured as sea level depression. (From top to bottom, references are: Shackleton, 1987; Cuffey *et al.*, 1995 and Grootes *et al.*, 1993; Cuffey and Clow, 1997; Chapellaz *et al.*, 1990; Brook *et al.*, 1996; Mayewski *et al.*, 1997; Saltzman *et al.*, 1997.)

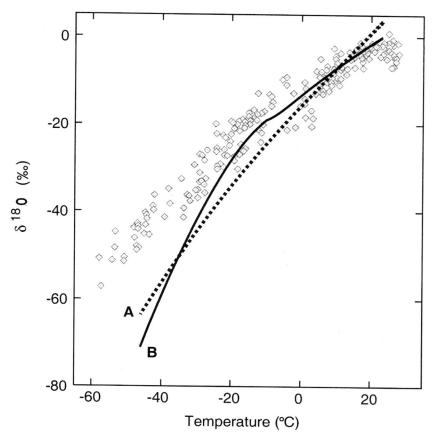

Fig. 18-23 Observed correlation of isotopic composition of precipitation with ground temperature (gray diamonds; Jouzel *et al.*, 1987), and predictions of simple isotopic models. A, prediction with constant α; B, prediction with temperature-dependent α.

Hendy, 1994) to involve a significant fraction of the globe. This particular change involved a sudden strengthening of the thermohaline circulation and consequent heat transport to the North Atlantic, and thus directly resulted from coupling of the oceans to other climate systems. A major question for research is whether such abrupt changes could be triggered by anthropogenic warming. There is at this point no reason to discount the possibility. Some modeling efforts (Stocker and Schmittner, 1997) have, for example, suggested that an increase in the freshwater flux to the North Atlantic from the continents could cause the thermohaline circulation to abruptly cease, plunging Europe into a cold climate unlike any in recent millennia.

18.4.3 Polar Amplification

A clear characteristic of global climate changes is their amplification in the polar regions. In glacial–interglacial cycling, temperature changes in the Arctic and Antarctic have been two to three times larger than the corresponding changes at low and mid-latitudes. This results, in part, from large albedo changes due to changing land and sea ice cover, and seasonal snowlines, and also from changes in ocean to atmosphere heat transfer as insulating sea-ice cover waxes and wanes. Polar amplification is very interesting because it implies the polar ice sheets (Greenland and West Antarctica) are particularly vulnerable to melt in a warmed climate, and because the melting of permafrost in tundra regions may

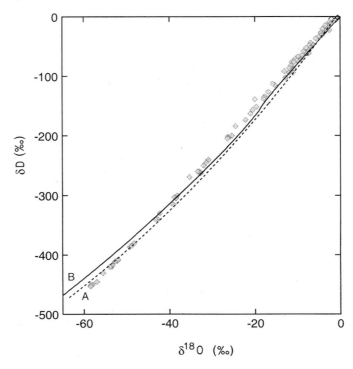

Fig. 18-24 Observed correlation (the Meteoric Water Line) of the two most important isotopic ratios in precipitation (gray diamonds; Jouzel *et al.*, 1987 and Dahe *et al.*, 1994), and predictions of simple isotopic models. A, prediction with constant α; B, prediction with temperature-dependent α.

release significant quantities of the powerful greenhouse gas CH_4. The increase of global mean temperature over the past century has, however, been only modestly amplified over the North Atlantic, Scandinavia and Greenland (Overpeck *et al.*, 1997), suggesting something is very different about the modern environment from past ones. Indeed there is now a substantial and negative anthropogenic radiative forcing at the mid-latitudes of the northern hemisphere, resulting from industry-generated particles and gases which form aerosols (see Chapter 7). Because these will change through time as technologies and economies change, it would be incorrect to view the modern Arctic as safe from amplified warming based on the past century.

18.4.4 Global Climate Sensitivity

One of the primary goals of "Earth systems" research is to learn how sensitive global climate

is to changes in forcings, for instance how much temperature change results from a given increase in atmospheric CO_2 content. A particular difficulty in learning the sensitivity to a given forcing is that the magnitude of climate change strongly depends on coincident changes in other forcings as a result of feedbacks and couplings. For the particular case of CO_2, an initial increase in temperature due to CO_2 change may increase atmospheric H_2O and decrease polar albedo, both of which are strong climate forcings themselves. Subsequent changes in biogenic emissions and ocean circulation may be very important too. Thus the net sensitivity due to CO_2 change is very difficult to quantify from physical models alone, due to the great complexity of system components and their couplings. Current models, for instance, do not attempt to predict changes in planetary albedo resulting from changes in marine biogenic emissions.

The   information   in   paleoenvironmental

records therefore is very valuable for better constraining sensitivities. The use of paleoenvironmental records is either through general circulation models (physical models of climate processes combined on a numerical imitation of our planet), or through robust but simplistic "state-variable" analyses.

18.4.5   *General Circulation Models (GCMs)*

Insofar as paleoenvironmental records reveal histories of both forcings and environments, the accuracy of GCMs may be tested by efforts to replicate these histories. Successful replication would suggest the models capture the essential behavior of the Earth and therefore have predictive ability. Further, GCM simulations of past climates may allow partitioning of net climate changes into components due to various forcings. GCM simulations of the ice-age Earth are very much works in progress.

Recent revisions to the boundary conditions (ice-sheet topography and sea surface temperatures) have added uncertainty to many of the GCM calculations of the past two decades. Moreover, all of these calculations use prescriptions for at least one central component of the climate system, generally oceanic heat transport and/or sea surface temperatures. This limits the predictive benefit of the models. Nonetheless, these models are the only appropriate way to integrate physical models of diverse aspects of the Earth systems into a unified climate prediction tool.

A consistent result from GCM simulations of ice-age climate is that the global cooling of 5 to 8°C can not result solely from the direct radiative forcings (Hansen *et al.*, 1984; Webb *et al.*, 1997) of insolation, greenhouse gas, ice cover and terrestrial albedo changes. To explain the ice ages, there must be a strong net feedback that magnifies temperature changes by a factor of 2 to 4. This feedback is most likely a combination of atmospheric water vapor and clouds. The global climate sensitivity implied by these models is at least $0.5°C/(W/m^2)$, and plausibly greater than $1°C/(W/m^2)$.

18.4.6   *State-Variable Analyses*

An alternative approach to assessing climate sensitivity is to assume that some or all important elements of the Earth's environmental systems are so strongly coupled (as one may infer from Figs 18-21 and 18-22) that the behavior of the whole system may be represented by the behavior of a few variables, which are related using feedback factors or simple dynamic models. Though such analyses cannot substitute for process models operating on realistic geographic domains, they distill the essential lessons of environmental history without creating an illusory aura of physical completeness and accuracy.

Lorius *et al.* (1990) performed a simple multivariate analysis in which they correlate the temperature changes of the past 160 kyr (as recorded in the Vostok δD record) with changes in five forcings: atmospheric CO_2 plus CH_4, ice volume, aerosol loading (dust and SO_4^{2-} separately), and insolation. These analyses establish that greenhouse gas variations correlate with temperature variations better than do any other forcings, and that these gas variations account for 40 to 60% of the climate change (and most likely 55 to 65%), in the correlative sense. This suggests that gas changes account for more than 2°C of the global deglacial warming. As the gas changes themselves can account directly for only 0.7°C , a feedback factor of greater than 3 is necessary (feedback factor being defined as the ratio of a temperature change to the temperature change due only to direct forcings).

A broader view was explored by Hoffert and Covey (1992), who used estimated forcings and geographically broad data from both the last glacial maximum and the mid Cretaceous to calculate two estimates of bulk climate sensitivity (change in temperature per change in radiative forcing) necessary to explain the paleotemperatures for these climates. Net forcings relative to the pre-industrial Holocene are $-6.7 W/m^2$ and $+15.7 W/m^2$ for the glacial and the Cretaceous, respectively, which yield sensitivities of $0.45°C/(W/m^2)$ and $0.57°C/(W/m^2)$. The recent revisions of last glacial maximum reconstructed temperatures (both low-latitude and polar) to lower values increases Hoffert and Covey's LGM sensitivity to $0.9°C/(W/m^2)$, and Lorius *et al.*'s feedback factor to 3.5 or 4. Regardless, all these sensitivities are comfortably within the range predicted for modern climate by GCMs. To infer a substantially lower sensitivity of Earth's climate to forcings, one would have to invoke physically implausible large additional forcings for both Cretaceous and glacial periods. It is thus a reasonable suggestion (Hoffert and Covey, 1992) that these paleoclimate records preclude a low global sensitivity.

18.4.7 Perspective on Anthropogenic Forcings

Finally, paleoenvironmental records demonstrate conclusively that anthropogenic changes of CO_2, CH_4, SO_4, NO_3, and N_2O are very large relative to their natural variability during the entire Holocene, and the modern polluted environment has no analog in at least the past 400 kyr. The magnitude of this human "experiment" is of the same magnitude as the natural experiment that buried the sites of Boston, Chicago, London and Stockholm beneath more than a kilometer of glacial ice, and produced the moraines and lakes that stimulated Agassiz's revelations.

References

Agassiz, L. (1840). "Etudes sur les glaciers." Privately published, Neuchatel.

Alley, R. B., Meese, D. A., Schuman, C. A. *et al.* (1993). Abrupt increase in Greenland snow accumulation at the end of the Younger Dryas event. *Nature* **362**, 527–528.

Alley, R. B., Finkel, R. C., Nishiizumi, K. *et al.* (1995). Changes in continental and sea-salt atmospheric loadings in central Greenland during the most recent deglaciation: model-based estimates. *J. Glaciol.* **41**, 503–514.

Alley, R. B., Mayewski, P. A., Sowers, T. *et al.* (1997). Holocene climatic instability: a prominent, widespread event 8200 years ago. *Geology* **25**, 483–486.

Archer, D. and Maier-Reimer, E. (1994). Effect of deep sea sedimentary calcite preservation on atmospheric CO_2 concentration. *Nature* **367**, 260–264.

Barnola, J. M., Raynaud, D., Korotkevitch, Y. S., and Lorius, C. (1987). Vostok ice core provides 160 000 year record of atmospheric CO_2. *Nature* **329**, 408–413.

Battle, M., Bender, M., Dowers, T. *et al.* (1996). Atmospheric gas concentrations over the past century measured in air from firn at the South Pole. *Nature* **383**, 231–235.

Bender, M., Sowers, T., Dickson, M.-L. *et al.* (1994). Climate connections between Greenland and Antarctica during the last 100 000 years. *Nature* **372**, 663–666.

Berner, R. A. (1990). Atmospheric carbon dioxide levels over Phanerozoic time. *Science* **249**, 1382–1386.

Biscaye, P. E., Grousset, F. E., Revel, M. *et al.* (1997). Asian provenance of glacial dust (stage 2) in the Greenland Ice Sheet Project 2 ice core, Summit, Greenland. *J. Geophys. Res.* **102**, 26765–26781.

Blunier, T., Chappellaz, J. A., Schwander, J. *et al.* (1993). Atmospheric methane record from a Greenland ice core over the last 1000 years. *Geophys. Res. Lett.* **20**, 2219–2222.

Blunier, T., Chappellaz, J., Schwander, J. *et al.* (1998). Asynchrony of Antarctic and Greenland climate change during the last glacial period. *Nature* **394**, 739–743.

Bond, G., Broecker, W., Johnsen, S. *et al.* (1993). Correlations between climate records from North Atlantic sediments and Greenland ice. *Nature* **365**, 143–147.

Boyle, E. A. (1988). The role of vertical chemical fractionation in controlling late Quaternary atmospheric carbon dioxide. *J. Geophys. Res.* **93**, 701–715.

Broecker, W. S. (1994). Massive iceberg discharges as triggers for global climate change. *Nature* **372**, 421–424.

Broecker, W. S. and Denton, G. H. (1989). The role of ocean-atmosphere reorganizations in glacial cycles. *Geochim. Cosmochim. Acta* **53**, 2465–2501.

Broecker, W. S. and Henderson, G. M. (1998). The sequence of events surrounding Termination II and their implications for the cause of glacial–interglacial CO_2 changes. *Paleoceanography* **13**, 352–364.

Brook, E. J., Sowers. T., and Orchardo, J. (1996). Rapid variations in atmospheric methane concentration during the past 110 000 years. *Science* **273**, 1087–1091.

Brook, E. J., Harder, S., Severinghaus, J., and Bender, M. (in press). Atmospheric methane during the past 50 000 years: trends, interpolar gradient, and rate of change. *In* "AGU Monograph on Mechanisms of Millennial Scale Climate Change" (P. Clark, R. Webb, and L. Keigwin, eds). American Geophysical Union, Washington DC.

Cerling, T., Wang, Y., and Quade, J. (1993). Expansion of C4 ecosystems as an indicator of global ecological change in the late Miocene. *Nature* **361**, 344–345.

Chappellaz, J., Barnola, J. M., Raynaud, D. *et al.* (1990). Atmospheric methane record over the last climatic cycle revealed by the Vostok ice core. *Nature* **345**, 127–131.

Chappellaz, J., Blunier, T., Raynaud, D. *et al.* (1993). Synchronous changes in atmospheric methane and Greenland climate between 40 and 8 kyr BP. *Nature* **366**, 443–445.

CLIMAP Project members (1976). The surface of the ice age Earth. *Science* **191**, 1131–1137.

COHMAP Members (1988). Climate changes of the last 18 000 years: Observations and model simulations. *Science* **241**, 1043–1052.

Craig, H. (1961). Isotopic variations in meteoric waters. *Science* **133**, 1702–1703.

Crowley, T. J. and North, G. R. (1991). "Paleoclimatology." Oxford University Press, New York.

Cuffey, K. M., Clow, G. D., Alley, R. B. *et al.* (1995). Large Arctic temperature change at the Wisconsin-Holocene glacial transition. *Science* **270**, 455–458.

Cuffey, K. M. and Clow, G. D. (1997). Temperature, accumulation and ice sheet elevation in central Greenland through the last deglacial transition. *J. Geophys. Res.* **102**, 26383–26396.

Curry, W. B., Duplessy, J. C., Labeyrie, L. D., and Shackleton, N. J. (1988). Changes in the distribution of $\delta^{13}C$ of deep water between the last glaciation and the Holocene. *Paleoceanography* **3**, 317–341.

Dahe, Q., Petit, J. R., Jouzel J., and Stievenard, M. (1994). Distribution of stable isotopes in surface snow along the route of the 1990 International Trans-Antarctica Expedition. *J. Glaciol.* **40**(134), 107–118.

Dahl-Jensen, D. and Johnsen, S. J. (1986). Paleotemperatures still exist in the Greenland ice sheet. *Nature* **320**, 250–252.

Dansgaard, W. (1964). Stable isotopes in precipitation. *Tellus* **16**, 436–468.

Dansgaard, W., Johnsen, S. J., Clausen, H. B. *et al.* (1993). Evidence for general instability of past climate from a 250-kyr ice-core record. *Nature* **364**, 218–220.

De Angelis, M., Barkov, N. I., and Petrov, V. N. (1987). Aerosol concentration over the last climatic cycle (160 kyr) from an Antarctic ice core. *Nature* **235**, 318–321.

Delmas, R. J. (1992). Environmental information from ice cores. *Rev. Geophys.* **30**, 1–21.

Delmotte, M., Chappellaz, J., Raynaud, D. *et al.* (1998). Atmospheric methane changes during the last 400 kyr revealed by the Vostok ice core. *EOS* **79**, F151.

Denton, G. H., Bickheim, J., Wilson, S. C., and Stuiver, M. (1989). Late Wisconsin and early Holocene glacial history, inner Ross Embayment, Antarctica. *Quatern. Res.* **31**, 151–182.

Denton, G. H. and Hendy, C. H. (1994). Younger Dryas age advance of Franz Josef Glacier in the Southern Alps of New Zealand. *Science* **264**, 1434–1437

Dlugokencky, K., Masarie, A., Lang, P. M., and Tans, P. P. (1998). Continuing decline in the growth rate of the atmospheric methane burden, *Nature* **393**, 447–450.

Emiliani, C. (1955). Pleistocene temperatures. *J. Geol.* **63**, 538–578.

Etheridge, D. M., Pearman, G. I., and Fraser, P. J. (1992). Changes in tropospheric methane between 1841 and 1978 from a high accumulation rate Antarctic ice core. *Tellus* **44B**, 282–294.

Etheridge, D. M., Steele, L. P., Lagenfelds, R. L. *et al.* (1996). Natural and anthropogenic changes in atmospheric CO_2 over the last 1000 years from air in Antarctic ice cores. *J. Geophys. Res.* **101**, 4115–4128.

Fuhrer, K. and Legrand, M. (1997). Continental biogenic species in the Greenland Ice Core Project ice core: Tracing back the biomass history of the North American continent. *J. Geophys. Res.* **102**(C12), 26735–26745.

Grootes, P. M., Stuiver, M., White, J. W. C. *et al.* (1993). Comparison of oxygen isotope records from the GISP2 and GRIP Greenland ice cores. *Nature* **366**, 552–554.

Guilderson, T. P., Fairbanks, R. G., and Rubenstone, J. L. (1994). Tropical temperature variations since 20 000 years ago: modulating interhemispheric climate change. *Science* **263**, 663–665.

Hammer, C. U., Clausen, H. B., Dansgaard, W. *et al.* (1987). Dating of Greenland ice cores by flow models, isotopes, volcanic debris, and continental dust. *J. Glaciol.* **20**, 3–26.

Hammer, C., Mayewski, P. A., Peel, D., and Stuiver, M. (eds) (1997). Greenland Summit ice cores. *J. Geophys. Res.* **102**(C12), 26317–26886.

Hansen, J., Lacis, A., Rind, D. *et al.* (1984). Climate sensitivity: analysis of feedback mechanisms. *Geophys. Monogr.* **20**, pp. 130–163. Am. Geophys. Union, Washington DC.

Hartman, D. (1994). "Global Physical Climatology." Academic Press, New York.

Hays, J. D., Imbrie, J., and Shackleton, N. J. (1976). Variations in the Earth's orbit: Pacemaker of the ice ages. *Science* **194**, 1121–1132.

Herron, M. M. and Langway, C. C. (1980). Firn densification: An empirical model. *J. Glaciol.* **25**, 373–385.

Herron, M. M. and Langway Jr., C. C. (1985). Chloride, nitrate and sulfate in the Dye 3 and Camp Century Greenland ice cores. *In* "Greenland Ice Core: Geophysics, Geochemistry and the Environment" (C. C. Langway Jr. *et al.*, eds), pp. 77–84. *Geophys. Monogr. Series* **33**, AGU, Washington DC.

Hoffert, M. I. and Covey, C. (1992). Deriving global climate sensitivity from paleoclimate reconstructions. *Nature* **360**, 573–576.

Horita, J. and Wesolowski, D. J. (1994). Liquid-vapor fractionation of oxygen and hydrogen isotopes of water from the freezing to the critical temperature. *Geochim. Cosmochim. Acta.* **58**(16), 3425–3437.

Johnsen, S. J., Dansgaard, W., Clausen, H. B., and Langway, C. C. (1972). Oxygen isotope profiles through the Antarctic and Greenland ice sheets. *Nature* **235**, 429–434.

Johnsen, S. J., Dansgaard, W., and White, J. W. C. (1989). The origin of Arctic precipitation under present and glacial conditions. *Tellus* **41**, 452–469.

Johnsen, S. J., Clausen, H. B., Dansgaard W. *et al.* (1997). The $\delta^{18}O$ record along the Greenland Ice Core Project deep ice core and the problem of possible Eemian climatic instability. *J. Geophys. Res.* **102**(C12), 26397–26410.

Jouzel, J., Merlivat, L., and Lorius, C. (1982). Deuterium excess in an East Antarctic ice core suggests higher relative humidity at the oceanic surface during the last glacial maximum. *Nature* **299**, 688–691.

Jouzel, J. and Merlivat, L. (1984). Deuterium and oxygen 18 in precipitation: modelling of the isotopic effects during snow formation. *J. Geophys. Res.* **89**(D7), 11749–11757.

Jouzel, J., Russell, G. L., Koster, R. D. *et al.* (1987). Simulations of the HDO and $H_2^{18}O$ atmospheric cycles using the NASA GISS general circulation model: the seasonal cycle for present-day conditions. *J. Geophys. Res.* **92**(D12), 14739–14760.

Jouzel, J., Barkov, N. I., Barnola, J. M. *et al.* (1993). Extending the Vostok ice-core record of paleoclimate to the penultimate glacial period. *Nature* **364**, 407–412.

Jouzel, J., Waelbroeck, C., Malaize, B. *et al.* (1996). Climatic interpretation of the recently extended Vostok ice records. *Clim. Dynam.* **12**, 513–521.

Kennett, J. P. and Ingram, B. L. (1995). A 20,000-year record of ocean circulation and climate change from the Santa Barbara Basin. *Nature* **377**, 510–514.

Legrand, M. (1987). Chemistry of Antarctic snow and ice. *J. Physique* **48**(3), *Colloque* C1, 77–86.

Legrand, M. R., Lorius, C., Barkov, N. I., and Petrov, V. N. (1988). Vostok (Antarctica) ice core: atmospheric chemistry changes over the last climatic cycle (160 000 years). *Atmos. Environ.* **22**(2), 317–331.

Legrand, M., Feniet-Saigne, C., Saltzman, E. S. *et al.* (1991). Ice core record of oceanic emissions of dimethyl sulphide during the last climate cycle. *Nature* **350**, 144–146.

Leuenberger, M. and Siegenthaler, U. (1992). Ice-age concentration nitrous oxide from an Antarctic ice core, *Nature* **360**, 449–451.

Lorius, C., Jouzel, J., Raynaud, D. *et al.* (1990). The ice-core record: climate sensitivity and future greenhouse warming. *Nature* **347**, 139–145.

MacAyeal, D. R., Firestone, J., and Waddington, E. (1993). Paleothermometry redux. *J. Glaciol.* **39**(132), 423–431.

Machida, T., Nakazawa, T., Fujii, Y. *et al.* (1995). Increase in atmospheric nitrous oxide concentration during the last 250 years. *Geophys. Res. Lett.* **22**, 2921–2924.

Majoube, M. (1971). Fractionnement en oxygene 18 et en deuterium entre l'eau et sa vapeur. *J. Chim. Phys.* **68**, 1423–1436.

Martin, J. H. (1990). Glacial–interglacial CO_2 change: The iron hypothesis. *Paleoceanography* **5**, 1–13.

Mayewski, P., Lyons, B., Spencer, M. J. *et al.* (1990). An ice core record of atmospheric response to anthropogenic sulphate and nitrate. *Nature* **346**, 554–556.

Mayewski, P. A., Meeker, L. D., Twickler, M. S. *et al.* (1993). Ice core sulphate from three northern hemisphere sites: Source and temperature forcing implications. *Atmos. Environ. A* **27**(17/18), 2915–2919.

Mayewski, P. A. *et al.* (1997). Major features and forcing of high-latitude northern hemisphere atmospheric circulation using a 110 000-year-long glaciochemical series. *J. Geophys. Res.* **102**(C12), 26345–26366.

Meeker, L. D., Mayewski, P. A., Twickler, M. S. *et al.* (1997). A 110 000-year history of change in continental biogenic emissions and related atmospheric circulation inferred from the Greenland Ice Sheet Project Ice Core. *J. Geophys. Res.* **102**(C12), 26489–26504.

Merlivat, L. and Nief, G. (1967). Fractionnement isotopique lors des changements d'etats solide-vapeur et liquide-vapeur de l'eau a des temperatures inferieures a 0°C. *Tellus* **19**(1), 122–127.

Merlivat, L. and Jouzel, J. (1979). Global climatic interpretation of the deterium-oxygen 18 relationship for precipitation. *J. Geophys. Res.* **84**, 5029–5033.

Molnar, P. and England, P. (1990). Late Cenozoic uplift of mountain ranges and global climate change: chicken or egg? *Nature* **346**, 29–34.

Muller, R. A. and MacDonald, G. J. (1997). Glacial cycles and astronomical forcing. *Science* **277**(5323), 215–218.

Neftel, A., Oeschger, H., Schwander, J. *et al.* (1982). Ice core sample measurements give atmospheric CO_2 content during the past 40 000 years, *Nature* **295**, 220–223.

Overpeck, J., Hughen, K., Hardy, D. *et al.* (1997). Arctic environmental change of the last four centuries. *Science* **278**, 1251–1256.

Paterson, W. S. B. (1994). "The Physics of Glaciers," 3rd edn. Pergamon, Tarrytown, NY.

Peltier, W. R. (1987). Glacial isostasy, mantle viscosity and Pleistocene climatic change. *In* North America and Adjacent Oceans during the last deglaciation. "The Geology of North America," Vol. K-3 (W. F. Ruddiman and H. E. Wright, Jr, eds), pp. 155–182. Geological Society of America, Boulder, CO.

Raisbeck, G. M., Yiou, F., Bourles, D. *et al.* (1987). Evidence for two intervals of enhanced ^{10}Be deposition in Antarctic ice during the last glacial period. *Nature* **326**, 273–277.

Raymo, M. E., Ruddiman, W. F., Backman, J. *et al.* (1989). Late Pliocene variation in northern hemisphere ice sheets and north Atlantic deep water circulation. *Paleoceanography* **4**, 413–446.

Raymo, M. E. and Ruddiman, W. F. (1992). Tectonic forcing of late Cenozoic climate. *Nature* **359**, 117–122.

Reeh, N. (1989). Dating by ice flow modeling: a useful tool or an exercise in applied mathematics? *In* "The Environmental Record in Glaciers and Ice Sheets" (H. Oeschger and C. C. Langway, eds), pp. 141–159. Wiley, New York.

Ruddiman, W. F. and Kutzbach, J. E. (1991). Plateau uplift and climatic change. *Scient. Am.* **264**, 66–75.

Salamatin, A. N., Lipenkov, V. Y., Barkov, N. I. *et al.* (1998). Ice core age dating and paleothermometer calibration based on isotope and temperature profiles from deep boreholes at Vostok Station (East Antarctica). *J. Geophys. Res.* **103**(D8), 8963–8977.

Saltzman, E. S., Whung, P.-Y., and Mayewski, P. A. (1997). Methanesulfonate in the Greenland Ice Sheet Project 2 Ice Core. *J. Geophys. Res.* **102**(C12), 26649–26658.

Saltzman, B. and Verbitsky, M. (1994). Late Pleistocene climatic trajectory in the phase space of global ice, ocean state, and CO_2: Observations and theory. *Paleoceanography* **9**(6), 767–779.

Schlesinger, W. H. (1996). Biogeochemistry: an analysis of global change. Academic Press.

Severinghaus, J. P., Sowers, T., Brook, E. J. *et al.* (1998).

Timing of abrupt climate change at the end of the Younger Dryas interval from thermally fractionated gases in polar ice. *Nature* **391**, 141–146.

Shackleton, N. J. (1967). Oxygen isotope analyses and Pleistocene temperatures re-assessed. *Nature* **215**, 15–17.

Shackleton, N. J. (1987). Oxygen isotopes, ice volume and sea level. *Quatern. Sci. Rev.* **6**, 183–190.

Shackleton, N. J., Berger, A., and Peltier, W. R. (1990). An alternative astronomical calibration of the lower Pleistocene timescale based on ODP Site 677. *Trans R. Soc. Edin. Earth Sci.* **81**, 251–261.

Shuman, C. A., Alley, R. B., Anandakrishnan, S. *et al.* (1995). Temperature and accumulation at the Greenland Summit: Comparison of high-resolution isotope profiles and passive microwave brightness temperature trends. *J. Geophys. Res.* **100**(D5), 9165–9177.

Sowers, T. *et al.* (1993). A 135,000 year Vostok-SPEC-MAP common temporal framework. *Paleoceanography* **8**, 737–766.

Sowers, T. A. and Bender, M. (1995). Climate records during the last deglaciation. *Science* **269**, 210–214.

Sowers, T., Brook, E. J., Etheridge, D. *et al.* (1997). An interlaboratory comparison of techniques for extracting and analyzing trapped gases in ice cores. *J. Geophys. Res.* **102**, 26527–26538.

Steig, E. J., Brook, E. J., White, J. W. C. *et al.* (1998). Synchronous climate changes in Antarctica and the North Atlantic. *Science* **282**, 92–95.

Stocker, T. F. and Schmittner, A. (1997). Influence of CO_2 emission rates on the stability of the thermo-haline circulation. *Nature* **388**(6645), 862–865.

Stute, M., Schlosser, P., Clark, J. F., and Broecker, W. S. (1992). Paleotemperatures in the southwestern United States derived from noble gases in groundwater. *Science* **256**, 1000–1003.

Stute, M., Schlosser, P., Talma, A. S. *et al.* (1995). Uniform cooling of the low latitude continents during the last glacial maximum. *EOS* **76**, F296.

Thompson, L. *et al.* (1995). Late glacial stage and Holocene tropical ice core records from Huascaran, Peru. *Science* **269**, 46–50.

Turekian, K. K. T. (1996). "Global Environmental Change." Prentice Hall, Upper Saddle River, NJ.

Webb, R. S., Rind, D. H., Lehman, S. J. *et al.* (1997). Influence of ocean heat transport on the climate of the Last Glacial Maximum. *Nature* **385**, 695–699.

Whung, P.-Y., Saltzman, E. S., Spencer, M. J. *et al.* (1994). A two hundred year record of biogenic sulfur in a south Greenland ice core (20D). *J. Geophys. Res.* **97**, 6023–6036.

19

Human Modification of the Earth System: Global Change

Robert J. Charlson, Gordon H. Orians, and Gordon V. Wolfe

Chapter 1 posed several examples of global, societally important environmental issues, the solution or management of which requires an understanding and an integrated treatment of the Earth systems. These issues appear frequently throughout the book, particularly in the references to anthropogenic fluxes of substances that significantly perturb individual elemental cycles. In this brief concluding chapter we revisit these issues to call attention to their biogeochemical and global nature.

19.1 Global Climate Change

Many biogeochemical and physical processes are involved in determining the climate of the Earth, and some of these are being significantly perturbed by human activity. Some of these perturbations can be assessed quantitatively, some have been discovered recently and can be described only qualitatively, and still others are yet to be discovered. As discussed in Chapter 17, there are many large questions regarding both physical and biogeochemical feedbacks within the Earth's climate system. Of particular importance is the atmosphere, through which all energy enters and leaves the Earth. The physical and chemical composition of the atmosphere determines the transmission, absorption, and reflection of incoming solar radiation, outgoing

terrestrial radiation, the dynamical distribution of energy with latitude. The resulting energy balance determines the surface temperature. The biogeochemical cycles of sulfur – a crucial component of clouds and most aerosols – and of carbon and nitrogen, which form radiatively important trace gases, are central to the radiative properties of the atmosphere. The cycles of these three elements are also severely perturbed by human activity.

Figure 19-1 sketches the interplay of these cycles and climate. In this figure we denote observable quantities by boxes and the processes that affect them by ovals. The figure illustrates the two major processes by which chemical cycles affect climate: the greenhouse effect and aerosol/cloud formation. We see for example the radiatively important natural atmospheric trace gases water vapor, carbon dioxide, methane, and nitrous oxide, as well as the radiatively important anthropogenic chlorofluorocarbons. This group of gases, produced from a variety of natural and human processes affecting the cycles of water, carbon, nitrogen, and halocarbons and absorb infrared radiation in the atmosphere, changing the global heat balance – the greenhouse effect (see Chapter 7, Section 7.11). The other important climate-affecting process, aerosol and cloud formation, appears to be dominated by the sulfur cycle: the production of sulfur gases

Earth System Science
ISBN 0-12-379370-X

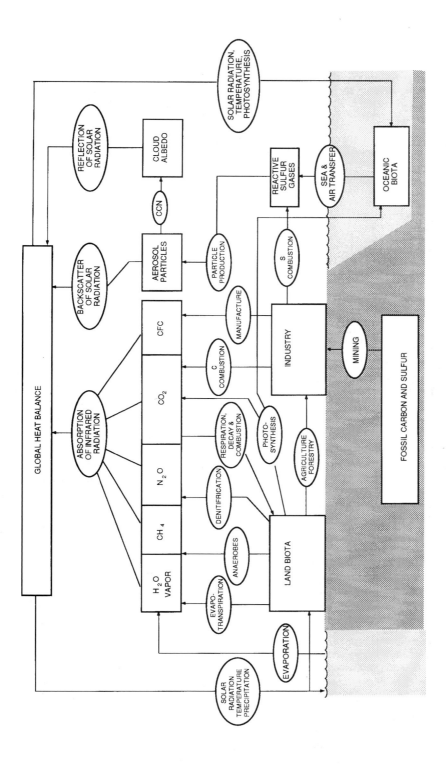

Fig. 19-1 Schematic of the processes that connect global biogeochemical cycles and climate. Boxes denote observables and ovals indicate processes that affect these.

that are oxidized to sulfuric acid in the atmosphere, forming new aerosol particles. Some of these have direct radiative effects and some may act as cloud condensation nuclei (CCN) to produce clouds and affect cloud albedo. Therefore, the sulfur cycle determines the shortwave radiation properties of the atmosphere, and the cycles of water, carbon, nitrogen, and trace halocarbons determine the longwave properties.

Of course, such a flow diagram cannot accurately portray the complete climate–biogeochemical cycle system. Rather, diagrams such as this one are intended to provide integration and, most especially, the definition of key quantities and processes to be observed and measured. Ultimately, our goal is to include a full quantitative model of the biogeochemical fluxes, their geographical variations and changes along with a complete biogeochemical and physical model of the climate.

Table 19-1 demonstrates that with the exception of water vapor, all of these cycles have been severely perturbed by human activity. Of course, all of these cycles are also linked in many ways. For example, the combustion of fossil fuel has increased the fluxes of carbon, sulfur, and nitrogen oxides to the atmosphere. Denitrification, the production of N_2O, is linked with the production of CO_2 during respiration and decay. And of course, other important cycles are involved which are not depicted here. Look back at Fig. 17-8, which sums up the climate *forcings* by the key agents.

19.2   Acid Precipitation

The combustion of fossil fuels, and the consequent oxidation of nitrogen in combustion air (see Chapters 5 and 12) has greatly modified the natural atmospheric cycles of C, N, and S, particularly in and near large regions of human population (e.g., the eastern US, Europe, and East Asia). Although the change of atmospheric CO_2 has little effect on the chemical composition of precipitation, sulfur- and nitrogen-containing acids have a major impact on the chemical composition of rain and snow. These acids in turn perturb the cycles of important minor elements, such as aluminum, through the weathering process in rocks and soils (Chapters 8 and 9). In developing a simple model of this phenomenon, both as a way to define key processes and to understand and quantify them, it is important to recognize the fundamental chemical nature of the acid–base balance of aqueous solutions, in this case, rainwater and melted snow water. As shown in Fig. 16-4, the pH of rainwater is especially sensitive to the addition of small amounts of strong acids because the natural system is like a titration near its endpoint.

We have only recently understood the phenomena that control rainwater pH in the natural, unpolluted environment. As pointed out in Section 16.2, these appear to be mainly the cycles of sulfur and nitrogen compounds. A model of the unperturbed system is necessary in order to understand and predict the changes that occur when strong sulfur- and nitrogen-acids are

Table 19-1   Radiatively important trace species in the atmosphere. Percent change in flux measured relative to the pre-industrial time

Cycle change	Species	Percent change
Longwave absorbers		
Water	H_2O (vapor)	Not known
Carbon	CO_2	+30
	CH_4	>140
Nitrogen	N_2O	+10
Halogens	Chlorofluorocarbons	$+\infty$
Shortwave reflectors		
Sulfur	SO_4^{2-}	+230

added, as well as to foresee the complex effects that perturbed rainwater pH has on other biogeochemical cycles and ecosystem processes.

19.3   Food Production

Few limitations have more profound implication for human welfare than the availability of food and water. Food production, whether by agriculture or hunting and gathering, is strongly influenced by climate, by the availability of key nutrients in adequate amounts during the growing season, by the presence of toxic materials, and by the physical, chemical, and microbiological properties of the soils. The biogeochemical cycles of C, N, P, and S are central to food production. A supply of minor elements is also important under some soil conditions.

Food production is influenced directly by biogeochemistry via precipitation chemistry and changes in soil properties induced by it. Potential benefits of increased NO_x in precipitation are an increased availability of nitrogen, a limiting nutrient in most agricultural regions. Potential negative consequences are increased soil acidity. The outcomes are certain to vary with soil types and climates. Better information on interrelationships between precipitation chemistry and soil chemistry will help agriculture adapt to changes induced by human perturbations of biogeochemical cycles.

Agriculture, the major human activity directly exposed to climate, has always been sensitive to climate change. Droughts and floods have depressed food production. Desperate farmers have fled from formerly productive areas to seek a better life. Despite these difficulties, agricultural scientists have developed many different strains of crops adapted to different climates that have helped to buffer agricultural productivity and keep pace with increasing demands for food. However, in poorer countries where the agricultural infrastructure is poorly developed, technical services are poor or lacking. Where climate change may occur, these countries are likely to experience severe difficulties in adapting to the change.

Predictions of changes in productivity are difficult in part because a major cause of probable climate change is increasing atmospheric carbon dioxide, which is also the raw material for photosynthesis. Enriching the atmosphere with CO_2 could potentially speed rates of photosynthesis and, thereby, increase agricultural production. Indeed, early laboratory experiments in which CO_2 concentrations were increased from 300 to 600 ppm, increased photosynthetic rates by 20% in maize and 60% in wheat (Akita and Moss, 1973), results that have been repeated successfully many times. Laboratory experiments also show that enrichment with CO_2 increases growth of root shoots and increases efficiency of water use.

What will happen under field conditions is, however, highly uncertain. Laboratory experiments have been conducted under conditions of abundant nutrients and water, ideal temperatures, and no competition among experimental plants. Such conditions are rare in the field. The best prediction is that, whereas increases similar to those found in the laboratory are unlikely under field conditions, increased concentrations of CO_2 are likely to ameliorate to some extent the detrimental effects of climate change. However, field-scale experiments under a variety of soils and climates and with several crop plants are needed to provide information on effects of climate change accompanied by higher concentrations of CO_2 on agricultural productivity.

19.4   Stratospheric Ozone Depletion

The Antarctic "ozone hole" is one of the most dramatic indications of anthropogenic environmental change. Depletion of stratospheric ozone via the catalytic mechanisms described in Chapters 5, 7, and 12 was first detected in a surprising fashion by British observers in Antarctica (Farman *et al.*, 1985). They measured increases of springtime solar UV radiation penetrating the atmosphere at wavelengths that normally are absorbed by O_3. The results showed almost a factor of two depletion of springtime ozone between 1956 and 1985, with most of the change occurring after 1976, as seen in Fig. 19-2. Subsequent reanalysis of satellite data revealed that most of this depletion was limited to a geographical area perhaps 1000 km across,

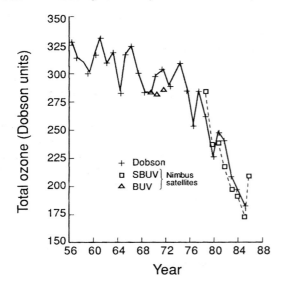

Fig. 19-2 Depletion of Antarctic ozone during October between 1956 and 1985. (Reprinted with permission from R. S. Stolarski (1988). Changes in ozone over the Antarctic. *In* F. S. Rowland and I. S. A. Isaksen, "The Changing Atmosphere," p. 112. John Wiley, Chichester.)

fortunately situated over an unpopulated place with no agriculture and very few people (Stolarski, 1988).

The chemical reactions in the oxygen-only mechanism, Sections 5.4.3 and 10.4 substantially underestimate the ozone destruction rate:

$$O_2 + h\nu \rightarrow O + O$$

$$O + O_2 + M \rightarrow O_3 + M$$

$$O_3 + h\nu \rightarrow O + O_2$$

$$O + O_3 \rightarrow 2\,O_2$$

Crutzen (1971) and Molina and Rowland (1974) showed that a second class of catalytic processes exists that result in destruction of ozone:

$$X + O_3 \rightarrow XO + O_2$$

$$\underline{XO + O \rightarrow X + O_2}$$

$$\text{net: } O_3 + O \rightarrow 2\,O_2$$

Here, X may be H, OH, NO, Cl, or Br. Cl from the photodissociation of chlorofluorocarbons such as CFC-12 (CCl_2F_2) is the main known catalyst currently acting in the so-called

"ozone hole" in the Antarctic spring. For this pioneering work, Crutzen, Molina, and Rowland were awarded the Nobel Prize in Chemistry in 1995.

It now appears that both the extreme magnitude and geographic limitations of the Antarctic ozone depletion are due to meteorologic patterns peculiar to the South Polar regions. The large decrease beyond the small reduction in the rest of the stratosphere apparently involves the circulation of the polar vortex, a complex interaction of Cl with oxides of nitrogen, their physical trapping in extremely cold ($T < -80°C$) clouds and preferential removal of some species by precipitation.

Although the unique circumstance of the Antarctic ozone hole does not extend to heavily populated parts of the world, there was disturbing evidence that stratospheric ozone levels worldwide also began to drop. Certainly the distribution of CFCs, which are extremely long-lived compounds, is global, and until recently their atmospheric burdens were ever increasing. The Montreal Protocol signed in 1987 has resulted in a nearly worldwide ban on the production and release of molecules like CFC-12 (CCl_2F_2), halting and now reversing the trend.

The depletion of stratospheric ozone is an example of the interaction of chemical cycles that was not predicted to occur when the supposedly inert CFCs were invented. It is in fact the extremely low reactivity of CFCs which made their use so popular and which is responsible for their very long lifetimes in the atmosphere. Therefore one lesson of the ozone hole has been that we must have a more complete understanding of the natural chemical cycles in order to better predict the effects and environmental implications of our novel compounds. Examples abound of our failure to predict the interactions of products of industrialization with natural chemical cycles: leaded gasolines, halogenated pesticides such as DDT, the biovolatilization of waste selenium and mercury, and so on. Currently we develop, license, and use hundreds of new compounds each year. The greater our understanding of biogeochemical cycles and the Earth system, the greater the chances that potentially damaging compounds will be recognized before they cause new environmental crises.

19.5   Large-scale Eutrophication

Inasmuch as the global cycles of several biologically essential elements have been substantially modified by human activity, it is appropriate to briefly review the consequences as they appear in large aquatic systems. As might be gathered from Figs 12-6 and 14-7, riverine fluxes of dissolved N (mostly NO_3^-) and both particulate and dissolved phosphorus have been greatly enhanced by human activity. Known sources, for example the application of fertilizers in agricultural areas, and urban sewage account for a substantial part of the fluxes carried to the ocean via the hydrologic process.

The consequences of addition of nutrients to aquatic systems are most evident in eutrophication of small systems such as lakes and estuaries where concentrations of key algal nutrient species are highest. Problems can occur even in large lakes like Lake Erie and Lake Geneva, where extreme conditions have occurred yielding sufficient production of organic matter to create anaerobic conditions in deeper parts of the lakes. When this occurs, the biological response is to utilize SO_4^{2-} as the oxidizing agent, yielding $S(-II)$ in compounds like H_2S. While local and regional scale effects of eutrophication due to N and P pollution have been recognized for decades, a more global view has revealed the existence of larger-scale phenomena in large estuaries and, in the vicinity of some intensely populated regions, the coastal ocean. Two examples close to each other are the North Sea and the Baltic Sea, both of which receive large fluxes of nutrients from European rivers (especially the Rhine). The consequences range from enhancement of blooms of certain phytoplankton species to the development of large deep-water areas that are anaerobic.

Counterexamples teach a lesson that these exaggerations of aquatic biological activity are highly idiosyncratic and depend on the fluxes of nutrients, the types of phytoplankton ecosystems that are involved, and – most importantly – the local and regional circulations of the aquatic system. For example, the Mediterranean Sea is landlocked and has many large pollution sources, but the large flux of nutrient-poor ("impoverished") water from the Atlantic Ocean prevents eutrophication. Much more water flows into the Mediterranean Sea than is required to replace evaporation from it. The excess, high salinity water exits Gibraltar below the water flowing in at the surface. Nutrients that enter the Mediterranean Sea from pollution sources are utilized by marine phytoplankton that sinks and exits with the outflow. Another example is that estuaries often have lower salinity or even freshwater at the surface with a denser saline layer at the bottom. An estuarine circulation occurs with nutrients being trapped in the saline bottom water.

19.6   Oxidative Capacity of the Global Troposphere

Oxidation processes are common in the atmosphere, surface waters and soils because of the abundance of molecular oxygen found in these parts of the Earth. Often, the reduced materials that enter into oxidation reactions are produced by biota, such as in photosynthesis. Because the atmospheric reservoir of O_2 is a major feature of this global redox system, its concentration and involvement in chemical reactions are important. Although no major changes in atmospheric O_2 as a result of human activity are forecast, changes in the oxidative capacity of the atmosphere are possible. Interestingly, there were alarmist statements and rumors in the 1970s to the effect that deforestation would cause a decrease in O_2. Signs and posters could be found admonishing people to not walk on the grass in order to save the planetary oxygen.

Ordinarily, the atmosphere is a self-cleansing system due to the abundance of O_3, OH, NO_2, and other reactive species. For example, hydrocarbon emissions from biota (such as terpenes) are oxidized in a matter of hours or days to CO and then on to CO_2. Alternatively, carboxylic acids may be formed and then transferred to the hydrosphere or pedosphere by rain. The atmosphere acts much like a low-temperature flame, converting numerous reduced compounds to oxidized ones that are more readily removed from the air. The limit to the rate of oxidation can be defined by the concentration of OH radicals and other trace species, such as hydrogen peroxide, H_2O_2, which is found in cloud and rainwater.

The increase of several reduced species in air (e.g., CH_4, CO, N_2O, SO_2, non-methane hydrocarbons) suggests that the oxidative capacity of the atmosphere may be decreasing. The increase in these reduced gases may even be the cause of a decrease in oxidative capacity, by lowering the steady-state concentrations of OH and other oxidizing agents. Rasmussen and Khalil (1984) describe the mysterious increase of CH_4 starting around A.D. 1800 and accelerating in recent years to almost 1% per year (Fig. 19-3e). This increase is still a mystery because there is no clear indication of its main cause. Studies of the presence of the reduced species in air strongly suggest that several of the major biogeochemical cycles are involved in controlling the oxidative capacity of the atmosphere. The nitrogen cycle (which is heavily perturbed by the human activities of combustion and fertilizer production) is a key factor through its role in atmospheric photochemistry. The sulfur cycle is involved through the presence of SO_2 from combustion of fossil sulfur in coal and oil. Carbon monoxide as a part of the carbon cycle also is implicated. It is clearly impossible to separate out any one chemical species as the single controlling factor; rather, it is necessary to consider the entire system as a unit.

19.7   Life and Biogeochemical Cycles

The evolution of life on Earth has depended on a sustained supply of nutrients provided by the physical environment. Life, in turn, has profoundly influenced the availability and cycling of these nutrients; hence the inclusion of *bio* in biogeochemical cycles. The involvement of the biosphere with biogeochemical cycles has been determined by the evolution of life's biochemical properties in the context of the physical and chemical properties of planet Earth.

Not surprisingly, only about 20 of the chemical elements found on Earth are used by living organisms (Chapters 3 and 8). Most of them are common elements. Rare elements are used, if at all, only at extremely low concentrations for specialized functions. An example of the latter is the use of molybdenum as an essential component of nitrogenase, the enzyme that catalyzes the fixation of elemental dinitrogen. Because they are composed of common elements, living organisms exert their most profound effects on the cycles of those elements.

Fig. 19-3   Illustration of the linked behavior of radiatively important trace species concentrations over different time scales. (a, b, c) Concentrations of methane, nss-SO_4^{2-}, and CO_2 over the past 160 kyr found in ice cores from Vostok, Antartica (temperature deduced from $^{18}O_2$ also shown). (d, e, f) Secular trends in nitrous oxide, methane, and CO_2 over the past 250 years. (g, h, i, j) Changes in October stratospheric ozone column burden over Antarctica, and chlorofluorocarbon, methane, sulfate, and nitrate from south Greenland ice, and carbon dioxide concentrations over the past 30 years. (Figures adapted with permission from: (a) J. Chappellaz *et al.* (1990). Ice-core record of atmospheric methane over the past 160 000 years. *Nature* **345**, 127–131, Macmillan Magazines; (b) M. R. Legrand, R. J. Delmas, and R. J. Charlson (1988). Climate forcing implications from Vostok ice core sulphate data. *Nature* **334**, 418–420, Macmillan Magazines; (c) J. M. Barnola, D. Raynaud, Y. S. Korotkevich, and C. Lorius (1987). Vostok ice core provides 160 000 year record of atmospheric CO_2. *Nature* **329**, 408–414, Macmillan Magazines; (d) M. A. K. Khalil and R. A. Rasmussen (1988). Nitrous oxide: trends and global mass balance. *Annal. Glaciol.* **10**, 73–39, International Glaciological Society, Cambridge; (e) B. Stauffer, G. Fischer, A. Neftel, and H. Oeschger (1985). Increase of atmospheric methane record in Antarctic ice core. *Science* **229**, 1386–1388, AAAS; (f) U. Siegenthaler and H. Oeschger (1987). Biospheric CO_2 emissions during the past 200 years reconstructed by deconvolution of ice core data. *Tellus* **39B**, 140–154, Swedish Geophysical Society, Stockholm; (g) R. S. Stolarski (1988). Changes in ozone over the Antarctic. *In* F. S. Rowland and I. S. A. Isaksen, "The Changing Atmosphere," p. 112, John Wiley, Chichester; (h) D. M. Cunold *et al.* (1986). Atmospheric lifetime and annual release estimates for $CHCl_3$ and CF_2Cl_2 from 5 years of ALE data. *J. Geophys. Res.* **91**, 10797–10817, American Geophysical Union; (i) D. R. Blake and F. S. Rowland (1988). Continuing worldwide increase in tropospheric methane 1978 to 1987. *Science* **239**, 1129–1131, AAAS; (j) C. D. Keeling (1989). A three dimensional model of atmospheric CO_2 transport based on observed winds: 1. Analysis of observational data. *Geophys. Monogr.* **55**, 165–236, American Geophysical Union.)

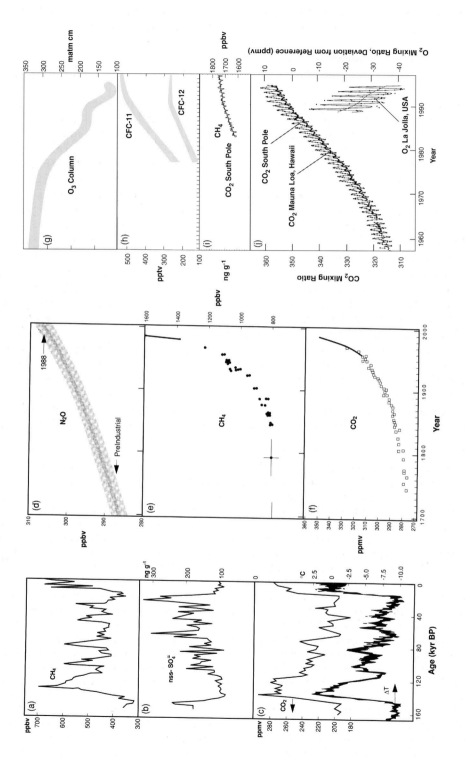

Fig. 19-3

Finding and extracting elements from the environment is a potentially costly process. Organisms have evolved powerful mechanisms for scavenging nutrients from rock, soil, water, and air. In addition, they re-use most materials extensively in more or less cyclic patterns. The more abundant elements leak from these cycles because organisms are unable to use everything present. Also, some elements are released rather than being recycled when the cost of recycling exceeds the cost of scavenging new molecules. Thus, amino acids and carbohydrates are discarded when deciduous leaves are dropped from plants because costs of breakdown and reabsorption of those molecules are too high. Nonetheless, the amount of leakage from ecosystems may be low because microorganisms and roots of plants quickly capture and re-incorporate the elements.

Organisms also evolved powerful detoxifying mechanisms that remove toxic materials or convert them to non-toxic forms or nutrients. Examples of alterations to non-toxic forms are the conversions of hydrogen sulfide to sulfate and nitrite to nitrate. The prime example of development of the ability to use a toxic substance is the evolution of aerobic metabolism, which converted a serious and widespread toxin, oxygen, into a major resource. This development, as we have seen, greatly increased the productivity of the biosphere and generated the oxygen-rich atmosphere of today's Earth.

How the biosphere would respond to human-induced climate changes is uncertain. Atmospheric concentrations of CO_2, which have been increasing for many decades, have apparently been accompanied by increased rates of photosynthesis at mid-latitudes and high latitudes during the summer (Fan *et al.*, 1998). Evidence for this is provided by the increased amplitude of the high latitude seasonal fluctuations in CO_2 concentrations. Whether or not the biosphere is, or will continue to be, a growing sink for CO_2 has profound implications for future climate change, productivity of agriculture and forestry, and the functioning of natural ecosystems. If increases in CO_2 concentrations are accompanied by modest and slow changes in climate, the biosphere may be able to respond by increasing productivity and sequestering more carbon.

However, if climate changes are rapid, the adaptive flexibility of long-lived plants may be exceeded, lowering productivity and, hence, storage of carbon. Rapid, or even modest, climate changes are likely to shift climates latitudinally and longitudinally faster than slow-growing trees with large seeds have migrated in the past. Implications for the biological productivity and survival of species in the face of such changes are profound, lending importance to the need to better understand and predict likely climate changes. Climate change also has ramifications for the animals that feed on the plants.

19.8   Conclusion

We have learned much about the individual parts and processes of the Earth's atmospheric, oceanic, continental, physical, chemical, and biological systems. However, we have just started to understand the linkages and feedbacks that make these systems function as a single entity. The degree to which the new discipline of Earth system science develops will depend upon the ways that world society chooses to respond to the many global processes that presently are changing. It would be speculative even to suggest that this new scientific discipline will emerge with a particular lexicon, focus, and practice. However, we do know that the global system *is* changing as a result of known processes, that it has changed continuously throughout the geologic past and in recent times, and that current changes are large compared to natural ones in the past.

The characteristics of the changes in several species discussed in this chapter are shown in Fig. 19-3. This figure depicts the changing composition of the atmosphere on three time scales. Figures 19-3a–c show the simultaneous variation of CO_2, CH_4, temperature, and SO_4^{2-} from the Vostok ice core (see also Fig. 1-2). These records also clearly demonstrate that the Earth functions as a coupled system. Temperature, CO_2 and CH_4 are positively correlated with one another, but each is negatively correlated with SO_4^{2-} (for reasons that are not yet known). This time period covers 160 000 years including the present interglacial climate, the last glacial period (known in the US as the Wisconsin ice age), the previous interglacial time and the last bit of the penultimate ice age.

Figures 19-3d–f depict both the recent data from ice cores and the contemporary records of N_2O, CH_4, and CO_2 during the most recent 250 years. These illustrate the profound changes that have occurred since the industrial revolution. Although the exact causes of the increases of N_2O and CH_4 are not yet fully agreed upon, there is no debate regarding the relationship of the increase of CO_2 to the burning of fossil carbon and deforestation. In the case of CH_4 and CO_2, there is also excellent agreement between the ice-core records and the records from direct sampling of the atmosphere, which began in 1957 for CO_2 and in ca. 1973 for CH_4.

Finally, Figs 19-3g–j illustrate the accelerated rate of change that has occurred for some of these atmospheric chemical variables for the three decades from 1957 to the present. The O_3 column data (g) are for the month of October at Halley Bay, Antarctica, while the remainder are global mean values.

We want to draw two final conclusions from these figures.

1. The oldest records (a–c) and Fig. 1-2 clearly show a strong degree of temporal correlation between three biologically involved atmospheric components and climate (as indicated by temperature). Because there is a sound physical basis for the involvement of all three in climatic processes, it is necessary to study, view, and understand these variables and climate as linked components of a system. They are all *dependent* variables and cannot be viewed as independent with climate being imposed as an exogenous factor.

2. Humans have so modified the main biogeochemical cycles of Earth that the chemical composition of the atmosphere exhibits differences that are approaching the magnitude of the changes that occurred between ice ages and interglacial periods. For example, the change in atmospheric CO_2 from the Wisconsin ice age to the pre-industrial value was

from 190 to 280 ppmv, an increase of about 50%. The increase from A.D. 1800 until now is from 280–365 ppmv or an additional 30%. It appears to be inevitable that CO_2 will continue to increase, doubling from the pre-industrial value by ca. AD 2050–2100. That the Earth's heat balance will change is also inevitable (see Chapter 17).

As much as we know about the increase in CO_2, the forecast of climatic response is unclear, largely because the current changes are unprecedented. It is likely that the near-term climatic future will be affected by human activity. Understanding the response of climate to our actions will depend upon an understanding of the biogeochemical functioning of the Earth as a linked system.

References

Akita, S. and Moss, D. N. (1973). Photosynthetic responses to carbon dioxide and light by maize and wheat leaves adjusted for constant stomatal apertures. *Crop Sci.* **13**, 234-237.

Crutzen, P. J. (1971). Ozone production rates in an oxygen, hydrogen, nitrogen oxide atmosphere. *J. Geophys. Res.* **76**, 7311–7327.

Fan, S., Gloor, M., Mahlman, J. *et al.* (1998). A large terrestrial carbon sink in North America implied by atmospheric and oceanic carbon dioxide data and models. *Science* **282**, 442–446.

Farman, J. C., Gardiner, B. G., and Shouklin, J. D. (1985). Large losses of total ozone in Antarctica reveal seasonal ClO_x/NO_x interaction. *Nature* **315**, 207–210.

Molina, M. J. and Rowland, F. S. (1974). Stratospheric sink for chlorofluoromethanes: chlorine-catalyzed destruction of ozone. *Nature* **249**, 810–812.

Rasmussen, R. A. and Khalil, M. A. K. (1984). Atmospheric methane in recent and ancient atmospheres: Concentrations, trends, and interhemispheric gradient. *J. Geophys. Res.* **89**, 11599–11604.

Stolarski, R. S. (1988). Changes in ozone over the Antarctic. *In* "The Changing Atmosphere" (F. S. Rowland and I. S. A. Isaksen, eds). Wiley, New York.

Answers

Answers are given for some of the numerical questions in the text.

Chapter 4

4-1   If steady state is assumed ($Q_1 + Q_2 = S$), Q_2 can be estimated as $S - Q_1 = 25$. If the uncertainties in S and Q_1 are independent, the uncertainty range of the Q_2 estimate is from -25 to $+75$.

4-2   Atmosphere: 3.4 yr; Surface water: 4.3 yr; Short-lived biota (land): 1 yr, etc.

4-3 $\tau_0 = \tau_r = 2\tau_a$

4-4   The response time of the system is 0.83 yr.

4-5   The response time is $1/k$. However, the turnover time τ_0 is $M_0/k(M_0 - M_1)$ and thus different from the response time. The turnover time depends on the steady-state content M_0. If M_0 is just a little bit larger than the threshold value, τ_0 will be very large.

Chapter 5

5-1   (a) 6.7×10^{-19}. (b) 1.5×10^{-10} bar. (c) 4.2×10^{-14}.

5-2   (b) -623.13 kJ/mol. (c) 0.0292 M/bar.

5.3   (a) 0.10 M. (b), (c) 0.20 M. (d) -0.10 M.

5-4   (a) 2027. (b) 1018 will condense, 1009 will remain in the atmosphere. (c) $-4.4‰$ for the precipitation, $-13.2‰$ for the remaining vapor.

5-5   0.106 V.

Chapter 10

10-1 τ_{water} (river input) = 43 750 yr; τ_{water} (hydrothermal output) = 1.4×10^7 yr; τ_{Mg} (river input) = 1.36×10^7 yr; τ_{Mg} (hydrothermal) = 1.4×10^7 yr.

10-2   (a) Dissolved oxygen will increase from the saturation value of 252 μM to 424 μM. You need the RKR ratio for N and O. (b) The flux is 6.88×10^{-6} mol/cm^2/s out of the ocean.

10-4   The activity of water would be 0.60. This is substantially different than the activity of water in seawater and thus could not be in equilibrium with water in seawater. The spontaneous reaction in seawater will be to the right.

Earth System Science
ISBN 0-12-379370-X

Index

Note: emboldened page numbers indicate chapters

International Geophysics Series

Edited by

JAMES R. HOLTON

Department of Atmospheric Sciences
University of Washington
Seattle, Washington

* Out of Print